Mathematica by Example

Mathematica by Example

Fifth Edition

Martha L. Abell
Georgia Southern University, Statesboro, USA

James P. Braselton
Georgia Southern University, Statesboro, USA

Academic Press is an imprint of Elsevier
125 London Wall, London EC2Y 5AS, United Kingdom
525 B Street, Suite 1800, San Diego, CA 92101-4495, United States
50 Hampshire Street, 5th Floor, Cambridge, MA 02139, United States
The Boulevard, Langford Lane, Kidlington, Oxford OX5 1GB, United Kingdom

Library of Congress Cataloging-in-Publication Data
A catalog record for this book is available from the Library of Congress

British Library Cataloguing-in-Publication Data
A catalogue record for this book is available from the British Library

ISBN: 978-0-12-812481-9

For information on all Academic Press publications
visit our website at https://www.elsevier.com/books-and-journals

Publisher: Katey Birtcher
Acquisition Editor: Graham Nisbet
Editorial Project Manager: Susan Ikeda
Production Project Manager: Paul Prasad Chandramohan
Designer: Mark Rogers

Typeset by VTeX

Contents

Preface

Mathematica by Example bridges the gap that exists between the very elementary handbooks available on Mathematica and those reference books written for the advanced Mathematica users. *Mathematica by Example* is an appropriate reference for all users of Mathematica and, in particular, for beginning users like students, instructors, engineers, business people, and other professionals first learning to use Mathematica. *Mathematica by Example* introduces the very basic commands and includes typical examples of applications of these commands. In addition, the text also includes commands useful in areas such as calculus, linear algebra, business mathematics, ordinary and partial differential equations, and graphics. In all cases, however, examples follow the introduction of new commands. Readers from the most elementary to advanced levels will find that the range of topics covered addresses their needs.

Taking advantage of Version 11 of Mathematica, *Mathematica by Example*, Fifth Edition, introduces the fundamental concepts of Mathematica to solve typical problems of interest to students, instructors, and scientists. The Fifth Edition is an extensive revision of the text. Features that make *Mathematica by Example*, Fifth Edition, as easy to use as a reference and as useful as possible for the beginner include the following.

1. **Version 11 Compatibility.** All examples illustrated in *Mathematica by Example*, Fifth Edition, were completed using Version 11 of Mathematica. Although many computations can continue to be carried out with earlier versions of Mathematica, we have taken advantage of the new features in Version 11 as much as possible.
2. **Applications.** New applications, many of which are documented by references, from a variety of fields, especially biology, physics, and engineering, are included throughout the text. Especially notice the new examples regarding series in Chapter 3 and manipulation of photographs in Chapters 2, 4, and 5.
3. **Detailed Table of Contents.** The table of contents includes all chapter, section, and subsection headings. Along with the comprehensive index, we hope that users will be able to locate information quickly and easily.
4. **Additional examples.** We have considerably expanded the topics in Chapters 1 through 6. The results should be more useful to instructors, students, business people, engineers, and other professionals using Mathematica on a variety of platforms.
5. **Comprehensive Index.** In the index, mathematical examples and applications are listed by topic, or name, as well as commands along with frequently used options: particular mathematical examples as well as examples illustrating how to use frequently used commands are easy to locate. In addition, commands in the index are cross-referenced with frequently used options. Functions available in the various packages are cross-referenced both by package and alphabetically.
6. As technology has changed, so has the publication of a book. When *Mathematica by Example* was first published in 1992, it was published as a single color book. Consequently, at that time, it was important to use various gray levels in plots to help distinguish them. Now, many of you will download an electronic copy of the text and print it on a high-resolution color printer with high-quality paper. The result will be outstanding. To illustrate the use of color, we have chosen from various universities and colleges throughout the United States. We *tried* to use the colors from at least one university or college in each state. Sometimes this was difficult to do because obtaining the color codes from some colleges was easier than from others. Of course, in the print version of the text, all images will still be in various levels of gray.

We began *Mathematica by Example* in 1990 and the first edition was published in 1991. Back then, we were on top of the world using Macintosh IIcx's with 8 megs of RAM and 40 meg hard drives. We tried to choose examples that we thought would be relevant to beginning users – typically in the context of mathematics encountered in the undergraduate curriculum. Those examples could also be carried out by Mathematica in a timely manner on a computer as powerful as a Macintosh IIcx.

When working on the Fifth edition, we are on the top of the world with iMacs with dual Intel processors complete with 8 gigs of RAM and 1 or 2 Terabyte hard drives. Now we are working with machines with more memory than we can comprehend and so fast we can't believe that computers will be faster but they will almost certainly be nearly obsolete by the time you are reading this. The examples presented in *Mathematica by Example* continue to be the ones that we think are most similar to the problems encountered by beginning users and are presented in the context of someone familiar with mathematics typically encountered by undergraduates. However, for this Fifth edition of *Mathematica by Example* we have taken the opportunity to expand on several of our favorite examples because the machines now have the speed and power to explore them in greater detail.

Other improvements to the Fifth edition include:

1. In Chapter 3, we have increased the number of examples relating to applications of series, particularly discussing "how is π approximated accurately now?"
2. Chapter 4, Introduction to Lists and Tables, contains several examples illustrating various techniques of how to quickly create plots of bifurcation diagrams, Julia sets, and the Mandelbrot set with new Mathematica functions that make visualizing these sets remarkably easy.
3. In Chapter 6, we have taken advantage of the `Manipulate` function to illustrate a variety of situations and expand on many examples throughout the chapter. For example, see Example 6.13 for a comparison of solutions of nonlinear equations to their corresponding linear approximations.
4. We have included references that we find particularly interesting in the **Bibliography**, even if they are not specific Mathematica-related texts. A comprehensive list of Mathematica-related publications can be found at the Wolfram web site.

http://store.wolfram.com/catalog/books/

Also, be sure to investigate, use, and support Wolfram's MathWorld – simply an amazing web resource for mathematics, Mathematica, and other information.

http://store.wolfram.com/catalog/books/

Martha L. Abell
James P. Braselton
Statesboro, GA, USA
June, 2017

Chapter 1

Getting Started

1.1 INTRODUCTION TO MATHEMATICA

Mathematica, first released in 1988 by Wolfram Research, Inc.,

`http://www.wolfram.com/`,

is a system for doing mathematics on a computer. Mathematica combines symbolic manipulation, numerical mathematics, outstanding graphics, and a sophisticated programming language. Because of its versatility, Mathematica has established itself as the computer algebra system of choice for many computer users. Among the over 1,000,000 users of Mathematica, 28% are engineers, 21% are computer scientists, 20% are physical scientists, 12% are mathematical scientists, and 12% are business, social, and life scientists. Two-thirds of the users are in industry and government with a small (8%) but growing number of student users. However, due to its special nature and sophistication, beginning users need to be aware of the special syntax required to make Mathematica perform in the way intended. You will find that calculations and sequences of calculations most frequently used by beginning users are discussed in detail along with many typical examples. In addition, the comprehensive index not only lists a variety of topics but also cross-references commands with frequently used options. *Mathematica By Example* serves as a valuable tool and reference to the beginning user of Mathematica as well as to the more sophisticated user, with specialized needs.

For information, including purchasing information, about Mathematica contact:

Corporate Headquarters:
Wolfram Research, Inc.
100 Trade Center Drive
Champaign, IL 61820
USA
telephone: 217-398-0700
fax: 217-398-0747
email: `info@wolfram.com`
web: `http://www.wolfram.com`

Europe:
Wolfram Research Europe Ltd.
10 Blenheim Office Park
Lower Road, Long Hanborough
Oxfordshire OX8 8LN
United Kingdom
telephone: +44-(0) 1993-883400
fax: +44-(0) 1993-883800
email: `info-europe@wolfram.com`

Asia:
Wolfram Research Asia Ltd.
Izumi Building 8F
3-2-15 Misaki-cho
Chiyoda-ku, Tokyo 101
Japan

Mathematica by Example. http://dx.doi.org/10.1016/B978-0-12-812481-9.00001-6

telephone: +81-(0)3-5276-0506
fax: +81-(0)3-5276-0509
email: `info-asia@wolfram.com`

A Note Regarding Different Versions of Mathematica

With the release of Version 10.4 of Mathematica, many new functions and features have been added to Mathematica. We encourage users of earlier versions of Mathematica to update to Version 11 as soon as they can. All examples in *Mathematica By Example*, fifth edition, were completed with Version 11. In most cases, the same results will be obtained if you are using Version 10.4 or later, although the appearance of your results will almost certainly differ from that presented here. However, particular features of Version 10.4 are used and in those cases, of course, these features are not available in earlier versions. If you are using an earlier or later version of Mathematica, your results may not appear in a form identical to those found in this book: some commands found in Version 11 are not available in earlier versions of Mathematica; in later versions some commands will certainly be changed, new commands added, and obsolete commands removed. For details regarding these changes, please refer to the **Documentation Center**. You can determine the version of Mathematica you are using during a given Mathematica session by entering either the command `$Version` or the command `$VersionNumber`. In this text, we assume that Mathematica has been correctly installed on the computer you are using. If you need to install Mathematica on your computer, please refer to the documentation that came with the Mathematica software package.

On-line help for upgrading older versions of Mathematica and installing new versions of Mathematica is available at the Wolfram Research, Inc. website:

`http://www.wolfram.com/.`

Details regarding what is different in Mathematica 11 from previous versions of Mathematica can be found at

`http://www.wolfram.com/mathematica/new-in-11/?source=frontpage-stripe.`

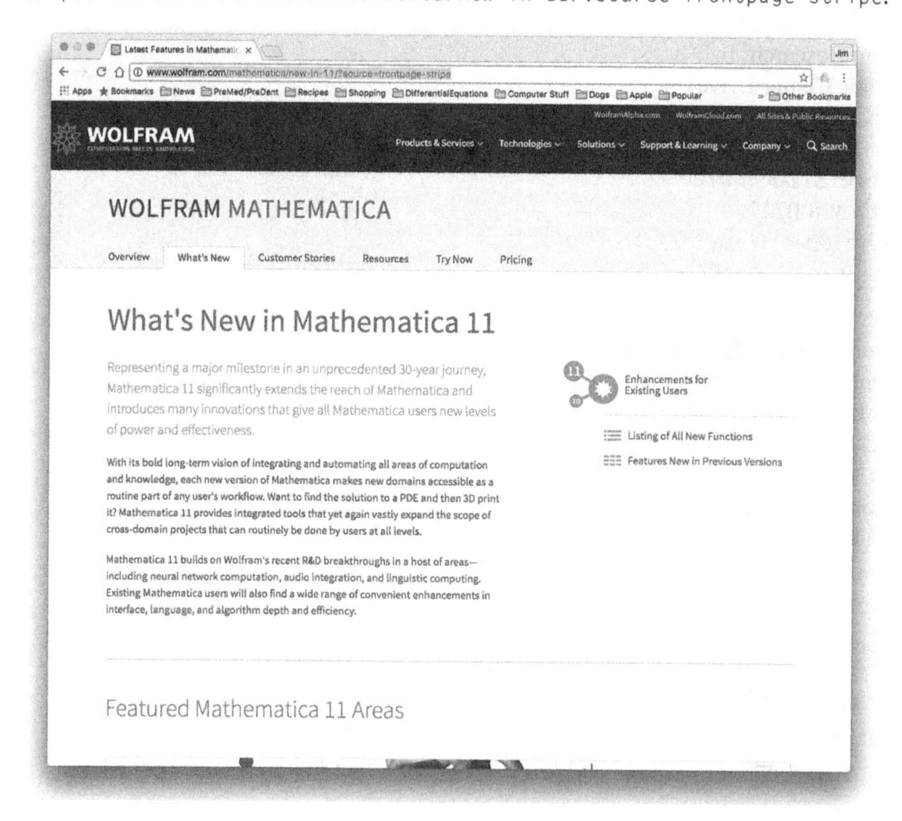

Also, when you go to the **Wolfram Documentation** center (under **Help** in the Mathematica menu) you can choose **Wolfram Documentation** to see the major differences. Also, the upper right hand corner of the main help page for each function will tell you if it is new in Version 11 or has been updated in Version 11.

1.1.1 Getting Started with Mathematica

We begin by introducing the essentials of Mathematica. The examples presented are taken from algebra, trigonometry, and calculus topics that most readers are familiar with to assist you in becoming acquainted with the Mathematica computer algebra system.

We assume that Mathematica has been correctly installed on the computer you are using. If you need to install Mathematica on your computer, please refer to the documentation that came with the Mathematica software package.

Start Mathematica on your computer system. Using Windows or Macintosh mouse or keyboard commands, start the Mathematica program by selecting the Mathematica icon or an existing Mathematica document (or notebook), and then clicking, double-clicking or right-clicking and selecting **Open** on the icon.

If you start Mathematica by selecting the Mathematica icon, Mathematica's startup window, "welcome screen," is displayed, as illustrated in the following screen shot.

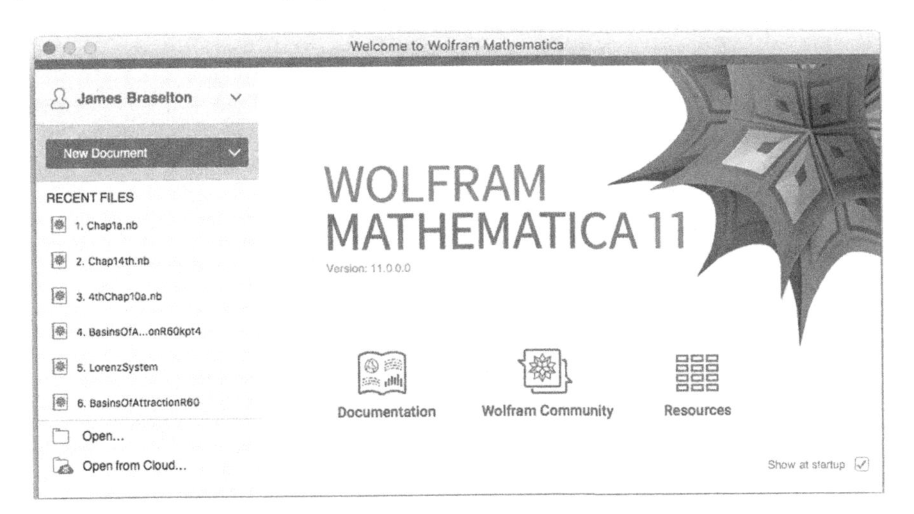

From the startup window, you can perform a variety of actions such as creating a new notebook. For example, selecting **New Document** generates a new Mathematica notebook.

Mathematica's online help facilities are spectacular. For beginning users, one of the more convenient features are the various **Palettes** that are available. The **Palettes** provide a variety of fill-in-the blank templates to perform a wide variety of action. To access a **Palette** go to the Mathematica menu, select **Palettes** and then select a given **Palette**. The following screen shots show the **Basic Math Assistant** and **Classroom Assistant** palettes.

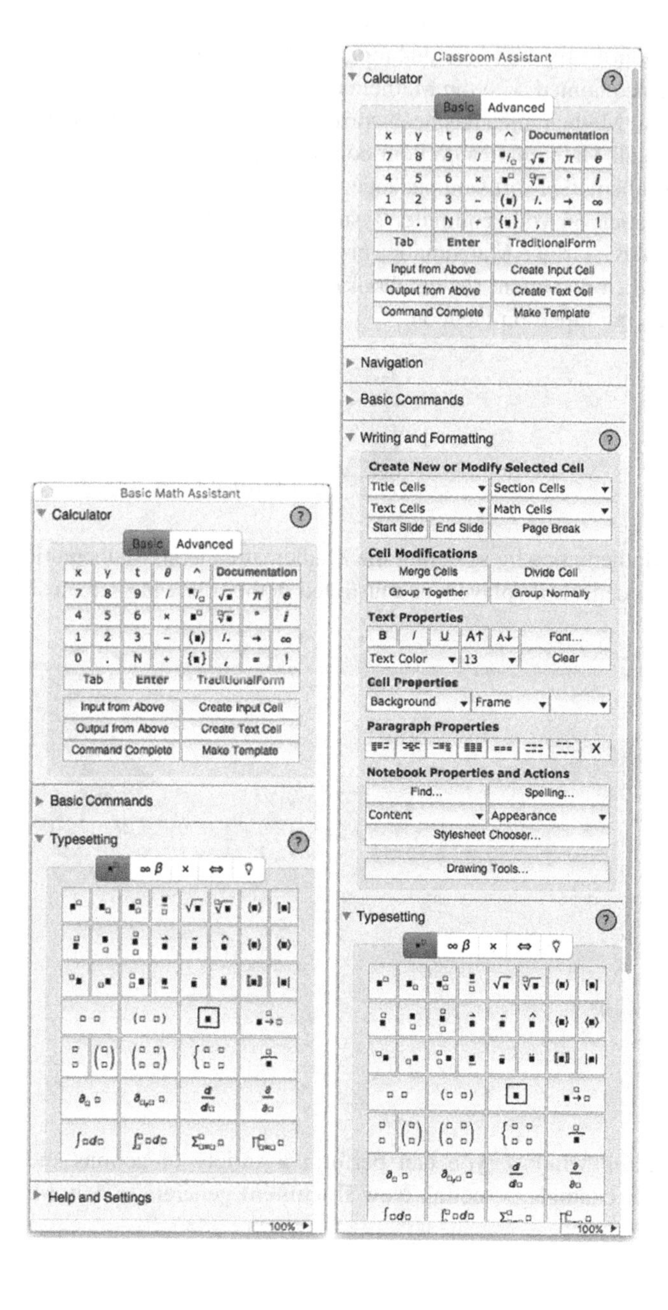

The standard Mathematica palettes are summarized in Fig. 1.5.

If you go further into the submenu and select **Other...**, you will find the **Algebraic Manipulation** palette, a slightly different **Basic Math Input** palette from that mentioned above, and the **Basic Typesetting** palette.

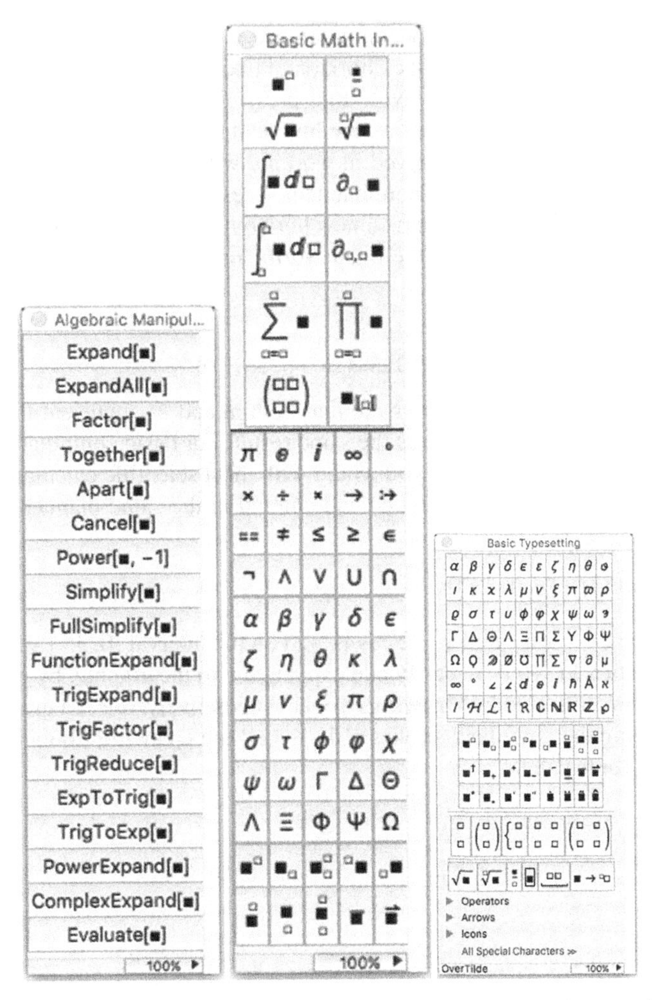

When you start typing in the new notebook created above, the thin black horizontal line near the top of the window is replaced by what you type.

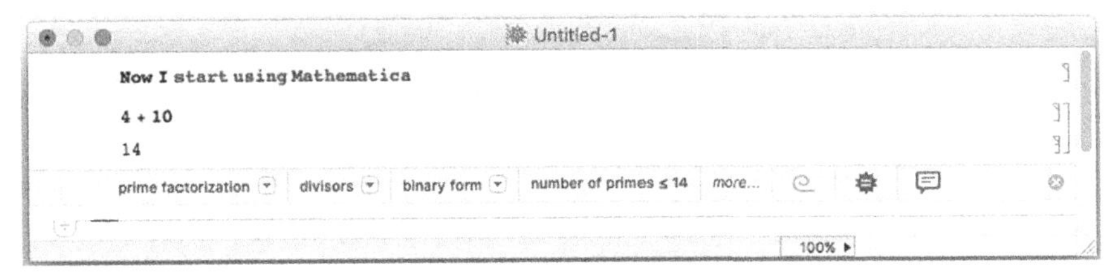

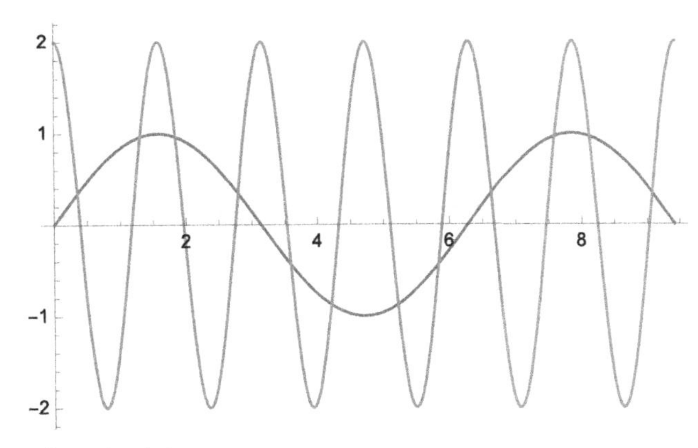

FIGURE 1.1 A two-dimensional plot.

With some operating systems, **Return** evaluates commands and **Shift-Return** yields a new line.

The **Basic MathInput** palette:

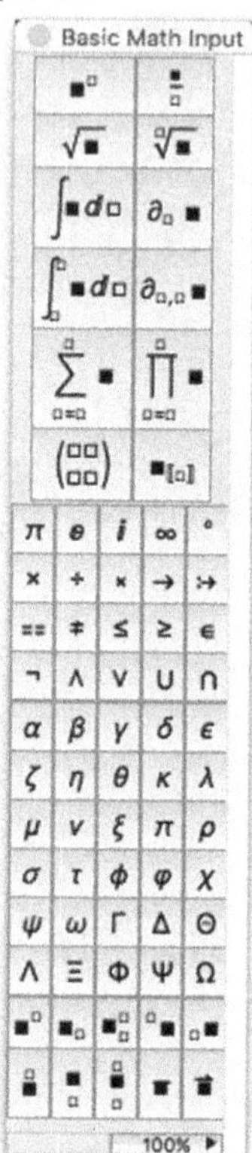

Once Mathematica has been started, computations can be carried out immediately. Mathematica commands are typed and the black horizontal line is replaced by the command, which is then evaluated by pressing **Enter**. Note that pressing **Enter** or **Shift-Return** evaluates commands and pressing **Return** yields a new line. Output is displayed below input. We illustrate some of the typical steps involved in working with Mathematica in the calculations that follow. In each case, we type the command and press **Enter**. Mathematica evaluates the command, displays the result, and inserts a new horizontal line after the result. For example, typing `N[`, then pressing the π key on the **Basic Math Input** palette, followed by typing `,50]` and pressing the enter key

```
N[π, 50]
3.1415926535897932384626433832795028841971693993751
```

returns a 50-digit approximation of π. Note that both π and `Pi` represent the mathematical constant π so entering `N[Pi,50]` returns the same result. For basic computations, enter them into Mathematica in the same way as you would with most scientific calculators.

The next calculation can then be typed and entered in the same manner as the first. For example, entering

Plot[{Sin[x], 2Cos[4x]}, {x, 0, 3π}]

graphs the functions $y = \sin x$ and $y = 2\cos 4x$ and on the interval $[0, 3\pi]$ shown in Fig. 1.1.

With Mathematica 11, you can easily add explanation to the graphic, by going to **Graphics** in the main menu, followed by **Drawings Tools**. Alternatively, select a graphic by clicking on it and then typing the command strokes ctrl-t to call the **Drawing Tools** palette. You can use the **Drawing Tools** palette

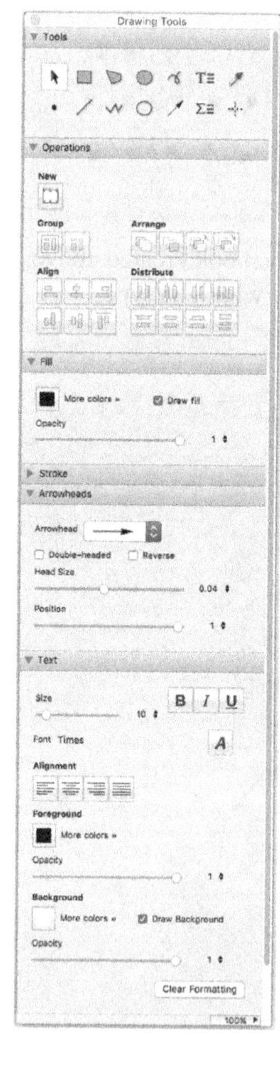

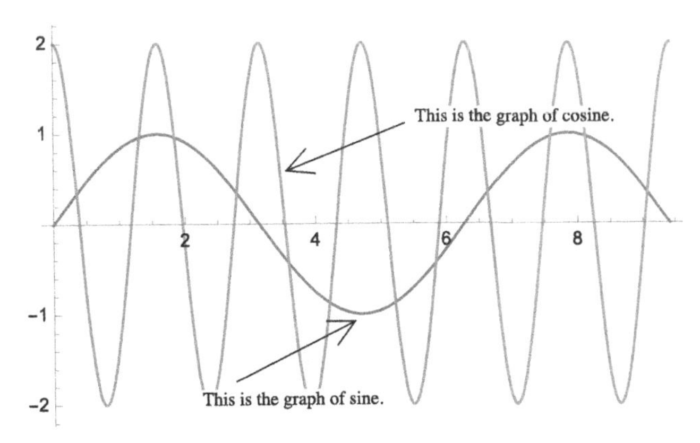

FIGURE 1.2 Using **Drawing Tools** to enhance a graphic.

to quickly enhance a graphic. In this case we use the arrow button and "T" (text) button twice to identify each curve shown in Fig. 1.2.

The various elements can be modified by clicking on them and moving and/or typing as needed. In particular, notice the "cross-hair" button in the second row.

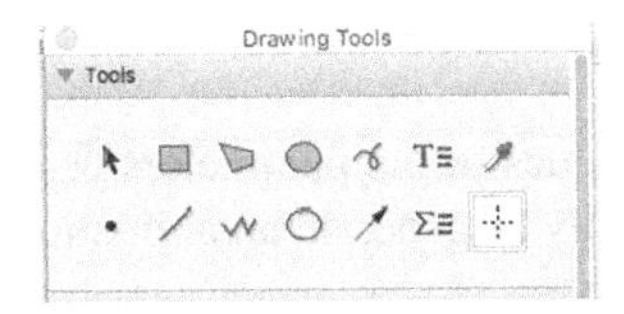

Use this button to identify coordinates in a plot. After selecting the graphic, selecting the button, then moving the cursor within the graphic will show you the coordinates.

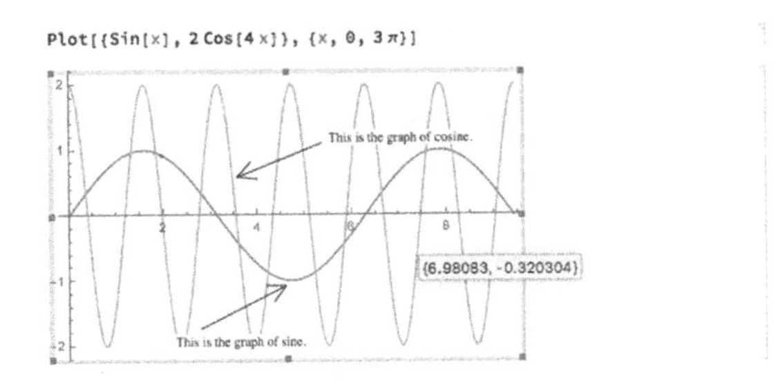

With Mathematica 11, you can use `Manipulate` to illustrate how changing various parameters affect a given function or functions. With the following command we illustrate how a and b affect the period of sine and cosine and c affects the amplitude of the cosine function.

```
Manipulate[Plot[{Sin[2Pi/ax], cCos[2Pi/bx]}, {x, 0, 4π}, PlotRange → {−4π/2, 4π/2},
  AspectRatio → 1], {{a, 2Pi, "Period for Sine"}, .1, 4Pi},
  {{b, 2Pi, "Period for Cosine"}, .1, 4Pi},
  {{d, 1, "Amplitude for Sine"}, .1, 5}, {{c, 1, "Amplitude for Cosine"}, .1, 5}]
```

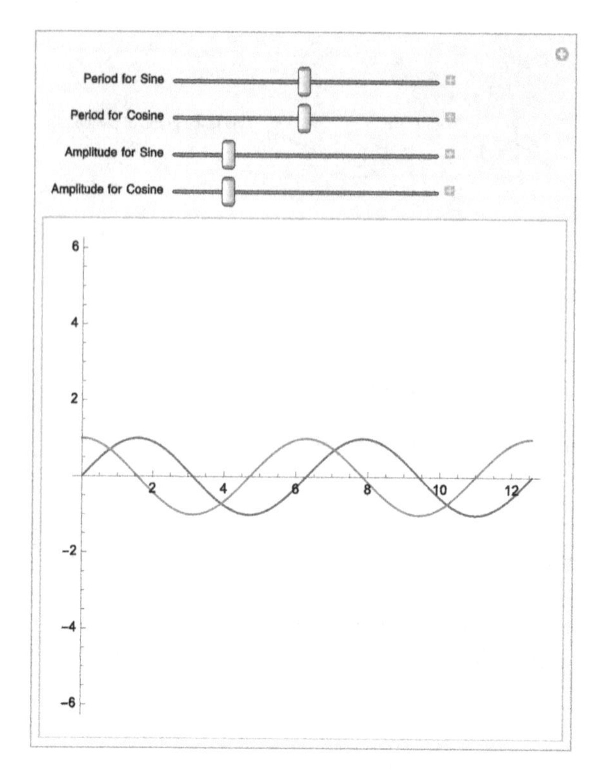

Use the slide bars to adjust the values of the parameters or click on the + button to expand the options to enter values explicitly or generate an animation to illustrate how changing the parameter values changes the problem.

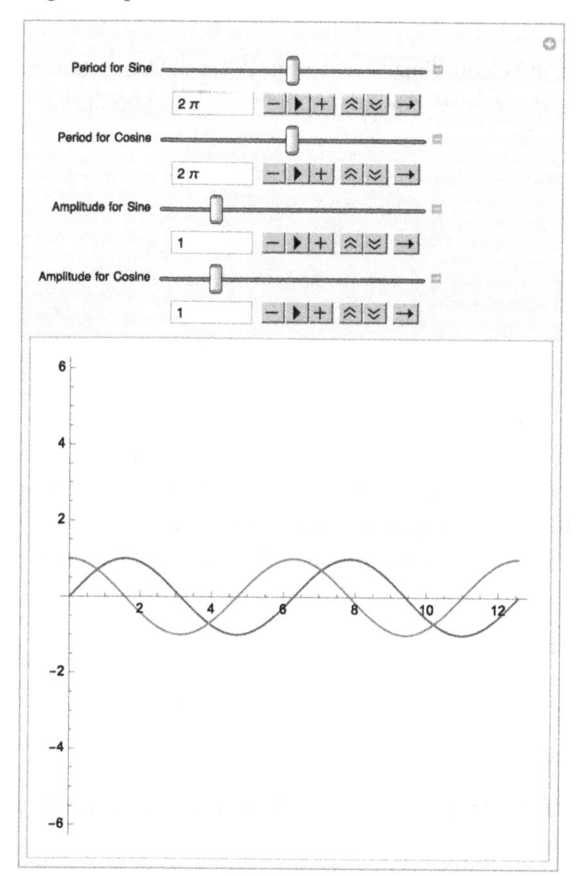

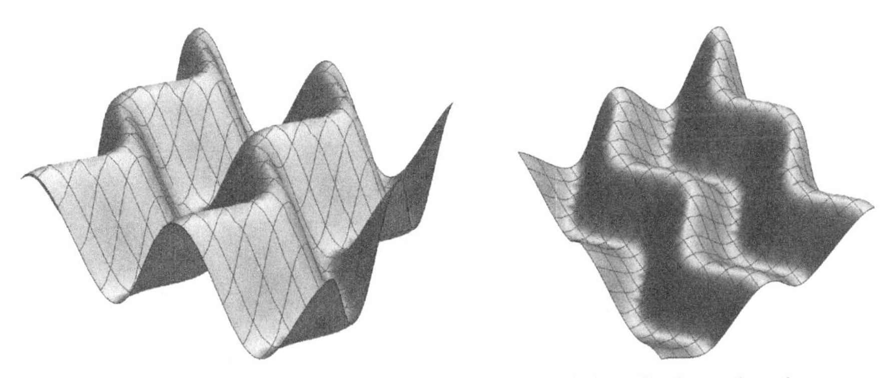

FIGURE 1.3 On the left, a three-dimensional plot. On the right, select the bounding box and use the cursor to move the graphic to the desired perspective.

Use `Plot3D` to generate basic three-dimensional plots. Entering

Plot3D[Sin[x + Cos[y]], {x, 0, 4π}, {y, 0, 4π}, Ticks → None, Boxed → False,

Axes → None]

graphs the function $z = \sin(x + \cos(y))$ for $0 \le x \le 4\pi, 0 \le y \le 4\pi$. Later we will learn other ways of changing the viewing perspective. For now, we note that by selecting the graphic, you can often use the cursor to move the graphic to the desired perspective or viewing angle as illustrated in Fig. 1.3.

Notice that every Mathematica command begins with capital letters and the argument is enclosed by square brackets `[...]`.

To print three-dimensional objects with your 3D printer or a 3D printing service, you need to generate an STL file. To create an STL file, an object must be *orientable*. This basically means that a three-dimensional object has an inside and an outside. Objects like the Mobius strip, Klein bottle, and the projective plane are not orientable so printing likenesses of them can be challenging. On the other hand, objects like spheres, toruses, and so on are orientable.

In the case of the previous plot, it has a top and bottom but neither an inside nor an outside. With Mathematica 11, provided that an object is orientable, you can use Mathematica code to generate an STL object and print it on either your own 3-dimensional printer or have it printed by one of the many 3-dimensional printing services.

For the previous example, there are multiple ways of proceeding. We also illustrate how to use Mathematica's extensive help facilities. From the **Welcome Screen**, select **Documentation**

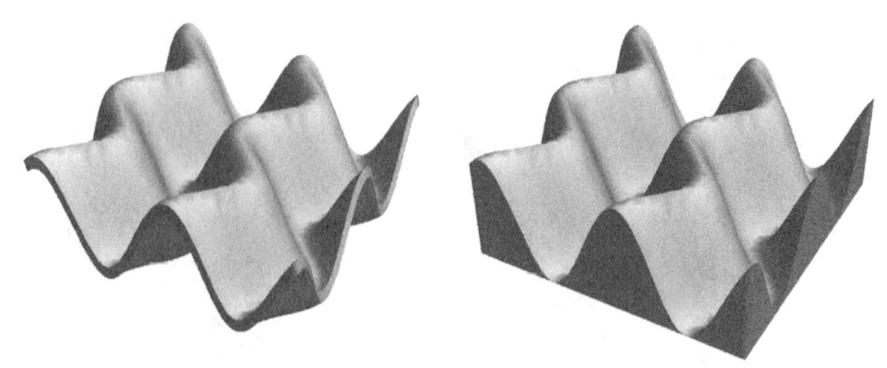

FIGURE 1.4 On the left, a "thickened" three-dimensional plot. On the right, a "filled" three-dimensional plot.

and then at the bottom of the screen select **New Features**. Scroll down to the area labeled **Geometry & 3D Printing**.

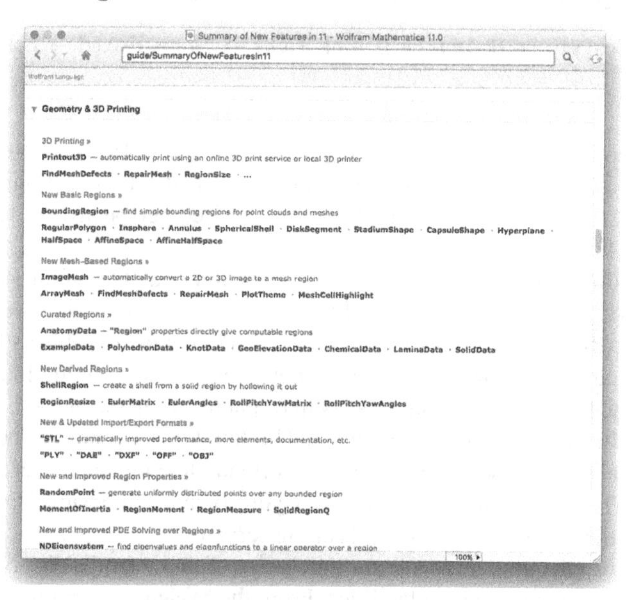

We see that there are two `Plot3D` options, `ThickSurface` and `FilledSurface` that will be able to generate STL files. Both approaches are illustrated as follows and illustrated in Fig. 1.4.

p1 = Plot3D[Sin[x + Cos[y]], {x, 0, 4π}, {y, 0, 4π}, Ticks → None, Boxed → False,

Axes → None, PlotTheme → "ThickSurface"]

p2 = Plot3D[Sin[x + Cos[y]], {x, 0, 4π}, {y, 0, 4π}, Ticks → None, Boxed → False,

Axes → None, PlotTheme → "FilledSurface"]

Show[GraphicsRow[{p1, p2}]]

We can now save the results as an STL file, print the result to our 3D printer, or print to a 3D printing service.
Entering

Printout3D[p1, "p1.stl"]

saves `p1` to an STL file.

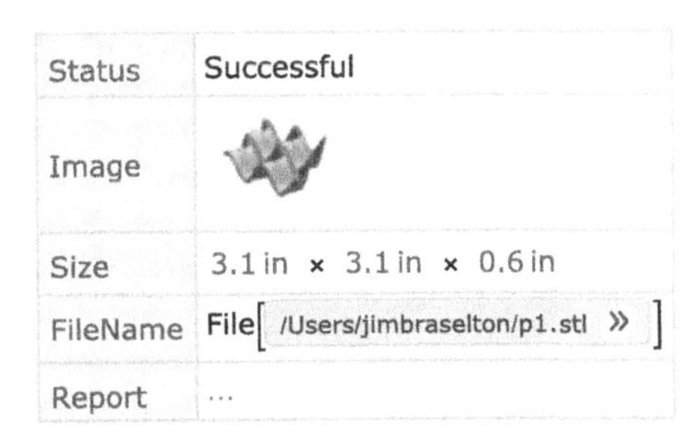

Status	Successful
Image	
Size	3.1 in × 3.1 in × 0.6 in
FileName	File[/Users/jimbraselton/p1.stl »]
Report	···

On the other hand, the following command sends the result directly to Sculpteo.

Printout3D[p2, "Sculpteo"];

Your browser window will open and you can adjust the image to your satisfaction before ordering (or not). Many printing services are supported. You can also use this command to print directly to your own 3D printer.

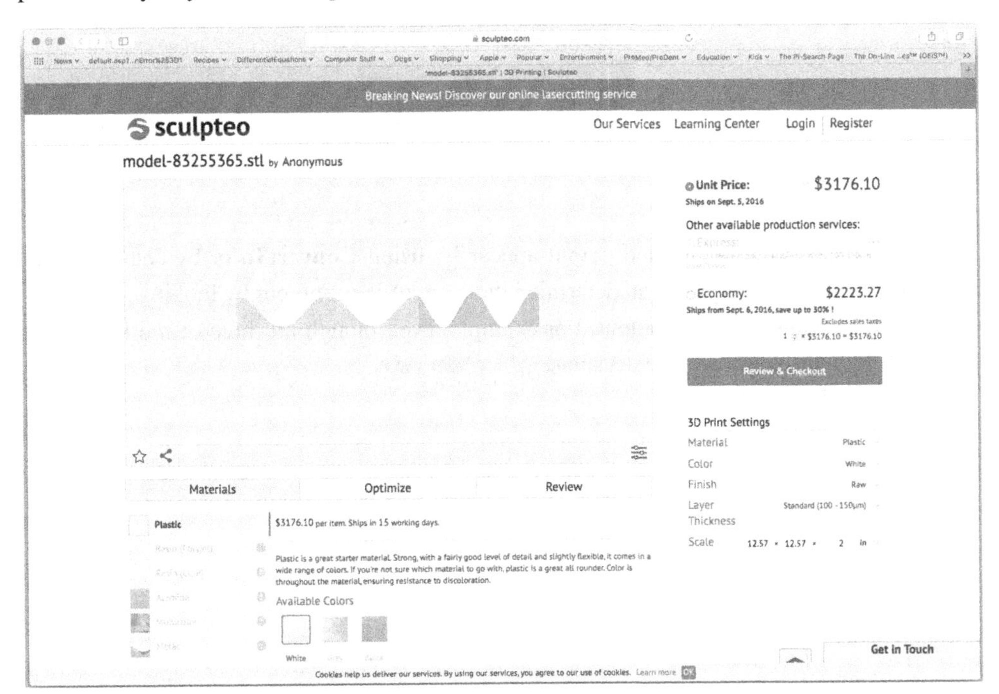

Notice that all three of the following commands

```
Solve[x^3 - 2 x + 1 == 0]
```

$$\{\{x\to 1\}, \{x\to \frac{1}{2}(-1-\sqrt{5})\}, \{x\to \frac{1}{2}(-1+\sqrt{5})\}\}$$

```
Solve[x³ - 2 x + 1 == 0]
```

$$\{\{x\to 1\}, \{x\to \frac{1}{2}(-1-\sqrt{5})\}, \{x\to \frac{1}{2}(-1+\sqrt{5})\}\}$$

$$\text{Solve}[x^3 - 2x + 1 == 0]$$

$$\{\{x\to 1\}, \{x\to \frac{1}{2}(-1-\sqrt{5})\}, \{x\to \frac{1}{2}(-1+\sqrt{5})\}\}$$

solve the equation $x^3 - 3x + 1 = 0$ for x.

In the first case, the input and output are in **StandardForm**, in the second case the input and output are in **InputForm**, and in the third case, the input and output are in **TraditionalForm**.

To type x^3 in Mathematica, press the ▪ on the **Basic Math Input** palette, type x in the base position, and then click (or tab to) the exponent position and type 3. Use the **esc** key, tab button, or mouse to help you place or remove the cursor from its current location.

To convert cells from one type to another, first select the cell, and then move the cursor to the Mathematica menu,

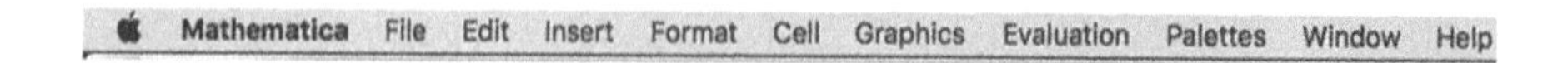

select **Cell**, and then **Convert To**, as illustrated in the following screen shot.

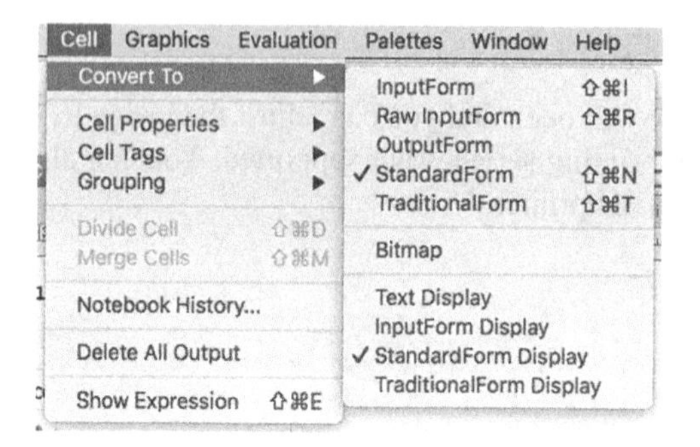

You can change how input and output appear by using **ConvertTo** or by changing the default settings. Moreover, you can determine the form of input/output by looking at the cell bracket that contains the input/output. For example, even though all three of the following commands look different, all three evaluate $\int_0^{2\pi} x^3 \sin x \, dx$.

```
Integrate[x^3 Sin[x], {x, 0, 2 Pi}]
```

$12\pi - 8\pi^3$

$\int_0^{2\pi} x^3 \text{Sin}[x] \, dx$

$12\pi - 8\pi^3$

$\int_0^{2\pi} x^3 \sin(x) \, dx$

$12\pi - 8\pi^3$

In the first calculation, the input is in **Input Form** and the output in **Output Form**, in the second the input and output are in **Standard Form**, and in the third the input and output are in **TraditionalForm**. Throughout *Mathematica By Example*, fifth edition, we display input and output using **Input Form** (for input) or **Standard Form** (for output), unless otherwise stated.

To enter code in **Standard Form**, we often take advantage of the **Basic Math Input** palette, which is accessed by going to **Palettes** under the Mathematica menu and then selecting **Basic Math Input**.

Use the buttons to create templates and enter special characters. Alternatively, you can access a complete list of typesetting shortcuts from Mathematica help at `guide/MathematicalTypesetting` in the **Documentation Center**.

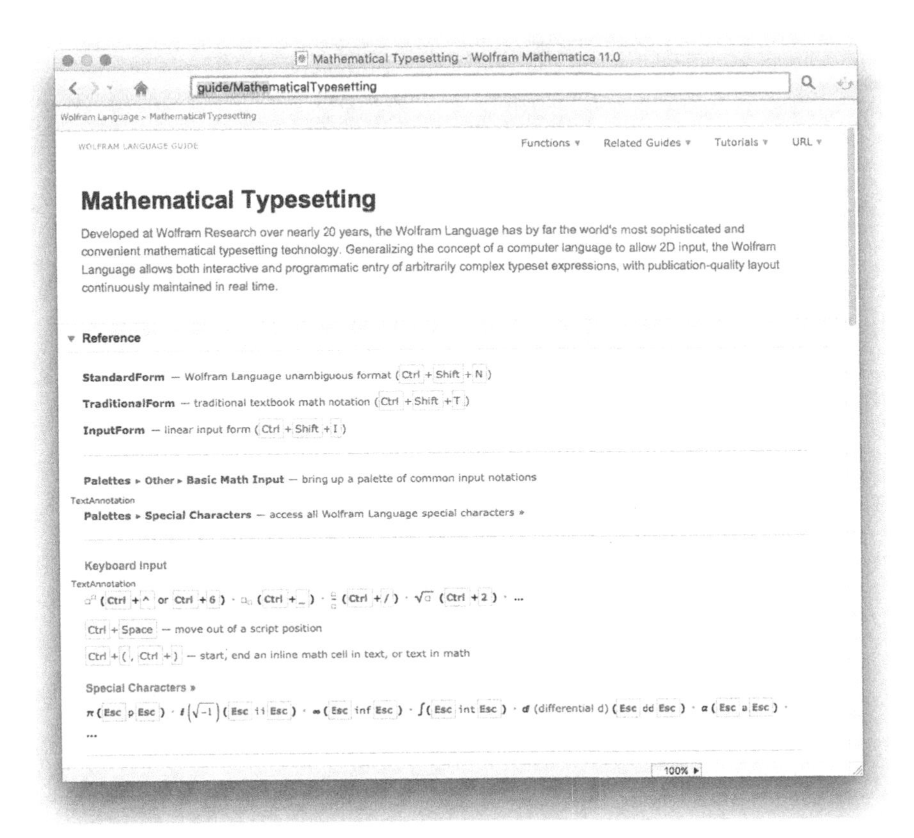

Mathematica sessions are terminated by entering `Quit[]` or by selecting **Quit** from the **File** menu, or by using a keyboard shortcut, like **command-Q**, as with other applications. They can be saved by referring to **Save** from the **File** menu.

Mathematica allows you to save notebooks (as well as combinations of cells) in a variety of formats, in addition to the standard Mathematica format. From the Mathematica menu, select **Save As...** and then select one of the following options.

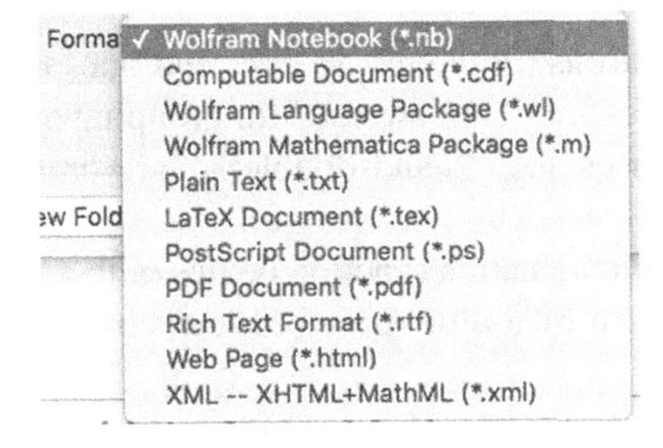

Remark 1.1. Input and text regions in notebooks can be edited. Editing input can create a notebook in which the mathematical output does not make sense in the sequence it appears. It is also possible to simply go into a notebook and alter input without doing any recalculation. This also creates misleading notebooks. Hence, common sense and caution should be used when editing the input regions of notebooks. Recalculating all commands in the notebook will clarify any confusion.

Preview

In order for the Mathematica user to take full advantage of this powerful software, an understanding of its syntax is imperative. The goal of *Mathematica By Example* is to introduce the reader to the Mathematica commands and sequences of commands most frequently used by beginning users of Mathematica. Although all of the rules of Mathematica syntax are far

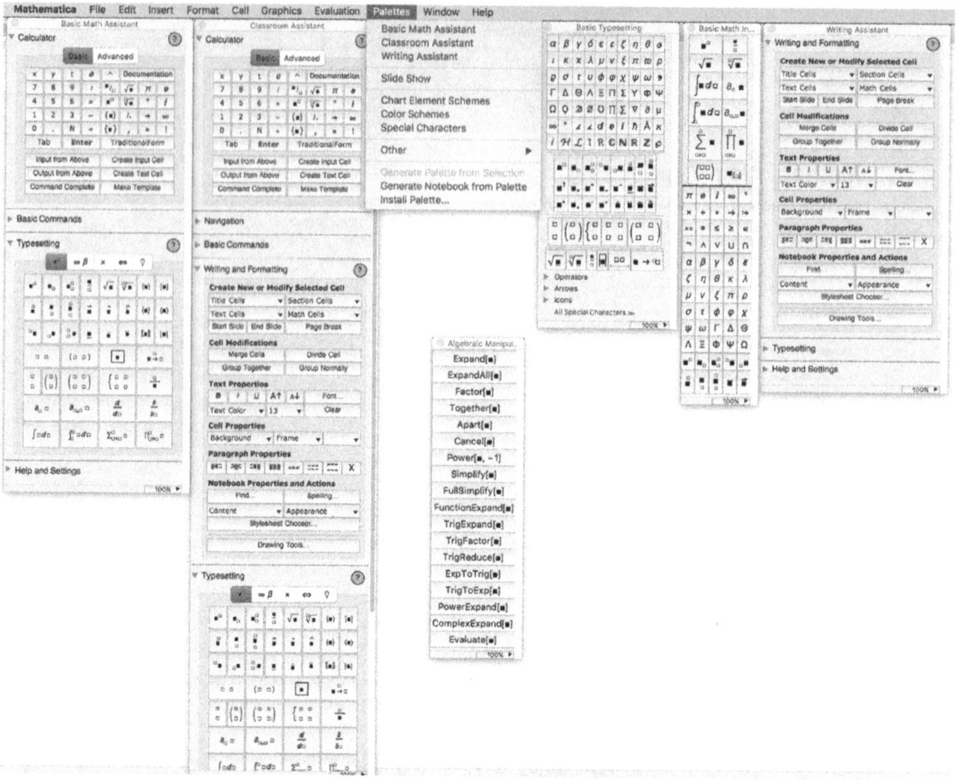

FIGURE 1.5 The standard Mathematica palettes.

too numerous to list here, knowledge of the following five rules equips the beginner with the necessary tools to start using the Mathematica program with little trouble.

Five Basic Rules of Mathematica Syntax

1. The arguments of *all* functions (both built-in ones and ones that you define) are given in brackets `[...]`. Parentheses are `(...)` are used for grouping operations; vectors, matrices, and lists are given in braces `{...}`; and double square brackets `[[...]]` are used for indexing lists and tables.
2. Every word of a built-in Mathematica function begins with a capital letter.
3. Multiplication is represented by a $*$ or space between characters. Enter `2*x*y` or `2x y` to evaluate $2xy$ *not* `2xy`.
4. Powers are denoted by a ^. Enter `(8*x^3)^(1/3)` to evaluate $(8x^3)^{1/3} = 8^{1/3}(x^3)^{1/3} = 2x$ instead of `8x^1/3`, which returns `8x/3`.
5. Mathematica follows the order of operations *exactly*. Thus, entering `(1+x)^1/x` returns $\dfrac{(1+x)^1}{x} = 1 + 1/x$ while `(1+x)^(1/x)` returns $(1+x)^{1/x}$. Similarly, entering `x^3x` returns $x^3 \cdot x = x^4$ while entering `x^(3x)` returns x^{3x}.

   Remark 1.2. If you get no response or an incorrect response, you may have entered or executed the command incorrectly. In some cases, the amount of memory allocated to Mathematica can cause a crash. Like people, Mathematica is not perfect and errors can occur.

6. Many calculations and tasks encountered by beginning users can be completed by filling in templates that are provided in the various palettes and accessed from the Mathematica menu. The standard Mathematica palettes are shown in Fig. 1.5.

1.2 GETTING HELP FROM MATHEMATICA

Becoming competent with Mathematica can take a serious investment of time. Hopefully, messages that result from syntax errors are viewed lightheartedly. Ideally, instead of becoming frustrated, beginning Mathematica users will find it challenging and fun to locate the source of errors. Frequently, Mathematica's error messages indicate where the error(s) has (have) occurred. In this process, it is natural that you will become more proficient with Mathematica. In addition to Mathematica's extensive help facilities, which are described next, a tremendous amount of information is available for all Mathematica users at the Wolfram Research website.

`http://www.wolfram.com/`

Not only can you get significant Mathematica help at the Wolfram website, you can also access outstanding *mathematical* resources at Wolfram's Mathematica resources that are accessed from the Welcome Screen followed by selecting **Resources**.

The exact URL for this address will vary depending upon your license number and resources that your license provides.

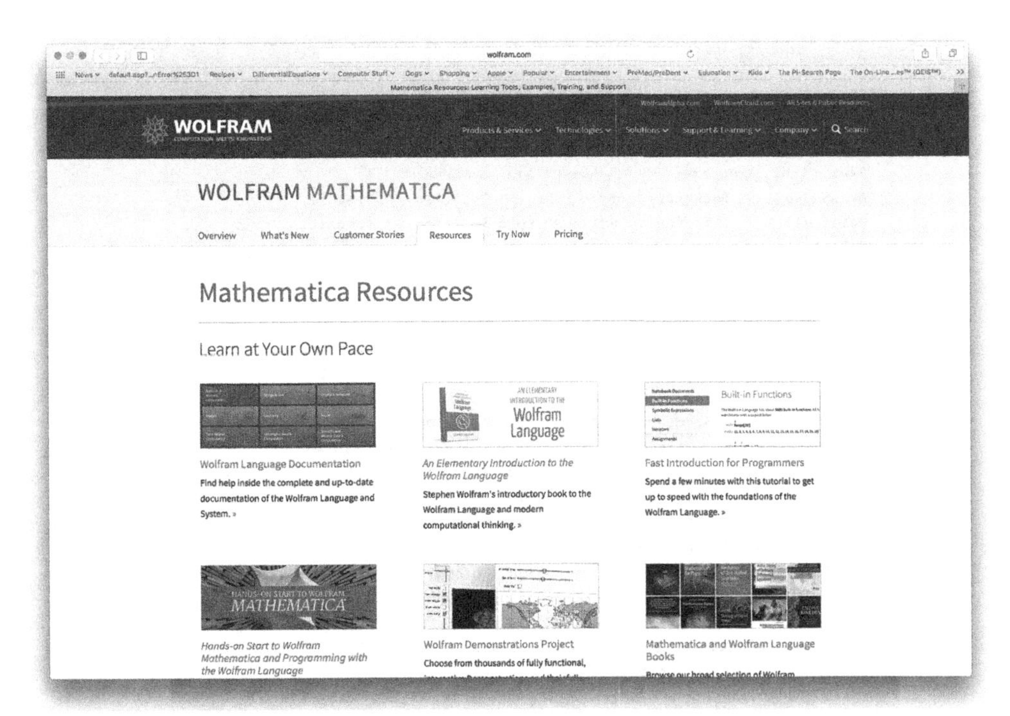

One way to obtain information about Mathematica commands and functions, including user-defined functions, is the command `?`. `?object` gives a basic description and syntax information of the Mathematica object `object`. `??object` yields detailed information regarding syntax and options for the object `object`. Equivalently, `Information[object]` yields the information on the Mathematica object `object` returned by both `?object` and `Options[object]` in addition to a list of attributes of object. Note that `object` may either be a user-defined object or a built-in Mathematica object, such as a built-in function or sequence of commands.

Example 1.1

Use ? and ?? to obtain information about the command `Plot`.

Solution. `?Plot` uses basic information about the `Plot` function

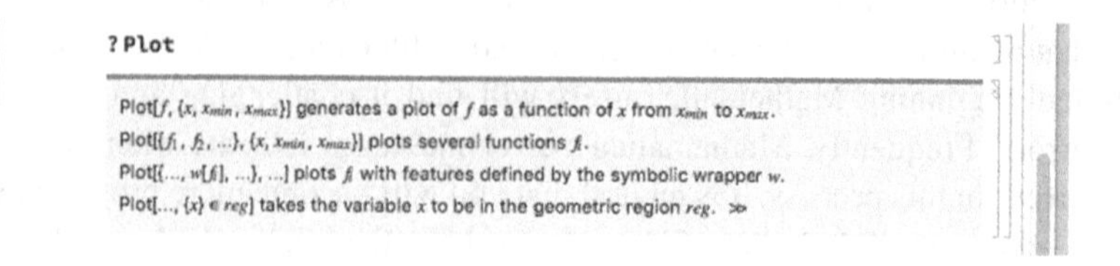

while `??Plot` includes basic information as well as a list of options and their default values.

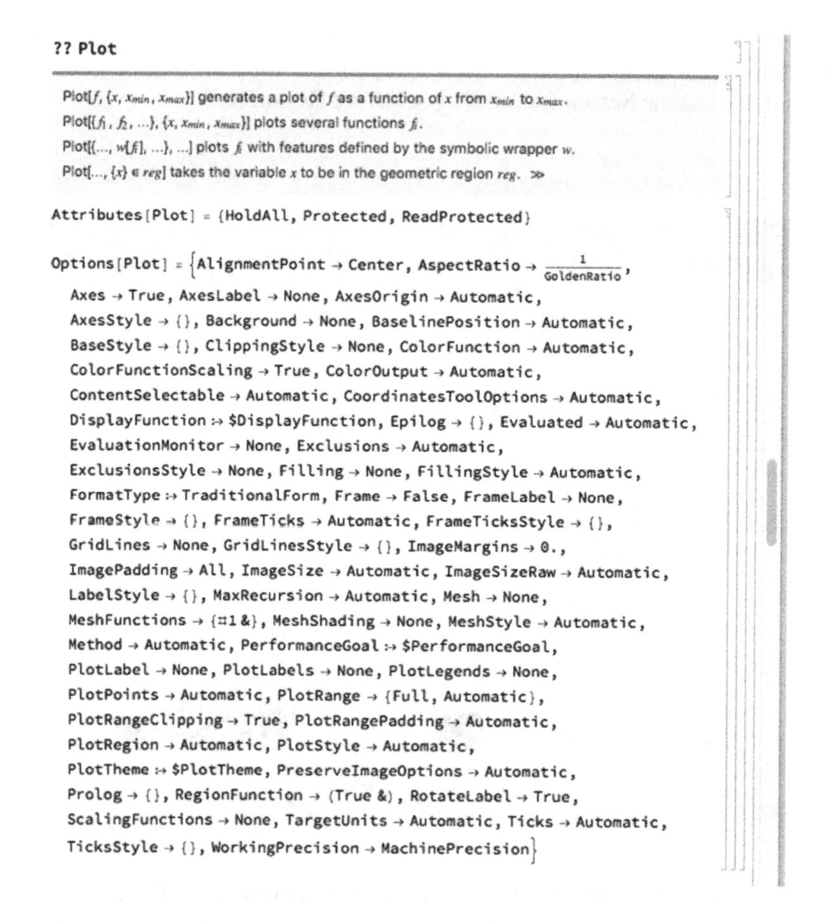

If you click on the >> button, Mathematica returns its extensive description of the function. Notice that the `Plot` function has been updated in Version 11. The

button shows that `Plot` has been updated. Click on **Show Changes** to see the changes in Version 11.

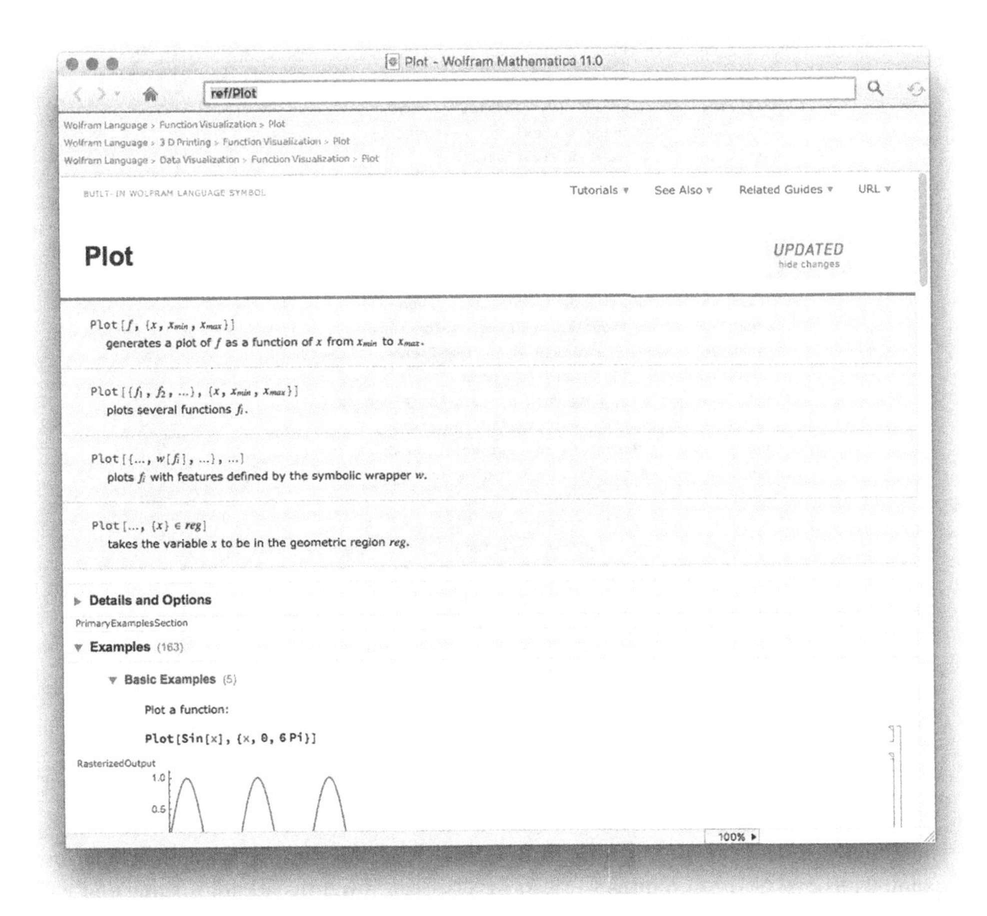

□

`Options[object]` returns a list of the available options associated with object along with their current settings. This is quite useful when working with a Mathematica command such as `ParametricPlot` which has many options. Notice that the default value (the value automatically assumed by Mathematica) for each option is given in the output.

Example 1.2

Use `Options` to obtain a list of the options and their current settings for the command `ParametricPlot`.

Solution. The command `Options[ParametricPlot]` lists all the options and their current settings for the command `ParametricPlot`.

```
Options[ParametricPlot]

{AlignmentPoint → Center, AspectRatio → Automatic, Axes → True,
 AxesLabel → None, AxesOrigin → Automatic, AxesStyle → {},
 Background → None, BaselinePosition → Automatic,
 BaseStyle → {}, BoundaryStyle → Automatic,
 ColorFunction → Automatic, ColorFunctionScaling → True,
 ColorOutput → Automatic, ContentSelectable → Automatic,
 CoordinatesToolOptions → Automatic,
 DisplayFunction :→ $DisplayFunction, Epilog → {},
 Evaluated → Automatic, EvaluationMonitor → None,
 Exclusions → Automatic, ExclusionsStyle → None,
 FormatType :→ TraditionalForm, Frame → Automatic,
 FrameLabel → None, FrameStyle → {}, FrameTicks → Automatic,
 FrameTicksStyle → {}, GridLines → None,
 GridLinesStyle → {}, ImageMargins → 0., ImagePadding → All,
 ImageSize → Automatic, ImageSizeRaw → Automatic,
 LabelStyle → {}, MaxRecursion → Automatic, Mesh → Automatic,
 MeshFunctions → Automatic, MeshShading → None,
 MeshStyle → Automatic, Method → Automatic,
 PerformanceGoal :→ $PerformanceGoal, PlotLabel → None,
 PlotLegends → None, PlotPoints → Automatic, PlotRange → Automatic,
 PlotRangeClipping → True, PlotRangePadding → Automatic,
 PlotRegion → Automatic, PlotStyle → Automatic,
 PlotTheme :→ $PlotTheme, PreserveImageOptions → Automatic,
 Prolog → {}, RegionFunction → (True &), RotateLabel → True,
 TargetUnits → Automatic, TextureCoordinateFunction → Automatic,
 TextureCoordinateScaling → Automatic, Ticks → Automatic,
 TicksStyle → {}, WorkingPrecision → MachinePrecision}
```

□

The command `Names["form"]` lists all objects that match the pattern defined in `form`. For example, `Names["Plot"]` returns `Plot`, `Names["*Plot"]` returns all objects that end with the string `Plot`, and `Names["Plot*"]` lists all objects that begin with the string `Plot`, and `Names["*Plot*"]` lists all objects that contain the string `Plot`. `Names["form",SpellingCorrection->True]` finds those symbols that match the pattern defined in `form` after a spelling correction.

Example 1.3

Create a list of all built-in functions beginning with the string `Plot`.

Solution. We use `Names` to find all objects that match the pattern `Plot`.

```
Names["Plot"]

{Plot}
```

Next, we use `Names` to create a list of all built-in functions beginning with the string `Plot`.

```
Names["Plot*"]

{Plot, Plot3D, Plot3Matrix, PlotDivision, PlotJoined, PlotLabel,
 PlotLabels, PlotLayout, PlotLegends, PlotMarkers, PlotPoints,
 PlotRange, PlotRangeClipping, PlotRangeClipPlanesStyle,
 PlotRangePadding, PlotRegion, PlotStyle, PlotTheme}
```

□

In the following, after using `?` to learn about the Mathematica function `ColorData`, we go to the Mathematica menu and select **Palettes** followed by **Color Schemes**.

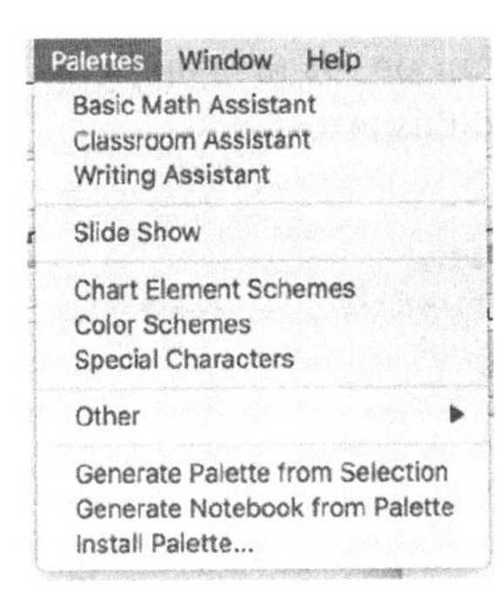

We are given a variety of choices. Using these choices is illustrated throughout *Mathematica By Example*.

Remember that on a computer running Mathematica or in the electronic version of this book, these graphics will appear in color rather than in black-and-white as seen in the printed version of this text.

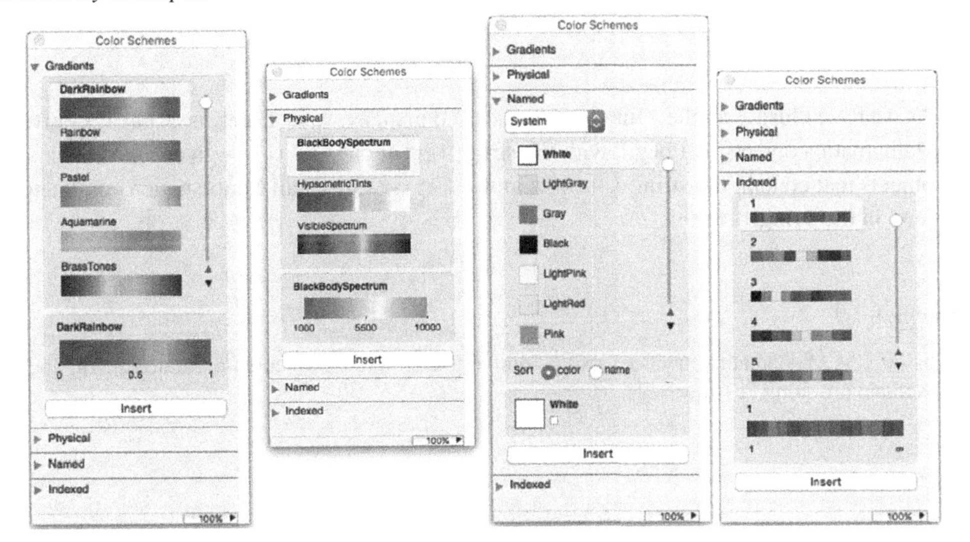

We then use the help facilities description of the **ColorData** function

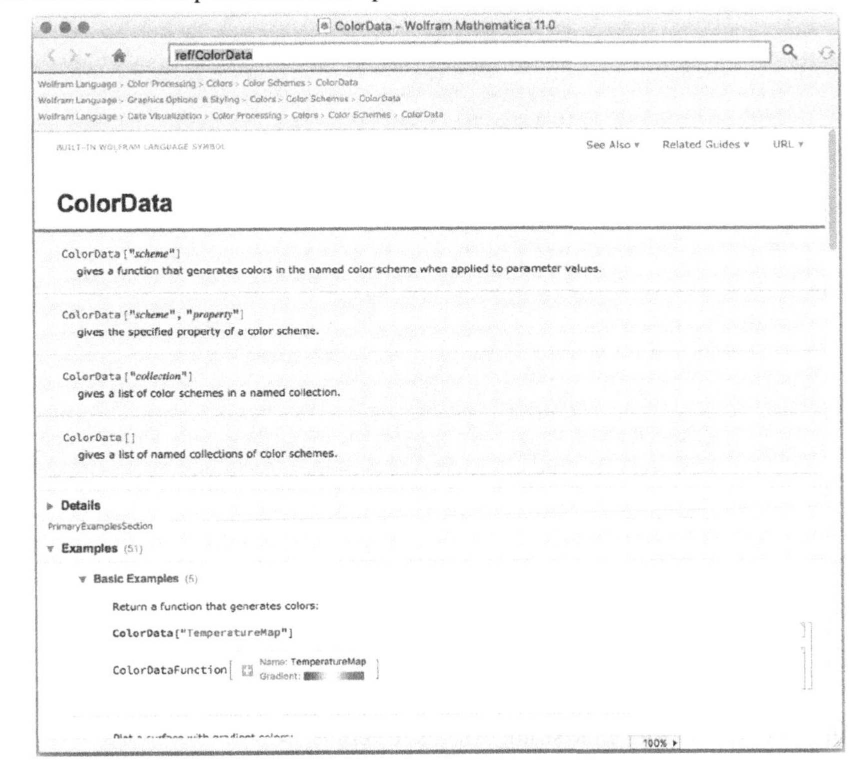

to help us generate a plot of $y = \sin x$ on the interval $[0, 2\pi]$ in deep red on our computer or that shown in the on-line version of this text.

Of course, the plot is dark gray in the black-and-white printed version of the text. We will illustrate other ways of controlling the color of graphics later in the text.

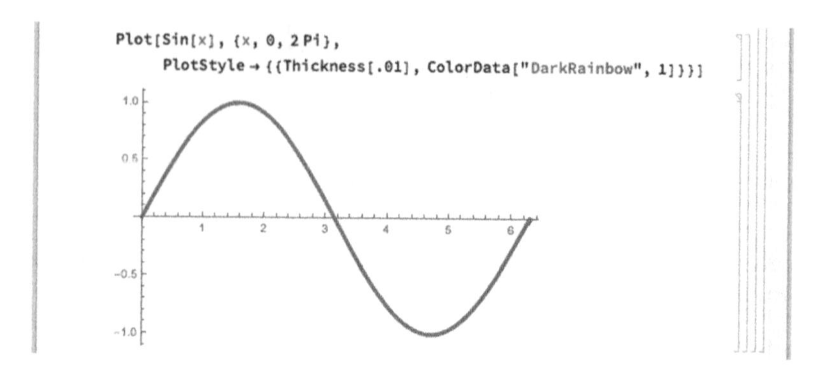

As we have illustrated, the `?` function can be used in many ways. Entering `?letters*` gives all Mathematica objects that begin with the string `letters`; `?*letters*` gives all Mathematica objects that contain the string `letters`; and `?*letters` gives all Mathematica commands that end in the string `letters`.

Example 1.4

What are the Mathematica functions that (a) end in the string `Cos`; (b) contain the string `Sin`; and (c) contain the string `Polynomial`?

Solution. Entering

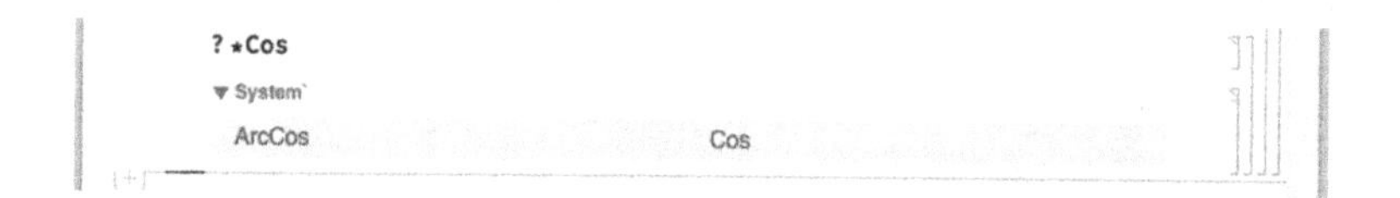

returns all functions ending with the string `Cos`, entering

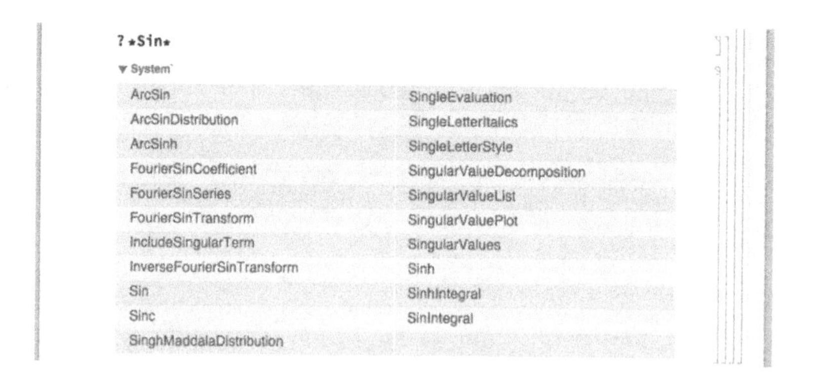

returns all functions containing the string `Sin`, and entering

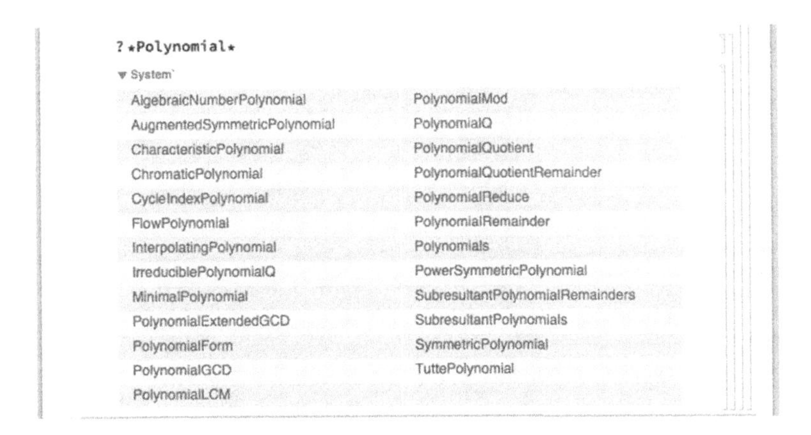

returns all functions that contain the string `Polynomial`. □

Mathematica Help

Additional help features are accessed from the Mathematica menu under **Help**. Although many of these topics have already been discussed, they are included again here for easy reference. For basic help information about Mathematica, go to the Mathematica menu, followed by **Help** and select **Wolfram Documentation**

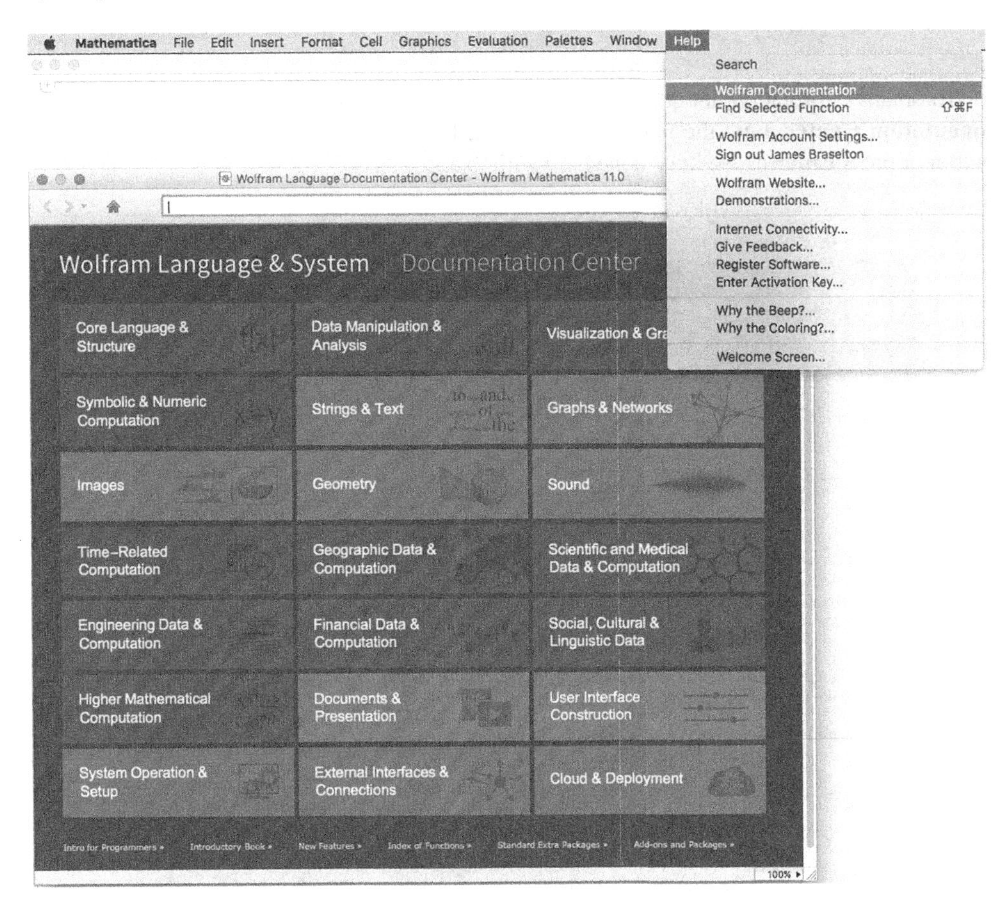

From this menu, you can choose a subject and then a sub-subject. For illustrative purposes, we select **Symbolic & Numeric Computation** followed by **Calculus**.

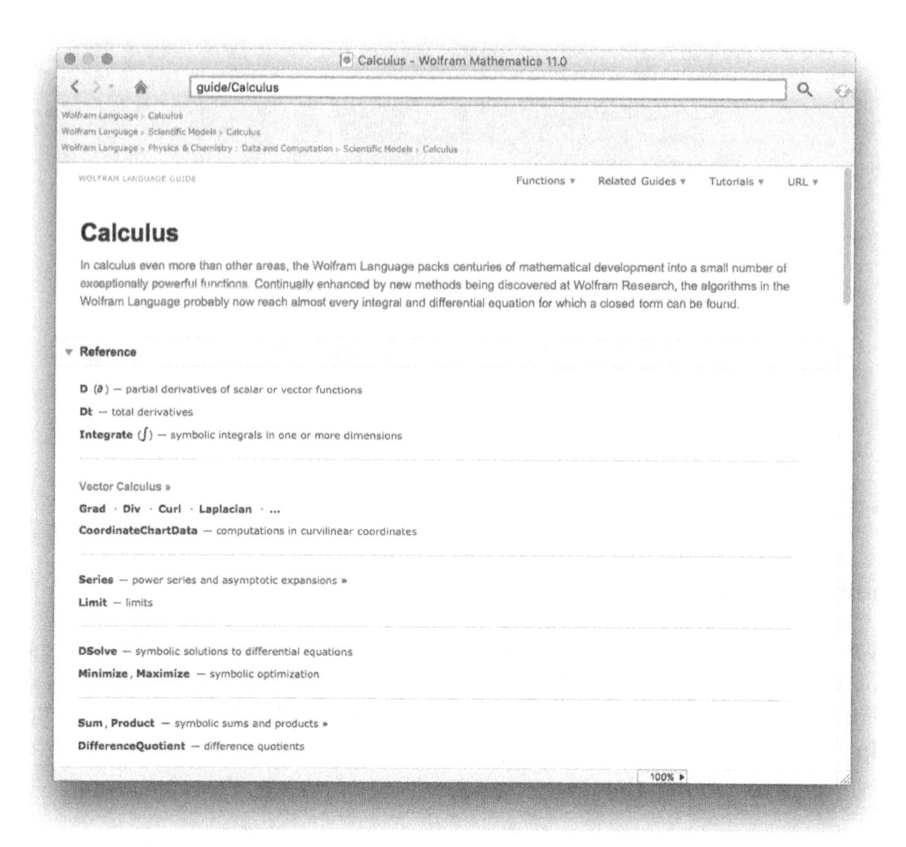

To obtain information about a particular Mathematica object or function, open the **Documentation Center**, type the name of the object, function, or topic and press the **Go** (>) button or press **Enter** as we have done here with `ExampleData`.

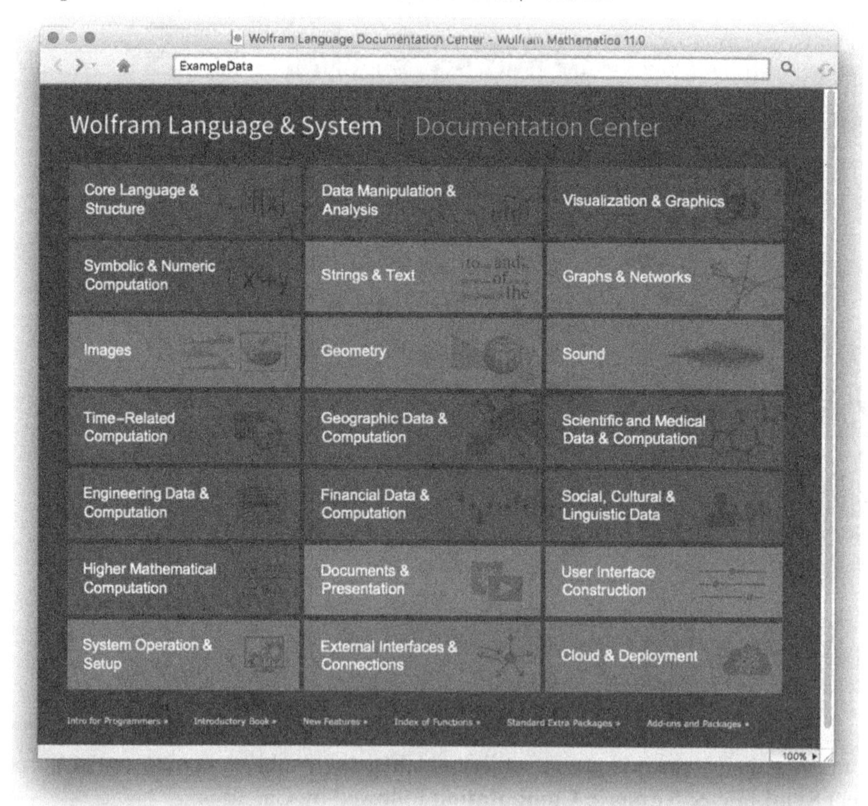

A typical help window not only contains a detailed description of the command and its options as well as hyperlinked cross-references to related commands and can be accessed by clicking on the appropriate links.

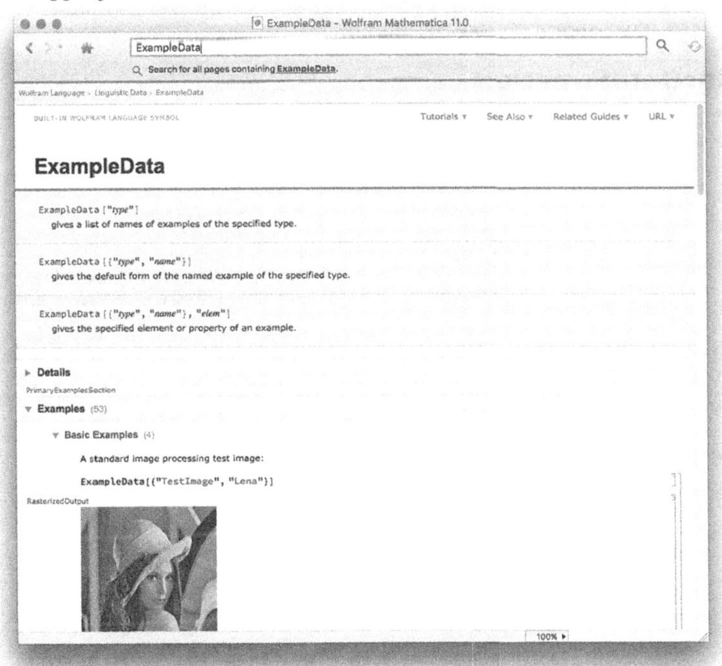

You can also use the **Documentation Center** to search for help regarding a particular topic. In this case we enter `color schemes` in the top line of the **Documentation Center** and then click on the > button (or press **Enter**) to see all the on-line help regarding "color schemes."

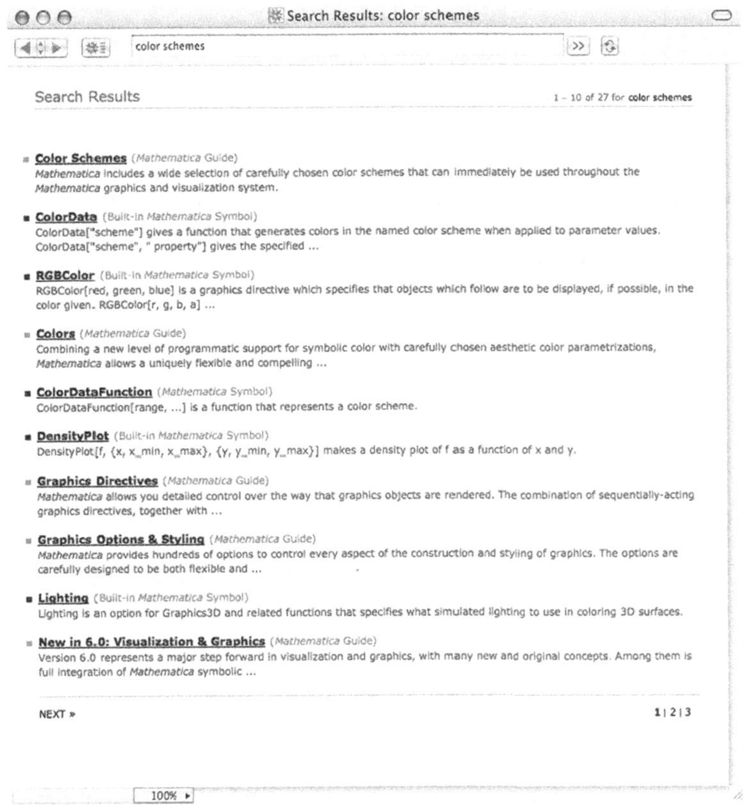

Clicking on the topic will take you to the documentation for the topic. Here is what we see when we select `Color Data`.

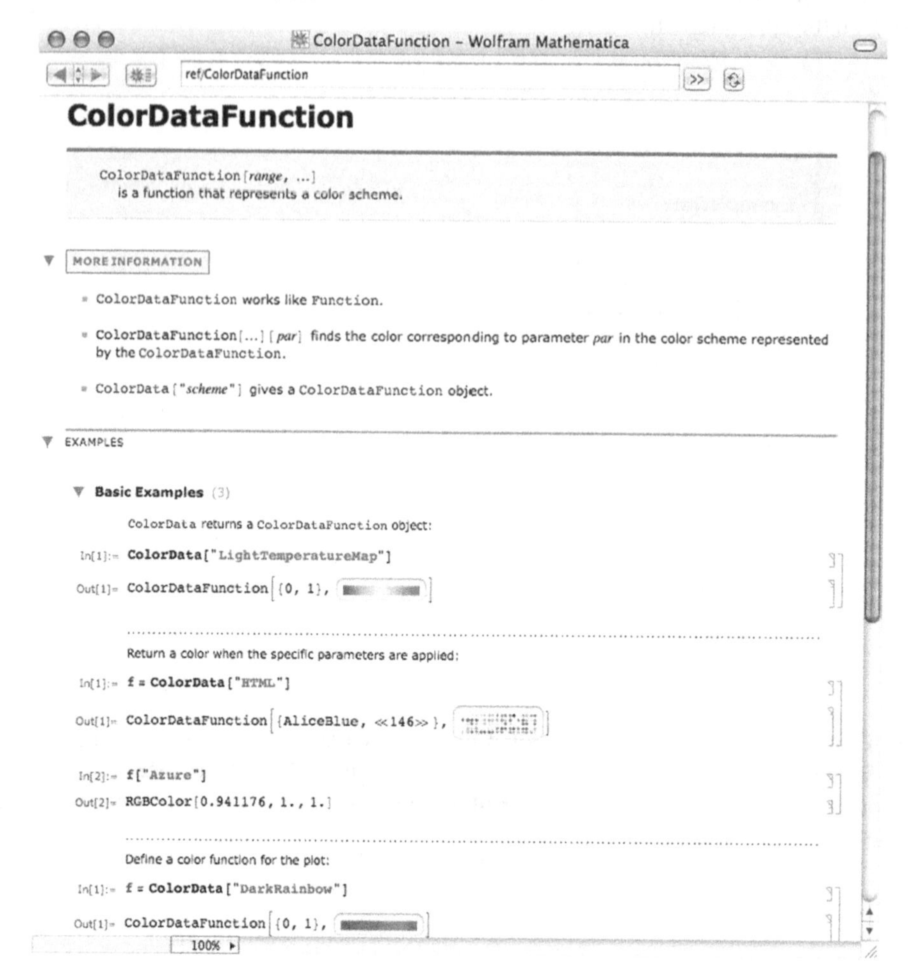

As you become more proficient with Mathematica, you will want to learn to take advantage of its extensive capabilities. Remember that Mathematica contains thousands of functions to perform many tasks. If you wish to perform a task that is not discussed here, go to the **Documentation Center** and type a few words related to what you want to do. For example, to investigate integer operations, we search for them.

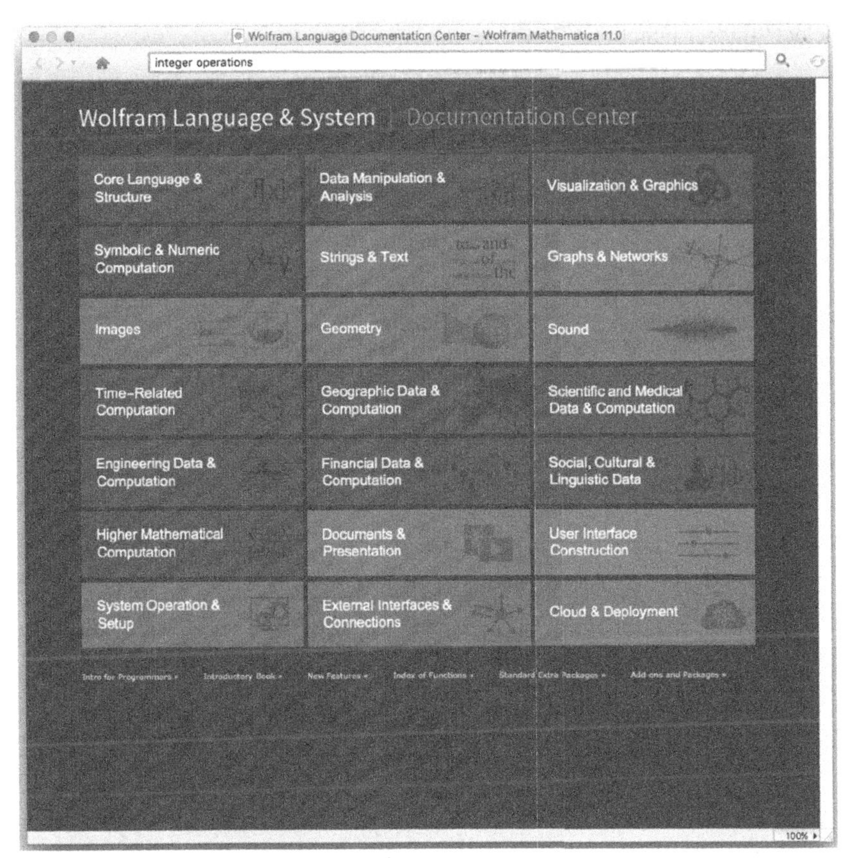

Here is the result of the search.

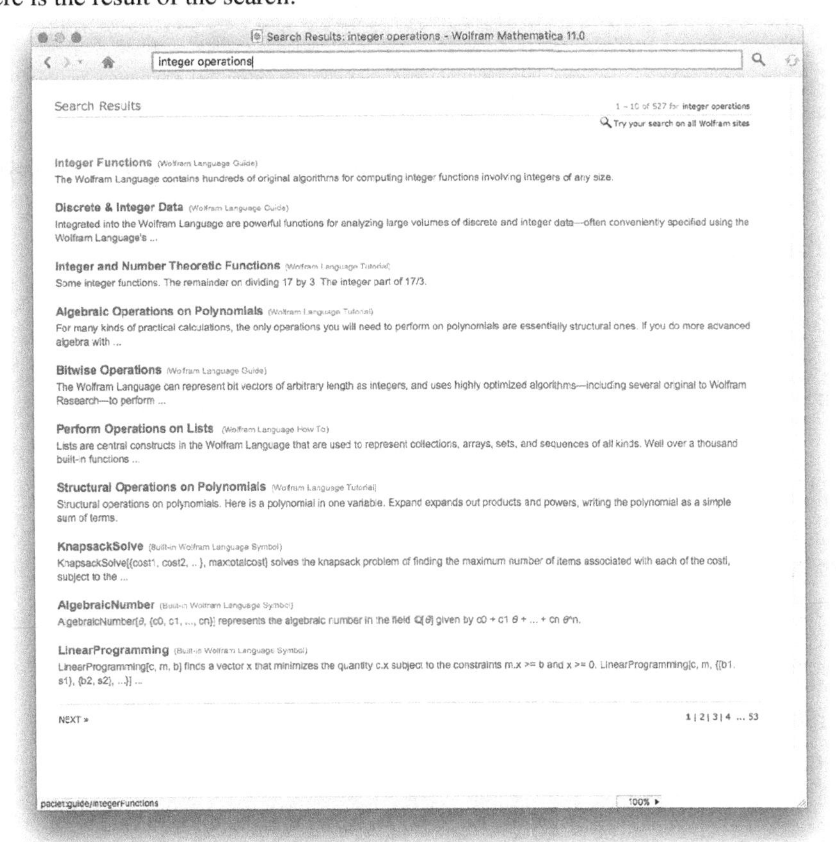

We select "Integer Functions."

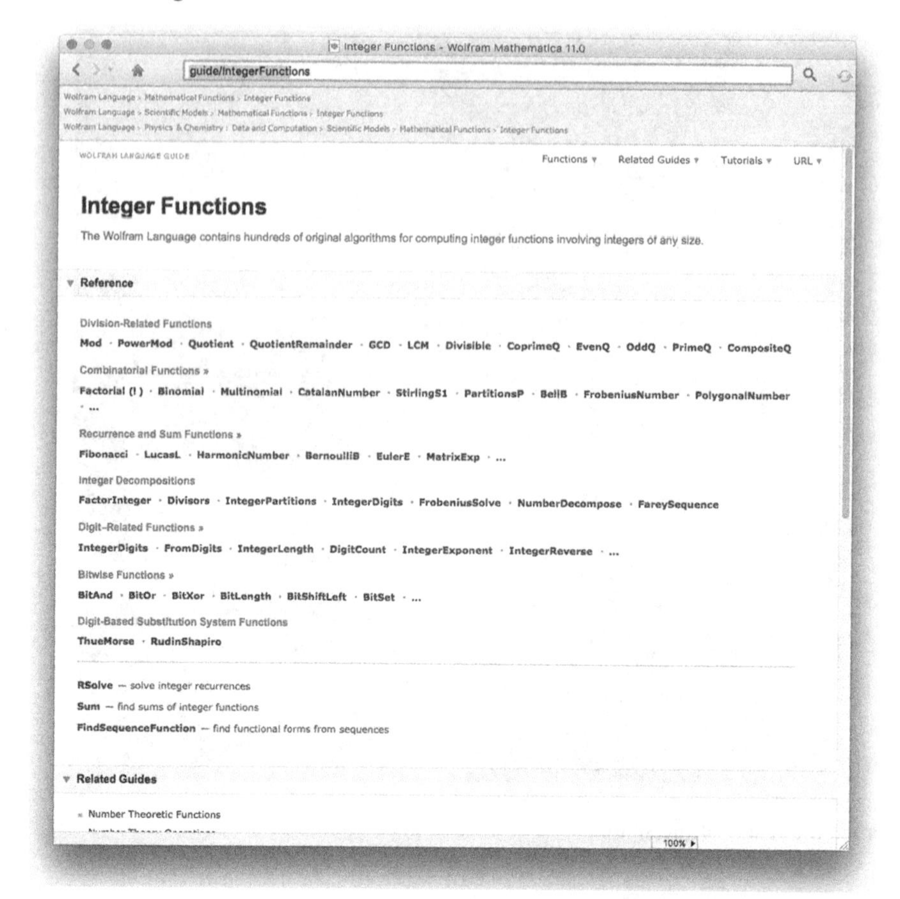

Example 1.5

In this example, we investigate digit operations. *Mathematica By Example*, Fifth Edition, is copyright in 2017, which has four digits.

IntegerDigits[2017]

{2, 0, 1, 7}

As a string, the number is

IntegerString[2017]

2017

In base 2, the copyright year is

IntegerString[2017, 2]

11111100001

On the other hand, with Roman numerals the copyright year for the Fourth edition is

IntegerString[2008, "Roman"]

MMVIII

Chapter 2

Basic Operations on Numbers, Expressions, and Functions

2.1 NUMERICAL CALCULATIONS AND BUILT-IN FUNCTIONS

2.1.1 Numerical Calculations

The basic arithmetic operations (addition, subtraction, multiplication, division, and exponentiation) are performed in the natural way with Mathematica. Whenever possible, Mathematica gives an exact answer and reduces fractions.

1. "a plus b," $a + b$, is entered as `a+b`;
2. "a minus b," $a - b$, is entered as `a-b`;
3. "a times b," ab, is entered as either `a*b` or `a b` (note the space between the symbols `a` and `b`);
4. "a divided by b," a/b, is entered as `a/b`. Executing the command `a/b` results in a fraction reduced to lowest terms; and
5. "a raised to the bth power," a^b, is entered as `a^b`.

Example 2.1

Calculate (a) $121 + 542$; (b) $3231 - 9876$; (c) $(-23)(76)$; (d) $(22341)(832748)(387281)$; and (e) $\frac{467}{31}$.

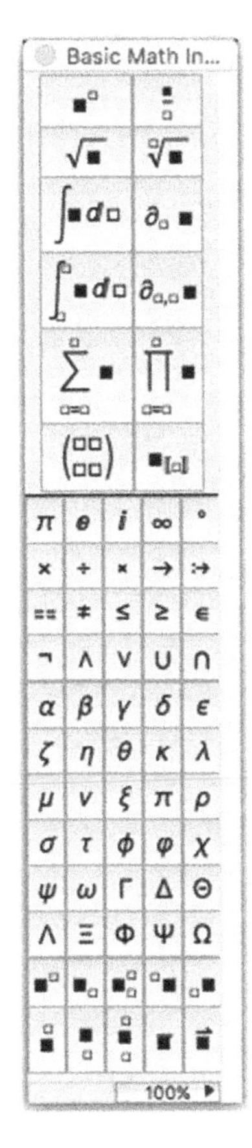

Solution. These calculations are carried out in the following screen shot. In each case, the input is typed and then evaluated by pressing **Enter**. In the last case, the **Basic Math** template is used to enter the fraction.

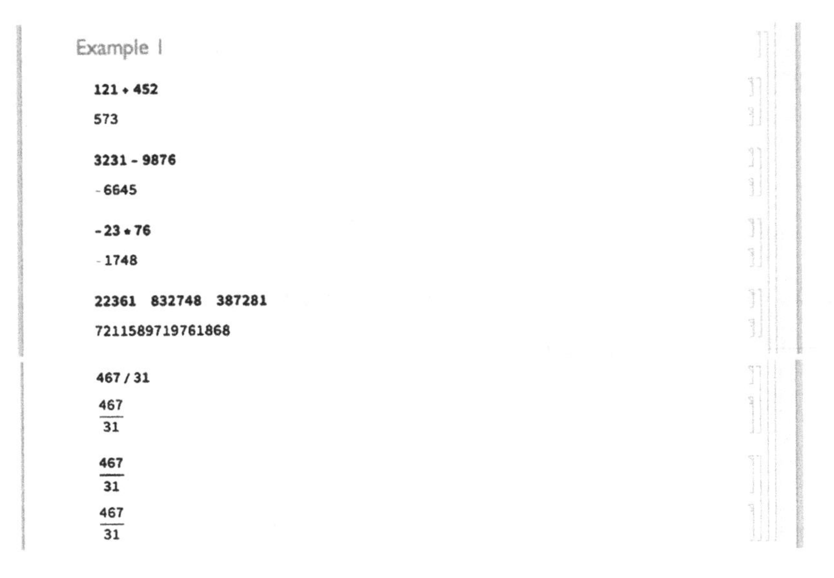

Mathematica by Example. http://dx.doi.org/10.1016/B978-0-12-812481-9.00002-8

The term $a^{n/m} = \sqrt[m]{a^n} = \left(\sqrt[m]{a}\right)^n$ is entered as `a^(n/m)`. For $n/m = 1/2$, the command `Sqrt[a]` can be used instead. Usually, the result is returned in unevaluated form but `N` can be used to obtain numerical approximations to virtually any degree of accuracy. With `N[expr,n]`, Mathematica yields a numerical approximation of `expr` to n digits of precision, if possible. At other times, `Simplify` can be used to produce the expected results.

Remark 2.1. If the expression b in a^b contains more than one symbol, be sure that the exponent is included in parentheses. Entering `a^n/m` computes $a^n/m = \frac{1}{m}a^n$ while entering `a^(n/m)` computes $a^{n/m}$.

Example 2.2

Compute (a) $\sqrt{27}$ and (b) $\sqrt[3]{8^2} = 8^{2/3}$.

Solution. (a) Mathematica automatically simplifies $\sqrt{27} = 3\sqrt{3}$. We use `N` to obtain an approximation of $\sqrt{27}$. (b) Mathematica automatically simplifies $8^{2/3}$.

`N[number]` and `number//N` return numerical approximations of `number`.

```
Sqrt[27]
3 √3

N[Sqrt[27]]
5.19615

8^(2/3)
4
```

□

When computing odd roots of negative numbers, Mathematica's results are surprising to the novice. Namely, Mathematica returns a complex number. We will see that this has important consequences when graphing certain functions.

Example 2.3

Calculate (a) $\frac{1}{3}\left(-\frac{27}{64}\right)^2$ and (b) $\left(-\frac{27}{64}\right)^{2/3}$.

Solution. (a) Because Mathematica follows the order of operations, `(-27/64)^2/3` first computes $(-27/64)^2$ and then divides the result by 3.

```
(-27/64)^2/3
243
────
4096
```

(b) On the other hand, `(-27/64)^(2/3)` raises $-27/64$ to the 2/3 power. Mathematica does not automatically simplify $\left(-\frac{27}{64}\right)^{2/3}$.

```
(-27/64)^(2/3)
9
── (-1)^(2/3)
16
```

However, when we use `N`, Mathematica returns the numerical version of the principal root of $\left(-\frac{27}{64}\right)^{2/3}$.

```
N[(-27/64)^(2/3)]
-0.28125 + 0.487139 i
```

To obtain the result

$$\left(-\frac{27}{64}\right)^{2/3} = \left(\sqrt[3]{\frac{-27}{64}}\right)^2 = \left(-\frac{3}{4}\right)^2 = \frac{9}{16},$$

which would be expected by most algebra and calculus students, we first square $-27/64$ and then take the third root.

```
((-27/64)^2)^(1/3)
9
──
16
```

The `Surd` function automatically returns the real-valued nth root of x that most beginning users of Mathematica expect.

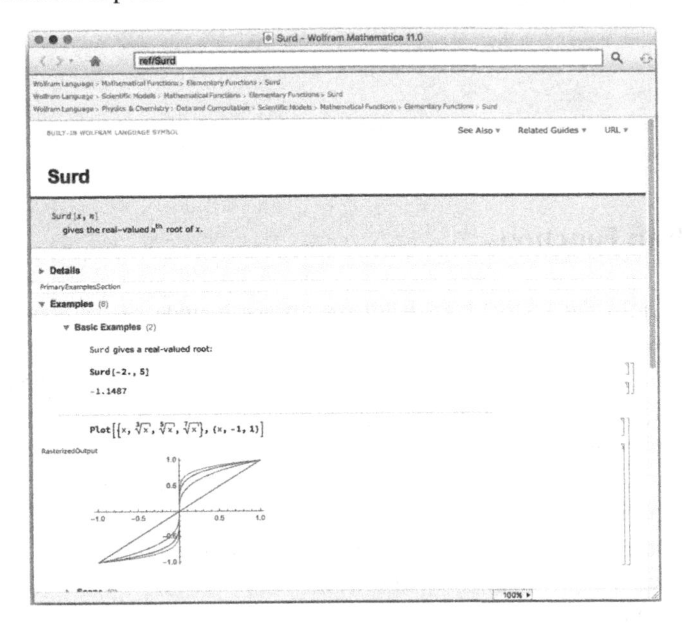

Then,

```
? Surd
Surd[x, n] gives the real-valued n th root of x. >>
Surd[(-27/64)^2, 3]
9
──
16
```

returns the result $9/16$. □

2.1.2 Built-In Constants

Mathematica has built-in definitions of nearly all commonly used mathematical constants and functions. To list a few, $e \approx 2.71828$ is denoted by `E`, $\pi \approx 3.14159$ is denoted by `Pi`, and $i = \sqrt{-1}$ is denoted by `I`. Usually, Mathematica performs complex arithmetic automatically.

Other built-in constants include ∞, denoted by `Infinity`, Euler's constant, $\gamma \approx 0.577216$, denoted by `EulerGamma`, Catalan's constant, approximately 0.915966, denoted by `Catalan`, and the golden ratio, $\frac{1}{2}\left(1+\sqrt{5}\right) \approx 1.61803$, denoted by `GoldenRatio`.

Example 2.4

Entering

$N[E, 50]$

2.7182818284590452353602874713526624977572470937000

returns a 50-digit approximation of e. Entering

N[π, 25]

3.141592653589793238462643

returns a 25-digit approximation of π. Entering

$\frac{3+I}{4-I}$

$\frac{11}{17}+\frac{7i}{17}$

performs the division $(3+i)/(4-i)$ and writes the result in standard form.

2.1.3 Built-In Functions

Functions frequently encountered by beginning users include the exponential function, `Exp[x]`; the natural logarithm, `Log[x]`; the absolute value function, `Abs[x]`; the trigonometric functions `Sin[x]`, `Cos[x]`, `Tan[x]`, `Sec[x]`, `Csc[x]`, and `Cot[x]`; the inverse trigonometric functions `ArcSin[x]`, `ArcCos[x]`, `ArcTan[x]`, `ArcSec[x]`, `ArcCsc[x]`, and `ArcCot[x]`; the hyperbolic trigonometric functions `Sinh[x]`, `Cosh[x]`, and `Tanh[x]`; and their inverses `ArcSinh[x]`, `ArcCosh[x]`, and `ArcTanh[x]`. Generally, Mathematica tries to return an exact value unless otherwise specified with `N`.

Several examples of the natural logarithm and the exponential functions are given next. Mathematica often recognizes the properties associated with these functions and simplifies expressions accordingly.

Example 2.5

Entering

E^{-5}

$\frac{1}{e^5}$

$E^{-5}//N$

0.00673795

returns an approximation of $e^{-5} = 1/e^5$. Entering

N[number] or number//N return approximations of number.

Exp[x] computes e^x. Enter E to compute $e \approx 2.718$.

Log[Exp[3]]

3

computes $\ln e^3 = 3$. Entering

Log[x] computes $\ln x$. $\ln x$ and e^x are inverse functions ($\ln e^x = x$ and $e^{\ln x} = x$) and Mathematica uses these properties when simplifying expressions involving these functions.

Exp[Log[Pi]]

π

computes $e^{\ln \pi} = \pi$. Entering

Abs[−5]

5

computes $|-5| = 5$. Entering

Abs[x] returns the absolute value of x, $|x|$.

Abs[(3 + 2*I*)/(2 − 9*I*)]

$\sqrt{\frac{13}{85}}$

computes $|(3+2i)/(2-9i)|$. Entering

Cos[Pi/12]

$\frac{1+\sqrt{3}}{2\sqrt{2}}$

***N*[Cos[Pi/12]]**

0.965926

computes the exact value of $\cos(\pi/12)$ and then an approximation. Although Mathematica cannot compute the exact value of tan 1000, entering

N[number] and number//N return approximations of number.

***N*[Tan[1000]]**

1.47032

returns an approximation of tan 1000. Similarly, entering

***N*[ArcSin[1/3]]**

0.339837

returns an approximation of $\sin^{-1}(1/3)$ and entering

ArcCos[2/3]//N

0.841069

returns an approximation of $\cos^{-1}(2/3)$.

Mathematica is able to apply many identities that relate the trigonometric and exponential functions using the functions `TrigExpand`, `TrigFactor`, `TrigReduce`, `TrigToExp`, and `ExpToTrig`.

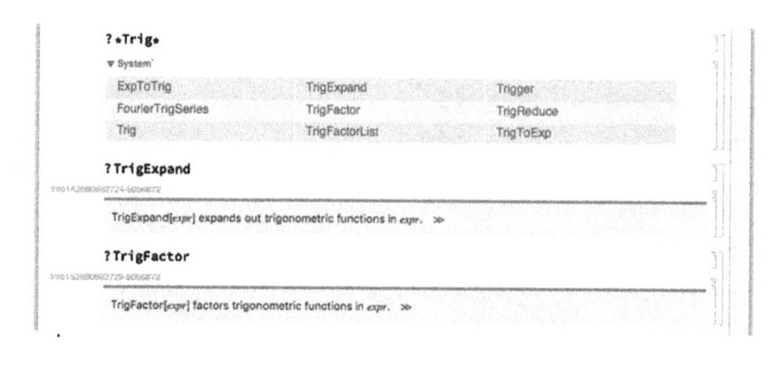

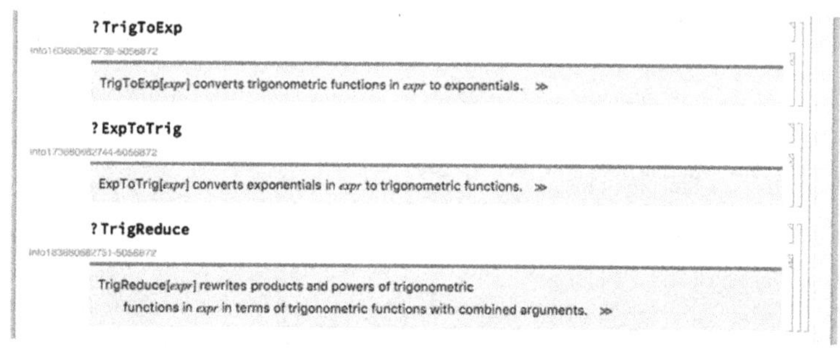

Example 2.6

Mathematica does not automatically apply the identity $\sin^2 x + \cos^2 x = 1$.

Cos[x]^2 + Sin[x]^2

$\text{Cos}[x]^2 + \text{Sin}[x]^2$

To apply the identity, we use `Simplify`. Generally, `Simplify[expression]` attempts to simplify `expression`.

Simplify[Cos[x]^2 + Sin[x]^2]

1

Use `TrigExpand` to multiply expressions or to rewrite trigonometric functions. In this case, entering

TrigExpand[Cos[3*x*]]

Cos[*x*]^3 − 3Cos[*x*]Sin[*x*]^2

writes $\cos 3x$ in terms of trigonometric functions with argument x. We use the `TrigReduce` function to convert products to sums.

Many of the algebraic manipulation commands can be accessed from the **AlgebraicManipulation** palette.

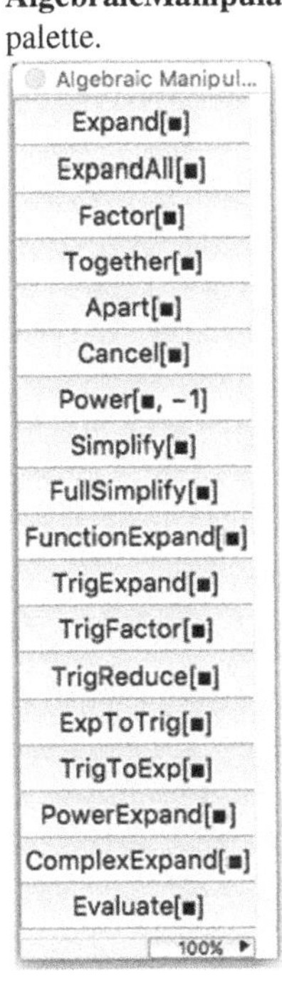

TrigReduce[Sin[3*x*]Cos[4*x*]]

$\frac{1}{2}$(−Sin[*x*] + Sin[7*x*])

We use `TrigExpand` to write

TrigExpand[Sin[3*x*]Cos[4*x*]]

TrigExpand[Cos[2*x*]]

$-\frac{\text{Sin}[x]}{2} + \frac{7}{2}\text{Cos}[x]^6\text{Sin}[x] - \frac{35}{2}\text{Cos}[x]^4\text{Sin}[x]^3 + \frac{21}{2}\text{Cos}[x]^2\text{Sin}[x]^5 - \frac{\text{Sin}[x]^7}{2}$

Cos[*x*]^2 − Sin[*x*]^2

in terms of trigonometric functions with argument x. We use `ExpToTrig` to convert exponential expressions to trigonometric expressions.

ExpToTrig[1/2**(Exp[**x**] + Exp[**−x**])]**
Cosh[x]

Similarly, we use `TrigToExp` to convert trigonometric expressions to exponential expressions.

TrigToExp[Sin[x**]]**
$\frac{1}{2}ie^{-ix} - \frac{1}{2}ie^{ix}$

Usually, you can use `Simplify` to apply elementary identities.

Simplify[Tan[x**]^**2 **+** 1**]**
Sec[x]^2

A Word of Caution

Remember that there are certain ambiguities in traditional mathematical notation. For example, the expression $\sin^2(\pi/6)$ is usually interpreted to mean "compute $\sin(\pi/6)$ and square the result." That is, $\sin^2(\pi/6) = \left[\sin(\pi/6)\right]^2$. The symbol sin is not being squared; the number $\sin(\pi/6)$ *is* squared. With Mathematica, we must be especially careful and follow the standard order of operations exactly, especially when using **InputForm**. We see that entering

```
Sin[Pi / 6] ^ 2
```

$\frac{1}{4}$

FIGURE 2.1 Visualizing the order in which Mathematica carries out a sequence of operations.

computes $\sin^2(\pi/6) = [\sin(\pi/6)]^2$ while

```
Sin^2[Pi/6]
```

$\text{Sin}^2[\frac{\pi}{6}]$

raises the symbol `Sin` to the power $2[\frac{\pi}{6}]$. Mathematica interprets $\sin^2\left(\frac{\pi}{6}\right)$ to be the product of the symbols sin^2 and $\pi/6$. However, using **TraditionalForm** we are able to evaluate $\sin^2(\pi/6) = [\sin(\pi/6)]^2$ with Mathematica using conventional mathematical notation.

$\sin^2\left(\frac{\pi}{6}\right)$

$\frac{1}{4}$

Be aware, however, that traditional mathematical notation does contain certain ambiguities and Mathematica may not return the result you expect if you enter input using **TraditionalForm** unless you are especially careful to follow the standard order of operations, as the following warning message indicates.

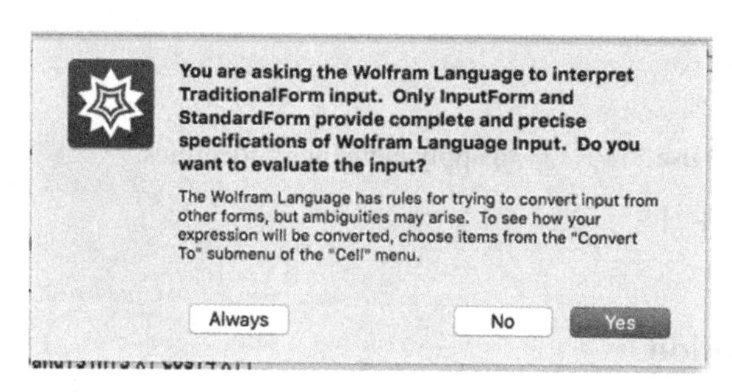

Example 2.7

As stated, Mathematica follows the order of operations exactly. To see how Mathematica performs a calculation, `TreeForm` presents the sequence graphically.

For example, for the calculation $\dfrac{(a+b)^2(c+d)^3}{x+y-z}$, `TreeForm` gives us the results shown in Fig. 2.1.

```
Clear[a, b, c, d, x, y, z]
TreeForm[(a + b)^2(c + d)^3/(x + y - z)]
```

2.2 EXPRESSIONS AND FUNCTIONS: ELEMENTARY ALGEBRA

2.2.1 Basic Algebraic Operations on Expressions

Expressions involving unknowns are entered in the same way as numbers. Mathematica performs standard algebraic operations on mathematical expressions. For example, the commands

1. `Factor[expression]` factors `expression`;
2. `Expand[expression]` multiplies `expression`;
3. `Together[expression]` writes `expression` as a single fraction; and
4. `Simplify[expression]` performs basic algebraic manipulations on `expression` and returns the simplest form it finds.

For basic information about any of these commands (or any other) enter `?command` as we do here for `Factor`.

Many of the algebraic manipulation commands can be accessed from the **AlgebraicManipulation** palette.

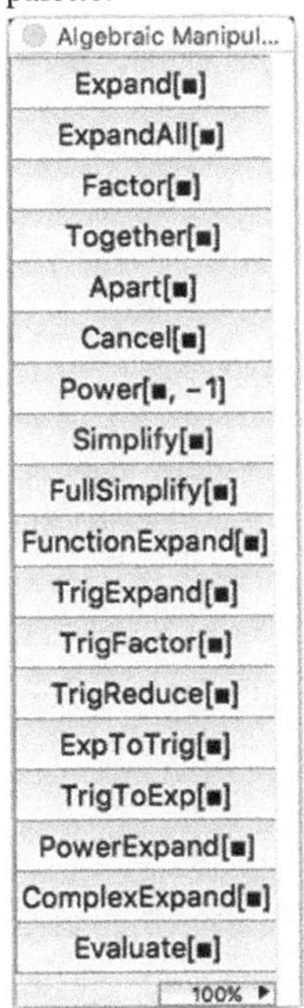

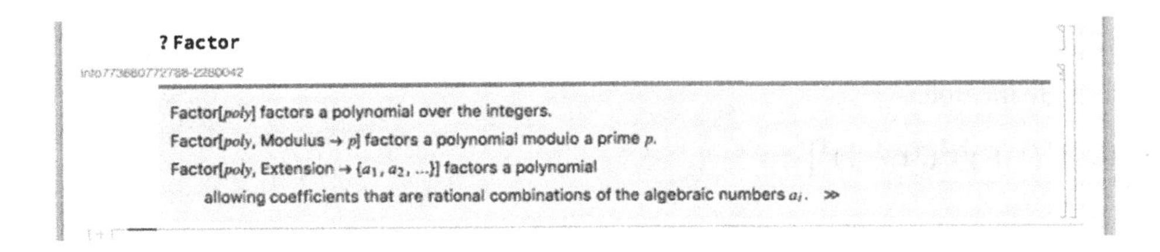

or access the **Documentation Center** as we do here for `Factor`.

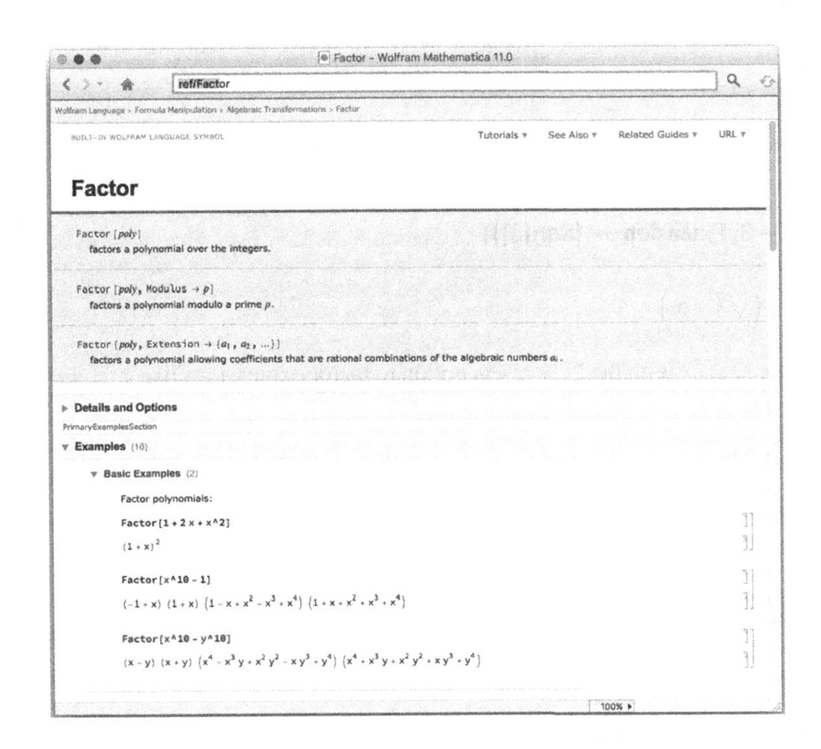

When entering expressions, be sure to include a space or * between variables to denote multiplication to avoid any ambiguity or confusion. For example, Mathematica interprets `4x` to be the product of x and 4. On the other hand, Mathematica interprets `x4` to be a new symbol with the name "x4." Similarly, xy is interpreted as a symbol with the name "xy" while x y and x*y are interpreted as the product of x and y, xy.

Example 2.8

(a) Factor the polynomial $12x^2+27xy-84y^2$. (b) Expand the expression $(x+y)^2(3x-y)^3$. (c) Write the sum $\frac{2}{x^2}-\frac{x^2}{2}$ as a single fraction.

Solution. The result obtained with Factor indicates that $12x^2+27xy-84y^2=3(4x-7y)(x+4y)$. When typing the command, be sure to include a space, or *, between the x and y terms to denote multiplication. xy represents an expression while x y or x*y denotes x multiplied by y.

Factor$\left[12x^2+27xy-84y^2\right]$

$3(4x-7y)(x+4y)$

We use Expand to compute the product $(x+y)^2(3x-y)^3$ and Together to express $\frac{2}{x^2}-\frac{x^2}{2}$ as a single fraction.

Expand$\left[(x+y)^2(3x-y)^3\right]$

$27x^5+27x^4y-18x^3y^2-10x^2y^3+7xy^4-y^5$

Together$\left[\frac{2}{x^2}-\frac{x^2}{2}\right]$

$\frac{4-x^4}{2x^2}$

□

Factor[x^2-3] returns x^2-3.

To factor an expression like $x^2-3=x^2-\left(\sqrt{3}\right)^2=\left(x-\sqrt{3}\right)\left(x+\sqrt{3}\right)$, use Factor with the Extension option.

Factor[x^2 − 3, Extension → {Sqrt[3]}]

$-\left(\sqrt{3}-x\right)\left(\sqrt{3}+x\right)$

Similarly, use Factor with the Extension option to factor expressions like $x^2+1=x^2-i^2=(x+i)(x-i)$.

Factor[x^2 + 1]

$1+x^2$

Factor[x^2 + 1, Extension → {I}]

$(-i+x)(i+x)$

Mathematica does not automatically simplify $\sqrt{x^2}$, to the expression x

Sqrt[x^2]

$\sqrt{x^2}$

because without restrictions on x, $\sqrt{x^2} = |x|$. The command PowerExpand[expression] simplifies expression assuming that all variables are positive. Alternatively, you can use Assumptions to tell Mathematica to assume that $x > 0$ or to assume that $x < 0$.

Simplify[Sqrt[*x*^2], Assumptions → *x* > 0]

x

Simplify[Sqrt[*x*^2], Assumptions → *x* < 0]

$-x$

Alternatively, PowerExpand will automatically make "appropriate" assumptions when simplifying expressions.

PowerExpand[Sqrt[*x*^2]]

x

Thus, entering

Simplify[Sqrt[*a*^2*b*^4]]

$\sqrt{a^2b^4}$

returns $\sqrt{a^2b^4}$ but entering

PowerExpand[Sqrt[*a*^2*b*^4]]

ab^2

Simplify[Sqrt[*a*^2*b*^4], Assumptions →

{*a* > 0, *b* > 0}]

ab^2

returns ab^2. However, if $a < 0$ and $b > 0$, $\sqrt{a^2b^4} = -ab^2$.

Simplify[Sqrt[*a*^2*b*^4], Assumptions →

{*a* < 0, *b* > 0}]

$-ab^2$

In general, a space is not needed between a number and a symbol to denote multiplication when a symbol follows a number. That is, 3dog means "3 times variable dog; dog3 is a variable with name dog3. Mathematica interprets 3 dog, dog*3, and dog 3 as "3 times variable dog. However, when multiplying two variables, either include a space or * between the variables.

1. cat dog means "variable cat times variable dog."
2. cat*dog means "variable cat times variable dog."
3. But, catdog is interpreted as a variable catdog.

The command Apart[expression] computes the partial fraction decomposition of expression; Cancel[expression] factors the numerator and denominator of expression then reduces expression to lowest terms.

Example 2.9

(a) Determine the partial fraction decomposition of $\frac{1}{(x-3)(x-1)}$. (b) Simplify $\frac{x^2-1}{x^2-2x+1}$.

Solution. `Apart` is used to see that $\frac{1}{(x-3)(x-1)} = \frac{1}{2(x-3)} - \frac{1}{2(x-1)}$. Then, `Cancel` is used to find that $\frac{x^2-1}{x^2-2x+1} = \frac{(x-1)(x+1)}{(x-1)^2} = \frac{x+1}{x-1}$. In this calculation, we have assumed that $x \neq 1$, an assumption made by `Cancel` but not by `Simplify`.

Apart$\left[\frac{1}{(x-3)(x-1)}\right]$

$\frac{1}{2(-3+x)} - \frac{1}{2(-1+x)}$

Cancel$\left[\frac{x^2-1}{x^2-2x+1}\right]$

$\frac{1+x}{-1+x}$ □

In addition, Mathematica has several built-in functions for manipulating parts of fractions.

1. `Numerator[fraction]` yields the numerator of `fraction`.
2. `ExpandNumerator[fraction]` expands the numerator of `fraction`.
3. `Denominator[fraction]` yields the denominator of `fraction`.
4. `ExpandDenominator[fraction]` expands the denominator of `fraction`.

Example 2.10

Given $\frac{x^3+2x^2-x-2}{x^3+x^2-4x-4}$, (a) factor both the numerator and denominator; (b) reduce $\frac{x^3+2x^2-x-2}{x^3+x^2-4x-4}$ to lowest terms; and (c) find the partial fraction decomposition of $\frac{x^3+2x^2-x-2}{x^3+x^2-4x-4}$.

Solution. The numerator of $\frac{x^3+2x^2-x-2}{x^3+x^2-4x-4}$ is extracted with `Numerator`. We then use `Factor` together with `%`, which is used to refer to the most recent output, to factor the result of executing the `Numerator` command.

Numerator$\left[\frac{x^3+2x^2-x-2}{x^3+x^2-4x-4}\right]$

$-2-x+2x^2+x^3$

Factor[%]

$(-1+x)(1+x)(2+x)$

Similarly, we use `Denominator` to extract the denominator of the fraction. Again, `Factor` together with `%` is used to factor the previous result, which corresponds to the denominator of the fraction.

Denominator$\left[\frac{x^3+2x^2-x-2}{x^3+x^2-4x-4}\right]$

$-4-4x+x^2+x^3$

Factor[%]

$(-2+x)(1+x)(2+x)$

`Cancel` is used to reduce the fraction to lowest terms.

Cancel$\left[\frac{x^3+2x^2-x-2}{x^3+x^2-4x-4}\right]$

$\frac{-1+x}{-2+x}$

Finally, `Apart` is used to find its partial fraction decomposition.

Apart$\left[\frac{x^3+2x^2-x-2}{x^3+x^2-4x-4}\right]$

$1+\frac{1}{-2+x}$ □

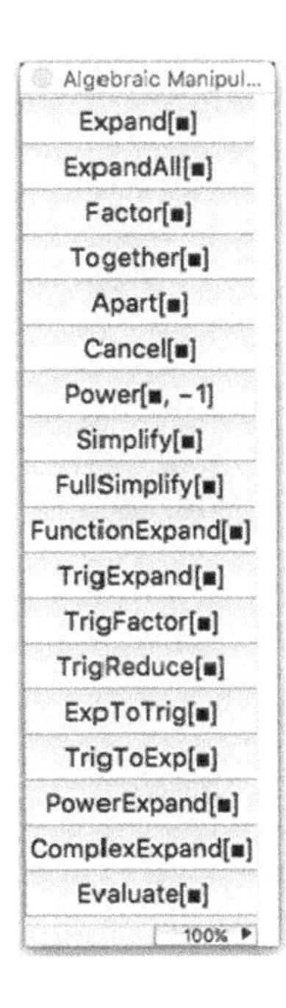

You can also take advantage of the **AlgebraicManipulation** palette, which is accessed by going to **Palettes** under the Mathematica menu, followed by **AlgebraicManipulation**, to evaluate expressions.

Example 2.11

Simplify $\frac{2(x-3)^2(x+1)}{3(x+1)^{4/3}}+2(x-3)(x+1)^{2/3}$.

Solution. First, we type the expression.

```
2 (x - 3) ^ 2 (x + 1) / (3 (x + 1) ^ (4 / 3)) + 2 (x - 3) (x + 1) ^ (2 / 3)
```

$$\frac{2\,(-3+x)^2}{3\,(1+x)^{1/3}}+2\,(-3+x)\,(1+x)^{2/3}$$

Then, select the expression.

$$\frac{2\,(-3+x)^2}{3\,(1+x)^{1/3}}+2\,(-3+x)\,(1+x)^{2/3}$$

Move the cursor to the palette and click on `Simplify`. Mathematica simplifies the expression.

$$\frac{8\,(-3+x)\,x}{3\,(1+x)^{1/3}}$$

□

2.2.2 Naming and Evaluating Expressions

In Mathematica, objects can be named. Naming objects is convenient: we can avoid typing the same mathematical expression repeatedly (as we did in Example 2.10) and named expressions can be referenced throughout a notebook or Mathematica session. Every Mathematica object can be named – expressions, functions, graphics, and so on can be named with Mathematica. Objects are named by using a single equals sign (=).

Because every built-in Mathematica function begins with a capital letter, we adopt the convention that *every* mathematical object we name in this text will begin with a *lowercase* letter. Consequently, we will be certain to avoid any possible ambiguity with any built-in Mathematica objects.

With Mathematica 11, the default option is to display *known* objects in black and unknown objects in blue. Thus, in the following screen shot,

```
Fraction x y
Apart
apart
2 Pi pi π
? Plot Expand Cancel
E e Exp
```

Fraction, x, y, apart, pi, and e are in blue; Apart, 2, Pi, π, ?, Plot, Expand, Cancel, Exp, and E are in black.

To automatically update named variables, `Dynamic[x]` returns the current value of x. Thus, `Dynamic[x]` returns `dog`.

```
x = dog
dog
Dynamic[x]
dog
```

However, when we enter $x = 7$ afterwards, `Dynamic[x]` is automatically updated to the new value of x.

Expressions are easily evaluated using `ReplaceAll`, which is abbreviated with `/.` and obtained by typing a backslash (/) followed by a period (.), together with `Rule`, which is abbreviated with `->` and obtained by typing a dash (minus sign) (-) followed by a greater than sign (>). For example, entering the command

```
x^2 /. x->3
```

returns the value of the expression x^2 if $x = 3$. Note, however, this does not assign the symbol x the value 3: entering `x=3` assigns x the value 3.

Example 2.12

Evaluate $\dfrac{x^3+2x^2-x-2}{x^3+x^2-4x-4}$ if $x=4$, $x=-3$, and $x=2$.

Solution. To avoid retyping $\frac{x^3+2x^2-x-2}{x^3+x^2-4x-4}$, we define `fraction` to be $\frac{x^3+2x^2-x-2}{x^3+x^2-4x-4}$.

Of course, you can simply copy and paste this expression if you neither want to name it nor retype it.

Clear[x]

fraction = $\frac{x^3+2x^2-x-2}{x^3+x^2-4x-4}$

$\frac{-2-x+2x^2+x^3}{-4-4x+x^2+x^3}$

`/.` is used to evaluate `fraction` if $x = 4$ and then if $x = -3$.

If you include a semi-colon (;) at the end of the command, the resulting output is suppressed.

fraction/.x->4

$\frac{3}{2}$

fraction/.x-> − 3

$\frac{4}{5}$

When we try to replace each x in `fraction` by 2, we see that the result is undefined: division by 0 is always undefined.

```
fraction /. x → -2
··· Power: Infinite expression 1/0 encountered.
··· Infinity: Indeterminate expression 0 ComplexInfinity encountered.
Indeterminate
```

However, when we use `Cancel` to first simplify and then use `ReplaceAll` to evaluate,

```
fraction2 = Cancel[fraction]
-1 + x
------
-2 + x

fraction2 /. x → -2
3
-
4
```

we see that the result is 3/4. The result indicates that $\lim_{x\to -2} \frac{x^3+2x^2-x-2}{x^3+x^2-4x-4} = \frac{3}{4}$. We confirm this result with `Limit`.

```
Limit[fraction, x → -2]
3
-
4
```

Generally, `Limit[f[x],x->a]` attempts to compute $\lim_{x\to a} f(x)$. The `Limit` function is discussed in more detail in the next chapter. □

2.2.3 Defining and Evaluating Functions

It is important to remember that functions, expressions, and graphics can be named anything that is not the name of a built-in Mathematica function or command. As previously indicated, every built-in Mathematica object begins with a capital letter so every user-defined function, expression, or other object in this text will be assigned a name using lowercase letters, exclusively. This way, the possibility of conflicting with a built-in Mathematica command or function is completely eliminated. Because definitions of functions and names of objects are frequently modified, we introduce the command `Clear`. `Clear[expression]` clears all definitions of expression, if any. You can see if a particular symbol has a definition by entering `?symbol`.

In Mathematica, an elementary function of a single variable, $y = f(x) = expression\ in\, x$, is typically defined using the form

`f[x_]=expression in x` or `f[x_]:=expression in x.`

Notice that when you first define a function, you must always enclose the argument in square brackets (`[...]`) and place an underline (or blank) "_" after the argument on the left-hand side of the equals sign in the definition of the function.

Example 2.13

Entering

f[x_]=x/(x^2+1)

$\frac{x}{1+x^2}$

defines and computes $f(x) = x/(x^2+1)$. Entering

f[3]

$\frac{3}{10}$

computes $f(3) = 3/(3^2+1) = 3/10$. Entering

f[a]

$\frac{a}{1+a^2}$

computes $f(a) = a/(a^2+1)$. Entering

f[3+h]

$\frac{3+h}{1+(3+h)^2}$

computes $f(3+h) = (3+h)/((3+h)^2+1)$. Entering

n1 = Simplify[(f[3+h] − f[3])/h]

$-\frac{8+3h}{10(10+6h+h^2)}$

computes and simplifies $\frac{f(3+h) - f(3)}{h}$ and names the result `n1`. Entering

n1/.h → 0

$-\frac{2}{25}$

evaluates `n1` if $h = 0$. Entering

n2 = Together[(f[a + h] − f[a])/h]

$$\frac{1-a^2-ah}{(1+a^2)(1+a^2+2ah+h^2)}$$

computes and simplifies $\frac{f(a+h)-f(a)}{h}$ and names the result n2. Entering

n2/.h → 0

$$\frac{1-a^2}{(1+a^2)^2}$$

evaluates n2 if $h = 0$.

Often, you will need to evaluate a function for the values in a **list**,

$$\texttt{list} = \{a_1, a_2, a_3, \ldots, a_n\}.$$

Once $f(x)$ has been defined, `Map[f,list]` returns the list

$$\{f(a_1), f(a_2), f(a_3), \ldots, f(a_n)\}$$

Also,

The `Table` function will be discussed in more detail as needed as well as in Chapters 4 and 5.

1. `Table[f[n],{n,n1,n2}]` returns the list

$$\{f(n_1), f(n_1+1), f(n_1+2), \ldots, f(n_2)\}$$

2. `Table[{n,f[n]},{n,n1,n2}]` returns the list of ordered pairs

$$\{(n_1, f(n_1)), (n_1+1, f(n_1+1)), (n_1+2, f(n_1+2)), \ldots, (n_2, f(n_2))\}$$

Example 2.14

Entering

Clear[h]

h[t_] = (1 + t)^(1/t)

h[1]

$$(1+t)^{\frac{1}{t}}$$

2

defines $h(t) = (1+t)^{1/t}$ and then computes $h(1) = 2$. Because division by 0 is always undefined, $h(0)$ is undefined.

```
h[0]
··· Power: Infinite expression 1/0 encountered.
··· Infinity: Indeterminate expression 1^ComplexInfinity encountered.
Indeterminate
```

However, $h(t)$ is defined for all $t > 0$. In the following, we use `RandomReal` together with `Table` to generate 6 random numbers "close" to 0 and name the resulting list `t1`. Because we are using `RandomReal`, your results will almost certainly differ from those here.

`RandomReal[{a,b}]` returns a random real number between a and b; `RandomReal[{a,b},n]` returns n random real numbers between a and b.

t1 = Table[RandomReal[{0, 10^(−n)}], {n, 0, 5}]

{0.399958, 0.0392505, 0.00674767, 0.00018339, 0.0000760418,

4.3836670778376435*^-6}

We then use Map to compute $h(t)$ for each of the values in the list t1.
From the result, we might correctly deduce that $\lim_{t\to 0^+}(1+t)^{1/t}=e$. We can compute the limit with the Limit function, which will be discussed in more detail in Chapter 3. For now, we remark that Limit[f[x],x->a] attempts to compute $\lim_{x\to a} f(x)$.

Limit[*h*[*t*], *t* → 0]

e

In each of these cases, do not forget to include the "blank" (or underline) (_) on the left-hand side of the equals sign and right bracket in the definition of each function. Remember to always include arguments of functions in square brackets.

Example 2.15

Including a semi-colon at the end of a command suppresses the resulting output because Mathematica names (or, assigns) the function a name but does not evaluate the function after the assignment.

Entering

Clear[*f*]

***f*[0] = 1;**

***f*[1] = 1;**

***f*[n_]:=*f*[*n* − 1] + *f*[*n* − 2]**

defines the recursively-defined function defined by $f(0)=1$, $f(1)=1$, and $f(n)=f(n-1)+f(n-2)$. For example, $f(2)=f(1)+f(0)=1+1=2$; $f(3)=f(2)+f(1)=2+1=3$. We use Table to create a list of ordered pairs $(n, f(n))$ for $n=0, 1, \ldots, 10$.

The f_n we have defined here return the **Fibonacci number** F_n. Fibonacci[n] also returns the nth Fibonacci number.

Table[{*n*, *f*[*n*]}, {*n*, 0, 10}]

{{0, 1}, {1, 1}, {2, 2}, {3, 3}, {4, 5}, {5, 8}, {6, 13}, {7, 21}, {8, 34}, {9, 55}, {10, 89}}

In the preceding examples, the functions were defined using each of the forms f[x_]:=... and f[x_]=.... As a practical matter, when defining "routine" functions with domains consisting of sets of real numbers and ranges consisting of sets of real numbers, either form can be used. Defining a function using the form f[x_]=... instructs Mathematica to define f and then compute and return f[x] (**immediate assignment**); defining a function using the form f[x_]:=... instructs Mathematica to define f. In this case, f[x] is not computed and, thus, Mathematica returns no output (**delayed assignment**). The form f[x_]:=... should be used when Mathematica cannot evaluate f[x] unless x is a particular value, as with recursively-defined functions or piecewise-defined functions which we will discuss shortly.

Generally, if attempting to define a function using the form f[x_]=... produces one or more error messages, use the form f[x_]:=... instead.

To define piecewise-defined functions, we usually use Condition (/;) as illustrated in the following example. In simple situations we take advantage of Piecewise.

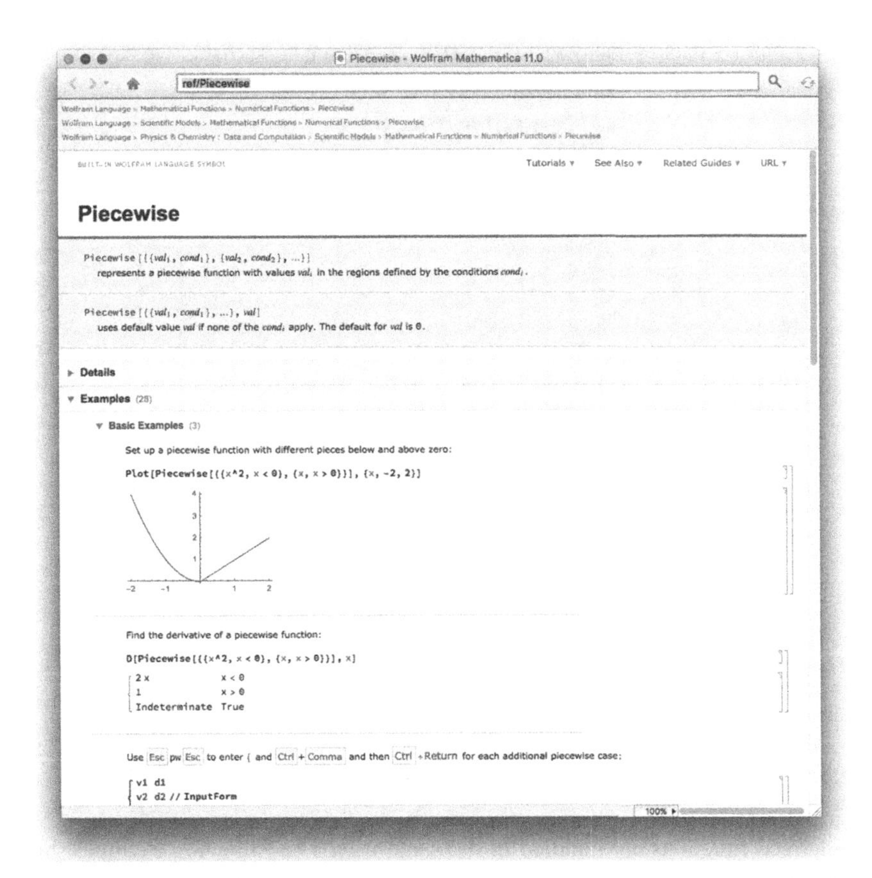

Example 2.16

Entering

```
Clear[f]
f[t_]:=Sin[1/t]/;t > 0
```

defines $f(t) = \sin(1/t)$ for $t > 0$. Entering

```
f[1/(10Pi)]
```

0

is evaluated because $1/(10\pi) > 0$. However, both of the following commands are returned unevaluated. In the first case, -1 is not greater than 0 ($f(t)$ is not defined for $t \leq 0$). In the second case, Mathematica does not know the value of a so it cannot determine if it is or is not greater than 0.

```
f[-1]
```

$f[-1]$

```
f[a]
```

$f[a]$

Entering

```
f[t_]:= - t/;t ≤ 0
```

defines $f(t) = -t$ for $t \leq 0$. Now, the domain of $f(t)$ is all real numbers. That is, we have defined the piecewise-defined function

$$f(t) = \begin{cases} \sin(1/t), & t > 0 \\ -t, & t \leq 0. \end{cases}$$

We can now evaluate $f(t)$ for any real number t.

```
f[2/(5Pi)]
```

1

```
f[0]
```

0

```
f[-10]
```

10

However, $f(a)$ still returns unevaluated because Mathematica does not know if $a \leq 0$ or if $a > 0$.

```
f[a]
```

$f[a]$

Recursively-defined functions are handled in the same way. The following example shows how to define a periodic function.

Example 2.17

Entering

```
Clear[g]
g[x_]:=x/;0 ≤ x < 1
g[x_]:=1/;1 ≤ x < 2
g[x_]:=3 − x/;2 ≤ x < 4
g[x_]:=g[x − 3]/;x ≥ 3
```

defines the recursively-defined function $g(x)$. For $0 \leq x < 3$, $g(x)$ is defined by

$$g(x) = \begin{cases} x, & 0 \leq x < 1 \\ 1, & 1 \leq x < 2 \\ 3 - x, & 2 \leq x < 3. \end{cases}$$

For $x \geq 3$, $g(x) = g(x-3)$. Entering

```
g[7]
```

1

computes $g(7) = g(4) = g(1) = 1$. We use `Table` to create a list of ordered pairs $(x, g(x))$ for 25 equally spaced values of x between 0 and 6.

```
Table[{x, g[x]}, {x, 0, 6, 6/24}]
{{0, 0}, {1/4, 1/4}, {1/2, 1/2}, {3/4, 3/4}, {1, 1}, {5/4, 1},
  {3/2, 1}, {7/4, 1}, {2, 1}, {9/4, 3/4}, {5/2, 1/2}, {11/4, 1/4}, {3, 0},
   {13/4, 1/4}, {7/2, 1/2}, {15/4, 3/4}, {4, 1}, {17/4, 1}, {9/2, 1},
   {19/4, 1}, {5, 1}, {21/4, 3/4}, {11/2, 1/2}, {23/4, 1/4}, {6, 0}}
```

We will discuss additional ways to define, manipulate, and evaluate functions as needed. However, Mathematica's extensive programming language allows a great deal of flexibility in defining functions, many of which are beyond the scope of this text. These powerful techniques are discussed in detail in texts like Wellin's *Programming with Mathematica: An Introduction* [18], Gray's *Mastering Mathematica: Programming Methods and Applications* [9], Wellin's *Essentials of Programming in Mathematica* [19] and Maeder's *The Mathematica Programmer II* and *Programming in Mathematica* [12,13].

2.3 GRAPHING FUNCTIONS, EXPRESSIONS, AND EQUATIONS

One of the best features of Mathematica is its graphics capabilities. In this section, we discuss methods of graphing functions, expressions, and equations and several of the options available to help graph functions.

2.3.1 Functions of a Single Variable

The commands

`Plot[f[x],{x,a,b}]` and `Plot[f[x],{x,a,x1,x2,...,xn,b}]`

graph the function $y = f(x)$ on the intervals $[a, b]$ and $[a, x_1) \cup (x_1, x_2) \cup \cdots \cup (x_n, b]$, respectively. Mathematica returns information about the basic syntax of the `Plot` command with `?Plot` or use the **Documentation Center** to obtain detailed information regarding `Plot`.

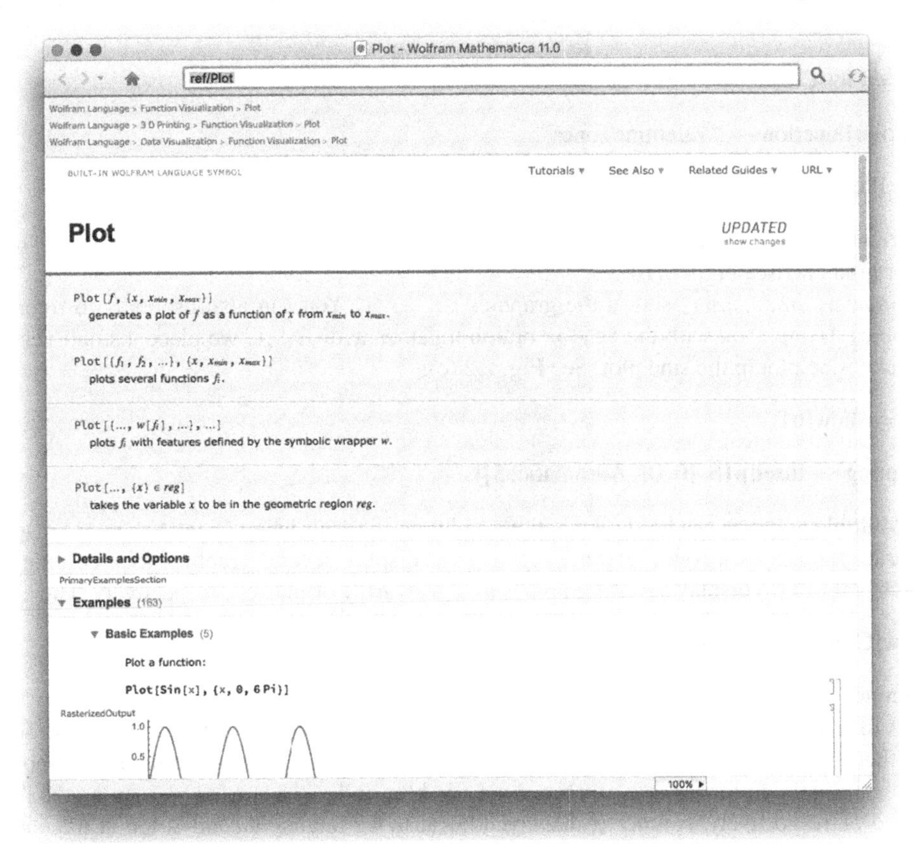

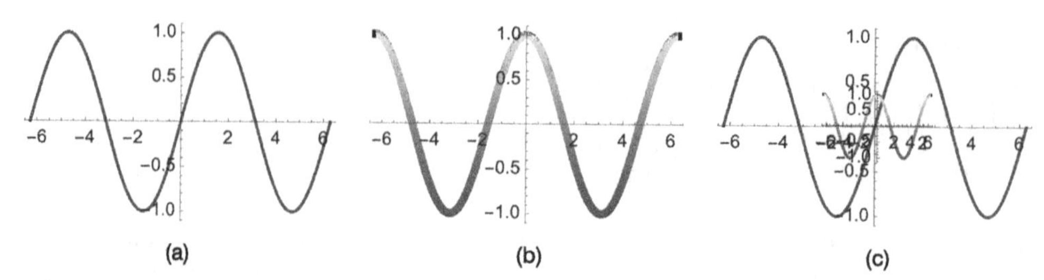

FIGURE 2.2 (a) $y = \sin x$ for $-2\pi \le x \le 2\pi$. (b) A "reddish" plot of $y = \cos x$ for $-2\pi \le x \le 2\pi$. (c) Combining two graphics with `Epilog` and `Inset`.

Remember that every Mathematica object can be assigned a name, including graphics. Use `Show` to combine graphics together. `Show[p1,p2,...,pn]` displays the graphics `p1`, `p2`, ..., `pn` together. `Show[GraphicsRow[{p1,p2,...,pn}]]` displays the graphics `p1`, `p2`, ..., `pn` in a row. `Show[GraphicsColumn[{p1,p2,...,pn}]]` displays the graphics `p1`, `p2`, ..., `pn` in a column.

```
Show[GraphicsGrid[{{p11,p12,...,p1m},{p21,p22,...,p2m},...,
    {pn1,pn2,...,pnm}}]]
```

displays the graphics `p11`, `pp12`, ..., `pnm` in an n-row by m-column array.

Example 2.18

Graph $y = \sin x$ and $y = \cos x$ for $-2\pi \le x \le 2\pi$.

Solution. Entering

p1 = Plot[Sin[x], {x, −2Pi, 2Pi}]

graphs $y = \sin x$ for $-2\pi \le x \le 2\pi$ and names the result `p1`. The plot is shown in Fig. 2.2 (a). With

p1b = Plot[Cos[x], {x, −2Pi, 2Pi},

ColorFunction → "ValentineTones",

PlotStyle → Thickness[.025]]

we create a slightly thicker plot of $y = \cos x$ and shade the plot using the `ValentineTones` color gradient. See Fig. 2.2 (b).

`Show[p1,p2,...,pn]` shows the graphics `p1`, ..., `pn`. You can also use `Show` to rerender graphics. Using `Show` with the `Epilog` option together with `Inset`, we place a small version of the cosine plot in the sine plot. See Fig. 2.2 (c).

p1c = Show[p1,

Epilog → Inset[p1b, {0, 0}, Automatic, 5]]

Multiple graphics can be shown in rows, columns, or grids using `GraphicsRow` (to display several graphics as a row), `GraphicsColumn` (to display several graphics as a column) or `GraphicsGrid` (to display several graphics as a grid, array or matrix), respectively. Thus,

Show[GraphicsRow[{p1, p1b, p1c}]]

generates Fig. 2.2.

A different way of approaching the problem would be to graph the two functions to scale and then show them together. Include a semi-colon (;) at the end of a command to suppress the result. In the following, we graph $y = \sin x$ in blue and $y = \cos x$ in red, naming the plots `p1` and `p2` respectively. Neither is displayed because we include a semi-colon at the end of

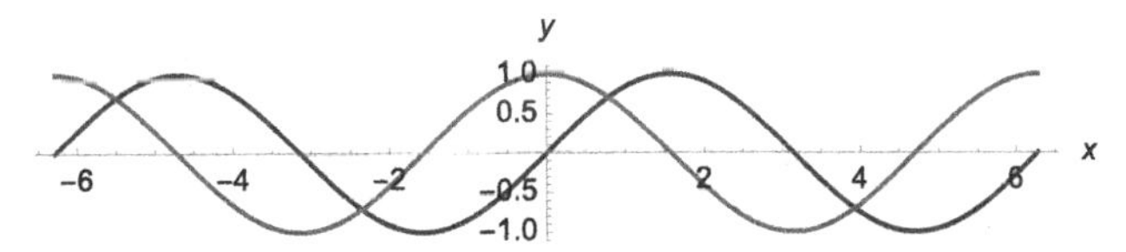

FIGURE 2.3 $y = \sin x$ in blue and $y = \cos x$ in red graphed to scale on the interval $[-2\pi, 2\pi]$. (For interpretation of the colors in this figure, the reader is referred to the web version of this chapter.)

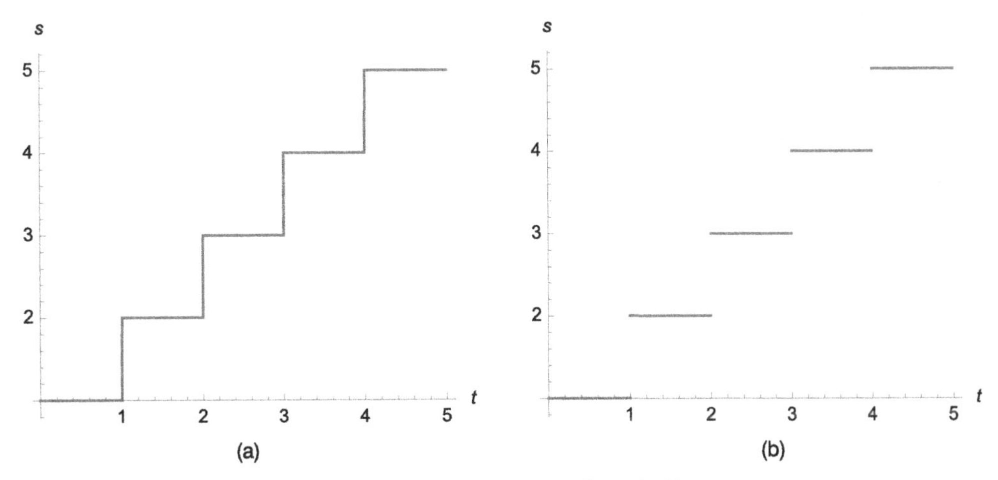

FIGURE 2.4 (a) $s(t) = 1 + s(t-1)$, $0 \le t \le 5$. (b) Catching the discontinuities.

each command. We then combine the graphics with `Show`. The result is shown to scale because the option `AspectRatio->Automatic` is included in the `Show` command (see Fig. 2.3).

```
p1 = Plot[Sin[x], {x, −2Pi, 2Pi}, PlotStyle → Blue];

p2 = Plot[Cos[x], {x, −2Pi, 2Pi}, PlotStyle → Red];

Show[p1, p2, PlotRange → {{−2Pi, 2Pi}, Automatic},

 AxesLabel → {x, y},

  AspectRatio → Automatic]
```

□

Be careful when graphing functions with discontinuities. Often Mathematica will catch discontinuities. In other cases, it doesn't and you might need to use the `Exclusions` option to generate a more accurate plot.

Example 2.19

Graph $s(t)$ for $0 \le t \le 5$ where $s(t) = 1$ for $0 \le t < 1$ and $s(t) = 1 + s(t-1)$ for $t \ge 1$.

Solution. After defining $s(t)$,

```
Clear[s]

s[t_]:=1/;0 ≤ t < 1;

s[t_]:=1 + s[t − 1]/;t ≥ 1;
```

we use `Plot` to graph $s(t)$ for $0 \le t \le 5$ in Fig. 2.4 (a).

```
p1 = Plot[s[t], {t, 0, 5}, AspectRatio → Automatic,

 AxesLabel → {t, s}]
```

Of course, Fig. 2.4 (a) is not completely precise: vertical lines are never the graphs of functions. In this case, discontinuities occur at $t = 1, 2, 3, 4$, and 5. If we were to redraw the figure by hand, we would erase the vertical line segments, and then for emphasis place open dots at $(1, 1)$, $(2, 2)$, $(3, 3)$, $(4, 4)$, and $(5, 5)$ and then closed dots at $(1, 2)$, $(2, 3)$, $(3, 4)$, $(4, 5)$, and $(5, 6)$. In cases like this where `Plot` does not automatically detect discontinuities you can specify them with `Exclusions`. See Fig. 2.4 (b).

```
p2 = Plot[s[t], {t, 0, 5}, AspectRatio → Automatic,
  Exclusions → {1, 2, 3, 4}, AxesLabel → {t, s}]

Show[GraphicsRow[{p1, p2}]]
```

To fine-tune graphics, use the **Drawing Tools** and **Graphics Inspector** palettes, which are accessed under **Graphics** in the menu.

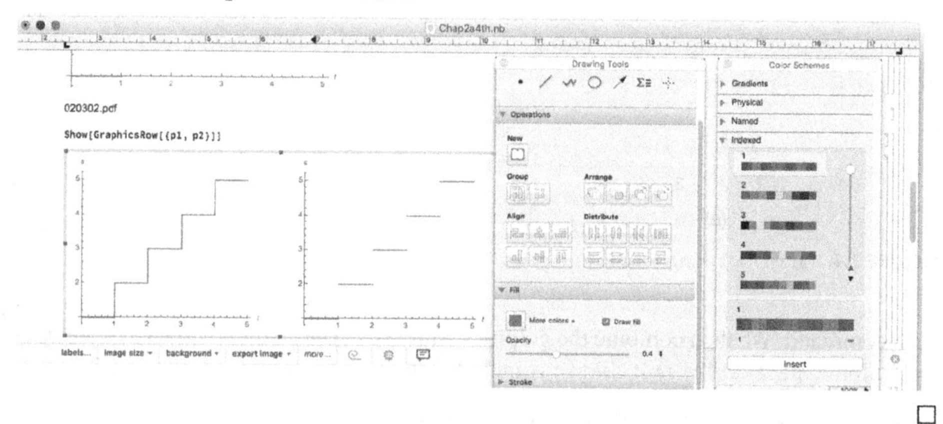

□

Entering `Options[Plot]` lists all `Plot` options and their default values. The most frequently used options include `PlotStyle`, `DisplayFunction`, `AspectRatio`, `PlotRange`, `PlotLabel`, and `AxesLabel`.

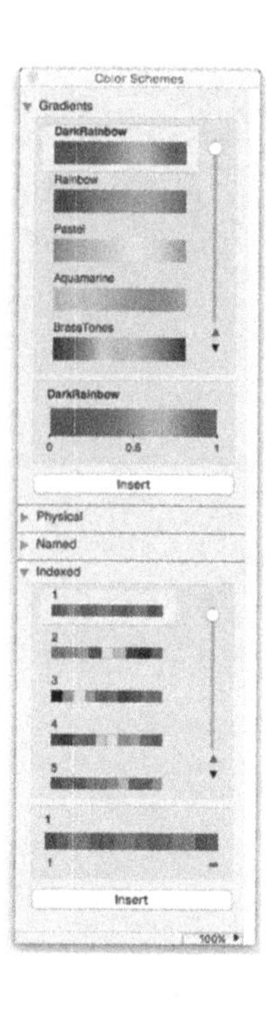

1. `PlotStyle` controls the color and thickness of a plot. `PlotStyle-> GrayLevel[w]`, where $0 \le w \le 1$ instructs Mathematica to generate the plot in `GrayLevel[w]`. `GrayLevel[0]` corresponds to black and `GrayLevel[1]` corresponds to white. Color plots can be generated using `RGBColor`. `RGBColor[1,0,0]` corresponds to red, `RGBColor[0,1,0]` corresponds to green, and `RGBColor[0,0,1]` corresponds to blue. You can also use any of the named colors listed on the **Color Schemes** palette. If you prefer CMYK color coding, use `CMYKColor[c,m,y,k]` to help achieve the desired effects in your graphics.
   `PlotStyle->Dashing[{a1,a2,...,an}]` indicates that successive segments be dashed with repeating lengths of $a_1, a_2, \ldots, a_n$. The thickness of the plot is controlled with `PlotStyle->Thickness[w]`, where w is the fraction of the total width of the graphic. For a single plot, the `PlotStyle` options are combined with `PlotStyle->{{option1, option2, ... , optionn}]`.
2. A plot is not displayed when the option `DisplayFunction-> Identity` is included or when a semi-colon (;) is included at the end of the command. Including the option `DisplayFunction->$ DisplayFunction` in `Show` or `Plot` commands instructs Mathematica to display graphics.
3. The ratio of height to width of a plot is controlled by `AspectRatio`. The default is `1/GoldenRatio`. Generally, a plot is drawn to scale when the option `AspectRatio-> Automatic` is included in the `Plot` or `Show` command.
4. `PlotRange` controls the horizontal and vertical axes. `PlotRange->{c,d}` specifies that the vertical axis displayed corresponds to the interval $c \le y \le d$ while `PlotRange->{{a,b}, {c,d}}` specifes that the horizontal axis displayed corresponds to the interval $a \le x \le b$ and that the vertical axis displayed corresponds to the interval $c \le y \le d$.

5. `PlotLabel->"titleofplot"` labels the plot `titleofplot`.
6. `AxesLabel->{"xaxislabel","yaxislabel"}` labels the x-axis with `xaxislabel` and the y-axis with `yaxislabel`.

Example 2.20

Graph $y = \sin x$, $y = \cos x$, and $y = \tan x$ together with their inverse functions.

Solution. In `p2` and `p3`, we use `Plot` to graph $y = \sin^{-1} x$ and $y = x$, respectively. Neither plot is displayed because we include a semi-colon at the end of the `Plot` commands. `p1`, `p2`, and `p3` are displayed together with `Show` in Fig. 2.5 (a). The plot is shown to scale; the graph of $y = \sin x$ is in black, $y = \sin^{-1} x$ is in gray, and $y = x$ is dashed.

```
p1 = Plot[Sin[x], {x, -2Pi, 2Pi},
  PlotStyle -> Black];
p2 = Plot[ArcSin[x], {x, -1, 1},
  PlotStyle -> Gray];
p3 = Plot[x, {x, -2Pi, 2Pi},
  PlotStyle -> {{Black, Dashing[{0.01}]}}];
t1a = Show[p1, p2, p3, PlotRange -> {-2Pi, 2Pi},
  AspectRatio -> Automatic, AxesLabel -> {x, y}]
```

The command `Plot[{f1[x],f2[x],...,fn[x]},{x,a,b}]` plots $f_1(x)$, $f_2(x)$, ..., $f_n(x)$ together for $a \le x \le b$. Simple `PlotStyle` options are incorporated with

```
PlotStyle->{option1,option2,...,optionn
```

where `optioni` corresponds to the plot of $f_i(x)$. Multiple options are incorporated using

```
PlotStyle->{{options1},{options2},...,{optionsn}}
```

where `optionsi` are the options corresponding to the plot of $f_i(x)$.

In the following, we use `Plot` to graph $y = \cos x$, $y = \cos^{-1} x$, and $y = x$ together. The plot in Fig. 2.5 (b) is shown to scale; the graph of $y = \cos x$ is in black, $y = \cos^{-1} x$ is in gray, and $y = x$ is dashed.

```
p1 = Plot[Cos[x], {x, 0, 2Pi},
  PlotStyle -> Black];
p2 = Plot[ArcCos[x], {x, -1, 1},
  PlotStyle -> Gray];
p3 = Plot[x, {x, -2Pi, 2Pi},
  PlotStyle -> {{Black, Dashing[{0.01}]}}];
t1b = Show[p1, p2, p3, PlotRange -> {{-Pi, 2Pi}, {-Pi, 2Pi}},
  AspectRatio -> Automatic, AxesLabel -> {x, y}]
```

We use the same approach to graph $y = \tan x$, $y = \tan^{-1} x$, and $y = x$ in Fig. 2.6.

```
p1 = Plot[Tan[x], {x, -Pi/2, Pi/2},
  PlotStyle -> Black];
```

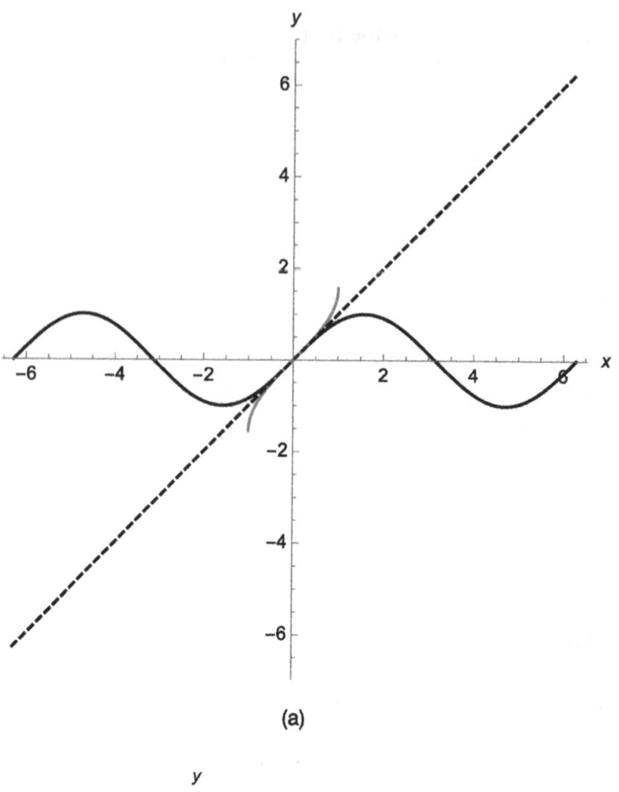

(a)

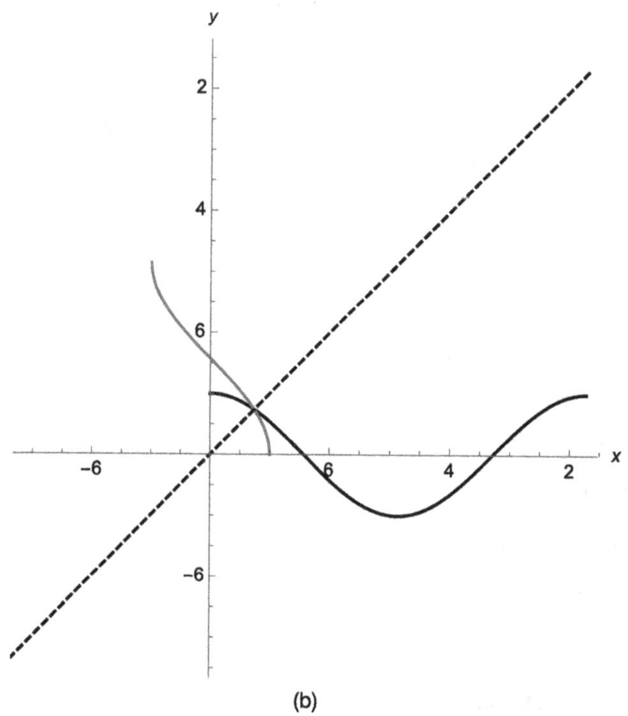

(b)

FIGURE 2.5 (a) $y = \sin x$, $y = \sin^{-1} x$, and $y = x$. (b) $y = \cos x$, $y = \cos^{-1} x$, and $y = x$.

```
p2 = Plot[ArcTan[x], {x, -2Pi, 2Pi},
  PlotStyle → Gray];
p3 = Plot[x, {x, -2Pi, 2Pi},
  PlotStyle → {{Black, Dashing[{0.01}]}}];
```

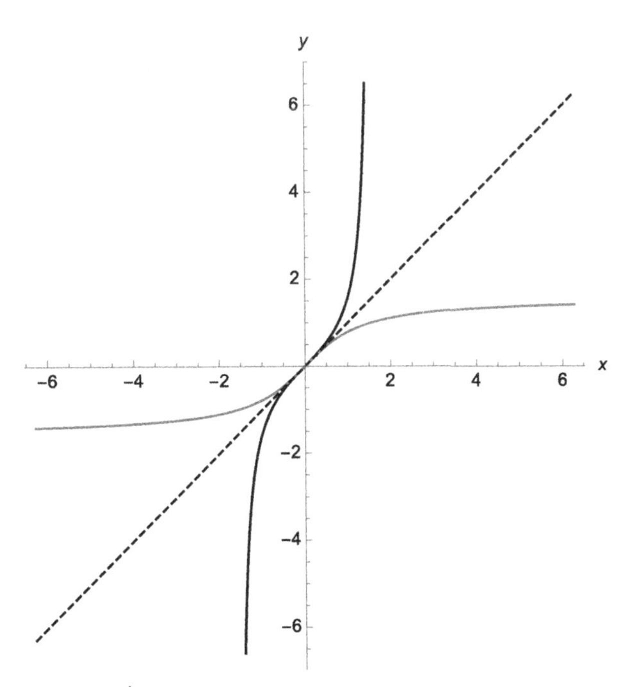

FIGURE 2.6 $y = \tan x$, $y = \tan^{-1} x$, and $y = x$.

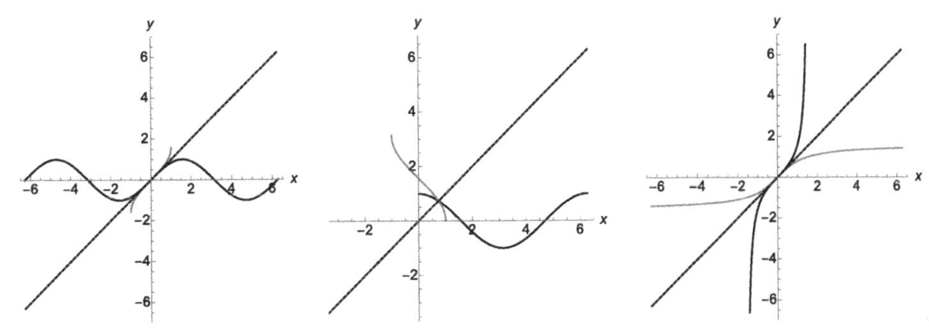

FIGURE 2.7 The elementary trigonometric functions and their inverses.

t1c = Show[p1, p2, p3, PlotRange → {{−2Pi, 2Pi}, {−2Pi, 2Pi}},

AspectRatio → Automatic, AxesLabel → {x, y}]

Use `Show` together with `GraphicsRow` to display graphics in rectangular arrays. Entering

Show[GraphicsRow[{p2, p3, t1c}]]

shows the three plots `p2`, `p3`, and `t1c` in a row as shown in Fig. 2.7. □

Example 2.21: Inverse Functions

$f(x)$ and $g(x)$ are **inverse functions** if

$$f(g(x)) = g(f(x)) = x.$$

If $f(x)$ and $g(x)$ are inverse functions, their graphs are symmetric about the line $y = x$. The command

```
Composition[f1,f2,f3,...,fn,x]
```

computes the composition

$$(f_1 \circ f_2 \circ \cdots \circ f_n)(x) = f_1(f_2(\cdots(f_n(x)))).$$

For two functions $f(x)$ and $g(x)$, it is usually easiest to compute the composition $f(g(x))$ with `f[g[x]]` or `f[x]//g`.

Show that

$$f(x)=e^x \quad \text{and} \quad g(x)=\ln x$$

are inverse functions.

Solution. When using Mathematica, `Exp[x]` represents the function $y = e^x$. Similarly, `Log[x]` represents the natural logarithm function, $y = \ln x$.

After defining $f(x)$ and $g(x)$,

$f(x)$ and $g(x)$ are not returned because a semi-colon is included at the end of each command.

```
f[x_] = Exp[x];
g[x_] = Log[x];
```

```
f[g[x]]
```

x

```
Simplify[g[f[x]]]
```

$\text{Log}\left[e^x\right]$

we compute and simplify the compositions $f(g(x))$ and $g(f(x))$. Because both results simplify to $f(g(x)) = g(f(x)) = x$, $f(x)$ and $g(x)$ are inverse functions. Observe that `Simplify` does not automatically reduce $\ln(e^x) = x$. To do so, we use `PowerExpand`.

```
PowerExpand[g[f[x]]]
```

x

To see that the graphs of $f(x)$ and $g(x)$ are symmetric about the line $y = x$, we use `Plot` to graph $f(x)$, $g(x)$, and $y = x$ together in Fig. 2.8. Because `Tooltip` is being applied to the set of functions being plotted, you can identify each curve by sliding the cursor over the curve: when the cursor is placed over a curve, Mathematica displays its definition. In the `Plot` command, observe how we are able to graph all three functions, $f(x) = e^x$, $g(x) = \ln x$, and $y = x$ together as well as control how each graph is plotted using the `PlotStyle` option. The range (y-axis) displayed is controlled using the `PlotRange` option. The option `AspectRatio->1` instructs Mathematica to display the axes in a $1-1$ (square) ratio. In this case, because the domains and ranges are the same, the resulting plot is drawn to scale. If the domains and ranges are not the same and you wish to have the plot displayed to scale, use the option `AspectRatio->Automatic`.

```
Plot[Tooltip[{f[x], g[x], x}], {x, -10, 10},
  PlotStyle → {{Black}, {Gray}, {LightGray, Dashing[.01]}},
  PlotRange → {-10, 10}, AspectRatio → 1]
```

In the plot, observe that the graphs of $f(x) = e^x$ and $g(x) = \ln x$ are symmetric about the line $y = x$. The plot also illustrates that the domain and range of a function and its inverse are interchanged: $f(x)$ has domain $(-\infty, \infty)$ and range $(0, \infty)$; $g(x)$ has domain $(0, \infty)$ and range $(-\infty, \infty)$. □

For repeated compositions of a function with itself, `Nest[f,x,n]` computes the composition

$$\underbrace{(f \circ f \circ f \circ \cdots \circ f)}_{n \text{ times}}(x) = \underbrace{(f(f(f \cdots)))}_{n \text{ times}}(x) = f^n(x).$$

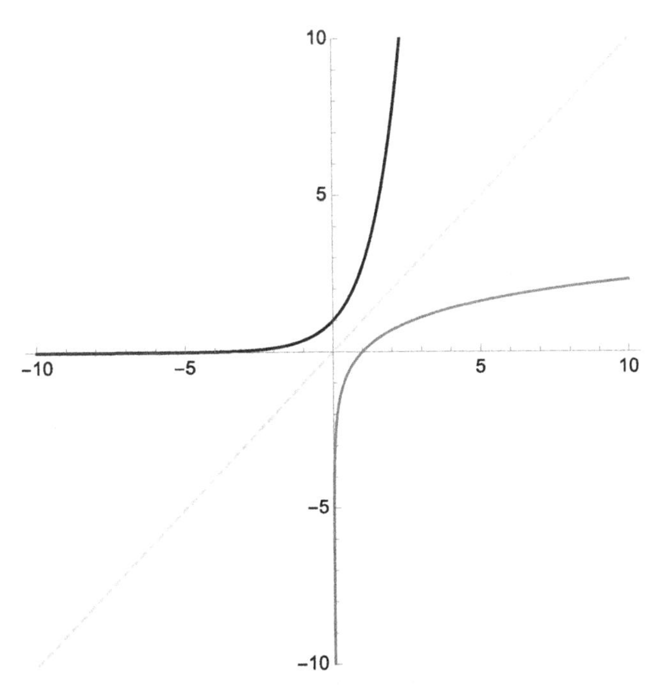

FIGURE 2.8 $f(x) = e^x$ in black, $g(x) = \ln x$ in gray, and $y = x$ dashed in light gray.

Example 2.22

Graph $f(x)$, $f^{10}(x)$, $f^{20}(x)$, $f^{30}(x)$, $f^{40}(x)$, and $f^{50}(x)$ if $f(x) = \sin x$ for $0 \le x \le 2\pi$.

Solution. After defining $f(x) = \sin x$,

```
f[x_] = Sin[x];
```

we graph $f(x)$ in `p1` with `Plot`

```
p1 = Plot[f[x], {x, 0, 2Pi}];
```

and then illustrate the use of `Nest` by computing $f^5(x)$.

```
Nest[f, x, 5]
```

Sin[Sin[Sin[Sin[Sin[x]]]]]

Next, we use `Table` together with `Nest` to create the list of functions

$$\left\{f^{10}(x),\ f^{20}(x),\ f^{30}(x),\ f^{40}(x),\ f^{50}(x)\right\}.$$

Because the resulting output is rather long, we include a semi-colon at the end of the `Table` command to suppress the resulting output.

`Table[f[i],{i,a,b,istep}]` computes $f(i)$ for i values from a to b using increments of *istep*.

```
toplot = Table[Nest[f, x, n], {n, 10, 50, 10}];
```

We then graph the functions in `toplot` on the interval $[0, 2\pi]$ with `Plot`, applying the `Tooltip` function to the list being plotted so they can easily be identified. Last, we use `Show` to display `p1` and `p2` together in Fig. 2.9.

```
p2 = Plot[Tooltip[Evaluate[toplot]], {x, 0, 2Pi}];

Show[p1, p2]
```

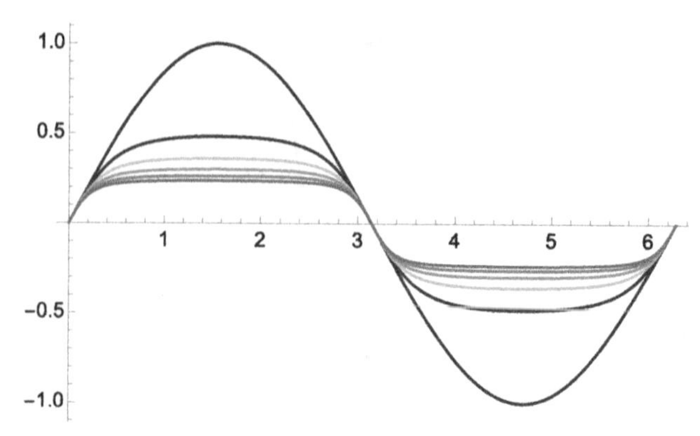

FIGURE 2.9 $f(x)$ in blue; the graphs of $f^{10}(x)$, $f^{20}(x)$, $f^{30}(x)$, $f^{40}(x)$, and $f^{50}(x)$ are in different colors. You can use `Tooltip` to identify each curve. (For interpretation of the references to color in this figure legend, the reader is referred to the web version of this chapter.)

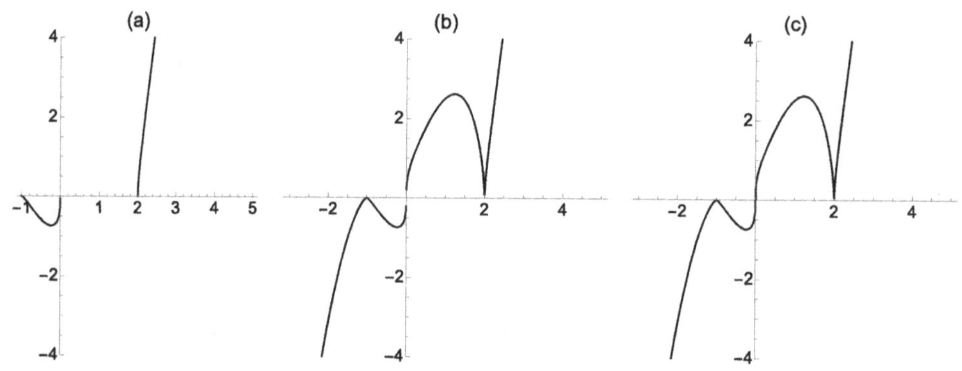

FIGURE 2.10 (a), (b) and (c) three plots of $y = x^{1/3}(x-2)^{2/3}(x+1)^{4/3}$.

In thc plot, we see that repeatedly composing sine with itself has a flattening effect on $y = \sin x$. □

When graphing functions involving odd roots, Mathematica's results may be surprising to the beginner. The key is to use the `Surd` function *or* remember that Mathematica follows the order of operations *exactly* and understand that without restrictions on x, $\sqrt{x^2} = |x|$.

Example 2.23

Graph $y = x^{1/3}(x-2)^{2/3}(x+1)^{4/3}$.

Solution. Entering

```
p1 = Plot[x^(1/3)(x − 2)^(2/3)(x + 1)^(4/3),
  {x, −3, 5}, PlotRange → {−4, 4},
  AspectRatio → Automatic, PlotStyle → Black,
   PlotLabel → "(a)"]
```

does not produce the graph we expect (see Fig. 2.10 (a)) because many of us consider $y = x^{1/3}(x-2)^{2/3}(x+1)^{4/3}$ to be a real-valued function with domain $(-\infty, \infty)$. Generally, Mathematica does return a real number when computing the odd root of a negative number. For example, $x^3 = -1$ has three solutions

`Solve` is discussed in more detail in the next section.

```
Clear[x]
```

s1 = Solve[x^3== − 1]

$\{\{x \to -1\}, \{x \to (-1)^{1/3}\}, \{x \to -(-1)^{2/3}\}\}$

***N*[s1]**

$\{\{x \to -1.\}, \{x \to 0.5 + 0.866025i\}, \{x \to 0.5 - 0.866025i\}\}$

When computing an odd root of a negative number, Mathematica has many choices (as illustrated above) and chooses a root with positive imaginary part – the result is not a real number.

(−1)^(1/3)//*N*

$0.5 + 0.866025i$

To obtain real values when computing odd roots of negative numbers, first let $sign(x) = \begin{cases} x/|x|, & \text{if } x \neq 0, \\ 0, & \text{if } x = 0. \end{cases}$ `Sign[x]` returns $sign(x)$. Then, for the reduced fraction n/m with m odd, $x^{n/m} = \begin{cases} sign(x)\,|x|^{n/m}, & \text{if } n \text{ is odd} \\ |x|^{n/m}, & \text{if } n \text{ is even} \end{cases}$. See Fig. 2.10 (b).

p2 = Plot[Sign[x]Abs[x]^(1/3)Abs[x − 2]^(2/3)Abs[x + 1]^(4/3),

{x, −3, 5}, PlotRange → {−4, 4}, PlotStyle → Black,

AspectRatio → Automatic, PlotLabel → “(b)”]

Alternatively, use the `Surd` function to obtain the desired results. `Surd[x,n]` returns the real-valued nth root of x.

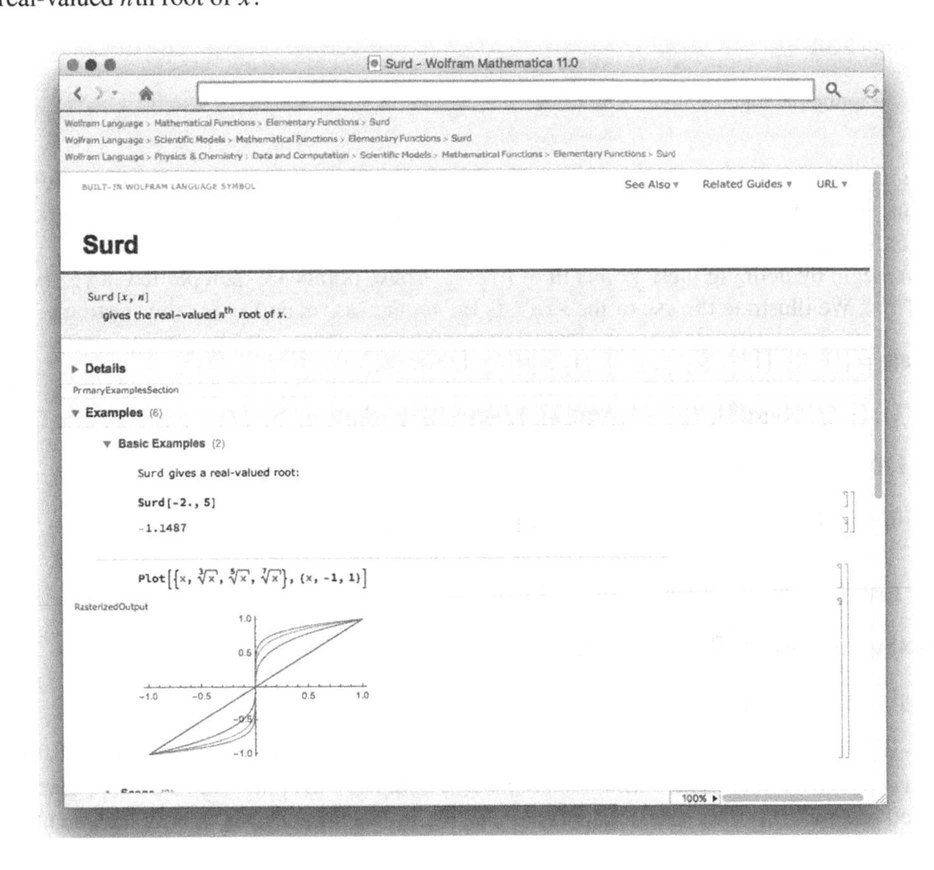

```
p3 = Plot[Surd[x, 3]Surd[(x − 2)^2, 3]Surd[(x + 1)^4, 3],
  {x, −3, 5}, PlotRange → {−4, 4},
  AspectRatio → Automatic, PlotStyle → Black,
   PlotLabel → "(c)"]
```

All three graphics are shown in Fig. 2.10 using `Show` together with `GraphicsRow`.

```
Show[GraphicsRow[{p1, p2, p3}]]
```

□

The command

`ListPlot[{{x1,y1},{x2,y2},...,{xn,yn}}]`

plots the list of points $\{(x_1, y_1), (x_2, y_2), \ldots, (x_n, y_n)\}$. The size of the points in the resulting plot is controlled with the option `PlotStyle->PointSize[w]`, where w is the fraction of the total width of the graphic. For two-dimensional graphics, the default value is 0.008. The command

`ListPlot[{y1,y2,...,yn}]`

plots the list of points $\{(1, y_1), (2, y_2), \ldots, (n, y_n)\}$. To connect successive points with line segments, use the command `ListLinePlot`.

`ListLinePlot[{x1,y1},{x2,y2},...,{xn,yn}}]`

connects the consecutive points (x_1, y_1), (x_2, y_2), ..., (x_n, y_n) with line segments.

Example 2.24

Plot the "principal trigonometric values."

Solution. The "principal trigonometric values" are usually interpreted to mean the values of the functions $\sin x$ and $\cos x$ at the angles $\theta = 0, \pi/6, \pi/4, \pi/3, \pi/2, 2\pi/3, 3\pi/4, 5\pi/6, \pi$, $7Pi/6, 5\pi/4, 4\pi/3, 3\pi/2, 5\pi/3, 7\pi/4, 11\pi/6$, and 2π.

We begin by defining these points in `prinvals`. These points are then plotted in `lp1` with `ListPlot`. We illustrate the use of the `PlotStyle`, `AspectRatio`, and `PlotLabel` options.

```
prinvals = {{1, 0}, {1/2, Sqrt[3]/2}, {1/Sqrt[2], 1/Sqrt[2]}, {Sqrt[3]/2, 1/2},
  {0, 1}, {−1/2, Sqrt[3]/2}, {−1/Sqrt[2], 1/Sqrt[2]}, {−Sqrt[3]/2, 1/2},
  {−1, 0}, {−1/2, −Sqrt[3]/2}, {−1/Sqrt[2], −1/Sqrt[2]}, {−Sqrt[3]/2, −1/2},
   {0, −1}, {1/2, −Sqrt[3]/2}, {1/Sqrt[2], −1/Sqrt[2]}, {Sqrt[3]/2, −1/2}};
```

```
lp1 = ListPlot[prinvals, AspectRatio → Automatic,
  PlotStyle → {Black, PointSize[.025]},
  PlotLabel → "(a)"]
```

Next, we define `prinlines` to be a set of ordered pairs that will draw line segments between connected points. These lines are graphed by using `Map` to apply `ListLinePlot` to each ordered pair in `prinlines`. Generally, `Map[f,list]` applies the function `f` to each element of list. We use Show and illustrate the use of the AspectRatio and PlotLabel options to graph these line segments in lp2.

```
prinlines = {{{1, 0}, {−1, 0}}, {{1/2, Sqrt[3]/2}, {−1/2, −Sqrt[3]/2}},
  {{1/Sqrt[2], 1/Sqrt[2]}, {−1/Sqrt[2], −1/Sqrt[2]}},
  {{Sqrt[3]/2, 1/2}, {−Sqrt[3]/2, −1/2}},
   {{0, 1}, {0, −1}}, {{−1/2, Sqrt[3]/2}, {1/2, −Sqrt[3]/2}},
  {{−1/Sqrt[2], 1/Sqrt[2]}, {1/Sqrt[2], −1/Sqrt[2]}},
   {{−Sqrt[3]/2, 1/2}, {Sqrt[3]/2, −1/2}}};

prinlines2 = Map[ListLinePlot[#,
  PlotStyle → {{GrayLevel[.5], Dashing[.01]}}]&, prinlines];

lp2 = Show[prinlines2, AspectRatio → Automatic,
  PlotLabel → "(b)"]
```

In lp3, we use Plot to graph the functions $y = \sqrt{1-x^2}$ and $y = -\sqrt{1-x^2}$. In the color version of this text, the CMYKColor code corresponds to a deep red. The graph is drawn to scale because we include the option AspectRatio->Automatic.

```
lp3 = Plot[{Sqrt[1 − x^2], −Sqrt[1 − x^2]}, {x, −1, 1},
  PlotStyle → CMYKColor[0, 0.89, 0.94, 0.28],
  AspectRatio → Automatic, PlotLabel → "(c)"]
```

Finally, we use Show to display all four graphics together in lp4.

```
lp4 = Show[lp1, lp2, lp3, PlotLabel → "(d)"]
```

The four images are displayed as a graphics grid using Show together with GraphicsGrid in Fig. 2.11.

```
Show[GraphicsGrid[{{lp1, lp2}, {lp3, lp4}}]]
```

□

A comprehensive discussion of Mathematica's extensive graphics capabilities cannot be reasonably covered in a single section of a text so our approach is to address issues that might of interest or present a different point of view to the beginning user of Mathematica. In the previous example, we saw that $x^3 + 1 = 0$ has three solutions, two of which are complex. To visualize this graphically, observe that the zeros of $z^3 + 1 = 0$ are the level curves of $f(x, y) = \left|(x + iy)^3 + 1\right|$ (x, y real) corresponding to 0. In a plot of $f(x, y)$, the solutions are the zeros. Shortly we will discuss ContourPlot and Plot3D. For now, we remark that

```
cp1 = ContourPlot[Abs[(x + I y)^3 + 1], {x, −2, 1}, {y, −3/2, 3/2},
  Contours → 30, Axes → True]

p13d = Plot3D[Abs[(x + I y)^3 + 1], {x, −2, 1}, {y, −3/2, 3/2},
  Axes → True, PlotRange → {0, 15}, MeshFunctions → (#3&),
  Mesh → 35]
```

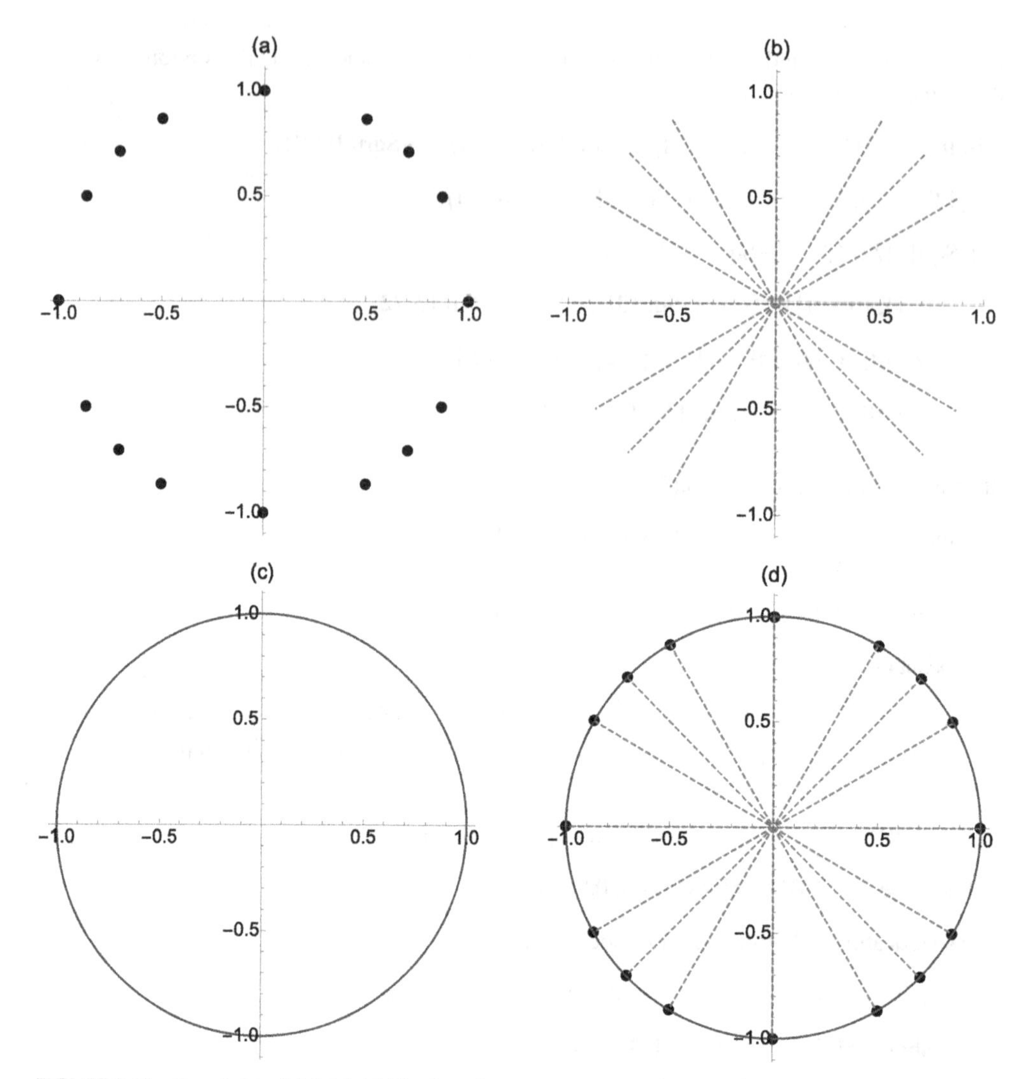

FIGURE 2.11 (from left to right) (a)–(d) The four plots `lp1`, `lp2`, `lp3`, and `lp4` combined into a single graphic.

Show[GraphicsRow[{cp1, p13d}]]

generates several level curves of $f(x, y)$ (Fig. 2.12 (a)) and a 3-d plot of $f(x, y)$ (Fig. 2.12 (b)) that help us see the zeros of the original equation. In the 3-d plot, note how we use the `MeshFunctions` option to generate contours on the 3-d plot.

2.3.2 Parametric and Polar Plots in Two Dimensions

For parametrically defined functions, $x = x(t)$, $y = y(t)$, $a \le t \le b$, use `ParametricPlot`. `ParametricPlot[{x[t],y[t]},{t,a,b}]` attempts to graph the parametrically defined functions $x = x(t)$, $y = y(t)$, $a \le t \le b$.
To graph the polar function, $r = f(\theta)$, use `PolarPlot`.

```
PolarPlot[f[theta],{theta,alpha,beta}]
```

attempts to graph $r = f(\theta)$ for $\alpha \le \theta \le \beta$.

To obtain the θ symbol with Mathematica on a Mac, press the esc key, type "theta", and then press the esc key again. Use the same process to obtain other symbols, such as α, β, and so on.

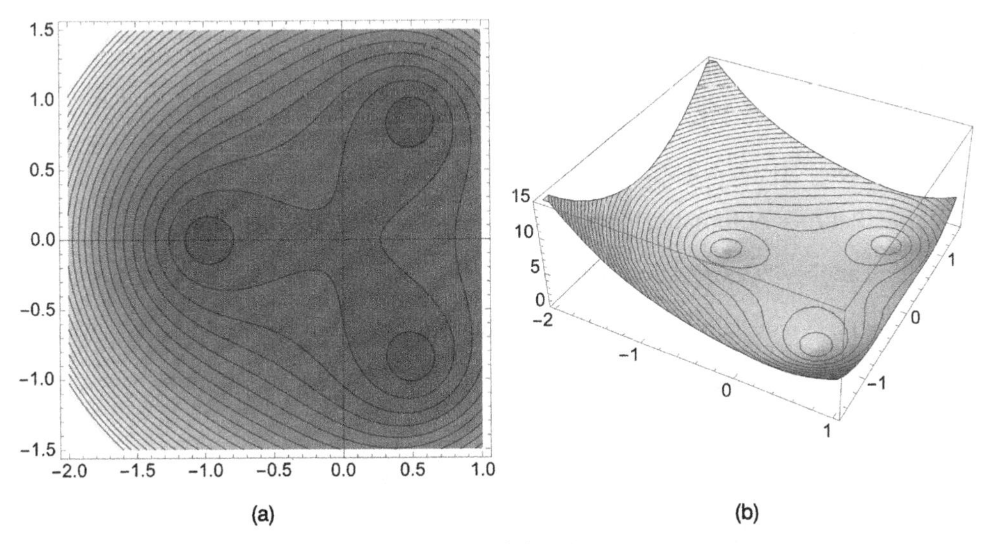

FIGURE 2.12 (a) Contour plot of $f(x, y)$. (b) 3-d plot of $f(x, y)$.

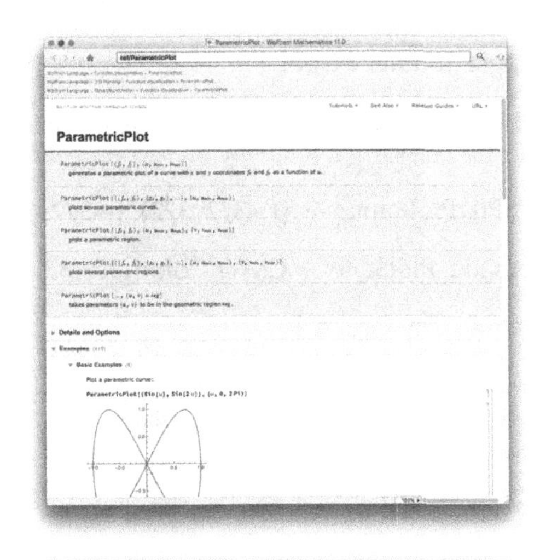

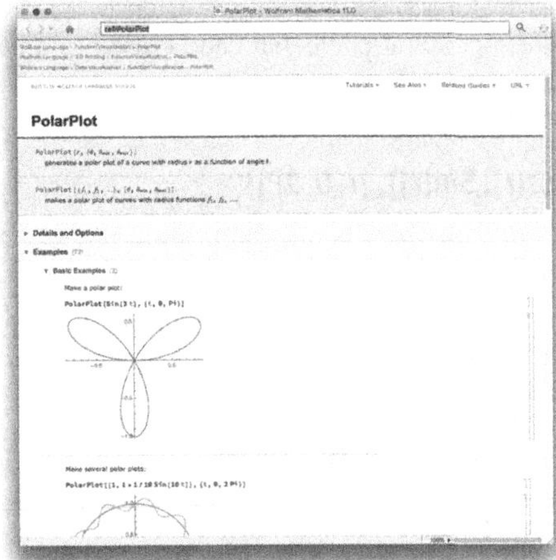

Example 2.25: The Unit Circle

The **unit circle** is the set of points (x, y) exactly 1 unit from the origin, $(0, 0)$, and, in rectangular coordinates, has equation $x^2 + y^2 = 1$. The unit circle is the classic example of a relation that is neither a function of x nor a function of y. The top half of the unit circle is given by $y=\sqrt{1-x^2}$ and the bottom half is given by $y=-\sqrt{1-x^2}$.

```
p1 = Plot[{Sqrt[1 - x^2], -Sqrt[1 - x^2]}, {x, -1, 1},
  PlotRange → {{-3/2, 3/2}, {-3/2, 3/2}},
  AspectRatio → Automatic, PlotStyle → CMYKColor[1.00, .70, 0, .75],
   PlotLabel → "Georgia Southern Blue"]
```

Each point (x, y) on the unit circle is a function of the angle, t, that subtends the x-axis, which leads to a parametric representation of the unit circle, $\begin{cases} x=\cos t, \\ y=\sin t, \end{cases}$ $0 \le t \le 2\pi$, which we graph with `ParametricPlot`.

```
p2 = ParametricPlot[{Cos[t], Sin[t]}, {t, 0, 2Pi},
  PlotRange → {{-3/2, 3/2}, {-3/2, 3/2}},
  AspectRatio → Automatic, PlotStyle → CMYKColor[.40, .43, .84, .08],
   PlotLabel → "Georgia Southern Gold"]
```

Using the change of variables $x=r\cos t$ and $y=r\sin t$ to convert from rectangular to polar coordinates, a polar equation for the unit circle is $r=1$. We use `PolarPlot` to graph $r=1$.

```
p3 = PolarPlot[1, {t, 0, 2Pi}, PlotRange → {{-3/2, 3/2}, {-3/2, 3/2}},
  AspectRatio → Automatic, PlotStyle → CMYKColor[.97, .33, .78, .24],
  PlotLabel → "Ohio University Green"]
```

We display `p1`, `p2`, and `p3` side-by-side using `Show` together with `GraphicsRow` in Fig. 2.13. Of course, they all look the same, just in different colors (or shades of gray if you are reading the printed version of this text).

```
Show[GraphicsRow[{p1, p2, p3}]]
```

To illustrate several other features of Mathematica's extensive graphics capability, we elaborate on the example and construct a unit circle that can be used as a handout in a trigonometry, pre-calculus, or calculus class.

In `p1`, we graph the unit circle.

```
p1 = ParametricPlot[{Cos[t], Sin[t]}, {t, 0, 2Pi},
  PlotStyle → Black]
```

Next, we use `ParametricPlot` to connect the principal value points on the unit circle with line segments by observing that the parametric equations $x=(1-t)a+tc$, $y=(1-t)b+td$, $0 \le t \le 1$ graphs a line segment connecting (a, b) and (c, d).

```
p2a = ParametricPlot[{{1/2, Sqrt[3]/2}(t - 1),
  {-1/2, -Sqrt[3]/2}(t - 1)},
  {t, 0, 1}, PlotStyle → {{Black, Dashing[{.01}]}},
   PlotLabel → "(b)"]
```

```
p2b = ParametricPlot[{{Sqrt[3]/2, 1/2}(t − 1),
    {−Sqrt[3]/2, −1/2}(t − 1)},
    {t, 0, 1}, PlotStyle → {{Black, Dashing[{.01}]}}]
p2c = ParametricPlot[{{1/Sqrt[2], 1/Sqrt[2]}(t − 1),
    {−1/Sqrt[2], −1/Sqrt[2]}(t − 1)},
    {t, 0, 1}, PlotStyle → {{Black, Dashing[{.01}]}}]
p2d = ParametricPlot[{{−1/2, Sqrt[3]/2}(t − 1),
    {1/2, −Sqrt[3]/2}(t − 1)},
    {t, 0, 1}, PlotStyle → {{Black, Dashing[{.01}]}}]
p2e = ParametricPlot[{{−Sqrt[3]/2, 1/2}(t − 1),
    {Sqrt[3]/2, −1/2}(t − 1)},
    {t, 0, 1}, PlotStyle → {{Black, Dashing[{.01}]}}]
p2f = ParametricPlot[{{−1/Sqrt[2], 1/Sqrt[2]}(t − 1),
    {1/Sqrt[2], −1/Sqrt[2]}(t − 1)},
    {t, 0, 1}, PlotStyle → {{Black, Dashing[{.01}]}}]
q1 = Show[p1, p2a, p2b, p2c, p2f, p2e, p2d]
```

Next, we use the graphics primitive `Text` to label the principal values on the unit circle.

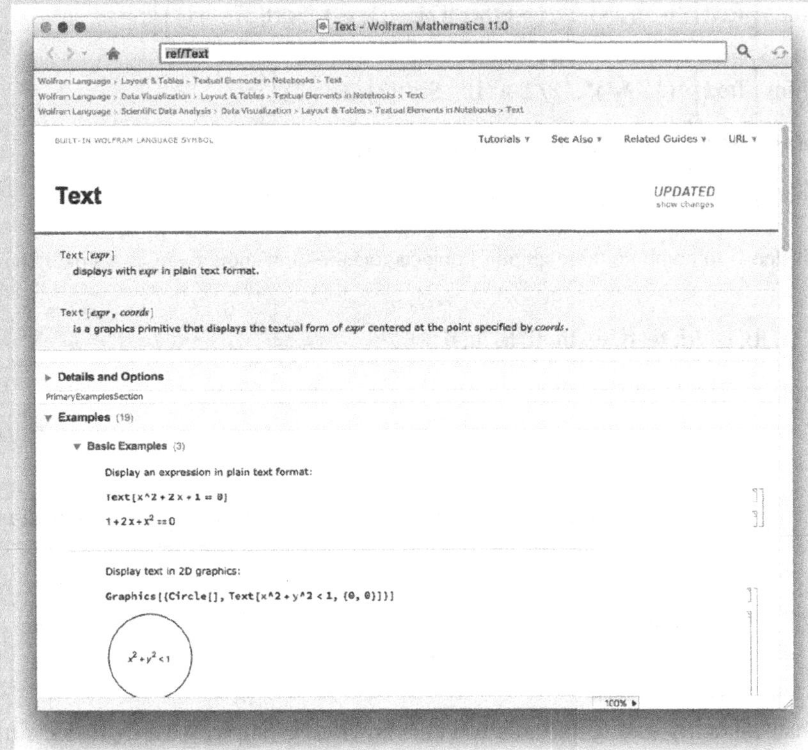

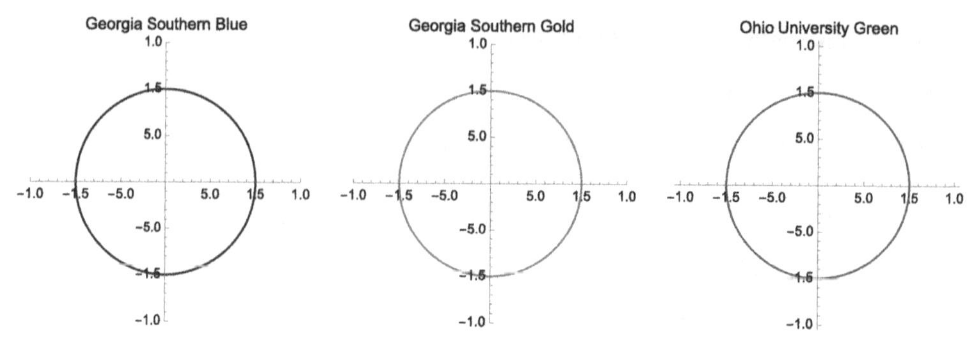

FIGURE 2.13 The unit circle generated with `Plot`, `ParametricPlot`, and `PolarPlot`.

```
ta = Graphics[Text["(1/2, √3/2)", {1/2 + .1, Sqrt[3]/2 + .1}]];
tb = Graphics[Text["(√3/2, 1/2)", {Sqrt[3]/2 + .1, 1/2 + .1}]];
tc = Graphics[Text["(1/√2, 1/√2)", {1/Sqrt[2] + .1, 1/Sqrt[2] + .1}]];

td = Graphics[Text["(-1/2, √3/2)", {−1/2 − .1, Sqrt[3]/2 + .1}]];
te = Graphics[Text["(-√3/2, 1/2)", {−Sqrt[3]/2 − .1, 1/2 + .1}]];
tf = Graphics[Text["(-1/√2, 1/√2)", {−1/Sqrt[2] − .1, 1/Sqrt[2] + .1}]];

tg = Graphics[Text["(-1/2,-√3/2)", {−1/2 − .1, −Sqrt[3]/2 − .1}]];
th = Graphics[Text["(-√3/2,-1/2)", {−Sqrt[3]/2 − .1, −1/2 − .1}]];
ti = Graphics[Text["(-1/√2,-1/√2)", {−1/Sqrt[2] − .1, −1/Sqrt[2] − .1}]];

tj = Graphics[Text["(1/2,-√3/2)", {1/2 + .1, −Sqrt[3]/2 − .1}]];
tk = Graphics[Text["(√3/2,-1/2)", {Sqrt[3]/2 + .1, −1/2 − .1}]];
tl = Graphics[Text["(1/√2,-1/√2)", {1/Sqrt[2] + .1, −1/Sqrt[2] − .1}]];
```

Our last step is to combine these graphics objects together and show them as a single image with `Show`.

```
Show[q1, ta, tb, tc, td, te, tf, tg, th, ti, tj, tk, tl,
  PlotLabel → "The Unit Circle"]
```

In the following example, the equations of the parametrically defined functions involve integrals.

Remark 2.2. Topics from calculus are discussed in Chapter 3. For now, we state that `Integrate[f[x],{x,a,b}]` attempts to evaluate $\int_a^b f(x)\,dx$.

Rather than explicitly stating the color as in the previous example, we use gradients from the **Color Scheme** palette to adjust the colors in the resulting graphics (see Fig. 2.14).

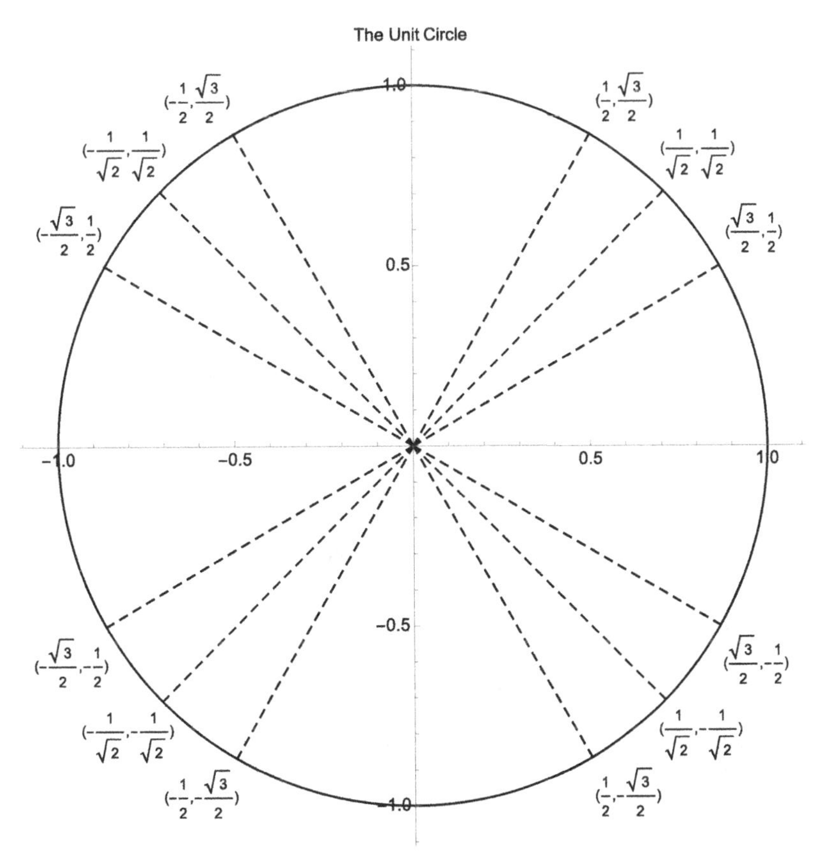

FIGURE 2.14 A labeled unit circle. Within Mathematica, you can increase the size of the graphic by clicking and dragging on the bounding box so that the result fills a page that when printed is suitable for distribution to classes such as trigonometry, pre-calculus, and calculus classes.

Example 2.26: Cornu Spiral

The **Cornu spiral** (or **clothoid**) (see [8] and [17]) has parametric equations

$$x(t) = \int_0^t \sin\left(\frac{1}{2}u^2\right)\,du \quad \text{and} \quad y(t) = \int_0^t \cos\left(\frac{1}{2}u^2\right)\,du.$$

Graph the Cornu spiral.

Solution. We begin by defining x and y. Notice that Mathematica can evaluate these integrals, even though the results are in terms of the `FresnelS` and `FresnelC` functions, which are defined in terms of integrals:

$$\texttt{FresnelS[t]} = \int_0^t \sin\left(\frac{\pi}{2}u^2\right)\,du \quad \text{and} \quad \texttt{FresnelC[t]} = \int_0^t \cos\left(\frac{\pi}{2}u^2\right)\,du.$$

```
x[t_] = Integrate[Sin[u^2/2], {u, 0, t}]
```

$\sqrt{\pi}\text{FresnelS}\left[\frac{t}{\sqrt{\pi}}\right]$

```
y[t_] = Integrate[Cos[u^2/2], {u, 0, t}]
```

$\sqrt{\pi}\text{FresnelC}\left[\frac{t}{\sqrt{\pi}}\right]$

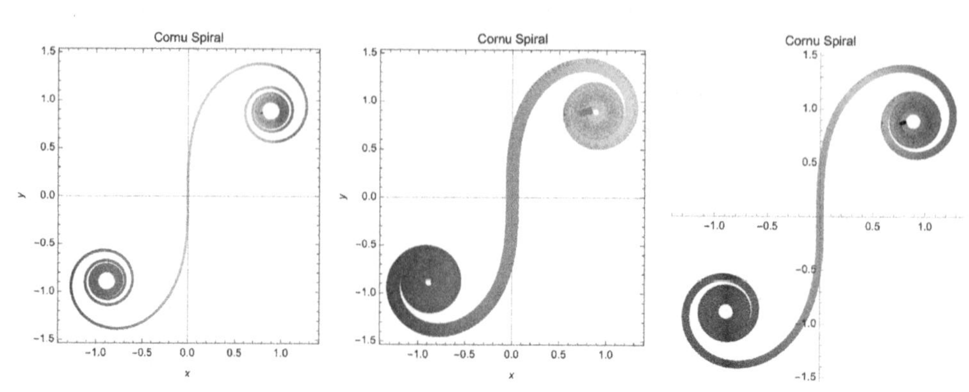

FIGURE 2.15 Graphing the Cornu Spiral using different color gradients and thickness. Note that in the printed version of this text, all images will be in gray.

We use `ParametricPlot` to graph the Cornu spiral in Fig. 2.15. The option `AspectRatio->Automatic` instructs Mathematica to generate the plot to scale; `PlotLabel->"Cornu spiral"` labels the plot. Rather than explicitly stating the color, we use gradients from the **Color Scheme** palette to adjust the color of the plot.

```
p1 = ParametricPlot[{x[t], y[t]}, {t, −10, 10}, AspectRatio → Automatic,
  PlotStyle → Thickness[.01], PlotLabel → "Cornu Spiral",
  Frame → True, FrameLabel → {x, y}, ColorFunction →
   (ColorData["SouthwestColors"][#1]&)]

p2 = ParametricPlot[{x[t], y[t]}, {t, −10, 10}, AspectRatio → Automatic,
  PlotStyle → Thickness[.05], PlotLabel → "Cornu Spiral",
  Frame → True, FrameLabel → {x, y}, ColorFunction →
   (ColorData["SolarColors"][#1]&)]

p3 = ParametricPlot[{x[t], y[t]}, {t, −10, 10}, AspectRatio → Automatic,
  PlotStyle → Thickness[.025], PlotLabel → "Cornu Spiral",
  ColorFunction →
   (ColorData["DarkRainbow"][#1]&)]
```

We use `Show` together with `GraphicsRow` to display the three graphics together in Fig. 2.15.

```
Show[GraphicsRow[{p1, p2, p3}]]
```
□

Observe that the graph of the polar equation $r(\theta) = f(\theta), \alpha \le \theta \le \beta$ is the same as the graph of the parametric equations

$$x(\theta) = f(\theta)\cos\theta \quad \text{and} \quad y(\theta) = f(\theta)\sin\theta, \quad \alpha \le \theta \le \beta$$

so both `ParametricPlot` and `PolarPlot` can be used to graph polar equations.

Example 2.27

Graph (a) $r = \sin(8\theta/7)$, $0 \le \theta \le 14\pi$; (b) $r = \theta\cos\theta$, $-19\pi/2 \le \theta \le 19\pi/2$; (c) ("The Butterfly") $r = e^{\cos\theta} - 2\cos 4\theta + \sin^5(\theta/12)$, $0 \le \theta \le 24\pi$; and (d) ("The Lituus") $r^2 = 1/\theta$, $0.1 \le \theta \le 10\pi$.

Solution. For (a) and (b) we use `ParametricPlot`. First define r and then use `ParametricPlot` to generate the graph of the polar curve. No graphics are displayed because we place a semicolon at the end of each command. We illustrate the use of various options, particularly the `PlotStyle` and `PlotLabel` options in each.

```
Clear[r]
r[θ_] = Sin[8θ/7];
pp1 = ParametricPlot[{r[θ]Cos[θ], r[θ]Sin[θ]},
  {θ, 0, 14Pi}, AspectRatio → Automatic,
  PlotStyle → CMYKColor[0, 1, .7, .1],
   PlotLabel → "(a)-University of Georgia Red"]
```

For (b), we use the option `PlotRange->{{-30,30},{-30,30}}` to indicate that the range displayed on both the vertical and horizontal axes corresponds to the interval $[-30, 30]$. To help assure that the resulting graphic appears "smooth", we increase the number of points that Mathematica samples when generating the graph by including the option `PlotPoints->200`.

```
Clear[r]
r[θ_] = θCos[θ];
pp2 = ParametricPlot[{r[θ]Cos[θ], r[θ]Sin[θ]},
  {θ, −19Pi/2, 19Pi/2}, AspectRatio → Automatic,
  PlotRange → {{−30, 30}, {−30, 30}},
  PlotStyle → CMYKColor[.03, 1, .63, .12],
   PlotLabel → "(b)-The Ohio State University Red"]
```

For (c) and (d), we use `PolarPlot`. Using standard mathematical notation, we know that $\sin^5(\theta/12) = (\sin(\theta/12))^5$. However, when defining r with Mathematica, be sure you use the form `Sin(θ/12)^5`, not `Sin^5[θ/12]`, which Mathematica will not interpret in the way intended.

```
Clear[r]
r[θ_] = Exp[Cos[θ]] − 2Cos[4θ] + Sin[θ/12]^5;
pp3 = PolarPlot[r[θ], {θ, 0, 24Pi}, PlotPoints → 200,
  PlotRange → {{−4, 5}, {−4.5, 4.5}},
  AspectRatio → Automatic,
  PlotLabel → "(c)-Oklahoma State Orange",
   PlotStyle → CMYKColor[0, .63, 1.0, 0]]

pp4 = PolarPlot[{Sqrt[1/θ], −Sqrt[1/θ]}, {θ, 0.1, 10Pi},
  AspectRatio → Automatic, PlotRange → All,
  PlotLabel → "(d)-Harvey Mudd Gold",
```

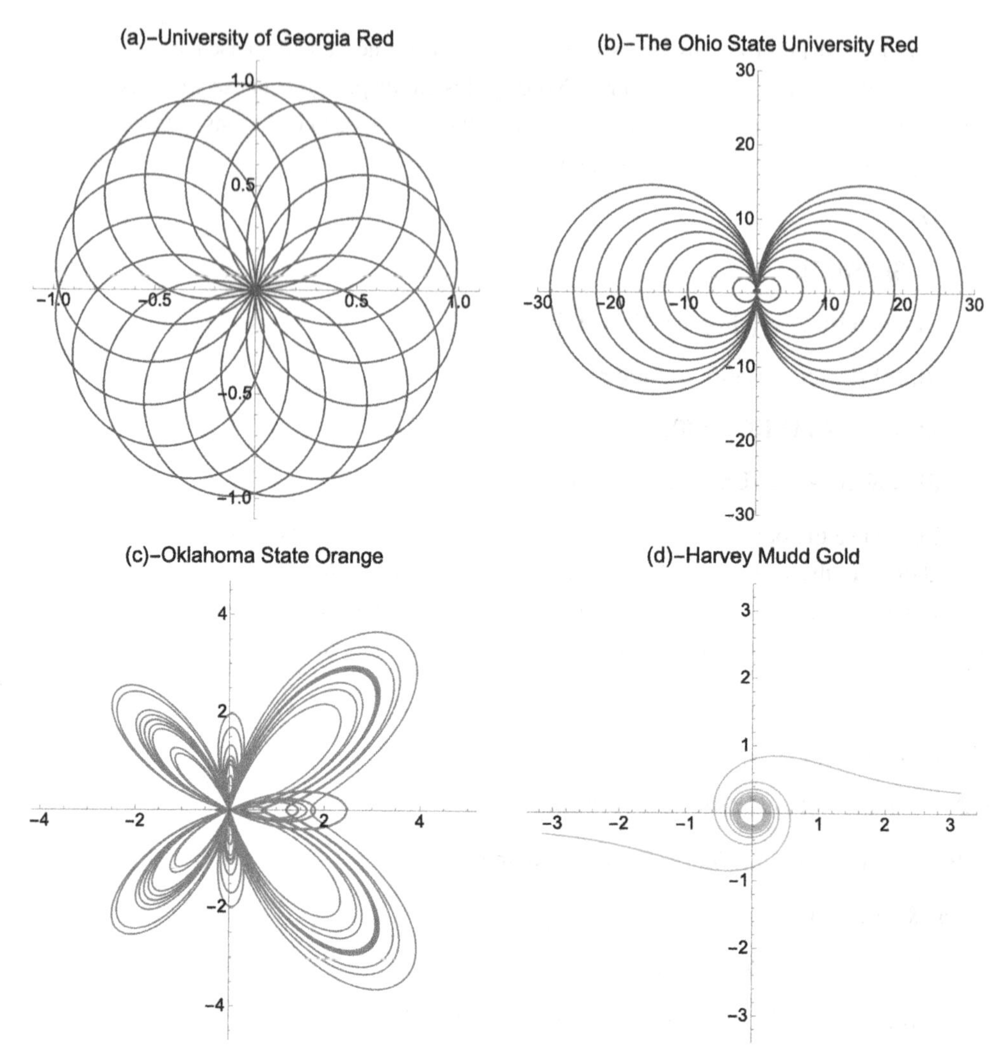

FIGURE 2.16 Graphs of four polar equations.

PlotStyle → CMYKColor[0, .3, 1., 0]]

Finally, we use `Show` together with `GraphicsGrid` to display all four graphs as a graphics array in Fig. 2.16. `pp1` and `pp2` are shown in the first row; `pp3` and `pp4` in the second.

Show[GraphicsGrid[{{pp1, pp2}, {pp3, pp4}}]] □

2.3.3 Three-Dimensional and Contour Plots; Graphing Equations

An elementary function of two variables, $z = f(x, y) = expression\, in\, x\, and\, y$, is typically defined using the form

```
f[x_,y_]=expression in x and y.
```

For delayed evaluation, use `f[x_,y_]:=...` rather than `f[x_,y_]=...` (immediate evaluation). Once a function has been defined, a basic graph is generated with `Plot3D`:

```
Plot3D[f[x,y],{x,a,b},{y,c,d}]
```

graphs $f(x, y)$ for $a \le x \le b$ and $c \le y \le d$.

For details regarding `Plot3D` and its options enter `?Plot3D` or `??Plot3D` or access the **Documentation Center** to obtain information about the `Plot3D` command, as we do here.

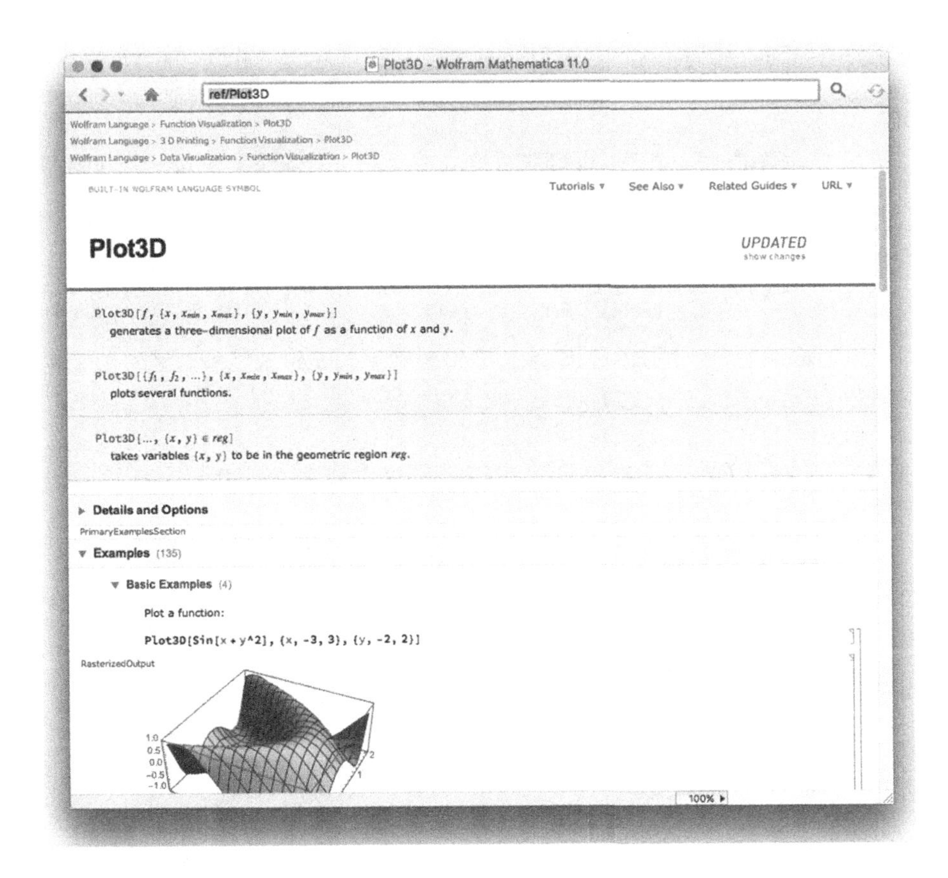

Graphs of several level curves of $z = f(x, y)$ are generated with

```
ContourPlot[f[x,y],{x,a,b},{y,c,d}].
```

A density plot of $z = f(x, y)$ is generated with

```
DensityPlot[f[x,y],{x,a,b},{y,c,d}].
```

For details regarding `ContourPlot` (`DensityPlot`) and its options enter `?ContourPlot` (`?DensityPlot`) or `??ContourPlot` (`??DensityPlot`) or access the **Documentation Center**.

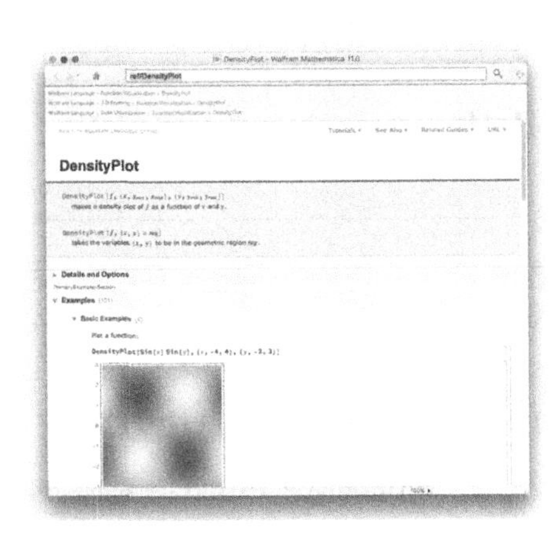

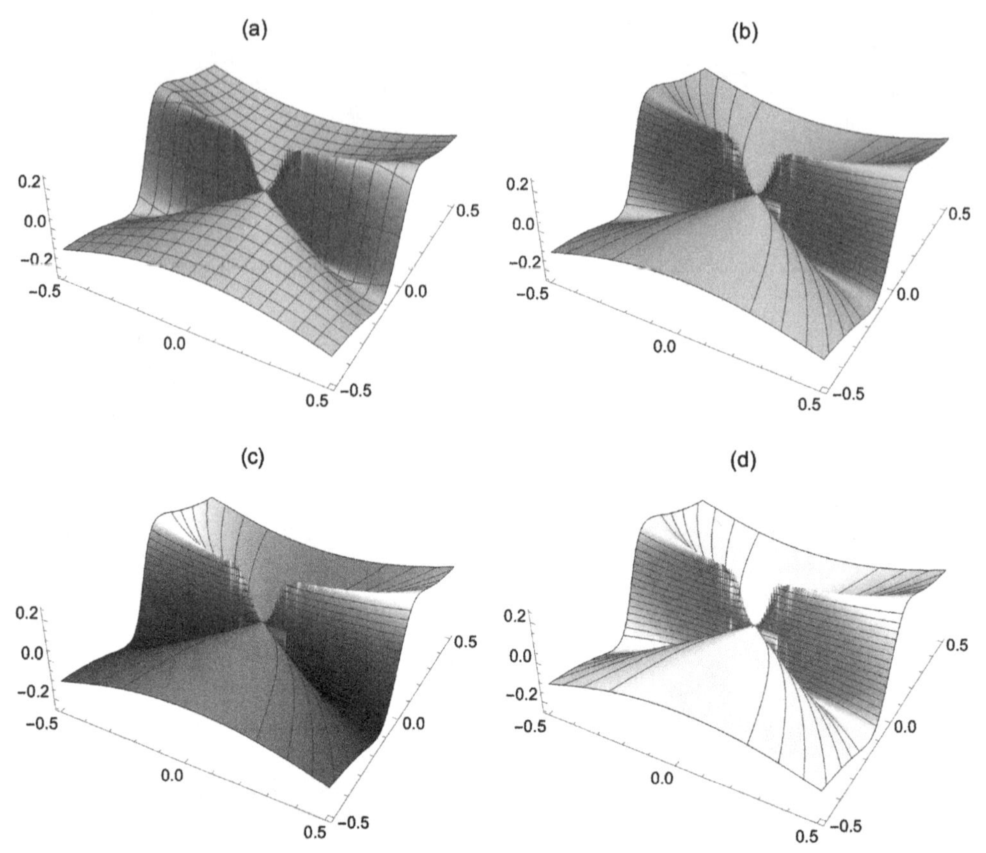

FIGURE 2.17 Three-dimensional plot of $f(x, y)$: Upper left is the basic plot generated with `Plot3D`; in upper right, we use contour lines to determine the mesh; in lower left, we use the `GrayTones` color gradient to shade the plot; in lower right, we create a transparent plot with `Opacity`.

Example 2.28

Let $f(x, y) = \dfrac{x^2 y}{x^4 + 4y^2}$. (a) Calculate $f(1, -1)$. (b) Graph $f(x, y)$ and several contour plots of $f(x, y)$ on a region containing the origin, $(0, 0)$.

Solution. After defining $f(x, y)$, we evaluate $f(1, -1) = -1/5$.

```
Clear[f]

f[x_, y_] = x^2y/(x^4 + 4y^2);

f[1, 1]
```

$\frac{1}{5}$

Next, we use `Plot3D` to graph $f(x, y)$ for $-1/2 \le x \le 1/2$ and $-1/2 \le y \le 1/2$ in Fig. 2.17. We illustrate the use of the `Axes`, `Boxed`, `PlotPoints`, `MeshFunctions`, `PlotStyle`, and `ColorFunction` options.

```
p1 = Plot3D[f[x, y], {x, -1/2, 1/2}, {y, -1/2, 1/2},
  Axes → Automatic, Boxed → False, PlotPoints → 60,
  PlotLabel → "(a)"]
```

Usc MeshFunctions to modify thc standard rcctangular grid. In Fig. 2.17 (b), we use the level curves of the function for the grid.

To adjust the viewing angle of three-dimensional graphics, select the graphic and drag to the desired viewing angle.

```
p2 = Plot3D[f[x, y], {x, −1/2, 1/2}, {y, −1/2, 1/2},
  Axes → Automatic, Boxed → False, PlotPoints → 60,
  MeshFunctions → (#3&), PlotLabel → "(b)"]
```

We use the GrayTones color gradient to shade the graph. (Fig. 2.17 (c))

```
p3 = Plot3D[f[x, y], {x, −1/2, 1/2}, {y, −1/2, 1/2},
  Axes → Automatic, Boxed → False, PlotPoints → 60,
  MeshFunctions → (#3&), PlotLabel → "(c)",
    ColorFunction → {ColorData["GrayTones"][#3]&}]
```

Use Opacity to make a "clear" plot. (See Fig. 2.17 (d).) We use Show together with GraphicsGrid to display all four plots together in Fig. 2.17.

```
p4 = Plot3D[f[x, y], {x, −1/2, 1/2}, {y, −1/2, 1/2},
  Axes → Automatic, Boxed → False, PlotPoints → 60,
  MeshFunctions → (#3&), PlotLabel → "(d)",
    ColorFunction → {Opacity[.5]}]

Show[GraphicsGrid[{{p1, p2}, {p3, p4}}]]
```

Four contour plots are generated with ContourPlot. The second through fourth illustrate the use of the PlotPoints, Frame, ContourShading, Axes, AxesOrigin, ColorFunction, and Contours options. (See Fig. 2.18.)

```
cp1 = ContourPlot[f[x, y], {x, −1/2, 1/2}, {y, −1/2, 1/2},
  Contours → 30, PlotPoints → 50,
  PlotLabel → "(a)"]

cp2 = ContourPlot[f[x, y], {x, −1/2, 1/2}, {y, −1/2, 1/2},
  Contours → 30, PlotPoints → 50,
  ColorFunction → ColorData["GrayTones"],
    PlotLabel → "(b)"]

cp3 = ContourPlot[f[x, y], {x, −1/2, 1/2}, {y, −1/2, 1/2},
  Contours → 30, PlotPoints → 50, ContourShading → False,
  Frame → False, Axes → Automatic, AxesOrigin → {0, 0},
  AxesLabel → {x, y}, ContourStyle → Black,
    PlotLabel → "(c)"]
```

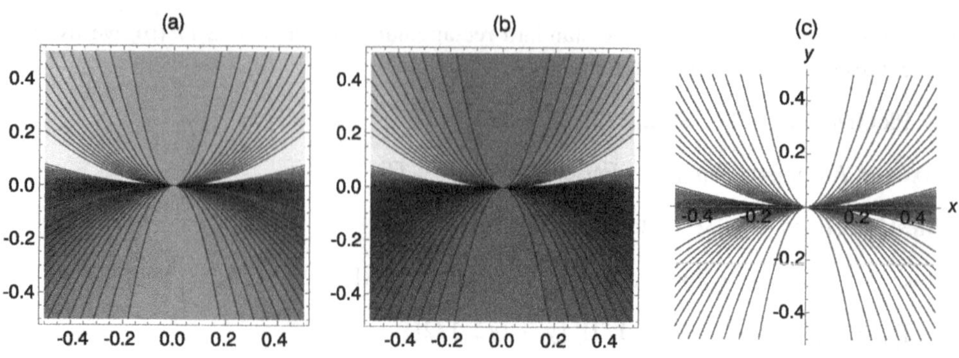

FIGURE 2.18 Three contour plots of $f(x, y)$.

Show[GraphicsRow[{cp1, cp2, cp3}]]

Fig. 2.18 shows the graphics array generated with the previous commands. With Mathematica 11, if you want to adjust your array, drag and move the objects within the graphic. □

With Mathematica 11, you can adjust the viewing angle of a three-dimensional graphics by selecting the graphic and dragging it to the desired position.

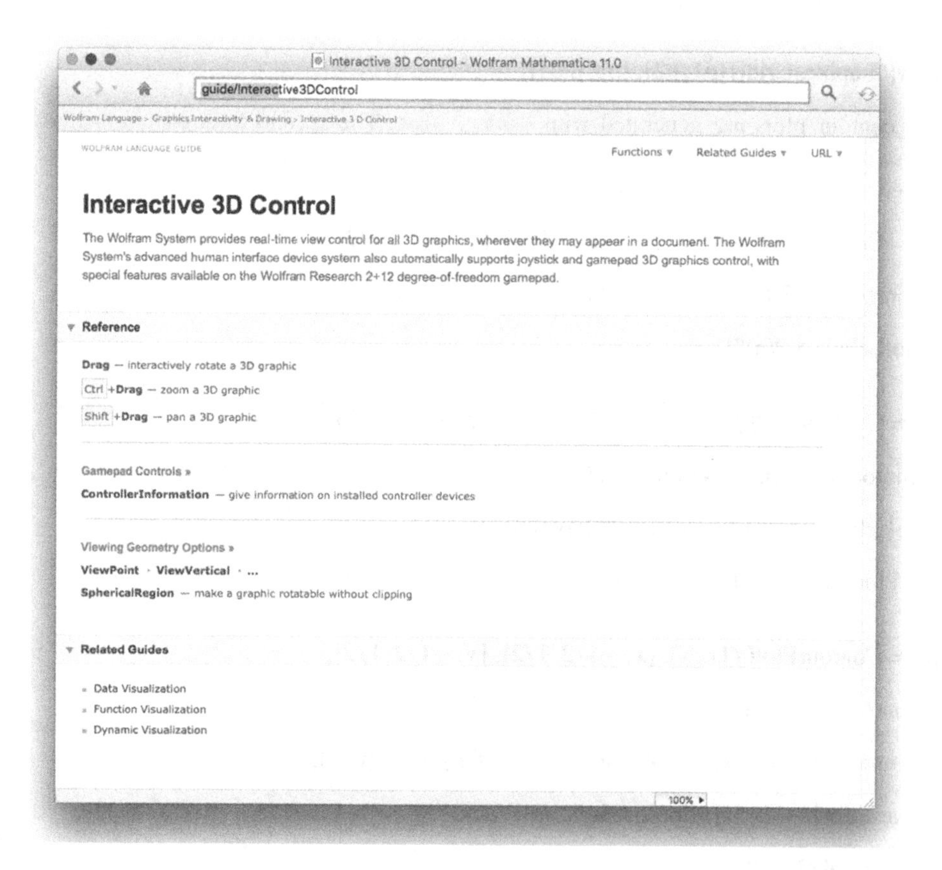

Manually, use the `ViewPoint` option.

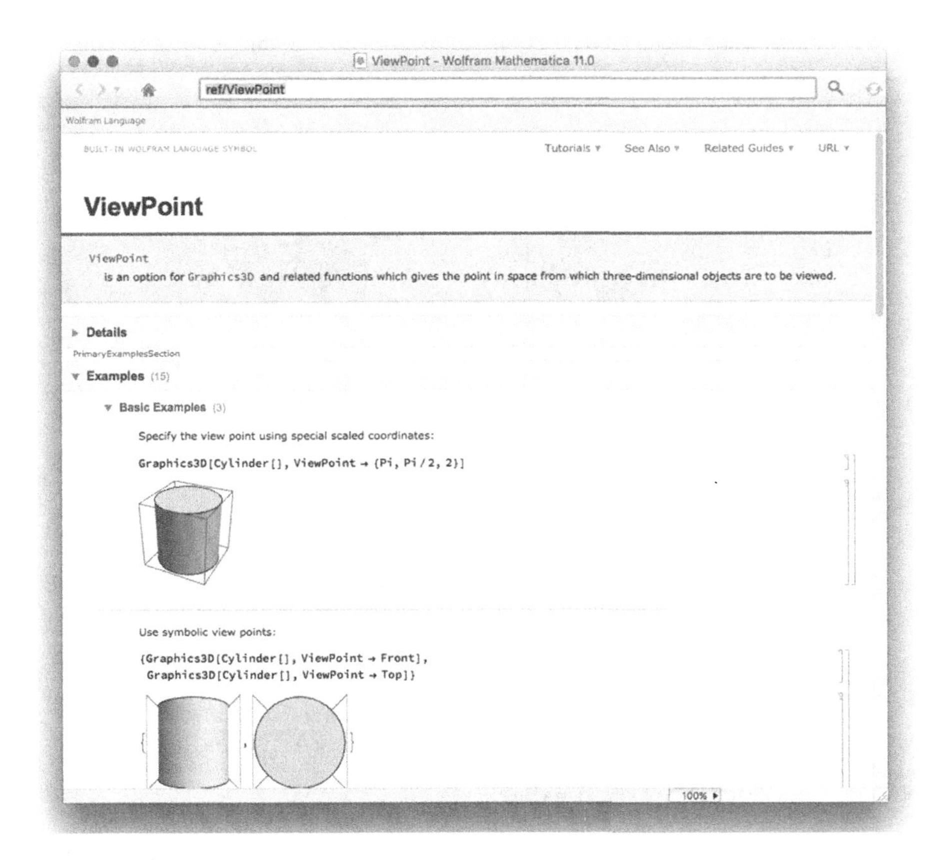

Fig. 2.19 shows four different views of the graph of $g(x, y) = x \sin y + y \sin x$ for $0 \le x \le 5\pi$ and $0 \le y \le 5\pi$. The options `AxesLabel`, `BoxRatios`, `ViewPoint`, `PlotPoints`, `Shading`, and `Mesh` are also illustrated.

```
Clear[g]

g[x_, y_] = xSin[y] + ySin[x];

p1 = Plot3D[g[x, y], {x, 0, 5Pi}, {y, 0, 5Pi},
  PlotPoints → 60, AxesLabel → {x, y, z},
  PlotLabel → "(a)"];

p2 = Plot3D[g[x, y], {x, 0, 5Pi}, {y, 0, 5Pi},
  PlotPoints → 60, ViewPoint->{−2.846, −1.813, 0.245},
  Boxed → False, BoxRatios → {1, 1, 1},
    AxesLabel → {x, y, z},
    PlotLabel → "(b)"];

p3 = Plot3D[g[x, y], {x, 0, 5π}, {y, 0, 5π},
  PlotPoints → 60, ViewPoint → {1.488, −1.515, 2.634},
  AxesLabel → {x, y, z}, ColorFunction → (White&),
```

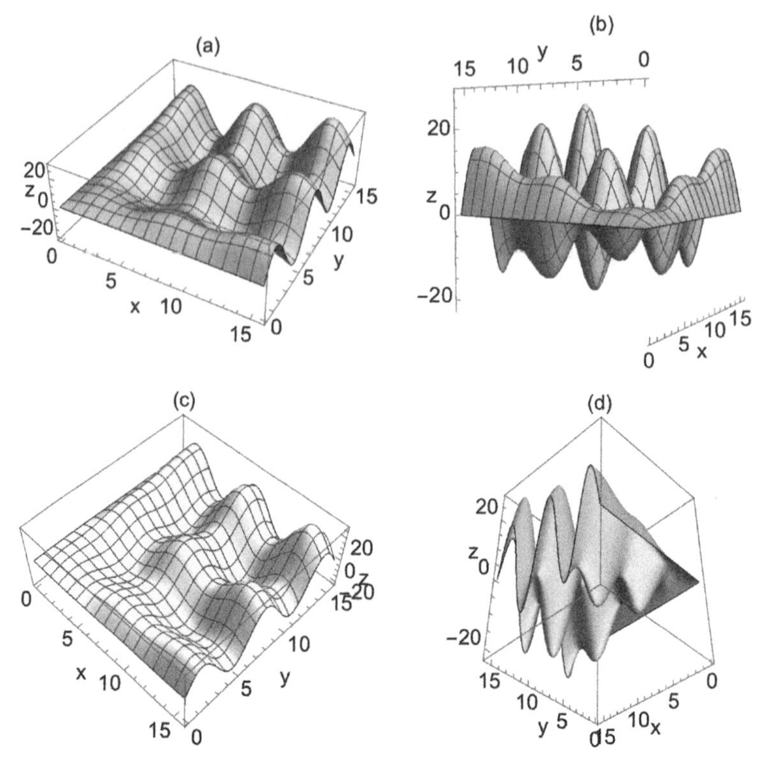

FIGURE 2.19 Four different plots of $g(x, y) = x \sin y + y \sin x$ for $0 \le x \le 5\pi$.

```
    PlotLabel → "(c)"];

p4 = Plot3D[g[x, y], {x, 0, 5Pi}, {y, 0, 5Pi},
   PlotPoints → 60, AxesLabel → {x, y, z},
   Mesh → False, BoxRatios → {2, 2, 3},
    ViewPoint->{−1.736, 1.773, −2.301},
    PlotLabel → "(d)"];

Show[GraphicsGrid[{{p1, p2}, {p3, p4}}]]
```

`ContourPlot` is especially useful when graphing equations. The graph of the equation $f(x, y) = C$, where C is a constant, is the same as the contour plot of $z = f(x, y)$ corresponding to C. That is, the graph of $f(x, y) = C$ is the same as the level curve of $z = f(x, y)$ corresponding to $z = C$.

Use `ContourPlot` to graph equations of the form $f(x, y) = g(x, y)$ with

```
ContourPlot[f[x,y]==g[x,y],{x,a,b},{y,c,d}].
```

Example 2.29

Graph the equation $y^2 - 2x^4 + 2x^6 - x^8 = 0$ for $-1.5 \le x \le 1.5$.

Solution. We define $f(x, y)$ to be the left-hand side of the equation $y^2 - 2x^4 + 2x^6 - x^8 = 0$ and then use `ContourPlot` to graph `eq` for $-1.5 \le x \le 1.5$ in Fig. 2.20.

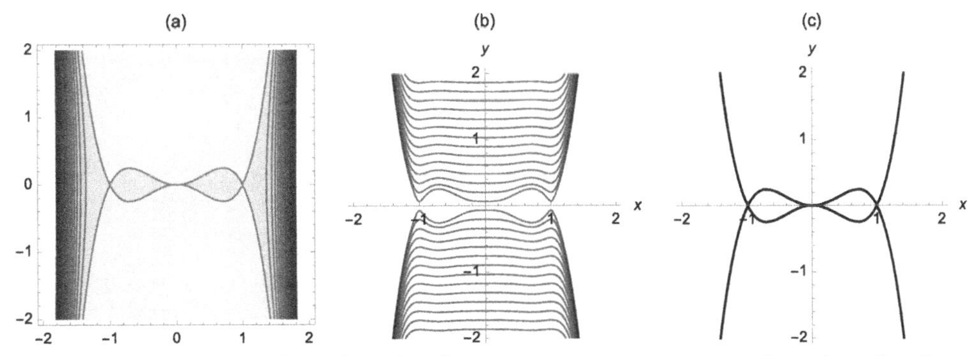

FIGURE 2.20 Three plots of $y^2 - 2x^4 + 2x^6 - x^8 = C$. In (c), the graph is the plot of $y^2 - 2x^4 + 2x^6 - x^8 = 0$.

```
Clear[x, y]

f[x_, y_] = y^2 − x^4 + 2x^6 − x^8;

cp1 = ContourPlot[f[x, y], {x, −2, 2}, {y, −2, 2},
  AspectRatio → Automatic, PlotLabel → "(a)",
  Contours → 30]
```

To graph the contour plots of $z = f(x, y)$ for particular values of C, create a list of the values of C for which you want the contour plots and then use the option `Contours->List of C values.`

For example, here we use `Table` to create a list of 30 equally spaced values of $f(0, y)$ for $-2 \le y \le 2$.

```
vals = Table[f[0, y], {y, −2, 2, 4/29}];
```

Next, we use `ContourPlot` to graph $f(x, y) = C$ for each C-value in `vals` and illustrate various options associated with the `ContourPlot` function.

```
cp2 = ContourPlot[lhseq, {x, −2, 2}, {y, −2, 2},
  AspectRatio → Automatic, Frame → False, Contours → vals,
  Axes → Automatic, AxesLabel → {x, y}, PlotLabel → "(b)",
    AxesOrigin → {0, 0}, ContourShading → False]

cp3 = ContourPlot[lhseq == 0, {x, −2, 2}, {y, −2, 2},
  AspectRatio → Automatic, Frame → False,
  Axes → Automatic, AxesLabel → {x, y}, PlotLabel → "(c)",
    AxesOrigin → {0, 0}, ContourStyle → Black]
```

Finally, we use `Show` together with `GraphicsRow` to display all three graphics side-by-side in Fig. 2.20.

```
Show[GraphicsRow[{cp1, cp2, cp3}]]
```

□

Equations can be plotted together, as with the commands `Plot` and `Plot3D`, with

```
ContourPlot[{eq1,eq2,...,eqn},{x,a,b},{y,c,d}].
```

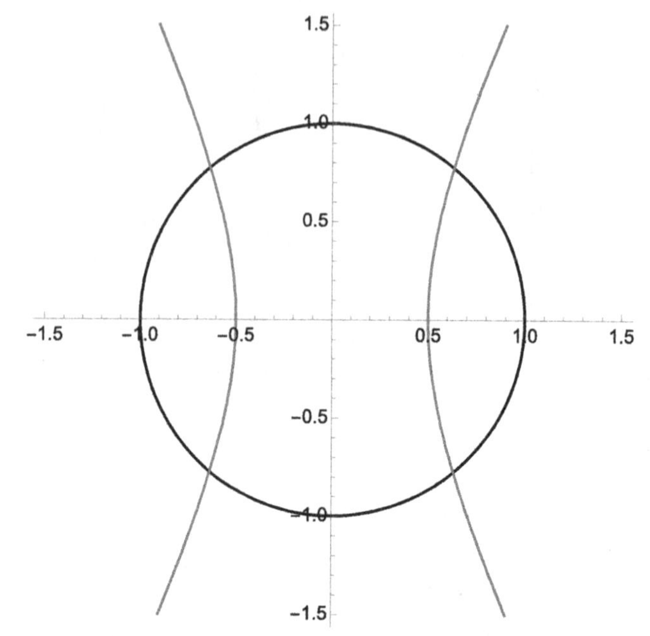

FIGURE 2.21 Plots of $x^2 + y^2 = 1$ and $4x^2 - y^2 = 1$.

Example 2.30

Graph the equations $x^2 + y^2 = 1$ and $4x^2 - y^2 = 1$ for $-1.5 \le x \le 1.5$.

Solution. We use `ContourPlot` to graph the equations together on the same axes in Fig. 2.21. The graph of $x^2 + y^2 = 1$ is the unit circle while the graph of $4x^2 - y^2 = 1$ is a hyperbola.

```
cp1 = ContourPlot[{x^2 + y^2 == 1, 4x^2 - y^2 == 1},
  {x, -3/2, 3/2}, {y, -3/2, 3/2}, Frame -> False,
  Axes -> Automatic, AxesOrigin -> {0, 0},
   ContourStyle -> {Black, Gray}]
```

From Fig. 2.21, we see four intersection points. To solve an equation of the form $f(x) = g(x)$ for x, use `Solve`. `Solve[f[x]==g[x],x]` attempts to solve the equation $f(x) = g(x)$ for x. For a system of two equations in two variables, try using `Solve[{f[x,y]==0, g[x,y]==0},{x,y}]` to solve the system $f(x, y) = 0$, $g(x, y) = 0$. `Solve` will be discussed in more detail in the next section.

```
Solve[{x^2 + y^2 == 1, 4x^2 - y^2 == 1}]
```

$$\left\{\left\{x \to -\sqrt{\tfrac{2}{5}}, y \to -\sqrt{\tfrac{3}{5}}\right\}, \left\{x \to -\sqrt{\tfrac{2}{5}}, y \to \sqrt{\tfrac{3}{5}}\right\}, \left\{x \to \sqrt{\tfrac{2}{5}}, y \to -\sqrt{\tfrac{3}{5}}\right\}, \left\{x \to \sqrt{\tfrac{2}{5}}, y \to \sqrt{\tfrac{3}{5}}\right\}\right\}$$

□

Example 2.31: Conic Sections

A **conic section** is a graph of the equation

$$Ax^2+Bxy+Cy^2+Dx+Ey+F=0.$$

Except when the conic is degenerate, the conic $Ax^2+Bxy+Cy^2+Dx+Ey+F=0$ is a (an)
1. **Ellipse** or **circle** if $B^2-4AC<0$;
2. **Parabola** if $B^2-4AC=0$; or
3. **Hyperbola** if $B^2-4AC>0$.

Graph the conic section $ax^2+bxy+cy^2=1$ for $-4\le x\le 4$ and for a, b, and c equal to all possible combinations of -1, 1, and 2.

Also see Example 2.34.

Solution. We begin by defining `conic` to be the equation $ax^2+bxy+cy^2=1$ and then use `Permutations` to produce all possible orderings of the list of numbers $\{-1,1,2\}$, naming the resulting output `vals`.

`Permutations[list]` returns a list of all possible orderings of the list `list`.

```
Clear[a, b, c, x, y, p]
conic = ax^2 + bxy + cy^2 == 1;
vals = Permutations[{-1, 1, 2}]
```

{{−1, 1, 2}, {−1, 2, 1}, {1, −1, 2}, {1, 2, −1}, {2, −1, 1}, {2, 1, −1}}

Next we define the function `p`. Given `a1`, `b1`, and `c1`, `p` defines `toplot` to be the equation obtained by replacing a, b, and c in `conic` by `a1`, `b1`, and `c1`, respectively. Then, `toplot` is graphed for $-4\le x\le 4$. The function `p` returns a graphics object.

```
Clear[p]
p[{a1_, b1_, c1_}]:=Module[{toplot},
toplot = Evaluate[conic/.{a → a1, b → b1, c → c1}];
  ContourPlot[Evaluate[toplot], {x, -5, 5},
  {y, -5, 5}, Frame → False, Axes → Automatic, Ticks → None]
]
```

We then use `Map` to compute `p` for each ordered triple in `vals`. The resulting output, named `graphs`, is a set of six graphics objects.

```
graphs = Map[p, vals];
```

`Partition` is then used to partition `graphs` into three element subsets. The resulting array of graphics objects named `toshow` is displayed with `Show` and `GraphicsGrid` in Fig. 2.22.

```
Show[GraphicsGrid[Partition[graphs, 3]]]
```
□

Studying how parameter values affect the conic is particularly well-suited to a `Manipulate` object as illustrated in the following commands and then in Fig. 2.23.

```
Manipulate[ContourPlot[ax^2 + bxy + cy^2 == 1,
  {x, -5, 5},
  {y, -5, 5}, Frame → False, Axes → Automatic, Ticks → None,
```

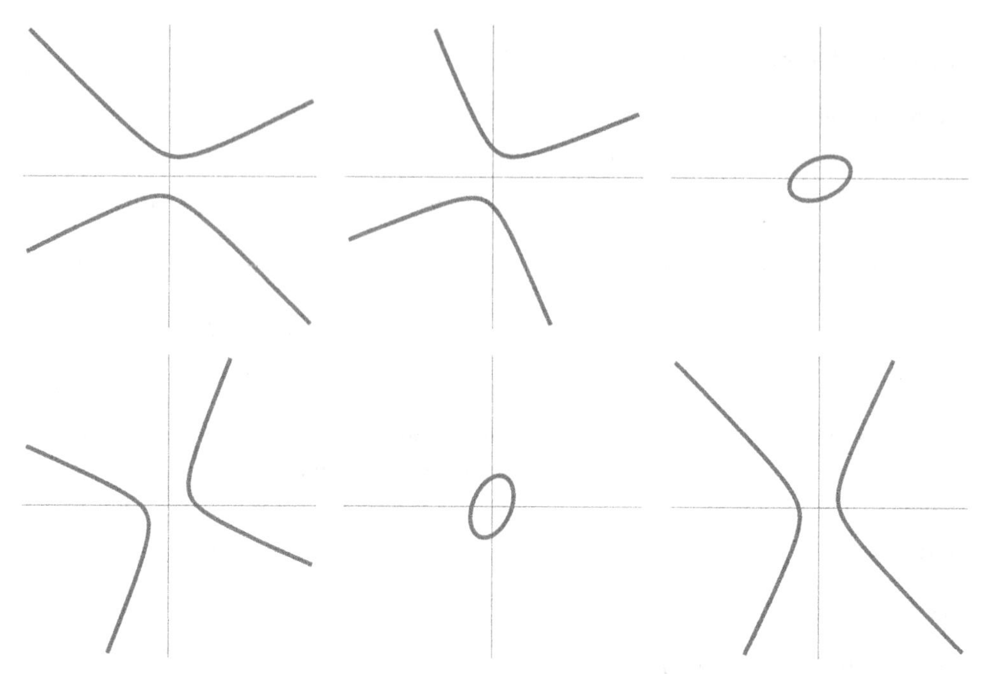

FIGURE 2.22 Plots of six conic sections.

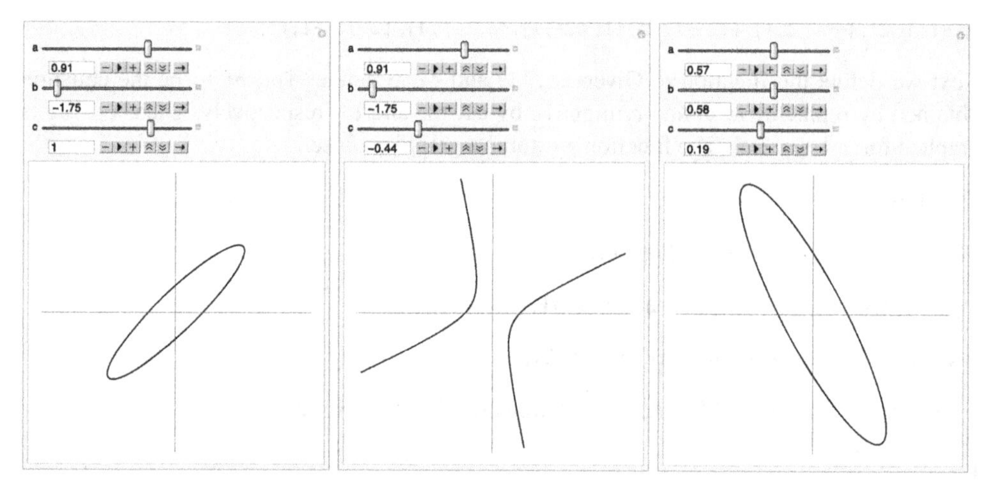

FIGURE 2.23 Using `Manipulate` to adjust parameter values.

```
ContourStyle → Black, PlotPoints->60], {{a, 1}, −2, 2},
{{b, 0}, −2, 2}, {{c, 1}, −2, 2}]
```

2.3.4 Parametric Curves and Surfaces in Space

The command

```
ParametricPlot3D[{x[t],y[t],z[t]},{t,a,b}]
```

generates the three-dimensional curve $\begin{cases} x = x(t), \\ y = y(t), \\ z = z(t), \end{cases}$ $a \le t \le b$ and the command

```
ParametricPlot3D[{x[u,v],y[u,v],z[u,v]},{u,a,b},{v,c,d}]
```

plots the surface $\begin{cases} x = x(u, v), \\ y = y(u, v), \\ z = z(u, v), \end{cases} \quad a \le u \le b, c \le v \le d.$

Entering `Information[ParametricPlot3D]` or `??ParametricPlot3D` returns a description of the `ParametricPlot3D` command along with a list of options and their current settings.

Example 2.32: Umbilic Torus NC

A parametrization of **umbilic torus NC** is given by $\mathbf{r}(s, t) = x(s, t)\mathbf{i} + y(s, t)\mathbf{j} + z(s, t)\mathbf{k}$, $-\pi \le s \le \pi$, $-\pi \le t \le \pi$, where

$$x = \left[7 + \cos\left(\frac{1}{3}s - 2t\right) + 2\cos\left(\frac{1}{3}s + t\right)\right]\sin s$$

$$y = \left[7 + \cos\left(\frac{1}{3}s - 2t\right) + 2\cos\left(\frac{1}{3}s + t\right)\right]\cos s$$

and

$$z = \sin\left(\frac{1}{3}s - 2t\right) + 2\sin\left(\frac{1}{3}s + t\right).$$

Graph the torus.

Solution. We define $x = x(s, t)$, $y = y(s, t)$, $z = z(s, t)$, and $\mathbf{r}(s, t) = \langle x(s, t), y(s, t), z(s, t)\rangle$.

```
c = 3;

a = 1;

x[s_, t_] = (7 + Cos[s/3 - 2t] + 2Cos[s/3 + t])Sin[s];

y[s_, t_] = (7 + Cos[s/3 - 2t] + 2Cos[s/3 + t])Cos[s];

z[s_, t_] = Sin[s/3 - 2t] + 2Sin[s/3 + t];

r[s_, t_] = {x[s, t], y[s, t], z[s, t]};
```

The torus is then graphed with `ParametricPlot3D`, `DensityPlot`, and `ContourPlot` in Fig. 2.24. In the plots, we illustrate the `Mesh`, `MeshFunctions`, `PlotPoints`, and `PlotRange` options. All four plots are shown together with `Show` and `GraphicsGrid`. Notice that `DensityPlot` and `ContourPlot` yield very similar results: a basic density plot is similar to a basic contour plot but without the contour lines.

```
threedp1uta = ParametricPlot3D[r[s, t], {s, -Pi, Pi},
  {t, -Pi, Pi}, PlotPoints->{30, 30},
  AspectRatio->1, AxesLabel->{x, y, z},
    PlotRange->{{-12, 12}, {-12, 12}, {-3, 3}},
  BoxRatios->{4, 4, 1}, Mesh → False, PlotStyle → Opacity[.9],
  PlotLabel → "(a)"]

threedp1utb = ParametricPlot3D[r[s, t], {s, -Pi, Pi},
  {t, -Pi, Pi}, PlotPoints->{50, 50},
  AspectRatio->1, AxesLabel->{x, y, z},
    PlotRange->{{-12, 12}, {-12, 12}, {-3, 3}},
```

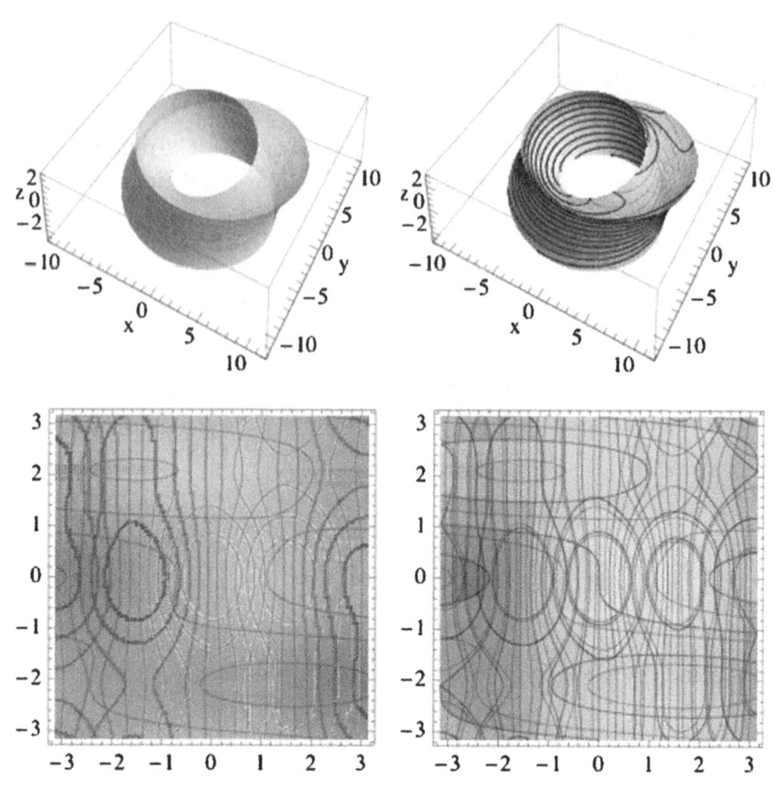

FIGURE 2.24 On the top row, two plots of Umbilic torus; on the bottom, comparing a density plot (on the left) to a contour plot (on the right).

```
    BoxRatios->{4, 4, 1},
    MeshFunctions->{#3&}, Mesh → 10,
      PlotLabel → "(b)"]

threedplutc = DensityPlot[r[s, t], {s, −Pi, Pi},
    {t, −Pi, Pi}, PlotPoints->{100, 100},
    AspectRatio->1, AxesLabel->{x, y, z},
      MeshFunctions->{#3&}, Mesh → 10,
    PlotLabel → "(c)"]

threedplutd = ContourPlot[r[s, t], {s, −Pi, Pi},
    {t, −Pi, Pi}, PlotPoints->{100, 100},
    AspectRatio->1, AxesLabel->{x, y, z},
      MeshFunctions->{#3&}, Mesh → 10, PlotLabel → "(d)"]

Show[GraphicsGrid[{{threedpluta, threedplutb},
    {threedplutc, threedplutd}}]]
```
□

Example 2.33: Gray's Torus Example

A parametrization of an **elliptical torus** is given by

$$x = (a + b\cos v)\cos u, \quad y = (a + b\cos v)\sin u, \quad z = c\sin v$$

For positive integers p and q, the curve with parametrization

$$x = (a + b\cos qt)\cos pt, \quad y = (a + b\cos qt)\sin pt, \quad z = c\sin qt$$

winds around the elliptical torus and is called a **torus knot**.

Plot the torus if $a = 8$, $b = 3$, and $c = 5$ and then graph the torus knots for $p = 2$ and $q = 5$, $p = 1$ and $q = 10$, and $p = 2$ and $q = 3$.

This example is explored in detail in Sections 8.2 and 11.4 of Gray's *Modern Differential Geometry of Curves and Surfaces*, [8], an indispensible reference for those who use Mathematica's graphics extensively. Note that since Dr. Gray's death, the text has been revised and updated in its third edition by Abbena and Salamon, [1].

Solution. We begin by defining `torus` and `torusknot`.

```
torus[a_, b_, c_][p_, q_][u_, v_]:=
  {(a + bCos[u])Cos[v], (a + bCos[u])Sin[v], cSin[u]}

torusknot[a_, b_, c_][p_, q_][t_]:=
  {(a + bCos[qt])Cos[pt], (a + bCos[qt])Sin[pt], cSin[qt]}
```

Next, we use `ParametricPlot3D` to generate all four graphs

```
pp1 = ParametricPlot3D[Evaluate[torus[8, 3, 5][2, 5][u, v]],
  {u, 0, 2Pi}, {v, 0, 2Pi}, PlotPoints → 60];

pp2 = ParametricPlot3D[Evaluate[torusknot[8, 3, 5][2, 5][t]],
  {t, 0, 3Pi}, PlotPoints → 200];

pp3 = ParametricPlot3D[Evaluate[torusknot[8, 3, 5][1, 10][t]],
  {t, 0, 3Pi}, PlotPoints → 200];

pp4 = ParametricPlot3D[Evaluate[torusknot[8, 3, 5][2, 3][t]],
  {t, 0, 3Pi}, PlotPoints → 200];
```

and show the result as a graphics array with `Show` and `GraphicsGrid` in Fig. 2.25.

```
Show[GraphicsGrid[{{pp1, pp2}, {pp3, pp4}}]]
```

If we take advantage of a few options, such as eliminating the mesh (`Mesh->False`) and increasing the opacity (`PlotStyle->Opacity[.4]`), we can produce a graphic of the knot on the torus. After using the `PlotStyle` option together with `Opacity`, we produce a nearly transparent torus. Then, each knot is plotted. To assure smooth plots, we increase the number of points plotted with `PlotPoints` and also increase the thickness of the curve with `Thickness`.

```
pp1 = ParametricPlot3D[Evaluate[torus[8, 3, 5][2, 5][u, v]],
  {u, 0, 2Pi}, {v, 0, 2Pi}, PlotPoints → 60,
  Mesh → False, PlotStyle → Opacity[.4],
   ColorFunction → "AlpineColors"];
```

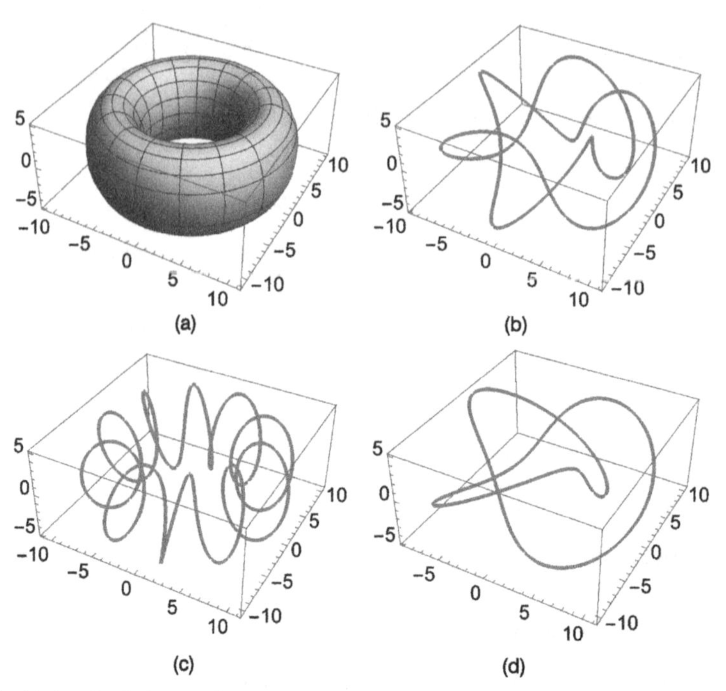

FIGURE 2.25 (a) An elliptical torus. (b) This knot is also known as the trefoil knot. (c) The curve generated by `torusknot[8,3,5][2,3][1,10]` is not a knot. (d) The torus knot with $p = 2$ and $q = 3$.

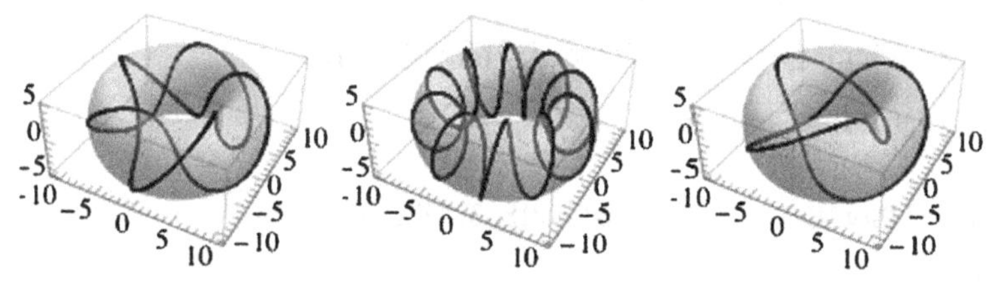

FIGURE 2.26 The knots in Fig. 2.25 on the torus.

```
pp2 = ParametricPlot3D[Evaluate[torusknot[8, 3, 5][2, 5][t]],
  {t, 0, 3Pi}, PlotPoints → 200, PlotStyle → {Black, {Thickness[.025]}}];
pp3 = ParametricPlot3D[Evaluate[torusknot[8, 3, 5][1, 10][t]],
  {t, 0, 3Pi}, PlotPoints → 200, PlotStyle → {Black, {Thickness[.025]}}];
pp4 = ParametricPlot3D[Evaluate[torusknot[8, 3, 5][2, 3][t]],
  {t, 0, 3Pi}, PlotPoints → 200, PlotStyle → {Black, {Thickness[.025]}}];
```

We use `Show` twice together with `GraphicsRow` to first display the torus with each knot and then display all three graphics side-by-side in Fig. 2.26.

```
Show[GraphicsRow[{Show[{pp1, pp2}], Show[{pp1, pp3}], Show[{pp1, pp4}]}]]
```
□

Example 2.34: Quadric Surfaces

The **quadric surfaces** are the three-dimensional objects corresponding to the conic sections in two dimensions. A **quadric surface** is a graph of

$$Ax^2 + By^2 + Cz^2 + Dxy + Exz + Fyz + Gx + Hy + Iz + J = 0,$$

where A, B, C, D, E, F, G, H, I, and J are constants.

Also see Example 2.31.

The intersection of a plane and a quadric surface is a conic section.

Several of the basic quadric surfaces, in standard form, and a parametrization of the surface are listed in the following table.

Name	Parametric Equations	
Ellipsoid		
$\frac{x^2}{a^2}+\frac{y^2}{b^2}+\frac{z^2}{c^2}=1$	$\begin{cases} x=a\cos t\cos r, \\ y=b\cos t\sin r, \\ z=c\sin t, \end{cases}$	$-\pi/2\le t\le\pi/2,\ -\pi\le r\le\pi$
Hyperboloid of One Sheet		
$\frac{x^2}{a^2}+\frac{y^2}{b^2}-\frac{z^2}{c^2}=1$	$\begin{cases} x=a\sec t\cos r, \\ y=b\sec t\sin r, \\ z=c\tan t, \end{cases}$	$-\pi/2< t<\pi/2,\ -\pi\le r\le\pi$
Hyperboloid of Two Sheets		
$\frac{x^2}{a^2}-\frac{y^2}{b^2}-\frac{z^2}{c^2}=1$	$\begin{cases} x=a\sec t, \\ y=b\tan t\cos r, \\ z=c\tan t\sin r, \end{cases}$	$-\pi/2< t<\pi/2$ or $\pi/2< t<3\pi/2,\ -\pi\le r\le\pi$

Graph the ellipsoid with equation $\frac{1}{16}x^2+\frac{1}{4}y^2+z^2=1$, the hyperboloid of one sheet with equation $\frac{1}{16}x^2+\frac{1}{4}y^2-z^2=1$, and the hyperboloid of two sheets with equation $\frac{1}{16}x^2-\frac{1}{4}y^2-z^2=1$.

Solution. A parametrization of the ellipsoid with equation $\frac{1}{16}x^2+\frac{1}{4}y^2+z^2=1$ is given by

$$x=4\cos t\cos r,\quad y=2\cos t\sin r,\quad z=\sin t,\quad -\pi/2\le t\le\pi/2,\ -\pi\le r\le\pi,$$

which is graphed with `ParametricPlot3D`.

```
Clear[x, y, z]
x[t_, r_] = 4Cos[t]Cos[r];
y[t_, r_] = 2Cos[t]Sin[r];
z[t_, r_] = Sin[t];
pp1 = ParametricPlot3D[{x[t, r], y[t, r], z[t, r]}, {t, -Pi/2, Pi/2}, {r, -Pi, Pi},
  PlotPoints -> 30, DisplayFunction -> Identity];
```

A parametrization of the hyperboloid of one sheet with equation $\frac{1}{16}x^2+\frac{1}{4}y^2-z^2=1$ is given by

$$x=4\sec t\cos r,\quad y=2\sec t\sin r,\quad z=\tan t,\quad -\pi/2< t<\pi/2,\ -\pi\le r\le\pi.$$

Because $\sec t$ and $\tan t$ are undefined if $t=\pm\pi/2$, we use `ParametricPlot3D` to graph these parametric equations on a subinterval of $[-\pi/2,\pi/2]$, $[-\pi/3,\pi/3]$.

```
Clear[x, y, z]
x[t_, r_] = 4Sec[t]Cos[r];
y[t_, r_] = 2Sec[t]Sin[r];
z[t_, r_] = Tan[t];
pp2 = ParametricPlot3D[{x[t, r], y[t, r], z[t, r]}, {t, -Pi/3, Pi/3}, {r, -Pi, Pi},
  PlotPoints -> 30, DisplayFunction -> Identity];
```

`pp1` and `pp2` are shown together in Fig. 2.27 using `Show` and `GraphicsRow`.

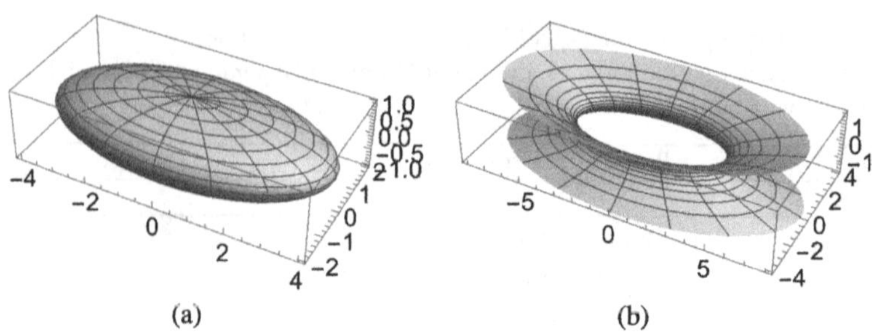

FIGURE 2.27 (a) Plot of $\frac{1}{16}x^2 + \frac{1}{4}y^2 + z^2 = 1$. (b) Plot of $\frac{1}{16}x^2 + \frac{1}{4}y^2 - z^2 = 1$.

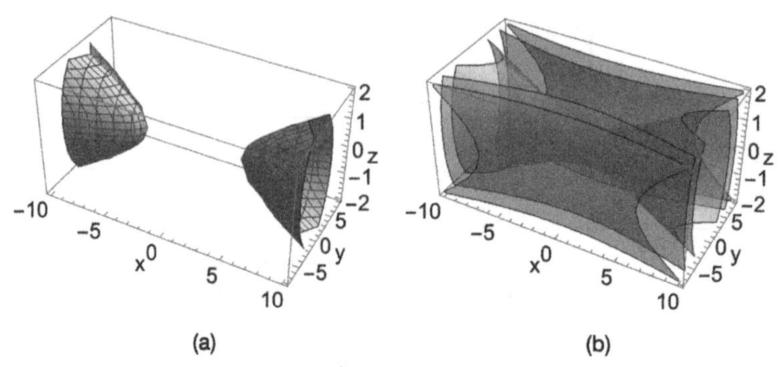

FIGURE 2.28 (a) Plot of $\frac{1}{16}x^2 - \frac{1}{4}y^2 - z^2 = 1$ generated with `ContourPlot3D`. (b) Several level surfaces of $f(x, y, z) = \frac{1}{16}x^2 - \frac{1}{4}y^2 - z^2$.

```
Show[GraphicsRow[{pp1, pp2}]]
```

For (c), we take advantage of the `ContourPlot3D` function:

```
ContourPlot3D[f[x,y,z],{x,a,b},{y,c,d},{z,u,v}]
```

graphs several level surfaces of $w = f(x, y, z)$.

We use `ContourPlot3D` to graph the equation $\frac{1}{16}x^2 - \frac{1}{4}y^2 - z^2 - 1 = 0$ in Fig. 2.28 (a), illustrating the use of the `PlotPoints`, `Axes`, `AxesLabel`, and `BoxRatios` options. In Fig. 2.28 (b), several level surfaces are drawn that illustrate the use of the `Opacity` function with the `ContourStyle` and `Mesh` options.

```
cp3d1 = ContourPlot3D[x^2/16 − y^2/4 − z^2 − 1 == 0,
  {x, −10, 10}, {y, −8, 8}, {z, −2, 2},
  PlotPoints → {8, 8, 8}, Axes → Automatic,
    AxesLabel → {x, y, z}, BoxRatios → {2, 1, 1}]

cp3d2 = ContourPlot3D[x^2/16 − y^2/4 − z^2 − 1,
  {x, −10, 10}, {y, −8, 8},
  {z, −2, 2}, PlotPoints → {8, 8, 8}, Axes → Automatic,
    AxesLabel → {x, y, z},
  BoxRatios → {2, 1, 1}, Mesh → False,
  ContourStyle → Opacity[.5]]

Show[GraphicsRow[{cp3d1, cp3d2}]]
```

□

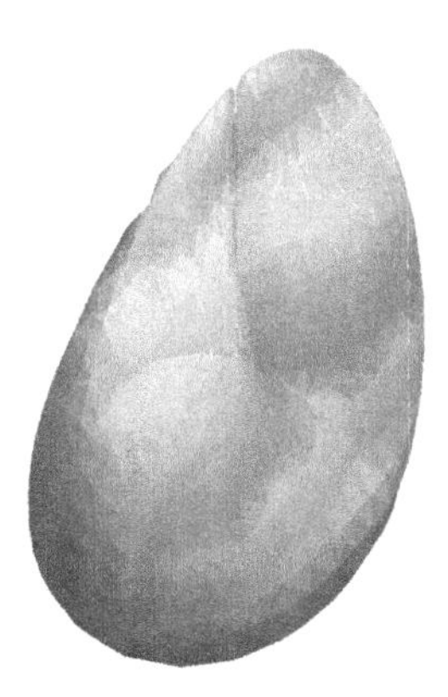

FIGURE 2.29 The Cross-Cap.

`ContourPlot3D` is especially useful in plotting equations involving three variables x, y, and z for which it is difficult to solve for one variable as a function of the other two.

Example 2.35: Cross-Cap

The **Cross-Cap** has equation

$$4x^2\left(x^2+y^2+z^2+z\right)+y^2\left(y^2+z^2-1\right).$$

We use `ContourPlot3D` to generate the plot of the Cross-Cap shown in Fig. 2.29

```
ContourPlot3D[4x^2(x^2+y^2+z^2+z)+
  y^2(y^2+z^2-1)==0, {x,-1,1}, {y,-1,1},
  {z,-1,1}, Mesh→False, Boxed→False,
   Axes→None, ContourStyle→Opacity[.8]]
```

Example 2.36

A homotopy from the **Roman surface** to the **Boy surface** is given by

$$x(u,v)=\frac{\sqrt{2}\cos(2u)\cos^2 v+\cos u\sin(2v)}{2-\alpha\sqrt{2}\sin(3u)\sin(2v)},$$
$$y(u,v)=\frac{\sqrt{2}\sin(2u)\cos^2 v+\sin u\sin(2v)}{2-\alpha\sqrt{2}\sin(3u)\sin(2v)},\text{ and}$$
$$z(u,v)=\frac{3\cos^2 v}{2-\alpha\sqrt{2}\sin(3u)\sin(2v)}.$$

If f and g are functions from X to Y, a **homotopy** from f to g is a continuous function H from $X\times[0,1]$ to Y satisfying $H(x,0)=f(x)$ and $H(x,1)=g(x)$.

Here, $\alpha=0$ gives the Roman surface and $\alpha=1$ gives the Boy surface.

To see the homotopy we first define x, y, z, and $\mathbf{r}=x(s,t)\mathbf{i}+y(s,t)\mathbf{j}+z(s,t)\mathbf{k}$.

Tables and lists are discussed in more detail in Chapters 4 and 5.

```
Clear[x, y, z, r]
x[α_][s_, t_] = (Sqrt[2]Cos[t]^2Cos[2s] + Cos[s]Sin[2t])/(2 - αSqrt[2]Sin[3s]Sin[2t]);
y[α_][s_, t_] = (Sqrt[2]Cos[t]^2Sin[2s] - Sin[s]Sin[2t])/(2 - αSqrt[2]Sin[3s]Sin[2t]);
```

```
z[α_][s_, t_] = 3Cos[t]^2/(2 - αSqrt[2]Sin[3s]Sin[2t]);
r[α_][s_, t_] = {x[α][s, t], y[α][s, t], z[α][s, t]};
```

We then use `Table` together with `ParametricPlot3D` to parametrically plot x, y, and z, $0 \le u \le 2\pi$, $0 \le v \le 2\pi$ for nine equally spaced values of α between 0 and 1. Note that if the semi-colon is omitted at the end of the command, the nine plots are displayed.

```
smalltable = Table[ParametricPlot3D[r[α][s, t],
  {s, 0, 2Pi}, {t, 0, 2Pi}, Boxed → False, Axes → None,
  PlotRange → {{-2, 5/2}, {-2, 2}, {0, 7/2}}],
   {α, 0, 1, 1/8}];
```

We then use `Partition` to partition `smalltable` into three element subsets. The resulting 3×3 array of graphics is shown as a grid with `Show` together with `GraphicsGrid` in Fig. 2.30.

To adjust the viewing angles of three dimensional plots, select the graphic and drag to the desired viewing angle.

```
Show[GraphicsGrid[Partition[smalltable, 3]]]
```

Another way of seeing the transformation is to use `Manipulate`. `Manipulate` is *very* powerful. In its most basic form, `Manipulate[f[x],{x,a,b}]` creates an interactive display of $f(x)$ for x values from a to b. Because the previous commands depended only on α, we combine the commands into a single `Manipulate` object that depends on α.

```
Manipulate[
Clear[x, y, z, r];
x[α_][s_, t_] = (Sqrt[2]Cos[t]^2Cos[2s] + Cos[s]Sin[2t])/
  (2 - αSqrt[2]Sin[3s]Sin[2t]);
y[α_][s_, t_] = (Sqrt[2]Cos[t]^2Sin[2s] - Sin[s]Sin[2t])/
  (2 - α  Sqrt[2]Sin[3s]Sin[2t]);
z[α_][s_, t_] = 3Cos[t]^2/(2 - αSqrt[2]Sin[3s]Sin[2t]);
r[α_][s_, t_] = {x[α][s, t], y[α][s, t], z[α][s, t]};
ParametricPlot3D[r[α][s, t],
  {s, 0, 2Pi}, {t, 0, 2Pi}, Boxed → False, Axes → None,
  PlotRange → {{-2, 5/2}, {-2, 2}, {0, 7/2}}],
   {{α, .25}, 0, 1}]
```

Several images from the result are shown in Fig. 2.31.

Manipulation of graphics is discussed in more detail in Section 5.6, **Matrices and Graphics**. Here, we simply illustrate a few quick ways to manipulate a basic jpeg that illustrate a few of the features of Mathematica 11.

Mathematica 11 provides numerous ways to adjust digital images.

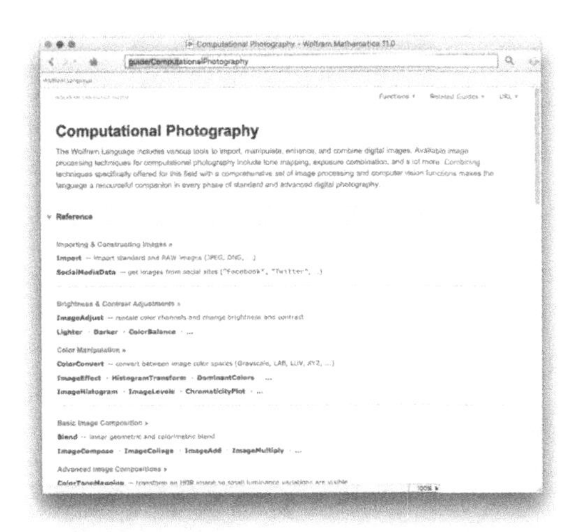

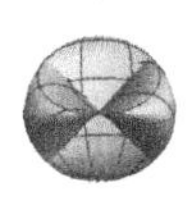

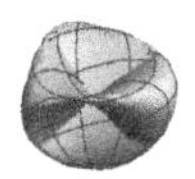

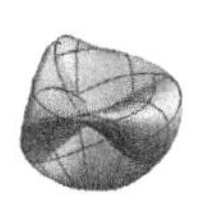
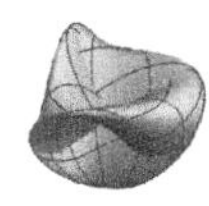
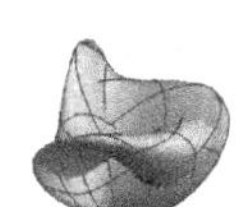

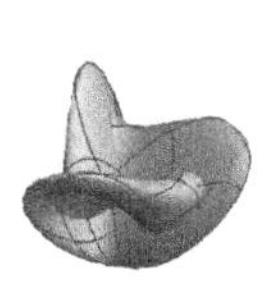
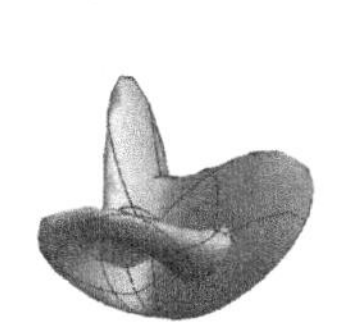
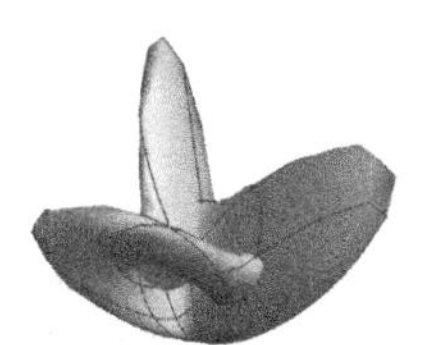

FIGURE 2.30 Seeing the Roman surface continuously transform to the Boy surface.

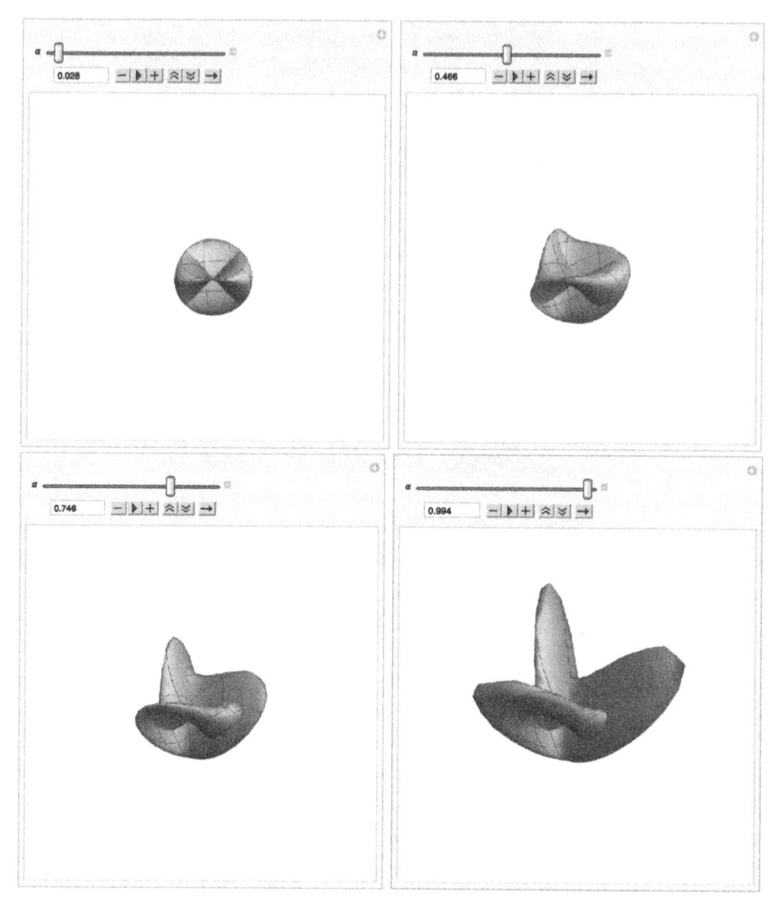

FIGURE 2.31 With `Manipulate` we can create an animation of the transformation of the Roman surface to the Boy surface or inspect the plot for various values of α.

FIGURE 2.32 Importing elementary graphics into Mathematica.

Generally, be sure the graphics file that you want to adjust is in the same folder/directory as the open Mathematica notebook. Use `Import["file"]` to import the file into your open Mathematica notebook. Keep in mind that these graphics can be assigned names just like any other Mathematica objects. Thus, use assign, =, to assign a name to each graphic as you make adjustments. Finally, use `Export`, `Export["file.ext",expr]` to export your modified graphic `expr` using the name `file.ext`, where `ext` is usually something similar to `jpg` or `pdf`. You can then move those graphics into your favorite photography or printing program so you can fine-tune the adjustments and then print the results.

Example 2.37

We use `Import` to import a few graphics into Mathematica. The four graphs are displayed in a row using `Show` and `GraphicsRow` in Fig. 2.32.

FIGURE 2.33 Combining images into a single collage.

FIGURE 2.34 Applying basic visual effects to an image.

```
p1 = Import["rome.jpg"];
p2 = Import["atlanta.jpg"];
p3 = Import["thehigh.jpg"];
p4 = Import["roundchurch.jpg"];

Show[GraphicsRow[{p1, p2, p3, p4}]]
```

Mathematica's tools for editing digital images are quite extensive. Here, we illustrate just a few of the basic available functions. You can combine graphics together with `ImageCollage`. The command

```
ImageCollage[{n1->image1,n2->image2,...,nm->imagem}]
```

creates a collage that weights the various images according to the n-value (see Fig. 2.33).

```
r1 = ImageCollage[{p1, p2, p3, p4}];
r2 = ImageCollage[{5 → p1, 2 → p2, 3 → p3, 1 → p4}];
Show[GraphicsRow[{r1, r2}]]
```

`ImageEffect` can be used to apply various effects to an image. Use `?ImageEffect` for a comprehensive discussion of the different effects that are available (see Fig. 2.34).

```
p1a = ImageEffect[p1, "Charcoal"];
p1b = ImageEffect[p1, "OilPainting"];
p1c = ImageEffect[p1, "Solarization"];
p1d = ImageEffect[p1, "Comics"];
Show[GraphicsRow[{p1a, p1b, p1c, p1d}]]
```

FIGURE 2.35 Approximating the image of The High with 2, 4, 5, and 6 colors.

FIGURE 2.36 A basic adjustment of color levels in an image.

`DominantColors` returns a list of the dominant colors in a graphic (see Fig. 2.35).

```
DominantColors[p3]
```

{■, ■, □, □, ■, ■, ■, ■, ■, ■, ■, □}

You can then use commands like `ColorQuantize` to approximate an image with *n* discrete colors. Later, we will use `ImageApply` to change the colors of individual pixels.

```
p3a = ColorQuantize[p3, 2];
p3b = ColorQuantize[p3, 4];
p3c = ColorQuantize[p3, 5];
p3d = ColorQuantize[p3, 6];
Show[GraphicsRow[{p3a, p3b, p3c, p3d}]]
```

`ColorToneMapping` transforms an image so that you can see small variations in color. On the other hand, `ImageAdjust` adjusts the color levels (see Fig. 2.36).

```
p2b = ColorToneMapping[p2];
p2c = ImageAdjust[p2, .5];
Show[GraphicsRow[{p2, p2b, p2c}]]
```

2.3.5 Miscellaneous Comments

Clearly, Mathematica's graphics capabilities are extensive and *volumes* could be written about them. You can see many commands that we haven't discussed here by using `?` to see those commands that contain the string `Plot`.

Be sure to take advantage of **MathWorld** for a huge number of resources related to graphics and Mathematica.

```
?*Plot*
▼ System`
AnatomyPlot3D                  MatrixPlot
ArrayPlot                      MaxPlotPoints
AudioPlot                      NicholsPlot
BodePlot                       NumberLinePlot
ChromaticityPlot               NyquistPlot
ChromaticityPlot3D             ParametricPlot
CommunityGraphPlot             ParametricPlot3D
ContourPlot                    Plot
ContourPlot3D                  Plot3D
DateListLogPlot                Plot3Matrix
DateListPlot                   PlotDivision
DateListStepPlot               PlotJoined
DensityPlot                    PlotLabel
DensityPlot3D                  PlotLabels
DiscretePlot                   PlotLayout
DiscretePlot3D                 PlotLegends
GeoListPlot                    PlotMarkers
GeoRegionValuePlot             PlotPoints
GraphPlot                      PlotRange
GraphPlot3D                    PlotRangeClipping
JuliaSetPlot                   PlotRangeClipPlanesStyle
LayeredGraphPlot               PlotRangePadding
LineIntegralConvolutionPlot    PlotRegion
ListContourPlot                PlotStyle
ListContourPlot3D              PlotTheme
ListCurvePathPlot              PolarPlot

ListCurvePathPlot              PolarPlot
ListDensityPlot                ProbabilityPlot
ListDensityPlot3D              ProbabilityScalePlot
ListLineIntegralConvolutionPlot QuantilePlot
ListLinePlot                   RegionPlot
ListLogLinearPlot              RegionPlot3D
ListLogLogPlot                 ReliefPlot
ListLogPlot                    RevolutionPlot3D
ListPlot                       RootLocusPlot
ListPlot3D                     RulePlot
ListPointPlot3D                SingularValuePlot
ListPolarPlot                  SliceContourPlot3D
ListSliceContourPlot3D         SliceDensityPlot3D
ListSliceDensityPlot3D         SliceVectorPlot3D
ListSliceVectorPlot3D          SphericalPlot3D
ListStepPlot                   StreamDensityPlot
ListStreamDensityPlot          StreamPlot
ListStreamPlot                 TimelinePlot
ListSurfacePlot3D              TreePlot
ListVectorDensityPlot          VectorDensityPlot
ListVectorPlot                 VectorPlot
ListVectorPlot3D               VectorPlot3D
LogLinearPlot                  WaveletImagePlot
LogLogPlot                     WaveletListPlot
LogPlot                        WaveletMatrixPlot
MandelbrotSetPlot              $PlotTheme
```

You can obtain detailed information regarding any of these commands from the **Documentation Center** by clicking on the command's name.

For now, we briefly mention a few of the ones that were not discussed previously. To plot lists of numbers or lists of ordered pairs, use `ListPlot` or `ListLinePlot`, which are discussed in more detail in Chapter 4. For matrices and other arrays use commands such as `MatrixPlot` or `ArrayPlot` that are discussed in more detail in Chapter 4.

Example 2.38: Cellular Automaton

Very loosely speaking, a **cellular automaton** is a discrete function that assigns values to subsequent rows based on the values of the cells in the previous row(s). For a concise discussion of cellular automaton refer to Weisstein (Weisstein, Eric W. "Cellular Automaton." From *MathWorld–A Wolfram Web Resource*. `http://mathworld.wolfram.com/CellularAutomaton.html`), `CellularAutomatan` is a powerful command that allows you to investigate (quite complicated) cellular automaton. In its simplest form,

```
CellularAutomaton[rule,initialvalues,n]
```

returns the first n generations of the cellular automaton following the specified rule and having the indicated initial values.

The simplest cellular automaton are called **elementary cellular automaton** (Weisstein, Eric W. "Elementary Cellular Automaton." From *MathWorld–A Wolfram Web Resource*. `http://mathworld.wolfram.com/ElementaryCellularAutomaton.html`). Based on basic counting principals, there are 256 elementary cellular automaton. They are cataloged by number. With

CellularAutomaton[146, {{1}, 0}, 5]

{{0, 0, 0, 0, 0, 1, 0, 0, 0, 0, 0}, {0, 0, 0, 0, 1, 0, 1, 0, 0, 0, 0}, {0, 0, 0, 1, 0, 0, 0, 1, 0, 0, 0},

{0, 0, 1, 0, 1, 0, 1, 0, 1, 0, 0}, {0, 1, 0, 0, 0, 0, 0, 0, 0, 1, 0}, {1, 0, 1, 0, 0, 0, 0, 0, 1, 0, 1}}

{{0, 0, 0, 0, 0, 1, 0, 0, 0, 0, 0}, {0, 0, 0, 0, 1, 0, 1, 0, 0, 0, 0},

{0, 0, 0, 1, 0, 0, 0, 1, 0, 0, 0}, {0, 0, 1, 0, 1, 0, 1, 0, 1, 0, 0},

{0, 1, 0, 0, 0, 0, 0, 0, 0, 1, 0}, {1, 0, 1, 0, 0, 0, 0, 0, 1, 0, 1}}

we calculate the first five generations of the elementary cellular automaton with a 1 at position 0 on generation 0 using Rule 146. To calculate the first 100 generations, we use `CellularAutomaton[146, {{1},0}, 100]`. The resulting array is rather large so we use `ArrayPlot` to visualize it in Fig. 2.37 (a). Using our color scheme, the cells with value 1 are shaded in red and those with 0 are in light green.

```
a1 = ArrayPlot[CellularAutomaton[146, {{1}, 0}, 100],
  ColorFunction → "NeonColors", AspectRatio → 1]
```

In this case the grid is initially spaced so that positions 1, 11, 21, 31, and 41 have the value one. The first three generations using Rule 146 are calculated.

```
CellularAutomaton[146,
   {SparseArray[{1 → 1, 11 → 1, 21 → 1, 31 → 1, 41 → 1}], 0}, 3]
{{0, 0, 0, 1, 0, 0, 0, 0, 0, 0, 0, 0, 0, 1, 0, 0, 0, 0, 0, 0,
  0, 0, 0, 1, 0, 0, 0, 0, 0, 0, 0, 0, 0, 1, 0, 0, 0, 0, 0, 0, 0, 0, 0, 1, 0, 0, 0},
   {0, 0, 1, 0, 1, 0, 0, 0, 0, 0, 0, 0, 1, 0, 1, 0, 0, 0, 0, 0, 0,
     0, 1, 0, 1, 0, 0, 0, 0, 0, 0, 0, 1, 0, 1, 0, 0, 0, 0, 0, 0, 1, 0, 1, 0, 0},
   {0, 1, 0, 0, 0, 1, 0, 0, 0, 0, 0, 1, 0, 0, 0, 1, 0, 0, 0, 0, 0,
  1, 0, 0, 0, 1, 0, 0, 0, 0, 0, 1, 0, 0, 0, 1, 0, 0, 0, 0, 0, 1, 0, 0, 0, 1, 0},
   {1, 0, 1, 0, 1, 0, 1, 0, 0, 0, 1, 0, 1, 0, 1, 0, 1, 0, 0, 0, 1,
     0, 1, 0, 1, 0, 1, 0, 0, 0, 1, 0, 1, 0, 1, 0, 1, 0, 0, 0, 1, 0, 1, 0, 1, 0, 1}}
```

To see how the situation evolves over 100 generations is more easily seen using `ArrayPlot`. See Fig. 2.37 (b)

```
a2 = ArrayPlot[CellularAutomaton[146,
  {SparseArray[{1 → 1, 11 → 1, 21 → 1, 31 → 1, 41 → 1}], 0}, 100],
  ColorFunction → "NeonColors", AspectRatio → 1]
Show[GraphicsRow[{a1, a2}]]
```

Of the 256 elementary cellular automaton, many are equivalent. To see that some of them are equivalent, we create a plot of the 256 elementary cellular automaton for 50 generations as done with Rule 146. All 256 plots are shown on the left in Fig. 2.38 (a). With `Union`, we remove and sort the ones that are identically equal. Those are shown on the right in Fig. 2.38 (b).

```
t1 = Table[ArrayPlot[CellularAutomaton[i, {{1}, 0}, 50]],
  {i, 0, 255}];
t2 = Partition[t1, 16];
p1 = Show[GraphicsGrid[t2]];
t3 = Union[t1];
t4 = Partition[t3, 12];
p2 = Show[GraphicsGrid[t4]];
Show[GraphicsRow[{p1, p2}]]
```

To see the plots together with the rule number, use `Table`. Each ordered pair returned consists of the rule number and the 50 generation plot. To display the ordered pairs in an organized fashion, we use `Grid`. Of course, the result is quite large so just a portion of the actual grid is displayed in Fig. 2.39.

```
t5 = Table[{i, ArrayPlot[CellularAutomaton[i, {{1}, 0}, 50]]},
  {i, 0, 255}];
t6 = Partition[t5, 16];
Grid[t6]
```

Note that `MatrixPlot` and `ArrayPlot` are discussed in more detail in Chapter 5.

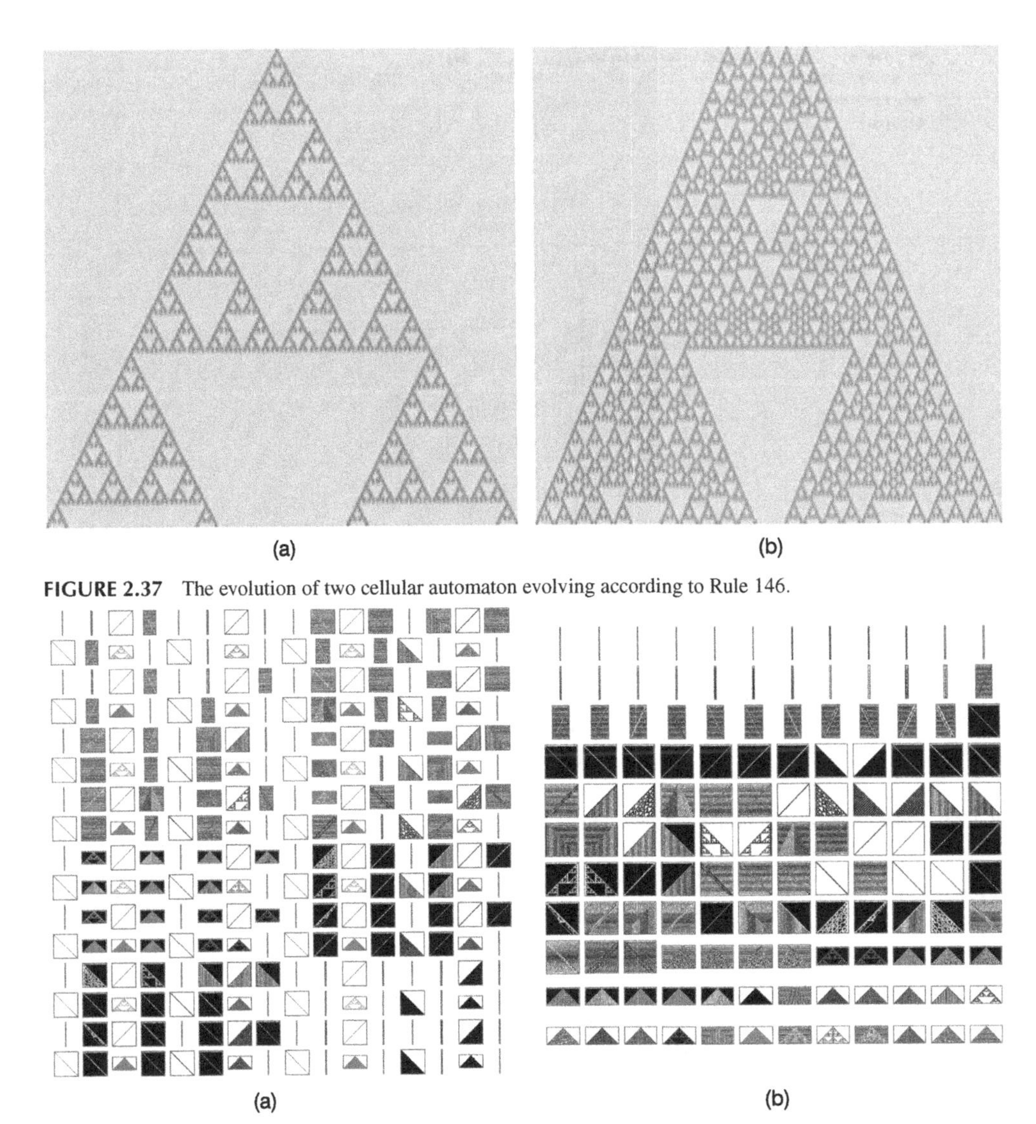

FIGURE 2.37 The evolution of two cellular automaton evolving according to Rule 146.

FIGURE 2.38 (a) The first 50 generations for the 256 elementary cellular automaton. (b) Removal of the identical ones.

For graphs that involve points or nodes or connecting them by edges (graph theory), you can use `GraphPlot` to help investigate some problems. For trees, use `TreePlot`.

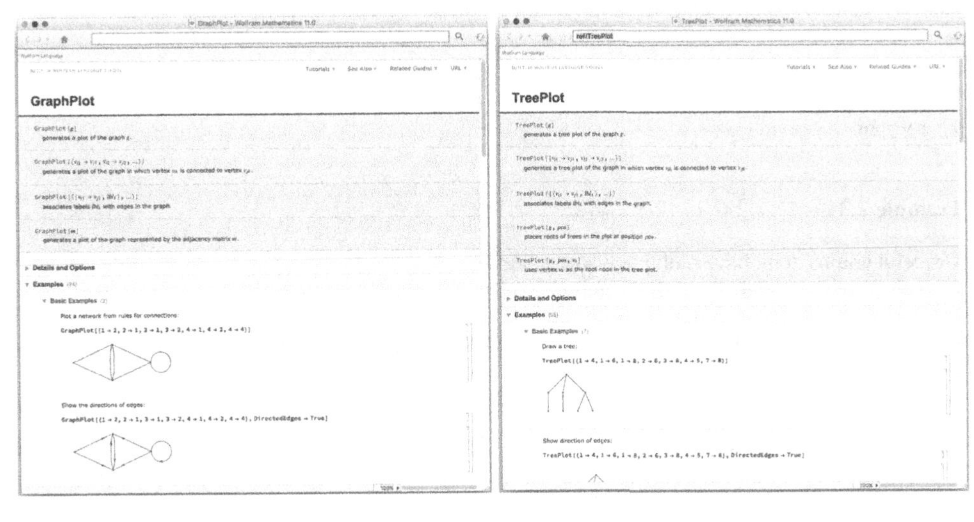

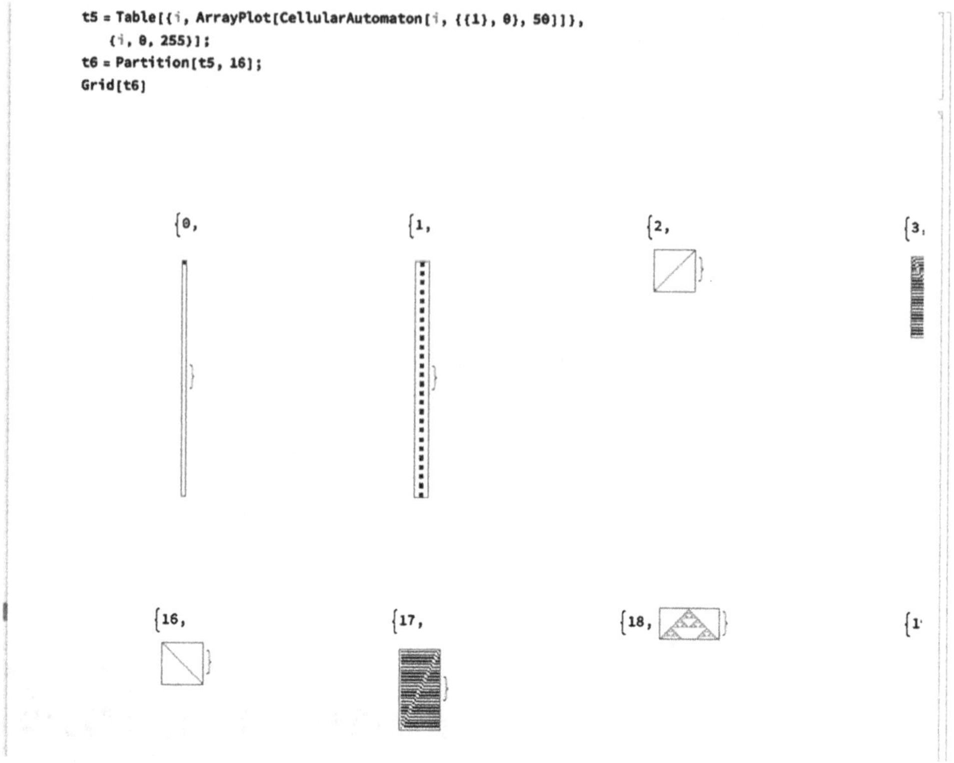

t5 = Table[{i, ArrayPlot[CellularAutomaton[i, {{1}, 0}, 50]]},
{i, 0, 255}];
t6 = Partition[t5, 16];
Grid[t6]
{0,
{1,
{2,
{3,
{16,
{17,
{18,
{1

FIGURE 2.39 Seeing the automaton together with its rule number.

1
9
12
6
0
10
2
8
11
9
1
9
6
12
10
(a)
0
4
3
2
(b)

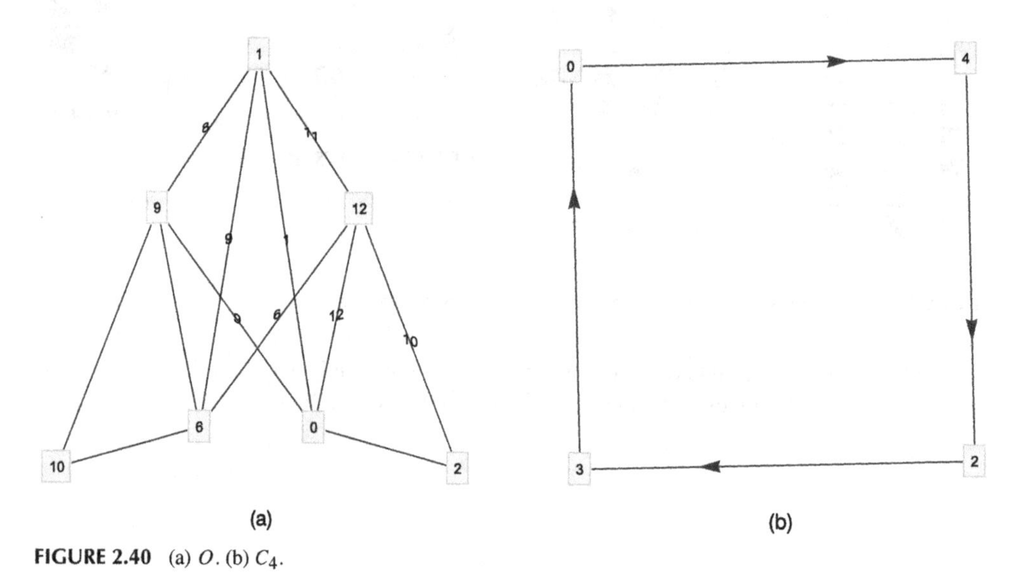

FIGURE 2.40 (a) O. (b) C_4.

Example 2.39

Graceful graphs don't have multiple edges or loops.

We generate O with `GraphPlot` and display the result in Fig. 2.40 (a).

gp1 = GraphPlot[{{0->12, "12"}, {12->1, "11"}, {1->0, "1"}, {0->9, "9"},
{1->9, "8"}, {1->6, "5"}, {12->6, "6"}, {2 → 12, "10"},
{0->2, "2"}, 6->9, 9 → 10, 10 → 6},

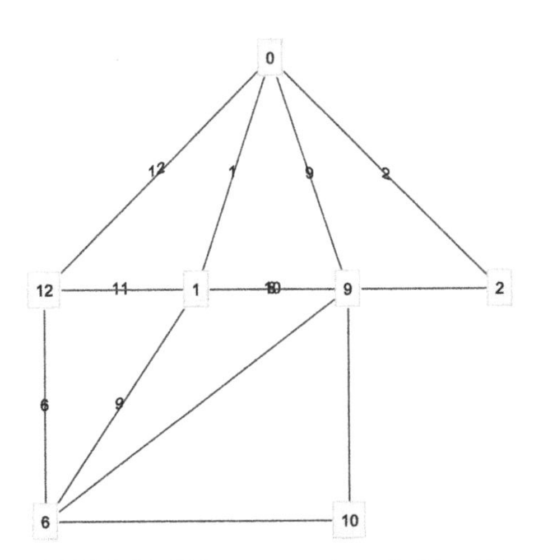

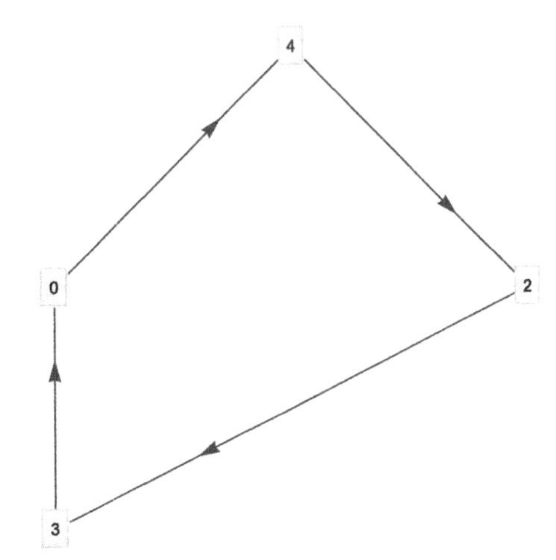

FIGURE 2.41 Using `TreePlot` instead of `GraphPlot`.

```
VertexLabeling → True, AspectRatio → 1]
```

C_4 is shown in Fig. 2.40 (b).

```
gp2 = GraphPlot[{0 → 4, 4 → 2, 2 → 3, 3 → 0}, DirectedEdges → True,
    VertexLabeling → True, AspectRatio → 1]
```

Replacing `GraphPlot` with `TreePlot` gives us Fig. 2.41.

```
tp1 = TreePlot[{{0->12, "12"}, {12->1, "11"}, {1->0, "1"}, {0->9, "9"},
  {1->9, "8"}, {1->6, "5"}, {12->6, "6"}, {2 → 12, "10"},
  {0->2, "2"}, 6->9, 9 → 10, 10 → 6},
    VertexLabeling → True, AspectRatio → 1]

tp2 = TreePlot[{0 → 4, 4 → 2, 2 → 3, 3 → 0}, DirectedEdges → True,
    VertexLabeling → True, AspectRatio → 1]
```

2.4 SOLVING EQUATIONS

2.4.1 Exact Solutions of Equations

Mathematica can find exact solutions to many equations and systems of equations, including exact solutions to polynomial equations of degree four or less. Because a single equals sign "=" is used to name objects and assign values in Mathematica, equations in Mathematica are of the form

`left-hand side==right-hand side.`

The "double-equals" sign "==" between the left-hand side and right-hand side specifies that the object is an equation. For example, to represent the equation $3x + 7 = 4$ in Mathematica, type `3x+7==4`. The command `Solve[lhs==rhs,x]` solves the equation `lhs` = `rhs` for x. If the only unknown in the equation `lhs` = `rhs` is x and Mathematica does not need to use inverse functions to solve for x, the command `Solve[lhs==rhs]` solves the equation `lhs` = `rhs` for x. Hence, to solve the equation $3x + 7 = 4$, both the commands `Solve[3x+7==4]` and `Solve[3x+7==4,x]` return the same result.

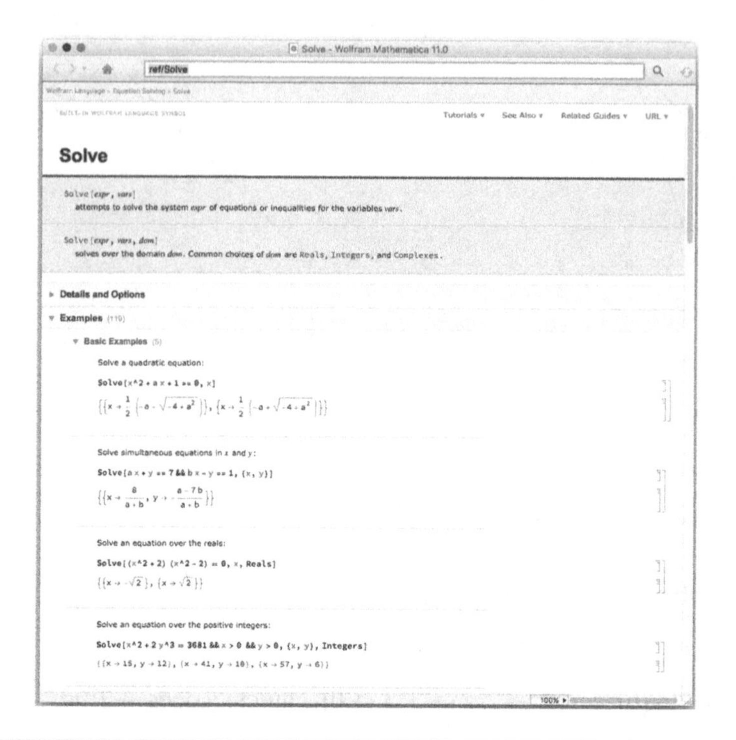

Example 2.40

Solve the equations $3x+7=4$, $(x^2-1)/(x-1)=0$, and $x^3+x^2+x+1=0$.

Solution. In each case, we use `Solve` to solve the indicated equation. Be sure to include the double equals sign "==" between the left and right-hand sides of each equation. Thus, the result of entering

Solve[3x + 7==4]

$\{\{x \to -1\}\}$

means that the solution of $3x+7=4$ is $x=-1$ and the result of entering

Solve$\left[\frac{x^2-1}{x-1}==0\right]$

$\{\{x \to -1\}\}$

means that the solution of $\dfrac{x^2-1}{x-1}=0$ is $x=-1$. On the other hand, the equation $x^3+x^2+x+1=0$ has two imaginary roots. We see that entering

Solve$\left[x^3+x^2+x+1==0\right]$

$\{\{x \to -1\}, \{x \to -i\}, \{x \to i\}\}$

yields all three solutions. Thus, the solutions of $x^3+x^2+x+1=0$ are $x=-1$ and $x=\pm i$. Remember that the Mathematica symbol `I` represents the complex number $i=\sqrt{-1}$. In general, Mathematica can find the exact solutions of any polynomial equation of degree four or less. □

Observe that the results of a `Solve` command are a **list**.

Lists and tables are discussed in more detail in Chapters 4 and 5.

Mathematica can also solve equations involving more than one variable for one variable (literal equations) in terms of other unknowns.

Example 2.41

(a) Solve the equation $v=\pi r^2/h$ for h. (b) Solve the equation $a^2+b^2=c^2$ for a.

Solution. These equations involve more than one unknown so we must specify the variable for which we are solving in the `Solve` commands. Thus, entering

Solve[v == Pir^2/h, h]

$$\left\{\left\{h\to \frac{\pi r^2}{v}\right\}\right\}$$

solves the equation $v=\pi r^2/h$ for h. (Be sure to include a space or $*$ between π and r.) Similarly, entering

Solve[a^2 + b^2 == c^2, a]

$$\left\{\left\{a\to -\sqrt{-b^2+c^2}\right\},\left\{a\to \sqrt{-b^2+c^2}\right\}\right\}$$

solves the equation $a^2+b^2=c^2$ for a. □

If Mathematica needs to use inverse functions to solve an equation, you must be sure to specify the variable(s) for which you want Mathematica to solve.

Example 2.42

Find a solution of $\sin^2 x - 2\sin x - 3 = 0$.

Solution. When the command `Solve[Sin[x]^2-2Sin[x]-3==0]` is entered, Mathematica solves the equation for `Sin[x]`. However, when the command

```
Solve[Sin[x]^2-2Sin[x]-3==0,x]
```

is entered, Mathematica attempts to solve the equation for x.

Solve[Sin[x]^2 − 2Sin[x] − 3==0]

$\{\{x\to \text{ConditionalExpression}\left[-\frac{\pi}{2}+2\pi C[1], C[1]\in \text{Integers}\right]\},$

$\{x\to \text{ConditionalExpression}\left[\frac{3\pi}{2}+2\pi C[1], C[1]\in \text{Integers}\right]\},$

$\{x\to \text{ConditionalExpression}[\pi-\text{ArcSin}[3]+2\pi C[1], C[1]\in \text{Integers}]\},$

$\{x\to \text{ConditionalExpression}[\text{ArcSin}[3]+2\pi C[1], C[1]\in \text{Integers}]\}\}$

Mathematica's result indicates that the solutions to the equation are $x=-\pi/2+2C\pi$ and $x=3\pi/2+2C\pi$, where C is any integer. $\sin^{-1}3$ is not a real number so in the context of solving the equation, we ignore the complex-valued solutions. Therefore this equation has infinitely many solutions of the form $x=\frac{1}{2}(4k-1)\pi$, $k=0, \pm1, \pm2, \ldots$; $\sin x = 3$ has no solutions. □

Example 2.43

Let $f(\theta) = \sin 2\theta + 2\cos\theta$, $0 \le \theta \le 2\pi$. (a) Solve $f'(\theta) = 0$. (b) Graph $f(\theta)$ and $f'(\theta)$.

Solution. After defining $f(\theta)$, we use `D` to compute $f'(\theta)$ and then use `Solve` to solve $f'(\theta) = 0$.

`D[f[x],x]` computes $f'(x)$; `D[f[x],{x,n}]` computes $f^{(n)}(x)$. Topics from calculus are discussed in more detail in Chapter 3.

f[θ_] = Sin[2θ] + 2Cos[θ];

df = f′[θ]

2Cos[2θ] − 2Sin[θ]

Solve[df == 0, θ]

$\{\{\theta \to \text{ConditionalExpression}\left[-\frac{\pi}{2} + 2\pi C[1], C[1] \in \text{Integers}\right]\},$

$\{\theta \to \text{ConditionalExpression}\left[\frac{\pi}{6} + 2\pi C[1], C[1] \in \text{Integers}\right]\},$

$\{\theta \to \text{ConditionalExpression}\left[\frac{5\pi}{6} + 2\pi C[1], C[1] \in \text{Integers}\right]\}\}$

As in the previous example, the solutions are given as a sequence of conditionals: $-\pi/2 + 2C\pi$, $\pi/6 + 2C\pi$ and $5\pi/6 + 2C\pi$, where C is an integer. Of these infinitely many solutions, $\theta = \pi/6$, $5\pi/6$, and $3\pi/2$ are in the interval $[0, 2\pi]$.
To verify by hand, we use the identity $\cos 2\theta = 1 - 2\sin^2\theta$ and factor:

$$\begin{aligned}
2\cos 2\theta - 2\sin\theta &= 0 \\
1 - 2\sin^2\theta - \sin\theta &= 0 \\
2\sin^2\theta + \sin\theta - 1 &= 0 \\
(2\sin\theta - 1)(\sin\theta + 1) &= 0
\end{aligned}$$

so $\sin\theta = 1/2$ or $\sin\theta = -1$. Because we are assuming that $0 \le \theta \le 2\pi$, we obtain the solutions $\theta = \pi/6$, $5\pi/6$, or $3\pi/2$. We perform the same steps with Mathematica.

`expression /. x->y` replaces all occurrences of x in *expression* by y.

s1 = TrigExpand[df]

$2\text{Cos}[\theta]^2 - 2\text{Sin}[\theta] - 2\text{Sin}[\theta]^2$

s2 = s1/.Cos[θ]^2 → 1 − Sin[θ]^2

$-2\text{Sin}[\theta] - 2\text{Sin}[\theta]^2 + 2\left(1 - \text{Sin}[\theta]^2\right)$

Factor[s2]

−2(1 + Sin[θ])(−1 + 2Sin[θ])

Finally, we graph $f(\theta)$ and $f'(\theta)$ with `Plot` in Fig. 2.42. Note that the plot is drawn to scale because we include the option `AspectRatio->Automatic`.

p1 = Plot[{f[θ], df}, {θ, 0, 2π}, AspectRatio → Automatic,

PlotStyle → {{CMYKColor[0, .09, .80, 0]}, {CMYKColor[1, 6.0, 0, .60]}},

PlotLabel → "University of Iowa Yellow and University of Michigan Blue"] □

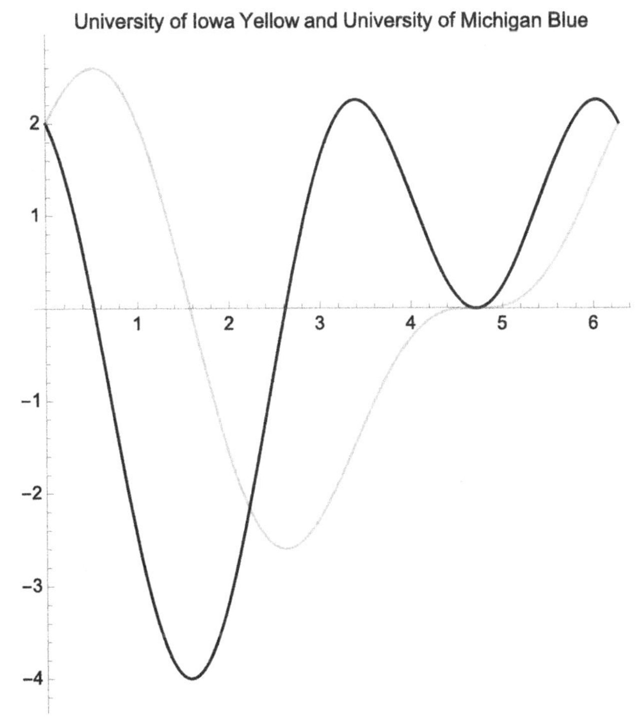

FIGURE 2.42 Graphs of $f(\theta)$ in University of Iowa yellow and $f'(\theta)$ in University of Michigan blue. (For interpretation of the colors in this figure, the reader is referred to the web version of this chapter.)

We can also use `Solve` to find the solutions, if any, of various types of systems of equations. Entering

```
Solve[{lhs1==rhs1,lhs2==rhs2},{x,y}]
```

solves a system of two equations for x and y while entering

```
Solve[{lhs1==rhs1,lhs2==rhs2}]
```

attempts to solve the system of equations for all unknowns. In general, `Solve` can find the solutions to a system of linear equations. In fact, if the systems to be solved are inconsistent or dependent, Mathematica's output indicates so.

Example 2.44

Solve each system:

(a) $\begin{cases} 3x - y = 4 \\ x + y = 2 \end{cases}$; (b) $\begin{cases} 2x - 3y + 4z = 2 \\ 3x - 2y + z = 0 \\ x + y - z = 1 \end{cases}$; (c) $\begin{cases} 2x - 2y - 2z = -2 \\ -x + y + 3z = 0 \\ -3x + 3y - 2z = 1 \end{cases}$; and (d) $\begin{cases} -2x + 2y - 2z = -2 \\ 3x - 2y + 2z = 2 \\ x + 3y - 3z = -3 \end{cases}$.

Solution. In each case we use `Solve` to solve the given system. For (a), the result of entering

Solve[{3*x* − *y*==4, *x* + *y*==2}, {*x*, *y*}]

$\left\{\left\{x \to \frac{3}{2}, y \to \frac{1}{2}\right\}\right\}$

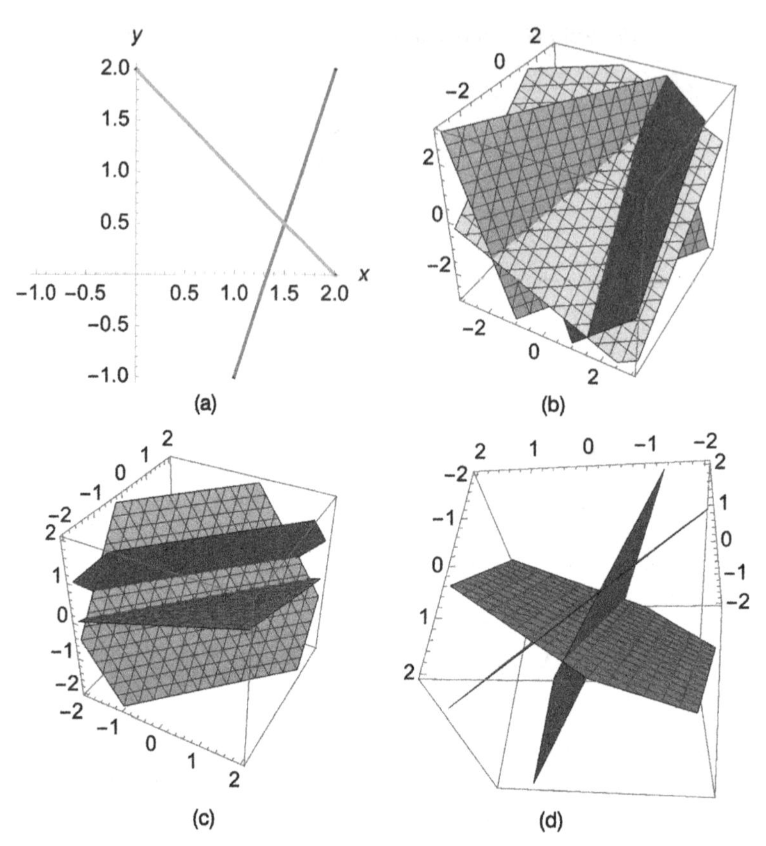

FIGURE 2.43 (a) Two intersecting lines. (b) Three planes that intersect in a single point. (c) These three planes have no point in common. (d) The intersection of these three planes is a line.

means that the solution of $\begin{cases} 3x - y = 4 \\ x + y = 2 \end{cases}$ is $(x, y) = (3/2, 1/2)$, which is the point of intersection of the lines with equations $3x - y = 4$ and $x + y = 2$. See Fig. 2.43 (a).

```
cp1 = ContourPlot[{3x − y == 4, x + y == 2},
  {x, −1, 2}, {y, −1, 2}, Frame → False,
    Axes → Automatic, AxesOrigin → {0, 0},
      AxesLabel → {x, y}]
```

(b) We can verify that the results returned by Mathematica are correct. First, we name the system of equations `sys` and then use `Solve` to solve the system of equations naming the result `sols`.

```
sys = {2x − 3y + 4z==2, 3x − 2y + z==0,

x + y − z==1};
```

```
sols = Solve[sys, {x, y, z}]
```

$\left\{\left\{x \to \frac{7}{10}, y \to \frac{9}{5}, z \to \frac{3}{2}\right\}\right\}$

We verify the result by substituting the values obtained with `Solve` back into `sys` with `ReplaceAll` (`/.`).

```
sys/.sols
```

{{True, True, True}}

means that the solution of $\begin{cases} 2x - 3y + 4z = 2 \\ 3x - 2y + z = 0 \\ x + y - z = 1 \end{cases}$ is $(x, y, z) = (7/10, 9/5, 3/2)$, which is the point of intersection of the planes with equations $2x - 3y + 4z = 2$, $3x - 2y + z = 0$, $x + y - z = 1$. See Fig. 2.43 (b).

```
cp2a = ContourPlot3D[{2x − 3y + 4z==2, 3x − 2y + z==0,
  x + y − z==1}, {x, −3, 3},
    {y, −3, 3}, {z, −3, 3}]
```

To better see the intersection point, click within the graphic and then drag to an appropriate viewing angle.

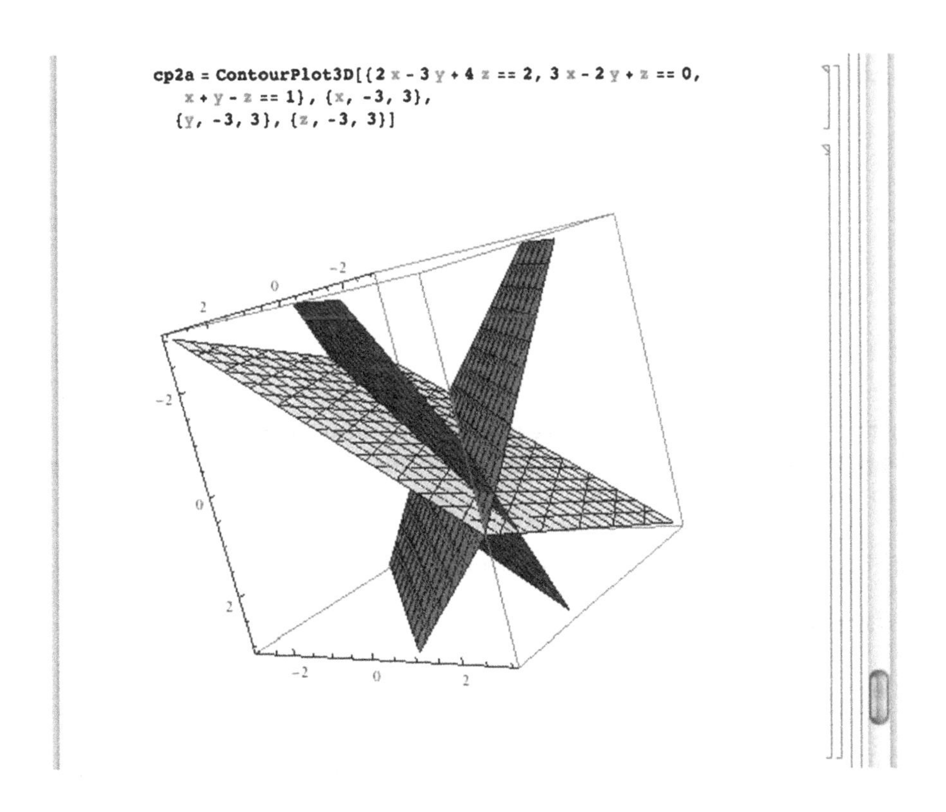

(c) When we use `Solve` to solve this system, Mathematica returns `{}`, which indicates that the system has no solution; the system is inconsistent.

```
Solve[{2x − 2y − 2z== − 2, −x + y + 3z==0,

  −3x + 3y − 2z==1}]
```

```
{}
```

To see that the planes with equations $2x - 2y - 2z = -2$, $-x + y + 3z = 0$, and $-3x + 3y - 2z = 1$ have no points in common, graph them with in Fig. 2.43 (c).

```
cp3a = ContourPlot3D[{2x − 2y − 2z == −2,
  −x + y + 3z == 0, −3x + 3y − 2z == 1}, {x, −2, 2},
    {y, −2, 2}, {z, −2, 2}]
```

To better see that the planes do not intersect, we click and drag the graphic to an appropriate viewing angle.

```
cp3a = ContourPlot3D[{2 x - 2 y - 2 z == -2,
   -x + y + 3 z == 0, -3 x + 3 y - 2 z == 1}, {x, -2, 2},
  {y, -2, 2}, {z, -2, 2}]
```

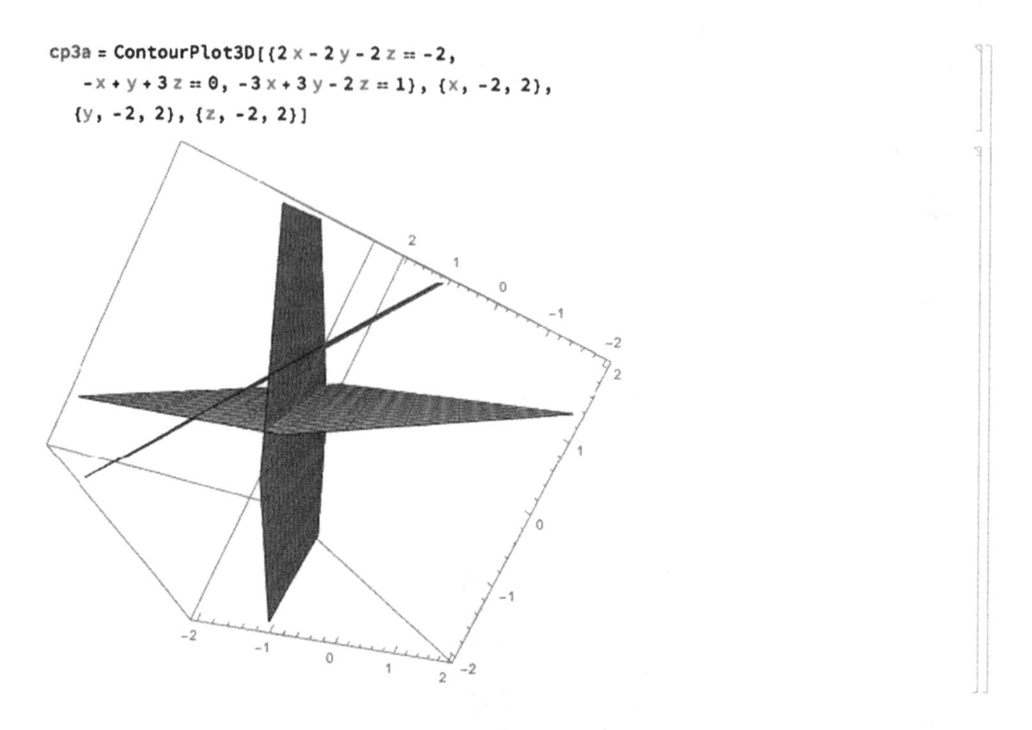

(d) On the other hand, when we use `Solve` to solve this system, Mathematica's result indicates that the system has infinitely many solutions. That is, all ordered triples of the form $\{(0, y, 1 + y) | y \text{ real}\}$ are solutions of the system.

```
Solve[{-2x + 2y - 2z == -2, 3x - 2y + 2z == 2,
  x + 3y - 3z == -3}]
```

$\{\{x \to 0, z \to 1 + y\}\}$

We see that the intersection of the three planes is a line with `ContourPlot3D`. See Fig. 2.43 (d).

```
cp3a = ContourPlot3D[{2x - 2y - 2z == -2,
  3x - 2y + 2z == 2, x + 3y - 3z == -3}, {x, -2, 2},
   {y, -2, 2}, {z, -2, 2}]
Show[GraphicsGrid[{{cp1, cp2a}, {cp3a, cp4a}}]]
```

□

We can often use `Solve` to find solutions of a nonlinear system of equations as well.

Example 2.45

Solve the systems (a) $\begin{cases} 4x^2 + y^2 = 4 \\ x^2 + 4y^2 = 4 \end{cases}$ and (b) $\begin{cases} \frac{1}{a^2}x^2 + \frac{1}{b^2}y^2 = 1 \\ y = mx \end{cases}$ (a, b greater than zero) for x and y.

Solution. The graphs of the equations are both ellipses. We use `ContourPlot` to graph each equation, naming the results `cp1` and `cp2`, respectively, and then use `Show` to display both graphs together in Fig. 2.44 (a). The solutions of the system correspond to the intersection

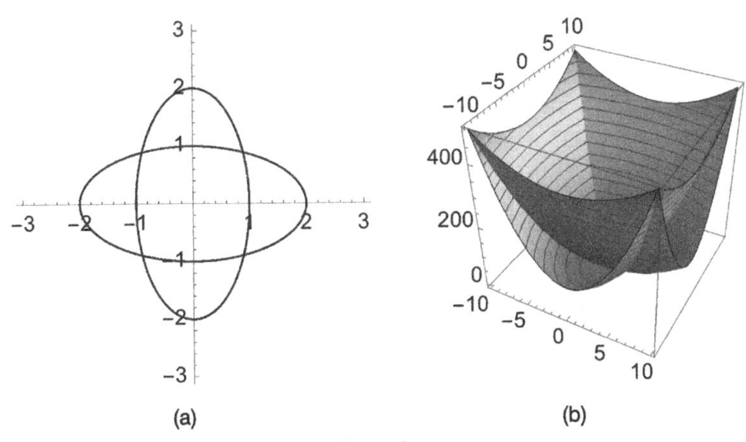

FIGURE 2.44 (a) Graphs of $4x^2 + y^2 = 4$ and $x^2 + 4y^2 = 4$. (b) 3-dimensional plots of $f(x, y)$ and $g(x, y)$ together with their level curves shown as contours.

points of the two graphs. Alternatively, the solutions of the system correspond to the intersection points of the level curves of $f(x, y) = 4x^2 + y^2 - 4$ and $g(x, y) = x^2 + 4y^2 - 4$ corresponding to 0. See Fig. 2.44 (b).

```
cp1 = ContourPlot[4x^2 + y^2 − 4, {x, −3, 3}, {y, −3, 3}, Contours → {0},
    ContourShading → False, PlotPoints → 60];
cp2 = ContourPlot[x^2 + 4y^2 − 4, {x, −3, 3}, {y, −3, 3}, Contours → {0},
    ContourShading → False, PlotPoints → 60];
cp3 = Show[cp1, cp2, Frame → False, Axes → Automatic, AxesOrigin → {0, 0}]
cp4 = Plot3D[{4x^2 + y^2 − 4, x^2 + 4y^2 − 4}, {x, −10, 10},
  {y, −10, 10}, BoxRatios → {1, 1, 1}, MeshFunctions->{#3&},
    ColorFunction → (ColorData["Rainbow"][#3]&),
      PlotStyle → {Opacity[.4], Opacity[.8]}]
Show[GraphicsRow[{cp3, cp4}]]
```

Finally, we use `Solve` to find the solutions of the system.

```
.1in Solve[{4x^2 + y^2==4, x^2 + 4y^2==4}]
```

$$\left\{\left\{x \to -\frac{2}{\sqrt{5}}, y \to -\frac{2}{\sqrt{5}}\right\}, \left\{x \to -\frac{2}{\sqrt{5}}, y \to \frac{2}{\sqrt{5}}\right\}, \left\{x \to \frac{2}{\sqrt{5}}, y \to -\frac{2}{\sqrt{5}}\right\}, \left\{x \to \frac{2}{\sqrt{5}}, y \to \frac{2}{\sqrt{5}}\right\}\right\}$$

For (b), we also use `Solve` to find the solutions of the system. However, because the unknowns in the equations are a, b, m, x, and y, we must specify that we want to solve for x and y in the `Solve` command.

```
Solve[{x^2/a^2 + y^2/b^2==1, y==mx}, {x, y}]
```

$$\left\{\left\{x \to -\frac{ab}{\sqrt{b^2+a^2m^2}}, y \to -\frac{abm}{\sqrt{b^2+a^2m^2}}\right\}, \left\{x \to \frac{ab}{\sqrt{b^2+a^2m^2}}, y \to \frac{abm}{\sqrt{b^2+a^2m^2}}\right\}\right\}$$ □

Although Mathematica can find the exact solution to every polynomial equation of degree four or less, exact solutions to some equations may not be meaningful. In those cases, Mathematica can provide approximations of the exact solutions using either the `N[expression]` or the `expression // N` commands.

Example 2.46

Approximate the solutions to the equations (a) $x^4 - 2x^2 = 1 - x$; and (b) $1 - x^2 = x^3$.

Solution. Each of these is a polynomial equation with degree less than five so `Solve` will find the exact solutions of each equation. However, the solutions are quite complicated so we use `N` to obtain approximate solutions of each equation. For (a), entering

```
N[Solve[x^4 - 2x^2==1 - x]]
{{x → 0.182777 - 0.633397i}, {x → 0.182777 + 0.633397i}, {x → -1.71064},
  {x → 1.34509}}
```

first finds the exact solutions of the equation $x^4 - 2x^2 = 1 - x$ and then computes approximations of those solutions. The resulting output is the list of approximate solutions. Approximating solutions of equations is discussed further in the next subsection.
For (b), entering

```
Solve[1 - x^2==x^3, x]//N
{{x → 0.754878}, {x → -0.877439 + 0.744862i},
  {x → -0.877439 - 0.744862i}}
    {{x → 0.754878}, {x → -0.877439 + 0.744862i},
      {x → -0.877439 - 0.744862i}}
```

first finds the exact solutions of the equation $1 - x^2 = x^3$ and then computes approximations of those solutions. The resulting output is the list of approximate solutions. □

2.4.2 Approximate Solutions of Equations

When solving an equation is either impractical or impossible, Mathematica provides several functions including `FindRoot`, `NRoots`, and `NSolve` to approximate solutions of equations. `NRoots` and numerically approximate the roots of any polynomial equation. The command `NRoots[poly1==poly2,x]` approximates the solutions of the polynomial equation `poly1==poly2`, where both `poly1` and `poly2` are polynomials in x. The syntax for `NSolve` is the same as the syntax of `NRoots` although `NSolve` is usually capable of approximating solutions of more complicated equations than `NRoots`. Often you will find that using `Solve` together with `N` will yield the exact same results as those obtained using `NSolve`. Regardless, it is relatively easy to ask Mathematica to approximate the solutions of a complicated equation and it cannot. In a general sense, solving and approximating solutions of equations in a general sense is a challenging problem.

`FindRoot` attempts to approximate a root to an equation provided that a "reasonable" guess of the root is given. `FindRoot` works on functions other than polynomials. The command

```
FindRoot[lhs==rhs,{x,firstguess}]
```

searches for a numerical solution to the equation `lhs==rhs`, starting with x = `firstguess`. To locate more than one root, `FindRoot` must be used several times. One way of obtaining `firstguess` (for real-valued solutions) is to graph both `lhs` and `rhs` with `Plot`, find the point(s) of intersection, and estimate the x-coordinates of the point(s) of intersection. Generally, `NRoots` is easier to use than `FindRoot` when trying to approximate the roots of a polynomial. For "simple" non-polynomial equations, `NSolve` may work well. If `NRoots` or `NSolve` do not yield the desired results, it may be best to use `FindRoot` to approximate each root (solution) individually.

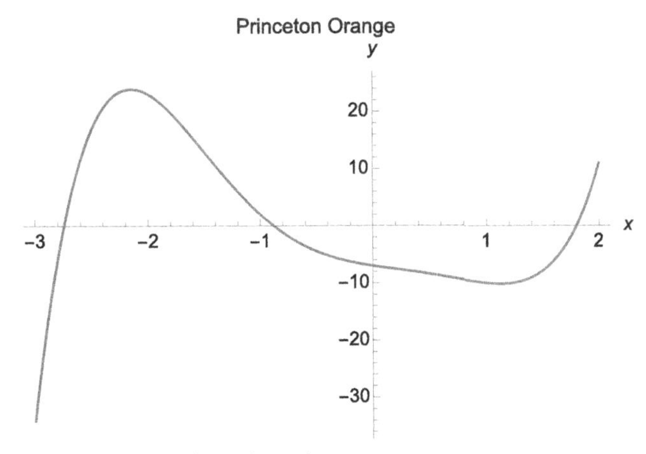

FIGURE 2.45 Graph of $f(x) = x^5 + x^4 - 4x^3 + 2x^2 - 3x - 7$ in Princeton orange.

Example 2.47

Approximate the solutions of $x^5 + x^4 - 4x^3 + 2x^2 - 3x - 7 = 0$.

Solution. Because $x^5 + x^4 - 4x^3 + 2x^2 - 3x - 7 = 0$ is a polynomial equation, we may use `NRoots` to approximate the solutions of the equation. Thus, entering

```
NRoots[x^5 + x^4 - 4x^3 + 2x^2 - 3x - 7==0, x]
x == -2.74463||x == -0.880858||x == 0.41452 - 1.19996i||
  x == 0.41452 + 1.19996i||x == 1.79645
  x == -2.74463||x == -0.880858||x == 0.41452 - 1.19996i||
    x == 0.41452 + 1.19996i||x == 1.79645
```

approximates the solutions of $x^5 + x^4 - 4x^3 + 2x^2 - 3x - 7 = 0$. The symbol || appearing in the result represents "or".

We obtain equivalent results with `NSolve`.

```
NSolve[x^5 + x^4 - 4x^3 + 2x^2 - 3x - 7==0, x]
{{x -> -2.74463}, {x -> -0.880858}, {x -> 0.41452 - 1.19996i},
    {x -> 0.41452 + 1.19996i}, {x -> 1.79645}}
  {{x -> -2.74463}, {x -> -0.880858}, {x -> 0.41452 - 1.19996i},
    {x -> 0.41452 + 1.19996i}, {x -> 1.79645}}
```

`FindRoot` may also be used to approximate each root of the equation. However, to use `FindRoot`, we must supply an initial approximation of the solution that we wish to approximate. The real solutions of $x^5 + x^4 - 4x^3 + 2x^2 - 3x - 7 = 0$ correspond to the values of x where the graph of $f(x) = x^5 + x^4 - 4x^3 + 2x^2 - 3x - 7$ intersects the x-axis. We use `Plot` to graph $f(x)$ in Fig. 2.45.

```
Plot[x^5 + x^4 - 4x^3 + 2x^2 - 3x - 7, {x, -3, 2},
  PlotStyle -> CMYKColor[{0, .62, .95, 0}],
  PlotLabel -> "Princeton Orange",
    AxesLabel -> {x, y}]
```

We see that the graph intersects the x-axis near $x \approx -2.5$, -1, and 1.5. We use these values as initial approximations of each solution. Thus, entering

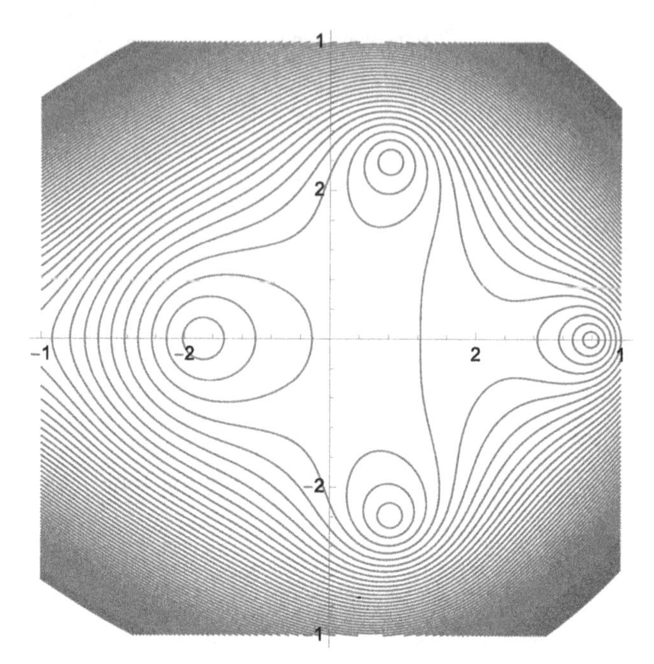

FIGURE 2.46 Level curves of $w = |f(x+iy)|$, $-2 \le x \le 2$, $-2 \le y \le 2$.

```
FindRoot[x^5 + x^4 - 4x^3 + 2x^2 - 3x - 7==0, {x, -2.5}]
{x → -2.74463}
```

approximates the solution near -2.5, entering

```
FindRoot[x^5 + x^4 - 4x^3 + 2x^2 - 3x - 7==0, {x, -1}]
{x → -0.880858}
```

approximates the solution near -1, and entering

```
FindRoot[x^5 + x^4 - 4x^3 + 2x^2 - 3x - 7==0, {x, 2}]
{x → 1.79645}
```

approximates the solution near 1.5, which is relatively close to $x = 2$, which is why we used $x = 2$ as our initial guess in the `FindRoot` command. Note that `FindRoot` may be used to approximate complex solutions as well. To obtain initial guesses, observe that the solutions of $f(z) = 0$, $z = x + iy$, x, y real, are the level curves of $w = |f(z)|$ that are points. In Fig. 2.46, we use `ContourPlot` to graph various level curves of $w = |f(x+iy)|$, $-2 \le x \le 2$, $-2 \le y \le 2$. In the plot, observe that the two complex solutions occur at $x \pm iy \approx 0.5 \pm 1.2i$.

```
f[z_] = z^5 + z^4 - 4z^3 + 2z^2 - 3z - 7;
ContourPlot[Abs[f[x + Iy]], {x, -2, 2}, {y, -2, 2},
  ContourShading → False, Contours → 60,
   PlotPoints → 200, Frame → False, Axes → Automatic,
    AxesOrigin → {0, 0}]
```

Thus, entering

```
FindRoot[x^5 + x^4 - 4x^3 + 2x^2 - 3x - 7==0, {x, 0.5 + I}]
{x → 0.41452 + 1.19996i}
```

approximates the solution near $x + iy \approx 0.5 + 1.2i$. For polynomials with real coefficients, complex solutions occur in conjugate pairs so the other complex solution is approximately $0.41452 - 1.19996i$.

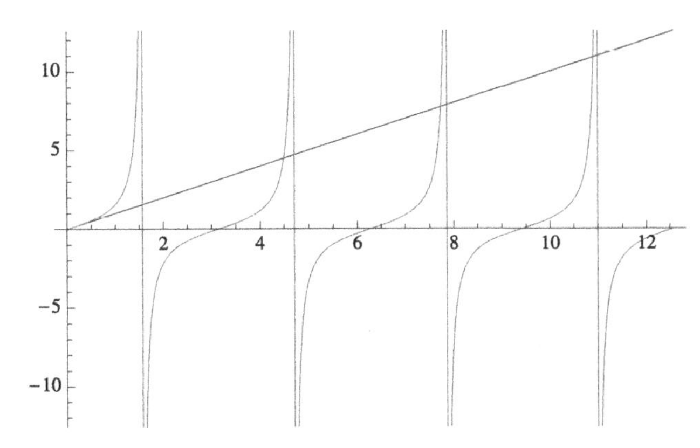

FIGURE 2.47 $y = x$ and $y = \tan x$.

Example 2.48

Find the first three nonnegative solutions of $x = \tan x$.

Solution. We attempt to solve $x = \tan x$ with `Solve`.

```
Solve[x == Tan[x], x]
··· Solve: This system cannot be solved with the methods available to Solve.
Solve[x == Tan[x], x]
```

We next graph $y = x$ and $y = \tan x$ together in Fig. 2.47.

```
Plot[Tooltip[{x, Tan[x]}], {x, 0, 4Pi},
    PlotRange → {−4Pi, 4Pi}]
```

□

In the graph, we see that $x = 0$ is a solution. This is confirmed with `FindRoot`.

Remember that vertical lines are never the graphs of functions. In this case, they represent the vertical asymptotes at odd multiples of $\pi/2$.

```
FindRoot[x == Tan[x], {x, 0}]
{x → 0.}
```

The second solution is near 4 while the third solution is near 7. Using `FindRoot` together with these initial approximations locates the second two solutions.

```
FindRoot[x == Tan[x], {x, 4}]
{x → 4.49341}
FindRoot[x == Tan[x], {x, 7}]
{x → 7.72525}
```

□

`FindRoot` can also be used to approximate solutions to systems of equations. (Although `NRoots` can solve a polynomial equation, `NRoots` cannot be used to solve a system of polynomial equations.) When approximations of solutions of systems of equations are desired, use either `Solve` and `N` together, when possible, or `FindRoot`.

Example 2.49

Approximate the solutions to the system of equations $\begin{cases} x^2 + 4xy + y^2 = 4 \\ 5x^2 - 4xy + 2y^2 = 8 \end{cases}$.

Solution. We begin by using `ContourPlot` to graph each equation in Fig. 2.48. From the resulting graph, we see that $x^2 + 4xy + y^2 = 4$ is a hyperbola, $5x^2 - 4xy + 2y^2 = 8$ is an ellipse, and there are four solutions to the system of equations.

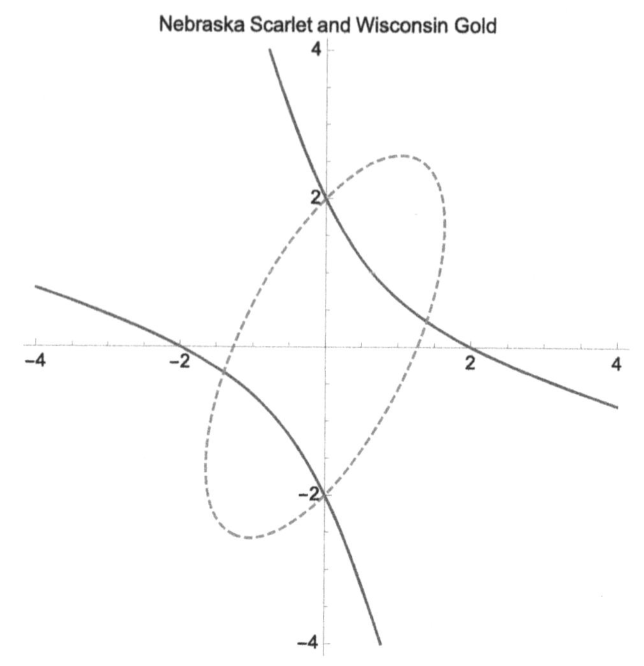

FIGURE 2.48 Graphs of $x^2 + 4xy + y^2 = 4$ (Nebraska scarlet) and $5x^2 - 4xy + 2y^2 = 8$ (Wisconsin gold).

```
cp1 = ContourPlot[x^2 + 4xy + y^2 - 4 == 0, {x, -4, 4}, {y, -4, 4},
  PlotPoints -> 60, ContourShading -> False,
   ContourStyle -> CMYKColor[{.002, 1, .85, .06}]];
cp2 = ContourPlot[5x^2 - 4xy + 2y^2 - 8 == 0, {x, -4, 4},
  {y, -4, 4}, PlotPoints -> 60,
  ContourShading -> False,
  ContourStyle -> {Dashing[{0.01}],
   CMYKColor[{.05, .26, 1, .27}]}];
Show[cp1, cp2, Frame -> False, Axes -> Automatic,
  AxesOrigin -> {0, 0},
  PlotLabel -> "Nebraska Scarlet and Wisconsin Gold"]
```

From the graph we see that possible solutions are $(0, 2)$ and $(0, -2)$. In fact, substituting $x = 0$ and $y = -2$ and $x = 0$ and $y = 2$ into each equation verifies that these points are both exact solutions of the equation. The remaining two solutions are approximated with `FindRoot`.

```
FindRoot[{x^2 + 4xy + y^2==4, 5x^2 - 4xy + 2y^2==8},
   {x, -1}, {y, -0.25}]
{x -> -1.39262, y -> -0.348155}
```

□

Chapter 3

Calculus

3.1 LIMITS AND CONTINUITY

One of the first topics discussed in calculus is that of limits. Mathematica can be used to investigate limits graphically and numerically. In addition, the Mathematica command `Limit[f[x],x->a]` attempts to compute the limit of $y = f(x)$ as x approaches a, $\lim_{x\to a} f(x)$, where a can be a finite number, ∞ (`Infinity`), or $-\infty$ (`-Infinity`). The arrow "`->`" is obtained by typing a minus sign "`-`" followed by a greater than sign ">".

Remark 3.1. To define a function of a single variable, $f(x) = expression\, in\, x$, enter `f[x_]=expression in x`. To generate a basic plot of $y = f(x)$ for $a \le x \le b$, enter `Plot[f[x],{x,a,b}]`.

3.1.1 Using Graphs and Tables to Predict Limits

Example 3.1

Use a graph and table of values to investigate $\lim_{x\to 0} \frac{\sin 3x}{x}$.

Solution. We clear all prior definitions of f, define $f(x) = (\sin 3x)/x$, and then graph $y = f(x)$ on the interval $[-\pi, \pi]$ with `Plot`.

`Clear[f]` clears all prior definitions of f, if any. Clearing function definitions before defining new ones helps eliminate any possible confusion and/or ambiguities.

```
Clear[f]
f[x_] = Sin[3x]/x;
Plot[f[x], {x, -2π, 2π},
  PlotStyle → {{Thickness[.01], CMYKColor[{.05, .26, 1, .27}]}},
  PlotRange → All]
```

From the graph shown in Fig. 3.1, we might, correctly, conclude that $\lim_{x\to 0} \frac{\sin 3x}{x} = 3$. Further evidence that $\lim_{x\to 0} \frac{\sin 3x}{x} = 3$ can be obtained by computing the values of $f(x)$ for values of x "near" $x = 0$. In the following, we use `RandomReal`. `RandomReal[{a,b}]` returns a "random" real number between a and b. Because we are generating "random" numbers, your results will differ from those obtained here, to define `xvals` to be a table of 6 "random" real numbers. The first number in `xvals` is between -1 and 1, the second between $-1/10$ and $1/10$, and so on.

Remark 3.2. Throughout *Mathematica by Example* we illustrate how different options such as those that affect the coloring and labeling of a graphic are used, which include options such as `PlotStyle`, `PlotLabel`, and `AxesLabel`.

```
xvals = Table[RandomReal[{-10^-n, 10^-n}], {n, 0, 5}]
```

Mathematica by Example. http://dx.doi.org/10.1016/B978-0-12-812481-9.00003-X

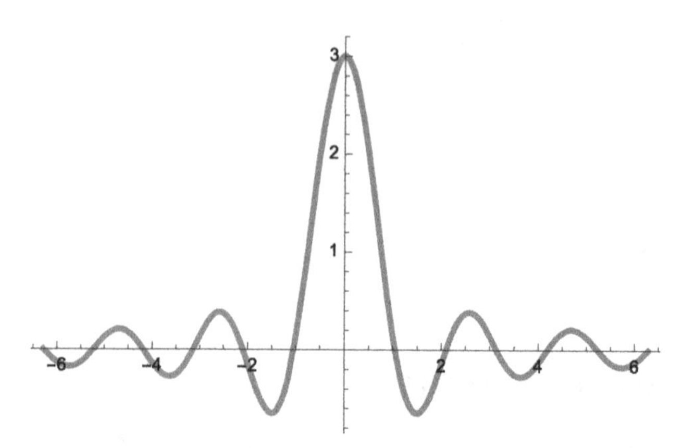

FIGURE 3.1 Graph of $f(x) = (\sin 3x)/x$ on the interval $[-2\pi, 2\pi]$ in Wisconsin gold.

{−0.710444, −0.0134926, −0.00161622, −0.000546629, 0.0000262461,

3.6298466289834744*^ − 6}

We then use `Map` to compute the value of $f(x)$ for each x in `xvals`. `Map[f,{x1,x2,...,xn}]` returns the list $\{f(x_1), f(x_2), ..., f(x_n)\}$. We use `Table` to display the results in tabular form. Generally, `list[[i]]` returns the ith element of `list` while

```
Table[f[i],{i,start,finish,stepsize}]
```

computes each value of $f(i)$ from `start` to `finish` in increments of `stepsize`. To create a table consisting of n equally spaced values, use `stepsize=(finish-start)/(n-1)`. `TableForm` attempts to display a table form in a standard format such as the row-and-column format that follows.

pairs = Table[{xvals[[*i*]], fvals[[*i*]]}, {*i*, 1, 6}];

TableForm[pairs]

−0.710444	1.19217
−0.0134926	2.99918
−0.00161622	2.99999
−0.000546629	3.
0.0000262461	3.
3.6298466289834744*^ − 6	3.

From these values, we might again correctly deduce that $\lim_{x\to 0} \frac{\sin 3x}{x} = 3$. Of course, these results do not prove that $\lim_{x\to 0} \frac{\sin 3x}{x} = 3$ but they are helpful in convincing us that $\lim_{x\to 0} \frac{\sin 3x}{x} = 3$. □

For piecewise-defined functions, you can either use Mathematica's *conditional command* (/;) to define the piecewise-defined function or use `Piecewise`.

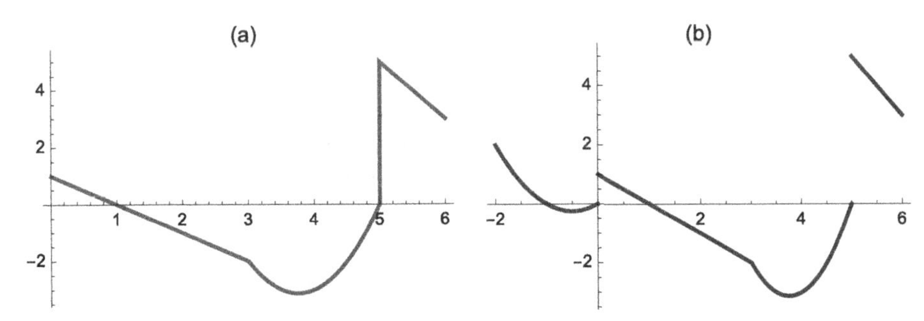

FIGURE 3.2 (a) `Plot` does not catch the breaks in the piecewise defined function (Badger red). (b) If you use `Piecewise`, `Plot` can catch jumps (Illinois blue).

Example 3.2

If $h(x) = \begin{cases} x^2 + x, \text{ if } x \le 0 \\ 1 - x, \text{ if } 0 < x < 3 \\ 2x^2 - 15x + 25, \text{ if } 3 \le x \le 5 \\ 15 - 2x, \text{ if } x > 5 \end{cases}$, compute the following limits: (a) $\lim_{x\to 0} h(x)$, (b) $\lim_{x\to 3} h(x)$, (c) $\lim_{x\to 5} h(x)$.

Solution. We use Mathematica's conditional command, `/;` to define h. We must use delayed evaluation (`:=`) because $h(x)$ cannot be computed unless Mathematica is given a particular value of x. The first line of the following defines $h(x)$ to be $x^2 + x$ for $x \le 0$, the second line defines $h(x)$ to be $1 - x$ for $0 < x < 3$ and so on. Notice that Mathematica accidentally connects (0, 0) to (0, 1) and then (5, 0) to (5, 5). (See Fig. 3.2 (a).) The delayed evaluation is also incompatible with Mathematica's `Limit` function.

The plots `p1` and `p2` are not displayed because a semi-colon is included at the end of each `Plot` command.

```
Clear[h]
h[x_]:=x^2 + x/;x ≤ 0
h[x_]:=1 − x/;0 < x < 3
h[x_]:=2x^2 − 15x + 25/;3 ≤ x ≤ 5
h[x_]:=15 − 2x/;x > 5
p1 = Plot[h[x], {x, 0, 6},
  PlotStyle → {{Thickness[.01], CMYKColor[{.002, 1, .85, .06}]}}]
```

To avoid these problems and see the discontinuities in our plots, we redefine h using Mathematica's `Piecewise` function as follows.

```
Clear[h]
h[x_]:=Piecewise[{{x^2 + x, x < 0}, {1 − x, 0 < x < 3}, {2x^2 − 15x + 25, 3<=x ≤ 5},
  {15 − 2x, x > 5}}];
p2 = Plot[h[x], {x, −2, 6}, PlotStyle → {{Thickness[.01], CMYKColor[1, .86, .24, .09]}},
  PlotRange → All];
```

```
Show[GraphicsRow[{p1, p2}]]
```

Notice that when we execute the `Plot` command, Mathematica "catches" the breaks between (0, 0) and (0, 1) and then at (5, 0) and (5, 5) shown in Fig. 3.2 (b).

From Fig. 3.2, we see that $\lim_{x\to 0} h(x)$ does not exist, $\lim_{x\to 3} h(x) = -2$, and $\lim_{x\to 5} h(x)$ does not exist. □

When limits exist, you can often use `Limit[f[x],x->a]` (where a may be ±`Infinity`) to compute $\lim_{x\to a} f(x)$. Thus, for the previous example we see that

```
Limit[h[x], x → 3]
```

```
-2
```

is correct. On the other hand

```
Limit[h[x], x → 5]
```

```
5
```

is incorrect. We check by computing the right hand limit, $\lim_{x\to 5^+} h(x)$, using the `Direction->-1` option in the `Limit` command and then the left limit, $\lim_{x\to 5^-} h(x)$, using the `Direction->1` in the `Limit` command.

```
Limit[h[x], x → 5, Direction → 1]
```

```
0
```

```
Limit[h[x], x → 5, Direction → -1]
```

```
5
```

We follow the same procedure for $x = 0$

```
Limit[h[x], x → 0]
```

```
1
```

```
Limit[h[x], x → 0, Direction → 1]
```

```
0
```

```
Limit[h[x], x → 0, Direction → -1]
1
```

3.1.2 Computing Limits

Some limits involving rational functions can be computed by factoring the numerator and denominator.

Example 3.3

Compute $\lim_{x\to -9/2} \dfrac{2x^2+25x+72}{72-47x-14x^2}$.

Solution. We define `frac1` to be the rational expression $\dfrac{2x^2+25x+72}{72-47x-14x^2}$. We then attempt to compute the value of `frac1` if $x = -9/2$ by using `ReplaceAll (/.)` to evaluate `frac1` if $x = -9/2$ but see that it is undefined.

```
frac1 = (2 x^2 + 25 x + 72)/(72 - 47 x - 14 x^2);

frac1 /. x → -9/2

... Power: Infinite expression 1/0 encountered.

... Infinity: Indeterminate expression 0 ComplexInfinity encountered.

Indeterminate
```

Factoring the numerator and denominator with `Factor`, `Numerator`, and `Denominator`, we see that

$$\lim_{x\to -9/2} \frac{2x^2+25x+72}{72-47x-14x^2} = \lim_{x\to -9/2} \frac{(x+8)(2x+9)}{(8-7x)(2x+9)} = \lim_{x\to -9/2} \frac{x+8}{8-7x}.$$

The fraction $(x+8)/(8-7x)$ is named `frac2` and the limit is evaluated by computing the value of `frac2` if $x = -9/2$

Factor[Numerator[frac1]]
Factor[Denominator[frac1]]

$(8+x)(9+2x)$

$-(9+2x)(-8+7x)$

frac2 = Cancel[frac1]

$\frac{-8-x}{-8+7x}$

frac2/.x → $-\frac{9}{2}$

$\frac{7}{79}$

or by using the `Limit` function on the original fraction.

Limit[frac1, x → −9/2]

$\frac{7}{79}$

We conclude that

$$\lim_{x\to -9/2} \frac{2x^2+25x+72}{72-47x-14x^2} = \frac{7}{79}. \qquad \square$$

As stated previously, `Limit[f[x],x->a]` attempts to compute $\lim_{x\to a} f(x)$, `Limit[f[x],x->a,Direction->1]` attempts to compute $\lim_{x\to a^-} f(x)$, and `Limit[f[x],x->a,Direction->-1]` attempts to compute $\lim_{x\to a^+} f(x)$. Generally, a can be a number, ±`Infinity` ($\pm\infty$), or another symbol.

Thus, entering

Limit$\left[\frac{2x^2+25x+72}{72-47x-14x^2}, x \to -\frac{9}{2}\right]$

$\frac{7}{79}$

computes $\lim_{x\to -9/2} \frac{2x^2+25x+72}{72-47x-14x^2} = 7/79$.

Example 3.4

Calculate each limit: (a) $\lim_{x\to -5/3} \frac{3x^2-7x-20}{21x^2+14x-35}$; (b) $\lim_{x\to 0} \frac{\sin x}{x}$; (c) $\lim_{x\to\infty} \left(1+\frac{z}{x}\right)^x$; (d) $\lim_{x\to 0} \frac{e^{3x}-1}{x}$; (e) $\lim_{x\to\infty} e^{-2x}\sqrt{x}$; and (f) $\lim_{x\to 1^+} \left(\frac{1}{\ln x}-\frac{1}{x-1}\right)$.

Solution. In each case, we use `Limit` to evaluate the indicated limit. Entering

```
Limit[(3x^2-7x-20)/(21x^2+14x-35), x -> -5/3]
17/56
```

computes $\lim_{x\to -5/3} \frac{3x^2-7x-20}{21x^2+14x-35} = \frac{17}{56}$; and entering

```
Limit[Sin[x]/x, x -> 0]
1
```

computes $\lim_{x\to 0} \frac{\sin x}{x} = 1$. Mathematica represents ∞ by `Infinity`. Thus, entering

```
Limit[(1 + z/x)^x, x -> Infinity]
e^z
```

computes $\lim_{x\to\infty} \left(1+\frac{z}{x}\right)^x = e^z$. Entering

```
Limit[(Exp[3x] - 1)/x, x -> 0]
3
```

computes $\lim_{x\to 0} \frac{e^{3x}-1}{x} = 3$. Entering

```
Limit[Exp[-2x]Sqrt[x], x -> Infinity]
0
```

computes $\lim_{x\to\infty} e^{-2x}\sqrt{x} = 0$, and entering

```
Limit[1/Log[x] - 1/(x - 1), x -> 1, Direction -> -1]
1/2
```

computes $\lim_{x\to 1^+} \left(\frac{1}{\ln x}-\frac{1}{x-1}\right) = \frac{1}{2}$. □

Because $\ln x$ is undefined for $x \leq 0$, a right-hand limit is mathematically necessary, even though Mathematica's `Limit` function computes the limit correctly without the distinction.

3.1.3 One-Sided Limits

As illustrated previously, Mathematica can compute certain one-sided limits. The command `Limit[f[x],x->a,Direction->1]` attempts to compute $\lim_{x\to a^-} f(x)$ while `Limit[f[x],x->a,Direction->-1]` attempts to compute $\lim_{x\to a^+} f(x)$.

Example 3.5

Compute (a) $\lim_{x\to 0^+} |x|/x$; (b) $\lim_{x\to 0^-} |x|/x$; (c) $\lim_{x\to 0^+} e^{-1/x}$; and (d) $\lim_{x\to 0^-} e^{-1/x}$.

Solution. Even though $\lim_{x\to 0} |x|/x$ does not exist, $\lim_{x\to 0^+} |x|/x = 1$ and $\lim_{x\to 0^-} |x|/x = -1$, as we see using `Limit` together with the `Direction->-1` and `Direction->1` options, respectively.

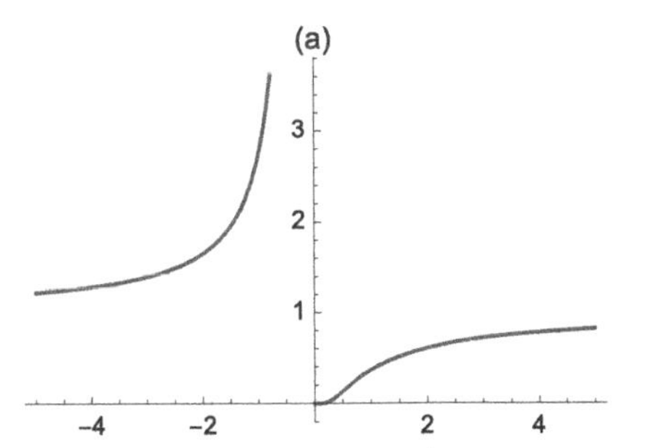

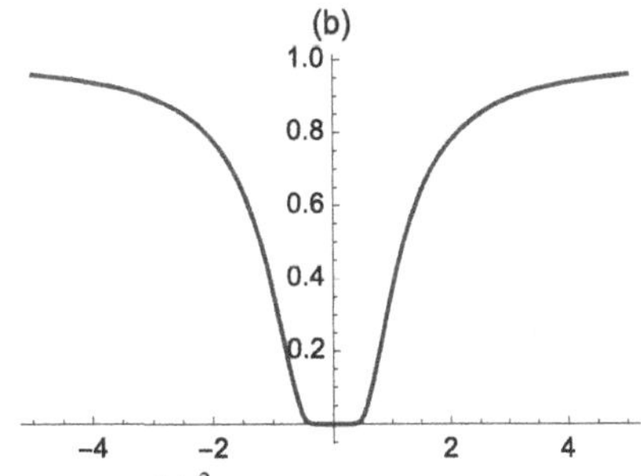

FIGURE 3.3 (a) Graph of $y = e^{-1/x}$ (Army green). (b) Graph of $y = e^{-1/x^2}$ (Navy blue).

```
Limit[Abs[x]/x, x → 0, Direction → 1]
Limit[Abs[x]/x, x → 0, Direction → -1]
-1
1
```

The `Direction->-1` and `Direction->1` options are used to calculate the correct values for (c) and (d), respectively. For (c), we have:

```
Limit[1/x, x → 0, Direction → -1]
∞
```

Technically, $\lim_{x\to 0} e^{-1/x}$ does not exist (see Fig. 3.3 (a)) so the following is incorrect.

```
Limit[Exp[-1/x], x → 0]
0
```

However, using `Limit` together with the `Direction` option gives the correct left and right limits.

```
Limit[Exp[-1/x], x → 0, Direction → 1]
∞
Limit[Exp[-1/x], x → 0, Direction → -1]
0
```

We confirm these results by graphing $y = e^{-1/x}$ with `Plot` in Fig. 3.3 (a). In (b), we also show the graph of $y = e^{-1/x^2}$ in Fig. 3.3 (b), which is further discussed in the exercises.

```
p1 = Plot[Exp[-1/x], {x, -5, 5}, PlotStyle → CMYKColor[.64, .47, 1, .4],
  PlotLabel → "(a)"];
p2 = Plot[Exp[-1/x^2], {x, -5, 5},
  PlotStyle → CMYKColor[1, .38, 0, .64],
  PlotLabel → "(b)"];
Show[GraphicsRow[{p1, p2}]]
```

□

The `Limit` command together and its options (`Direction->1` and `Direction->-1`) are "fragile" and should be used with caution because the results can be unpredictable. It is wise to check or confirm results using a different technique for nearly all problems encountered.

3.1.4 Continuity

Definition 3.1. The function $y = f(x)$ is **continuous** at $x = a$ if

1. $\lim_{x\to a} f(x)$ exists;
2. $f(a)$ exists; and
3. $\lim_{x\to a} f(x) = f(a)$.

Note that the third item in the definition means that both (1) and (2) are satisfied. But, if either of (1) or (2) is not satisfied the function is not continuous at the number in question. The function $y = f(x)$ is **continuous** on the open interval I if $f(x)$ is continuous at each number a contained in the interval I. Loosely speaking, the "standard" set of functions (polynomials, rational, trigonometric, etc...) are continuous on their domains.

Remark 3.3. Be careful with regard to this. For example, since $\lim_{x\to 0^-} \sqrt{x}$ does not exist, many would say that $f(x) = \sqrt{x}$ is *right continuous* at $x = 0$.

Example 3.6

For what value(s) of x, if any, are each of the following functions continuous? (a) $f(x) = x^3 - 8x$; (b) $f(x) = \sin 2x$; (c) $f(x) = (x-1)/(x+1)$; (d) $f(x) = \sqrt{(x-1)/(x+1)}$.

Solution. (a) Polynomial functions are continuous for all real numbers. In interval notation, $f(x)$ is continuous on $(-\infty, \infty)$. (b) Because the sine function is continuous for all real numbers, $f(x) = \sin 2x$ is continuous for all real numbers. In interval notation, $f(x)$ is continuous on $(-\infty, \infty)$. (c) The rational function $f(x) = (x-1)/(x+1)$ is continuous for all $x \neq -1$. In interval notation, $f(x)$ is continuous on $(-\infty, -1) \cup (-1, \infty)$. (d) $f(x) = \sqrt{(x-1)/(x+1)}$ is continuous if the radicand is nonnegative. In interval notation, $f(x)$ is strictly continuous on $(-\infty, -1) \cup (1, \infty)$ but some might say that $f(x)$ is continuous on $(-\infty, -1) \cup [1, \infty)$, where it is understood that $f(x)$ is *right continuous* at $x = 1$. We see this by graphing each function with the following commands. See Fig. 3.4. Note that in `p3`, the vertical line is *not* a part of the graph of the function—it is a vertical asymptote. If you were to redraw the figure by hand, the vertical line would *not* be a part of the graph.

```
p1 = Plot[x^3 − 8x, {x, −5, 5}, PlotLabel → "(a)",
  PlotStyle → CMYKColor[.82, 0, .64, .7]];
p2 = Plot[Sin[2x], {x, −5, 5}, PlotLabel → "(b)",
  PlotStyle → CMYKColor[.56, .47, .47, .15]];
p3 = Plot[((x − 1)/(x + 1)), {x, −5, 5}, PlotLabel → "(c)",
  PlotStyle → CMYKColor[.25, .97, .76, .18]];
p4 = Plot[Sqrt[(x − 1)/(x + 1)], {x, −5, 5}, PlotLabel → "(d)",
  PlotStyle → CMYKColor[0, .61, .97, 0]];
Show[GraphicsGrid[{{p1, p2}, {p3, p4}}]]
```

□

Computers are finite state machines so handling "interesting" functions can be problematic, especially when one must distinguish between rational and irrational numbers. We assume that if $x = p/q$ is a *rational* number (p and q integers), p/q is a reduced fraction. One way of tackling these sorts of problems is to view rational numbers as ordered pairs, $\{a, b\}$. If a and b are integers, Mathematica automatically reduces a/b so `Denominator[a/b]`

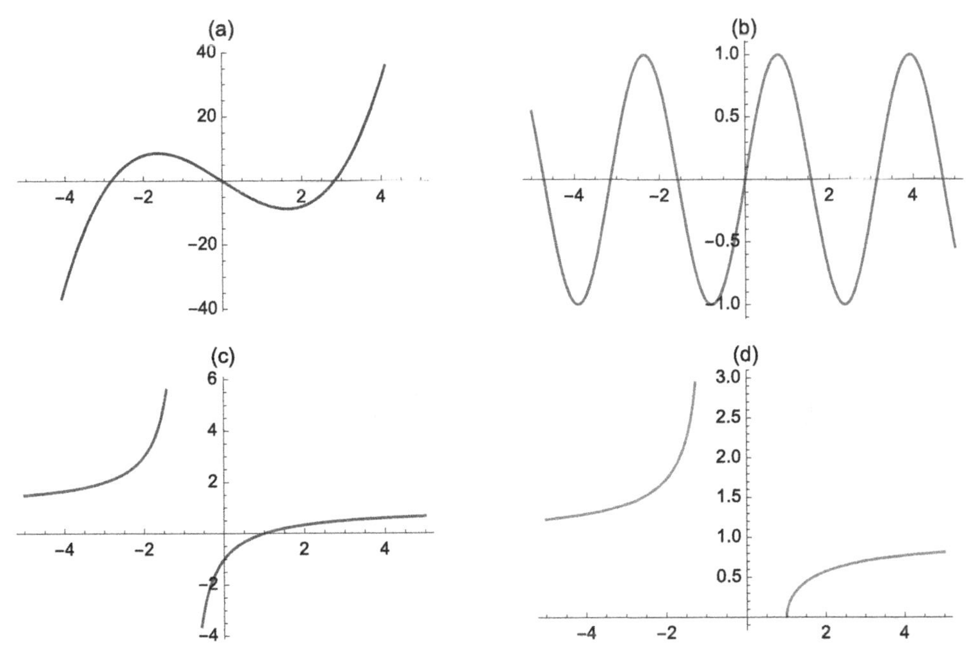

FIGURE 3.4 (a) Polynomials (Michigan State green), (b) trigonometric (Ohio State gray), (c) rational (Alabama crimson), and (d) root functions (Auburn orange) are usually continuous on their domains.

or `a/b//Denominator` returns the denominator of the reduced fraction; `Numerator[a/b]` or `a/b//Numerator` returns the numerator of the reduced fraction. If you want to see the points $(x, f(x))$ for which x is rational, we use `ListPlot`.

Example 3.7

Let $f(x) = \begin{cases} 1/q, & \text{if } x = p/q \text{ is rational} \\ 0, & \text{if } x \text{ is irrational} \end{cases}$. Create a representative graph of $f(x)$.

Solution. You cannot see points: the measure of the rational numbers is 0, and the measure of the irrational numbers is the **continuum**, C. A true graph of $f(x)$ would look like the graph of $y = 0$. In the context of the example, we want to see how the graph of $f(x)$ looks for rational values of x. We use a few points to illustrate the technique by using `Table` and `Flatten` to generate a set of ordered pairs.

`Flatten[list,n]` flattens `list` to level `n`.
In Mathematica, an ordered pair (a, b) is represented by $\{a, b\}$.

```
t1 = Flatten[Table[{n, m}, {n, 1, 5}, {m, 1, 5}], 1]
{{1, 1}, {1, 2}, {1, 3}, {1, 4}, {1, 5}, {2, 1}, {2, 2},
  {2, 3}, {2, 4}, {2, 5}, {3, 1}, {3, 2}, {3, 3}, {3, 4}, {3, 5}, {4, 1},
    {4, 2}, {4, 3}, {4, 4}, {4, 5}, {5, 1},
      {5, 2}, {5, 3}, {5, 4}, {5, 5}}
```

Next, we defined a function f. Assuming that a and b are integers, given an ordered pair $\{a, b\}$, $f(\{a, b\})$ returns the point $\{a/b, 1/(\text{Reduced denominator of } a/b)\}$

```
f[{a_, b_}]:={a/b, 1/(a/b//Denominator)}
```

We use `Map` to compute the value of f for each ordered pair in `t1`. The resulting list is named `t2`.

```
t2 = Map[f, t1]
```
$\{\{1, 1\}, \{\frac{1}{2}, \frac{1}{2}\}, \{\frac{1}{3}, \frac{1}{3}\}, \{\frac{1}{4}, \frac{1}{4}\}, \{\frac{1}{5}, \frac{1}{5}\}, \{2, 1\}, \{1, 1\},$
$\{\frac{2}{3}, \frac{1}{3}\}, \{\frac{1}{2}, \frac{1}{2}\}, \{\frac{2}{5}, \frac{1}{5}\}, \{3, 1\}, \{\frac{3}{2}, \frac{1}{2}\}, \{1, 1\},$

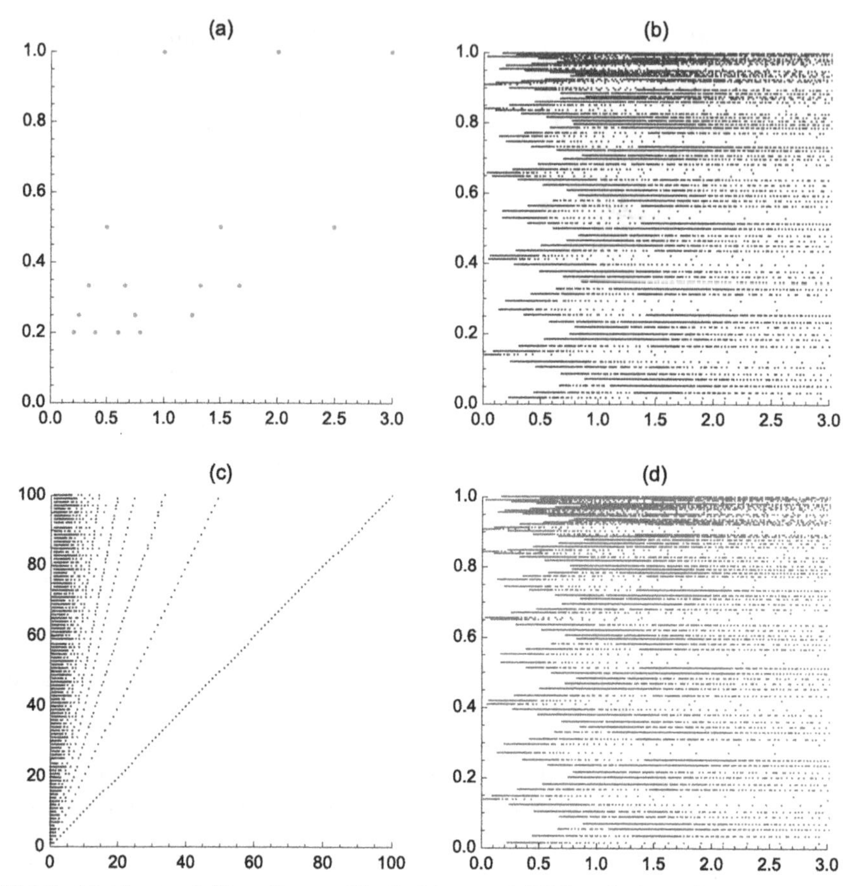

FIGURE 3.5 (a) After step 1 (Notre Dame gold). (b) After step 2 (USC cardinal). (c) Examining the numerator rather than the denominator (Oklahoma crimson). (d) The sine of the numerator (Texas burnt orange).

$\left\{\frac{3}{4}, \frac{1}{4}\right\}, \left\{\frac{3}{5}, \frac{1}{5}\right\}, \{4, 1\}, \{2, 1\}, \left\{\frac{4}{3}, \frac{1}{3}\right\}, \{1, 1\},$
$\left\{\frac{4}{5}, \frac{1}{5}\right\}, \{5, 1\}, \left\{\frac{5}{2}, \frac{1}{2}\right\}, \left\{\frac{5}{3}, \frac{1}{3}\right\}, \left\{\frac{5}{4}, \frac{1}{4}\right\}, \{1, 1\}\}$

Notice that `t2` contains duplicate entries. We can remove them using `Flatten` but doing so does not affect the plot shown in Fig. 3.5 (a).

```
p1 = ListPlot[t2, PlotRange → {{0, 3}, {0, 1}}, AspectRatio → 1,
  PlotStyle->CMYKColor[.06, .27, 1, .12],
  PlotLabel → "(a)"];
```

To generate a "prettier" plot, we repeat the procedure using more points. After entering each command the results are not displayed because we include a semicolon (;) at the end of each. See Fig. 3.5 (b).

```
t3 = Flatten[Table[{n, m}, {n, 1, 300}, {m, 1, 200}], 1];
t4 = Map[f, t3];
p2 = ListPlot[t4, PlotRange → {{0, 3}, {0, 1}}, AspectRatio → 1,
  PlotStyle->CMYKColor[.07, 1, .65, .32],
  PlotLabel → "(b)"];
```

This function is interesting because it is continuous at the irrationals and discontinuous at the rationals.

We can consider other functions in similar contexts. In the following the y-coordinate is the numerator rather than the denominator. See Fig. 3.5 (c).

```
Clear[f]
f[{a_, b_}]:={a/b, a/b//Numerator};
t3 = Flatten[Table[{n, m}, {n, 1, 100}, {m, 1, 100}], 1];
t4 = Map[f, t3];
p3 = ListPlot[t4, PlotRange → {{0, 100}, {0, 100}}, AspectRatio → 1,
  PlotStyle → CMYKColor[0, 1, .65, .34],
  PlotLabel → "(c)"];
```

With Mathematica, we can modify commands to investigate how changing parameters affect a given situation. In the following, we compute the sine of p if $x = p/q$. See Fig. 3.5 (d).

```
Clear[f]
f[{a_, b_}]:={a/b, Sin[(a/b//Numerator)]};
t5 = Flatten[Table[{n, m}, {n, 1, 300}, {m, 1, 200}], 1];
t6 = Map[f, t5];
p4 = ListPlot[t6, PlotRange → {{0, 3}, {0, 1}}, AspectRatio → 1,
  PlotStyle → CMYKColor[0, .65, 1, .09],
  PlotLabel → "(d)"];

Show[GraphicsGrid[{{p1, p2}, {p3, p4}}]]
```

□

3.2 DIFFERENTIAL CALCULUS

3.2.1 Definition of the Derivative

Definition 3.2. The **derivative** of $y = f(x)$ is

$$y' = f'(x) = \frac{dy}{dx} = \lim_{h \to 0} \frac{f(x+h) - f(x)}{h}, \tag{3.1}$$

provided the limit exists.

Assuming the derivative exists, as h approaches 0 the secants approach the tangent. Hence, if the limit exists the derivative gives us the slope of a function at that particular value of x.

The `Limit` command can be used along with `Simplify` to compute the derivative of a function using the definition of the derivative.

Example 3.8

Use the definition of the derivative to compute the derivative of (a) $f(x) = x + 1/x$, (b) $g(x) = 1/\sqrt{x}$ and (c) $f(x) = \sin 2x$.

Solution. For (a), we first define f, compute the difference quotient, $(f(x+h) - f(x))/h$, simplify the difference quotient with `Simplify`, and use `Limit` to calculate the derivative.

```
f[x_]=x+1/x;
```

```
step1 = (f[x+h] - f[x])/h
step2 = Simplify[step1]
Limit[step2, h → 0]
```

$$\frac{h - \frac{1}{x} + \frac{1}{h+x}}{h}$$

$$\frac{-1+hx+x^2}{x(h+x)}$$

$$1 - \frac{1}{x^2}$$

For (b), we use the same approach as in (a) but use `Together` rather than `Simplify` to reduce the complex fraction.

```
g[x_] = 1/Sqrt[x];
```

```
step1 = (g[x+h] - g[x])/h
step2 = Together[step1]
Limit[step2, h → 0]
```

$$\frac{-\frac{1}{\sqrt{x}} + \frac{1}{\sqrt{h+x}}}{h}$$

$$\frac{\sqrt{x}-\sqrt{h+x}}{h\sqrt{x}\sqrt{h+x}}$$

$$-\frac{1}{2x^{3/2}}$$

For (c), observe that `Simplify` instructs Mathematica to apply elementary trigonometric identities.

```
f[x_] = Sin[2x];
```

```
step1 = (f[x+h] - f[x])/h
step2 = Simplify[step1]
Limit[step2, h → 0]
```

$$\frac{-\mathrm{Sin}[2x]+\mathrm{Sin}[2(h+x)]}{h}$$

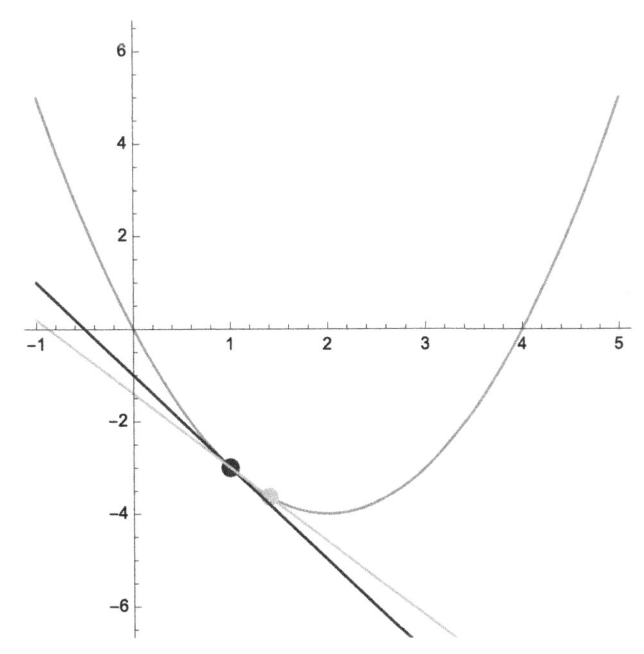

FIGURE 3.6 Plots of $y = x^2 - 4x$, the tangent at $(1, f(1))$, and the secant containing $(1, f(1))$ and $(1+h, f(1+h))$ if $h = 0.4$ (Georgia Tech colors).

$\frac{2\text{Cos}[h+2x]\text{Sin}[h]}{h}$

2Cos[2x] □

If the derivative of $y = f(x)$ exists at $x = a$, a geometric interpretation of $f'(a)$ is that $f'(a)$ is the slope of the line tangent to the graph of $y = f(x)$ at the point $(a, f(a))$.

To motivate the definition of the derivative, many calculus texts choose a value of x, $x = a$, and then draw the graph of the secant line passing through the points $(a, f(a))$ and $(a + h, f(a+h))$ for "small" values of h to show that as h approaches 0, the secant line approaches the tangent line. An equation of the secant line passing through the points $(a, f(a))$ and $(a+h, f(a+h))$ is given by

$$y - f(a) = \frac{f(a+h) - f(a)}{(a+h) - a}(x - a) \quad \text{or} \quad y = \frac{f(a+h) - f(a)}{h}(x - a) + f(a).$$

Example 3.9

If $f(x) = x^2 - 4x$, graph $f(x)$ together with the secant line containing $(1, f(1))$ and $(1+h, f(1+h))$ for various values of h.

Solution. We begin by considering a particular h value. We choose $h = 0.4$. We then define $f(x) = x^2 - 4x$. In `p1`, we graph $f(x)$ in black on the interval $[-1, 5]$, in `p2` we place a blue point at $(1, f(1))$ and a green point at $(1.4, f(1.4)$, in `p3` we graph the tangent to $y = f(x)$ at $(1, f(1))$ in red, in `p4` we graph the secant containing $(1, f(1))$ and $(1.4, f(1.4)$ in purple, and finally we show all four graphics together with `Show` in Fig. 3.6.

Remember that when a semi-colon is placed at the end of a command, the resulting output is not displayed. The names of the colors that Mathematica knows are listed in the **ColorSchemes** palette followed by "Known" and then "System."

```
f[x_] = x^2 - 4x;

p1 = Plot[f[x], {x, -1, 5}, PlotStyle → CMYKColor[.08, .3, 1, 0]];
```

```
p2 = Graphics[{PointSize[.03], CMYKColor[1, .88, .39, .42], Point[{1, f[1]}],
  CMYKColor[.04, .14, .67, 0], Point[{1 + .4, f[1 + .4]}]}];
p3 = Plot[f'[1](x - 1) + f[1], {x, -1, 5}, PlotStyle -> CMYKColor[1, .88, .39, .42]];
p4 = Plot[(f[1 + .4] - f[1])/.4(x - 1) + f[1], {x, -1, 5},
  PlotStyle -> CMYKColor[.04, .14, .67, 0]];
Show[p1, p2, p3, p4, PlotRange -> {{-1, 5}, {-6, 6}}, AspectRatio -> 1]
```

We now generalize the previous set of commands for arbitrary $h \neq 0$ values. $g(h)$ shows plots of $y = x^2 - 4x$, the tangent at $(1, f(1))$, and the secant containing $(1, f(1))$ and $(1 + h, f(1 + h))$.

```
Clear[f, g];
f[x_] = x^2 - 4x;
g[h_]:=Module[{p1, p2, p3, p4},
p1 = Plot[f[x], {x, -1, 5}, PlotStyle -> CMYKColor[.08, .3, 1, 0]];
p2 = Graphics[{PointSize[.03], CMYKColor[1, .88, .39, .42], Point[{1, f[1]}],
  CMYKColor[.04, .14, .67, 0], Point[{1 + h, f[1 + h]}]}];
p3 = Plot[f'[1](x - 1) + f[1], {x, -1, 5}, PlotStyle -> CMYKColor[1, .88, .39, .42]];
p4 = Plot[(f[1 + h] - f[1])/h(x - 1) + f[1], {x, -1, 5},
  PlotStyle->CMYKColor[.04, .14, .67, 0]];
Show[p1, p2, p3, p4, PlotRange -> {{-1, 5}, {-6, 6}}, AspectRatio -> 1]]
```

`Table[f[x],{x,start,stop,stepsize}]` creates a table of $f(x)$ values beginning with *start* and ending with *stop* using increments of *stepsize*. Given a table, `Partition[table,n]` partitions the table into n element subgroups. Thus if a table, `t1` has 9 elements, `Partition[t1, 3]` creates a 3×3 grid; three sets of three elements each.

Using `Table` followed by `GraphicsGrid`, we can create an table of graphics for various values of h like that shown in Fig. 3.7. With `Table`, the dimensions of the grid displayed on your computer are based on the size of the active Mathematica window. To control the dimensions of the grid, we use `GraphicsGrid` together with `Partition` and `Show`.

```
t1 = Table[g[k], {k, 1, .0001, -(1 - .0001)/8}];
Show[GraphicsGrid[Partition[t1, 3]]]
```

`Animate` works in the same way as `Table` and is very similar to `Manipulate`. Entering

```
Animate[g[k], {k, 1, .0001, -(1 - .0001)/99}]
```

generates an animation of $g(k)$ and 100 equally spaced values of k starting with $k = 1$ and ending with $k = 0.0001$. To animate the result, use the toolbar in the animate graphic. To vary g continuously and not use a specific stepsize enter `Animate[g[k],{k,1,.0001}]`.

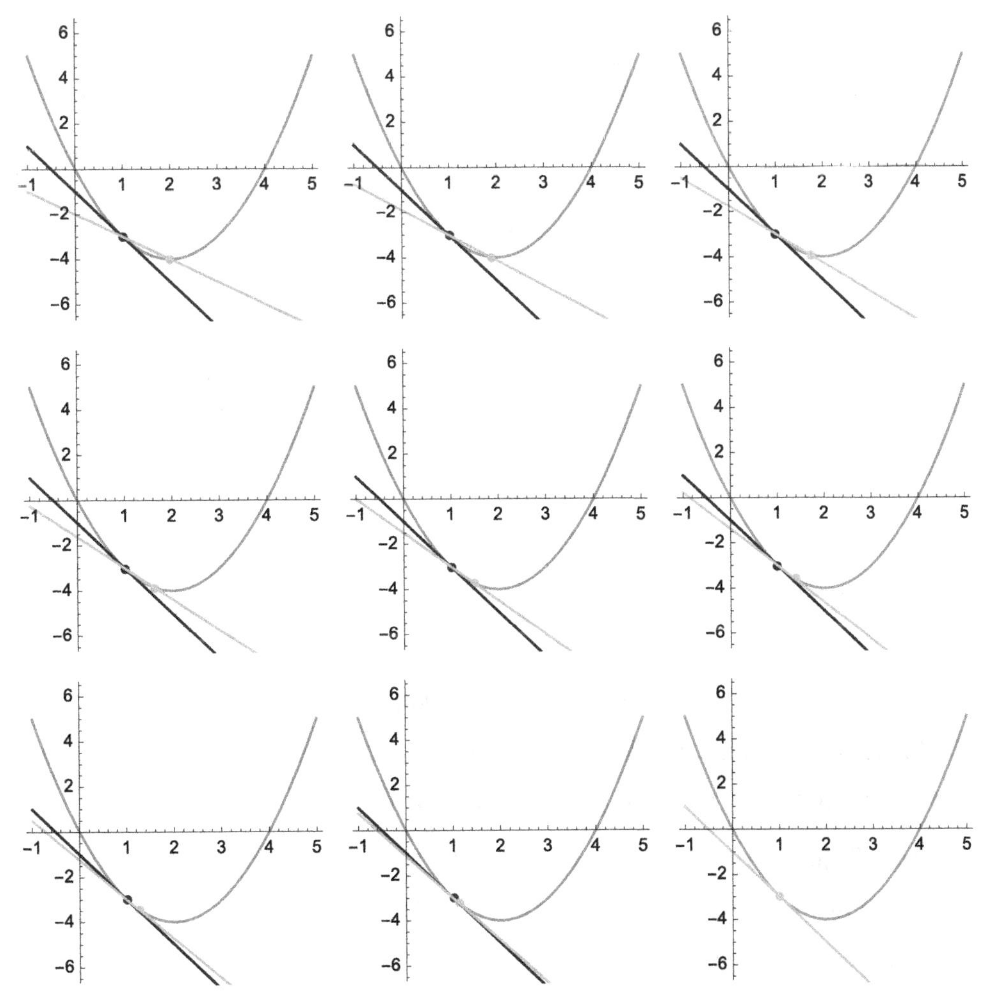

FIGURE 3.7 Plots of $y = x^2 - 4x$, the tangent at $(1, f(1))$, and the secant containing $(1, f(1))$ and $(1+h, f(1+h))$ for various values of h.

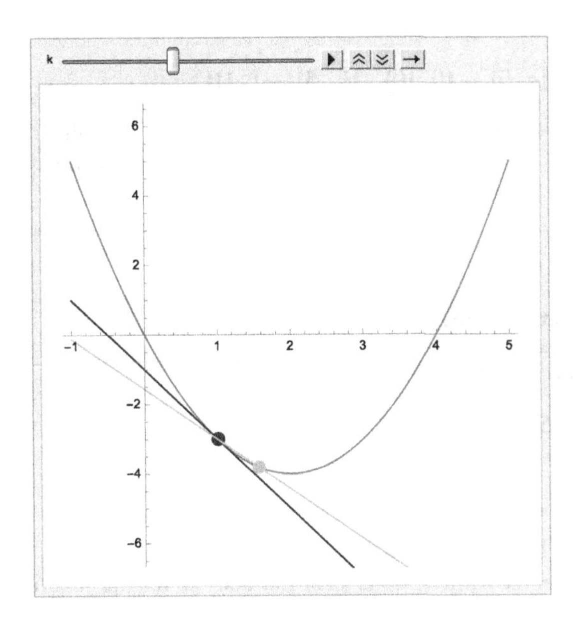

After animating the selection, you can control the animation (speed, direction, pause, and so on) with the buttons at the top of the animation window.

With Mathematica 11, you can use `Manipulate` to help generate animations and images that you can adjust based on changing parameter values.

To illustrate how to do so, we begin by redefining f and then defining $m(a, h)$. Given a and h values, $m(a, h)$ plots $f(x)$ for $-10 \le x \le 10$ (`p1`), plots a blue point at $(a, f(a))$ and a green point at $(a+h, f(a+h))$ (`p2`), plots $f'(a)(x-a)+f(a)$ (the tangent to the graph of $f(x)$ at $(a, f(a))$) for $-1 \le x \le 5$ in red (`p3`), the secant containing $(a, f(a))$ and $(a+h, f(a+h))$ for $-10 \le x \le 10$ in purple (`p4`) and finally displays all four graphics together with `Show`. Using `PlotRange`, we indicate that horizontal axis displays x values between -10 and 10, the vertical axis displays y values between -10 and 10; `AspectRatio->1` means that the ratio of the lengths of the x to y axes is 1. Thus, the plot scaling is correct. Note that when we use `Module` to define m, `p1`, `p2`, `p3`, and `p4` are *local* to the function m. This means that if you have such objects defined elsewhere in your Mathematica notebook, those objects are not affected when you compute m.

```
Clear[m, f];
f[x_] = x^2 - 4x;
m[a_, h_]:=Module[{p1, p2, p3, p4},
p1 = Plot[f[x], {x, -10, 10}, PlotStyle -> CMYKColor[.08, .3, 1, 0]];
p2 = Graphics[{PointSize[.03], CMYKColor[1, .88, .39, .42], Point[{a, f[a]}],
   CMYKColor[.04, .14, .67, 0], Point[{a + h, f[a + h]}]}];
p3 = Plot[f'[a](x - a) + f[a], {x, -1, 5}, PlotStyle -> CMYKColor[1, .88, .39, .42]];
p4 = Plot[(f[a + h] - f[a])/h(x - a) + f[a], {x, -10, 10},
   PlotStyle -> CMYKColor[.04, .14, .67, 0]];
Show[p1, p2, p3, p4, PlotRange -> {{-10, 10}, {-10, 10}}, AspectRatio -> 1]]
```

Now we use `Manipulate` to create a "mini" program. The sliders (centered at $a = 0$ and $h = .5$ with range from -10 to 10 an -1 to 1, respectively) allow you to see how changing a and h affects the plot. See Fig. 3.8.

```
Manipulate[m[a, h], {{a, 0}, -10, 10}, {{h, .5}, -1, 1}]
```

Fig. 3.8 illustrates the special case when $f(x) = x^2 - 4x$. To illustrate the same concept using a "standard" set of functions (polynomials, rational, root, and trig), we first define the functions

```
quad[x_] = (x + 2)^2 - 2;
cubic[x_] = -1/10x(x^2 - 25);
rational[x_] = 50/((x + 5)(x - 5));
root[x_] = 3Sqrt[x + 5];
sin[x_] = 5Sin[x];
```

and then we adjust `m` by defining a few of these "standard" and then defining the function `mmore`, which performs the same actions as m but does so for the function selected. We then

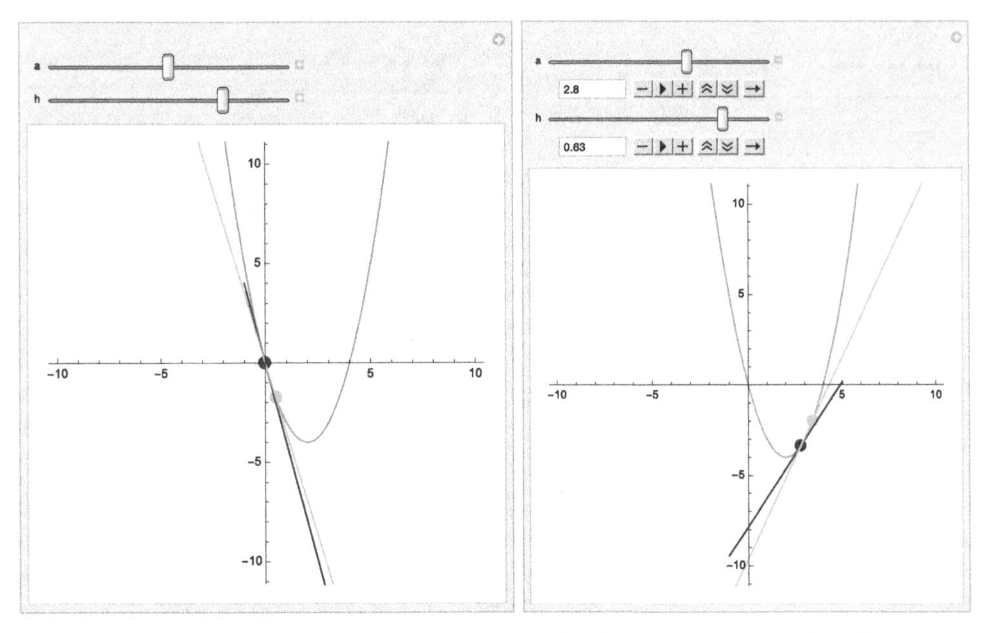

FIGURE 3.8 With `Manipulate`, we can perform animations and see how a function changes depending on parameter values. When you press the + button, the bar expands so that you can control the `Manipulate` object.

use `Manipulate` to create an object that shows the secant (in yellow), the tangent (in blue) for the selected function, a value, and h value. See Fig. 3.9.

```
Clear[mmore];
mmore[f_, a_, h_]:=Module[{p1, p2, p3, p4},
p1 = Plot[f[x], {x, −10, 10}, PlotStyle → CMYKColor[.08, .3, 1, 0]];
p2 = Graphics[{PointSize[.03], CMYKColor[1, .88, .39, .42], Point[{a, f[a]}],
  CMYKColor[.04, .14, .67, 0], Point[{a + h, f[a + h]}]}];
p3 = Plot[f'[a](x − a) + f[a], {x, −10, 10}, PlotStyle → CMYKColor[1, .88, .39, .42]];
p4 = Plot[(f[a + h] − f[a])/h(x − a) + f[a], {x, −10, 10},
  PlotStyle → CMYKColor[.04, .14, .67, 0]];
Show[p1, p2, p3, p4, PlotRange → {{−10, 10}, {−10, 10}}, AspectRatio → 1]]

Manipulate[mmore[f, a, h], {{f, quad}, {quad, cubic, rational, root, sin}},
  {{a, 0}, −10, 10}, {{h, 1}, −2, 2}]
```

□

3.2.2 Calculating Derivatives

The functions `D` and `'` are used to differentiate functions. Assuming that $y = f(x)$ is differentiable,

1. `D[f[x],x]` computes and returns $f'(x) = df/dx$,
2. `f'[x]` computes and returns $f'(x) = df/dx$,

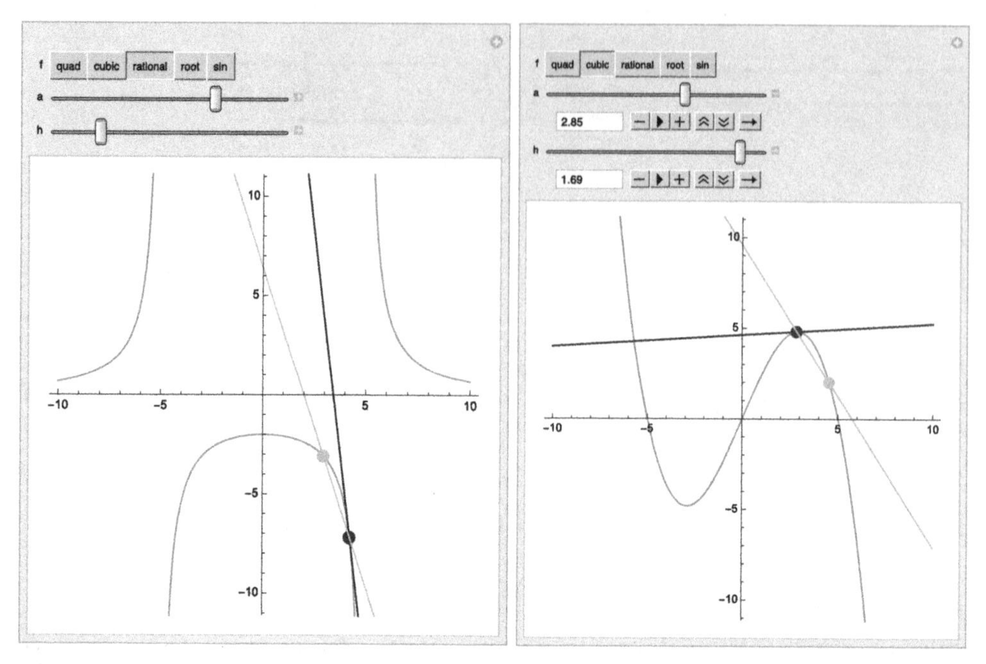

FIGURE 3.9 With this `Manipulate` object, we see how various functions, a values, and h values affect the secant to $y = f(x)$ passing through $(a, f(a))$ and $(a + h, f(a + h))$ and the tangent to $y = f(x)$ at $(a, f(a))$. (For interpretation of the colors in this figure, the reader is referred to the web version of this chapter.)

```
Clear[f, g]
f[x_] = x^3 Exp[-2 x];
g[x_] = x ArcTan[x];

f'[x]
3 e^(-2 x) x^2 - 2 e^(-2 x) x^3

D[f[x], x]
3 e^(-2 x) x^2 - 2 e^(-2 x) x^3

∂x f[x]
3 e^(-2 x) x^2 - 2 e^(-2 x) x^3

g'[x]
x/(1 + x^2) + ArcTan[x]

D[g[x], x]
x/(1 + x^2) + ArcTan[x]

∂x g[x]
x/(1 + x^2) + ArcTan[x]
```

FIGURE 3.10 You can use ′, `D`, and ∂ to compute derivatives of functions.

3. `f''[x]` computes and returns $f^{(2)}(x) = d^2 f(x)/dx^2$, and
4. `D[f[x],{x,n}]` computes and returns $f^{(n)}(x) = d^n f(x)/dx^n$.
5. You can use the ∂▪ button located on the **Basic Math Assistant** and **Basic Math Input** palettes to create templates to compute derivatives.

Fig. 3.10 illustrates various ways of computing derivatives using the ′ symbol, `D`, and the ∂ symbol.

Mathematica knows the numerous differentiation rules, including the product, quotient, and chain rules. Thus, entering

Clear[f, g]

D[f[x]g[x], x]

$g[x]f'[x] + f[x]g'[x]$

shows us that $\dfrac{d}{dx}(f(x)\cdot g(x)) = f'(x)g(x) + f(x)g'(x)$; entering

Together[D[f[x]/g[x], x]]

$\frac{g[x]f'[x]-f[x]g'[x]}{g[x]^2}$

shows us that

Remark 3.4. Throughout the text, input is in `Bold` and output is `Not`; output follows input $\dfrac{d}{dx}(f(x)/g(x)) = (f'(x)g(x) - f(x)g'(x))/(g(x))^2$; and entering

D[f[g[x]], x]

$f'[g[x]]g'[x]$

shows us that $\dfrac{d}{dx}(f\,(g(x))) = f'\,(g(x))\,g'(x)$.

Example 3.10

Compute the first and second derivatives of (a) $y = x^4 + \frac{4}{3}x^3 - 3x^2$, (b) $f(x) = 4x^5 - \frac{5}{2}x^4 - 10x^3$, (c) $y = \sqrt{e^{2x} + e^{-2x}}$, and (d) $y = (1 + 1/x)^x$.

Solution. For (a), we use `D`.

D[x^4 + 4/3x^3 − 3x^2, x]

$-6x + 4x^2 + 4x^3$

For (b), we first define f and then use $'$ together with `Factor` to calculate and factor $f'(x)$ and $f''(x)$.

f[x_] = 4x^5 − 5/2x^4 − 10x^3;

Factor[f'[x]]

$10x^2(1 + x)(-3 + 2x)$

Factor[f''[x]]

$10x\left(-6 - 3x + 8x^2\right)$

For (c), we use `Simplify` together with `D` to calculate and simplify y' and y''.

D[Sqrt[Exp[2x] + Exp[−2x]], x]

$$\frac{-2e^{-2x}+2e^{2x}}{2\sqrt{e^{-2x}+e^{2x}}}$$

D[Sqrt[Exp[2x] + Exp[−2x]], {x, 2}]//Simplify

$$\frac{\sqrt{e^{-2x}+e^{2x}}\left(1+6e^{4x}+e^{8x}\right)}{\left(1+e^{4x}\right)^2}$$

By hand, (d) would require logarithmic differentiation. The second derivative would be particularly difficult to compute by hand. Mathematica quickly computes and simplifies each derivative.

Simplify[D[(1 + 1/x)^x, x]]

$$\frac{\left(1+\frac{1}{x}\right)^x\left(-1+(1+x)\text{Log}\left[1+\frac{1}{x}\right]\right)}{1+x}$$

Simplify[D[(1 + 1/x)^x, {x, 2}]]

$$\frac{\left(1+\frac{1}{x}\right)^x\left(-1+x-2x(1+x)\text{Log}\left[1+\frac{1}{x}\right]+x(1+x)^2\text{Log}\left[1+\frac{1}{x}\right]^2\right)}{x(1+x)^2}$$

□

`Map` and operations on lists are discussed in more detail in Chapter 4.

The command `Map[f,list]` applies the function `f` to each element of the list `list`. Thus, if you are computing the derivatives of a large number of functions, you can use `Map` together with `D`.

A built-in Mathematica function is **threadable** if `f[list]` returns the same result as `Map[f,list]`. Many familiar functions like `D` and `Integrate` are threadable.

Example 3.11

Compute the first and second derivatives of $\sin x$, $\cos x$, $\tan x$, $\sin^{-1} x$, $\cos^{-1} x$, and $\tan^{-1} x$.

Solution. Notice that lists are contained in braces. Thus, entering

Map[D[#, x]&, {Sin[x], Cos[x], Tan[x],

ArcSin[x], ArcCos[x], ArcTan[x]}]

$$\left\{\text{Cos}[x], -\text{Sin}[x], \text{Sec}[x]^2, \frac{1}{\sqrt{1-x^2}}, -\frac{1}{\sqrt{1-x^2}}, \frac{1}{1+x^2}\right\}$$

computes the first derivative of the three trigonometric functions and their inverses. In this case, we have applied a *pure function* to the list of trigonometric functions and their inverses. Given an argument `#`, `D[#,x]&` computes the derivative of `#` with respect to x. The `&` symbol is used to mark the end of a *pure* function. Similarly, entering

Map[D[#, {x, 2}]&, {Sin[x], Cos[x], Tan[x],

ArcSin[x], ArcCos[x], ArcTan[x]}]

$$\left\{-\text{Sin}[x], -\text{Cos}[x], 2\text{Sec}[x]^2\text{Tan}[x], \frac{x}{\left(1-x^2\right)^{3/2}}, -\frac{x}{\left(1-x^2\right)^{3/2}}, -\frac{2x}{\left(1+x^2\right)^2}\right\}$$

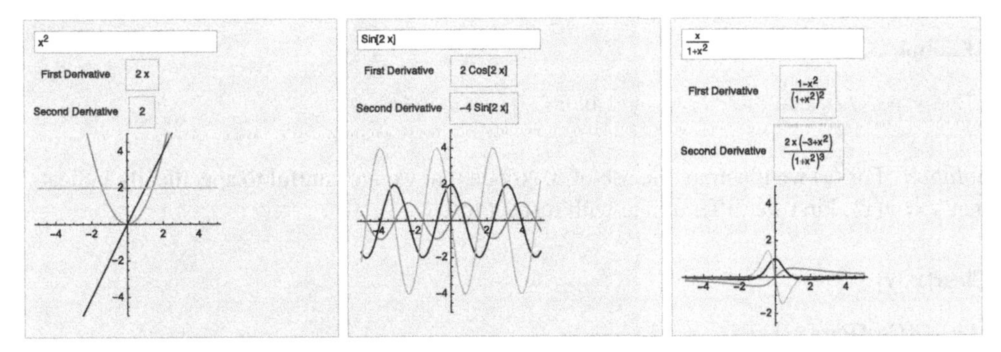

FIGURE 3.11 Seeing the relationship between the first and second derivative of a function and the original function (Mercer University colors).

computes the second derivative of the three trigonometric functions and their inverses. Because D is threadable, the same results are obtained with the following commands.

```
D[{Sin[x], Cos[x], Tan[x],
  ArcSin[x], ArcCos[x], ArcTan[x]}, x]
```

$$\left\{\text{Cos}[x], -\text{Sin}[x], \text{Sec}[x]^2, \frac{1}{\sqrt{1-x^2}}, -\frac{1}{\sqrt{1-x^2}}, \frac{1}{1+x^2}\right\}$$

```
D[{Sin[x], Cos[x], Tan[x],
  ArcSin[x], ArcCos[x], ArcTan[x]}, {x, 2}]
```

$$\left\{-\text{Sin}[x], -\text{Cos}[x], 2\text{Sec}[x]^2\text{Tan}[x], \frac{x}{(1-x^2)^{3/2}}, -\frac{x}{(1-x^2)^{3/2}}, -\frac{2x}{(1+x^2)^2}\right\}$$ □

With DynamicModule, we create a simple dynamic that lets you compute the first and second derivatives of basic functions and plot them on a standard viewing window, $[-5,5]\times[-5,5]$. The layout of Fig. 3.11 is primarily determined by Panel, Column, and Grid. The default function is $y = x^2$. To compute and graph the first and second derivatives of a different function, simply type over x^2 with the desired function.

```
Panel[DynamicModule[{f = x^2},
   Column[{InputField[Dynamic[f]], Grid[{{"First Derivative",
  Panel[Dynamic[D[f, x]//Simplify]]},
   {"Second Derivative", Panel[Dynamic[D[f, {x, 2}]//Simplify]]}}],
   Dynamic[Plot[Evaluate[Tooltip[{f, D[f, x], D[f, {x, 2}]}]],
   {x, -5, 5}, PlotRange → {-5, 5},
   PlotStyle → {{CMYKColor[0, .65, 1, .04]}, {Black}, {CMYKColor[.01, .05, .23, .03]}},
   AspectRatio → Automatic]]}]], ImageSize → {300, 300}]
```

3.2.3 Implicit Differentiation

If an equation contains two variables, x and y, implicit differentiation can be carried out by explicitly declaring y to be a function of x, $y = y(x)$, and using D or by using the Dt command.

Example 3.12

Find $y' = dy/dx$ if (a) $\cos(e^{xy}) = x$ and (b) $\ln(x/y) + 5xy = 3y$.

Solution. For (a) we illustrate the use of `D`. Notice that we are careful to specifically indicate that $y = y(x)$. First we differentiate with respect to x

Clear[x, y]

s1 = D[Cos[Exp[xy[x]]] − x, x]

$$-1 - e^{xy[x]}\text{Sin}\left[e^{xy[x]}\right]\left(y[x] + xy'[x]\right)$$

and then we solve the resulting equation for $y' = dy/dx$ with `Solve`.

Solve[s1 == 0, y'[x]]

$$\left\{\left\{y'[x] \to -\frac{e^{-xy[x]}\left(\text{Csc}\left[e^{xy[x]}\right] + e^{xy[x]}y[x]\right)}{x}\right\}\right\}$$

For (b), we use `Dt`. When using `Dt`, we interpret `Dt[x]`= 1 and `Dt[y]`= $y' = dy/dx$. Thus, entering

s2 = Dt[Log[x/y] + 5xy − 3y]

$$5y\text{Dt}[x] - 3\text{Dt}[y] + 5x\text{Dt}[y] + \frac{y\left(\frac{\text{Dt}[x]}{y} - \frac{x\text{Dt}[y]}{y^2}\right)}{x}$$

s3 = s2/.{Dt[x] → 1, Dt[y] → dydx}

$$-3\text{dydx} + 5\text{dydx}x + 5y + \frac{\left(-\frac{\text{dydx}x}{y^2} + \frac{1}{y}\right)y}{x}$$

and solving for `dydx` with `Solve`

Solve[s3 == 0, dydx]

$$\left\{\left\{\text{dydx} \to -\frac{y(1+5xy)}{x(-1-3y+5xy)}\right\}\right\}$$

shows us that if $\ln(x/y) + 5xy = 3y$, $y' = \dfrac{dy}{dx} = -\dfrac{(1+5xy)y}{(5xy-3y-1)x}$.

To graph each equation, we use `ContourPlot`. Generally, given an equation of the form $f(x, y) = g(x, y)$, the command

```
ContourPlot[f[x,y]==g[x,y],{x,a,b},{y,c,d}]
```

attempts to plot the graph of $f(x, y) = g(x, y)$ on the rectangle $[a, b] \times [c, d]$. Using `Show` together with `GraphicsRow`, we show the two graphs side-by-side in Fig. 3.12.

cp1 = ContourPlot[Cos[Exp[xy]]==x, {x, −2, 2}, {y, −4, 4}, PlotPoints → 120,

Frame → False, Axes → Automatic, AxesOrigin → {0, 0}, Contours → 40,

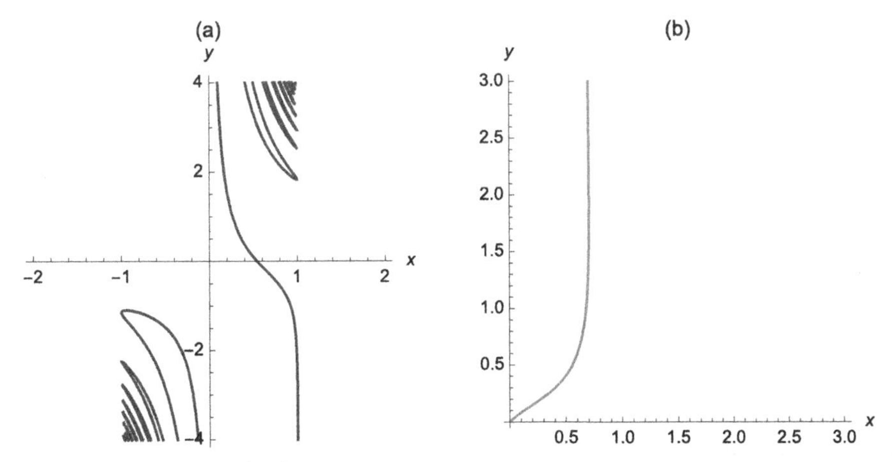

FIGURE 3.12 On the left, $\cos(e^{xy}) = x$ for $-2 \le x \le 2$ and $-4 \le y \le 4$; on the right, $\ln(x/y) + 5xy = 3y$ for $.01 \le x \le 3$ and $.01 \le y \le 3$ (MIT colors).

```
  ContourStyle → CMYKColor[.24, 1, .78, .17], AxesLabel → {x, y},
  PlotLabel → "(a)"];
cp2 = ContourPlot[Log[x/y] + 5xy == 3y, {x, .01, 3}, {y, .01, 3}, PlotPoints → 120,
  AxesLabel → {x, y}, PlotLabel → "(b)",
  ContourStyle → CMYKColor[.48, .39, .39, .04],
  Frame → False, Axes → Automatic, AxesOrigin → {0, 0}];
Show[GraphicsRow[{cp1, cp2}]]
```

□

3.2.4 Tangent Lines

If $f'(a)$ exists, $f'(a)$ is interpreted to be the slope of the line tangent to the graph of $y = f(x)$ at the point $(a, f(a))$. In this case, an equation of the tangent is given by

$$y - f(a) = f'(a)(x - a) \quad \text{or} \quad y = f'(a)(x - a) + f(a).$$

Example 3.13

Find an equation of the line tangent to the graph of $f(x) = \sin x^{1/3} + \cos^{1/3} x$ at the point with x-coordinate $x = 5\pi/3$.

Solution. Recall that when computing odd roots of negative numbers, Mathematica returns the value with the largest imaginary part. However, for our purposes, we need the real-valued root. To obtain the real-valued root, use `Surd`: `Surd[x,n]` returns the real-valued root of x if n is odd. Because we will be graphing a function involving odd roots of negative numbers, we define $f(x)$ using the `Surd` function and then compute $f'(x)$.

```
f[x_] = Sin[Surd[x, 3]] + Surd[Cos[x], 3];
```

```
f'[x]
```

$$\frac{\text{Cos}[\sqrt[3]{x}]}{3\sqrt[3]{x}^2} - \frac{\text{Sin}[x]}{3\sqrt[3]{\text{Cos}[x]}^2}$$

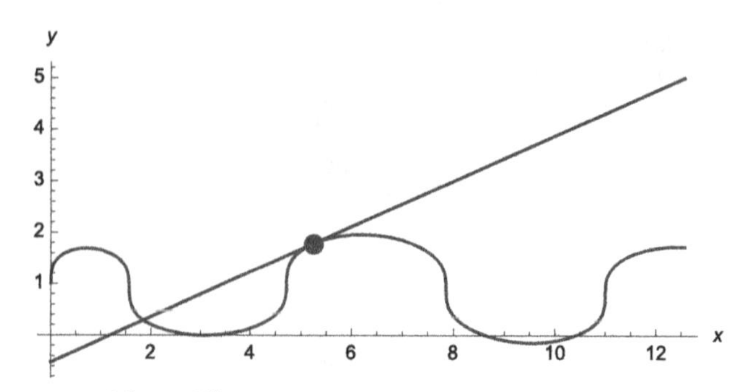

FIGURE 3.13 $f(x) = \sin x^{1/3} + \cos^{1/3} x$ together with its tangent at the point $(5\pi/3, f(5\pi/3))$ (Stanford colors).

Then, the slope of the line tangent to the graph of $f(x)$ at the point with x-coordinate $x = 5\pi/3$ is

```
f'[5Pi/3]
```

$$\frac{1}{2^{1/3}\sqrt{3}} + \frac{\mathrm{Cos}\left[\left(\frac{5\pi}{3}\right)^{1/3}\right]}{3^{1/3}(5\pi)^{2/3}}$$

```
f'[5Pi/3]//N
```

0.440013

while the y-coordinate of the point is

```
f[5Pi/3]
```

$$\frac{1}{2^{1/3}} + \mathrm{Sin}\left[\left(\frac{5\pi}{3}\right)^{1/3}\right]$$

```
f[5Pi/3]//N
```

1.78001

Thus, an equation of the line tangent to the graph of $f(x)$ at the point with x-coordinate $x = 5\pi/3$ is

$$y - \left(\frac{1}{\sqrt[3]{2}} + \sin\sqrt[3]{5\pi/3}\right) = \left(\frac{\cos\sqrt[3]{5\pi/3}}{\sqrt[3]{3}\sqrt[3]{25\pi^2}} + \frac{1}{\sqrt[3]{2}\sqrt{3}}\right)\left(x - \frac{5\pi}{3}\right),$$

as shown in Fig. 3.13.

```
p1 = Plot[f[x], {x, 0, 4Pi}, PlotStyle → CMYKColor[0, 1, .65, .34]];
p2 = ListPlot[{{5Pi/3, f[5Pi/3]}//N}, PlotStyle → {PointSize[.03],
   CMYKColor[.48, .36, .24, .66]}];
p3 = Plot[f'[5Pi/3](x − 5Pi/3) + f[5Pi/3], {x, 0, 4Pi},
   PlotStyle->CMYKColor[.48, .36, .24, .66]];
```

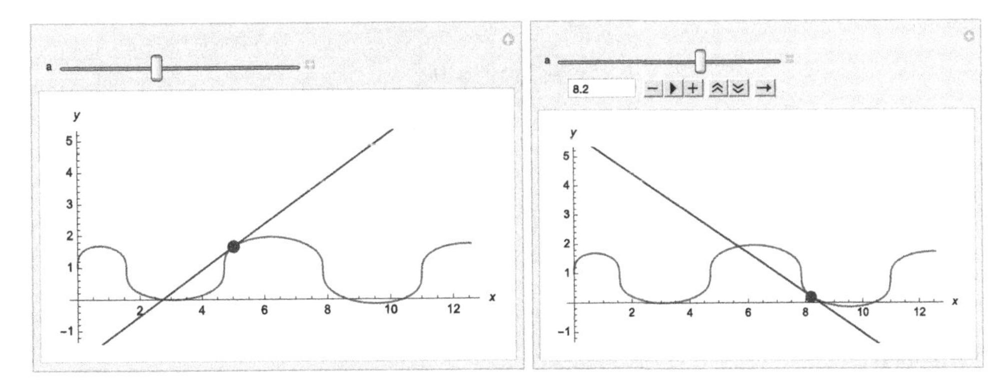

FIGURE 3.14 Using `Manipulate` to plot $f(x)$ together with various tangents.

```
Show[p1, p2, p3, AspectRatio->Automatic, PlotRange → All,
  AxesLabel → {x, y}]
```

We use `Manipulate` to plot different tangents or animate the tangents as shown in Fig. 3.14.

```
Manipulate[
p1 = Plot[f[x], {x, 0, 4Pi}, PlotStyle → CMYKColor[0, 1, .65, .34]];
p2 = ListPlot[{{a, f[a]}//N}, PlotStyle → {PointSize[.03],
  CMYKColor[.48, .36, .24, .66]}];
p3 = Plot[f'[a](x − a) + f[a], {x, 0, 4Pi}, PlotStyle->CMYKColor[.48, .36, .24, .66]];
Show[p1, p2, p3, AspectRatio->Automatic, PlotRange → {−1, 5},
  AxesLabel → {x, y}], {{a, 5}, .1, 4Pi − .1}]
```

□

Remark 3.5. Sometimes using `Surd` may be cumbersome. Therefore, it is worth noting that if you desire a real root to $x^{n/m}$ and there is a real root, `x^(n/m)=Abs[x]^(n/m)` if n is even and `x^(n/m)=Sign[x]Abs[x]^(n/m)` if n is odd. Of course, remember that if m is even and x is negative, $x^{n/m}$ is not a real number. Note that `Sign[x]` returns 1 if x is positive and -1 if x is negative.

Tangent Lines of Implicit Functions

Example 3.14

Find equations of the tangent line and normal line to the graph of $x^2y - y^3 = 8$ at the point $(-3, 1)$. Find and simplify $y'' = d^2y/dx^2$.

Solution. We evaluate $y' = dy/dx$ if $x = -3$ and $y = 1$ to determine the slope of the tangent line at the point $(-3, 1)$. Note that we cannot (easily) solve $x^2y - y^3 = 8$ for y so we use implicit differentiation to find $y' = dy/dx$:

$$\frac{d}{dx}\left(x^2y - y^3\right) = \frac{d}{dx}(8)$$

If the line ℓ has slope m_1 the line perpendicular (or, normal) to ℓ has slope $-1/m_1$.
By the product and chain rules, $\frac{d}{dx}(x^2y) = \frac{d}{dx}(x^2)y + x^2\frac{d}{dx}(y) = 2x \cdot y + x^2 \cdot \frac{dy}{dx} = 2xy + x^2y'$.

$$2xy + x^2y' - 3y^2y' = 0$$
$$y' = \frac{-2xy}{x^2 - 3y^2}.$$

eq = x^2y − y^3 == 8

$x^2y - y^3 == 8$

s1 = Dt[eq]

$2xy\text{Dt}[x] + x^2\text{Dt}[y] - 3y^2\text{Dt}[y] == 0$

s2 = s1/.Dt[x] → 1

$2xy + x^2\text{Dt}[y] - 3y^2\text{Dt}[y] == 0$

s3 = Solve[s2, Dt[y]]

$\left\{\left\{\text{Dt}[y] \to -\frac{2xy}{x^2-3y^2}\right\}\right\}$

Lists are discussed in more detail in Chapter 4.

Notice that `s3` is a **list.** The formula for $y' = dy/dx$ is the second part of the first part of the first part of `s3` and extracted from `s3` with

s3[[1, 1, 2]]

$-\frac{2xy}{x^2-3y^2}$

We then use `ReplaceAll` (`/.`) to find that the slope of the tangent at $(-3, 1)$ is

s3[[1, 1, 2]]/.{x → −3, y → 1}

1

The slope of the normal is $-1/1 = -1$. Equations of the tangent and normal are given by

$$y - 1 = 1(x + 3) \quad \text{and} \quad y - 1 = -1(x + 3),$$

respectively. See Fig. 3.15.

```
cp1 = ContourPlot[x^2y − y^3 − 8, {x, −5, 5}, {y, −5, 5}, Contours → {0},
  ContourShading → False, PlotPoints → 200,
  ContourStyle → {Thickness[.01], CMYKColor[1, .45, 0, .66]}];
p1 = ListPlot[{{−3, 1}}, PlotStyle → {Black, PointSize[.03]}];
p2 = Plot[{(x + 3) + 1, −(x + 3) + 1}, {x, −5, 5},
  PlotStyle → {{CMYKColor[0, .3, .93, 0], Thickness[.01]}}];
```

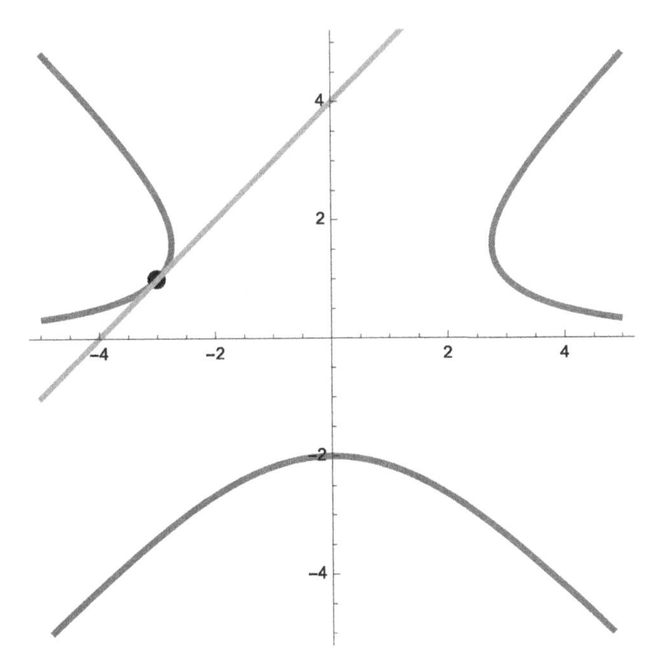

FIGURE 3.15 Graphs of $x^2y - y^3 = 8$ and the tangent and normal at $(-3, 1)$ (University of California Berkeley colors).

```
Show[cp1, p1, p2, Frame → False, Axes → Automatic, AxesOrigin → {0, 0},
  AspectRatio → Automatic, DisplayFunction → $DisplayFunction]
```

To find $y'' = d^2y/dx^2$, we proceed as follows.

```
s4 = Dt[s3[[1, 1, 2]]]//Simplify
```

$$-\frac{2(x^2+3y^2)(-y\text{Dt}[x]+x\text{Dt}[y])}{(x^2-3y^2)^2}$$

```
s5 = s4/.Dt[x] → 1/.s3[[1]]//Simplify
```

$$\frac{6y(x^2-y^2)(x^2+3y^2)}{(x^2-3y^2)^3}$$

The result means that

$$y'' = \frac{d^2y}{dx^2} = \frac{6\left(x^2y - y^3\right)\left(x^2+3y^2\right)}{\left(x^2-3y^2\right)^3}.$$

Because $x^2y - y^3 = 8$, the second derivative is further simplified to

$$y'' = \frac{d^2y}{dx^2} = \frac{48\left(x^2+3y^2\right)}{\left(x^2-3y^2\right)^3}.$$

□

Parametric Equations and Polar Coordinates

For the parametric equations $\{x = f(t), y = g(t)\}, t \in I$,

$$y' = \frac{dy}{dx} = \frac{dy/dt}{dx/dt} = \frac{g'(t)}{f'(t)}$$

and

$$y'' = \frac{d^2y}{dx^2} = \frac{d}{dx}\frac{dy}{dx} = \frac{d/dt(dy/dx)}{dx/dt}.$$

If $\{x = f(t), y = g(t)\}$ has a tangent line at the point $(f(a), g(a))$, parametric equations of the tangent are given by

$$x = f(a) + tf'(a) \qquad \text{and} \qquad y = g(a) + tg'(a). \tag{3.2}$$

If $f'(a)$, $g'(a) \neq 0$, we can eliminate the parameter from (3.2)

$$\frac{x - f(a)}{f'(a)} = \frac{y - g(a)}{g'(a)}$$

$$y - g(a) = \frac{g'(a)}{f'(a)}(x - f(a))$$

and obtain an equation of the tangent line in point-slope form.

```
l = Solve[x[a] + tx'[a] == cx, t]
r = Solve[y[a] + ty'[a] == cy, t]
```

$\left\{\left\{t \to \frac{cx - x[a]}{x'[a]}\right\}\right\}$

$\left\{\left\{t \to \frac{cy - y[a]}{y'[a]}\right\}\right\}$

Example 3.15: The Cycloid

The **cycloid** has parametric equations

$$x = t - \sin t \qquad \text{and} \qquad y = 1 - \cos t.$$

Graph the cycloid together with the line tangent to the graph of the cycloid at the point $(x(a), y(a))$ for various values of a between -2π and 4π.

Solution. After defining x and y we use ' to compute dy/dt and dx/dt. We then compute $dy/dx = (dy/dt)/(dx/dt)$ and d^2y/dx^2.

```
x[t_] = t - Sin[t];
y[t_] = 1 - Cos[t];
dx = x'[t]
dy = y'[t]
dydx = dy/dx
```

1 − Cos[t]

Sin[t]

$\frac{\text{Sin}[t]}{1 - \text{Cos}[t]}$

```
dypdt = Simplify[D[dydx, t]]
```

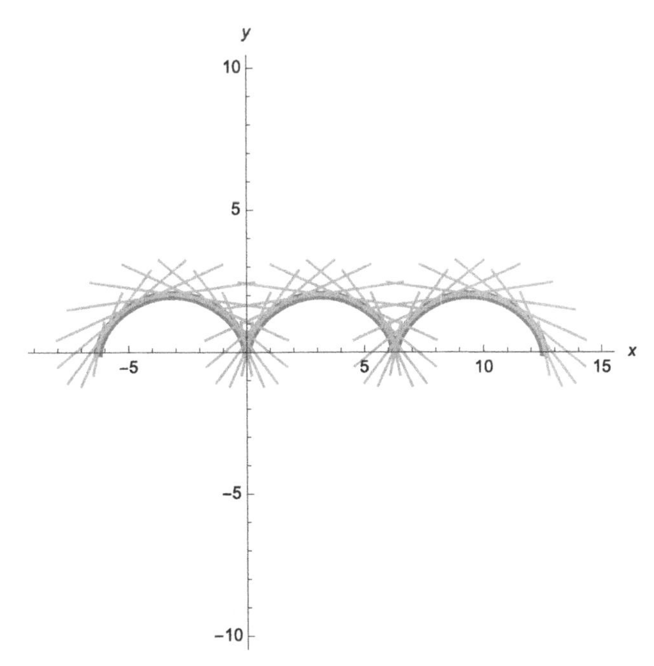

FIGURE 3.16 The cycloid with various tangents (California Institute of Technology colors).

$\frac{1}{-1+\text{Cos}[t]}$

```
seconddderiv = Simplify[dypdt/dx]
```

$-\frac{1}{(-1+\text{Cos}[t])^2}$

We then use `ParametricPlot` to graph the cycloid for $-2\pi \le t \le 4\pi$, naming the resulting graph `p1`.

```
p1 = ParametricPlot[{x[t], y[t]}, {t, -2Pi, 4Pi},
   PlotStyle → {{CMYKColor[0, .68, .98, 0], Thickness[.015]}}];
```

Next, we use `Table` to define `toplot` to be 40 tangent lines (3.2) using equally spaced values of a between -2π and 4π. We then graph each line `toplot` and name the resulting graph `p2`. Finally, we show `p1` and `p2` together with the `Show` function. The resulting plot is shown to scale because the lengths of the x and y-axes are equal and we include the option `AspectRatio->1`. In the graphs, notice that on intervals for which dy/dx is defined, dy/dx is a decreasing function and, consequently, $d^2y/dx^2 < 0$. (See Fig. 3.16.)

```
toplot = Table[{x[a] + tx'[a], y[a] + ty'[a]}, {a, -2Pi, 4Pi, 6Pi/39}];
p2 = ParametricPlot[Evaluate[toplot], {t, -2, 2}, PlotStyle → CMYKColor[.35, .28, .35, 0]];
Show[p1, p2, AspectRatio → 1, PlotRange → {-3Pi, 3Pi},
   AxesLabel → {x, y}]
```

With `Manipulate`, you can animate the tangents. (See Fig. 3.17.)

```
Manipulate[x[t_] = t - Sin[t];
```

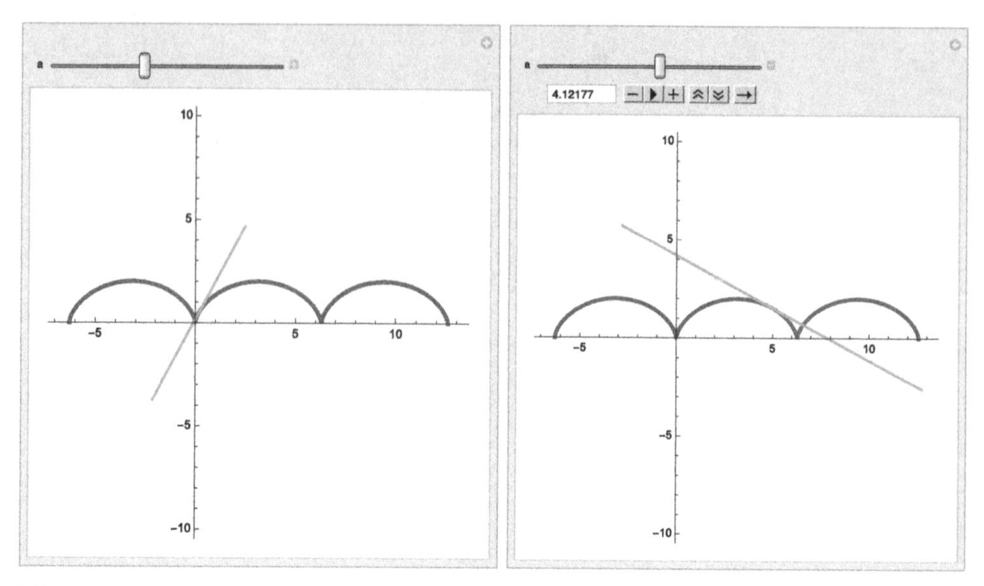

FIGURE 3.17 With `Manipulate` or `Animate` you can animate the tangents.

```
y[t_] = 1 − Cos[t];
y[t_] = Module[{p1, p2}, p1 = ParametricPlot[{x[t], y[t]}, {t, −2Pi, 4Pi},
PlotStyle → {{CMYKColor[0, .68, .98, 0], Thickness[.01]}}];
p2 = ParametricPlot[{x[a] + tx'[a], y[a] + ty'[a]}, {t, −5, 5},
  PlotStyle → {{CMYKColor[.35, .28, .35, 0], Thickness[.0075]}}];
Show[p1, p2, AspectRatio → 1, PlotRange → {{−2Pi, 4Pi}, {−3Pi, 3Pi}}]],
  {{a, 1}, −2Pi, 4Pi}]
```

□

Example 3.16: Orthogonal Curves

Two lines L_1 and L_2 with slopes m_1 and m_2, respectively, are **orthogonal** if their slopes are negative reciprocals: $m_1 = -1/m_2$.
Extended to curves, we say that the curves C_1 and C_2 are **orthogonal** at a point of intersection if their respective tangent lines to the curves at that point are orthogonal.
Show that the family of curves with equation $x^2+2xy-y^2=C$ is orthogonal to the family of curves with equation $y^2+2xy-x^2=C$.

Solution. We begin by defining `eq1` and `eq2` to be equations $x^2+2xy-y^2=C$ and $y^2+2xy-x^2=C$, respectively. Then, use `Dt` to differentiate and `Solve` to find $y'=dy/dx$.

```
eq1 = x^2 + 2xy − y^2==c;
eq2 = y^2 + 2xy − x^2==c;
```

```
Simplify[Solve[Dt[eq1, x], Dt[y, x]]/.Dt[c, x] → 0]
```

$\left\{\left\{\mathrm{Dt}[y,x]\to -\frac{x+y}{x-y}\right\}\right\}$

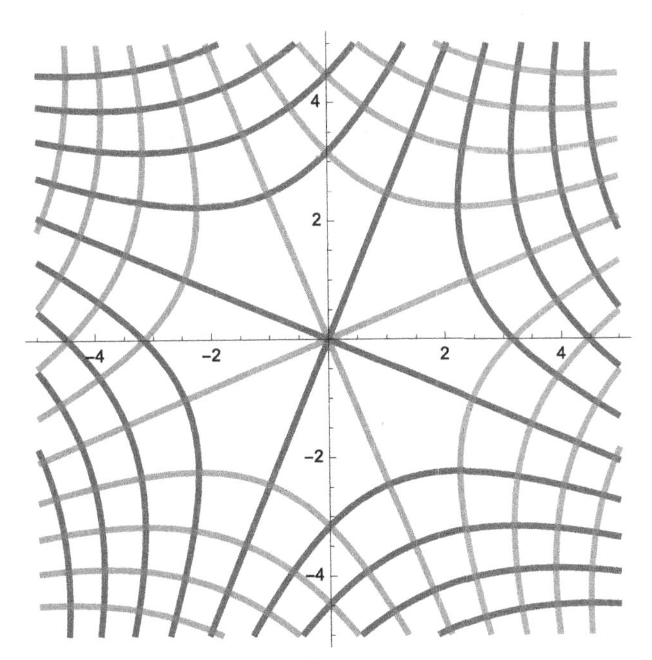

FIGURE 3.18 $x^2+2xy-y^2=C$ and $y^2+2xy-x^2=C$ for various values of C (University of Illinois colors).

```
Simplify[Solve[Dt[eq2, x], Dt[y, x]]/.Dt[c, x] → 0]
```

$$\left\{\left\{\text{Dt}[y,x]\to\frac{x-y}{x+y}\right\}\right\}$$

Because the derivatives are negative reciprocals, we conclude that the curves are orthogonal. We confirm this graphically by graphing several members of each family with `ContourPlot` and showing the results together. (See Fig. 3.18.)

```
cp1 = ContourPlot[x^2 + 2xy − y^2, {x, −5, 5}, {y, −5, 5}, ContourShading → False,
  ContourStyle → {{Thickness[.01], CMYKColor[1, .9, .1, .77]}}];

cp2 = ContourPlot[y^2 + 2xy − x^2, {x, −5, 5}, {y, −5, 5}, ContourShading → False,
  ContourStyle → {{Thickness[.01], CMYKColor[0, .76, 1.0]}}];

Show[cp1, cp2, Frame → False, Axes → Automatic, AxesOrigin → {0, 0}]
```
□

Theorem 3.1 (The Mean-Value Theorem for Derivatives). *If $y=f(x)$ is continuous on $[a,b]$ and differentiable on (a,b) then there is at least one value of c between a and b for which*

$$f'(c)=\frac{f(b)-f(a)}{b-a}\quad \textit{or, equivalently,}\quad f(b)-f(a)=f'(c)(b-a). \tag{3.3}$$

Example 3.17

Find all number(s) c that satisfy the conclusion of the Mean-Value Theorem for $f(x)=x^2-3x$ on the interval $[0,7/2]$.

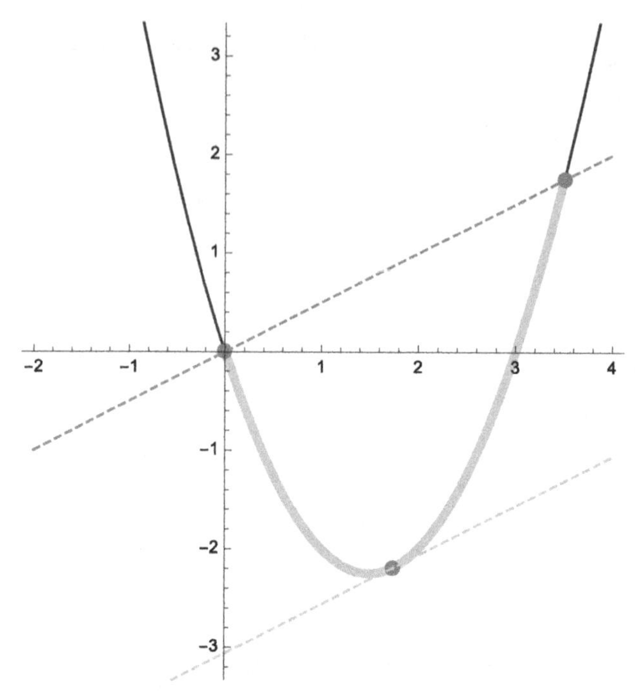

FIGURE 3.19 Graphs of $f(x) = x^2 - 3x$, the secant containing $(0, f(0))$ and $(7/2, f(7/2))$, and the tangent at $(7/4, f(7/4))$ (University of Michigan colors).

Solution. By the power rule, $f'(x) = 2x - 3$. The slope of the secant containing $(0, f(0))$ and $(7/2, f(7/2))$ is

$$\frac{f(7/2) - f(0)}{7/2 - 0} = \frac{1}{2}.$$

Solving $2x - 3 = 1/2$ for x gives us $x = 7/4$.

f[x_]=x^2−3x

$-3x + x^2$

Solve[f'[x]==0,x]

$\left\{\left\{x \to \frac{3}{2}\right\}\right\}$

Solve[f'[x]==(f[7/2]−f[0])/(7/2−0)]

$\left\{\left\{x \to \frac{7}{4}\right\}\right\}$

$x = 7/4$ satisfies the conclusion of the Mean-Value Theorem for $f(x) = x^2 - 3x$ on the interval $[0, 7/2]$, as shown in Fig. 3.19.

p1=Plot[f[x],{x,−1,4},PlotStyle→CMYKColor[1,.6,0,.6]];

p2=Plot[f[x],{x,0,7/2},PlotStyle→{{Thickness[.015],CMYKColor[0,.18,1,0]}}];

p3=ListPlot[{{0,f[0]},{7/4,f[7/4]},{7/2,f[7/2]}},

```
  PlotStyle → {{CMYKColor[.2, .14, .12, .4], PointSize[.025]}}];
p4 = Plot[{f'[7/4](x − 7/4) + f[7/4], (f[7/2] − f[0])/(7/2 − 0)x + f[0]},
  {x, −2, 4}, PlotStyle → {{Dashing[{.01}], CMYKColor[.06, .14, .38, .08]},
  {Dashing[{.01}], CMYKColor[.21, .15, .54, .31]}}];
Show[p1, p2, p3, p4, DisplayFunction → $DisplayFunction, AspectRatio → Automatic,
  PlotRange → {−3, 3}]
```

□

3.2.5 The First Derivative Test and Second Derivative Test

Example 3.15 illustrates the following properties of the first and second derivative.

Theorem 3.2. *Let $y = f(x)$ be continuous on $[a, b]$ and differentiable on (a, b).*

1. *If $f'(x) = 0$ for all x in (a, b), then $f(x)$ is constant on $[a, b]$.*
2. *If $f'(x) > 0$ for all x in (a, b), then $f(x)$ is increasing on $[a, b]$.*
3. *If $f'(x) < 0$ for all x in (a, b), then $f(x)$ is decreasing on $[a, b]$.*

For the second derivative, we have the following theorem.

Theorem 3.3. *Let $y = f(x)$ have a second derivative on (a, b).*

1. *If $f''(x) > 0$ for all x in (a, b), then the graph of $f(x)$ is concave up on (a, b).*
2. *If $f''(x) < 0$ for all x in (a, b), then the graph of $f(x)$ is concave down on (a, b).*

The **critical points** correspond to those points on the graph of $y = f(x)$ where the tangent line is horizontal or vertical; the number $x = a$ is a **critical number** if $f'(a) = 0$ or $f'(x)$ does not exist if $x = a$. The **inflection points** correspond to those points on the graph of $y = f(x)$ where the graph of $y = f(x)$ is neither concave up nor concave down. Theorems 3.2 and 3.3 help establish the first derivative test and second derivative test.

Theorem 3.4 (First Derivative Test). *Let $x = a$ be a critical number of a function $y = f(x)$ continuous on an open interval I containing $x = a$. If $f(x)$ is differentiable on I, except possibly at $x = a$, $f(a)$ can be classified as follows.*

1. *If $f'(x)$ makes a simple change in sign from positive to negative at $x = a$, then $f(a)$ is a **relative maximum**.*
2. *If $f'(x)$ makes a simple change in sign from negative to positive at $x = a$, then $f(a)$ is a **relative minimum**.*

Theorem 3.5 (Second Derivative Test). *Let $x = a$ be a critical number of a function $y = f(x)$ and suppose that $f''(x)$ exists on an open interval containing $x = a$.*

1. *If $f''(a) < 0$, then $f(a)$ is a relative maximum.*
2. *If $f''(a) > 0$, then $f(a)$ is a relative minimum.*

Example 3.18

Graph $f(x) = 3x^5 - 5x^3$.

Solution. We begin by defining $f(x)$ and then computing and factoring $f'(x)$ and $f''(x)$.

```
f[x_] = 3x^5 − 5x^3;
d1 = Factor[f'[x]]
```

d2 = Factor[f''[x]]

$15(-1+x)x^2(1+x)$

$30x\left(-1+2x^2\right)$

By inspection, we see that the critical numbers are $x = 0$, 1, and -1 while $f''(x) = 0$ if $x = 0$, $1/\sqrt{2}$, or $-1/\sqrt{2}$. Of course, these values can also be found with `Solve` as done next in `cns` and `ins`, respectively.

cns = Solve[d1 == 0]

ins = Solve[d2 == 0]

$\{\{x \to -1\}, \{x \to 0\}, \{x \to 0\}, \{x \to 1\}\}$

$\left\{\{x \to 0\}, \left\{x \to -\frac{1}{\sqrt{2}}\right\}, \left\{x \to \frac{1}{\sqrt{2}}\right\}\right\}$

We find the critical and inflection points by using `/.` (`Replace All`) to compute $f(x)$ for each value of x in `cns` and `ins`, respectively. The result means that the critical points are $(0,0)$, $(1,-2)$ and $(-1,2)$; the inflection points are $(0,0)$, $(1/\sqrt{2}, -7\sqrt{2}/8)$, and $(-1/\sqrt{2}, 7\sqrt{2}/8)$. We also see that $f''(0) = 0$ so Theorem 3.5 cannot be used to classify $f(0)$. On the other hand, $f''(1) = 30 > 0$ and $f''(-1) = -30 < 0$ so by Theorem 3.5, $f(1) = -2$ is a relative minimum and $f(-1) = 2$ is a relative maximum.

cps = {x, f[x]}/.cns

$\{\{-1,2\}, \{0,0\}, \{0,0\}, \{1,-2\}\}$

f''[x]/.cns

$\{-30, 0, 0, 30\}$

ips = {x, f[x]}/.ins

$\left\{\{0,0\}, \left\{-\frac{1}{\sqrt{2}}, \frac{7}{4\sqrt{2}}\right\}, \left\{\frac{1}{\sqrt{2}}, -\frac{7}{4\sqrt{2}}\right\}\right\}$

We can graphically determine the intervals of increase and decrease by noting that if $f'(x) > 0$ ($f'(x) < 0$), $a|f'(x)|/f'(x) = a$ ($a|f'(x)|/f'(x) = -a$). Similarly, the intervals for which the graph is concave up and concave down can be determined by noting that if $f''(x) > 0$ ($f''(x) < 0$), $a|f''(x)|/f''(x) = a$ ($a|f''(x)|/f''(x) = -a$). We use `Plot` to graph $|f'(x)|/f'(x)$ and $2|f''(x)|/f''(x)$ (different values are used so we can differentiate between the two plots) in Fig. 3.20.

```
Plot[{Abs[d1]/d1, 2Abs[d2]/d2}, {x, -2, 2}, PlotRange → {-3, 3},
  PlotStyle → {{CMYKColor[0, 1., .79, .2]}, {CMYKColor[.72, .66, .65, 72]}}]
```

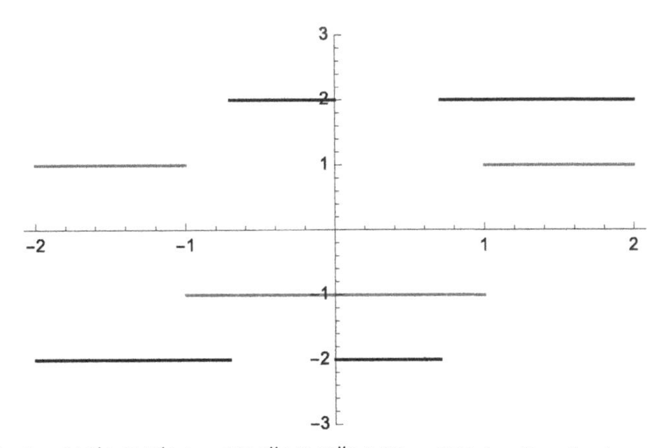

FIGURE 3.20 Graphs of $|f'(x)|/f'(x)$ and $2|f''(x)|/f''(x)$ (Cornell University colors).

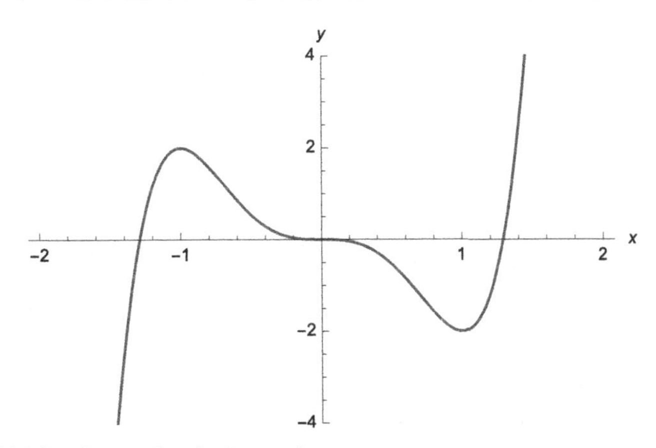

FIGURE 3.21 $f(x)$ for $-2 \le x \le 2$ and $-4 \le y \le 4$.

From the graph, we see that $f'(x) > 0$ for x in $(-\infty, -1) \cup (1, \infty)$, $f'(x) < 0$ for x in $(-1, 1)$, $f''(x) > 0$ for x in $(-1/\sqrt{2}, 0) \cup (1/\sqrt{2}, \infty)$, and $f''(x) < 0$ for x in $(-\infty, -1/\sqrt{2}) \cup (0, 1/\sqrt{2})$. Thus, the graph of $f(x)$ is

- increasing and concave down for x in $(-\infty, -1)$,
- decreasing and concave down for x in $(-1, -1/\sqrt{2})$,
- decreasing and concave up for x in $(-1/\sqrt{2}, 0)$,
- decreasing and concave down for x in $(0, 1\sqrt{2})$,
- decreasing and concave up for x in $(1/\sqrt{2}, 1)$, and
- increasing and concave up for x in $(1, \infty)$.

We also see that $f(0) = 0$ is neither a relative minimum nor maximum. To see all points of interest, our domain must contain -1 and 1 while our range must contain -2 and 2. We choose to graph $f(x)$ for $-2 \le x \le 2$; we choose the range displayed to be $-4 \le y \le 4$. (See Fig. 3.21.)

```
Plot[f[x], {x, −2, 2}, PlotRange → {−4, 4},
  PlotStyle->CMYKColor[0, 1., .79, .2],
  AxesLabel → {x, y}]
```
□

Remember to be especially careful when working with functions that involve odd roots. If n is odd and x is even, use `Surd[x,n]` to obtain the real-valued nth root of x.

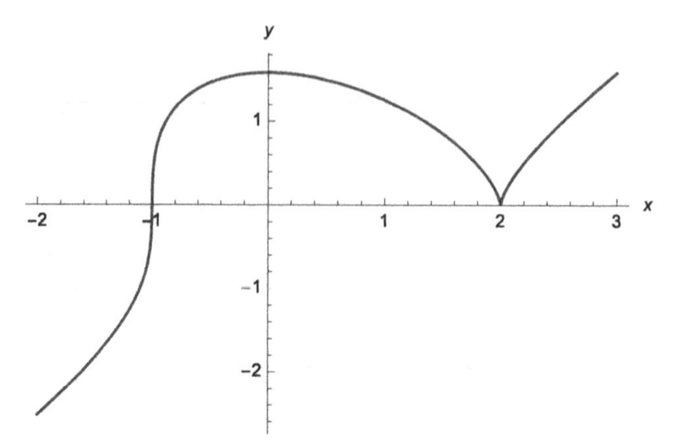

FIGURE 3.22 $f(x)$ for $-2 \le x \le 3$ (Carnegie Mellon colors).

Example 3.19

Graph $f(x) = (x-2)^{2/3}(x+1)^{1/3}$.

Solution. We begin by defining $f(x)$ and then computing and simplifying $f'(x)$ and $f''(x)$ with $'$ and `Simplify`.

```
Clear[f]
f[x_] = Surd[(x − 2)^2, 3]Surd[(x + 1), 3];
d1 = Simplify[f'[x]]
d2 = Simplify[f"[x]]
```

$$\frac{(-2+x)x}{\sqrt[3]{(-2+x)^2}^2\sqrt[3]{1+x}^2}$$

$$-\frac{2}{(1+x)\sqrt[3]{(-2+x)^2}^2\sqrt[3]{1+x}^2}$$

By inspection, we see that the critical numbers are $x = 0$, 2, and -1. We cannot use Theorem 3.5 to classify $f(2)$ and $f(-1)$ because $f''(x)$ is undefined if $x = 2$ or -1. On the other hand, $f''(0) < 0$ so $f(0) = 2^{2/3}$ is a relative maximum. By hand, we make a sign chart to see that the graph of $f(x)$ is

- increasing and concave up on $(-\infty, -1)$,
- increasing and concave down on $(-1, 0)$,
- decreasing and concave down on $(0, 2)$, and
- increasing and concave down on $(2, \infty)$.

Hence, $f(-1) = 0$ is neither a relative minimum nor maximum while $f(2) = 0$ is a relative minimum by Theorem 3.4. We use `Plot` to graph $f(x)$ for $-2 \le x \le 3$ in Fig. 3.22.

```
f[0]
Plot[f[x], {x, −2, 3},
  PlotStyle → CMYKColor[.07, 1, .82, .26],
```

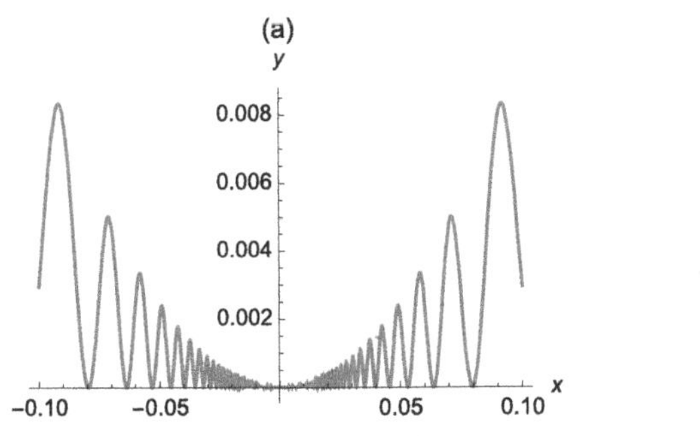

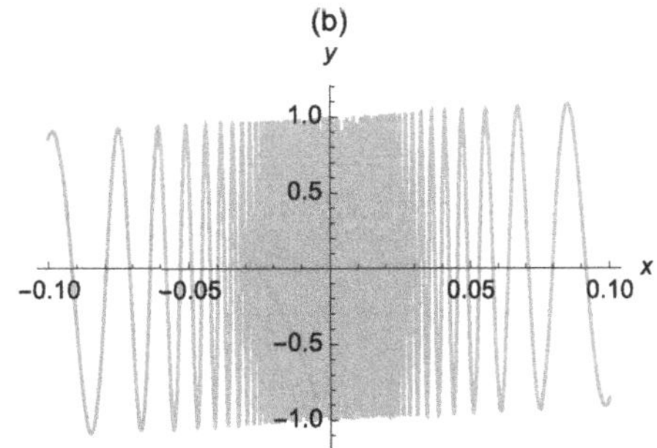

FIGURE 3.23 $f(x) = \left[x \sin\left(\frac{1}{x}\right)\right]^2$ and $f'(x)$ for $-0.1 \le x \le 0.1$ (Carnegie Mellon colors).

```
AxesLabel → {x, y}]
```

$2^{2/3}$

□

The previous examples illustrate that if $x = a$ is a critical number of $f(x)$ and $f'(x)$ makes a *simple change in sign* from positive to negative at $x = a$, then $(a, f(a))$ is a relative maximum. If $f'(x)$ makes a simple change in sign from negative to positive at $x = a$, then $(a, f(a))$ is a relative minimum. Mathematica is especially useful in investigating interesting functions for which this may not be the case.

Example 3.20

Consider

$$f(x) = \begin{cases} x^2 \sin^2\left(\dfrac{1}{x}\right), x \neq 0 \\ 0, x = 0, \end{cases}$$

$x = 0$ is a critical number because $f'(x)$ does not exist if $x = 0$. The point $(0, 0)$ is both a relative and absolute minimum, even though $f'(x)$ does not make a simple change in sign at $x = 0$, as illustrated in Fig. 3.23.

```
f[x_] = (xSin[1/x])^2;
f'[x]//Factor
```

$-2\text{Sin}\left[\frac{1}{x}\right]\left(\text{Cos}\left[\frac{1}{x}\right] - x\text{Sin}\left[\frac{1}{x}\right]\right)$

```
p1 = Plot[f[x], {x, -0.1, 0.1},
  PlotStyle → CMYKColor[.92, .02, 1, .12],
  AxesLabel → {x, y}, PlotLabel → "(a)"];
p2 = Plot[f'[x], {x, -0.1, 0.1},
  PlotStyle → CMYKColor[0, .32, 1, 0],
  AxesLabel → {x, y}, PlotLabel → "(b)"];
Show[GraphicsRow[{p1, p2}]]
```

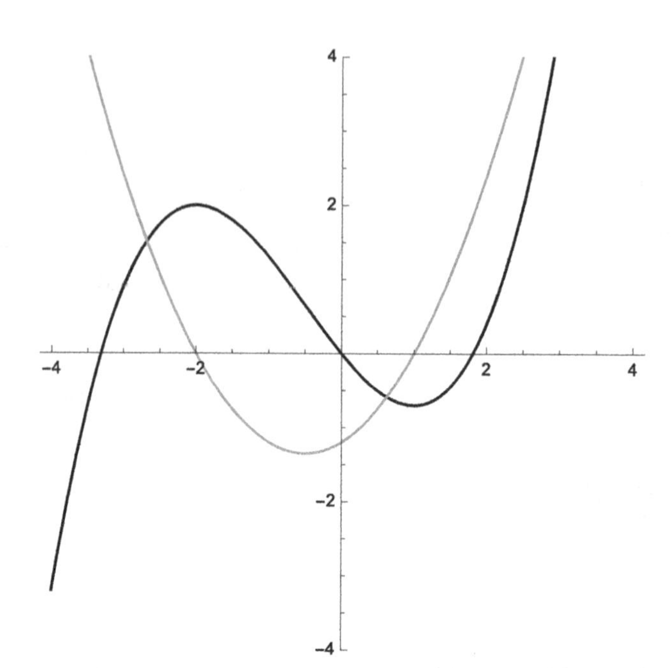

FIGURE 3.24 $f(x)$ has one relative maximum and one relative minimum but no absolute extreme values (Purdue University colors).

Notice that the derivative "oscillates" infinitely many times near $x = 0$, so the first derivative test cannot be used to classify $(0, 0)$.

The functions `Maximize` and `Minimize` can be used to assist with finding extreme values. For a function of a single variable `Maximize[f[x],x]` (`Minimize[f[x],x]`) attempts to find the maximum (minimum) values of $f(x)$; `Maximize[{f[x],a<=x<=b},x]` (`Minimize[{f[x],a<=x<=b},x]`) attempts to find the maximum (minimum) values of $f(x)$ on $[a, b]$.

Example 3.21

Consider $f(x) = \frac{1}{10}\left(-12x + 3x^2 + 2x^3\right)$. After defining $f(x)$, we plot $f(x)$ and $f'(x)$ together in Fig. 3.24.

```
f[x_]=1/10(-12x+3x^2+2x^3);
Plot[Tooltip[{f[x], f'[x]}], {x, -4, 4}, PlotRange -> {-4, 4},
  AspectRatio -> Automatic, PlotStyle -> {{Black}, {CMYKColor[.06, .27, 1, .12]}}]
```

With `Maximize`, we see that $f(x)$ does not have a maximum on its domain. However, when we restrict the interval to $-3 \le x \le 2$, `Maximize` finds the relative maximum at $x = -2$.

```
Maximize[f[x], x]
```

Maximize: The maximum is not attained at any point satisfying the given constraints.

$\{\infty, \{x \to \infty\}\}$

```
Maximize[{f[x], -3 ≤ x ≤ 2}, x]
```

$\{2, \{x \to -2\}\}$

Similarly, with `Minimize` we see that the $f(x)$ does not have a minimum value on its domain but find the relative minimum when we restrict the interval to $-3 \le x \le 2$.

Minimize[f[x], x]

Minimize: The minimum is not attained at any point satisfying the given constraints.

$\{-\infty, \{x \to -\infty\}\}$

Minimize[{f[x], −3 ≤ x ≤ 2}, x]

$\left\{-\frac{7}{10}, \{x \to 1\}\right\}$

However, with `Solve`, we easily find the two zeros of $f'(x)$ that we see in Fig. 3.24

Solve[f′[x] == 0, x]

$\{\{x \to -2\}, \{x \to 1\}\}$

When using `Maximize` or `Minimize` you should verify your results using another method.

Example 3.22

The function $f(x) = x/(x^2+1)$ is continuous on $(-\infty, \infty)$ and $\lim_{x\to\pm\infty} f(x) = 0$. Thus, $f(x)$ has an absolute minimum and maximum value on its domain. In this case,

Maximize[x/(x^2 + 1), x]

$\left\{\frac{1}{2}, \{x \to 1\}\right\}$

Minimize[x/(x^2 + 1), x]

$\left\{-\frac{1}{2}, \{x \to -1\}\right\}$

gives us the absolute maximum and minimum values of $f(x)$ and the x-values where they occur. On the other hand, $f(x) = x^4 - x^2$ is continuous on $(-\infty, \infty)$ and $\lim_{x\to\pm\infty} f(x) = \infty$. Thus, $f(x)$ has an absolute minimum on its domain. Because the derivative of a fourth degree polynomial is a third degree polynomial, we know that $f'(x)$ has three zeros, two of which probably correspond to relative minimums. Because the graph of $f(x)$ is symmetric with respect to the y-axis, we further suspect that the absolute minimum is obtained twice—at each relative minimum. `Maximize` and `Minimize` give us the following results.

A polynomial of degree n has n zeros (counting multiplicity).

Maximize[x^4 − x^2, x]

Maximize: The maximum is not attained at any point satisfying the given constraints.

$\{\infty, \{x \to -\infty\}\}$

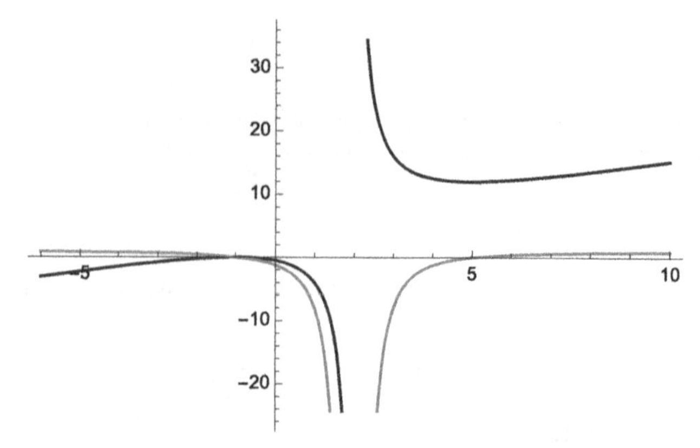

FIGURE 3.25 A function for which a relative minimum has a function value greater than the function value of a relative maximum (University of California–San Diego colors).

```
Minimize[x^4 - x^2, x]
```

$\left\{-\frac{1}{4}, \left\{x \to -\frac{1}{\sqrt{2}}\right\}\right\}$

Note that the result returned by `Maximize` is correct. Similarly, the result returned by `Minimize` is correct, but a complete answer would indicate that the absolute minimum value occurs at both $x = -1/\sqrt{2}$ and $x = 1/\sqrt{2}$.

Example 3.23

The function $f(x) = (x+1)^2/(x-2)$ has a vertical asymptote at $x = 2$. From the derivative,

```
f[x_] = (x + 1)^2/(x - 2);
d1 = Simplify[f'[x]]
cns = Solve[f'[x] == 0, x]
```

$\frac{(-5+x)(1+x)}{(-2+x)^2}$

$\{\{x \to -1\}, \{x \to 5\}\}$

```
f[x]/.cns
```

$\{0, 12\}$

we find two critical numbers, one of which is a relative maximum and one is a relative minimum. See Fig. 3.25.

```
Plot[Tooltip[{f[x], f'[x]}], {x, -6, 10},
  PlotStyle -> {{CMYKColor[1, .86, .42, .42]},
  {CMYKColor[{.06, .35, .99, .18}]}}]
```

On the other hand, `Maximize` and `Minimize` return confusing results because the function is undefined if $x = 2$. The function has relative extreme values but not absolute extreme values.

Maximize[f[x], x]

Maximize: The maximum is not attained at any point satisfying the given constraints.

$\{\infty, \{x \to 2\}\}$

Minimize[f[x], x]

Minimize: The minimum is not attained at any point satisfying the given constraints.

$\{-\infty, \{x \to 2\}\}$

From the graph in Fig. 3.25, we see that $\lim_{x\to 2^+} f(x) = +\infty$ while $\lim_{x\to 2^-} f(x) = -\infty$.

For periodic functions, such as sine and cosine, `Maximize` and `Minimize` generally don't indicate *all* extreme values.

Maximize[Sin[x], x]

$\left\{1, \left\{x \to \frac{\pi}{2}\right\}\right\}$

Minimize[Cos[x], x]

$\{-1, \{x \to \pi\}\}$

3.2.6 Applied Max/Min Problems

Mathematica can be used to assist in solving maximization/minimization problems encountered in a differential calculus course.

Example 3.24

A woman is located on one side of a body of water 4 miles wide. Her position is directly across from a point on the other side of the body of water 16 miles from her house, as shown in the following figure.

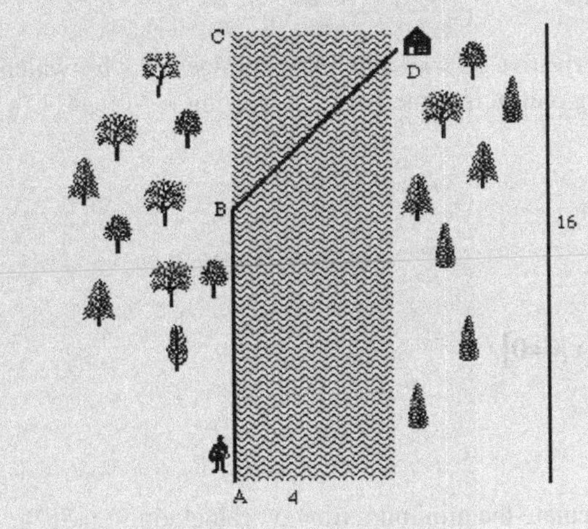

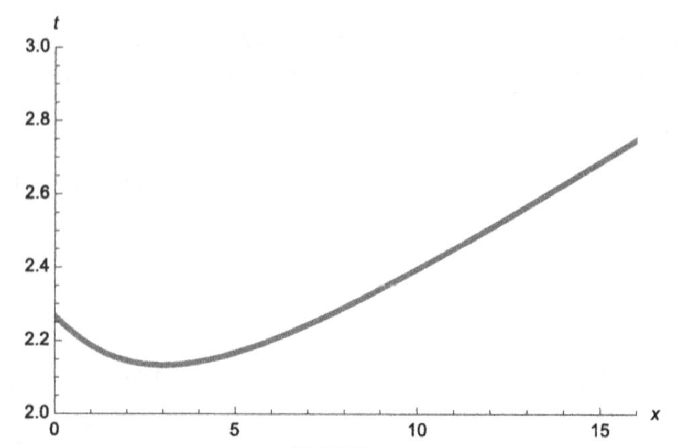

FIGURE 3.26 Plot of $time(x) = \frac{1}{10}(16-x) + \frac{1}{6}\sqrt{x^2+16}$, $0 \le x \le 16$ (The University of Texas at Austin colors).

If she can move across land at a rate of 10 miles per hour and move over water at a rate of 6 miles per hour, find the least amount of time for her to reach her house.

Solution. From the figure, we see that the woman will travel from A to B by land and then from B to D by water. We wish to find the least time for her to complete the trip.

Let x denote the distance BC, where $0 \le x \le 16$. Then, the distance AB is given by $16 - x$ and, by the Pythagorean theorem, the distance BD is given by $\sqrt{x^2+4^2}$. Because rate × time = distance, time = distance/rate. Thus, the time to travel from A to B is $\frac{1}{10}(16-x)$, the time to travel from B to D is $\frac{1}{6}\sqrt{x^2+16}$, and the total time to complete the trip, as a function of x, is

$$time(x) = \frac{1}{10}(16-x) + \frac{1}{6}\sqrt{x^2+16}, \quad 0 \le x \le 16.$$

We must minimize the function *time*. First, we define `time` and then verify that `time` has a minimum by graphing `time` on the interval [0, 16] in Fig. 3.26.

```
Clear[time]
time[x_] = (16-x)/10 + 1/6 Sqrt[x^2 + 16];
Plot[time[x], {x, 0, 16}, PlotRange -> {{0, 16}, {2, 3}},
  PlotStyle -> {Thickness[.01], CMYKColor[0, .65, 1, .09]},
  AxesLabel -> {x, t}]
```

Next, we compute the derivative of `time` and find the values of x for which the derivative is 0 with `Solve`. The resulting output is named `critnums` using `ReplaceAll (\.)`.

```
Together[time'[x]]
```

$$\frac{5x-3\sqrt{16+x^2}}{30\sqrt{16+x^2}}$$

```
critnums = Solve[time'[x]==0]
```

$\{\{x \to 3\}\}$

At this point, we can calculate the minimum time by calculating `time[3]`.

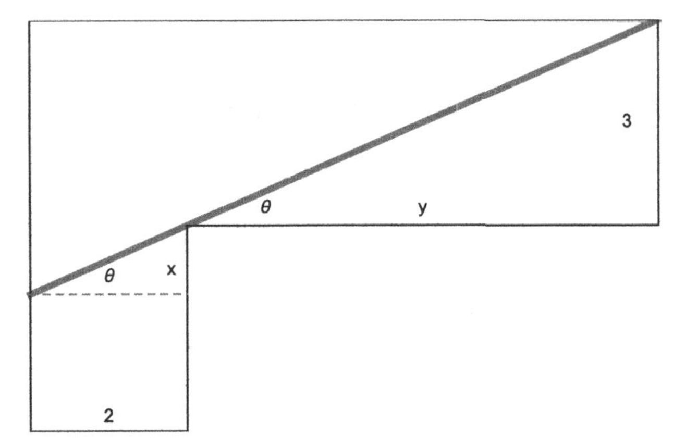

FIGURE 3.27 The length of the beam is found using similar triangles.

time[3]

$\frac{32}{15}$

Alternatively, we demonstrate how to find the value of `time[x]` for the value(s) listed in `critnums`.

time[x]/.$x \to 3$

$\frac{32}{15}$

Regardless, we see that the minimum time to complete the trip is 32/15 hours. □

One of the more interesting applied max/min problems is the *beam problem*. We present two solutions.

Example 3.25: The Beam Problem

Find the exact length of the longest beam that can be carried around a corner from a hallway 2 feet wide to a hallway that is 3 feet wide. (See Fig. 3.27.)

Solution. We assume that the beam has negligible thickness. Our first approach is algebraic. Using Fig. 3.27, which is generated with

```
f[x_] = x + 2;
p1 = Plot[f[x], {x, 0, 4}, PlotStyle → {Thickness[.01], CMYKColor[1, .31, .08, .42]},
   PlotRange->{0, 6}];

p2 =
   Graphics[{CMYKColor[.65, .43, .26, .78],
   Line[{{1, 0}, {1, f[1]}, {4, f[1]}, {4, f[4]}, {4, f[4]}, {0, f[4]},
   {0, 0}, {1, 0}}]}];
```

```
p3 = Graphics[{Text["2", {.5, .2}], {Text["3", {3.8, 4.5}]}}];

p4 = Graphics[{CMYKColor[.62, .19, .45, .5], Dashing[{0.01, 0.01}],
  Line[{{0, f[0]}, {1, f[0]}}]}];

p5 = Graphics[{Text["θ", {.5, 2.25}], Text["θ", {1.5, 3.25}]}];

p6 = Graphics[{Text[x, {.9, 2.35}], Text[y, {2.5, 3.25}]}];

Show[p1, p2, p3, p4, p5, p6, Axes->None]
```

and the Pythagorean theorem, the total length of the beam is

$$L = \sqrt{2^2 + x^2} + \sqrt{y^2 + 3^2}.$$

By similar triangles,

$$\frac{y}{3} = \frac{2}{x} \quad \text{so} \quad y = \frac{6}{x}$$

and the length of the beam, L, becomes

$$L(x) = \sqrt{4 + x^2} + \sqrt{9 + \frac{36}{x^2}}, \quad 0 < x < \infty.$$

Observe that the length of the longest beam is obtained by *minimizing* L. (Why?)

We ignore negative and imaginary values because length must be nonnegative real number.

```
Clear[l];
  l[x_] = Sqrt[2^2 + x^2] + Sqrt[y^2 + 3^2]/.y->6/x
```

$\sqrt{9 + \frac{36}{x^2}} + \sqrt{4 + x^2}$

We use two different methods to solve $L'(x) = 0$. Differentiating

```
l'[x]
```

$-\frac{36}{\sqrt{9+\frac{36}{x^2}}x^3} + \frac{x}{\sqrt{4+x^2}}$

```
36^2
```

1296

```
Solve[-12√(4 + x^2) + x^4 √((4+x^2)/x^2)==0, x]
```

$\{\{x \to -2i\}, \{x \to 2i\}, \{x \to -2^{2/3}3^{1/3}\}, \{x \to 2^{2/3}3^{1/3}\}\}$

```
p1 = x^8(9 + 36/x^2) - 1296(4 + x^2)//Expand//Factor
```

$9\left(4 + x^2\right)\left(-12 + x^3\right)\left(12 + x^3\right)$

and solving $L'(x)=0$ gives us

Solve[p1==0, *x*]

$\{\{x\to -2i\},\{x\to 2i\},\{x\to -(-3)^{1/3}2^{2/3}\},\{x\to (-3)^{1/3}2^{2/3}\},\{x\to -(-2)^{2/3}3^{1/3}\},$

$\{x\to (-2)^{2/3}3^{1/3}\},\{x\to -2^{2/3}3^{1/3}\},\{x\to 2^{2/3}3^{1/3}\}\}$

$N[2^{2/3}3^{1/3}]$

2.28943

$l[2^{2/3}3^{1/3}]$

$\sqrt{9+3\,2^{2/3}3^{1/3}}+\sqrt{4+2\,2^{1/3}3^{2/3}}$

***N*[%]**

7.02348

2(3/2)^(1/3)Sqrt[1 + (4/9)^(1/3)] + 3Sqrt[1 + (4/9)^(1/3)]//*N*

7.02348

It follows that the length of the beam is $L(2^{2/3}3^{1/3}) = \sqrt{9+3\cdot 2^{2/3}\cdot 3^{1/3}} + \sqrt{4+2\cdot 2^{1/3}\cdot 3^{2/3}} = \sqrt{13+9\cdot 2^{2/3}\cdot 3^{1/3}+6\cdot 2^{1/3}\cdot 3^{2/3}} \approx 7.02$. See Fig. 3.28.

Plot[*l*[*x*], {*x*, 0, 20}, PlotRange->{0, 20}, AspectRatio->Automatic,

PlotStyle → {Thickness[.01], CMYKColor[.62, .19, .45, .5]}, AxesLabel->{x, y}]

Our second approach uses right triangle trigonometry. In terms of θ, the length of the beam is given by

$$L(\theta)=2\csc\theta+3\sec\theta,\quad 0<\theta<\pi/2.$$

Differentiating gives us

$$L'(\theta)=-2\csc\theta\cot\theta+3\sec\theta\tan\theta.$$

To avoid typing the θ symbol, we define L as a function of t.

Clear[*l*]

***l*[t_] = 2Csc[*t*] + 3Sec[*t*]**

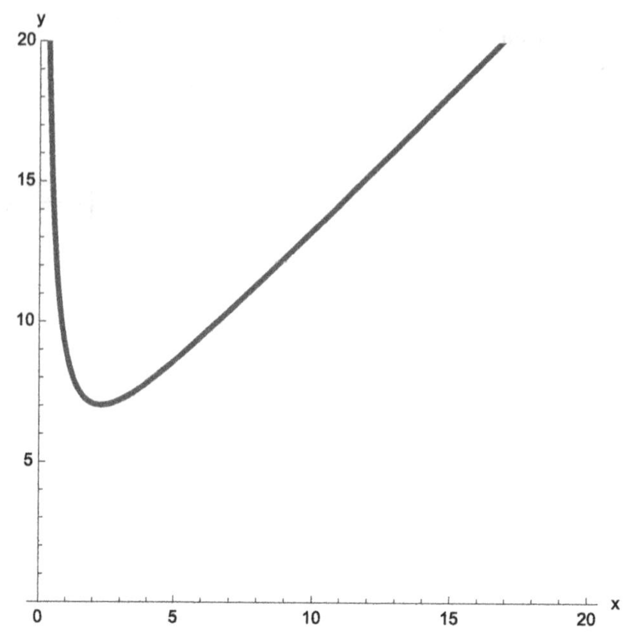

FIGURE 3.28 Graph of $L(x)$.

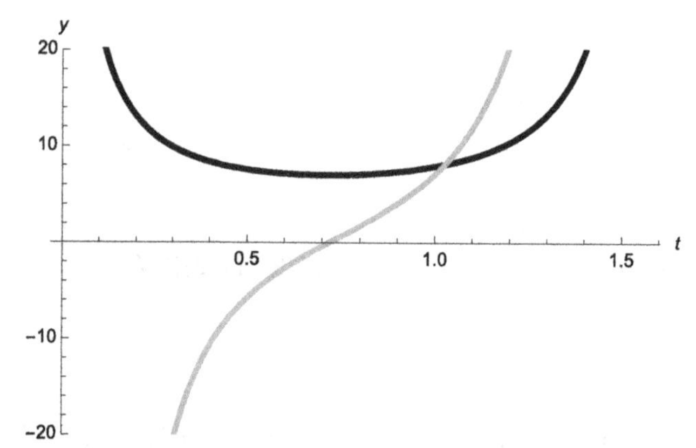

FIGURE 3.29 Graph of $L(\theta)$ and $L'(\theta)$.

2Csc[t] + 3Sec[t]

We now solve $L'(\theta) = 0$. First multiply through by $\sin\theta$ and then by $\tan\theta$.

$$\begin{aligned} 3\sec\theta\tan\theta &= 2\csc\theta\cot\theta \\ \tan^2\theta &= \frac{2}{3}\cot\theta \\ \tan^3\theta &= \frac{2}{3} \\ \tan\theta &= \sqrt[3]{\frac{2}{3}}. \end{aligned}$$

In this case, observe that we cannot compute θ exactly. However, we do not need to do so. Let $0 < \theta < \pi/2$ be the unique solution of $\tan\theta = \sqrt[3]{2/3}$. See Fig. 3.29. Using the identity $\tan^2\theta + 1 = \sec^2\theta$, we find that $\sec\theta = \sqrt{1 + \sqrt[3]{4/9}}$. Similarly, because $\cot\theta = \sqrt[3]{3/2}$ and

$\cot^2\theta + 1 = \csc^2\theta$, $\csc\theta = \sqrt[3]{3/2}\sqrt{1+\sqrt[3]{4/9}}$. Hence, the length of the beam is

$$L(\theta) = 2\sqrt[3]{\frac{3}{2}}\sqrt{1+\sqrt[3]{\frac{4}{9}}} + 3\sqrt{1+\sqrt[3]{\frac{4}{9}}} \approx 7.02.$$

```
Plot[Tooltip[{l[t], l'[t]}], {t, 0, Pi/2}, PlotRange->{−20, 20},
  PlotStyle->{{Thickness[.01], CMYKColor[.52, .59, .45, .9]},
  {Thickness[.01], CMYKColor[.03, .04, .14, .08]}},
   AxesLabel → {t, y}]
```

□

When you use Tooltip, scrolling the cursor over the plot will identify the plot for you.

In the next two examples, the constants do not have specific numerical values.

Example 3.26

Find the volume of the right circular cone of maximum volume that can be inscribed in a sphere of radius R.

Solution. Try to avoid three-dimensional figures unless they are absolutely necessary. For this problem, a cross-section of the situation is sufficient. See Fig. 3.30, which is created with

```
p1 = ParametricPlot[{Cos[t], Sin[t]}, {t, 0, 2Pi},
  PlotStyle → {{Thickness[.01], CMYKColor[0, .05, 1, 0]}}];
p2 = Graphics[{CMYKColor[.75, .35, 0, .07], Thickness[.005],
  Line[{{0, 1}, {Cos[4Pi/3], Sin[4Pi/3]}, {Cos[5Pi/3], Sin[5Pi/3]}, {0, 1}}],
  PointSize[.02], Point[{0, 0}], Line[{{Cos[4Pi/3], Sin[4Pi/3]}, {0, 0}, {0, 1}}],
  Line[{{0, 0}, {0, Sin[4Pi/3]}}]}];
p3 = Graphics[{Text[R, {−.256, −.28}], Text[R, {−.04, .5}], Text[y, {−.04, −.5}],
  Text[x, {−.2, −.8}]}];
Show[p1, p2, p3, AspectRatio->Automatic, Ticks->None, Axes->None]
```

The volume, V, of a right circular cone with radius r and height h is $V = \frac{1}{3}\pi r^2 h$. Using the notation in Fig. 3.30, the volume is given by

$$V = \frac{1}{3}\pi x^2 (R + y). \tag{3.4}$$

However, by the Pythagorean theorem, $x^2 + y^2 = R^2$ so $x^2 = R^2 - y^2$ and equation (3.4) becomes

$$V = \frac{1}{3}\pi\left(R^2 - y^2\right)(R + y) = \frac{1}{3}\pi\left(R^3 + R^2 y - R y^2 - y^3\right), \tag{3.5}$$

```
s1 = Expand[(r^2 − y^2)(r + y)]
```

$r^3 + r^2 y - r y^2 - y^3$

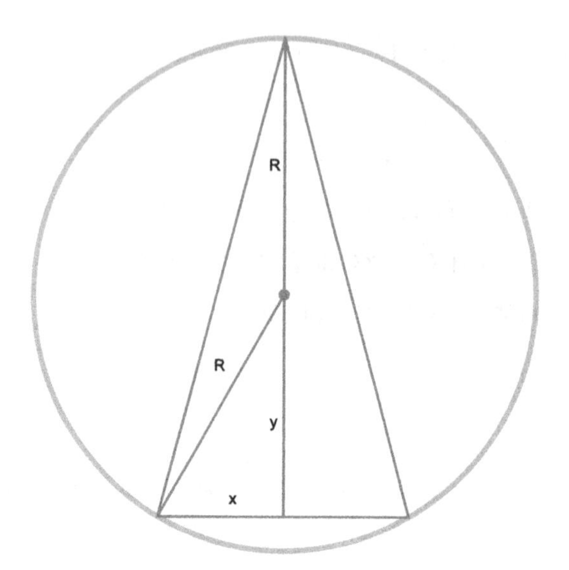

FIGURE 3.30 Cross-section of a right circular cone inscribed in a sphere (UCLA colors).

where $0 \le y \le R$. $V(y)$ is continuous on $[0, R]$ so it will have a minimum and maximum value on this interval. Moreover, the minimum and maximum values either occur at the end-points of the interval or at the critical numbers on the interior of the interval. Differentiating equation (3.5) with respect to y gives us

Remember that $R > 0$ is a constant.

$$\frac{dV}{dy} = \frac{1}{3}\pi\left(R^2 - 2Ry - 3y^2\right) = \frac{1}{3}\pi(R-3y)(R+y)$$

s2 = *D*[s1, y]

$r^2 - 2ry - 3y^2$

and we see that $dV/dy = 0$ if $y = \frac{1}{3}R$ or $y = -R$.

Factor[s2]

$(r - 3y)(r + y)$

Solve[s2 == 0, y]

$\{\{y \to -r\}, \{y \to \frac{r}{3}\}\}$

We ignore $y = -R$ because $-R$ is not in the interval $[0, R]$. Note that $V(0) = V(R) = 0$. The maximum volume of the cone is

$$V\left(\frac{1}{3}R\right) = \frac{1}{3}\pi \cdot \frac{32}{27}R^3 = \frac{32}{81}\pi R^2 \approx 1.24R^3.$$

s3 = s1/.y->r/3//Together

$\frac{32r^3}{27}$

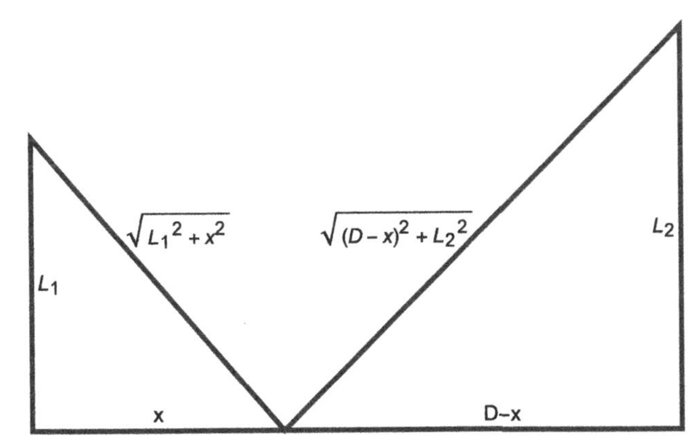

FIGURE 3.31 When the wire is stayed to minimize the length, the result is two similar triangles (Texas A & M colors).

s3 ∗ 1/3Pi

$\frac{32\pi r^3}{81}$

N[%]

$1.24112r^3$ □

Example 3.27: The Stayed-Wire Problem

Two poles D feet apart with heights L_1 feet and L_2 feet are to be stayed by a wire as shown in Fig. 3.31. Find the minimum amount of wire required to stay the poles, as illustrated in Fig. 3.31, which is generated with

```
p1 = Graphics[{CMYKColor[.15, 1, .39, .69], Thickness[.0075],
  Line[{{0, 0}, {0, 4}, {3.5, 0}, {9, 5.5}, {9, 0}, {0, 0}}]}];
p2 = Graphics[{Text["L1", {.2, 2}], Text["L2", {8.8, 2.75}], Text[x, {1.75, .2}],
  Text[x, {1.75, .2}], Text["√(L1^2 + x^2)", {1.75, 2.75}],
  Text["√((D − x)^2 + L2^2)", {5.5, 2.75}], Text["D-x", {6.5, .2}]}];
Show[p1, p2]
```

Solution. Using the notation in Fig. 3.31, the length of the wire, L, is

$$L(x) = \sqrt{L_1^2 + x^2} + \sqrt{L_2^2 + (D-x)^2}, \qquad 0 \le x \le D. \tag{3.6}$$

In the special case that $L_1 = L_2$, the length of the wire to stay the beams is minimized when the wire is placed halfway between the two beams, at a distance $D/2$ from each beam. Thus, we assume that the lengths of the beams are different; we assume that $L_1 < L_2$, as illustrated in Fig. 3.31. We compute $L'(x)$ and then solve $L'(x) = 0$. We use `PowerExpand` because `PowerExpand[expr]` expands out all products and powers assuming the variables are real and positive. That is, with `PowerExpand` we obtain that $\sqrt{x^2} = x$ rather than $\sqrt{x^2} = |x|$.

Clear[l]

l[x_] = Sqrt[x^2 + l1^2] + Sqrt[(d − x)^2 + l1^2]

$\sqrt{l1^2 + (d - x)^2} + \sqrt{l1^2 + x^2}$

l'[x]//Together

$\frac{-d\sqrt{l1^2+x^2}+x\sqrt{l1^2+x^2}+x\sqrt{d^2+l1^2-2dx+x^2}}{\sqrt{l1^2+x^2}\sqrt{d^2+l1^2-2dx+x^2}}$

Solve[l'[x]==0, x]

$\{\{x \to \frac{d}{2}\}\}$

The result indicates that $x = L_1 D/(L_1 + L_2)$ minimizes $L(x)$. (Note that we ignore the other value because $L_1 - L_2 < 0$.) Moreover, the triangles formed by minimizing L are similar triangles.

Clear[l]

l[x_] = Sqrt[x^2 + l1^2] + Sqrt[(d − x)^2 + l1^2]

$\sqrt{l1^2 + (d - x)^2} + \sqrt{l1^2 + x^2}$

l'[x]//Together

$\frac{-d\sqrt{l1^2+x^2}+x\sqrt{l1^2+x^2}+x\sqrt{d^2+l1^2-2dx+x^2}}{\sqrt{l1^2+x^2}\sqrt{d^2+l1^2-2dx+x^2}}$

l[0]//PowerExpand

$l1 + \sqrt{d^2 + l1^2}$

l[d]//PowerExpand

$l1 + \sqrt{d^2 + l1^2}$

Solve[l'[x]==0, x]

$\{\{x \to \frac{d}{2}\}\}$

l[d/2]//Together//PowerExpand

$\sqrt{d^2 + 4l1^2}$

$\boldsymbol{\sqrt{d^2 + 4l1^2}}$

$\sqrt{d^2 + 4l1^2}$

l1 $/\left(\frac{d\, l1}{l1+l2}\right)$ **//Simplify**

$\frac{l1+l2}{d}$

l2 $/\left(d-\frac{d\, l1}{l1+l2}\right)$ **//Simplify**

$\frac{l1+l2}{d}$

□

3.2.7 Antidifferentiation

3.2.7.1 Antiderivatives

$F(x)$ is an **antiderivative** of $f(x)$ if $F'(x)=f(x)$. The symbols

$$\int f(x)\,dx$$

mean "find all antiderivatives of $f(x)$." Because all antiderivatives of a given function differ by a constant, we usually find an antiderivative, $F(x)$, of $f(x)$ and then write

$$\int f(x)\,dx = F(x)+C,$$

where C represents an arbitrary constant. The command

```
Integrate[f[x],x]
```

attempts to find an antiderivative, $F(x)$, of $f(x)$. Instead of using `Integrate`, you might prefer to use the $\int \blacksquare d\square$ button on the **Basic Math Input** or **Basic Math Assistant** palettes to help you evaluate antiderivatives. Mathematica does not include the "$+C$" that we include when writing $\int f(x)\,dx = F(x)+C$. In the same way as `D` can differentiate many functions, `Integrate` can antidifferentiate many functions. However, antidifferentiation is a fundamentally difficult procedure so it is not difficult to find functions $f(x)$ for which the command `Integrate[f[x],x]` returns unevaluated.

Example 3.28

Evaluate each of the following antiderivatives: (a) $\int \frac{1}{x^2}e^{1/x}\,dx$, (b) $\int x^2\cos x\,dx$, (c) $\int x^2\sqrt{1+x^2}\,dx$, (d) $\int \frac{x^2-x+2}{x^3-x^2+x-1}\,dx$, and (e) $\int \frac{\sin x}{x}\,dx$.

Solution. Entering

Integrate[1/*x*^2Exp[1/*x*], *x*]

$-e^{\frac{1}{x}}$

shows us that $\int \frac{1}{x^2}e^{1/x}\,dx = -e^{1/x}+C$. To use the $\int \blacksquare d\square$ button, first click on the button, fill in the blanks, and press Enter.

```
∫□ d□
□^2/2
∫1/x^2 Exp[1/x] d□
e^(1/x) □/x^2
∫1/x^2 Exp[1/x] dx
-e^(1/x)
∫1/x^2 Exp[1/x] dx
-e^(1/x)
```

Notice that Mathematica does not automatically include the arbitrary constant, C. When computing several antiderivatives, you can use `Map` to apply `Integrate` to a list of antiderivatives. However, because `Integrate` is threadable,

```
Map[Integrate[#,x]&,list]
```

returns the same result as `Integrate[list,x]`, which we illustrate to compute (b), (c), and (d).

Integrate[{x^2Cos[x], x^2Sqrt[1 + x^2],

(x^2 − x + 2)/(x^3 − x^2 + x − 1)}, x]

$\left\{2x\text{Cos}[x] + \left(-2 + x^2\right)\text{Sin}[x], \frac{1}{8}\left(\sqrt{1 + x^2}\left(x + 2x^3\right) - \text{ArcSinh}[x]\right),\right.$

$\left.-\text{ArcTan}[x] + \text{Log}[1 - x]\right\}$

For (e), we see that there is not a "closed form" antiderivative of $\int \frac{\sin x}{x} dx$ and the result is given in terms of a definite integral, the **sine integral function**:

$$Si(x) = \int_0^x \frac{\sin t}{t} dt.$$

Integrate[Sin[x]/x, x]

SinIntegral[x] □

u-Substitutions

Usually, the first antidifferentiation technique discussed is the method of **u-substitution**. Suppose that $F(x)$ is an antiderivative of $f(x)$. Given

$$\int f(g(x))\, g'(x)\, dx,$$

we let $u = g(x)$ so that $du = g'(x)\,dx$. Then,

$$\int f(g(x))\,g'(x)\,dx = \int f(u)\,du = F(u) + C = F(g(x)) + C,$$

where $F(x)$ is an antiderivative of $f(x)$. After mastering u-substitutions, the **integration by parts formula**,

$$\int u\,dv = uv - \int v\,du, \tag{3.7}$$

is introduced.

Example 3.29

Evaluate $\int 2^x \sqrt{4^x - 1}\,dx$.

Solution. We use `Integrate` to evaluate the antiderivative.

i1 = Integrate[2^xSqrt[4^x − 1], x]

$$\frac{2^x\sqrt{-1+4^x} - \text{Log}\left[2^x + \sqrt{-1+4^x}\right]}{\text{Log}[4]}$$

Proceeding by hand, we let $u = 2^x$. Then, $du = 2^x \ln 2\,dx$ or, equivalently, $\dfrac{1}{\ln 2}du = 2^x\,dx$

D[2^x, x]

$2^x\text{Log}[2]$

so $\int 2^x\sqrt{4^x-1}\,dx = \dfrac{1}{\ln 2}\int\sqrt{u^2-1}\,du$. We now use `Integrate` to evaluate $\frac{1}{\ln 2}\int\sqrt{u^2-1}\,du$

i2 = 1/Log[2]Integrate[Sqrt[u^2 − 1], u]

$$\frac{\frac{1}{2}u\sqrt{-1+u^2} - \frac{1}{2}\text{Log}\left[u+\sqrt{-1+u^2}\right]}{\text{Log}[2]}$$

Simplify[i2]

$$\frac{u\sqrt{-1+u^2} - \text{Log}\left[u+\sqrt{-1+u^2}\right]}{\text{Log}[4]}$$

and then `/.` (`ReplaceAll`)/ to replace u with 2^x.

i3 = i2/.u → 2^x

$$\frac{2^{-1+x}\sqrt{-1+2^{2x}} - \frac{1}{2}\text{Log}\left[2^x+\sqrt{-1+2^{2x}}\right]}{\text{Log}[2]}$$

Observe that the result we obtained by hand is the same as the result obtained by `Integrate` directly. Sometimes, the results will *look* different and have slightly different forms. To verify that they are equivalent, subtract the two, and simplify the result. If the result is a constant, the two antiderivatives are equivalent. If not, they aren't. □

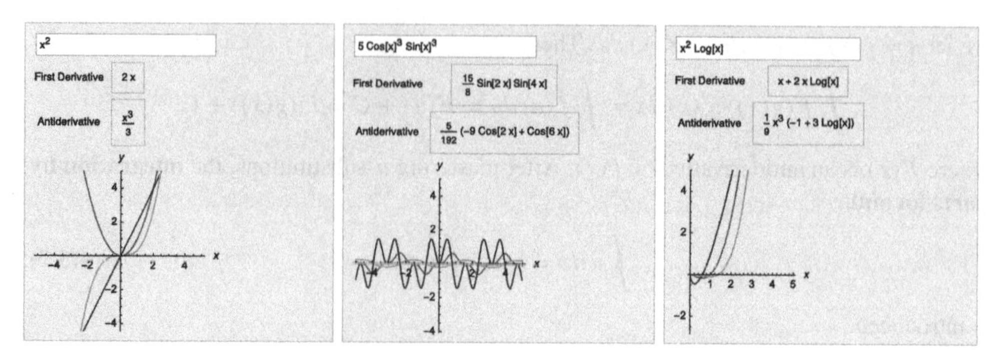

FIGURE 3.32 Seeing the relationship between the derivative and antiderivative of a function and the original function (Texas A & M colors).

As we did with derivatives, with `DynamicModule`, we create a simple dynamic that lets you compute the derivative and antiderivative of basic functions and plot them on a standard viewing window, $[-5, 5] \times [-5, 5]$. The layout of Fig. 3.32 is primarily determined by `Panel`, `Column`, and `Grid`.

```
Panel[DynamicModule[{f = x^2},
  Column[{InputField[Dynamic[f]], Grid[{{"First Derivative",
Panel[Dynamic[D[f, x]//Simplify]]},
  {"Antiderivative",
Panel[Dynamic[Integrate[f, x]//Simplify]]}}],
  Dynamic[Plot[Evaluate[Tooltip[{f, D[f, x],
  Integrate[f, x]}]], {x, -5, 5}, PlotRange → {-5, 5},
  AspectRatio → Automatic, AxesLabel → {x, y},
  PlotStyle → {{CMYKColor[.15, 1, .39, .69]}, {CMYKColor[1, .48, .09, .46]},
  {CMYKColor[.46, .23, .84, .68]}}]]}]], ImageSize → {300, 300}]
```

3.3 INTEGRAL CALCULUS

3.3.1 Area

In integral calculus courses, the definite integral is frequently motivated by investigating the area under the graph of a positive continuous function on a closed interval. Let $y = f(x)$ be a nonnegative continuous function on an interval $[a, b]$ and let n be a positive integer. If we divide $[a, b]$ into n subintervals of equal length and let $\left[x_{k-1}, x_k\right]$ denote the kth subinterval, the length of each subinterval is $(b-a)/n$ and $x_k = a + k\frac{b-a}{n}$. The area bounded by the graphs of $y = f(x)$, $x = a$, $x = b$, and the y-axis can be approximated with the sum

$$\sum_{k=1}^{n} f\left(x_k{}^*\right)\frac{b-a}{n}, \tag{3.8}$$

where $x_k{}^* \in [x_{k-1}, x_k]$. Typically, we take $x_k{}^* = x_{k-1} = a + (k-1)\frac{b-a}{n}$ (the left endpoint of the kth subinterval), $x_k{}^* = x_{k-1} = a + k\frac{b-a}{n}$ (the right endpoint of the kth subinterval),

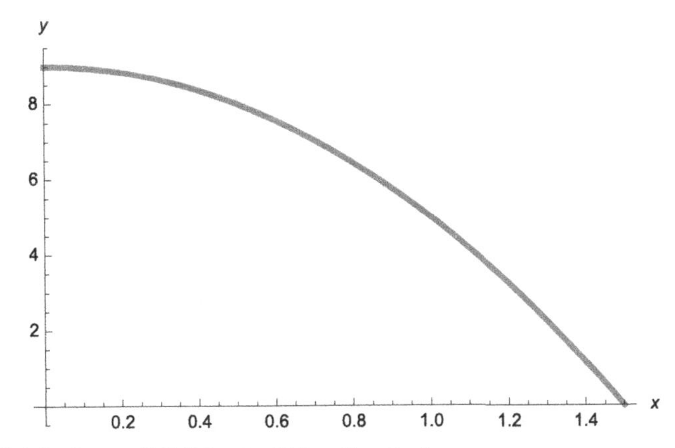

FIGURE 3.33 $f(x)$ for $0 \le x \le 3/2$ (Princeton University colors).

or $x_k{}^* = \frac{1}{2}(x_{k-1} + x_k) = a + \frac{1}{2}(2k-1)\frac{b-a}{n}$ (the midpoint of the kth subinterval). For these choices of $x_k{}^*$, (3.8) becomes

$$\frac{b-a}{n}\sum_{k=1}^{n} f\left(a + (k-1)\frac{b-a}{n}\right) \tag{3.9}$$

$$\frac{b-a}{n}\sum_{k=1}^{n} f\left(a + k\frac{b-a}{n}\right), \text{ and} \tag{3.10}$$

$$\frac{b-a}{n}\sum_{k=1}^{n} f\left(a + \frac{1}{2}(2k-1)\frac{b-a}{n}\right), \tag{3.11}$$

respectively. If $y = f(x)$ is increasing on $[a, b]$, (3.9) is an under approximation and (3.10) is an upper approximation: (3.9) corresponds to an approximation of the area using n inscribed rectangles; (3.10) corresponds to an approximation of the area using n circumscribed rectangles. If $y = f(x)$ is decreasing on $[a, b]$, (3.10) is an under approximation and (3.9) is an upper approximation: (3.10) corresponds to an approximation of the area using n inscribed rectangles; (3.9) corresponds to an approximation of the area using n circumscribed rectangles.

In the following example, we define the functions `leftsum[f[x],a,b,n]`, `middlesum[f[x],a,b,n]`, and `rightsum[f[x],a,b,n]` to compute (3.9), (3.11), and (3.10), respectively, and `leftbox[f[x],a,b,n]`, `middlebox[f[x], a,b,n]`, and `rightbox[f[x],a, b,n]` to generate the corresponding graphs. After you have defined these functions, you can use them with functions $y = f(x)$ that you define.

Remark 3.6. To define a function of a single variable, $f(x) = expression\, in\, x$, enter `f[x_]=expression in x`. To generate a basic plot of $y = f(x)$ for $a \le x \le b$, enter `Plot[f[x],{x,a,b}]`.

Example 3.30

Let $f(x) = 9 - 4x^2$. Approximate the area bounded by the graph of $y = f(x)$, $x = 0$, $x = 3/2$, and the y-axis using (a) 100 inscribed and (b) 100 circumscribed rectangles. (c) What is the exact value of the area?

Solution. We begin by defining and graphing $y = f(x)$ in Fig. 3.33.

```
f[x_]=9-4x^2;
Plot[f[x], {x, 0, 3/2},
```

```
AxesLabel → {x, y},
PlotStyle → {{CMYKColor[0, .61, .97, 0], Thickness[.01]}}]
```

The first derivative, $f'(x) = -8x$ is negative on the interval so $f(x)$ is decreasing on $[0, 3/2]$. Thus, an approximation of the area using 100 inscribed rectangles is given by (3.10) while an approximation of the area using 100 circumscribed rectangles is given by (3.9). After defining `leftsum`, `rightsum`, and `middlesum`, these values are computed using `leftsum` and `rightsum`. The use of `middlesum` is illustrated as well. Approximations of the sums are obtained with `N`.

`N[number]` returns a numerical approximation of `number`.

```
leftsum[f_, a_, b_, n_]:=Module[{},
    (b − a)/nSum[f/.x->a + (k − 1)(b − a)/n, {k, 1, n}]];
rightsum[f_, a_, b_, n_]:=Module[{},
    (b − a)/nSum[f/.x->a + k(b − a)/n, {k, 1, n}]];
middlesum[f_, a_, b_, n_]:=Module[{},
    (b − a)/nSum[f/.x->a + 1/2(2k − 1)(b − a)/n, {k, 1, n}]];
```

```
l100 = leftsum[f[x], 0, 3/2, 100]
N[%]
r100 = rightsum[f[x], 0, 3/2, 100]
N[%]
m100 = middlesum[f[x], 0, 3/2, 100]
N[%]
```

$\frac{362691}{40000}$

9.06728

$\frac{357291}{40000}$

8.93228

$\frac{720009}{80000}$

9.00011

Observe that these three values appear to be close to 9. In fact, 9 is the exact value of the area of the region bounded by $y = f(x)$, $x = 0$, $x = 3/2$, and the y-axis. To help us see why this is true, we define `leftbox`, `middlebox`, and `rightbox`, and then use these functions to visualize the situation using $n = 4$, 16, and 32 rectangles in Fig. 3.34.

It is not important that you understand the syntax of these three functions at this time. Once you have entered the code, you can use them to visualize the process for your own functions, $y = f(x)$.

```
leftbox[f_, a_, b_, n_, opts___]:=Module[{z, p1, recs, ls},
    z[k_] = a + (b − a)k/n;
```

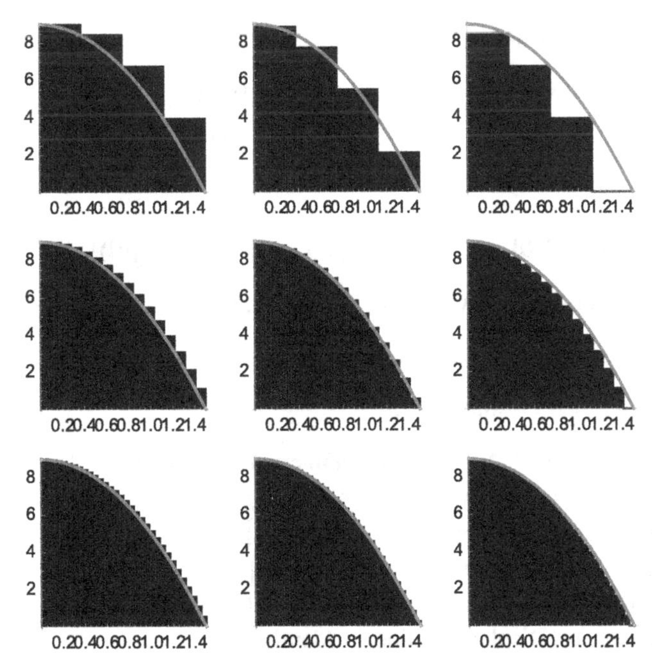

FIGURE 3.34 $f(x)$ with 4, 16, and 32 rectangles.

```
p1 = Plot[f, {x, a, b}, PlotRange → All,
  PlotStyle->{{Thickness[.02], CMYKColor[0, .61, .97, 0]}}];
  recs = Table[Rectangle[{z[k − 1], 0}, {z[k], f/.x->z[k − 1]}], {k, 1, n}];
  ls = Table[Line[{{z[k − 1], 0}, {z[k − 1], f/.x->z[k − 1]}, {z[k], f/.x->z[k − 1]},
  {z[k], 0}}], {k, 1, n}];
  Show[Graphics[{Black, recs}], Graphics[ls], p1, opts, Axes->Automatic,
  AspectRatio → 1]]

rightbox[f_, a_, b_, n_, opts___]:=Module[{z, p1, recs, ls},
    z[k_] = a + (b − a)k/n;
p1 = Plot[f, {x, a, b}, PlotRange → All,
  PlotStyle->{{Thickness[.02], CMYKColor[0, .61, .97, 0]}}];
  recs = Table[Rectangle[{z[k − 1], 0}, {z[k], f/.x->z[k]}], {k, 1, n}];
  ls = Table[Line[{{z[k − 1], 0}, {z[k − 1], f/.x->z[k]}, {z[k], f/.x->z[k]},
  {z[k], 0}}], {k, 1, n}];
  Show[Graphics[{Black, recs}], Graphics[ls], p1, opts,
  Axes->Automatic, AspectRatio → 1]]

middlebox[f_, a_, b_, n_, opts___]:=Module[{z, p1, recs, ls},
    z[k_] = a + (b − a)k/n;
```

```
p1 = Plot[f, {x, a, b}, PlotRange → All,
PlotStyle->{{Thickness[.02], CMYKColor[0, .61, .97, 0]}}];
recs = Table[Rectangle[{z[k − 1], 0}, {z[k], f/.x->1/2(z[k − 1] + z[k])}],
{k, 1, n}];
ls = Table[Line[{{z[k − 1], 0}, {z[k − 1], f/.x->1/2(z[k − 1] + z[k])},
   {z[k], f/.x->1/2(z[k − 1] + z[k])}, {z[k], 0}}], {k, 1, n}];
   Show[Graphics[{Black, recs}], Graphics[ls], p1, opts,
   Axes->Automatic, AspectRatio → 1]]
```

```
somegraphs = {{leftbox[f[x], 0, 3/2, 4, DisplayFunction → Identity],
   middlebox[f[x], 0, 3/2, 4, DisplayFunction → Identity],
   rightbox[f[x], 0, 3/2, 4, DisplayFunction → Identity]},
   {leftbox[f[x], 0, 3/2, 16, DisplayFunction → Identity],
   middlebox[f[x], 0, 3/2, 16, DisplayFunction → Identity],
   rightbox[f[x], 0, 3/2, 16, DisplayFunction → Identity]},
   {leftbox[f[x], 0, 3/2, 32, DisplayFunction → Identity],
   middlebox[f[x], 0, 3/2, 32, DisplayFunction → Identity],
   rightbox[f[x], 0, 3/2, 32, DisplayFunction → Identity]}};
Show[GraphicsGrid[somegraphs]]
```

```
somegraphs = {{leftbox[f[x], 0, 3/2, 4],
middlebox[f[x], 0, 3/2, 4],
rightbox[f[x], 0, 3/2, 4]},
{leftbox[f[x], 0, 3/2, 16],
middlebox[f[x], 0, 3/2, 16],
rightbox[f[x], 0, 3/2, 16]},
{leftbox[f[x], 0, 3/2, 32],
middlebox[f[x], 0, 3/2, 32],
rightbox[f[x], 0, 3/2, 32]}};
Show[GraphicsGrid[somegraphs]]
```

Notice that as n increases, the under approximations increase to the value of the area while the upper approximations decrease to the value of the area. In the limit as $n \to \infty$, if the two limits are equal, we can conclude that the area is the value of the limits.
These graphs help convince us that the limit of the sum as $n \to \infty$ of the areas of the inscribed and circumscribed rectangles is the same. We compute the exact value of (3.9) with `leftsum`, evaluate and simplify the sum with `Simplify`, and compute the limit as $n \to \infty$ with `Limit`. We see that the limit is 9.

```
ls = leftsum[f[x], 0, 3/2, n]

ls2 = Simplify[ls]
```

Limit[ls2, n->Infinity]

$$\frac{9(-1+3n+4n^2)}{4n^2}$$

$$\frac{9(-1+3n+4n^2)}{4n^2}$$

9

Similar calculations are carried out for (3.10) and again we see that the limit is 9. We conclude that the exact value of the area is 9.

rs = rightsum[f[x], 0, 3/2, n]

rs2 = Simplify[rs]

Limit[rs2, n->Infinity]

$$\frac{9(-1-3n+4n^2)}{4n^2}$$

$$\frac{9(-1-3n+4n^2)}{4n^2}$$

9

For illustrative purposes, we confirm this result with `middlesum`.

ms = middlesum[f[x], 0, 3/2, n]

ms2 = Simplify[ms]

Limit[ms2, n->Infinity]

$$\frac{9(1+8n^2)}{8n^2}$$

$$9+\frac{9}{8n^2}$$

9

As illustrated earlier, with `Manipulate`, you can experiment with different functions and different n values. First, we define a set of "typical" functions.

quad[x_] = 100 − x^2;

cubic[x_] = 4/9x^3 − 49/9x^2 + 100;

rational[x_] = 100/(x^2 + 1);

root[x_] = Sqrt[10 − x];

sin[x_] = 75Sin[Pix/5];

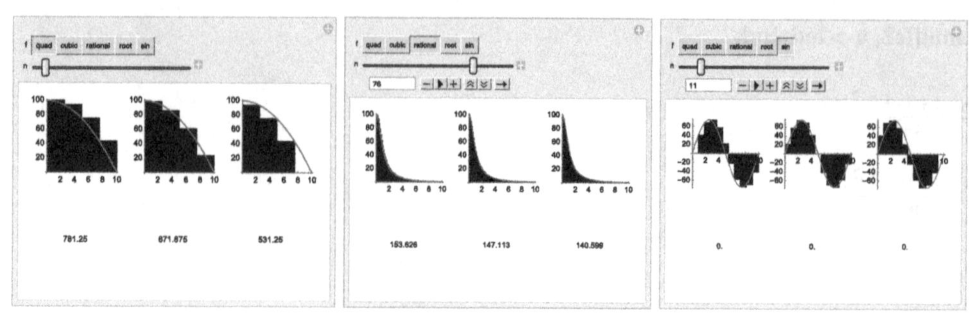

FIGURE 3.35 With `Manipulate`, we can investigate Riemann sum approximations and their graphical representations for various functions.

Next, we use `Manipulate` to create an object that allows us to experiment with how "typical" functions react to changes in n using left, middle, and right-hand endpoint approximations for computations of Riemann sums. In the resulting `Manipulate` object, $n = 4$ rectangles is the default; you can choose n-values from 0 to 100. The value of the corresponding Riemann sum is shown below the graphic. See Fig. 3.35.

How does the `Manipulate` object change if you remove `Transpose` from the command?

```
Manipulate[Show[GraphicsGrid[{{leftbox[f[x], 0, 10, n],
  Graphics[{Inset[leftsum[f[x], 0, 10, n]//N, {0, 0}]}]},
  {middlebox[f[x], 0, 10, n],
  Graphics[{Inset[middlesum[f[x], 0, 10, n]//N, {0, 0}]}]},
   {rightbox[f[x], 0, 10, n],
   Graphics[{Inset[rightsum[f[x], 0, 10, n]//N, {0, 0}]}]}}//
   Transpose]], {{f, quad}, {quad, cubic, rational, root, sin}},
  {{n, 4}, 0, 100, 1}]
```

□

3.3.2 The Definite Integral

In integral calculus courses, we formally learn that the **definite integral** of the function $y = f(x)$ from $x = a$ to $x = b$ is

$$\int_a^b f(x)\,dx = \lim_{|P| \to 0} \sum_{k=1}^{n} f\left(x_k{}^*\right) \Delta x_k, \tag{3.12}$$

provided that the limit exists. In equation (3.12), $P = \{a = x_0 < x_1 < x_2 < \cdots < x_n = b\}$ is a partition of $[a, b]$, $|P|$ is the **norm** of P,

$$|P| = \max\{x_k - x_{k-1} | k = 1, 2, \ldots, n\},$$

$\Delta x_k = x_k - x_{k-1}$, and $x_k{}^* \in \left[x_{k-1}, x_k\right]$.

The Fundamental Theorem of Calculus provides the fundamental relationship between differentiation and integration.

Theorem 3.6 (The Fundamental Theorem of Calculus). *Assume that $y = f(x)$ is continuous on $[a, b]$.*

1. *If $F(x) = \int_a^x f(t)\,dt$, then $F(x)$ is an antiderivative of $f(x)$: $F'(x) = f(x)$ or, equivalently,*

$$\frac{d}{dx}\left(\int_a^x f(t)\,dt\right) = f(x).$$

2. *If $G(x)$ is any antiderivative of $f(x)$, then $\int_a^b f(x)\,dx = G(b) - G(a)$.*

By the Fundamental Theorem of Calculus and the Chain Rule, it follows that

$$\frac{d}{dx}\left(\int_{g(x)}^{h(x)} f(t)\,dt\right) = f(h(x)) \cdot h'(x) - f(g(x)) \cdot g'(x).$$

Mathematica's `Integrate` command can compute many definite integrals. The command

```
Integrate[f[x],{x,a,b}]
```

attempts to compute $\int_a^b f(x)\,dx$ while

```
Integrate[f[x],x]
```

attempts to find an antiderivative of $f(x)$. Remember that Mathematica does not include the "$+C$" when computing antiderivatives. Because integration is a fundamentally difficult procedure, it is easy to create integrals for which the exact value cannot be found explicitly. In those cases, use `N` to obtain an approximation of its value or obtain a numerical approximation of the integral directly with

```
NIntegrate[f[x],{x,a,b}].
```

In the same way as you use the $\int \blacksquare\, d\square$ button to compute antiderivatives, you can use the $\int_\square^\square \blacksquare\, d\square$ button to compute definite integrals. If the result returned is unevaluated, use `N` to obtain a numerical approximation of the value of the integral or use `NIntegrate`.

Example 3.31

Evaluate (a) $\int_1^4 \left(x^2+1\right)/\sqrt{x}\,dx$; (b) $\int_0^{\sqrt{\pi/2}} x\cos x^2\,dx$; (c) $\int_0^{\pi} e^{2x}\sin^2 2x\,dx$; (d) $\int_0^1 \frac{2}{\sqrt{\pi}}e^{-x^2}\,dx$; and (e) $\int_{-1}^0 \sqrt[3]{u}\,du$.

Solution. We evaluate (a)–(c) directly with `Integrate`.

Integrate[(x^2 + 1)/Sqrt[x], {x, 1, 4}]

$\frac{72}{5}$

Integrate[xCos[x^2], {x, 0, Sqrt[Pi/2]}]

$\frac{1}{2}$

Integrate[Exp[2x]Sin[2x]^2, {x, 0, Pi}]

$\frac{1}{5}\left(-1 + e^{2\pi}\right)$

For (d), the result returned is in terms of the **error function**, `Erf[x]`, which is defined by the integral

$$\text{Erf[x]} = \frac{2}{\sqrt{\pi}} \int_0^x e^{-t^2}\, dt.$$

Integrate[2/Sqrt[Pi]Exp[−*x*^2], {*x*, 0, 1}]

Erf[1]

We use `N` to obtain an approximation of the value of the definite integral.

Integrate[2/Sqrt[Pi]Exp[−*x*^2], {*x*, 0, 1}]//*N*

0.842701

(e) Recall that Mathematica does not return a real number when we compute odd roots of negative numbers so the following result would be surprising to many students in an introductory calculus course because it is complex.

Integrate[*u*^(1/3), {*u*, −1, 0}]

$\frac{3}{4}(-1)^{1/3}$

Use `Surd[u,n]` to return the real nth root of u, $\sqrt[n]{u}$, if n is odd.

Integrate[Surd[*u*, 3], {*u*, −1, 0}]

$-\frac{3}{4}$ □

Improper integrals are computed using `Integrate` in the same way as with definite integrals.

Example 3.32

Evaluate (a) $\int_0^1 \frac{\ln x}{\sqrt{x}}\,dx$; (b) $\int_{-\infty}^{\infty} \frac{2}{\sqrt{\pi}} e^{-x^2}\,dx$; (c) $\int_1^{\infty} \frac{1}{x\sqrt{x^2-1}}\,dx$; (d) $\int_0^{\infty} \frac{1}{x^2+x^4}\,dx$; (e) $\int_2^4 \frac{1}{\sqrt[3]{(x-3)^2}}\,dx$; and (f) $\int_{-\infty}^{\infty} \frac{1}{x^2+x-6}\,dx$.

Solution. (a) This is an improper integral because the integrand is discontinuous on the interval [0, 1] but we see that the improper integral converges to −4.

Integrate[Log[*x*]/Sqrt[*x*], {*x*, 0, 1}]

−4

(b) This is an improper integral because the interval of integration is infinite but we see that the improper integral converges to 2.

Integrate[2/Sqrt[Pi]Exp[−x^2], {x, −Infinity, Infinity}]

2

(c) This is an improper integral because the integrand is discontinuous on the interval of integration and because the interval of integration is infinite but we see that the improper integral converges to $\pi/2$.

Integrate[1/(xSqrt[x^2 − 1]), {x, 1, Infinity}]

$\frac{\pi}{2}$

(d) As with (c), this is an improper integral because the integrand is discontinuous on the interval of integration and because the interval of integration is infinite but we see that the improper integral diverges to ∞.

```
Integrate[1/(x^2+x^4), {x, 0, Infinity}]
```

··· Integrate: Integral of $\frac{1}{x^2+x^4}$ does not converge on {0, ∞}.

$$\int_0^\infty \frac{1}{x^2+x^4}\,dx$$

(e) Recall that Mathematica does not return a real number when we compute odd roots of negative numbers so the following result would be surprising to many students in an introductory calculus course because it contains imaginary numbers.

Integrate[1/(x − 3)^(2/3), {x, 2, 4}]

$3 - 3(-1)^{1/3}$

Therefore, we use `Surd` to obtain the real-valued third root of $x - 3$.

Integrate[Surd[1/(x − 3)^2, 3], {x, 2, 4}]

6

(f) In this case, Mathematica warns us that the improper integral diverges.

```
s1 = Integrate[1/(x^2+x-6), {x, -Infinity, Infinity}]
```

··· Integrate: Integral of $\frac{1}{-6+x+x^2}$ does not converge on {−∞, ∞}.

$$\int_{-\infty}^{\infty} \frac{1}{-6+x+x^2}\,dx$$

To help us understand why the improper integral diverges, we note that $\frac{1}{x^2+x-6} = \frac{1}{5}\left(\frac{1}{x-2} - \frac{1}{x+3}\right)$ and

$$\int \frac{1}{x^2+x-6}\,dx = \int \frac{1}{5}\left(\frac{1}{x-2} - \frac{1}{x+3}\right)dx = \frac{1}{5}\ln\left(\frac{x-2}{x+3}\right) + C.$$

Integrate[1/(x^2+x−6), x]

$\frac{1}{5}$Log[2 − x] − $\frac{1}{5}$Log[3 + x]

Hence the integral is improper because the interval of integration is infinite and because the integrand is discontinuous on the interval of integration so

$$\begin{aligned}\int_{-\infty}^{\infty}\frac{1}{x^2+x-6}\,dx &= \int_{-\infty}^{-4}\frac{1}{x^2+x-6}\,dx+\int_{-4}^{-3}\frac{1}{x^2+x-6}\,dx\\ &+\int_{-3}^{0}\frac{1}{x^2+x-6}\,dx+\int_{0}^{2}\frac{1}{x^2+x-6}\,dx\\ &+\int_{2}^{3}\frac{1}{x^2+x-6}\,dx+\int_{3}^{\infty}\frac{1}{x^2+x-6}\,dx\end{aligned}\tag{3.13}$$

```
Integrate[1/(x^2+x-6), {x, -4, -3}]
```

··· Integrate: Integral of $\frac{1}{-6+x+x^2}$ does not converge on {−4, −3}.

$\int_{-4}^{-3}\frac{1}{-6+x+x^2}\,dx$

```
Integrate[1/(x^2+x-6), {x, -3, 0}]
```

··· Integrate: Integral of $\frac{1}{-6+x+x^2}$ does not converge on {−3, 0}.

$\int_{-3}^{0}\frac{1}{-6+x+x^2}\,dx$

```
Integrate[1/(x^2+x-6), {x, 0, 2}]
```

··· Integrate: Integral of $\frac{1}{-6+x+x^2}$ does not converge on {0, 2}.

$\int_{0}^{2}\frac{1}{-6+x+x^2}\,dx$

Evaluating each of these integrals,

```
Integrate[1/(x^2+x-6), {x, 2, 3}]
```

··· Integrate: Integral of $\frac{1}{-6+x+x^2}$ does not converge on {2, 3}.

$\int_{2}^{3}\frac{1}{-6+x+x^2}\,dx$

```
Integrate[1/(x^2+x-6), {x, 3, Infinity}]
```

$\frac{\text{Log}[6]}{5}$

```
Integrate[2 Pi f[x] Sqrt[1+f'[x]^2], {x, 1, Infinity}]
```

··· Integrate: Integral of $\frac{2\pi\sqrt{1+x^4}}{x^3}$ does not converge on {1, ∞}.

$\int_{1}^{\infty}\frac{2\pi\sqrt{1+\frac{1}{x^4}}}{x}\,dx$

we conclude that the improper integral diverges because at least one of the improper integrals in (3.13) diverges. □

In many cases, Mathematica can help illustrate the steps carried out when computing integrals using standard methods of integration like u-substitutions and integration by parts.

Example 3.33

Evaluate (a) $\int_e^{e^3} \frac{1}{x\sqrt{\ln x}}\,dx$ and (b) $\int_0^{\pi/4} x \sin 2x\,dx$.

Solution. (a) We let $u = \ln x$. Then, $du = 1/x\,dx$ so

$$\int_e^{e^3} \frac{1}{x\sqrt{\ln x}}\,dx = \int_1^3 \frac{1}{\sqrt{u}}du = \int_1^3 u^{-1/2}du,$$

which we evaluate with `Integrate`.

The new lower limit of integration is 1 because if $x = e$, $u = \ln e = 1$. The new upper limit of integration is 3 because if $x = e^3$, $u = \ln e^3 = 3$.

```
Integrate[1/Sqrt[u], {u, 1, 3}]
```

$2\left(-1+\sqrt{3}\right)$

To evaluate (b), we let $u = x \Rightarrow du = dx$ and $dv = \sin 2x\,dx \Rightarrow v = -\frac{1}{2}\cos 2x$.

```
u = x;
dv = Sin[2x];
du = D[x, x]
v = Integrate[Sin[2x], x]
```

1

$-\frac{1}{2}\text{Cos}[2x]$

The results mean that

$$\begin{aligned}\int_0^{\pi/4} x \sin 2x\,dx &= -\frac{1}{2}x\cos 2x\Big]_0^{\pi/4} + \frac{1}{2}\int_0^{\pi/4} \cos 2x\,dx \\ &= 0 + \frac{1}{2}\int_0^{\pi/4} \cos 2x\,dx.\end{aligned}$$

The resulting indefinite integral is evaluated with `Integrate`.

```
Integrate[xSin[2x], x]
```

$-\frac{1}{2}x\text{Cos}[2x] + \frac{1}{4}\text{Sin}[2x]$

Alternatively, we can illustrate the integration by parts calculation, $\int u\,dv = uv - v\int du$.

```
uv - Integrate[vdu, x]
```

$-\frac{1}{2}x\text{Cos}[2x] + \frac{1}{4}\text{Sin}[2x]$

We use `Integrate` to evaluate the definite integral.

Integrate[xSin[2x], {x, 0, Pi/4}]

$\frac{1}{4}$

□

3.3.3 Approximating Definite Integrals

Because integration is a fundamentally difficult procedure to produce *exact* answers or results, Mathematica is unable to compute a "closed form" of the value of many definite integrals. In these cases, numerical integration can be used to obtain an approximation of the definite integral using `N` together with `Integrate` or `NIntegrate`:

`NIntegrate[f[x],{x,a,b}]`

attempts to approximate $\int_a^b f(x)\,dx$.

Example 3.34

Evaluate $\int_0^{\sqrt[3]{\pi}} e^{-x^2} \cos x^3\, dx$.

We use the [button] button to complete the `Integrate` command.

Solution. In this case, Mathematica is unable to evaluate the integral with `Integrate`.

i1 = Integrate[Exp[−x^2]Cos[x^3], {x, 0, Pi^(1/3)}]

$\int_0^{\pi^{1/3}} e^{-x^2}\text{Cos}\left[x^3\right] dx$

An approximation is obtained with `N`.

***N*[i1]**

0.701566

Instead of using `Integrate` followed by `N`, you can use `NIntegrate` to numerically evaluate the integral.

NIntegrate[Exp[−x^2]Cos[x^3], {x, 0, Pi^(1/3)}]

0.701566

returns the same result as that obtained using `Integrate` followed by `N`. □

In some cases, you may wish to investigate particular numerical methods that can be used to approximate integrals. To implement numerical methods like Simpson's rule or the trapezoidal rule, redefine the function `leftsum` (`middlesum` or `rightsum`) discussed previously to perform the calculation for the desired method.

3.3.4 Area

Suppose that $y = f(x)$ and $y = g(x)$ are continuous on $[a, b]$ and that $f(x) \geq g(x)$ for $a \leq x \leq b$. The **area** of the region bounded by the graphs of $y = f(x)$, $y = g(x)$, $x = a$, and

$x = b$ is

$$A = \int_a^b \left[f(x) - g(x)\right] dx. \tag{3.14}$$

Sometimes determining the "greater" function and the "lower" function can be difficult. Equation (3.14) in its more general form tells us that the region bounded by the graphs of $y = f(x)$, $y = g(x)$, $x = a$, and $x = b$ is

$$A = \int_a^b |f(x) - g(x)| \, dx. \tag{3.15}$$

Example 3.35

Find the area between the graphs of $y = \sin x$ and $y = \cos x$ on the interval $[0, 2\pi]$.

Solution. We graph $y = \sin x$ and $y = \cos x$ on the interval $[0, 2\pi]$ in Fig. 3.36 with `Plot`. The graph of $y = \cos x$ is dashed. Observe that including the option `Filling->{1->{2}}` fills the region *between* the two plots.

```
Plot[{Sin[x], Cos[x]}, {x, 0, 2π},
  PlotStyle → {{CMYKColor[.03, 1, .66, .12], Thickness[.01]},
  {Thickness[.01], Dashing[{0.025}], CMYKColor[.05, .26, 1, .27]}}, Filling → {1 → {2}},
  AspectRatio → Automatic]
```

To find the upper and lower limits of integration, we must solve the equation $\sin x = \cos x$ for x. Observe that Mathematica returns all solutions to the equation that it can find. The results show us that there are infinitely many solutions to the equation.

```
Solve[Sin[x]==Cos[x], x]
```

$\left\{\left\{x \to \text{ConditionalExpression}\left[-\frac{3\pi}{4} + 2\pi C[1], C[1] \in \text{Integers}\right]\right\},\right.$
$\left.\left\{x \to \text{ConditionalExpression}\left[\frac{\pi}{4} + 2\pi C[1], C[1] \in \text{Integers}\right]\right\}\right\}$

For us the solutions of interest are valid for $0 \le x \le 2\pi$, which are $x = \pi/4$ and $x = 5\pi/4$. We check that these are valid solutions of $\sin x = \cos x$ with ==; in each case the returned result is `True`.

```
Sin[π/4]==Cos[π/4]
Sin[5π/4]==Cos[5π/4]
```

True

True

Hence, the area of the region between the graphs is given by

$$A = \int_0^{\pi/4} [\cos x - \sin x] \, dx + \int_{\pi/4}^{5\pi/4} [\sin x - \cos x] \, dx + \int_{5\pi/4}^{2\pi} [\cos x - \sin x] \, dx. \tag{3.16}$$

FIGURE 3.36 $y = \sin x$ and $y = \cos x$ on the interval $[0, 2\pi]$ (University of Wisconsin colors).

Notice that if we take advantage of symmetry we can simplify (3.16) to

$$A = 2\int_{\pi/4}^{5\pi/4} [\sin x - \cos x]\, dx. \tag{3.17}$$

We evaluate (3.17) with `Integrate` to see that the area of the region between the two graphs is $4\sqrt{2}$.

$$\int_0^{\frac{\pi}{4}} (\text{Cos}[x] - \text{Sin}[x])\, dx + \int_{\frac{\pi}{4}}^{\frac{5\pi}{4}} (\text{Sin}[x] - \text{Cos}[x])\, dx + \int_{\frac{5\pi}{4}}^{2\pi} (\text{Cos}[x] - \text{Sin}[x])\, dx$$

$4\sqrt{2}$ □

In cases when we cannot calculate the points of intersection of two graphs exactly, we can frequently use `FindRoot` to approximate the points of intersection.

Example 3.36

Let

$$p(x) = \frac{3}{10}x^5 - 3x^4 + 11x^3 - 18x^2 + 12x + 1$$

and

$$q(x) = -4x^3 + 28x^2 - 56x + 32.$$

Approximate the area of the region bounded by the graphs of $y = p(x)$ and $y = q(x)$.

Solution. After defining p and q, we graph them on the interval $[-1, 5]$ in Fig. 3.37 to obtain an initial guess of the intersection points of the two graphs.

When you use `Tooltip`, you can slide your cursor over a plot and the function being graphed is displayed.

```
Clear[p, q]
p[x_] = 3x^5/10 − 3x^4 + 11x^3 − 18x^2 + 12x + 1;
q[x_] = −4x^3 + 28x^2 − 56x + 32;
Plot[Tooltip[{p[x], q[x]}], {x, −1, 5},
  PlotStyle → {{Thickness[.01], CMYKColor[0, .91, .76, .06]},
  {Thickness[.01], CMYKColor[0, .15, .94, 0]}},
  AxesLabel → {x, y}, AspectRatio → 1]
```

The x-coordinates of the three intersection points are the solutions of the equation $p(x) = q(x)$. Although Mathematica can solve this equation exactly, approximate solutions are more useful for the problem and obtained with `NSolve`.

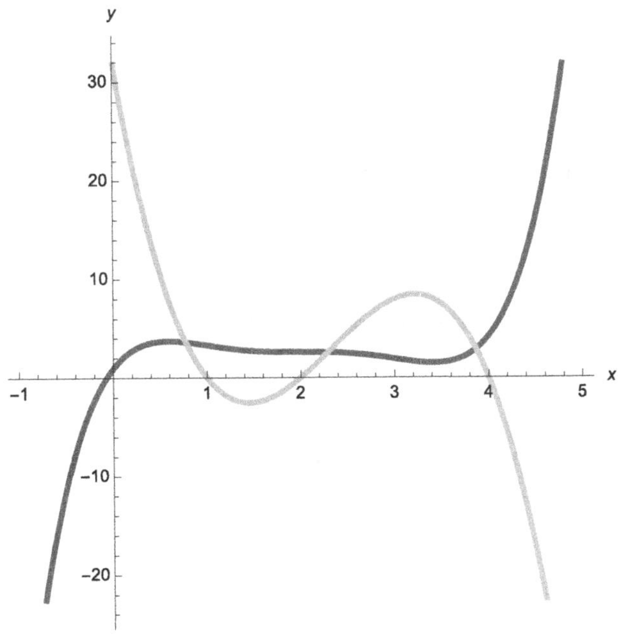

FIGURE 3.37 p and q on the interval $[-1, 5]$ (University of Maryland colors).

intpts = NRoots[p[x]==q[x], x]

$x == 0.772058 \| x == 1.5355 - 3.57094i \| x == 1.5355 + 3.57094i \| x == 2.29182 \| x ==$

3.86513

The numbers are extracted from the list with `Part` (`[[...]]`). For example, 0.772058 is the second part of the first part of `intpts`. Counting from left to right, 2.29182 is the second part of the fourth part of `intpts`.

x1 = intpts[[1, 2]]

x2 = intpts[[4, 2]]

x3 = intpts[[5, 2]]

0.772058

2.29182

3.86513

Using the roots to the equation $p(x) = q(x)$ and the graph we see that $p(x) \geq q(x)$ for $0.772 \leq x \leq 2.292$ and $q(x) \geq p(x)$ for $2.292 \leq x \leq 3.865$. Hence, an approximation of the area bounded by $p(x)$ and $q(x)$ is given by the sum

$$\int_{0.772}^{2.292} \left[p(x) - q(x)\right] dx + \int_{2.292}^{3.865} \left[q(x) - p(x)\right] dx.$$

These two integrals are computed with `Integrate` and `NIntegrate`. As expected, the two values are the same.

$\int_{x1}^{x2}(p[x]-q[x])\,dx+\int_{x2}^{x3}(q[x]-p[x])\,dx$

12.1951

```
NIntegrate[p[x] − q[x], {x, x1, x2}] + NIntegrate[q[x] − p[x], {x, x2, x3}]
```

12.1951

We conclude that the area is approximately 12.195 units^2. □

Parametric Equations

If the curve, C, defined parametrically by $x=x(t)$, $y=y(t)$, $a \le t \le b$ is a nonnegative continuous function of x and $x(a)<x(b)$ the area under the graph of C and above the x-axis is

$$\int_{x(a)}^{x(b)} y\,dx = \int_a^b y(t)x'(t)dt.$$

Graphically, y is a function of x, $y=y(x)$, if the graph of $y=y(x)$ passes the vertical line test.

Example 3.37: The Astroid

Find the area enclosed by the **astroid** $x=\sin^3 t$, $y=\cos^3 t$, $0\le t\le 2\pi$.

Solution. We begin by defining x and y and then graphing the astroid with `ParametricPlot` in Fig. 3.38.

```
x[t_] = Sin[t]^3;
y[t_] = Cos[t]^3;
ParametricPlot[{x[t], y[t]}, {t, 0, 2Pi}, AspectRatio->Automatic,
  PlotStyle → {Thickness[.01], CMYKColor[.07, .94, .65, .25]}]
```

Observe that $x(0)=0$ and $x(\pi/2)=1$ and the graph of the astroid in the first quadrant is given by $x=\sin^3 t$, $y=\cos^3 t$, $0\le t\le \pi/2$. Hence, the area of the astroid in the first quadrant is given by

$$\int_0^{\pi/2} y(t)x'(t)\,dt = 3\int_0^{\pi/2} \sin^2 t\cos^4 t\,dt$$

and the total area is given by

$$A = 4\int_0^{\pi/2} y(t)x'(t)\,dt = 12\int_0^{\pi/2} \sin^2 t\cos^4 t\,dt = \frac{3}{8}\pi \approx 1.178,$$

which is computed with `Integrate` and then approximated with `N`.

```
area = 4Integrate[y[t]x'[t], {t, 0, Pi/2}]
```

$\frac{3\pi}{8}$

```
N[area]
```

1.1781 □

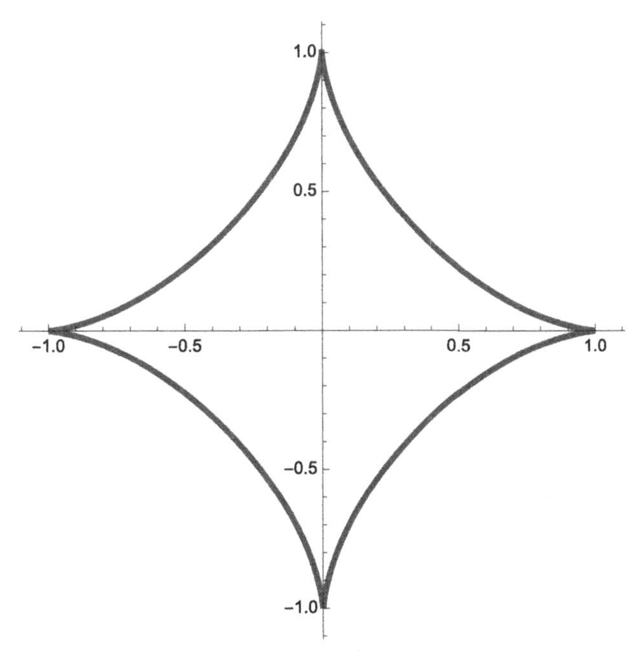

FIGURE 3.38 The astroid $x = \sin^3 t$, $y = \cos^3 t$, $0 \le t \le 2\pi$ (Harvard University colors).

Polar Coordinates

For problems involving "circular symmetry" it is often easier to work in polar coordinates. The relationship between (x, y) in rectangular coordinates and (r, θ) in polar coordinates is given by

$$x = r\cos\theta \qquad y = r\sin\theta$$

and

$$r^2 = x^2 + y^2 \qquad \tan\theta = \frac{y}{x}.$$

If $r = f(\theta)$ is continuous and nonnegative for $\alpha \le \theta \le \beta$, then the **area** A of the region enclosed by the graphs of $r = f(\theta)$, $\theta = \alpha$, and $\theta = \beta$ is

$$A = \frac{1}{2}\int_\alpha^\beta [f(\theta)]^2 \, d\theta = \frac{1}{2}\int_\alpha^\beta r^2 \, d\theta.$$

Example 3.38: Lemniscate of Bernoulli

The **lemniscate of Bernoulli** is given by

$$\left(x^2+y^2\right)^2 = a^2\left(x^2-y^2\right),$$

where a is a constant. (a) Graph the lemniscate of Bernoulli if $a = 2$. (b) Find the area of the region bounded by the lemniscate of Bernoulli.

Solution. This problem is much easier solved in polar coordinates so we first convert the equation from rectangular to polar coordinates with `ReplaceAll` (`/.`) and then solve for r with `Solve`.

```
lofb = (x^2 + y^2)^2==a^2(x^2 − y^2);

topolar = lofb/.{x->rCos[t], y->rSin[t]}
```

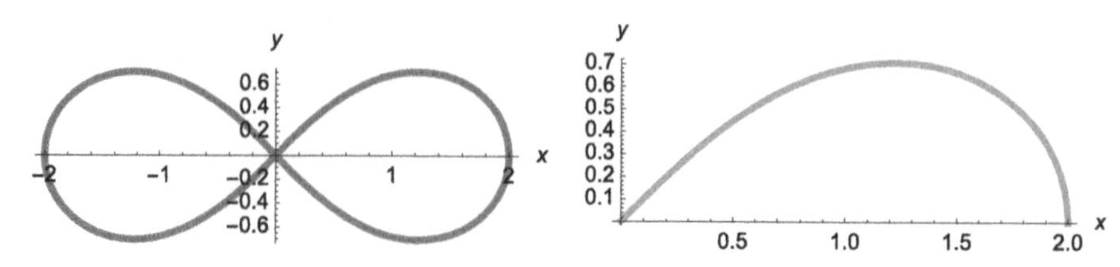

FIGURE 3.39 On the left, the lemniscate and on the portion of the lemniscate in quadrant 1 (Harvard University colors).

$$\left(r^2\text{Cos}[t]^2 + r^2\text{Sin}[t]^2\right)^2 == a^2\left(r^2\text{Cos}[t]^2 - r^2\text{Sin}[t]^2\right)$$

Solve[topolar, *r*]//Simplify

$$\left\{\{r \to 0\}, \{r \to 0\}, \left\{r \to -\sqrt{a^2\text{Cos}[2t]}\right\}, \left\{r \to \sqrt{a^2\text{Cos}[2t]}\right\}\right\}$$

These results indicate that an equation of the lemniscate in polar coordinates is $r^2 = a^2 \cos 2\theta$. The graph of the lemniscate is then generated in Fig. 3.39 (top figure) using `PolarPlot`. The portion of the lemniscate in quadrant one is obtained by graphing $r = 2\cos 2\theta$, $0 \leq \theta \leq \pi/4$.

```
p1 = PolarPlot[{-2Sqrt[Cos[2t]], 2Sqrt[Cos[2t]]}, {t, 0, 2Pi},
  PlotStyle → {Thickness[.015], CMYKColor[.16, .23, .23, .44]},
  AxesLabel → {x, y}];
p2 = PolarPlot[2Sqrt[Cos[2t]], {t, 0, Pi/4},
  PlotStyle → {Thickness[.015], CMYKColor[.33, .15, .14, .17]},
  AxesLabel → {x, y}];
Show[GraphicsRow[{p1, p2}]]
```

Then, taking advantage of symmetry, the area of the lemniscate is given by

$$A = 2 \cdot \frac{1}{2}\int_{-\pi/4}^{\pi/4} r^2\, d\theta = 2\int_0^{\pi/4} r^2\, d\theta = 2\int_0^{\pi/4} a^2 \cos 2\theta\, d\theta = a^2,$$

which we calculate with `Integrate`.

Integrate[2*a*^2Cos[2*t*], {*t*, 0, Pi/4}]

a^2

□

3.3.5 Arc Length

Let $y = f(x)$ be a function for which $f'(x)$ is continuous on an interval $[a, b]$. Then the **arc length** of the graph of $y = f(x)$ from $x = a$ to $x = b$ is given by

$$L = \int_a^b \sqrt{\left(\frac{dy}{dx}\right)^2 + 1}\, dx. \tag{3.18}$$

The resulting definite integrals used for determining arc length are usually difficult to compute because they involve a radical. In these situations, Mathematica is helpful with approximating solutions to these types of problems.

Example 3.39

Find the length of the graph of $y = \frac{x^4}{8} + \frac{1}{4x^2}$ from (a) $x = 1$ to $x = 2$ and from (b) $x = -2$ to $x = -1$.

Solution. With no restrictions on the value of x, $\sqrt{x^2} = |x|$. Generally, Mathematica does not automatically algebraically simplify $\sqrt{(dy/dx)^2 + 1}$ because Mathematica does not know if x is positive or negative.

```
y[x_] = x^4/8 + 1/(4x^2);
i1 = Factor[y'[x]^2 + 1]
```

$\frac{(1+x^2)^2(1-x^2+x^4)^2}{4x^6}$

```
i2 = PowerExpand[Sqrt[i1]]
```

$\frac{(1+x^2)(1-x^2+x^4)}{2x^3}$

In fact, for (b), x is negative so $\frac{1}{2}\sqrt{\frac{(x^6+1)^2}{x^6}} = -\frac{1}{2}\frac{x^6+1}{x^3}$. Mathematica simplifies $\frac{1}{2}\sqrt{\frac{(x^6+1)^2}{x^6}} = \frac{1}{2}\frac{x^6+1}{x^3}$ and correctly evaluates the arc length integral (3.18) for (a).

`PowerExpand[expr]` simplifies radicals in the expression `expr` assuming that all variables are positive.

```
Integrate[Sqrt[y'[x]^2 + 1], {x, 1, 2}]
```

$\frac{33}{16}$

For (b), we compute the arc length integral (3.18).

```
Integrate[Sqrt[y'[x]^2 + 1], {x, -2, -1}]
```

$\frac{33}{16}$

As we expect, both values are the same. □

Parametric Equations

If the smooth curve, C, defined parametrically by $x = x(t)$, $y = y(t)$, $t \in [a, b]$ is traversed exactly once as t increases from $t = a$ to $t = b$, the arc length of C is given by

$$L = \int_a^b \sqrt{\left(\frac{dx}{dt}\right)^2 + \left(\frac{dy}{dt}\right)^2}\, dt. \tag{3.19}$$

C is **smooth** if both $x'(t)$ and $y'(t)$ are continuous on (a, b) and not simultaneously zero for $t \in (a, b)$.

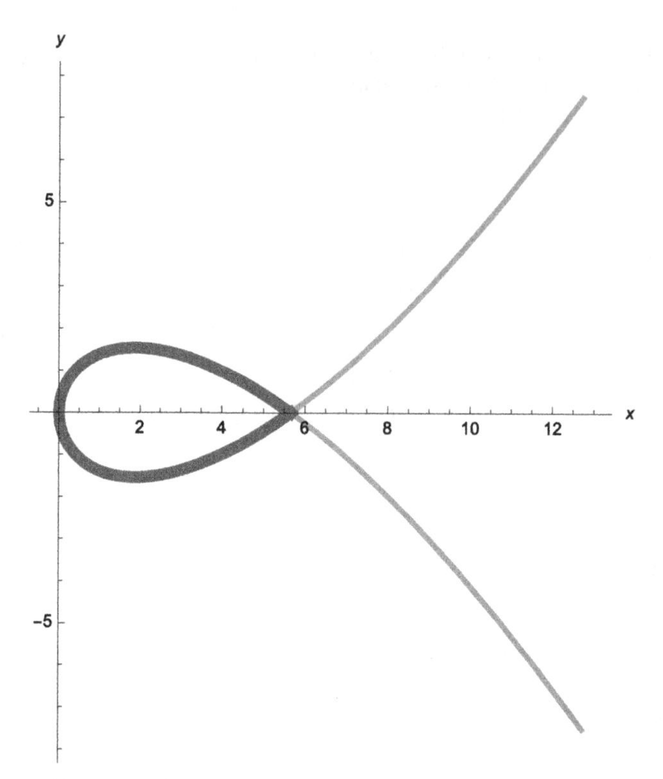

FIGURE 3.40 $x = \sqrt{2}t^2$, $y = 2t - \frac{1}{2}t^3$ (Pennsylvania State University colors).

Example 3.40

Find the length of the graph of $x = \sqrt{2}t^2$, $y = 2t - \frac{1}{2}t^3$, $-2 \le t \le 2$.

Solution. For illustrative purposes, we graph $x = \sqrt{2}t^2$, $y = 2t - \frac{1}{2}t^3$ for $-3 \le t \le 3$ and $-2 \le t \le 2$ (thickened) in Fig. 3.40.

```
x[t_] = t^2Sqrt[2]; y[t_] = 2t − 1/2t^3;
p1 = ParametricPlot[{x[t], y[t]}, {t, −3, 3},
   PlotStyle → {Thickness[.01], CMYKColor[.75, .2, 0, 0]}];
p2 = ParametricPlot[{x[t], y[t]}, {t, −2, 2},
   PlotStyle → {CMYKColor[.75, .66, 0, 0], Thickness[.02]}];
Show[p1, p2, PlotRange->All, AxesLabel → {x, y}]
```

Mathematica is able to compute the exact value of the arc length (3.19) although the result is quite complicated. For length considerations, the result of entering the `i1` command are not displayed here.

```
Factor[x'[t]^2 + y'[t]^2]
```

$\frac{1}{4}\left(4 - 4t + 3t^2\right)\left(4 + 4t + 3t^2\right)$

i1 = Intcgratc[2Sqrt[x'[t]^2 + y'[t]^2], {t, 0, 2}]

$$\int_0^2 2\sqrt{8t^2+\left(2-\frac{3t^2}{2}\right)^2}\,dt$$

A more meaningful approximation is obtained with `N` or using `NIntegrate`.

***N*[i1]**

13.7099

NIntegrate[2Sqrt[x'[t]^2 + y'[t]^2], {t, 0, 2}]

13.7099

We conclude that the arc length is approximately 13.71.
Observe that Mathematica 11 cannot evaluate the definite integral directly. However, if we first compute an anti-derivative with integrate in `i2`,

i2 = Integrate[2Sqrt[x'[t]^2 + y'[t]^2], t]

$$\left(27\sqrt{\frac{i}{i-2\sqrt{2}}}t\left(16+8t^2+9t^4\right)-16\left(i+2\sqrt{2}\right)\sqrt{\frac{4i-8\sqrt{2}+9it^2}{i-2\sqrt{2}}}\sqrt{\frac{4i+8\sqrt{2}+9it^2}{i+2\sqrt{2}}}\right.$$
$$\text{EllipticE}\left[i\text{ArcSinh}\left[\frac{3}{2}\sqrt{\frac{i}{i-2\sqrt{2}}}t\right],\frac{i-2\sqrt{2}}{i+2\sqrt{2}}\right]+32\left(-4i+\sqrt{2}\right)\sqrt{\frac{4i-8\sqrt{2}+9it^2}{i-2\sqrt{2}}}\sqrt{\frac{4i+8\sqrt{2}+9it^2}{i+2\sqrt{2}}}$$
$$\left.\text{EllipticF}\left[i\text{ArcSinh}\left[\frac{3}{2}\sqrt{\frac{i}{i-2\sqrt{2}}}t\right],\frac{i-2\sqrt{2}}{i+2\sqrt{2}}\right]\right)/\left(81\sqrt{\frac{i}{i-2\sqrt{2}}}\sqrt{16+8t^2+9t^4}\right)$$

and then apply the Fundamental Theorem of Calculus by subtracting the value of `i2` if $t = 0$ from the value of `i2` if $t = 2$,

ul = i2/.$t \to 2$

ll = i2/.$t \to 0$

val = ul − ll

$$\frac{1}{648\sqrt{\frac{3i}{i-2\sqrt{2}}}}(10368\sqrt{\frac{i}{i-2\sqrt{2}}}-16\sqrt{\frac{40i-8\sqrt{2}}{i-2\sqrt{2}}}(i+2\sqrt{2})\sqrt{\frac{40i+8\sqrt{2}}{i+2\sqrt{2}}}$$
$$\text{EllipticE}[i\text{ArcSinh}[3\sqrt{\frac{i}{i-2\sqrt{2}}}],\frac{i-2\sqrt{2}}{i+2\sqrt{2}}]+32\sqrt{\frac{40i-8\sqrt{2}}{i-2\sqrt{2}}}(-4i+\sqrt{2})\sqrt{\frac{40i+8\sqrt{2}}{i+2\sqrt{2}}}$$
$$\text{EllipticF}[i\text{ArcSinh}[3\sqrt{\frac{i}{i-2\sqrt{2}}}],\frac{i-2\sqrt{2}}{i+2\sqrt{2}}])$$

0

$$\frac{1}{648\sqrt{\frac{3i}{i-2\sqrt{2}}}}(10368\sqrt{\frac{i}{i-2\sqrt{2}}}-16\sqrt{\frac{40i-8\sqrt{2}}{i-2\sqrt{2}}}(i+2\sqrt{2})\sqrt{\frac{40i+8\sqrt{2}}{i+2\sqrt{2}}}$$
$$\text{EllipticE}[i\text{ArcSinh}[3\sqrt{\frac{i}{i-2\sqrt{2}}}],\frac{i-2\sqrt{2}}{i+2\sqrt{2}}]+32\sqrt{\frac{40i-8\sqrt{2}}{i-2\sqrt{2}}}(-4i+\sqrt{2})\sqrt{\frac{40i+8\sqrt{2}}{i+2\sqrt{2}}}$$
$$\text{EllipticF}[i\text{ArcSinh}[3\sqrt{\frac{i}{i-2\sqrt{2}}}],\frac{i-2\sqrt{2}}{i+2\sqrt{2}}])$$

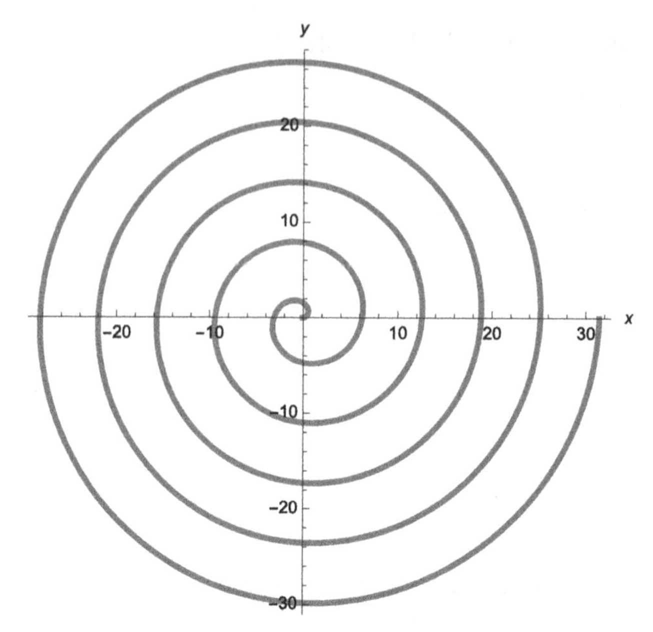

FIGURE 3.41 $r=\theta$ for $0 \le \theta \le 10\pi$ (Pennsylvania State University colors).

N[val]

13.7099 + 1.7763568394002505*^ − 15i

N[val]//Chop

13.7099

we obtain the same result. □

Polar Coordinates

If the smooth polar curve C given by $r = f(\theta)$, $\alpha \le \theta \le \beta$ is traversed exactly once as θ increases from α to β, the arc length of C is given by

$$L = \int_{\alpha}^{\beta} \sqrt{\left(\frac{dr}{d\theta}\right)^2 + r^2}\, d\theta \tag{3.20}$$

Example 3.41

Find the length of the graph of $r=\theta$, $0 \le \theta \le 10\pi$.

Solution. We begin by defining r and then graphing r with `PolarPlot` in Fig. 3.41.

```
r[t_]=t;
PolarPlot[r[t], {t, 0, 10Pi}, AspectRatio->Automatic,
  PlotStyle → {CMYKColor[0, .75, .6, 0], Thickness[.01]},
  AxesLabel → {x, y}]
```

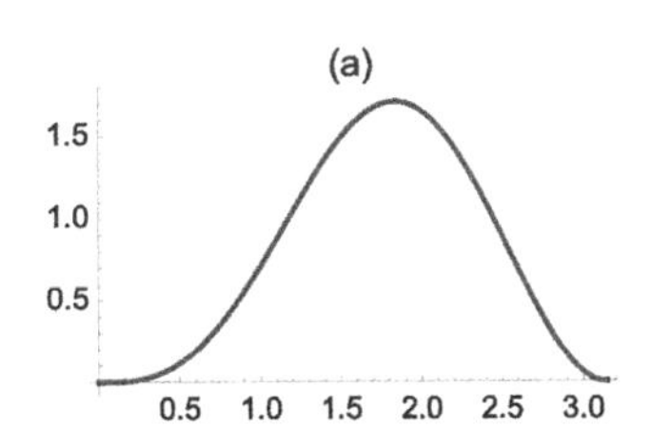

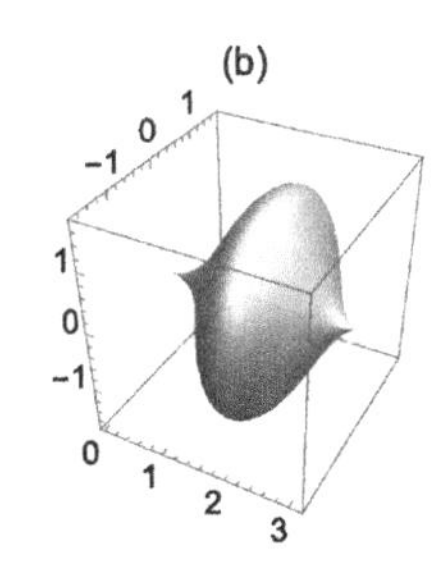

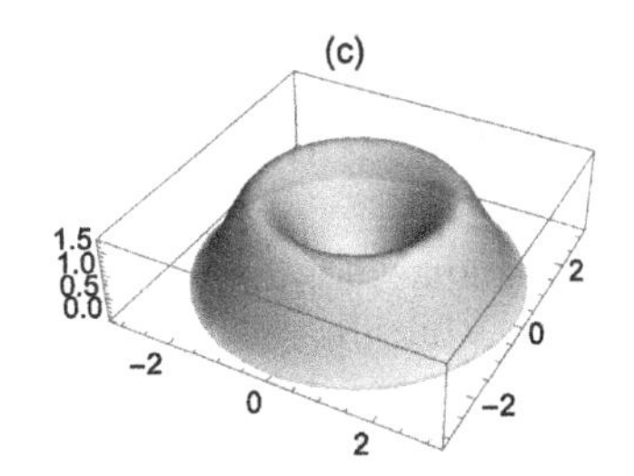

FIGURE 3.42 (a) $g(x)$ for $0 \le x \le \pi$. (b) $g(x)$ revolved about the x-axis. (c) $g(x)$ revolved about the y-axis (University of California–Santa Barbara colors).

Using (3.20), the length of the graph of r is given by $\int_0^{10\pi} \sqrt{1+\theta^2}\, d\theta$. The exact value is computed with `Integrate`

ev = Integrate[Sqrt[*r*′[*t*]^2 + *r*[*t*]^2], {*t*, 0, 10Pi}]

$5\pi\sqrt{1+100\pi^2} + \frac{1}{2}\text{ArcSinh}[10\pi]$

and then approximated with `N`.

***N*[ev]**

495.801

We conclude that the length of the graph is approximately 495.8 units. □

3.3.6 Solids of Revolution

Volume

Let $y = f(x)$ be a nonnegative continuous function on $[a, b]$. The **volume** of the solid of revolution obtained by revolving the region bounded by the graphs of $y = f(x)$, $x = a$, $x = b$, and the x-axis about the x-axis is given by

$$V = \pi \int_a^b [f(x)]^2\, dx. \tag{3.21}$$

If $0 \le a < b$, the **volume** of the solid of revolution obtained by revolving the region bounded by the graphs of $y = f(x)$, $x = a$, $x = b$, and the x-axis about the y-axis is given by

$$V = 2\pi \int_a^b x\, f(x)\, dx. \tag{3.22}$$

Example 3.42

Let $g(x) = x\sin^2 x$. Find the volume of the solid obtained by revolving the region bounded by the graphs of $y = g(x)$, $x = 0$, $x = \pi$, and the x-axis about (a) the x-axis; and (b) the y-axis.

Solution. After defining g, we graph g on the interval $[0, \pi]$ in Fig. 3.42 (a).

With Mathematica 11, for three dimensional graphics, you can adjust the viewpoint by clicking on the three-dimensional graphics object and dragging to the desired viewing angle.

g[x_] = *x*Sin[*x*]^2;

p1 = Plot[*g*[*x*], {*x*, 0, Pi}, AspectRatio->Automatic,

```
PlotLabel → "(a)",
PlotStyle → {{Thickness[.01], CMYKColor[1, .82, .17, .04]}}];
```

The volume of the solid obtained by revolving the region about the x-axis is given by equation (3.21) while the volume of the solid obtained by revolving the region about the y-axis is given by equation (3.22). These integrals are computed with `Integrate` and named `xvol` and `yvol`, respectively. We use `N` to approximate each volume.

```
xvol = Integrate[Pig[x]^2, {x, 0, Pi}]
N[xvol]
```

$\frac{1}{64}\pi^2(-15+8\pi^2)$

9.86295

```
yvol = Integrate[2Pixg[x], {x, 0, Pi}]
N[yvol]
```

$\frac{1}{6}\pi^2(-3+2\pi^2)$

27.5349

We can use `ParametricPlot3D` to visualize the resulting solids by parametrically graphing the equations given by

$$\begin{cases} x = r\cos t \\ y = r\sin t \\ z = g(r) \end{cases}$$

for r between 0 and π and t between $-\pi$ and π to visualize the graph of the solid obtained by revolving the region about the y-axis and by parametrically graphing the equations given by

$$\begin{cases} x = r \\ y = g(r)\cos t \\ z = g(r)\sin t \end{cases}$$

for r between 0 and π and t between $-\pi$ and π to visualize the graph of the solid obtained by revolving the region about the x-axis. (See Figs. 3.42 (b) and 3.42 (c).) In this case, we identify the z-axis as the y-axis. Notice that we are simply using polar coordinates for the x and y-coordinates, and the height above the x, y-plane is given by $z = g(r)$ because r is replacing x in the new coordinate system.

```
p2 = ParametricPlot3D[{r, g[r]Cos[t], g[r]Sin[t]}, {r, 0, Pi}, {t, 0, 2Pi},
  PlotPoints->{30, 30}, PlotLabel → "(b)",
  PlotStyle → {CMYKColor[0, .18, .89, 0], Specularity[White, 10]},
  Mesh → False];
```

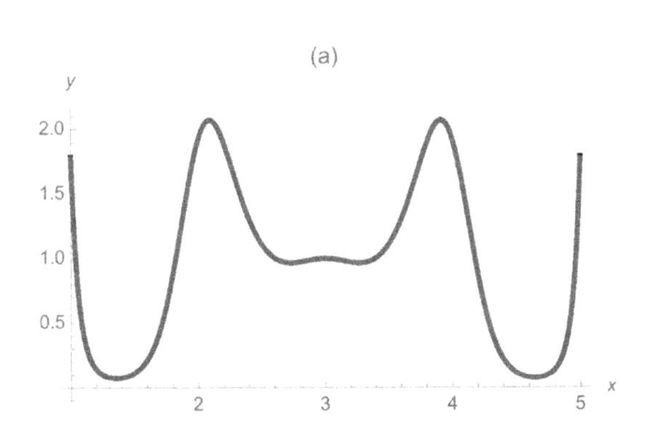

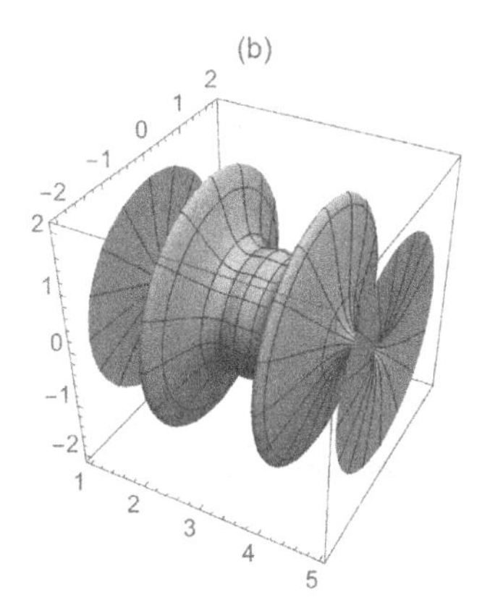

FIGURE 3.43 (a) $f(x)$ for $1 \leq x \leq 5$. (b) $f(x)$ revolved abut the x-axis (University of Southern California colors).

```
p3 = ParametricPlot3D[{rCos[t], rSin[t], g[r]}, {r, 0, Pi}, {t, 0, 2Pi},
  PlotPoints->{30, 30}, PlotLabel → "(c)",
  PlotStyle → {CMYKColor[1, .86, .36, .28],
  Opacity[.9], Specularity[White, 1]},
  Mesh → False];
```

p1, p2, and p3 are shown together side-by-side in Fig. 3.42 using `Show` together with `GraphicsRow`.

```
Show[GraphicsRow[{p1, p2, p3}]]
```

□

We now demonstrate a volume problem that requires the method of disks.

Example 3.43

Let $f(x) = e^{-(x-3)^2 \cos[4(x-3)]}$. Approximate the volume of the solid obtained by revolving the region bounded by the graphs of $y = f(x)$, $x = 1$, $x = 5$, and the x-axis about the x-axis.

Solution. Proceeding as in the previous example, we first define and graph f on the interval $[1, 5]$ in Fig. 3.43 (a).

```
f[x_] = Exp[-(x - 3)^2Cos[4(x - 3)]];
p1 = Plot[f[x], {x, 1, 5}, AspectRatio->Automatic,
  PlotLabel → "(a)", AxesLabel → {x, y},
  PlotStyle → {Thickness[.01], CMYKColor[.07, 1, .65, .32]}];
```

In this case, an approximation is desired so we use `NIntegrate` to approximate the integral $V = \int_1^5 \pi \left[f(x)\right]^2 dx$.

```
NIntegrate[Pi f[x]^2, {x, 1, 5}]
```

16.0762

In the same manner as before, `ParametricPlot3D` can be used to visualize the resulting solid by graphing the set of equations given parametrically by

$$\begin{cases} x = r \\ y = f(r)\cos t \\ z = f(r)\sin t \end{cases}$$

for r between 1 and 5 and t between 0 and 2π. In this case, polar coordinates are used in the y, z-plane with the distance from the x-axis given by $f(x)$. Because r replaces x in the new coordinate system, $f(x)$ becomes $f(r)$ in these equations. See Fig. 3.43 (b).

```
p2 = ParametricPlot3D[{r, f[r]Cos[t], f[r]Sin[t]}, {r, 1, 5}, {t, 0, 2Pi},
  PlotPoints->{45, 35}, PlotLabel → "(b)",
  PlotStyle → {CMYKColor[0, .27, 1, 0]}];
```

```
Show[GraphicsRow[{p1, p2}]]
```

□

When revolving a curve about the y-axis, you can use `RevolutionPlot3D` rather than the parametrization given previously.

Example 3.44

Let $f(x) = \exp\left(-2(x-2)^2\right) + \exp\left(-(x-4)^2\right)$ for $0 \le x \le 6$. (a) Find the minimum and maximum values of $f(x)$ on $[0, 6]$. Let R be the region bounded by $y = f(x)$, $x = 0$, $x = 6$, and the y-axis. (b) Find the volume of the solid obtained by revolving R about the y-axis. (c) Find the volume of the solid obtained by revolving R about the x-axis.

Solution. (a) Although `Maximize` and `Minimize` cannot find the exact maximum and minimum values, using `N` or `NMaximize` and `NMinimize` give accurate approximations.

`NMaximize` and `NMinimize` work in the same way as `Maximize` and `Minimize` but return approximations rather than exact results.

```
f[x_] = Exp[-2(x - 2)^2] + Exp[-(x - 4)^2];
```

```
Maximize[f[x], x]
```

$\{e^{-(-4+\text{Root}[\{-4+2\#1+e^{\#1^2}(-\frac{4}{e^8}+\frac{\#1}{e^8})\&,2.0196244769513376905\}])^2}$

$\times\, e^{-2(-2+\text{Root}[\{-4+2\#1+e^{\#1^2}(-\frac{4}{e^8}+\frac{\#1}{e^8})\&,2.0196244769513376905\}])^2}$

$(e^{(-4+\text{Root}[\{-4+2\#1+e^{\#1^2}(-\frac{4}{e^8}+\frac{\#1}{e^8})\&,2.0196244769513376905\}])^2}$

$+e^{2(-2+\text{Root}[\{-4+2\#1+e^{\#1^2}(-\frac{4}{e^8}+\frac{\#1}{e^8})\&,2.0196244769513376905\}])^2}), \{x \to \text{Root}[\{-4+2\#1$

$+e^{\#1^2}(-\frac{4}{e^8}+\frac{\#1}{e^8})\&, 2.0196244769513376905\}]\}\}$

```
Maximize[f[x], x]//N
```

{1.01903, {x → 2.01962}}

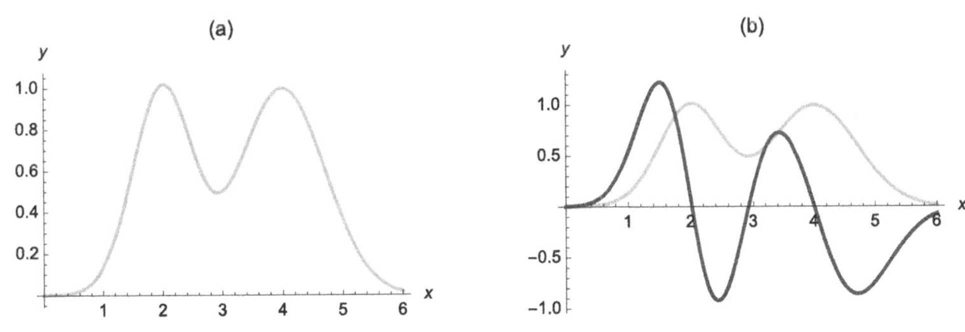

FIGURE 3.44 We use the graph of $f'(x)$ to help us estimate the initial values to approximate the critical numbers with `FindRoot` (University of Minnesota colors).

```
NMaximize[{f[x], 0 ≤ x ≤ 6}, x]
```

{1.00034, {x → 3.99864}}

```
Minimize[{f[x], 0 ≤ x ≤ 6}, x]
```

$\{\frac{e^8+e^{16}}{e^{24}}, \{x \to 0\}\}$

```
Minimize[{f[x], 0 ≤ x ≤ 6}, x]//N
```

{0.000335575, {x → 0.}}

```
NMinimize[{f[x], 0 ≤ x ≤ 6}, x]//N
```

{0.000335575, {x → 0.}}

We double check these results by graphing $f(x)$ and $f'(x)$ in Fig. 3.44 and then using `FindRoot` to approximate the critical numbers.

```
pf1 = Plot[f[x], {x, 0, 6}, PlotLabel → "(a)",
  AxesLabel → {x, y}, PlotStyle → {{Thickness[.01], CMYKColor[0, .16, 1, 0]}}]

pf2 = Plot[Tooltip[{f[x], f'[x]}], {x, 0, 6},
PlotStyle → {{Thickness[.01], CMYKColor[0, .16, 1, 0]},
{Thickness[.01], CMYKColor[0, 1, .63, .29]}},
PlotLabel → "(b)", AxesLabel → {x, y}]

Show[GraphicsRow[{pf1, pf2}]]

Map[FindRoot[f'[x] == 0, {x, #1}]&, {2, 3, 4}]
```

{{x → 2.01962}, {x → 2.92167}, {x → 3.99864}}

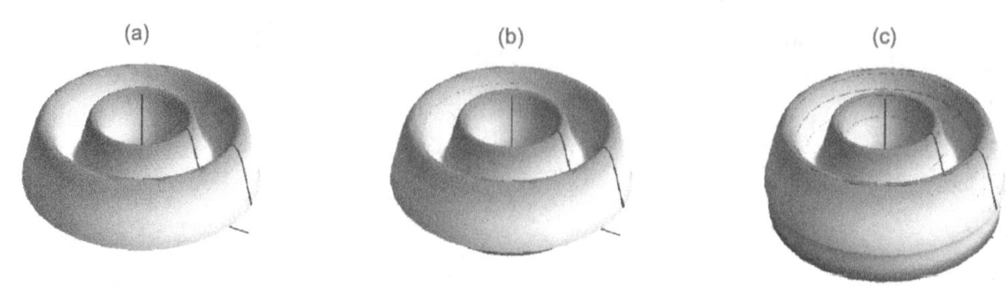

FIGURE 3.45 (a) The solid. (b) The solid with a "typical" shell. (c) Several shells.

(b) Mathematica finds the exact volume of the solids although the results are expressed in terms of the **Error function**, `Erf`.

Integrate[Pi*xf**[*x*], {*x*, 0, 6}]**

$$\frac{\pi(-1+2e^{16}+e^{24}-2e^{28}+2e^{32}\sqrt{\pi}(4\text{Erf}[2]+4\text{Erf}[4]+\sqrt{2}(\text{Erf}[2\sqrt{2}]+\text{Erf}[4\sqrt{2}])))}{4e^{32}}$$

NIntegrate[Pi*xf**[*x*], {*x*, 0, 6}]**

30.0673

Integrate[Pif**[*x*]^2, {*x*, 0, 6}]**

$$\frac{1}{12e^{8/3}}\pi^{3/2}(3e^{8/3}(\text{Erf}[4]+\text{Erf}[8]+\sqrt{2}(\text{Erf}[2\sqrt{2}]+\text{Erf}[4\sqrt{2}]))+4\sqrt{3}(\text{Erf}[\tfrac{8}{\sqrt{3}}]+\text{Erf}[\tfrac{10}{\sqrt{3}}]))$$

NIntegrate[Pif**[*x*]^2, {*x*, 0, 6}]**

7.1682

To visualize the solid revolved about the y-axis, we use `RevolutionPlot3D` in `p1`. We generate the curve in `p2`, a set of axes, and a representative "slice" of the curve. See Fig. 3.45 (a). After, we show the solid together with a representative shell. See Fig. 3.45 (b).

```
p1 = RevolutionPlot3D[f[x], {x, 1, 5},
  BoxRatios → {2, 2, 1}, PlotRange → {{−5, 5}, {−5, 5}, {0, 5/4}},
  Mesh → None, PlotStyle → Opacity[.4],
  ColorFunction → "LightTemperatureMap"];
p2 = ParametricPlot3D[{x, 0, Exp[−2(x − 2)^2] + Exp[−(x − 4)^2]},
  {x, 1, 5}, {t, 0, 2Pi},
  PlotStyle → Thickness[.05], BoxRatios → {2, 2, 1},
  Axes → Automatic, Boxed → False];
p3 = ParametricPlot3D[{x, 0, 0}, {x, −5, 5}, {t, 0, 2Pi},
  PlotStyle → {Gray, Thickness[.075]}, BoxRatios → {2, 2, 1},
```

```
  Axes → Automatic, Boxed → False];
p4 = ParametricPlot3D[{0, 0, x}, {x, 0, 5/4}, {t, 0, 2Pi},
  PlotStyle → {Gray, Thickness[.1]}, BoxRatios → {2, 2, 1},
  Axes → Automatic, Boxed → False];
p5 = Graphics3D[{Gray, Thickness[.01], Line[{{3.6, 0, 0},
  {3.6, 0, Exp[−2(3.6 − 2)^2] + Exp[−(3.6 − 4)^2]}}]}];
p6 = ParametricPlot3D[{3.6Cos[t], 3.6Sin[t], z}, {t, 0, 2Pi},
  {z, 0, Exp[−2(3.6 − 2)^2] + Exp[−(3.6 − 4)^2]}, Mesh → None,
  PlotStyle → Opacity[.8], ColorFunction → "TemperatureMap"];

g1 = Show[p1, p2, p3, p4, p5, Boxed → False, Axes → None]

g2 = Show[p1, p2, p3, p4, p5, p6, Boxed → False, Axes → None]

Show[GraphicsRow[{g1, g2}]]
```

Finally, we show the solid together with several shells in Fig. 3.45 (c).

```
sp = Table[j//N, {j, 1.1, 4.9, 3.8/14}];

p7 = Table[ParametricPlot3D[{sp[[i]]Cos[t], sp[[i]]Sin[t], z},
  {t, 0, 2Pi},
  {z, 0, Exp[−2(sp[[i]] − 2)^2] + Exp[−(sp[[i]] − 4)^2]},
  Mesh → None,
  PlotStyle → Opacity[.8],
  ColorFunction → "TemperatureMap"], {i, 1, Length[sp]}];

g3 = Show[p1, p2, p3, p4, p5, p7, Boxed → False, Axes → None]

Show[GraphicsRow[{g1, g2, g3}]]
```

You can use `Animate` to animate the process as illustrated in Fig. 3.46.

```
Animate[
p8 = ParametricPlot3D[{iCos[t], iSin[t], z}, {t, 0, 2Pi},
  {z, 0, Exp[−2(i − 2)^2] + Exp[−(i − 4)^2]}, Mesh → None,
  PlotStyle → Opacity[.8], ColorFunction → "TemperatureMap"];
  Show[p1, p2, p3, p4, p5, p8, Boxed → False, Axes → None],
  {i, 1.1, 4.9}]
```

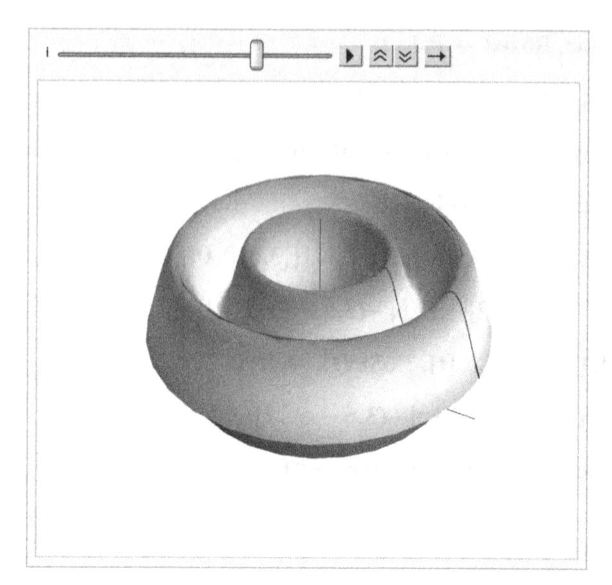

FIGURE 3.46 Using Animate to illustrate the finding the volume of a solid using the method of shells.

For revolving $f(x)$ about the x-axis, we proceed in much the same way. First, we plot $f(x)$ with a set of axes in three-space.

```
f[x_] = Exp[-2(x - 2)^2] + Exp[-(x - 4)^2];
p1 = ParametricPlot3D[{x, 0, f[x]}, {x, 0, 6}, PlotStyle → {Thick, Black},
    PlotRange → {{0, 6}, {-3/2, 3/2}, {-3/2, 3/2}}, BoxRatios → {1, 1, 1}];

p1b = ParametricPlot3D[{x, 0, -f[x]}, {x, 0, 6}, PlotStyle → {Thick, Black},
    PlotRange → {{0, 6}, {-3/2, 3/2}, {-3/2, 3/2}}, BoxRatios → {1, 1, 1}];

p2 = ParametricPlot3D[{x, 0, 0}, {x, 0, 6}, {t, 0, 2Pi},
  PlotStyle → {Gray, Thickness[.075]},
    PlotRange → {{0, 6}, {-3/2, 3/2}, {-3/2, 3/2}}, BoxRatios → {1, 1, 1}];

p3 = ParametricPlot3D[{0, 0, x}, {x, -3/2, 3/2}, {t, 0, 2Pi},
  PlotStyle → {Gray, Thickness[.1]},
    PlotRange → {{0, 6}, {-3/2, 3/2}, {-3/2, 3/2}}, BoxRatios → {1, 1, 1}];

Show[p1, p1b, p2, p3]
```

(c) Next, we generate a basic plot of the solid in p4 and then a set of disks inside the solid in t3d.

```
f[x_] = Exp[-2(x - 2)^2] + Exp[-(x - 4)^2];

p1 = ParametricPlot3D[{x, 0, f[x]}, {x, 0, 6}, PlotStyle → {Thick, Black},

  PlotRange → {{0, 6}, {-3/2, 3/2}, {-3/2, 3/2}}, BoxRatios → {1, 1, 1}];

p1b = ParametricPlot3D[{x, 0, -f[x]}, {x, 0, 6}, PlotStyle → {Thick, Black},

  PlotRange → {{0, 6}, {-3/2, 3/2}, {-3/2, 3/2}}, BoxRatios → {1, 1, 1}];

p2 = ParametricPlot3D[{x, 0, 0}, {x, 0, 6}, {t, 0, 2Pi},

  PlotStyle → {Gray, Thickness[.075]}, PlotRange → {{0, 6}, {-3/2, 3/2}, {-3/2, 3/2}},
```

```
    BoxRatios → {1, 1, 1}];
p3 = ParametricPlot3D[{0, 0, x}, {x, −3/2, 3/2}, {t, 0, 2Pi},
    PlotStyle → {Gray, Thickness[.1]}, PlotRange → {{0, 6}, {−3/2, 3/2}, {−3/2, 3/2}},
    BoxRatios → {1, 1, 1}];
Show[p1, p1b, p2, p3]

p4 = ParametricPlot3D[{r, f[r]Cos[t], f[r]Sin[t]},
    {r, 0, 6}, {t, 0, 2Pi}, PlotRange → {{0, 6}, {−3/2, 3/2}, {−3/2, 3/2}},
    BoxRatios → {1, 1, 1}]

t3d = Table[ParametricPlot3D[{x, rf[x]Cos[t], rf[x]Sin[t]}, {r, 0, 1}, {t, 0, 2Pi},
    PlotRange → {{0, 6}, {−3/2, 3/2}, {−3/2, 3/2}},
    BoxRatios → {1, 1, 1}, ColorFunction → "TemperatureMap", Mesh → 5], {x, 0, 6, 6/14}];
```

Two variations of the solid are plotted in `p5` and `p6`. In each case, we use `MeshFunctions` to have the contour lines (mesh) correspond to $f(x)$ values rather than the rectangular default mesh. In `p6` the solid is made transparent with the `Opacity` option.

```
p5 = ParametricPlot3D[{r, f[r]Cos[t], f[r]Sin[t]},
    {r, 0, 6}, {t, 0, 2Pi}, PlotRange → {{0, 6}, {−3/2, 3/2}, {−3/2, 3/2}},
    BoxRatios → {1, 1, 1}, MeshFunctions->{#1&}, Mesh → 60]

p6 = ParametricPlot3D[{r, f[r]Cos[t], f[r]Sin[t]},
    {r, 0, 6}, {t, 0, 2Pi}, PlotRange → {{0, 6}, {−3/2, 3/2},
    {−3/2, 3/2}},
    BoxRatios → {1, 1, 1}, MeshFunctions->{#1&}, Mesh → 25,
    PlotStyle → Opacity[.2],
    MeshStyle → {Gray, Thick}];
Show[p1, p1b, p2, p3, p6]
```

Several combinations of the images are shown in Figs. 3.47 and 3.48.

```
Show[GraphicsRow[{Show[p1, p1b, p2, p3, p6], Show[p1, p1b, p2, p3, p6, t3d]}]]

Show[GraphicsGrid[{{Show[p1, p1b, p2, p3], p4},
{p5, Show[p1, p1b, p2, p3, p6]}}]]
```

□

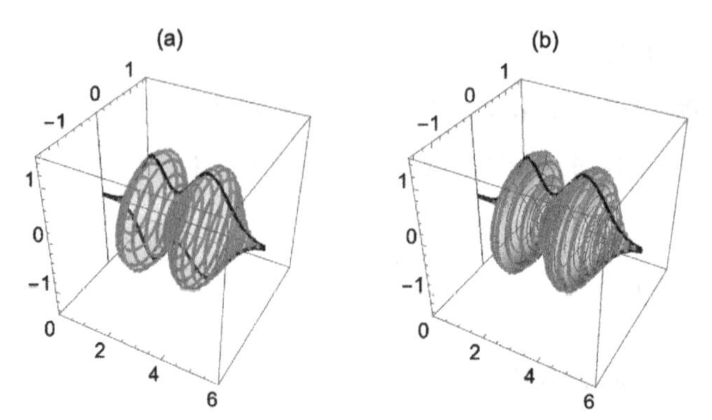

FIGURE 3.47 (a) Seeing $f(x)$ on the solid. (b) Disks in the solid.

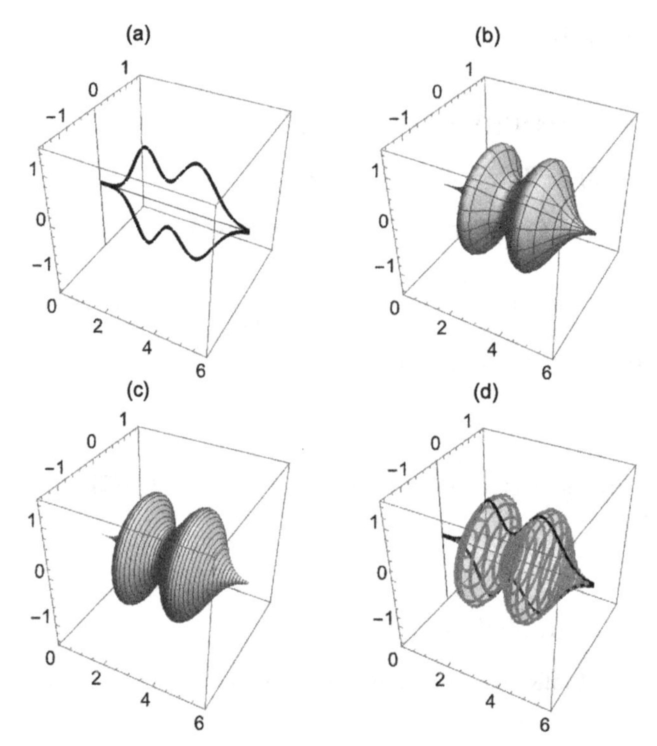

FIGURE 3.48 (a) $f(x)$ in space. (b) The basic solid. (c) Contours based on $f(x)$ values. (d) Seeing $f(x)$ on the solid.

To help identify regions, `RegionPlot[constraints,{x,a,b},{y,a,b}]` attempts to shade the region in the rectangle $[a, b] \times [c, d]$ that satisfies the constraints in `constraints`.

Example 3.45

Let $g(x) = \sqrt{x}$, $h(x) = x^2$, and R the region bounded by the graphs of $g(x)$ and $h(x)$. Find the volume of the solid obtained by revolving R about (a) the x-axis and (b) the y-axis.

Solution. We illustrate the use of `RegionPlot` to help us see R. See Fig. 3.49

```
g[x_] = Sqrt[x];

h[x_] = x^2;

p1a = Plot[Tooltip[{g[x], h[x]}], {x, 0, 2}, PlotRange → {{0, 2}, {0, 2}},
```

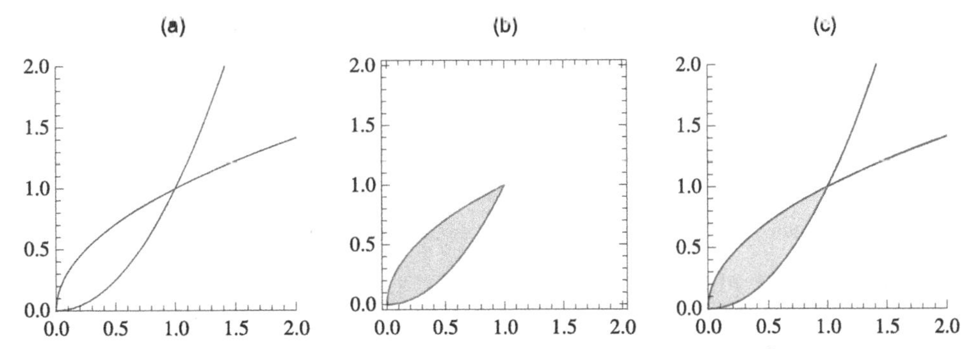

FIGURE 3.49 (a) Graphs of $f(x)$ and $g(x)$. (b) The region in $[0, 2] \times [0, 2]$ for which $x^2 \leq y \leq \sqrt{x}$. (c) The two plots displayed together.

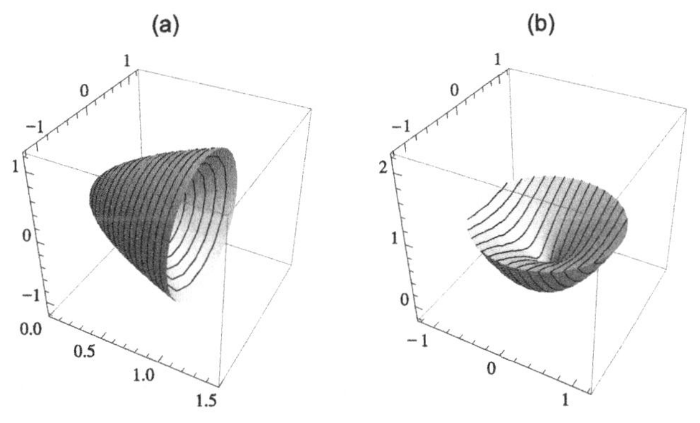

FIGURE 3.50 (a) The solid formed by revolving R about the x-axis. (b) The solid formed by revolving R about the y-axis (Northwestern University colors).

```
    AspectRatio → Automatic, PlotLabel → "(a)",
    PlotStyle → {{CMYKColor[.85, 1, 0, .15]}, {CMYKColor[.5, .5, .5, 1]}}];
p1b = RegionPlot[x^2 ≤ y ≤ Sqrt[x], {x, 0, 2}, {y, 0, 2}, PlotStyle → Purple,
    PlotLabel → "(b)"];
Show[GraphicsRow[{p1a, p1b, Show[{p1a, p1b}, PlotLabel → "(c)"]}]]
```

We plot the solids with `ParametricPlot3D` and contour lines along the function values using the `MeshFunctions` option in Fig. 3.50.

```
p4 = ParametricPlot3D[{{r, g[r]Cos[t], g[r]Sin[t]},
    {r, h[r]Cos[t], h[r]Sin[t]}}, {r, 0, 1},
    {t, 0, 2Pi}, PlotRange → {{0, 3/2}, {−5/4, 5/4}, {−5/4, 5/4}},
    BoxRatios → {1, 1, 1}, MeshFunctions → {#1&}, PlotStyle → {Purple, Opacity[.1]},
    PlotLabel → "(a)"];

p5 = ParametricPlot3D[{{rCos[t], rSin[t], g[r]},
    {rCos[t], rSin[t], h[r]}}, {r, 0, 1},
    {t, 0, 2Pi}, PlotRange → {{−5/4, 5/4}, {−5/4, 5/4}, {−1/4, 9/4}},
```

```
BoxRatios → {1, 1, 1}, MeshFunctions → {#1&}, PlotStyle → {Purple, Opacity[.1]},
PlotLabel → "(b)"];
```

```
Show[GraphicsRow[{p4, p5}]]
```

The volume of each solid is then found with `Integrate` and approximated with `N`.

```
Integrate[Pi(g[x]^2 - h[x]^2), {x, 0, 1}]
```

$\frac{3\pi}{10}$

```
N[%]
```

0.942478

```
Integrate[Pix(g[x] - h[x]), {x, 0, 1}]
```

$\frac{3\pi}{20}$

```
N[%]
```

0.471239 □

Surface Area

Let $y = f(x)$ be a nonnegative function for which $f'(x)$ is continuous on an interval $[a, b]$. Then the **surface area** of the solid of revolution obtained by revolving the region bounded by the graphs of $y = f(x)$, $x = a$, $x = b$, and the x-axis about the x-axis is given by

$$SA = 2\pi \int_a^b f(x)\sqrt{1 + \left[f'(x)\right]^2}\, dx. \tag{3.23}$$

Example 3.46: Gabriel's Horn

Gabriel's horn is the solid of revolution obtained by revolving the area of the region bounded by $y = 1/x$ and the x-axis for $x \geq 1$ about the x-axis. Show that the surface area of Gabriel's horn is infinite but that its volume is finite.

Solution. After defining $f(x) = 1/x$, we use `ParametricPlot3D` to visualize a portion of Gabriel's horn in Fig. 3.51.

```
f[x_] = 1/x;
```

```
ParametricPlot3D[{r, f[r]Cos[t], f[r]Sin[t]}, {r, 1, 10}, {t, 0, 2Pi},
  PlotPoints->{40, 40}, ViewPoint->{-1.509, -2.739, 1.294}]
```

Using equation (3.23), the surface area of Gabriel's horn is given by the improper integral

$$SA = 2\pi \int_1^\infty \frac{1}{x}\sqrt{1 + \frac{1}{x^4}}\, dx = 2\pi \lim_{L\to\infty} \int_1^L \frac{1}{x}\sqrt{1 + \frac{1}{x^4}}\, dx.$$

```
f'[x]
```

$-\frac{1}{x^2}$

```
step1 = Integrate[2Pi f[x]Sqrt[1 + f'[x]^2], {x, 1, capl}]
```

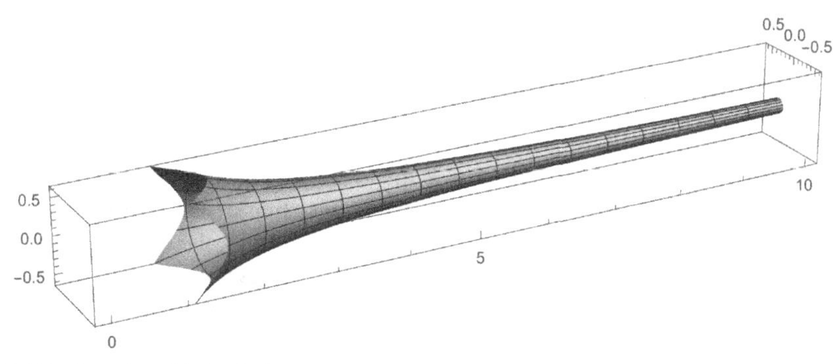

FIGURE 3.51 A portion of Gabriel's horn.

$$\text{ConditionalExpression}[2\pi(\tfrac{1}{2}(\sqrt{2}-\text{ArcSinh}[1])+\frac{\sqrt{1+\frac{1}{\text{capl}^4}}(-\sqrt{1+\text{capl}^4}+\text{capl}^2\text{ArcSinh}[\text{capl}^2])}{2\sqrt{1+\text{capl}^4}}),$$
$$\text{Re}[\tfrac{1}{\text{capl}^4}]\geq -1\&\&(.((\frac{(-1+\text{Re}[\text{capl}])^4}{\text{Im}[\text{capl}]^4}==-1\|\frac{(-1+\text{Re}[\text{capl}])^4}{\text{Im}[\text{capl}]^4}\geq -1)\&\&\text{capl}\notin\text{Reals})\|\text{Re}[\text{capl}]$$
$$\geq 0)\&\&((\text{Im}[\text{capl}]+\text{Re}[\text{capl}]==0\&\&(0<\text{Re}[\text{capl}]<1\|\text{Im}[\text{capl}]>0\|\text{Re}[\text{capl}]>1))\|-$$
$$1<\text{Im}[\text{capl}]<0\|(\text{Re}[\text{capl}]\geq 0\&\&-1<\text{Im}[\text{capl}]\leq 0)\|\text{Im}[\text{capl}]<-1)\&\&$$
$$(\frac{\text{Im}[\text{capl}](\text{Im}[\text{capl}]+\text{Re}[\text{capl}])}{1+\text{Im}[\text{capl}]}==0\|\frac{\text{Im}[\text{capl}]^2(\text{Im}[\text{capl}]+\text{Re}[\text{capl}])^2}{(1+\text{Im}[\text{capl}])^2}\leq 0\|\frac{1}{1+\text{Im}[\text{capl}]}==1\|\frac{1}{1+\text{Im}[\text{capl}]}\geq$$
$$1\|\text{Im}[\text{capl}]<-1)]$$

Limit[step1, capl->Infinity]

∞

On the other hand, using equation (3.21) the volume of Gabriel's horn is given by the improper integral

$$V=2\pi\int_1^\infty \frac{1}{x^2}\,dx=\pi\lim_{L\to\infty}\int_1^L\frac{1}{x^2}\,dx,$$

which converges to π.

step1 = Integrate[Pi *f*[*x*]^2, {*x*, 1, capl}]

$\text{ConditionalExpression}[(1-\frac{1}{\text{capl}})\pi, \text{Re}[\text{capl}]>0\|\text{capl}\notin\text{Reals}]$

Limit[step1, capl->Infinity]

π

Integrate[Pi *f*[*x*]^2, {*x*, 1, Infinity}]

π □

3.4 INFINITE SEQUENCES AND SERIES

3.4.1 Introduction to Sequences

Sequences and series are usually discussed in the third quarter or second semester of introductory calculus courses. Most students find that it is one of the most difficult topics covered in calculus. A **sequence** is a function with domain consisting of the positive integers. The **terms** of the sequence $\{a_n\}$ are $a_1, a_2, a_3, \ldots$. The nth term is a_n; the $(n+1)$st term is a_{n+1}. If $\lim_{n\to\infty} a_n = L$, we say that $\{a_n\}$ **converges** to L. If $\{a_n\}$ does not converge, $\{a_n\}$ **diverges**. We can sometimes prove that a sequence converges by applying the following theorem.

Theorem 3.7. *Every bounded monotonic sequence converges.*

A sequence $\{a_n\}$ is monotonic if $\{a_n\}$ is increasing ($a_{n+1} \geq a_n$ for all n) or decreasing ($a_{n+1} \leq a_n$ for all n).

In particular, Theorem 3.7 gives us the following special cases.

1. If $\{a_n\}$ has positive terms and is eventually decreasing, $\{a_n\}$ converges.
2. If $\{a_n\}$ has negative terms and is eventually increasing $\{a_n\}$ converges.

After you have defined a sequence, use `Table` to compute the first few terms of the sequence.

1. `Table[a[n],{n,1,m}]` returns the list $\{a_1, a_2, a_3, \ldots, a_m\}$.
2. `Table[a[n],{n,k,m}]` returns $\{a_k, a_{k+1}, a_{k+2}, \ldots, a_m\}$.

The command `ListPlot[listofpoints]` plots the list of points `listofpoints` while `ListLinePlot[listofpoints]` plots the list of points `listofpoints` and connects consecutive points with line segments.

Remark 3.7. An extensive database of integer sequences can be found at the **On-Line Encyclopedia of Integer Sequences,**

`http://www.research.att.com/~njas/sequences/Seis.html.`

Example 3.47

If $a_n = \dfrac{50^n}{n!}$, show that $\lim_{n\to\infty} a_n = 0$.

Solution. We remark that the symbol $n!$ in the denominator of a_n represents the **factorial sequence**:

$$n! = n \cdot (n-1) \cdot (n-2) \cdot \cdots \cdot 2 \cdot 1.$$

We begin by defining a_n and then computing the first few terms of the sequence with `Table`.

```
Clear[a]

a[n_]:=50^n/n!;

afewterms = Table[a[n], {n, 1, 10}]
```

$$\left\{50, 1250, \frac{62500}{3}, \frac{781250}{3}, \frac{7812500}{3}, \frac{195312500}{9}, \frac{9765625000}{63}, \frac{61035156250}{63}, \frac{3051757812500}{567}, \frac{15258789062500}{567}\right\}$$

```
N[afewterms]
```

$\{50., 1250., 20833.3, 260417., 2.60417\times 10^6, 2.17014\times 10^7, 1.5501\times 10^8, 9.68812\times 10^8, 5.38229\times 10^9, 2.69114\times 10^{10}\}$

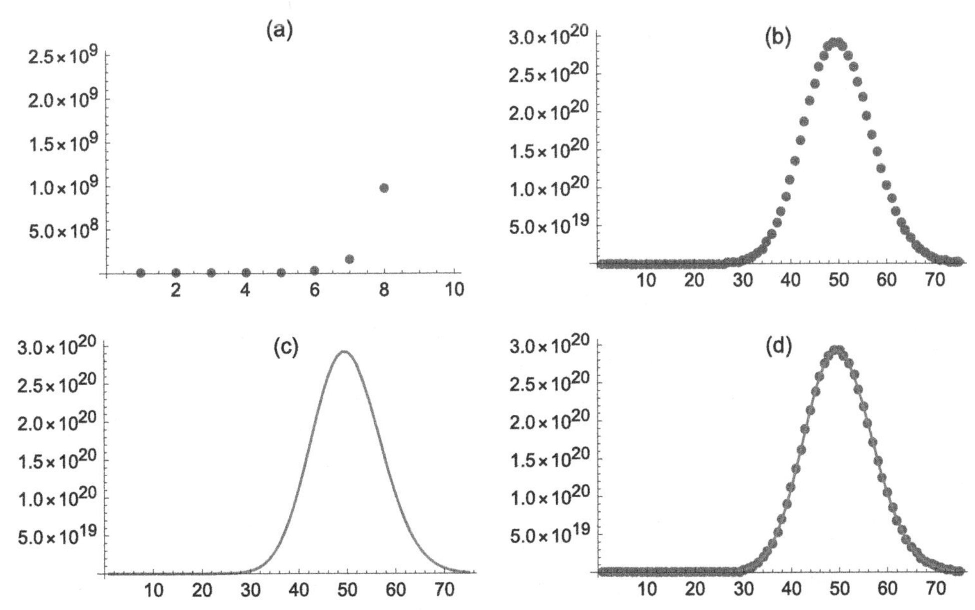

FIGURE 3.52 (a) The first few terms of a_n. (b) The first 75 terms of a_n. (c) The first 75 terms of a_n connected with line segments. (d) Displaying the points together with the first 75 terms of a_n connected with line segments (The Ohio State University colors).

The first few terms increase in magnitude. In fact, this is further confirmed by graphing the first few terms of the sequence with `ListPlot` in Fig. 3.52 (a). Based on the graph and the values of the first few terms we might incorrectly conclude that the sequence diverges.

```
p1 = ListPlot[afewterms, PlotStyle → {PointSize[.025], CMYKColor[.03, 1, .63, .12]},
  PlotLabel → "(a)"];
```

However, notice that $a_{n+1} = \dfrac{50}{n+1} a_n \Rightarrow \dfrac{a_{n+1}}{a_n} = \dfrac{50}{n+1}$. Because $50/(n+1) < 1$ for $n > 49$, we conclude that the sequence is decreasing for $n > 49$. Because it has positive terms, it is bounded below by 0 so the sequence converges by Theorem 3.7. Let $L = \lim_{n\to\infty} a_n$. Then,

$$\begin{aligned} \lim_{n\to\infty} a_{n+1} &= \lim_{n\to\infty} \frac{50}{n+1} a_n \\ L &= \lim_{n\to\infty} \frac{50}{n+1} \cdot L \\ L &= 0. \end{aligned}$$

When we graph a larger number of terms, it is clear that the limit is 0. (See Fig. 3.52 (b).) It is a good exercise to show that for any real value of x, $\lim_{n\to\infty} \dfrac{x^n}{n!} = 0$.

Use `ListLinePlot` to connect consecutive points with line segments.

```
p2 = ListPlot[Evaluate[Table[a[k], {k, 1, 75}]],
  PlotStyle → {PointSize[.025], CMYKColor[.03, 1, .63, .12]},
  PlotLabel → "(b)"];
p3 = ListLinePlot[Evaluate[Table[a[k], {k, 1, 75}]],
  PlotStyle → CMYKColor[.56, .47, .47, .15],
  PlotLabel → "(c)"];
```

```
p4 = Show[p2, p3, PlotLabel → "(d)"];
Show[GraphicsGrid[{{p1, p2}, {p3, p4}}]]
```
□

Example 3.48: The Rational Numbers and the Calkin–Wilf Sequence

A set S is **countably infinite** if there is a one-to-one correspondence between the elements of S and the natural numbers. In other words, S is countably infinite if it can be written as a sequence. Students in introductory set theory classes often learn that the set of rational numbers, $\mathbb{Q} = \left\{\frac{n}{m} : n, m \neq 0 \text{ integers}\right\}$ is countably infinite using *Cantor's Diagnalization Process*.
In 2000, Neil Calkin and Herbert Wilf published a paper introducing the **Calkin–Wilf sequence**. The Calkin–Wilf sequence is a recursively defined sequence that gives a one-to-one correspondence between the natural numbers and the positive rational numbers. Stating a specific sequence that yields the positive rational numbers proving the countability of the rationals provides an interesting contrast to the traditional method of using *Cantor's Diagnalization Process*.
The **Calkin–Wilf sequence** is defined by

$$q_1 = 1$$
$$q_i = \frac{1}{2\lfloor q_{i-1} \rfloor - q_{i-1} + 1} \text{ for } i > 1, \tag{3.24}$$

where $\lfloor q_{i-1} \rfloor$ represents the integer part of q_{i-1}. The Mathematica command `IntegerPart[x]` returns the integer part of x. When defining q, observe that we use the form `q[i_]:=q[i]=` so that Mathematica "remembers" the computed values of q_i. Thus, to compute q_n, Mathematica need only have computed q_{n-1} rather than re-computing the previous $n-1$ q_i values. Compute and plot the first n terms of the Calkin–Wilf sequence with $n = 50$ and $n = 500$.

Solution.

```
Clear[q]
q[1] = 1;
q[i_]:=q[i] = 1/(2IntegerPart[q[i − 1]] − q[i − 1] + 1)
```

In `q1` we compute the first 50 terms of the Calkin–Wilf sequence while in `q2` we compute the first 500 terms of the Calkin–Wilf but do not display the result.

```
q1 = Table[q[i], {i, 1, 50}]
```

$\{1, \frac{1}{2}, 2, \frac{1}{3}, \frac{3}{2}, \frac{2}{3}, 3, \frac{1}{4}, \frac{4}{3}, \frac{3}{5}, \frac{5}{2}, \frac{2}{5}, \frac{5}{3}, \frac{3}{4}, 4, \frac{1}{5}, \frac{5}{4}, \frac{4}{7}, \frac{7}{3}, \frac{3}{8}, \frac{8}{5}, \frac{5}{7}, \frac{7}{2}, \frac{2}{7}, \frac{7}{5}, \frac{5}{8}, \frac{8}{3}, \frac{3}{7}, \frac{7}{4}, \frac{4}{5}, 5, \frac{1}{6}, \frac{6}{5}, \frac{5}{9}, \frac{9}{4}, \frac{4}{11}, \frac{11}{7}, \frac{7}{10}, \frac{10}{3}, \frac{3}{11}, \frac{11}{8}, \frac{8}{13}, \frac{13}{5}, \frac{5}{12}, \frac{12}{7}, \frac{7}{9}, \frac{9}{2}, \frac{2}{9}, \frac{9}{7}, \frac{7}{12}\}$

```
q2 = Table[q[i], {i, 1, 500}];
```

We then graph both sets of points with `ListPlot` and show the results in Fig. 3.53.

```
p1 = ListPlot[q1, PlotStyle → CMYKColor[0, 1, 1, 0],
  PlotLabel → "(a)"]
```

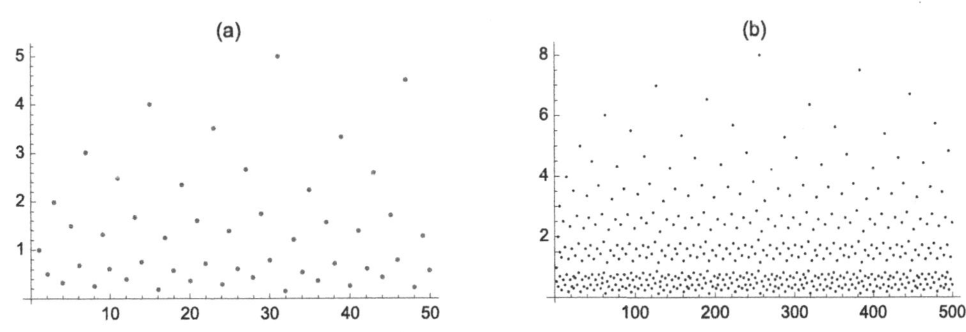

FIGURE 3.53 (a) The first 50 terms of the Calkin–Wilf sequence. (b) The first 500 terms of the Calkin–Wilf sequence (Rensselaer Polytechnic Institute colors).

```
p2 = ListPlot[q2, PlotStyle → CMYKColor[0, 0, 0, .61],
  PlotLabel → "(b)"]

Show[GraphicsRow[{p1, p2}]]
```

□

You can use Mathematica to define quite complex sequences. In the following example, we define a sequence that depends on four other sequences. Observe that when defining the terms, we consistently use the form `a[n_]:=a[n]=...` so that Mathematica "remembers" sequence values computed and need not recompute them to compute subsequent terms.

Example 3.49: The Gauss–Legendre Algorithm

One algorithm to approximate π is the **Gauss–Legendre algorithm**. Although it is memory intensive, 25 iterations approximate π to around 45 million digits. After defining a_k, b_k, t_k, and p_k, π is approximated with $\pi \approx \frac{(a_n + b_n)^2}{4t_n}$. Approximate π to 100 decimal places.

Solution.

```
Remove[a, b, t, p, pi]
a[0] = 1;
b[0] = 1/Sqrt[2];
t[0] = 1/4;
p[0] = 1;

a[n_]:=a[n] = N[(a[n − 1] + b[n − 1])/2, 10000];
b[n_]:=b[n] = N[Sqrt[a[n − 1]b[n − 1]], 10000];
t[n_]:=t[n] = N[t[n − 1] − p[n − 1](a[n − 1] − a[n])^2, 10000];
p[n_]:=p[n] = N[2p[n − 1], 10000];
```

π is then approximated with

```
pi[n_]:=N[(a[n] + b[n])^2/(4t[n]), 100]
```

To increase the accuracy of your approximation, increase 10000 and 100 to your desired values. However, even on relatively fast machines, if you replace all these numbers with a number such as 1000000, the computation time will increase considerably.

For illustrate purposes, we approximate π with $\pi \approx \dfrac{(a_6+b_6)^2}{4t_6}$. Observe that the result is accurate to at least 100 decimal places.

pi[6]

3.14159265358979323846264338327950288419716939937510582097494459230781640
628620899862803482534211 7068

***N*[Pi, 100]**

3.14159265358979323846264338327950288419716939937510582097494459230781640
628620899862803482534211 7068 □

3.4.2 Introduction to Infinite Series

An **infinite series** is a series of the form

$$\sum_{k=1}^{\infty} a_k = a_1 + a_2 + a_3 + \cdots \tag{3.25}$$

where $\{a_n\}$ is a sequence. The nth **partial sum** of (3.25) is

$$s_n = \sum_{k=1}^{n} a_k = a_1 + a_2 + \cdots + a_n. \tag{3.26}$$

Notice that the partial sums of the series (3.25) form a sequence $\{s_n\}$. Hence, we say that the infinite series (3.25) **converges** to L if the sequence of partial sums $\{s_n\}$ converges to L and write

$$\sum_{k=1}^{\infty} a_k = L.$$

The infinite series (3.25) **diverges** if the sequence of partial sums diverges. Given the infinite series (3.25),

```
Sum[a[k],{k,1,n}]
```

calculates the nth partial sum (3.26). In *some* cases, if the infinite series (3.25) converges,

```
Sum[a[k],{k,1,Infinity}]
```

can compute the value of the infinite sum. In addition to using `Sum` to compute finite and infinite sums, you can use the $\sum$▪ button on the **Basic Math Input** palette to calculate sums. You should think of the `Sum` function as a "fragile" command and be certain to carefully examine its results and double check their validity.

Example 3.50

Determine whether each series converges or diverges. If the series converges, find its sum. (a) $\sum_{k=1}^{\infty}(-1)^{k+1}$; (b) $\sum_{k=2}^{\infty}\frac{2}{k^2-1}$; and (c) $\sum_{k=0}^{\infty}ar^k$.

Solution. For (a), we compute the nth partial sum (3.26) in `sn` with `Sum`.

sn = Sum[(−1)^(k + 1), {k, 1, n}]

$\frac{1}{2}\left(1+(-1)^{1+n}\right)$

Notice that the odd partial sums are 1: $s_{2n+1}=\frac{1}{2}\left((-1)^{2n+1+1}+1\right)=\frac{1}{2}(1+1)=1$ while the even partial sums are 0: $s_{2n}=\frac{1}{2}\left((-1)^{2n+1}+1\right)=\frac{1}{2}(-1+1)=0$. We confirm that the limit of the partial sums does not exist with `Limit`. Mathematica's result indicates that it cannot determine the limit. The series diverges.

Limit[sn, n → Infinity]

$\frac{1}{2}\left(1+e^{2i\,\text{Interval}[\{0,\pi\}]}\right)$

Similarly, when we attempt to compute the infinite sum with `Sum`, Mathematica is able to determine that the partial sums diverge, which means that the infinite series diverges.

```
Sum[(-1)^(k+1), {k, 1, Infinity}]
··· Sum: Sum does not converge.
```

$\sum_{k=1}^{\infty}(-1)^{1+k}$

For (b), we have a *telescoping series*. Using partial fractions,

$$\begin{aligned}\sum_{k=2}^{\infty}\frac{2}{k^2-1}&=\sum_{k=2}^{\infty}\left(\frac{1}{k-1}-\frac{1}{k+1}\right)\\&=\left(1-\frac{1}{3}\right)+\left(\frac{1}{2}-\frac{1}{4}\right)+\left(\frac{1}{3}-\frac{1}{5}\right)+\cdots+\left(\frac{1}{n-2}-\frac{1}{n}\right)\\&\quad+\left(\frac{1}{n-1}-\frac{1}{n+1}\right)+\ldots\end{aligned}$$

we see that the nth partial sum is given by

$$s_n=\frac{3}{2}-\frac{1}{n}-\frac{1}{n+1}$$

and $s_n\to 3/2$ as $n\to\infty$ so the series converges to $3/2$:

$$\sum_{k=2}^{\infty}\frac{2}{k^2-1}=\frac{3}{2}.$$

We perform the same steps with Mathematica using `Sum`, `Apart`, and `Limit`.

`Apart` computes the partial fraction decomposition of a rational expression.

sn = Sum[1/(k − 1) − 1/(k + 1), {k, 2, n}]

$\frac{-2-n+3n^2}{2n(1+n)}$

Apart[sn]

$\frac{3}{2}-\frac{1}{n}-\frac{1}{1+n}$

Limit[sn, $n \to$ Infinity]

$\frac{3}{2}$

(c) A series of the form $\sum_{k=0}^{\infty} ar^k$ is called a **geometric series**. We compute the nth partial sum of the geometric series with `Sum`.

sn = Sum[ar^k, {k, 0, n}]

$\frac{a(-1+r^{1+n})}{-1+r}$

When using `Limit` to determine the limit of s_n as $n \to \infty$, we see that Mathematica returns the limit unevaluated because Mathematica does not know the value of r.

Limit[sn, $n \to$ Infinity]

$\text{Limit}\left[\frac{a(-1+r^{1+n})}{-1+r}, n \to \infty\right]$

In fact, the geometric series diverges if $|r| \geq 1$ and converges if $|r| < 1$. Observe that if we simply compute the sum with `Sum`, Mathematica returns $a/(1-r)$ which is correct if $|r| < 1$ but incorrect if $|r| \geq 1$.

Sum[ar^k, {k, 0, Infinity}]

$\frac{a}{1-r}$

However, the result of entering

```
Sum[(-5/3)^k, {k, 0, Infinity}]
··· Sum: Sum does not converge.
```

$\sum_{k=0}^{\infty}\left(-\frac{5}{3}\right)^k$

is correct because the series $\sum_{k=0}^{\infty}\left(-\frac{5}{3}\right)^k$ is geometric with $|r| = 5/3 \geq 1$ and, consequently, diverges. Similarly,

Sum[9(1/10)^k, {k, 1, Infinity}]

1

is correct because $\sum_{k=1}^{\infty} 9\left(\frac{1}{10}\right)^k$ is geometric with $a = 9/10$ and $r = 1/10$ so the series $0.999\ldots$ converges to

$$\frac{a}{1-r}=\frac{9/10}{1-1/10}=1.$$ □

3.4.3 Convergence Tests

Frequently used convergence tests for infinite series are stated in the following theorems. Note that the infinite series $\sum_{k=1}^{\infty} a_k$ **converges absolutely** means that $\sum_{k=1}^{\infty} |a_k|$ converges. If an infinite series converges absolutely, it converges. If a series converges, but does not converge absolutely, we say that the series **conditionally converges**.

The **alternating harmonic series**, $\sum_{k=1}^{\infty}(-1)^{k+1}1/k$ is an example of a series that converges conditionally. It is a good exercise to show that the alternating harmonic series converges to $\ln 2$. On the other hand, the harmonic series, $\sum_{k=1}^{\infty} 1/k$ diverges to $+\infty$.

Theorem 3.8 (The Divergence Test). *Let $\sum_{k=1}^{\infty} a_k$ be an infinite series. If $\lim_{k\to\infty} a_k \neq 0$, then $\sum_{k=1}^{\infty} a_k$ diverges.*

Theorem 3.9 (The Integral Test). *Let $\sum_{k=1}^{\infty} a_k$ be an infinite series with positive terms. If $f(x)$ is a decreasing continuous function for which $f(k)=a_k$ for all k, then $\sum_{k=1}^{\infty} a_k$ and $\int_1^{\infty} f(x)\,dx$ either both converge or both diverge.*

Theorem 3.10 (The Ratio Test). *Let $\sum_{k=1}^{\infty} a_k$ be an infinite series and let $\rho = \lim_{k\to\infty} \left|\frac{a_{k+1}}{a_k}\right|$.*

1. *If $\rho < 1$, $\sum_{k=1}^{\infty} a_k$ converges absolutely.*
2. *If $\rho > 1$, $\sum_{k=1}^{\infty} a_k$ diverges.*
3. *If $\rho = 1$, the Ratio Test is inconclusive.*

Theorem 3.11 (The Root Test). *Let $\sum_{k=1}^{\infty} a_k$ be an infinite series and let $\rho = \lim_{k\to\infty} \sqrt[k]{|a_k|}$.*

1. *If $\rho < 1$, $\sum_{k=1}^{\infty} a_k$ converges absolutely.*
2. *If $\rho > 1$, $\sum_{k=1}^{\infty} a_k$ diverges.*
3. *If $\rho = 1$, the Root Test is inconclusive.*

Theorem 3.12 (The Limit Comparison Test). *Let $\sum_{k=1}^{\infty} a_k$ and $\sum_{k=1}^{\infty} b_k$ be infinite series with positive terms and let $L = \lim_{k\to\infty} \frac{a_k}{b_k}$. If $0 < L < \infty$, then either both series converge or both series diverge.*

Example 3.51

Determine whether each series converges or diverges. (a) $\sum_{k=1}^{\infty}\left(1+\frac{1}{k}\right)^k$; (b) $\sum_{k=1}^{\infty}\frac{1}{k^p}$; (c) $\sum_{k=1}^{\infty}\frac{k}{3^k}$; (d) $\sum_{k=1}^{\infty}\frac{(k!)^2}{(2k)!}$; (e) $\sum_{k=1}^{\infty}\left(\frac{k}{4k+1}\right)^k$; (f) $\sum_{k=1}^{\infty}\frac{2\sqrt{k}+1}{(\sqrt{k}+1)(2k+1)}$.

Solution. (a) Using `Limit`, we see that the limit of the terms is $e \neq 0$ so the series diverges by the Divergence Test, Theorem 3.8.

```
Limit[(1 + 1/k)^k, k → Infinity]
```

e

It is a very good exercise to show that the limit of the terms of the series is e by hand. Let $L = \lim_{k\to\infty}(1+1/k)^k$. Take the logarithm of each side of this equation and apply

L'Hôpital's rule:

$$\ln L = \lim_{k\to\infty} \ln\left(1+\frac{1}{k}\right)^k$$
$$\ln L = \lim_{k\to\infty} k\ln\left(1+\frac{1}{k}\right)$$
$$\ln L = \lim_{k\to\infty} \frac{\ln\left(1+\frac{1}{k}\right)}{\frac{1}{k}}$$
$$\ln L = \lim_{k\to\infty} \frac{\frac{1}{1+\frac{1}{k}}\cdot -\frac{1}{k^2}}{-\frac{1}{k^2}}$$
$$\ln L = 1.$$

Exponentiating yields $L = e^{\ln L} = e^1 = e$.

(b) A series of the form $\sum_{k=1}^{\infty} \frac{1}{k^p}$ $(p > 0)$ is called a **p-series**. Let $f(x) = x^{-p}$. Then, $f(x)$ is continuous and decreasing for $x \geq 1$, $f(k) = k^{-p}$ and

$$\int_1^\infty x^{-p}dx = \begin{cases} \infty, \text{ if } p \leq 1 \\ 1/(p-1), \text{ if } p > 1 \end{cases}$$

so the p-series converges if $p > 1$ and diverges if $p \leq 1$. If $p = 1$, the series $\sum_{k=1}^{\infty} \frac{1}{k}$ is called the **harmonic series**. Observe that Mathematica's result is given in terms of a "conditional expression." In this example, the interpretation is that $\int_1^\infty x^{-p}\,dx = 1/(p-1)$ provided that the real part of p is greater than 1.

```
Clear[x, p]

s1 = Integrate[x^(-p), {x, 1, Infinity}]
```

ConditionalExpression$\left[\frac{1}{-1+p}, \text{Re}[p] > 1\right]$

(c) Let $f(x) = x \cdot 3^{-x}$. Then, $f(k) = k \cdot 3^{-k}$ and $f(x)$ is decreasing for $x > 1/\ln 3$ because $f'(x) < 0$ for $x > 1/\ln 3$.

```
f[x_] = x3^(-x);

Factor[f'[x]]
```

$-3^{-x}(-1 + x\text{Log}[3])$

```
Solve[-1 + xLog[3] == 0, x]
```

$\left\{\left\{x \to \frac{1}{\text{Log}[3]}\right\}\right\}$

Using `Integrate`, we see that the improper integral $\int_1^{\infty} f(x)\,dx$ converges.

ival = Integrate[f[x], {x, 1, Infinity}]

***N*[ival]**

$\frac{1+\text{Log}[3]}{3\text{Log}[3]^2}$

0.579592

Thus, by the Integral Test, Theorem 3.9, we conclude that the series converges. Note that when applying the Integral Test, if the improper integral converges its value is *not* the value of the sum of the series. In this case, we see that Mathematica is able to evaluate the sum with `Sum` and the series converges to $3/4$.

Sum[k3^(−k), {k, 1, Infinity}]

$\frac{3}{4}$

(d) If a_k contains factorial functions, the Ratio Test is often a good first test to try. After defining a_k we compute

$$\begin{aligned}\lim_{k\to\infty}\frac{a_{k+1}}{a_k} &= \lim_{k\to\infty}\frac{\dfrac{[(k+1)!]^2}{[2(k+1)]}}{\dfrac{(k!)^2}{(2k)!}} \\ &= \lim_{k\to\infty}\frac{(k+1)!\cdot(k+1)!}{k!\cdot k!}\frac{(2k)!}{(2k+2)!} \\ &= \lim_{k\to\infty}\frac{(k+1)^2}{(2k+2)(2k+1)} = \lim_{k\to\infty}\frac{(k+1)}{2(2k+1)} = \frac{1}{4}.\end{aligned}$$

Because $1/4 < 1$, the series converges by the Ratio Test. We confirm these results with Mathematica.

Remark 3.8. Use `FullSimplify` instead of `Simplify` to simplify expressions involving factorials.

a[k_] = (k!)^2/(2k)!;

s1 = FullSimplify[a[k + 1]/a[k]]

$\frac{1+k}{2+4k}$

Limit[s1, k → Infinity]

$\frac{1}{4}$

We illustrate that we can evaluate the sum using `Sum` and approximate it with `N` as follows.

ev = Sum[a[k], {k, 1, Infinity}]

$\frac{1}{27}\left(9+2\sqrt{3}\pi\right)$

N[ev]

0.7364

(e) Because

$$\lim_{k\to\infty}\sqrt[k]{\left(\frac{k}{4k+1}\right)^k}=\lim_{k\to\infty}\frac{k}{4k+1}=\frac{1}{4}<1,$$

the series converges by the Root Test.

```
a[k_] = (k/(4k + 1))^k;
Limit[a[k]^(1/k), k → Infinity]
```

$\frac{1}{4}$

As with (d), we can approximate the sum with `N` and `Sum`.

```
ev = Sum[a[k], {k, 1, Infinity}]
```

$\sum_{k=1}^{\infty}\left(\frac{k}{1+4k}\right)^k$

N[ev]

0.265757

(f) We use the Limit Comparison Test and compare the series to $\sum_{k=1}^{\infty}\frac{\sqrt{k}}{k\sqrt{k}}=\sum_{k=1}^{\infty}\frac{1}{k}$, which diverges because it is a p-series with $p=1$. Because

$$0<\lim_{k\to\infty}\frac{\frac{2\sqrt{k}+1}{(\sqrt{k}+1)(2k+1)}}{\frac{1}{k}}=1<\infty$$

and the harmonic series diverges, the series diverges by the Limit Comparison Test.

```
a[k_] = (2Sqrt[k] + 1)/((Sqrt[k] + 1)(2k + 1));
b[k_] = 1/k;
Limit[a[k]/b[k], k → Infinity]
```

1

□

3.4.4 Alternating Series

An **alternating series** is a series of the form

$$\sum_{k=1}^{\infty}(-1)^k a_k \quad \text{or} \quad \sum_{k=1}^{\infty}(-1)^{k+1} a_k \tag{3.27}$$

where $\{a_k\}$ is a sequence with positive terms.

Theorem 3.13 (Alternating Series Test). *If* $\{a_k\}$ *is decreasing and* $\lim_{k\to\infty} a_k = 0$, *the alternating series* (3.27) *converges.*

Definition 3.3. The alternating series (3.27) **converges absolutely** if $\sum_{k=1}^{\infty} a_k$ converges.

Theorem 3.14. *If the alternating series* (3.27) *converges absolutely, it converges.*

Definition 3.4. If the alternating series (3.27) converges but does not converge absolutely, we say that the alternating series, (3.27), **conditionally converges**.

Example 3.52

Determine whether each series converges or diverges. If the series converges, determine whether the convergence is conditional or absolute. (a) $\sum_{k=1}^{\infty}\frac{(-1)^{k+1}}{k}$; (b) $\sum_{k=1}^{\infty}(-1)^{k+1}\frac{(k+1)!}{4^k(k!)^2}$; (c) $\sum_{k=1}^{\infty}(-1)^{k+1}\left(1+\frac{1}{k}\right)^k$.

Solution. (a) Because $\{1/k\}$ is decreasing and $1/k \to 0$ as $k \to \infty$, the series converges. The series does not converge absolutely because the harmonic series diverges. Hence, $\sum_{k=1}^{\infty}\frac{(-1)^{k+1}}{k}$, which is called the **alternating harmonic series**, converges conditionally. We see that this series converges to $\ln 2$ with `Sum`.

```
a[k_] = (-1)^(k + 1)/k;
Sum[a[k], {k, 1, Infinity}]
```

Log[2]

(b) We test for absolute convergence first using the Ratio Test. Because

$$\lim_{k\to\infty}\frac{\dfrac{((k+1)+1)!}{4^{k+1}[(k+1)!]^2}}{\dfrac{(k+1)!}{4^k(k!)^2}} = \lim_{k\to\infty}\frac{k+2}{4(k+1)^2} = 0 < 1,$$

```
a[k_] = (k + 1)!/(4^k(k!)^2);
s1 = FullSimplify[a[k + 1]/a[k]]
Limit[s1, k → Infinity]
```

$\frac{2+k}{4(1+k)^2}$

0

the series converges absolutely by the Ratio Test. Absolute convergence implies convergence so the series converges.

(c) Because $\lim_{k\to\infty}\left(1+\frac{1}{k}\right)^k = e$, $\lim_{k\to\infty}(-1)^{k+1}\left(1+\frac{1}{k}\right)^k$ does not exist, so the series diverges by the Divergence Test. We confirm that the limit of the terms is not zero with `Limit`.

Sum[(−1)^(k + 1)a[k], {k, 1, Infinity}]

$-\frac{3-4e^{1/4}}{4e^{1/4}}$

```
a[k_] = (-1) ^ (k + 1) (1 + 1 / k) ^ k;
Sum[a[k], {k, 1, Infinity}]
```

··· Sum: Sum does not converge.

$\sum_{k=1}^{\infty}(-1)^{1+k}\left(1+\frac{1}{k}\right)^k$

```
Limit[a[k], k → Infinity]
```

$e^{1+2\,i\,\text{Interval}[\{0,\pi\}]}$

Limit[a[k], k → Infinity]

$e^{1+2i\text{Interval}[\{0,\pi\}]}$

□

3.4.5 Power Series

Let x_0 be a number. A **power series** in $x - x_0$ is a series of the form

$$\sum_{k=0}^{\infty} a_k (x-x_0)^k. \tag{3.28}$$

A fundamental problem is determining the values of x, if any, for which the power series converges, the **interval of convergence**.

Theorem 3.15. *For the power series (3.28), exactly one of the following is true.*

1. *The power series converges absolutely for all values of x. The interval of convergence is $(-\infty,\infty)$.*
2. *There is a positive number r so that the series converges absolutely if $x_0 - r < x < x_0 + r$. The series may or may not converge at $x = x_0 - r$ and $x = x_0 + r$. The interval of convergence will be one of $(x_0 - r, x_0 + r)$, $[x_0 - r, x_0 + r)$, $(x_0 - r, x_0 + r]$, or $[x_0 - r, x_0 + r]$.*
3. *The series converges only if $x = x_0$. The interval of convergence is $\{x_0\}$.*

Example 3.53

Determine the interval of convergence for each of the following power series. (a) $\sum_{k=0}^{\infty}\frac{(-1)^k}{(2k+1)!}x^{2k+1}$; (b) $\sum_{k=0}^{\infty}\frac{k!}{1000^k}(x-1)^k$; (c) $\sum_{k=1}^{\infty}\frac{2^k}{\sqrt{k}}(x-4)^k$.

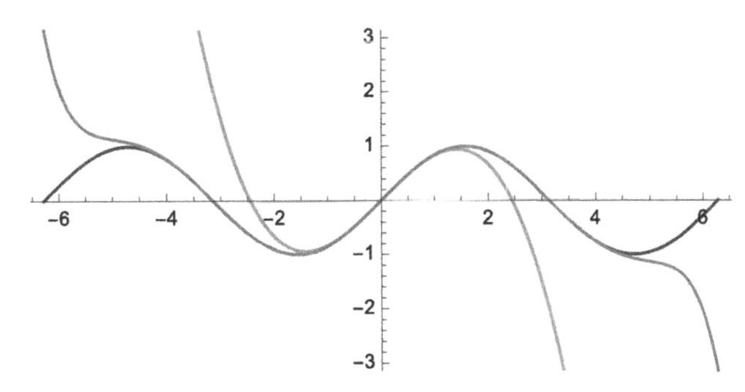

FIGURE 3.54 $y = \sin x$ together with the graphs of $p_1(x)$, $p_5(x)$, and $p_{10}(x)$.

Solution. (a) We test for absolute convergence first using the Ratio Test. Because

$$\lim_{k\to\infty}\left|\frac{\dfrac{(-1)^{k+1}}{(2(k+1)+1)!}x^{2(k+1)+1}}{\dfrac{(-1)^k}{(2k+1)!}x^{2k+1}}\right| = \lim_{k\to\infty}\frac{1}{2(k+1)(2k+3)}x^2 = 0 < 1$$

a[x_, k_] = (−1)^k/(2k + 1)!x^(2k + 1);

s1 = FullSimplify[a[x, k + 1]/a[x, k]]

Limit[s1, k → Infinity]

$-\frac{x^2}{6+10k+4k^2}$

0

for all values of x, we conclude that the series converges absolutely for all values of x; the interval of convergence is $(-\infty, \infty)$. In fact, we will see later that this series converges to $\sin x$:

$$\sin x = \sum_{k=0}^{\infty}\frac{(-1)^{k+1}}{(2k+1)!}x^{2k+1} = x - \frac{1}{3!}x^3 + \frac{1}{5!}x^5 - \frac{1}{7!}x^7 + \ldots,$$

which means that the partial sums of the series converge to $\sin x$. Graphically, we can visualize this by graphing partial sums of the series together with the graph of $y = \sin x$. Note that the partial sums of a series are a recursively defined function: $s_n = s_{n-1} + a_n$, $s_0 = a_0$. We use this observation to define `p` to be the nth partial sum of the series. We use the form `p[x_,n_]:=p[x,n]=...` so that Mathematica "remembers" the partial sums computed. That is, once `p[x,3]` is computed, Mathematica need not recompute `p[x,3]` when computing `p[x,4]`.

In Fig. 3.54 we graph $p_n(x) = \sum_{k=0}^{n}\frac{(-1)^k}{(2k+1)!}x^{2k+1}$ together with $y = \sin x$ for $n = 1, 5,$ and 10. In the graphs, notice that as n increases, the graphs of $p_n(x)$ more closely resemble the graph of $y = \sin x$.

When you use `Tooltip`, placing the cursor over the plot shows you the function being plotted.

Clear[p]

p[x_, 0] = a[x, 0];

p[x_, n_]:=p[x, n] = p[x, n − 1] + a[x, n]

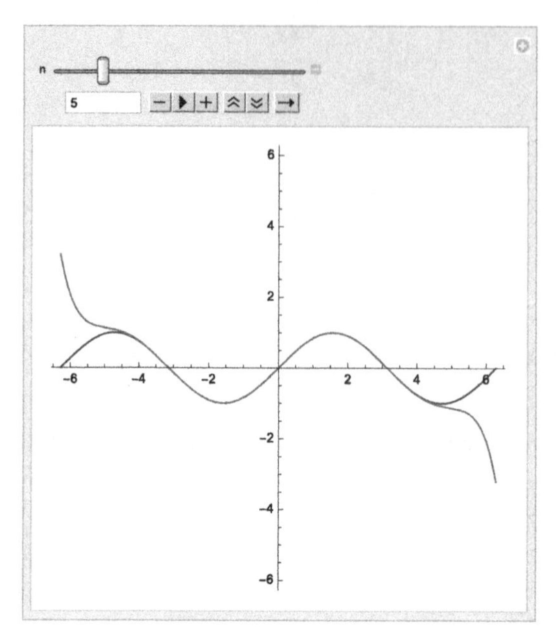

FIGURE 3.55 Using `Manipulate` to investigate the situation.

```
p[x, 2]
```

$x-\frac{x^3}{6}+\frac{x^5}{120}$

```
p1 = Plot[Tooltip[{Sin[x], p[x, 1], p[x, 5]}], {x, −2Pi, 2Pi},
  PlotStyle → {{CMYKColor[1, .7, .07, .45]}, {CMYKColor[0, .63, 1, 0]},
  {CMYKColor[.96, .09, .32, .29]}},
  PlotRange → {−Pi, Pi}, AspectRatio → Automatic]
```

We use `Manipulate` to investigate how n affects the situation with (see Fig. 3.55)

```
p2 = Manipulate[Plot[Tooltip[{Sin[x], p[x, n]}],
  {x, −2Pi, 2Pi},
  PlotStyle → {{CMYKColor[1, .7, .07, .45]}, {CMYKColor[0, .63, 1, 0]}},
  PlotRange → {−2Pi, 2Pi}, AspectRatio → Automatic], {{n, 5}, 1, 25, 1}]
```

(b) As in (a), we test for absolute convergence first using the Ratio Test:

$$\lim_{k\to\infty}\left|\frac{\frac{(k+1)k!}{1000^{k+1}}(x-1)^{k+1}}{\frac{k!}{1000^k}(x-1)^k}\right|=\frac{1}{1000}(k+1)|x-1|=\begin{cases}0, \text{ if } x=1\\ \infty, \text{ if } x\neq 1.\end{cases}$$

```
a[x_, k_] = k!/1000^k(x − 1)^k;
s1 = FullSimplify[a[x, k + 1]/a[x, k]]
Limit[s1, k → Infinity]
```

$\frac{(1+k)(-1+x)}{1000}$

$(-1+x)\infty$

Be careful of your interpretation of the result of the `Limit` command because Mathematica does not consider the case $x=1$ separately: if $x=1$ the limit is 0. Because $0<1$ the series converges by the Ratio Test.

The series converges only if $x=1$; the interval of convergence is $\{1\}$. You should observe that if you graph several partial sums for "small" values of n, you might incorrectly conclude that the series converges. (c) Use the Ratio Test to check absolute convergence first:

$$\lim_{k\to\infty}\left|\frac{\dfrac{2^{k+1}}{\sqrt{k+1}}(x-4)^{k+1}}{\dfrac{2^k}{\sqrt{k}}(x-4)^k}\right| = \lim_{k\to\infty} 2\sqrt{\frac{k}{k+1}}|x-4| = 2|x-4|.$$

By the Ratio Test, the series converges absolutely if $2|x-4|<1$. We solve this inequality for x with `Reduce` to see that $2|x-4|<1$ if $7/2<x<9/2$.

```
Clear[a, s1, k]
a[x_, k_] = 2^k/Sqrt[k](x - 4)^k;
s1 = Simplify[Abs[a[x, k + 1]/a[x, k]]]
Limit[s1, k → Infinity]
```

$2\text{Abs}\left[\sqrt{\frac{k}{1+k}}(-4+x)\right]$

$2\text{Abs}[-4+x]$

Use `Reduce` to solve the inequality.

```
Reduce[2Abs[x - 4] < 1, x]
```

$\frac{7}{2}<\text{Re}[x]<\frac{9}{2}\&\&-\frac{1}{2}\sqrt{-63+32\text{Re}[x]-4\text{Re}[x]^2}<\text{Im}[x]$

$<\frac{1}{2}\sqrt{-63+32\text{Re}[x]-4\text{Re}[x]^2}$

From the output, we see that for real values of x, the inequality is satisfied for $7/2<x<9/2$. We check $x=7/2$ and $x=9/2$ separately. If $x=7/2$, the series becomes $\sum_{k=1}^{\infty}(-1)^k\frac{1}{\sqrt{k}}$, which converges conditionally.

```
Simplify[a[x, k]/.x → 7/2]
```

$\frac{(-1)^k}{\sqrt{k}}$

On the other hand, if $x=9/2$,

Simplify[a[x, k]/.x → 9/2]

$\frac{1}{\sqrt{k}}$

the series is $\sum_{k=1}^{\infty} \frac{1}{\sqrt{k}}$, which diverges. We conclude that the interval of convergence is $[7/2, 9/2)$. □

3.4.6 Taylor and Maclaurin Series

Let $y = f(x)$ be a function with derivatives of all orders at $x = x_0$. The **Taylor series** for $f(x)$ about $x = x_0$ is

$$\sum_{k=0}^{\infty} \frac{f^{(k)}(x_0)}{k!}(x - x_0)^k. \tag{3.29}$$

The **Maclaurin series** for $f(x)$ is the Taylor series for $f(x)$ about $x = 0$. If $y = f(x)$ has derivatives up to at least order n at $x = x_0$, the nth degree **Taylor polynomial** for $f(x)$ about $x = x_0$ is

$$p_n(x) = \sum_{k=0}^{n} \frac{f^{(k)}(x_0)}{k!}(x - x_0)^k. \tag{3.30}$$

The nth degree **Maclaurin polynomial** for $f(x)$ is the nth degree Taylor polynomial for $f(x)$ about $x = 0$. Generally, finding Taylor and Maclaurin series using the definition is a tedious task at best.

Example 3.54

Find the first few terms of (a) the Maclaurin series and (b) the Taylor series about $x = \pi/4$ for $f(x) = \tan x$.

Use `Short` to obtain an abbreviated result. Many terms will be missing, but with `Short`, you will see the beginning and end of your result.

Solution. (a) After defining $f(x) = \tan x$, we use `Table` together with `/.` and `D` to compute $f^{(k)}(0)/k!$ for $k = 0, 1, \ldots, 8$.

```
Clear[f]
f[x_] = Tan[x];
t1 = Table[{k, D[f[x], {x, k}], D[f[x], {x, k}]/.x → 0}, {k, 0, 8}];
Short[t1]
```

{{0, Tan[x], 0}, ⟨⟨7⟩⟩, {8, 7936Sec[x]^8Tan[x] + 24576⟨⟨1⟩⟩^6⟨⟨1⟩⟩^3 + ⟨⟨1⟩⟩ + 128Sec[x]^2 Tan[x]^7, 0}}

To see these results in tabular form, enter

```
t1//TableForm
```

For length considerations, the resulting output is not shown here. Another way of approaching the problem is to use `Manipulate`. See Fig. 3.56.

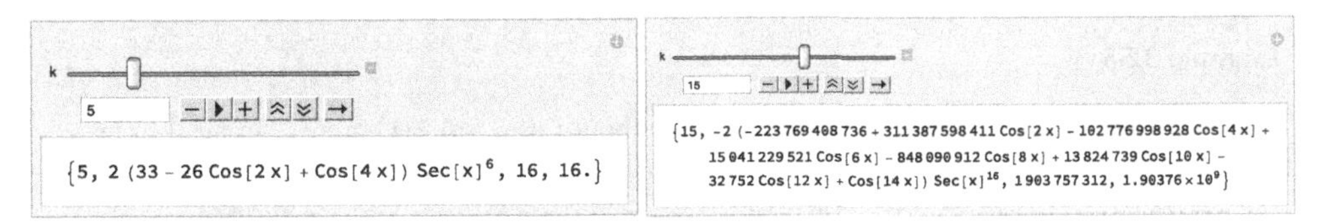

FIGURE 3.56 With `Manipulate`, we can adjust the function and function values.

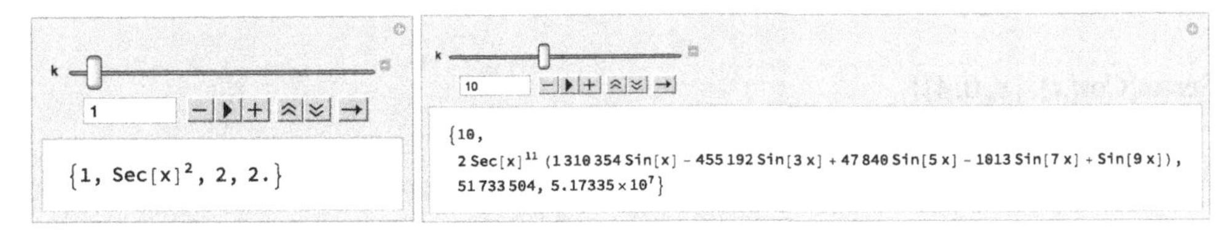

FIGURE 3.57 We use `Manipulate` to investigate series for the tangent function.

Manipulate[{k, D[f[x], {x, k}]//FullSimplify, D[f[x], {x, k}]/.$x \to$ 0,

D[f[x], {x, k}]/.$x \to$ 0//N}, {{k, 5}, 0, 25, 1}]

Using the values in the table or from the `Manipulate` object, we apply the definition to see that the Maclaurin series is

$$\sum_{k=0}^{\infty} \frac{f^{(k)}(0)}{k!} x^k = x + \frac{1}{3}x^3 + \frac{2}{15}x^5 + \frac{17}{315}x^7 + \ldots$$

For (b), we repeat (a) using $x = \pi/4$ instead of $x = 0$

Manipulate[{k, D[f[x], {x, k}]//FullSimplify, D[f[x], {x, k}]/.$x \to$ Pi/4,

D[f[x], {x, k}]/.$x \to$ Pi/4//N}, {{k, 1}, 0, 25, 1}]

and then apply the definition to see that the Taylor series about $x = \pi/4$ is

$$\sum_{k=0}^{\infty} \frac{f^{(k)}(x_0)}{k!}(x - x_0)^k = 1 + 2\left(x - \frac{\pi}{4}\right) + 2\left(x - \frac{\pi}{4}\right)^2 + \frac{8}{3}\left(x - \frac{\pi}{4}\right)^3 +$$
$$\frac{10}{3}\left(x - \frac{\pi}{4}\right)^4 + \frac{64}{15}\left(x - \frac{\pi}{4}\right)^5 + \frac{244}{45}\left(x - \frac{\pi}{4}\right)^6 + \ldots$$

From the series, we can see various Taylor and Maclaurin polynomials. For example, the third Maclaurin polynomial is

$$p_3(x) = x + \frac{1}{3}x^3$$

and the 4th degree Taylor polynomial about $x = \pi/4$ is

$$p_4(x) = 1 + 2\left(x - \frac{\pi}{4}\right) + 2\left(x - \frac{\pi}{4}\right)^2 + \frac{8}{3}\left(x - \frac{\pi}{4}\right)^3 + \frac{10}{3}\left(x - \frac{\pi}{4}\right)^4.$$

See Fig. 3.57. □

The command `Series[f[x],{x,x0,n}]` computes (3.29) to (at least) order $n - 1$. Because of the O-term in the result that represents the terms that are omitted from the power series for $f(x)$ expanded about the point $x = x_0$, the result of entering a `Series` command is not a function that can be evaluated if x is a particular number. We remove the remainder (O-) term of the power series `Series[f[x],{x,x0,n}]` with the command `Normal` and can then evaluate the resulting polynomial for particular values of x.

Example 3.55

Find the first few terms of the Taylor series for $f(x)$ about $x = x_0$. (a) $f(x) = \cos x$, $x = 0$; (b) $f(x) = 1/x^2$, $x = 1$.

Solution. Entering

Series[Cos[x], {x, 0, 4}]

$$1 - \frac{x^2}{2} + \frac{x^4}{24} + O[x]^5$$

computes the Maclaurin series to order 4. Entering

Series[Cos[x], {x, 0, 14}]

$$1 - \frac{x^2}{2} + \frac{x^4}{24} - \frac{x^6}{720} + \frac{x^8}{40320} - \frac{x^{10}}{3628800} + \frac{x^{12}}{479001600} - \frac{x^{14}}{87178291200} + O[x]^{15}$$

computes the Maclaurin series to order 14. In this case, the Maclaurin series for $\cos x$ converges to $\cos x$ for all real x. To graphically see this, we define the function `p`. Given n, `p[n]` returns the Maclaurin polynomial of degree n for $\cos x$.

```
p[n_]:=Series[Cos[x], {x, 0, n}]//Normal
```

```
p[3]
```

$$1 - \frac{x^2}{2}$$

We then graph $\cos x$ together with the Maclaurin polynomial of degree $n = 2, 4, 8$, and 16 on the interval $[-3\pi/2, 3\pi/2]$ in Fig. 3.58. Notice that as n increases, the graph of the Maclaurin polynomial more closely resembles the graph of $\cos x$. We would see the same pattern if we increased the length of the interval and the value of n.

```
Manipulate[Plot[Evaluate[Tooltip[{Cos[x], p[n]}]], {x, −3Pi/2, 3Pi/2},
  PlotStyle → {{CMYKColor[0, 1, .65, .34]}, {CMYKColor[1, .65, 0, .3]}},
  PlotRange → {{−3Pi/2, 3Pi/2}, {−3Pi/2, 3Pi/2}}, AspectRatio → Automatic],
  {{n, 4}, {2, 4, 8, 16, 32, 64}, ControlType → Setter}]
```

(b) After defining $f(x) = 1/x^2$, we compute the first 10 terms of the Taylor series for $f(x)$ about $x = 1$ with `Series`.

```
f[x_] = 1/x^2;
p10 = Series[f[x], {x, 1, 10}]
```

$$1 - 2(x-1) + 3(x-1)^2 - 4(x-1)^3 + 5(x-1)^4 - 6(x-1)^5 + 7(x-1)^6 - 8(x-1)^7 + 9(x-1)^8 - 10(x-1)^9 + 11(x-1)^{10} + O[x-1]^{11}$$

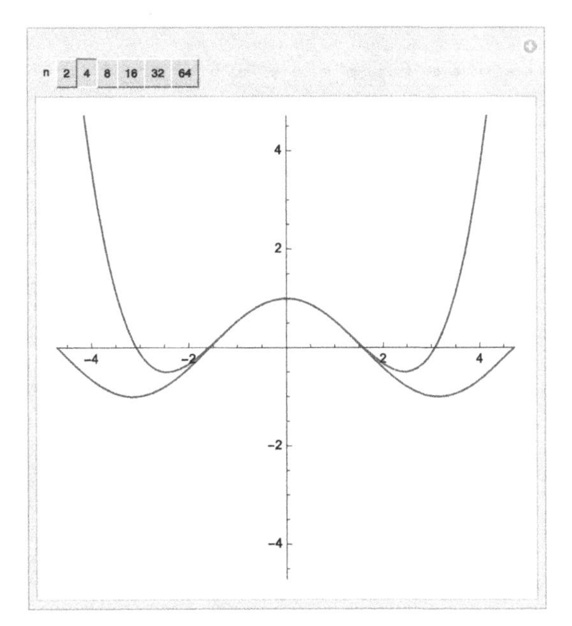

FIGURE 3.58 Using `Manipulate` to investigate graphs of $y = \cos x$ together with plots of several of its Maclaurin polynomials (University of Pennsylvania colors).

In this case, the pattern for the series is relatively easy to see: the Taylor series for $f(x)$ about $x = 1$ is

$$\sum_{k=0}^{\infty}(-1)^k(k+1)(x-1)^k.$$

This series converges absolutely if

$$\lim_{k\to\infty}\left|\frac{(-1)^{k+1}(k+2)(x-1)^{k+1}}{(-1)^k(k+1)(x-1)^k}\right| = |x-1| < 1$$

or $0 < x < 2$. The series diverges if $x = 0$ and $x = 2$. In this case, the series converges to $f(x)$ on the interval $(0, 2)$.

```
a[x_, k_] = (-1)^k(k + 1)(x - 1)^k;
s1 = FullSimplify[Abs[a[x, k + 1]/a[x, k]]]
```

$\text{Abs}\left[\frac{(2+k)(-1+x)}{1+k}\right]$

```
s2 = Limit[s1, k -> Infinity]
```

Abs[$-1 + x$]

```
Reduce[s2 < 1, x]
```

$0 < \text{Re}[x] < 2\&\& -\sqrt{2\text{Re}[x] - \text{Re}[x]^2} < \text{Im}[x] < \sqrt{2\text{Re}[x] - \text{Re}[x]^2}$

To see this, we use `Manipulate` to graph $f(x)$ together with the Taylor polynomial for $f(x)$ about $x = 1$ of degree n for large n. Regardless of the size of n, the graphs of $f(x)$ and the

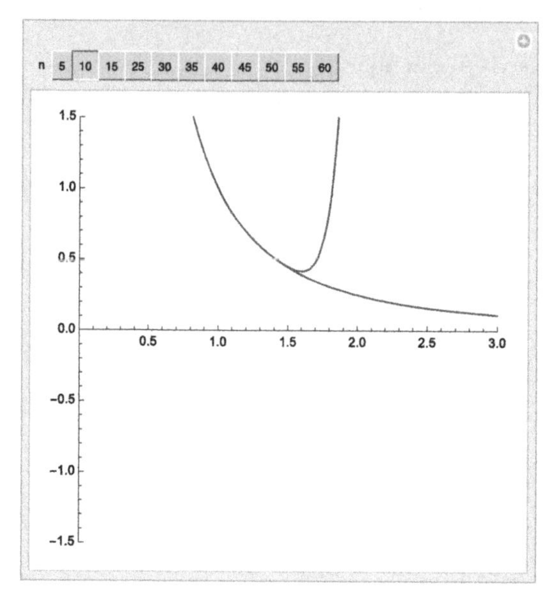

FIGURE 3.59 Graphs of $f(x)$ together with the various Taylor polynomials about $x = 1$.

Taylor polynomial closely resemble each other on the interval (0, 2) – but not at the endpoints or outside the interval. (See Fig. 3.59.)

```
p[n_]:=Series[f[x], {x, 1, n}]//Normal

Manipulate[Plot[Evaluate[Tooltip[{f[x], p[n]}]],
  {x, 0, 3}, PlotStyle → {{CMYKColor[0, 1, .65, .34]}, {CMYKColor[1, .65, 0, .3]}},
  PlotRange → {{0, 3}, {−3/2, 3/2}}, AspectRatio → Automatic],
  {{n, 10}, {5, 10, 15, 25, 30, 35, 40, 45, 50, 55, 60}, ControlType → Setter}]
```

□

3.4.7 Taylor's Theorem

Taylor's theorem states the relationship between $f(x)$ and the Taylor series for $f(x)$ about $x = x_0$.

Theorem 3.16 (Taylor's Theorem). *Let $y = f(x)$ have (at least) $n + 1$ derivatives on an interval I containing $x = x_0$. Then, for every number $x \in I$, there is a number z between x and x_0 so that*

$$f(x) = p_n(x) + R_n(x),$$

where $p_n(x)$ is given by equation (3.30) *and*

$$R_n(x) = \frac{f^{(n+1)}(z)}{(n+1)!}(x - x_0)^{n+1}. \tag{3.31}$$

Example 3.56

Use Taylor's theorem to show that

$$\sin x = \sum_{k=0}^{\infty} \frac{(-1)^k}{(2k+1)!} x^{2k+1}$$

Solution. Let $f(x) = \sin x$. Then, for each value of x, there is a number z between 0 and x so that $\sin x = p_n(x) + R_n(x)$ where $p_n(x) = \sum_{k=0}^{n} \frac{f^{(k)}(0)}{k!} x^k$ and $R_n(x) = \frac{f^{(n+1)}(z)}{(n+1)!} x^{n+1}$. Regardless of the value of n, $f^{(n+1)}(z)$ is one of $\sin z$, $-\sin z$, $\cos z$, or $-\cos z$, which are all bounded by 1. Then,

$$|\sin x - p_n(x)| = \left| \frac{f^{(n+1)}(z)}{(n+1)!} x^{n+1} \right|$$
$$|\sin x - p_n(x)| \le \frac{1}{(n+1)!} |x|^{n+1}$$

and $x^n/n! \to 0$ as $n \to \infty$ for all real values of x.

You should remember that the number z in $R_n(x)$ is guaranteed to exist by Taylor's theorem. However, from a practical point of view, you would rarely (if ever) need to compute the z value for a particular x value.

For illustrative purposes, we show the difficulties. Suppose we wish to approximate $\sin \pi/180$ using the Maclaurin polynomial of degree 4, $p_4(x) = x - \frac{1}{6}x^3$, for $\sin x$. The fourth remainder is $R_4(x) = \frac{1}{120} \cos z \, x^5$.

The Maclaurin polynomial of degree 4 for $\sin x$ is $\sum_{k=0}^{4} \frac{f^{(k)}(0)}{k!} x^4 = 0 + x + 0 \cdot x^2 + \frac{-1}{3!} x^3 + 0 \cdot x^4$.

```
Clear[f]
f[x_] = Sin[x];
r5 = D[f[z], {z, 5}]/5!x^5
```

$\frac{1}{120} x^5 \text{Cos}[z]$

If $x = \pi/180$ there is a number z between 0 and $\pi/180$ so that

$$\left| R_4\left(\frac{\pi}{180}\right) \right| = \frac{1}{120} \cos z \left(\frac{\pi}{180}\right)^5$$
$$\le \frac{1}{120} \left(\frac{\pi}{180}\right)^5 \approx 0.135 \times 10^{-10},$$

which shows us that the maximum the error can be is $\frac{1}{120} \left(\frac{\pi}{180}\right)^5 \approx 0.135 \times 10^{-10}$.

```
maxerror = N[1/120 * (Pi/180)^5]
```

1.349601623163255*^ − 11

Abstractly, the exact error can be computed. By Taylor's theorem, z satisfies

$$f\left(\frac{\pi}{180}\right) = p_4\left(\frac{\pi}{180}\right) + R_4\left(\frac{\pi}{180}\right)$$
$$\sin \frac{\pi}{180} = \frac{1}{180}\pi - \frac{1}{34992000}\pi^3 + \frac{1}{22674816000000}\pi^5 \cos z$$
$$0 = \frac{1}{180}\pi - \frac{1}{34992000}\pi^3 + \frac{1}{22674816000000}\pi^5 \cos z - \sin \frac{\pi}{180}.$$

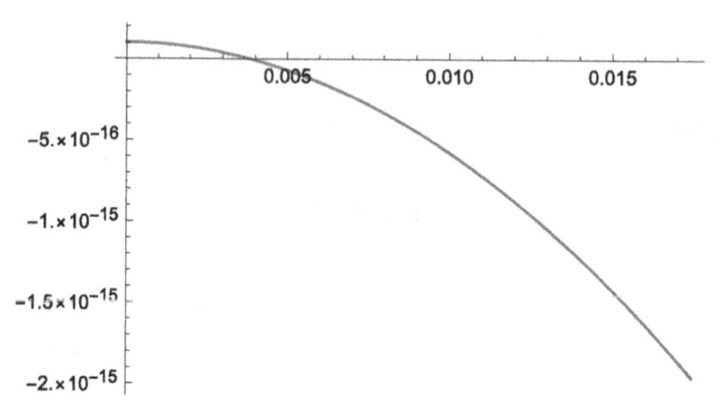

FIGURE 3.60 Finding z (Observe that Mathematica's default color is similar to Columbia University's blue).

We graph the right-hand side of this equation with `Plot` in Fig. 3.60. The exact value of z is the z-coordinate of the point where the graph intersects the z-axis.

p4 = Series[f[x], {x, 0, 4}]//Normal

$x - \frac{x^3}{6}$

exval = Sin[Pi/180]

p4b = p4/.x → Pi/180

r5b = r5/.x → Pi/180

$\text{Sin}\left[\frac{\pi}{180}\right]$

$\frac{\pi}{180} - \frac{\pi^3}{34992000}$

$\frac{\pi^5 \text{Cos}[z]}{22674816000000}$

toplot = r5b + p4b − exval;

Plot[toplot, {z, 0, Pi/180}]

We can use `FindRoot` to approximate z, if we increase the number of digits carried in floating point calculations with `WorkingPrecision`.

exz = FindRoot[toplot == 0, {z, 0, .004}, WorkingPrecision → 32]

$\{z \to 0.0038086149165541606417429516417308\}$

Alternatively, we can compute the exact value of z with `Solve`

cz = Solve[toplot == 0, z]

$\{\{z \to \text{ConditionalExpression}[-\text{ArcCos}[\frac{648000(-194400\pi + \pi^3 + 34992000\text{Sin}[\frac{\pi}{180}])}{\pi^5}] + 2\pi C[1],$

$C[1] \in \text{Integers}]\}, \{z \to \text{ConditionalExpression}[\text{ArcCos}[\frac{648000(-194400\pi + \pi^3 + 34992000\text{Sin}[\frac{\pi}{180}])}{\pi^5}]$

$+2\pi C[1], C[1] \in \text{Integers}]\}\}$

and then approximate the result with `N`.

N[cz]

$\{\{z \to \text{ConditionalExpression}[-0.00384232 + 6.28319C[1], C[1] \in \text{Integers}]\},$

$\{z \to \text{ConditionalExpression}[0.00384232 + 6.28319C[1], C[1] \in \text{Integers}]\}\}$ □

Example 3.57: Newton's Approximation of π.

Newton's Approximation of π is a fascinating combination of calculus, infinite series, and geometry. Newton began the approximation after proving the **Binomial Theorem**.

Theorem 3.17 (Binomial Theorem). *For $-1 < x < 1$,*

$$(1+x)^r = 1 + \sum_{k=1}^{\infty} \binom{r}{k} x^k, \tag{3.32}$$

where $\binom{r}{1} = 1$, $\binom{r}{2} = \frac{r(r-1)}{2!}$, *and* $\binom{r}{k} = \frac{r(r-1)(r-2)\cdots(r-k+1)}{k!}$ *for* $k \geq 3$.

We use `Series` to find the first few terms of the Binomial series.

Series[(1 + x)^r, {x, 0, 5}]

$1 + rx + \frac{1}{2}(-1+r)rx^2 + \frac{1}{6}(-2+r)(-1+r)rx^3 + \frac{1}{24}(-3+r)(-2+r)(-1+r)rx^4 + \frac{1}{120}(-4+r)(-3+r)(-2+r)(-1+r)rx^5 + O[x]^6$

Newton then used the Binomial series to find the first few terms of the series for $\sqrt{1-x} = (1-x)^{1/2}$.

f[x_] = Series[Sqrt[1 − x], {x, 0, 20}]//Normal

$1 - \frac{x}{2} - \frac{x^2}{8} - \frac{x^3}{16} - \frac{5x^4}{128} - \frac{7x^5}{256} - \frac{21x^6}{1024} - \frac{33x^7}{2048} - \frac{429x^8}{32768} - \frac{715x^9}{65536} - \frac{2431x^{10}}{262144} - \frac{4199x^{11}}{524288} - \frac{29393x^{12}}{4194304} - \frac{52003x^{13}}{8388608} - \frac{185725x^{14}}{33554432} - \frac{334305x^{15}}{67108864} - \frac{9694845x^{16}}{2147483648} - \frac{17678835x^{17}}{4294967296} - \frac{64822395x^{18}}{17179869184} - \frac{119409675x^{19}}{34359738368} - \frac{883631595x^{20}}{274877906944}$

In Fig. 3.61, observe that the series quickly converges to $\sqrt{1-x}$ if x is small. He used this observation to use the series to approximate $\sqrt{3}$ by observing

$$\sqrt{3} = \sqrt{4 \cdot \frac{3}{4}} = 2\sqrt{1 - \frac{1}{4}}$$

and substituting $x = 1/4$ into the first few terms of the series.

Plot[{f[x], Sqrt[1 − x]}, {x, −5, 5}, PlotRange → {−5, 5}, AspectRatio → 1,

PlotStyle → {{Thickness[.01], CMYKColor[.4, 1, .5, .15, .5]},

{Thickness[.01], CMYKColor[0, .65, .9, 0], Dashing[.02]}}, AxesLabel → {x, y}]

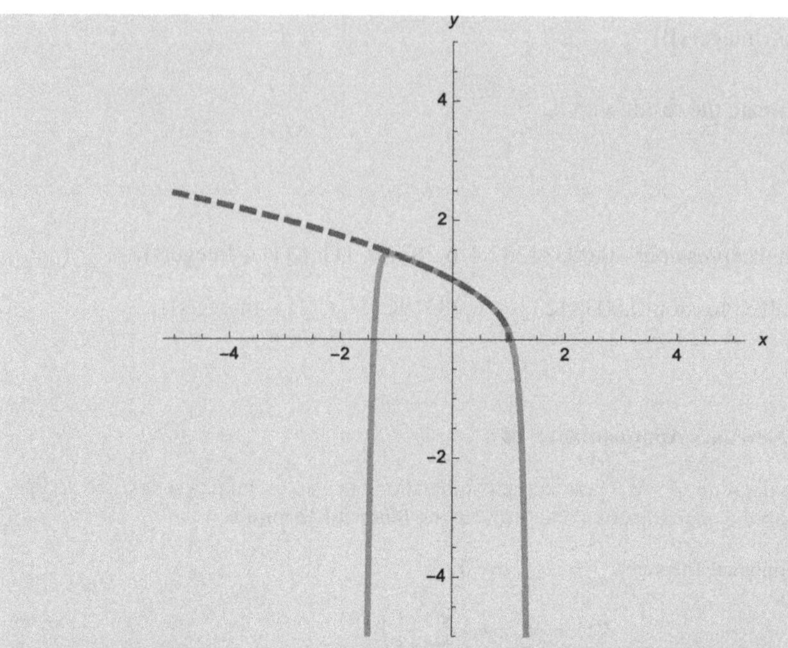

FIGURE 3.61 The Maclaurin series for $\sqrt{1-x}$ converges quickly to $\sqrt{1-x}$ when x is small (Virginia Polytechnic Institute & State University colors).

Next, Newton looked at the circle $x^2 - x + y^2 = (x - 1/2)^2 + y^2 = (1/2)^2$ with center at $(1/2, 0)$ and radius $1/2$. Refer to Fig. 3.62.

```
p1 = PolarPlot[Cos[r], {r, 0, 2Pi}, PlotStyle → {Thickness[.01],
   CMYKColor[.4, 1, .5, .15]}];
p2 = Graphics[{Thickness[.01], CMYKColor[0, .65, .9, 0],
   Line[{{1/2, 0}, {1/4, Sqrt[3]/4}}], Line[{{1/4, 0}, {1/4, Sqrt[3]/4}}],
   CMYKColor[0, .18, 1, .27],
   Line[{{1/4, -Sqrt[3]/4}, {3/4, Sqrt[3]/4}}],
   Line[{{1/2, 0}, {3/4, -Sqrt[3]/4}}]}];
p3 = Plot[Sqrt[x - x^2], {x, 0, 1/4},
   PlotStyle → {Thickness[.01], CMYKColor[.4, 1, .5, .15]},
   Filling → Bottom, FillingStyle → CMYKColor[.43, 0, .14, .21]];
Show[p1, p2, p3, AxesLabel → {x, y}, PlotLabel → "Newton's Circle"]
```

Newton understood that the area of a circle with radius r is πr^2 so it followed that the area of Newton's circle is $\pi/4$. Moreover, the area of the sector formed by the green region and the orange triangle is 1/6th the area of the circle so the area of the sector is $A_{sector} = \pi/24$. The orange triangle has width $1/4$ and height $\sqrt{3}/4$ so the orange triangle has area

$$A_{triangle} = \frac{1}{2} \cdot \frac{1}{4} \cdot \frac{\sqrt{3}}{4} = \frac{\sqrt{3}}{32}.$$

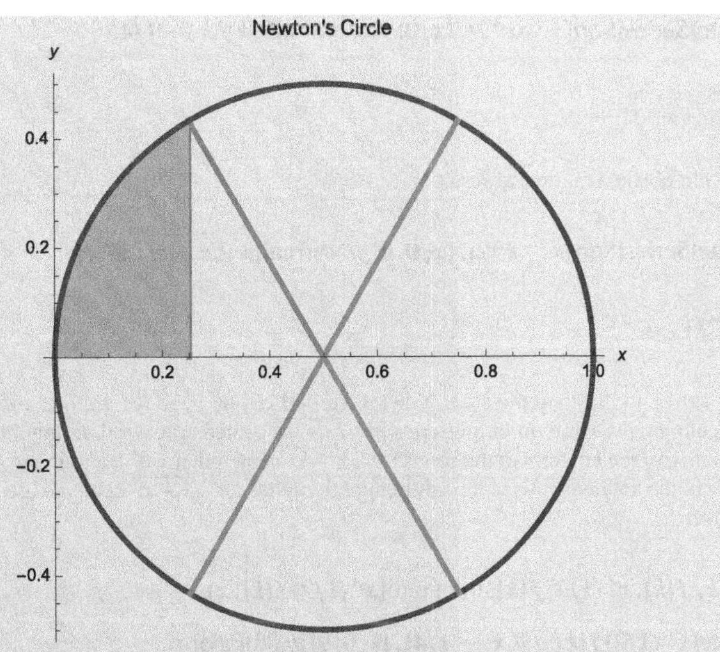

FIGURE 3.62 Newton's circle. The area of the green region is $\int_0^{1/4} \sqrt{x - x^2}\,dx$.

Because the top half of the circle is given by $y = \sqrt{x - x^2}$, the area of the green region, R, is $A_R = \int_0^{1/4} \sqrt{x - x^2}\,dx$. Newton then added and solved for π.

$$A_{sector} = A_R + A_{triangle}$$
$$\frac{\pi}{24} = \int_0^{1/4} \sqrt{x - x^2}\,dx + \frac{1}{32}\sqrt{3}$$
$$\pi = 24\left(\int_0^{1/4} \sqrt{x - x^2}\,dx + \frac{1}{32}\sqrt{3}\right).$$

Newton already had an excellent approximation of $\sqrt{3}$ as described above. Thus, to obtain an excellent approximation of π he needed only to find an excellent approximation of $\int_0^{1/4} \sqrt{x - x^2}\,dx$. To obtain it, he started with his power series for $\sqrt{1-x}$ and then observed that for $0 \le x \le 1$, $\sqrt{x - x^2} = \sqrt{x}\sqrt{x-1} = x^{1/2}(1-x)^{1/2}$. With that in mind, he decided to take his *power* series for $\sqrt{1-x}$ and multiply it by $x^{1/2}$ to obtain a series that he predicted would converge to $\sqrt{x - x^2}$. *Newton was right and it worked!*

```
p1 = Series[Sqrt[x - x^2], {x, 0, 9}]//Normal
```

$$\sqrt{x} - \frac{x^{3/2}}{2} - \frac{x^{5/2}}{8} - \frac{x^{7/2}}{16} - \frac{5x^{9/2}}{128} - \frac{7x^{11/2}}{256} - \frac{21x^{13/2}}{1024} - \frac{33x^{15/2}}{2048} - \frac{429x^{17/2}}{32768}$$

He then integrated his series (which was no longer a power series) term-by-term

```
Integrate[p1, x]//Simplify
```

$$\frac{2x^{3/2}}{3} - \frac{x^{5/2}}{5} - \frac{x^{7/2}}{28} - \frac{x^{9/2}}{72} - \frac{5x^{11/2}}{704} - \frac{7x^{13/2}}{1664} - \frac{7x^{15/2}}{2560} - \frac{33x^{17/2}}{17408} - \frac{429x^{19/2}}{311296}$$

and evaluated it if $x = 1/4$ to obtain his approximation of $\int_0^{1/4} \sqrt{x - x^2}\,dx$.

There is no need to show the evaluation of the series if $x = 0$ because the value of the series if $x = 0$ is 0.

```
Integrate[Series[Sqrt[x - x^2], {x, 0, 9}]//Normal, x]/.x → 1/4
```

$\frac{1919013728784979}{24995910798802944}$

We use N to obtain a numerical result.

```
Integrate[Series[Sqrt[x - x^2], {x, 0, 9}]//Normal, x]/.x → 1/4//N
```

0.0767731

We use Table to illustrate the steps Newton carried out by hand. In the first column, k, in the second column, the kth term of the series for $\sqrt{x-x^2}$ centered at $x=0$, in the third column, an antiderivative of the kth term of the series for $\sqrt{x-x^2}$ centered at $x=0$, and in the fourth column, the value of the antiderivative of the kth term of the series for $\sqrt{x-x^2}$ centered at $x=0$ evaluated at $x=1/4$.

```
Table[{k, f[k], x^(1/2)f[k], Integrate[x^(1/2)f[k], x],
Integrate[x^(1/2)f[k], x]/.x → 1/4}, {k, 0, 9}]//TableForm
```

0	1	$\sqrt{x}$	$\frac{2x^{3/2}}{3}$	$\frac{1}{12}$
1	$-\frac{x}{2}$	$-\frac{x^{3/2}}{2}$	$-\frac{x^{5/2}}{5}$	$-\frac{1}{160}$
2	$-\frac{x^2}{8}$	$-\frac{x^{5/2}}{8}$	$-\frac{x^{7/2}}{28}$	$-\frac{1}{3584}$
3	$-\frac{x^3}{16}$	$-\frac{x^{7/2}}{16}$	$-\frac{x^{9/2}}{72}$	$-\frac{1}{36864}$
4	$-\frac{5x^4}{128}$	$-\frac{5x^{9/2}}{128}$	$-\frac{5x^{11/2}}{704}$	$-\frac{5}{1441792}$
5	$-\frac{7x^5}{256}$	$-\frac{7x^{11/2}}{256}$	$-\frac{7x^{13/2}}{1664}$	$-\frac{7}{13631488}$
6	$-\frac{21x^6}{1024}$	$-\frac{21x^{13/2}}{1024}$	$-\frac{7x^{15/2}}{2560}$	$-\frac{7}{83886080}$
7	$-\frac{33x^7}{2048}$	$-\frac{33x^{15/2}}{2048}$	$-\frac{33x^{17/2}}{17408}$	$-\frac{33}{2281701376}$
8	$-\frac{429x^8}{32768}$	$-\frac{429x^{17/2}}{32768}$	$-\frac{429x^{19/2}}{311296}$	$-\frac{429}{163208757248}$
9	$-\frac{715x^9}{65536}$	$-\frac{715x^{19/2}}{65536}$	$-\frac{715x^{21/2}}{688128}$	$-\frac{715}{1443109011456}$

Thus, Newton's approximation of π, $\pi = 24\left(\int_0^{1/4}\sqrt{x-x^2}\,dx + \frac{1}{32}\sqrt{3}\right)$ using his approximations of $\int_0^{1/4}\sqrt{x-x^2}\,dx$ was given by

```
24(Sqrt[3]/32 + Integrate[Series[Sqrt[x - x^2], {x, 0, 9}]//Normal, x]/.x → 1/4)
```

$24\left(\frac{1919013728784979}{24995910798802944} + \frac{\sqrt{3}}{32}\right)$

We use N to approximate $\sqrt{3}$.

N[24(Sqrt[3]/32 + Integrate[Series[Sqrt[x − x^2], {x, 0, 9}]//Normal, x]/.x → 1/4), 20]

3.1415926683631729578

When we compare Newton's approximation of π to our present day approximation of π, it is truly remarkable that Newton's approximation was accurate to seven decimal places (to the right of the decimal point).

N[Pi, 20]

3.1415926535897932385

There are volumes of literature about the history of π, its applications, and how to efficiently and accurately approximate π (as well as other irrational and transcendental numbers).

Example 3.58

Determine whether the following series converge or diverge: (a) $\sum_{k=0}^{\infty} \frac{1}{16^k}\left(\frac{4}{8k+1} - \frac{2}{8k+4} - \frac{1}{8k+5} - \frac{1}{8k+6}\right)$; (b) $\sum_{k=0}^{\infty} \frac{2^k (k!)^2}{(2k+1)!}$; (c) $\frac{2\sqrt{2}}{9801}\sum_{k=0}^{\infty} \frac{(4k)!(1103+26390k)}{(k!)^4 396^{4k}}$; (d) $12\sum_{k=0}^{\infty} \frac{(-1)^k (6k)!(135910409+545140134k)}{(3k)!(k!)^3 640320^{3k+3/2}}$.

Solution. To determine whether each series converges or diverges, we use the Ratio Test. For (a), we see that $\lim_{k\to\infty} \left|\frac{a_{k+1}}{a_k}\right| = \frac{1}{16}$. Because $1/16 < 1$, the series converges (absolutely).

a[k_] = 1/16^k(4/(8k + 1) − 2/(8k + 4) − 1/(8k + 5) − 1/(8k + 6))

$16^{-k}\left(\frac{4}{1+8k} - \frac{2}{4+8k} - \frac{1}{5+8k} - \frac{1}{6+8k}\right)$

s1 = FullSimplify[a[k + 1]/a[k]]

$\frac{(1+2k)(3+4k)(1+8k)(5+8k)(318+k(391+120k))}{16(3+2k)(7+4k)(9+8k)(13+8k)(47+k(151+120k))}$

Limit[s1, k → Infinity]

$\frac{1}{16}$

In fact, in 1997, David Bailey, Peter Borwein, and Simon Plouffe (BBP) proved that

$$\pi = \sum_{k=0}^{\infty} \frac{1}{16^k}\left(\frac{4}{8k+1} - \frac{2}{8k+4} - \frac{1}{8k+5} - \frac{1}{8k+6}\right).$$

Notice that the second partial sum is accurate to three decimal places,

N[Sum[a[k], {k, 0, 1}], 10]

3.141422466

and the seventh partial sum is accurate to ten decimal places.

N[Sum[a[k], {k, 0, 6}], 15]

3.14159265357288

N[Pi, 15]

3.14159265358979

Interestingly, in 2013, Fabrice Bellard improved on the BBP equation and proved that

$$\pi = \frac{1}{2^6}\sum_{k=0}^{\infty}\frac{(-1)^k}{2^{10k}}\left(-\frac{2^5}{4k+1}-\frac{1}{4k+3}+\frac{2^8}{10k+1}-\frac{2^6}{10k+3}-\frac{2^2}{10k+5}\right.$$
$$\left.-\frac{2^2}{10k+7}+\frac{1}{10k+9}\right).$$

(b) Using the Ratio Test, we see that the series converges because the limit of the ratio of successive terms is $1/2 < 1$: $\lim_{k\to\infty}\left|\frac{a_{k+1}}{a_k}\right| = \frac{1}{2}$.

a[k_] = 2^k(k!)^2/(2k + 1)!

$\frac{2^k(k!)^2}{(1+2k)!}$

s1 = FullSimplify[a[k + 1]/a[k]]

$\frac{1+k}{3+2k}$

Limit[s1, k → Infinity]

$\frac{1}{2}$

Newton proved that this series converges to $\pi/2$:

$$\frac{\pi}{2} = \sum_{k=0}^{\infty}\frac{2^k(k!)^2}{(2k+1)!}.$$

Observe that the "patterns" of the terms of this series are much easier to see than in the example illustrating Newton's approximation of π using integral calculus, infinite series, and Euclidean geometry. You can use the 31st partial sum of this series to approximate $\pi/2$ to 10 decimal places.

N[Sum[a[k], {k, 0, 30}], 10]

1.570796327

N[Pi/2, 10]

1.570796327

(c) Using the Ratio Test, we see that

a[k_] = 2Sqrt[2]/9801(4k)!(1103 + 26390k)/((k!)^4396^(4k))

$$\frac{2^{\frac{3}{2}-8k}99^{-2-4k}(1103+26390k)(4k)!}{(k!)^4}$$

p1 = FullSimplify[a[k + 1]/a[k]]

$$\frac{(1+2k)(1+4k)(3+4k)(27493+26390k)}{3073907232(1+k)^3(1103+26390k)}$$

Limit[p1, k → Infinity]

$$\frac{1}{96059601}$$

$\lim_{k\to\infty}\left|\frac{a_{k+1}}{a_k}\right| = \frac{1}{96059601} < 1$ so the series converges absolutely. This series that was introduced by Srinivasa Ramanujan around 1910 converges to $1/\pi$:

$$\frac{1}{\pi} = \frac{2\sqrt{2}}{9801}\sum_{k=0}^{\infty}\frac{(4k)!(1103+26390k)}{(k!)^4 396^{4k}}.$$

As with the other series in this example, the convergence is quick.

N[Sum[a[k], {k, 0, 2}], 24]

0.318309886183790671537767

N[1/Pi, 24]

0.318309886183790671537768

With just the third partial sum of the series, we see that it approximates $1/\pi$ to 23 decimal places.

(d) Using the Ratio Test, we see that the series converges. This formulation of π was introduced by brothers David Chudnosky and Gregory Chudnosky in 1989:

$$\frac{1}{\pi} = 12\sum_{k=0}^{\infty}\frac{(-1)^k(6k)!(135910409+545140134k)}{(3k)!(k!)^3 640320^{3k+3/2}}.$$

a[k_] = 12(−1)^k(6k)!(13591409 + 545140134k)/((3k)!(k!)^3640320^(3k + 3/2))

$$\frac{(-1)^k 3^{-\frac{1}{2}-3k}4^{-\frac{7}{2}-9k}3335^{-\frac{3}{2}-3k}(13591409+545140134k)(6k)!}{(k!)^3(3k)!}$$

p1 = Abs[FullSimplify[a[k + 1]/a[k]]]

$$\frac{\mathrm{Abs}\left[\frac{(1+2k)(1+6k)(5+6k)(558731543+545140134k)}{(1+k)^3(13591409+545140134k)}\right]}{10939058860032000}$$

Limit[p1, k → Infinity]

$\frac{1}{151931373056000}$

The series converges very quickly. Using the third partial sum, we see that it yields an approximation of $1/\pi$ accurate to 42 decimal places.

```
N[Sum[a[k], {k, 0, 2}], 42]
```

0.318309886183790671537767526745028724068919

```
N[1/Pi, 42]
```

0.318309886183790671537767526745028724068919

To use the Chudnosky's series to approximate π, we first rewrite it as

$$\frac{1}{\pi} = \frac{1}{42680\sqrt{10005}} \sum_{k=0}^{\infty} \frac{(-1)^k (6k)!(13591409 + 545140134k)}{(3k)!(k!)^3 640320^{3k}}. \tag{3.33}$$

Now let $A = \sum_{k=0}^{\infty} a_k = \sum_{k=0}^{\infty} \frac{(-1)^k (6k)!}{(3k)!(k!)^3 640320^{3k}}$ and $B = \sum_{k=0}^{\infty} b_k = \sum_{k=0}^{\infty} \frac{(-1)^k (6k)! k}{(3k)!(k!)^3 640320^{3k}}$. Then (3.33) becomes

$$\frac{1}{\pi} = \frac{1351409A + 545140134B}{42680\sqrt{10005}} \tag{3.34}$$

and reciprocating gives us

$$\pi = \frac{42680\sqrt{10005}}{1351409A + 545140134B}. \tag{3.35}$$

Observe that $a_k = \frac{(-1)^k (6k)!}{(3k)!(k!)^3 640320^{3k}}$, $b_k = k \cdot a_k$, and

$$\frac{a_k}{a_{k-1}} = -\frac{(6k-1)(6k-5)(2k-1)}{10939058860032000k^2}$$
$$a_k = -\frac{(6k-1)(6k-5)(2k-1)}{10939058860032000k^2} a_{k-1}.$$

Then, if we let $A_n = \sum_{k=0}^{n} a_k$,

$$A_n = \sum_{k=0}^{n-1} a_k - \frac{(6n-1)(6n-5)(2n-1)}{10939058860032000(n-1)n^2} a_{n-1}$$
$$= A_{n-1} - \frac{(6n-1)(6n-5)(2n-1)}{10939058860032000(n-1)n^2} a_{n-1}$$

and if we let $B_n = \sum_{k=0}^{n} b_k$, $B_n = \sum_{k=0}^{n} k a_k = \sum_{k=0}^{n-1} k a_k + n a_n = B_{n-1} + n a_n$. Then, we approximate π with $\pi \approx= \frac{42680\sqrt{10005}}{1351409A_n + 545140134B_n}$.

```
Remove[a, b, pi, capa, capb]

a[0] = 1;

a[n_]:=a[n] = N[-((-1+2n)(-5+6n)(-1+6n))/(10939058860032000n^3) a[n - 1], 250];

b[n_]:=b[n] = na[n]
```

capa[0] = a[0];

capb[0] = b[0];

capa[n_]:=capa[n] = capa[n − 1] + a[n]

capb[n_]:=capb[n] = capb[n − 1] + b[n]

pi[n_]:=N[426880Sqrt[10005]/(13591409capa[n] + 545140134capb[n]), 250]

We illustrate the quick convergence using A_{10} and B_{10}. Observe that the approximation gives us an approximation of π accurate to at least 155 digits.

N[pi[10], 155]

3.1415926535897932384626433832795028841971693993751058209749

445923078164062862089986280348253421170679821480865132823066

47093844609550582231725359408128481 1

N[Pi, 155]

3.1415926535897932384626433832795028841971693993751058209749

4459230781640628620899862803482534211706798214808651328230664

709384460955058223172535940812848111

□

Closely related to these series is the **Madhava–Leibniz** series, $\pi = \sqrt{12}\sum_{k=0}^{\infty}\frac{(-1)^k}{2k+1}3^k$. It is a good exercise to verify that the first 21 terms of the **Madhava–Leibniz** series compute an approximation of π accurate to eleven decimal places.

3.4.8 Other Series

In calculus, we learn that the power series $f(x) = \sum_{k=0}^{\infty} a_k (x - x_0)^k$ is differentiable and integrable on its interval of convergence. However, for series that are not power series this result is not generally true. For example, in more advanced courses, we learn that the function

$$f(x) = \sum_{k=0}^{\infty} \frac{1}{2^k} \sin\left(3^k x\right)$$

is continuous for all values of x but nowhere differentiable. We can use Mathematica to help us see why this function is not differentiable. Let

$$f_n(x) = \sum_{k=0}^{n} \frac{1}{2^k} \sin\left(3^k x\right).$$

Notice that $f_n(x)$ is defined recursively by $f_0(x) = \sin x$ and $f_n(x) = f_{n-1}(x) + \frac{1}{2^n}\sin(3^n x)$. We use Mathematica to recursively define $f_n(x)$.

Clear[f]

f[0] = Sin[x];

f[k_]:=f[k] = f[k − 1] + $\frac{\text{Sin}[3^k x]}{2^k}$

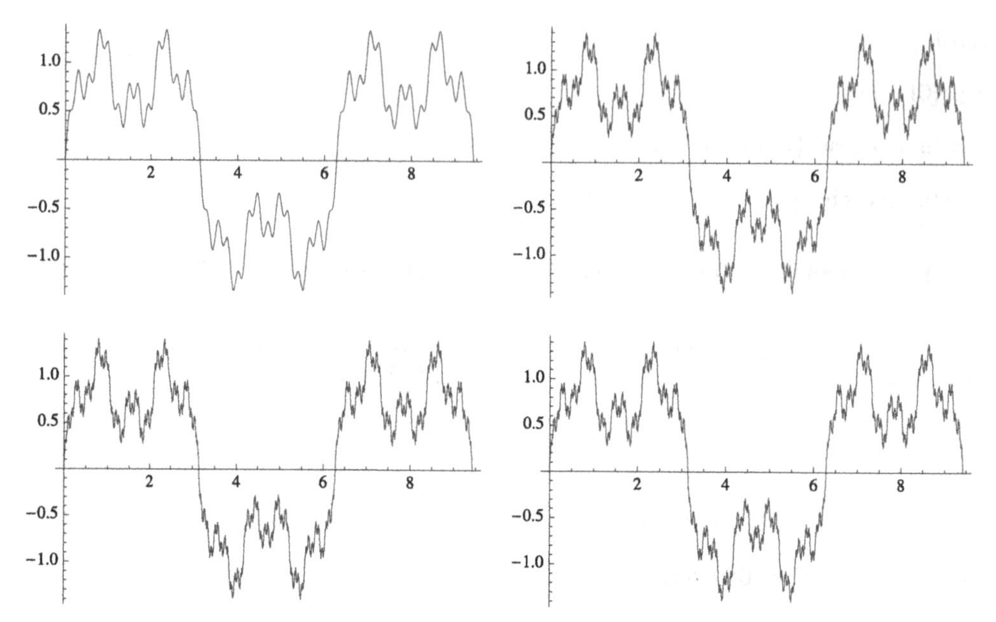

FIGURE 3.63 Approximating a function that is continuous everywhere but nowhere differentiable (Duke University colors).

We define $f_n(x)$ using the form

```
f[n_]:=f[n]=...
```

so that Mathematica "remembers" the values it computes. Thus, to compute `f[5]`, Mathematica uses the previously computed values, namely `f[4]`, to compute `f[5]`. Note that we can produce the same results by defining $f_n(x)$ with the command

```
f[n_]:=...
```

However, the disadvantage of defining $f_n(x)$ in this manner is that Mathematica does not "remember" the previously computed values and thus takes longer (and uses more memory) to compute $f_n(x)$ for larger values of n.

Next, we use `Table` to generate $f_3(x)$, $f_6(x)$, $f_9(x)$, and $f_{12}(x)$.

tograph = Table[*f*[*n*], {*n*, 3, 12, 3}]

We now graph each of these functions and show the results as a graphics array with `GraphicsGrid` in Fig. 3.63. (Note that you do not need to include the option `DisplayFunction->Identity` to suppress the resulting output unless you forget to include the semi-colon at the end of the command.)

graphs = Table[Plot[Evaluate[tograph[[*i*]]], {*x*, 0, 3π},

PlotStyle → CMYKColor[1, .75, .06, .24]], {*i*, 1, 4}];

toshow = Partition[graphs, 2];

Show[GraphicsGrid[toshow]]

From these graphs, we see that for large values of n, the graph of $f_n(x)$, although actually smooth, appears "jagged" and thus we might suspect that $f(x) = \lim_{n\to\infty} f_n(x) = \sum_{k=0}^{\infty} \frac{1}{2^k} \sin\left(3^k x\right)$ is indeed continuous everywhere but nowhere differentiable.

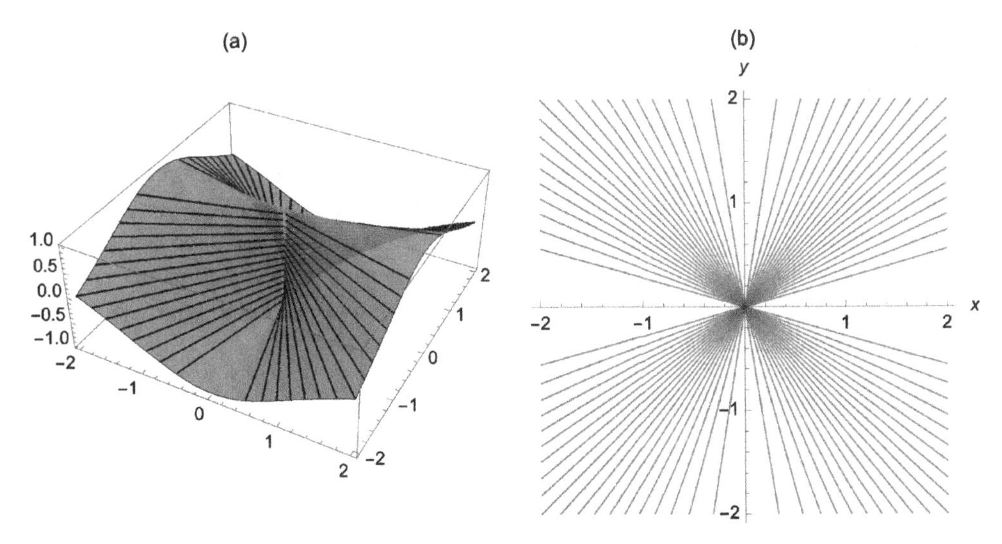

FIGURE 3.64 (a) Three-dimensional and (b) contour plots of $f(x, y)$ (North Carolina State University colors).

3.5 MULTI-VARIABLE CALCULUS

Mathematica is useful in investigating functions involving more than one variable. In particular, the graphical analysis of functions that depend on two (or more) variables is enhanced with the help of Mathematica's graphics capabilities.

3.5.1 Limits of Functions of Two Variables

Mathematica's graphics and numerical capabilities are helpful in investigating limits of functions of two variables.

Example 3.59

Show that the limit $\lim_{(x,y)\to(0,0)} \frac{x^2-y^2}{x^2+y^2}$ does not exist.

Solution. We begin by defining $f(x, y) = \left(x^2 - y^2\right) / \left(x^2 + y^2\right)$. Next, we use `Plot3D` to graph $z = f(x, y)$ for $-1/2 \le x \le 1/2$ and $-1/2 \le y \le 1/2$. `ContourPlot` is used to graph several level curves on the same rectangle. (See Fig. 3.64.) (To define a function of two variables, $f(x, y) = expression\,in\,x\,and\,y$, enter `f[x_,y_]=expression in x and y`. `Plot3D[f[x,y],{a,x,b},{y,c,d}]` generates a basic graph of $z = f(x, y)$ for $a \le x \le b$ and $c \le y \le d$)

```
Clear[f]

f[x_, y_] = (x^2 - y^2)/(x^2 + y^2);

p1 = Plot3D[f[x, y], {x, -2, 2}, {y, -2, 2}, PlotPoints → 40,
  PlotStyle → {Opacity[.8], CMYKColor[0, 1, .81, .04]},
  MeshFunctions → {#3&}, MeshStyle → {Black, Thickness[.005]}];

c1 = ContourPlot[f[x, y], {x, -2, 2}, {y, -2, 2}, ContourShading → False,
  Axes → Automatic, AxesOrigin → {0, 0}, PlotPoints → 60, Contours → 20,
  ContourStyle → CMYKColor[0, 1, .81, .04], Frame → False,
```

```
AxesLabel → {x, y}];
Show[GraphicsRow[{p1, c1}]]
```

When you slide the cursor over the contours in the contour plot, the contour values are displayed.

From the graph of the level curves, we suspect that the limit does not exist because we see that near $(0, 0)$, $z = f(x, y)$ attains many different values. We obtain further evidence that the limit does not exist by computing the value of $z = f(x, y)$ for various points chosen randomly near $(0, 0)$. We use `Table` and `RandomReal` to generate 10 ordered pairs (x, y) for x and y "close to" 0. Because `RandomReal` is included in the calculation, your results will almost certainly be different from those here.

```
pts = Table[RandomReal[{-10^-i, 10^-i}], {i, 1, 10}, {2}]
```

{{−0.0240696, 0.0964425}, {0.00929171, 0.00582256}, {−0.000875533, −0.000861975},

{4.93995041959814*^ − 6, 0.0000585256}, {−5.7785374509223095*^ − 6,

6.7869167280744185*^ − 6}, {−6.964772344758519*^ − 7, −9.809015465572932*^−8},

{1.669922073323435*^ − 8, −7.416883164217774*^ − 8},

{4.256718340841544*^ − 9, 2.2211732111688833*^ − 10},

{4.8322758801790404*^ − 11, −4.063639702523537*^ − 10},

{2.96738846034934*^ − 11, −9.412273092884795*^ − 11}}

Next, we define a function g that given an ordered pair (x, y) (`{x,y}` in Mathematica), $g((x, y))$ returns the ordered triple $(x, y, f(x, y))$ (`{x,y,f[x,y]}` in Mathematica).

```
g[{x_, y_}] = {x, y, f[x, y]}
```

$$\left\{x, y, \frac{x^2-y^2}{x^2+y^2}\right\}$$

We then use `Map` to apply g to the list `pts`.

```
Map[g, pts]//TableForm
```

−0.0240696	0.0964425	−0.882729
0.00929171	0.00582256	0.436083
−0.000875533	−0.000861975	0.0156052
4.93995041959814*^ − 6	0.0000585256	−0.985852
−5.7785374509223095*^ − 6	6.7869167280744185*^ − 6	−0.159473
−6.964772344758519*^ − 7	−9.809015465572932*^ − 8	0.961101
1.669922073323435*^ − 8	−7.416883164217774*^ − 8	−0.903505
4.256718340841544*^ − 9	2.2211732111688833*^ − 10	0.994569
4.8322758801790404*^ − 11	−4.063639702523537*^ − 10	−0.972113
2.96738846034934*^ − 11	−9.412273092884795*^ − 11	−0.819184

From the third column, we see that $z = f(x, y)$ does not appear to approach any particular value for points chosen randomly near $(0, 0)$. In fact, along the line $y = mx$ we see that $f(x, y) = f(x, mx) = \left(1 - m^2\right) / \left(1 + m^2\right)$. Hence as $(x, y) \to (0, 0)$ along $y = mx$, $f(x, y) = f(x, mx) \to \dfrac{1 - m^2}{1 + m^2}$. Thus, $f(x, y)$ does not have a limit as $(x, y) \to (0, 0)$.

We choose lines of the form $y = mx$ because near $(0, 0)$ the level curves of $z = f(x, y)$ look like lines of the form $y = mx$.

v1 = Simplify[*f*[*x*, *mx*]]

$\frac{1-m^2}{1+m^2}$

v1/.*m* → 0

v1/.*m* → 1

v1/.*m* → 1/2

1

0

$\frac{3}{5}$

□

In some cases, you can establish that a limit does not exist by converting to polar coordinates. For example, in polar coordinates, $f(x, y) = \dfrac{x^2 - y^2}{x^2 + y^2}$ becomes $f(r\cos\theta, r\sin\theta) = 2\cos^2\theta - 1$

Simplify[*f*[*r*Cos[*t*], *r*Sin[*t*]]]

Cos[2*t*]

and

$$\lim_{(x,y)\to(0,0)} f(x, y) = \lim_{r\to 0} f(r\cos\theta, r\sin\theta) = \lim_{r\to 0} 2\cos^2\theta - 1 = 2\cos^2\theta - 1 = \cos 2\theta$$

depends on θ.

3.5.2 Partial and Directional Derivatives

Partial derivatives of functions of two or more variables are computed with Mathematica using `D`. For $z = f(x, y)$,

1. `D[f[x,y],x]` computes $\frac{\partial f}{\partial x} = f_x(x, y)$,
2. `D[f[x,y],y]` computes $\frac{\partial f}{\partial y} = f_y(x, y)$,
3. `D[f[x,y],{x,n}]` computes $\frac{\partial^n f}{\partial x^n}$,
4. `D[f[x,y],y,x]` computes $\frac{\partial^2 f}{\partial y \partial x} = f_{xy}(x, y)$, and
5. `D[f[x,y],{x,n},{y,m}]` computes $\frac{\partial^{n+m} f}{\partial^n x \partial^m y}$.
6. You can use the $\partial_\square \blacksquare$ button located on the **BasicMathInput** palette to create templates to compute partial derivatives.

The calculations are carried out similarly for functions of more than two variables.

Example 3.60

Calculate $f_x(x,y)$, $f_y(x,y)$, $f_{xy}(x,y)$, $f_{yx}(x,y)$, $f_{xx}(x,y)$, and $f_{yy}(x,y)$ if $f(x,y)=\sin\sqrt{x^2+y^2+1}$.

Solution. After defining $f(x,y)=\sin\sqrt{x^2+y^2+1}$,

f[x_, y_] = Sin[Sqrt[x^2 + y^2 + 1]];

we illustrate the use of D to compute the partial derivatives. Entering

D[f[x, y], x]

$$\frac{x\mathrm{Cos}\left[\sqrt{1+x^2+y^2}\right]}{\sqrt{1+x^2+y^2}}$$

computes $f_x(x,y)$. Entering

D[f[x, y], y]

$$\frac{y\mathrm{Cos}\left[\sqrt{1+x^2+y^2}\right]}{\sqrt{1+x^2+y^2}}$$

computes $f_y(x,y)$. Entering

D[f[x, y], x, y]//Together

$$-\frac{xy\left(\mathrm{Cos}\left[\sqrt{1+x^2+y^2}\right]+\sqrt{1+x^2+y^2}\mathrm{Sin}\left[\sqrt{1+x^2+y^2}\right]\right)}{(1+x^2+y^2)^{3/2}}$$

computes $f_{yx}(x,y)$. Entering

D[f[x, y], y, x]//Together

$$-\frac{xy\left(\mathrm{Cos}\left[\sqrt{1+x^2+y^2}\right]+\sqrt{1+x^2+y^2}\mathrm{Sin}\left[\sqrt{1+x^2+y^2}\right]\right)}{(1+x^2+y^2)^{3/2}}$$

computes $f_{xy}(x,y)$. Remember that under appropriate assumptions, $f_{xy}(x,y)=f_{yx}(x,y)$. Entering

D[f[x, y], {x, 2}]//Together

$$\frac{\mathrm{Cos}\left[\sqrt{1+x^2+y^2}\right]+y^2\mathrm{Cos}\left[\sqrt{1+x^2+y^2}\right]-x^2\sqrt{1+x^2+y^2}\mathrm{Sin}\left[\sqrt{1+x^2+y^2}\right]}{(1+x^2+y^2)^{3/2}}$$

computes $f_{xx}(x,y)$. Entering

D[f[x, y], {y, 2}]//Together

$$\frac{\mathrm{Cos}\left[\sqrt{1+x^2+y^2}\right]+x^2\mathrm{Cos}\left[\sqrt{1+x^2+y^2}\right]-y^2\sqrt{1+x^2+y^2}\mathrm{Sin}\left[\sqrt{1+x^2+y^2}\right]}{(1+x^2+y^2)^{3/2}}$$

computes $f_{yy}(x,y)$. □

The **directional derivative** of $z = f(x, y)$ in the direction of the unit vector $\mathbf{u} = \cos\theta\,\mathbf{i} + \sin\theta\,\mathbf{j}$ is

$$D_{\mathbf{u}} f(x, y) = f_x(x, y)\cos\theta + f_y(x, y)\sin\theta,$$

provided that $f_x(x, y)$ and $f_y(x, y)$ both exist.

The vectors **i** and **j** are defined by $\mathbf{i} = \langle 1, 0\rangle$ and $\mathbf{j} = \langle 0, 1\rangle$.

If $f_x(x, y)$ and $f_y(x, y)$ both exist, the **gradient** of $f(x, y)$ is the vector-valued function

$$\nabla f(x, y) = f_x(x, y)\mathbf{i} + f_y(x, y)\mathbf{j} = \left\langle f_x(x, y), f_y(x, y)\right\rangle.$$

Calculus of vector-valued functions is discussed in more detail in Chapter 5.

Notice that if $\mathbf{u} = \langle \cos\theta, \sin\theta\rangle$,

$$D_{\mathbf{u}} f(x, y) = \nabla f(x, y) \cdot \langle \cos\theta, \sin\theta\rangle.$$

Let (x_0, y_0) be a point in the domain of the differentiable function $z = f(x, y)$. At (x_0, y_0), $f(x, y)$ increases most rapidly in the direction of the gradient, $\nabla f(x, y)$, evaluated at (x_0, y_0). Similarly, $f(x, y)$ decreases most rapidly in the direction of $-\nabla f(x, y)$. Consequently, the gradient is perpendicular to the level curves of $f(x, y)$.

```
StreamPlot[{f(x,y),g(x,y)},{x,a,b},{y,c,d}]
```

generates a stream plot of the vector field $< f(x, y), g(x, y) >$. Thus,

```
StreamPlot[{D[f[x,y],x],D[f[x,y],y]},{x,a,b},{y,c,d}]
```

generates a stream plot of the gradient of $f(x, y)$.

Vectors and vector-valued functions will be discussed in more detail in Chapters 4 and 5.

Example 3.61

Let $f(x, y) = 6x^2y - 3x^4 - 2y^3$. (a) Find $D_{\mathbf{u}} f(x, y)$ in the direction of $\mathbf{v} = \langle 3, 4\rangle$. (b) Compute $D_{(3/5,4/5)} f\left(\frac{1}{3}\sqrt{9+3\sqrt{3}}, 1\right)$. (c) Find an equation of the line tangent to the graph of $6x^2y - 3x^4 - 2y^3 = 0$ at the point $\left(\frac{1}{3}\sqrt{9+3\sqrt{3}}, 1\right)$.

Solution. After defining $f(x, y) = 6x^2y - 3x^4 - 2y^3$, we graph $z = f(x, y)$ with `Plot3D` in Fig. 3.65, illustrating the `PlotPoints`, `PlotRange`, and `ViewPoint` options.

```
f[x_, y_] = 6x^2y - 3x^4 - 2y^3;
Plot3D[f[x, y], {x, -2, 2}, {y, -2, 3}, PlotPoints -> 50,
  PlotRange -> {{-2, 2}, {-2, 3}, {-2, 2}},
  PlotStyle -> {CMYKColor[1, .93, .28, .22], Opacity[.5]},
  BoxRatios -> {1, 1, 1}, ViewPoint -> {1.887, 2.309, 1.6},
  ClippingStyle -> None]
```

(a) A unit vector, **u**, in the same direction as **v** is

$$\mathbf{u} = \left\langle \frac{3}{\sqrt{3^2+4^2}}, \frac{4}{\sqrt{3^2+4^2}}\right\rangle = \left\langle \frac{3}{5}, \frac{4}{5}\right\rangle.$$

```
v = {3, 4};
u = v/Sqrt[v.v]
```

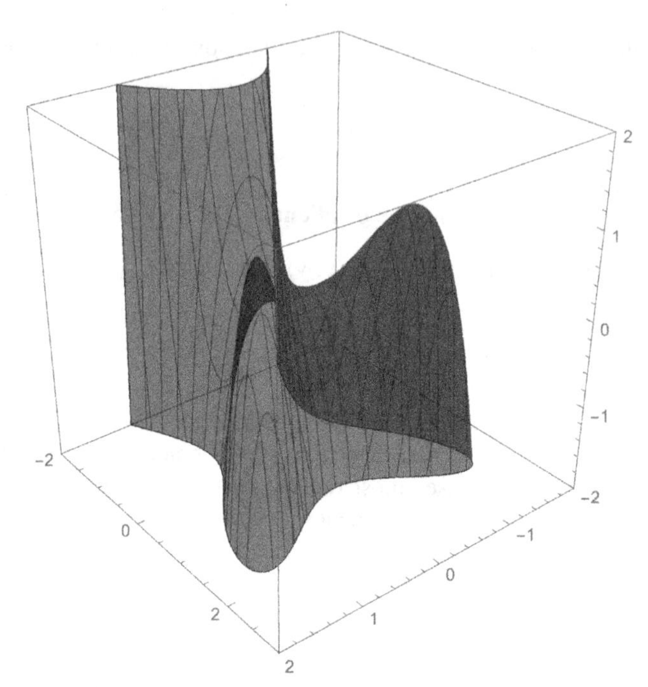

FIGURE 3.65 $f(x, y) = 6x^2y - 3x^4 - 2y^3$ for $-2 \le x \le 2$ and $-2 \le y \le 3$ (Rice University colors).

$\left\{\frac{3}{5}, \frac{4}{5}\right\}$

Then, $D_{\mathbf{u}}f(x, y) = \langle f_x(x, y), f_y(x, y)\rangle \cdot \mathbf{u}$, calculated in du.

gradf = {D[f[x, y], x], D[f[x, y], y]}

$\{-12x^3 + 12xy, 6x^2 - 6y^2\}$

du = Simplify[gradf.u]

$-\frac{12}{5}\left(-2x^2 + 3x^3 - 3xy + 2y^2\right)$

(b) $D_{\langle 3/5,4/5\rangle}f\left(\frac{1}{3}\sqrt{9+3\sqrt{3}}, 1\right)$ is calculated by evaluating du if $x = \frac{1}{3}\sqrt{9+3\sqrt{3}}$ and $y = 1$.

du1 = du/.{x → 1/3Sqrt[9 + 3Sqrt[3]], y → 1}//Simplify

$\frac{4}{5}\left(2\sqrt{3} - 3\sqrt{3+\sqrt{3}}\right)$

(c) The gradient is evaluated if $x = \frac{1}{3}\sqrt{9+3\sqrt{3}}$ and $y = 1$.

nvec = gradf/.{x → 1/3Sqrt[9 + 3Sqrt[3]], y → 1}//Simplify

$\left\{-4\sqrt{3+\sqrt{3}}, 2\sqrt{3}\right\}$

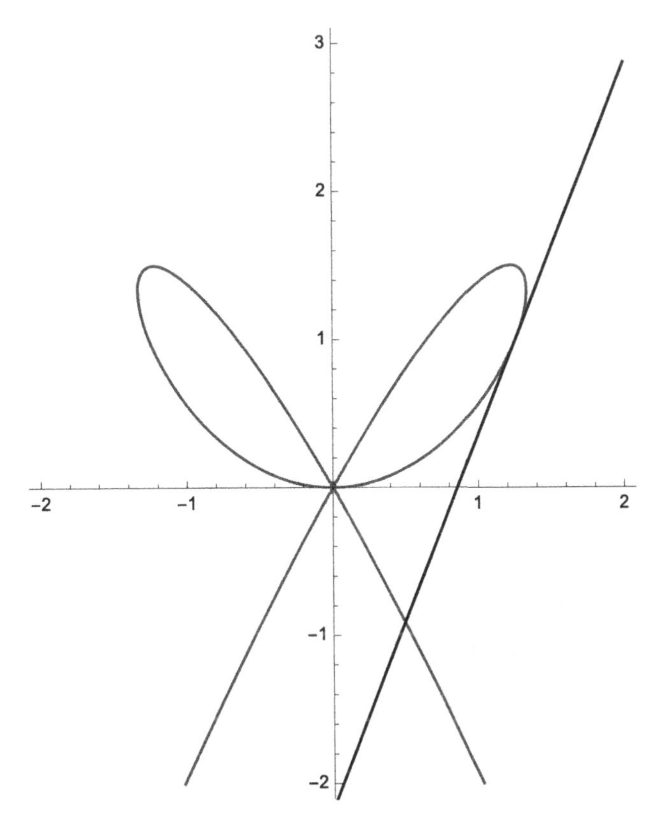

FIGURE 3.66 Level curve of $f(x, y)$ together with a tangent line.

Generally, $\nabla f(x, y)$ is perpendicular to the level curves of $z = f(x, y)$, so

$$\texttt{nvec} = \nabla f\left(\frac{1}{3}\sqrt{9+3\sqrt{3}}, 1\right) = \left\langle f_x\left(\frac{1}{3}\sqrt{9+3\sqrt{3}}, 1\right), f_y\left(\frac{1}{3}\sqrt{9+3\sqrt{3}}, 1\right)\right\rangle$$

is perpendicular to $f(x, y) = 0$ at the point $\left(\frac{1}{3}\sqrt{9+3\sqrt{3}}, 1\right)$. Thus, an equation of the line tangent to the graph of $f(x, y) = 0$ at the point $\left(\frac{1}{3}\sqrt{9+3\sqrt{3}}, 1\right)$ is

An equation of the line L containing (x_0, y_0) and perpendicular to $\mathbf{n} = \langle a, b\rangle$ is $a(x - x_0) + b(y - y_0) = 0$.

$$f_x\left(\frac{1}{3}\sqrt{9+3\sqrt{3}}, 1\right)\left(x - \frac{1}{3}\sqrt{9+3\sqrt{3}}\right) + f_y\left(\frac{1}{3}\sqrt{9+3\sqrt{3}}, 1\right)(y-1) = 0,$$

which we solve for y with `Solve`. We confirm this result by graphing $f(x, y) = 0$ using `ContourPlot` in `conf` and then graphing the tangent line in `tanplot`. `tanplot` and `conf` are shown together with `Show` in Fig. 3.66.

```
conf = ContourPlot[f[x, y] == 0, {x, −2, 2}, {y, −2, 2}, PlotPoints → 60,
  ContourShading → False, Frame → False, Axes → Automatic,
  ContourStyle->CMYKColor[1, .93, .28, .22],
  AxesOrigin → {0, 0}];
tanline = Solve[nvec[[1]](x − 1/3Sqrt[9 + 3Sqrt[3]]) + nvec[[2]](y − 1) == 0, y]
```

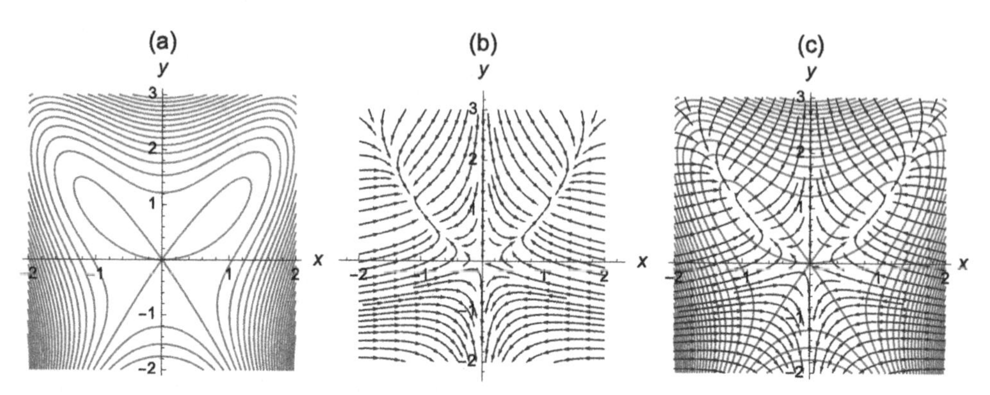

FIGURE 3.67 (a) Level curves of $z = f(x, y)$. (b) Gradient field of $z = f(x, y)$. (c) The gradient together with several level curves.

$$\left\{\left\{y \to -\frac{2+\sqrt{3}-2\sqrt{3+\sqrt{3}x}}{\sqrt{3}}\right\}\right\}$$

```
tanplot = Plot[Evaluate[y/.tanline], {x, −2, 2},
  PlotStyle->CMYKColor[0, 0, 0, .77]];
  Show[conf, tanplot, PlotRange → {{−2, 2}, {−2, 3}}, AspectRatio → Automatic]
```

More generally, we use `ContourPlot` together with the `StreamPlot` function to illustrate that the gradient vectors are perpendicular to the level curves of $z = f(x, y)$ in Fig. 3.67.

```
p1 = ContourPlot[f[x, y], {x, −2, 2}, {y, −2, 3},
  Contours → 25, ContourStyle → CMYKColor[0, 0, 0, .77],
  ContourShading → False, Frame → False, Axes → Automatic,
  AxesOrigin → {0, 0}, AxesLabel → {x, y},
  PlotLabel → "(a)"]

p2 = StreamPlot[Evaluate[{D[f[x, y], x], D[f[x, y], y]}],
  {x, −2, 2}, {y, −2, 3}, StreamStyle → CMYKColor[1, .93, .28, .22],
  Frame → False, Axes → Automatic, AxesLabel → {x, y},
  PlotLabel → "(b)"]

p3 = Show[p1, p2, PlotLabel → "(c)"]

Show[GraphicsRow[{p1, p2, p3}]]
```

In Figs. 3.67 (b) and (c), observe that if the gradient vectors are all converging to a point, $\nabla f(x, y) = 0$ and the point will be a relative maximum. At $(0, 0)$, $\nabla f(x, y) = 0$ but at this point, some gradient vectors are pointing towards the origin and some away so $(0, 0)$ is a saddle. Compare to Fig. 3.65. Note that if $\nabla f(x, y) = 0$ and all the gradient vectors were pointing away from the point, the point would be a relative minimum. □

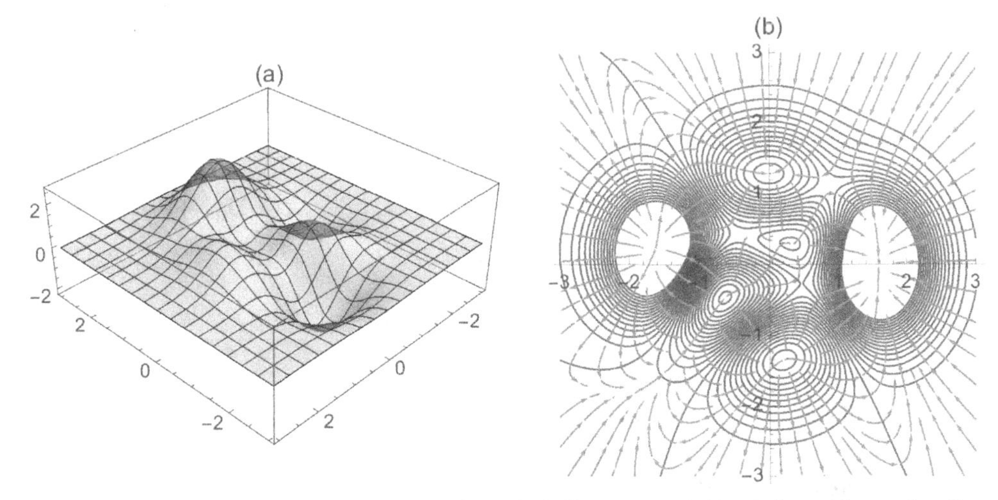

FIGURE 3.68 (a) $f(x, y)$ for $-3 \leq x \leq 3$ and $-3 \leq y \leq 2$. (b) Contour plot of $f(x, y)$ along with several gradient vectors (University of Washington colors).

Example 3.62

Let

$$f(x, y) = (y-1)^2 e^{-(x+1)^2-y^2} - \frac{10}{3}\left(-x^5 + \frac{1}{5}y - y^3\right)e^{-x^2-y^2} - \frac{1}{9}e^{-x^2-(y+1)^2}.$$

Calculate $\nabla f(x, y)$ and then graph $\nabla f(x, y)$ together with several level curves of $f(x, y)$.

Solution. We begin by defining and graphing $z = f(x, y)$ with `Plot3D` in Fig. 3.68 (a).

```
Clear[f]
f[x_, y_] = (y − 1)^2Exp[−(x + 1)^2 − y^2]−
  10/3(−x^5 + 1/5y − y^3)Exp[−x^2 − y^2]−
    1/9Exp[−x^2 − (y + 1)^2];

p1 = Plot3D[f[x, y], {x, −3, 3}, {y, −3, 3},
  PlotStyle → {CMYKColor[0, .13, .43, .13], Opacity[.5]},
  ViewPoint → {−1.99, 2.033, 1.833},
  PlotRange → All, PlotLabel → "(a)"];
conf = ContourPlot[f[x, y], {x, −3, 3}, {y, −3, 3},
  PlotPoints → 60, Contours → 30, ContourShading → False,
  Frame → False, Axes → Automatic, ContourStyle → CMYKColor[.93, 1, .18, .21],
  AxesOrigin → {0, 0}];
```

In the three-dimensional plot, notice that z appears to have six relative extrema: three relative maxima and three relative minima. We also graph several level curves of $f(x, y)$ with `ContourPlot` and name the resulting graphic `conf`.

Next we calculate $f_x(x, y)$ and $f_y(x, y)$ using `Simplify` and `D`. The gradient is the vector-valued function $\langle f_x(x, y), f_y(x, y)\rangle$.

```
gradf = {D[f[x, y], x], D[f[x, y], y]}//Simplify
```

$\{-\frac{2}{9}e^{-2x-x^2-(1+y)^2}(-e^{2x}x + 9e^{2y}(1+x)(-1+y)^2 + 3e^{1+2x+2y}x(-25x^3 + 10x^5 - 2y + 10y^3)), -\frac{2}{9}e^{-2x-x^2-(1+y)^2}(-e^{2x}(1+y) + 9e^{2y}(1-2y^2+y^3) + e^{1+2x+2y}(3 + 30x^5y - 51y^2 + 30y^4))\}$

We use `StreamPlot` to graph the gradient naming the resulting graphic `gradfplot`. `gradfplot` and `conf` are displayed together using `Show` in Fig. 3.68 (b).

```
gradfplot = StreamPlot[gradf, {x, −3, 3}, {y, −3, 3},
  StreamStyle → CMYKColor[.3, .35, .6, 0]
];

Show[GraphicsRow[{p1, Show[conf, gradfplot, PlotLabel → "(b)"]}]]
```

In the result (see Fig. 3.68 (b)), notice that the gradient is perpendicular to the level curves; the gradient is pointing in the direction of maximal increase of $z = f(x, y)$. □

Classifying Critical Points

Let $z = f(x, y)$ be a real-valued function of two variables with continuous second-order partial derivatives. A **critical point** of $z = f(x, y)$ is a point (x_0, y_0) in the interior of the domain of $z = f(x, y)$ for which

$$f_x(x_0, y_0) = 0 \qquad \text{and} \qquad f_y(x_0, y_0) = 0,$$

or, equivalently, $\nabla f(x_0, y_0) = \mathbf{0}$. Critical points are classified by the *Second Derivatives* (or *Partials*) *test*.

Theorem 3.18 (Second Derivatives Test). *Let (x_0, y_0) be a critical point of a function $z = f(x, y)$ of two variables and let*

$$d = f_{xx}(x_0, y_0)\, f_{yy}(x_0, y_0) - \left[f_{xy}(x_0, y_0)\right]^2. \tag{3.36}$$

1. *If $d > 0$ and $f_{xx}(x_0, y_0) > 0$, then $z = f(x, y)$ has a **relative** (or **local**) **minimum** at (x_0, y_0).*
2. *If $d > 0$ and $f_{xx}(x_0, y_0) < 0$, then $z = f(x, y)$ has a **relative** (or **local**) **maximum** at (x_0, y_0).*
3. *If $d < 0$, then $z = f(x, y)$ has a **saddle point** at (x_0, y_0).*
4. *If $d = 0$, no conclusion can be drawn and (x_0, y_0) is called a **degenerate critical point**.*

Example 3.63

Find the relative maximum, relative minimum, and saddle points of $f(x, y) = -2x^2 + x^4 + 3y - y^3$.

Solution. After defining $f(x, y)$, the critical points are found with `Solve` and named `critpts`.

```
f[x_, y_] = −2x^2 + x^4 + 3y − y^3;
```

```
critpts = Solve[{D[f[x, y], x] == 0, D[f[x, y], y] == 0}, {x, y}]
{{x → −1, y → −1}, {x → 0, y → −1}, {x → 1, y → −1}, {x → −1, y → 1}, {x → 0, y →
1}, {x → 1, y → 1}}
```

We then define `dfxx`. Given (x_0, y_0), `dfxx` (x_0, y_0) returns the ordered quadruple x_0, y_0, equation (3.36) evaluated at (x_0, y_0), and $f_{xx}(x_0, y_0)$.

```
dfxx[x0_, y0_] = {x0, y0,
   D[f[x, y], {x, 2}]D[f[x, y], {y, 2}] − D[f[x, y], x, y]^2/.
   {x → x0, y → y0}, D[f[x, y], {x, 2}]/.{x → x0, y → y0}}
```

$\{x0, y0, -6(-4 + 12x0^2)\, y0, -4 + 12x0^2\}$

For example,

```
dfxx[0, 1]
{0, 1, 24, −4}
```

shows us that a relative maximum occurs at (0, 1). We then use `/.` (`ReplaceAll`) to substitute the values in each element of `critpts` into `dfxx`.

```
dfxx[x, y]/.critpts//TableForm
```

−1	−1	48	8
0	−1	−24	−4
1	−1	48	8
−1	1	−48	8
0	1	24	−4
1	1	−48	8

From the result, we see that (0, 1) results in a relative maximum, (0, −1) results in a saddle, (1, 1) results in a saddle, (1, −1) results in a relative minimum, (−1, 1) results in a saddle, and (−1, −1) results in a relative minimum. We confirm these results graphically with a three-dimensional plot generated with `Plot3D` and a contour plot together with the gradient generated with `ContourPlot` and `StreamPlot` in Fig. 3.69.

```
p1 = Plot3D[f[x, y], {x, −3/2, 3/2}, {y, −3/2, 3/2}, PlotPoints → 40,
   PlotStyle → {CMYKColor[0, .7, 1, 0], Opacity[.4]},
   PlotLabel → "(a)"];
p2 = ContourPlot[f[x, y], {x, −3/2, 3/2}, {y, −3/2, 3/2},
```

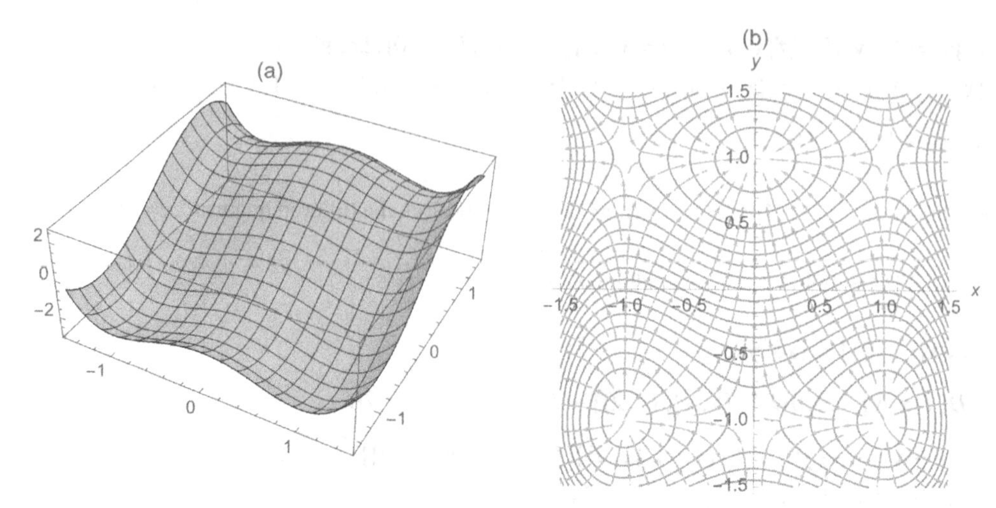

FIGURE 3.69 (a) Three-dimensional and (b) contour plots of $f(x, y)$ (University of Florida colors).

```
    PlotPoints → 40, ContourShading → False, Contours → 20,
    Frame → False, Axes → Automatic, AxesOrigin → {0, 0},
    ContourStyle → CMYKColor[1, .6, 0, .2],
    AxesLabel → {x, y}];
p3 = StreamPlot[Evaluate[{D[f[x, y], x], D[f[x, y], y]}], {x, −3/2, 3/2},
    {y, −3/2, 3/2},
    StreamStyle → CMYKColor[.1, .2, .4, .1]];
Show[GraphicsRow[{p1, Show[p2, p3, PlotLabel → "(b)"]}]]
```

In the contour plot, notice that near relative extrema, the level curves look like circles while near saddles they look like hyperbolas. Further, at the maximums, the gradient is directed towards the point while at the minimums it is directed away from the point. □

If the Second Derivatives Test fails, graphical analysis is especially useful.

Example 3.64

Find the relative extrema and saddle points of $f(x, y) = x^2 + x^2y^2 + y^4$.

Solution. Initially we proceed in the exact same manner as in the previous example: we define $f(x, y)$ and compute the critical points. Several complex solutions are returned, which we ignore.

```
f[x_, y_] = x^2 + x^2y^2 + y^4;
critpts = Solve[{D[f[x, y], x] == 0, D[f[x, y], y] == 0}, {x, y}]
```

$\{\{x \to 0, y \to 0\}, \{x \to 0, y \to 0\}, \{x \to -\sqrt{2}, y \to -i\}, \{x \to \sqrt{2}, y \to -i\}, \{x \to -\sqrt{2}, y \to i\}, \{x \to \sqrt{2}, y \to i\}\}$

We then compute the value of (3.36) at the real critical point, and the value of $f_{xx}(x, y)$ at this critical point.

```
dfxx[x0_, y0_] = {x0, y0,
D[f[x, y], {x, 2}]D[f[x, y], {y, 2}] − D[f[x, y], x, y]^2/.
{x → x0, y → y0}, D[f[x, y], {x, 2}]/.{x → x0, y → y0}}
```

$\{x0, y0, -16x0^2y0^2 + (2+2y0^2)(2x0^2+12y0^2), 2+2y0^2\}$

```
dfxx[0, 0]
```

$\{0, 0, 0, 2\}$

The result shows us that the Second Derivatives Test fails at (0, 0).

```
p1 = Plot3D[f[x, y], {x, −1, 1}, {y, −1, 1}, BoxRatios → Automatic,
  PlotStyle → {CMYKColor[1, .56, 0, .34], Opacity[.3]},
  MeshFunctions->{#3&}, MeshStyle → {CMYKColor[0, .18, 1, .15], Thickness[.0075]},
  PlotLabel → "(a)"];
p2 = ContourPlot[f[x, y], {x, −1, 1}, {y, −1, 1}, PlotPoints → 40,
  Contours → 20, ContourShading → False, Frame → False,
  Axes → Automatic, AxesOrigin → {0, 0}, ContourStyle →
  {{CMYKColor[0, .18, 1, .15], Thickness[.0075]}}];
p3 = StreamPlot[Evaluate[{D[f[x, y], x], D[f[x, y], y]}], {x, −1, 1},
  {y, −1, 1},
  StreamStyle → CMYKColor[.03, 1, .7, .12]];
Show[GraphicsRow[{p1, Show[p2, p3, PlotLabel → b]}]]
```

However, the contour plot and the stream plot of the gradient of $f(x, y)$ near (0, 0) indicates that an extreme value occurs at (0, 0). Because the gradient vectors are all pointing away from the origin, (0, 0) is a relative minimum which is confirmed in the three-dimensional plot. It is also an absolute minimum. (See Fig. 3.70.) □

Tangent Planes

Let $z = f(x, y)$ be a real-valued function of two variables. If both $f_x(x_0, y_0)$ and $f_y(x_0, y_0)$ exist, then an equation of the plane tangent to the graph of $z = f(x, y)$ at the point $(x_0, y_0, f(x_0, y_0))$ is given by

$$f_x(x_0, y_0)(x - x_0) + f_y(x_0, y_0)(y - y_0) - (z - z_0) = 0, \tag{3.37}$$

where $z_0 = f(x_0, y_0)$. Solving for z yields the function (of two variables)

$$z = f_x(x_0, y_0)(x - x_0) + f_y(x_0, y_0)(y - y_0) + z_0. \tag{3.38}$$

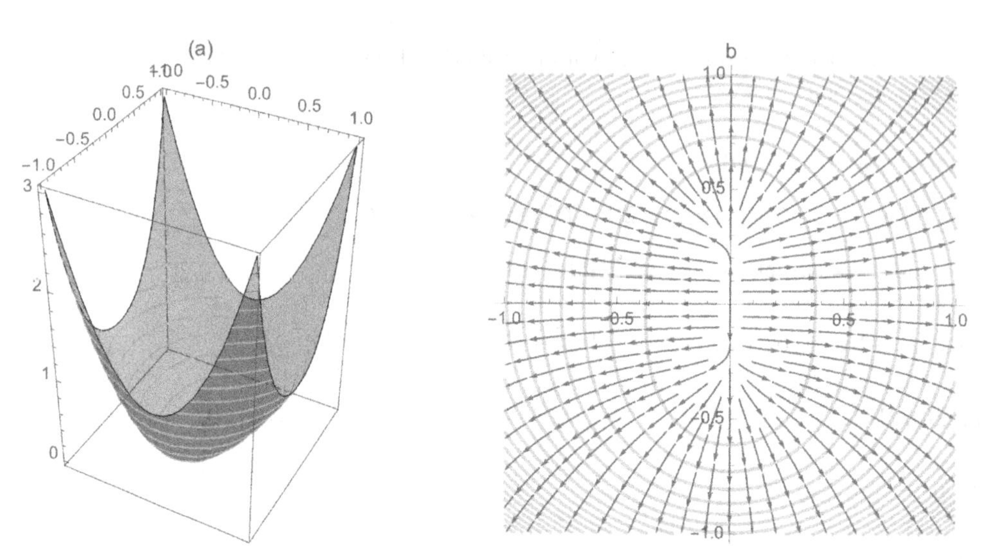

FIGURE 3.70 (a) Three-dimensional and (b) contour plots of $f(x, y)$ (University of California-Davis colors).

Symmetric equations of the line perpendicular to the surface $z = f(x, y)$ at the point (x_0, y_0, z_0) are given by

$$\frac{x - x_0}{f_x(x_0, y_0)} = \frac{y - y_0}{f_y(x_0, y_0)} = \frac{z - z_0}{-1} \tag{3.39}$$

and parametric equations are

$$\begin{cases} x = x_0 + f_x(x_0, y_0)\,t \\ y = y_0 + f_y(x_0, y_0)\,t \\ z = z_0 - t. \end{cases} \tag{3.40}$$

The plane tangent to the graph of $z = f(x, y)$ at the point $(x_0, y_0, f(x_0, y_0))$ is the "best" linear approximation of $z = f(x, y)$ near $(x, y) = (x_0, y_0)$ in the same way as the line tangent to the graph of $y = f(x)$ at the point $(x_0, f(x_0))$ is the "best" linear approximation of $y = f(x)$ near $x = x_0$.

Example 3.65

Find an equation of the plane tangent and normal line to the graph of $f(x, y) = 4 - \frac{1}{4}\left(2x^2 + y^2\right)$ at the point $(1, 2, 5/2)$.

Solution. We define $f(x, y)$ and compute $f_x(1, 2)$ and $f_y(1, 2)$.

```
f[x_,y_]=4 − 1/4(2x^2 + y^2);
f[1,2]
dx = D[f[x, y], x]/.{x → 1, y → 2}
dy = D[f[x, y], y]/.{x → 1, y → 2}
```

$\frac{5}{2}$

-1

-1

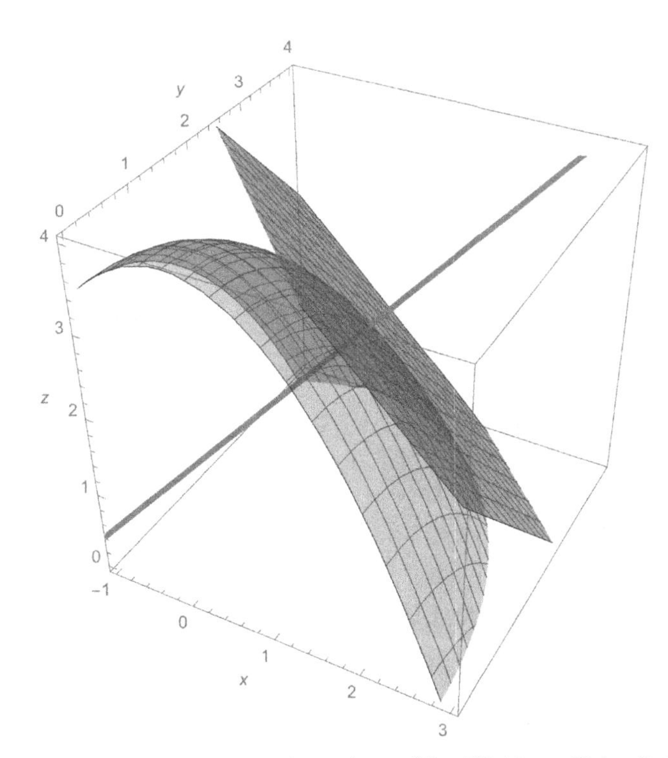

FIGURE 3.71 Graph of $f(x, y)$ with a tangent plane and normal line (Washington University in St. Louis colors).

Using (3.38), an equation of the tangent plane is $z = -1(x-1) - 1(y-2) + f(1,2)$. Using (3.40), parametric equations of the normal line are $x = 1 - t$, $y = 2 - t$, $z = f(1,2) - t$. We confirm the result graphically by graphing $f(x, y)$ together with the tangent plane in `p1` using `Plot3D`. We use `ParametricPlot3D` to graph the normal line in `p2` and then display `p1` and `p2` together with `Show` in Fig. 3.71.

```
p1 = Plot3D[f[x, y], {x, −1, 3}, {y, 0, 4},
  PlotStyle → {CMYKColor[0, 1, .59, .24], Opacity[.3]}];
p2 = Plot3D[dx(x − 1) + dy(y − 2) + f[1, 2], {x, −1, 3}, {y, 0, 4},
  PlotStyle → {CMYKColor[.59, .41, .42, .14], Opacity[.6]}];
p3 = ParametricPlot3D[{1 + dxt, 2 + dyt, f[1, 2] − t}, {t, −4, 4},
  PlotStyle → {CMYKColor[1, 0, .6, .4], Thickness[.01]}];
Show[p1, p2, p3, PlotRange → {{−1, 3}, {0, 4}, {0, 4}},
  BoxRatios → Automatic, AxesLabel → {x, y, z}]
```

Because $z = -1(x-1) - 1(y-2) + f(1,2)$ is the "best" linear approximation of $f(x, y)$ near $(1, 2)$, the graphs are very similar near $(1, 2)$ as shown in the three-dimensional plot. We also expect the level curves of each near $(1, 2)$ to be similar, which is confirmed with `ContourPlot` in Fig. 3.72.

```
p4 = ContourPlot[f[x, y], {x, 0.75, 1.25}, {y, 1.75, 2.25},
  Contours → 20,
```

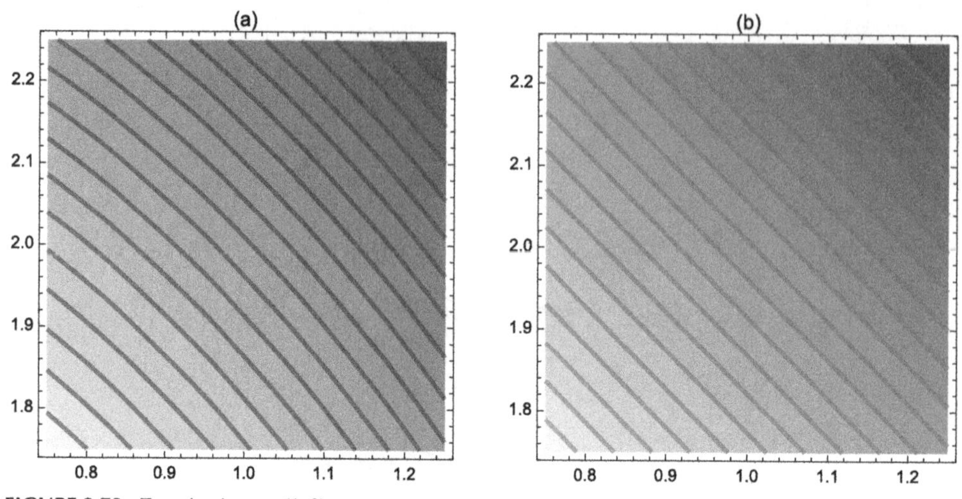

FIGURE 3.72 Zooming in near (1, 2).

```
    ContourStyle → {{CMYKColor[0, 1, .59, .24], Thickness[.01]}},
    PlotLabel → "(a)"];
p5 = ContourPlot[dx(x − 1) + dy(y − 2) + f[1, 2], {x, 0.75, 1.25}, {y, 1.75, 2.25},
    Contours → 20,
    ContourStyle → {{CMYKColor[.59, .41, .42, .14], Thickness[.01]}},
    PlotLabel → "(b)"];
Show[GraphicsRow[{p4, p5}]]
```

□

Lagrange Multipliers

Certain types of optimization problems can be solved using the method of *Lagrange multipliers* that is based on the following theorem.

Theorem 3.19 (Lagrange's Theorem). *Let $z = f(x, y)$ and $z = g(x, y)$ be real-valued functions with continuous partial derivatives and let $z = f(x, y)$ have an extreme value at a point (x_0, y_0) on the smooth constraint curve $g(x, y) = 0$. If $\nabla g(x_0, y_0) \neq \mathbf{0}$, then there is a real number λ satisfying*

$$\nabla f(x_0, y_0) = \lambda \nabla g(x_0, y_0). \tag{3.41}$$

Graphically, the points (x_0, y_0) at which the extreme values occur correspond to the points where the level curves of $z = f(x, y)$ are tangent to the graph of $g(x, y) = 0$.

Example 3.66

Find the maximum and minimum values of $f(x, y) = xy$ subject to the constraint $\frac{1}{4}x^2 + \frac{1}{9}y^2 = 1$.

Solution. For this problem, $f(x, y) = xy$ and $g(x, y) = \frac{1}{4}x^2 + \frac{1}{9}y^2 - 1$. Observe that parametric equations for $\frac{1}{4}x^2 + \frac{1}{9}y^2 = 1$ are $x = 2\cos t$, $y = 3\sin t$, $0 \le t \le 2\pi$. In Fig. 3.73 (a), we use `ParametricPlot3D` to parametrically graph $g(x, y) = 0$ and $f(x, y)$ for x and

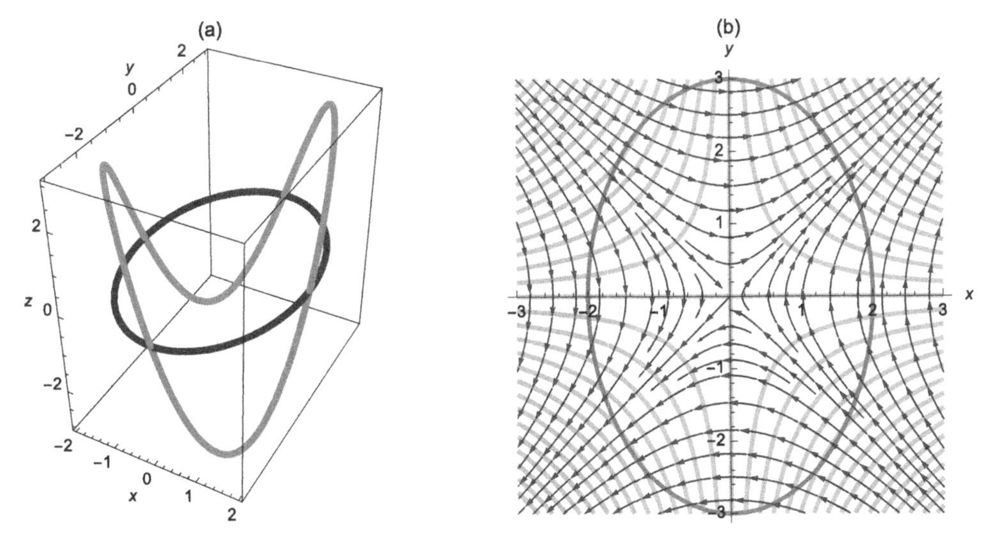

FIGURE 3.73 (a) $f(x, y)$ on $g(x, y) = 0$. (b) Level curves of $f(x, y)$ together with $g(x, y) = 0$ (Yale University colors).

y-values on the curve $g(x, y) = 0$ by graphing

$$\begin{cases} x = 2\cos t \\ y = 3\sin t \\ z = 0 \end{cases} \quad \text{and} \quad \begin{cases} x = 2\cos t \\ y = 3\sin t \\ z = x \cdot y = 6\cos t \sin t \end{cases}$$

for $0 \le t \le 2\pi$. Our goal is to find the minimum and maximum values in Fig. 3.73 (a) and the points at which they occur.

```
f[x_, y_] = xy;
g[x_, y_] = x^2/4 + y^2/9 − 1;

s1 = ParametricPlot3D[{2Cos[t], 3Sin[t], 0}, {t, 0, 2Pi},
  PlotStyle → {{CMYKColor[1, .75, .08, .4], Thickness[.02]}}];
s2 = ParametricPlot3D[{2Cos[t], 3Sin[t], 6Cos[t]Sin[t]}, {t, 0, 2Pi},
  PlotStyle → {{CMYKColor[.42, .4, .44, .04], Thickness[.02]}}];
plot1 = Show[s1, s2, BoxRatios → Automatic, PlotRange → All,
  PlotLabel → "(a)", AxesLabel → {x, y, z}];
```

To implement the method of Lagrange multipliers, we compute $f_x(x, y)$, $f_y(x, y)$, $g_x(x, y)$, and $g_y(x, y)$ with `D`.

```
fx = D[f[x, y], x]
fy = D[f[x, y], y]
gx = D[g[x, y], x]
gy = D[g[x, y], y]
```

y

x

$\frac{x}{2}$

$\frac{2y}{9}$

`Solve` is used to solve the system of equations (3.41):

$$f_x(x, y) = \lambda g_x(x, y)$$
$$f_y(x, y) = \lambda g_y(x, y)$$
$$g(x, y) = 0$$

for x, y, and λ.

vals = Solve[{fx == λgx, fy == λgy, g[x, y] == 0}, {x, y, λ}]

$\{\{x \to -\sqrt{2}, y \to -\frac{3}{\sqrt{2}}, \lambda \to 3\}, \{x \to -\sqrt{2}, y \to \frac{3}{\sqrt{2}}, \lambda \to -3\}, \{x \to \sqrt{2}, y \to -\frac{3}{\sqrt{2}}, \lambda \to -3\}, \{x \to \sqrt{2}, y \to \frac{3}{\sqrt{2}}, \lambda \to 3\}\}$

The corresponding values of $f(x, y)$ are found using `ReplaceAll` (`/.`).

n1 = {x, y, f[x, y]}/.vals//TableForm

$-\sqrt{2}$	$-\frac{3}{\sqrt{2}}$	3
$-\sqrt{2}$	$\frac{3}{\sqrt{2}}$	-3
$\sqrt{2}$	$-\frac{3}{\sqrt{2}}$	-3
$\sqrt{2}$	$\frac{3}{\sqrt{2}}$	3

***N*[n1]**

−1.41421	−2.12132	3.
−1.41421	2.12132	−3.
1.41421	−2.12132	−3.
1.41421	2.12132	3.

We conclude that the maximum value $f(x, y)$ subject to the constraint $g(x, y) = 0$ is 3 and occurs at $\left(\sqrt{2}, \frac{3}{2}\sqrt{2}\right)$ and $\left(-\sqrt{2}, -\frac{3}{2}\sqrt{2}\right)$. The minimum value is -3 and occurs at $\left(-\sqrt{2}, \frac{3}{2}\sqrt{2}\right)$ and $\left(\sqrt{2}, -\frac{3}{2}\sqrt{2}\right)$. We graph several level curves of $f(x, y)$ and the graph of $g(x, y) = 0$ with `ContourPlot` and show the graphs together with `Show`. The minimum and maximum values of $f(x, y)$ subject to the constraint $g(x, y) = 0$ occur at the points where the level curves of $f(x, y)$ are tangent to the graph of $g(x, y) = 0$ as illustrated in Fig. 3.73 (b).

```
gradfplot = StreamPlot[Evaluate[{D[f[x, y], x], D[f[x, y], y]}], {x, −3, 3},
  {y, −3, 3}, StreamStyle → CMYKColor[1, .75, .08, .4]];
cp1 = ContourPlot[f[x, y], {x, −3, 3}, {y, −3, 3}, Contours->30,
  ContourShading->False, PlotPoints->40,
  ContourStyle → {{CMYKColor[.42, .4, .44, .04], Thickness[.01]}}];
cp2 = ContourPlot[g[x, y]==0, {x, −3, 3}, {y, −3, 3}, ContourStyle->Thickness[0.01],
  ContourShading->False];
plot2 = Show[cp1, cp2, gradfplot, Frame → False,
  Axes → Automatic, AxesOrigin → {0, 0}, AxesLabel → {x, y},
  PlotLabel → "(b)"];

Show[GraphicsRow[{plot1, plot2}]]
```

Observe that the maximum and minimum values occur where the gradient vectors of $z = f(x, y)$ are parallel to the gradient vectors of $z = g(x, y)$ on the equation $g(x, y) = 0$. □

3.5.3 Iterated Integrals

The `Integrate` command, used to compute single integrals, is used to compute iterated integrals. The command

```
Integrate[f[x,y],{y,c,d},{x,a,b}]
```

attempts to compute the iterated integral

$$\int_c^d \int_a^b f(x, y)\,dx\,dy. \tag{3.42}$$

If Mathematica cannot compute the exact value of the integral, it is returned unevaluated, in which case numerical results may be more useful. The iterated integral (3.42) is numerically evaluated with the command `N` or

```
NIntegrate[f[x,y],{y,c,d},{x,a,b}]
```

Example 3.67

Evaluate each integral: (a) $\int_2^4 \int_1^2 \left(2xy^2 + 3x^2y\right) dx\,dy$; (b) $\int_0^2 \int_{y^2}^{2y} \left(3x^2 + y^3\right) dx\,dy$; (c) $\int_0^\infty \int_0^\infty xye^{-x^2-y^2}\,dy\,dx$; (d) $\int_0^\pi \int_0^\pi e^{\sin xy}\,dx\,dy$.

Solution. (a) First, we compute $\int\int \left(2xy^2 + 3x^2y\right) dx\,dy$ with `Integrate`. Second, we compute $\int_2^4 \int_1^2 \left(2xy^2 + 3x^2y\right) dx\,dy$ with `Integrate`.

```
Integrate[2xy^2 + 3x^2y, y, x]
```

$\frac{1}{6}x^2y^2(3x + 2y)$

Integrate[2xy^2 + 3x^2y, {y, 2, 4}, {x, 1, 2}]

98

(b) We illustrate the same commands as in (a), except we are integrating over a nonrectangular region.

Integrate[3x^2 + y^3, {x, y^2, 2y}]

$8y^3 + 2y^4 - y^5 - y^6$

Integrate[3x^2 + y^3, y, {x, y^2, 2y}]

$2y^4 + \frac{2y^5}{5} - \frac{y^6}{6} - \frac{y^7}{7}$

Integrate[3x^2 + y^3, {y, 0, 2}, {x, y^2, 2y}]

$\frac{1664}{105}$

(c) Improper integrals can be handled in the same way as proper integrals.

Integrate[x yExp[−x^2 − y^2], x, y]

$\frac{1}{4}e^{-x^2-y^2}$

Integrate[xyExp[−x^2 − y^2], {x, 0, Infinity}, {y, 0, Infinity}]

$\frac{1}{4}$

(d) In this case, Mathematica cannot evaluate the integral exactly so we use `NIntegrate` to obtain an approximation.

Integrate[Exp[Sin[xy]], y, x]

$\int\int e^{\text{Sin}[xy]}dxdy$

NIntegrate[Exp[Sin[xy]], {y, 0, Pi}, {x, 0, Pi}]

15.5092 □

Area, Volume, and Surface Area

Typical applications of iterated integrals include determining the area of a planar region, the volume of a region in three-dimensional space, or the surface area of a region in three-dimensional space. The area of the planar region R is given by

$$A = \iint_R dA. \tag{3.43}$$

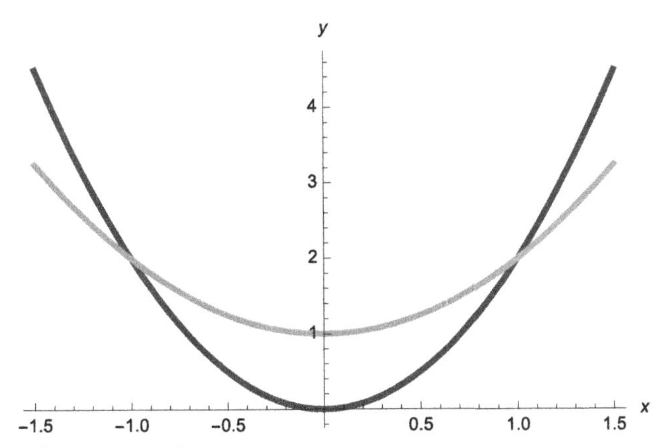

FIGURE 3.74 $y = 2x^2$ and $y = 1 + x^2$ for $-3/2 \le x \le 3/2$ (Michigan State University colors).

If $z = f(x, y)$ has continuous partial derivatives on a closed region R, then the surface area of the portion of the surface that projects onto R is given by

$$SA = \iint_R \sqrt{\left(\frac{\partial f}{\partial x}\right)^2 + \left(\frac{\partial f}{\partial y}\right)^2 + 1}\, dA. \tag{3.44}$$

If $f(x, y) \ge g(x, y)$ on R, the volume of the region between the graphs of $f(x, y)$ and $g(x, y)$ is

$$V = \iint_R (f(x, y) - g(x, y))\, dA. \tag{3.45}$$

Example 3.68

Find the area of the region R bounded by the graphs of $y = 2x^2$ and $y = 1 + x^2$.

Solution. We begin by graphing $y = 2x^2$ and $y = 1 + x^2$ with `Plot` in Fig. 3.74. The x-coordinates of the intersection points are found with `Solve`.

```
Plot[Tooltip[{2x^2, 1 + x^2}], {x, -3/2, 3/2},
  PlotStyle → {{CMYKColor[.82, 0, .64, .7], Thickness[.01]},
  {CMYKColor[.43, .3, .33, 0], Thickness[.01]}},
  AxesLabel → {x, y}]
```

```
Solve[2x^2 == 1 + x^2]
```

$\{\{x \to -1\}, \{x \to 1\}\}$

Using (3.43) and taking advantage of symmetry, the area of R is given by

$$A = \iint_R dA = 2 \int_0^1 \int_{2x^2}^{1+x^2} dy\, dx,$$

which we compute with `Integrate`.

```
2Integrate[1, {x, 0, 1}, {y, 2x^2, 1 + x^2}]
```

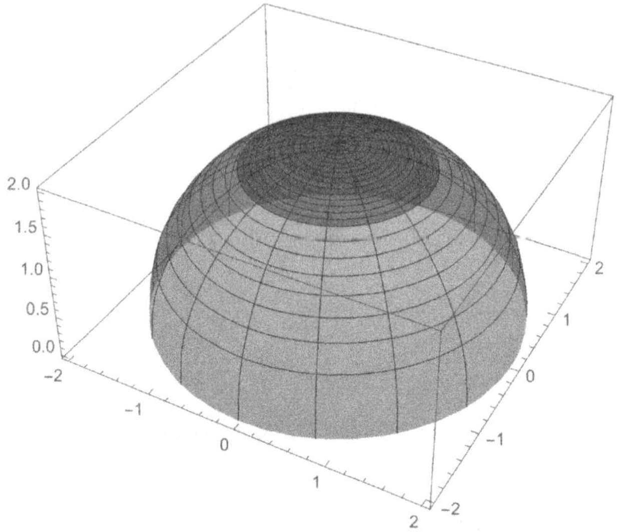

FIGURE 3.75 The portion of the graph of $f(x, y)$ above R (Iowa State University colors).

$\frac{4}{3}$

We conclude that the area of R is $4/3$. □

If the problem exhibits "circular symmetry," changing to polar coordinates is often useful. If $R = \{(r, \theta)\,|a \le r \le b, \alpha \le \theta \le \beta\}$, then

$$\iint_R f(x, y)\, dA = \int_\alpha^\beta \int_a^b f(r\cos\theta, r\sin\theta)\, r\, dr\, d\theta.$$

Example 3.69

Find the surface area of the portion of

$$f(x, y) = \sqrt{4 - x^2 - y^2}$$

that lies above the region $R = \{(x, y)\,|x^2 + y^2 \le 1\}$.

Solution. First, observe that the domain of $f(x, y)$ is

$$\left\{(x, y)\middle| -\sqrt{4 - y^2} \le x \le \sqrt{4 - y^2}, -2 \le y \le 2\right\} = \{(r, \theta)|0 \le r \le 2, 0 \le \theta \le 2\pi\}.$$

Similarly,

$$R = \left\{(x, y)\middle| -\sqrt{1 - y^2} \le x \le \sqrt{1 - y^2}, -1 \le y \le 1\right\} = \{(r, \theta)|0 \le r \le 1, 0 \le \theta \le 2\pi\}.$$

With this observation, we use `ParametricPlot3D` to graph $f(x, y)$ in `p1` and the portion of the graph of $f(x, y)$ above R in `p2` and show the two graphs together with `Show`. We wish to find the area of the black region in Fig. 3.75.

```
f[x_, y_] = Sqrt[4 − x^2 − y^2];
```

```
p1 = ParametricPlot3D[{rCos[t], rSin[t], f[rCos[t], rSin[t]]}, {r, 0, 2},
```

```
    {t, 0, 2Pi}, PlotPoints → 45, PlotStyle → {{CMYKColor[.02, 1, .85, .06],
    Opacity[.4]}}];
p2 = ParametricPlot3D[{rCos[t], rSin[t], f[rCos[t], rSin[t]]}, {r, 0, 1},
    {t, 0, 2Pi}, PlotPoints → 45, PlotStyle → {{CMYKColor[.31, .38, .75, .76],
    Opacity[.4]}}];
Show[p1, p2, BoxRatios → Automatic]
```

We compute $f_x(x, y)$, $f_y(x, y)$ and $\sqrt{[f_x(x, y)]^2 + [f_y(x, y)]^2 + 1}$ with `D` and `Simplify`.

```
fx = D[f[x, y], x]
fy = D[f[x, y], y]
```

$-\frac{x}{\sqrt{4-x^2-y^2}}$

$-\frac{y}{\sqrt{4-x^2-y^2}}$

```
s1 = Simplify[Sqrt[1 + fx^2 + fy^2]]
```

$2\sqrt{-\frac{1}{-4+x^2+y^2}}$

Then, using (3.44), the surface area is given by

$$\begin{aligned} SA &= \iint_R \sqrt{\left(\frac{\partial f}{\partial x}\right)^2 + \left(\frac{\partial f}{\partial y}\right)^2 + 1}\, dA \\ &= \iint_R \frac{2}{\sqrt{4-x^2-y^2}}\, dA \\ &= \int_{-1}^{1} \int_{-\sqrt{1-y^2}}^{\sqrt{1-y^2}} \frac{2}{\sqrt{4-x^2-y^2}}\, dx\, dy. \end{aligned} \tag{3.46}$$

However, notice that in polar coordinates,

$$R = \{(r, \theta)\, | 0 \le r \le 1, 0 \le \theta \le 2\pi\}$$

so in polar coordinates the surface area is given by

$$SA = \int_0^{2\pi} \int_0^1 \frac{2}{\sqrt{4-r^2}}\, r\, dr\, d\theta,$$

```
s2 = Simplify[s1/.{x → rCos[t], y → rSin[t]}]
```

$2\sqrt{\frac{1}{4-r^2}}$

which is much easier to evaluate than (3.46). We evaluate the iterated integral with `Integrate`

```
s3 = Integrate[rs2, {t, 0, 2Pi}, {r, 0, 1}]
```

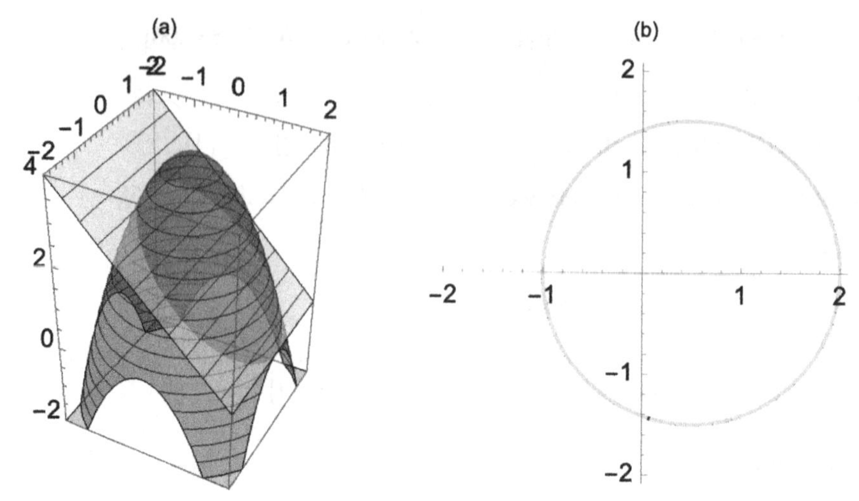

FIGURE 3.76 (a) $z = 4 - x^2 - y^2$ and $z = 2 - x$ for $-2 \leq x \leq 2$ and $-2 \leq y \leq 2$. (b) Graph of $4 - x^2 - y^2 = 2 - x$ (University of California-Irvine colors).

$$-4\left(-2+\sqrt{3}\right)\pi$$

```
N[s3]
```

3.36715

and conclude that the surface area is $\left(8 - 4\sqrt{3}\right)\pi \approx 3.367$. □

Example 3.70

Find the volume of the region between the graphs of $z = 4 - x^2 - y^2$ and $z = 2 - x$.

Solution. We begin by graphing $z = 4 - x^2 - y^2$ and $z = 2 - x$ together with `Plot3D` in Fig. 3.76 (a). The region of integration, R, is determined by graphing $4 - x^2 - y^2 = 2 - x$ with `ContourPlot` in Fig. 3.76 (b).

```
p1 = Plot3D[{4 − x^2 − y^2, 2 − x}, {x, −2, 2}, {y, −2, 2},
  PlotRange → {{−2, 2}, {−2, 2}, {−2, 4}},
  PlotStyle → {{CMYKColor[.93, .62, .09, .01], Opacity[.4]},
  {{CMYKColor[.01, .16, 1, 0], Opacity[.6]}}},
  MeshFunctions->{#3&},
  BoxRatios → Automatic];
p2 = ContourPlot[4 − x^2 − y^2 − (2 − x) == 0, {x, −2, 2}, {y, −2, 2},
  PlotPoints → 50, Frame → False, Axes → Automatic, AxesOrigin → {0, 0},
  ContourStyle → {{CMYKColor[.01, .16, 1, .0], Thickness[.01]}}];
Show[GraphicsRow[{p1, p2}]]
```

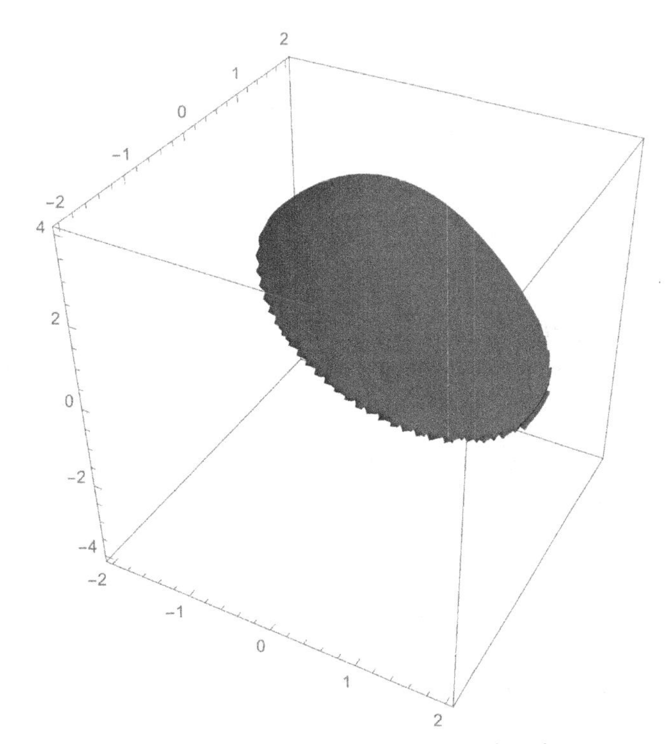

FIGURE 3.77 Using `RegionPlot3D` to see the region $2 - x \le z \le 4 - x^2 - y^2$ (University of California-Irvine colors).

Another way to see the situation illustrated in Fig. 3.77 is to use `RegionPlot3D`, which works in the same way as `RegionPlot` but in three dimensions.

```
RegionPlot3D[2 − x ≤ z<=4 − x^2 − y^2, {x, −2, 2}, {y, −2, 2},
  {z, −4, 4}, PlotPoints → 75, Mesh → False,
  PlotStyle->{CMYKColor[.93, .62, .09, .01], Opacity[.4]}]
```

Completing the square shows us that

$$\begin{aligned} R &= \left\{ (x, y) \left| \left(x - \frac{1}{2} \right)^2 + y^2 \le \frac{9}{4} \right. \right\} \\ &= \left\{ (x, y) \left| \frac{1}{2} - \frac{1}{2}\sqrt{9 - 4y^2} \le x \le \frac{1}{2} + \frac{1}{2}\sqrt{9 - 4y^2}, -\frac{3}{2} \le y \le \frac{3}{2} \right. \right\}. \end{aligned}$$

Thus, using (3.45), the volume of the solid is given by

$$\begin{aligned} V &= \iint_R \left[\left(4 - x^2 - y^2\right) - (2 - x) \right] dA \\ &= \int_{-\frac{3}{2}}^{\frac{3}{2}} \int_{\frac{1}{2}-\frac{1}{2}\sqrt{9-4y^2}}^{\frac{1}{2}+\frac{1}{2}\sqrt{9-4y^2}} \left[\left(4 - x^2 - y^2\right) - (2 - x) \right] dx\, dy, \end{aligned}$$

which we evaluate with `Integrate`.

```
i1 = Integrate[(4 − x^2 − y^2) − (2 − x), {y, −3/2, 3/2},
{x, 1/2 − 1/2Sqrt[9 − 4y^2], 1/2 + 1/2Sqrt[9 − 4y^2]}]
```

$\frac{81\pi}{32}$

N[i1]

7.95216

We conclude that the volume is $\frac{81}{32}\pi \approx 7.952$. □

Triple Iterated Integrals

Triple iterated integrals are calculated in the same manner as double iterated integrals.

Example 3.71

Evaluate

$$\int_0^{\pi/4}\int_0^y\int_0^{y+z}(x+2z)\sin y\,dx\,dz\,dy.$$

Solution. Entering

i1 = Integrate[(x + 2z)Sin[y], {y, 0, Pi/4}, {z, 0, y}, {x, 0, y + z}]

$-\frac{17(384-96\pi-12\pi^2+\pi^3)}{384\sqrt{2}}$

calculates the triple integral exactly with `Integrate`.

An approximation of the exact value is found with `N`.

N[i1]

0.157206 □

We illustrate how triple integrals can be used to find the volume of a solid when using spherical coordinates.

Example 3.72

Find the volume of the torus with equation in spherical coordinates $\rho = \sin\phi$.

Solution. We proceed by graphing the torus with `SphericalPlot3D` in Fig. 3.79 (see Fig. 3.78 for the help feature associated with this command).

SphericalPlot3D[Sin[Phi], {Phi, 0, Pi}, {theta, 0, 2Pi}, PlotPoints → 40]

In general, the volume of the solid region D is given by

$$V = \iiint_D dV.$$

Thus, the volume of the torus is given by the triple iterated integral

$$V = \int_0^{2\pi}\int_0^{\pi}\int_0^{\sin\phi}\rho^2\sin\phi\,d\rho\,d\phi\,d\theta,$$

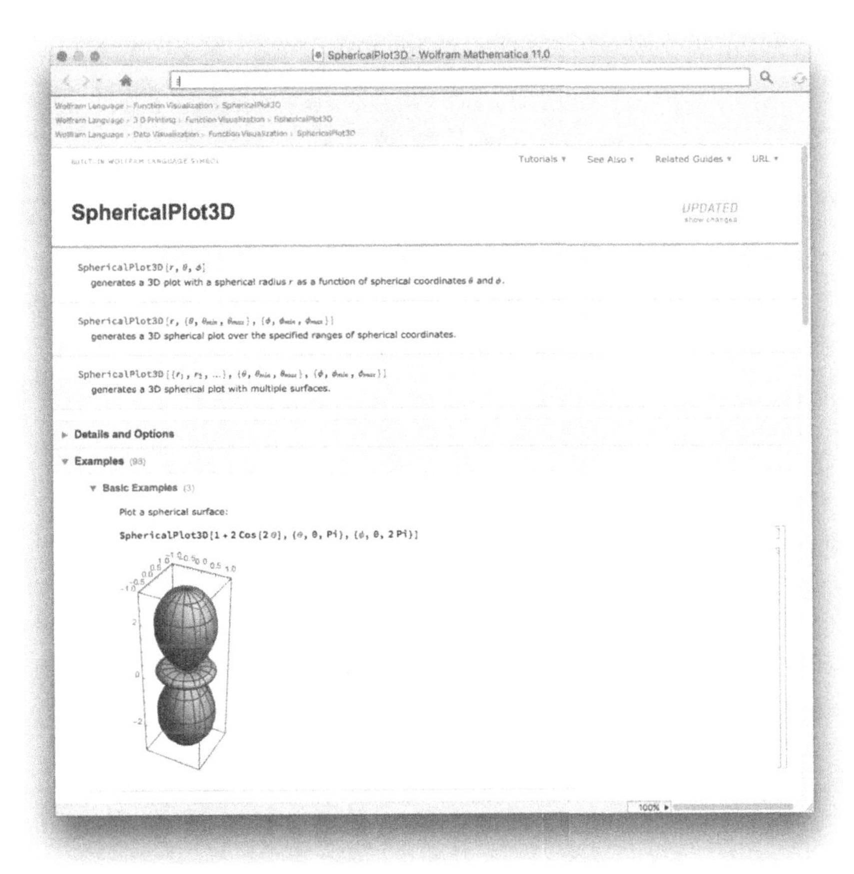

FIGURE 3.78 Mathematica's help for `SphericalPlot3D`.

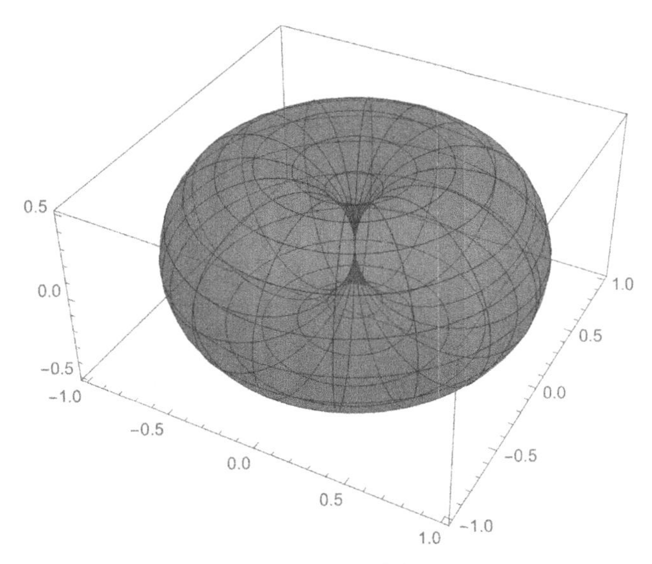

FIGURE 3.79 A graph of the torus (University of Arizona colors).

i1 = Integrate[rho^2Sin[phi], {theta, 0, 2Pi},

{phi, 0, Pi}, {rho, 0, Sin[phi]}]

$\frac{\pi^2}{4}$

```
N[i1]
```

2.4674

which we evaluate with `Integrate`. We conclude that the volume of the torus is $\frac{1}{4}\pi^2 \approx 2.467$. □

Chapter 4

Introduction to Lists and Tables

4.1 LISTS AND LIST OPERATIONS

4.1.1 Defining Lists

A **list** of n elements is a Mathematica object of the form

```
list={a1,a2,a3,...,an}.
```

The ith element of the list is extracted from `list` with `list[[i]]` or `Part[list,i]`.

Elements of a list are separated by commas. Lists are always enclosed in braces `{...}` and each element of a list may be (almost any) Mathematica object, even other lists. Because lists are Mathematica objects, they can be named. For easy reference, we will usually name lists.

Lists can be defined in a variety of ways: they may be completely typed in, imported from other programs and text files, or they may be created with either the `Table` or `Array` commands. Given a function $f(x)$ and a number n, the command

`Table` and `Manipulate` have nearly identical syntax. With `Manipulate`, you can create an interactive dynamic application; `Table` returns non-adjustable results.

1. `Table[f[i],{i,n}]` creates the list `{f[1],...,f[n]}`;
2. `Table[f[i],{i,0,n}]` creates the list `{f[0],...,f[n]}`;
3. `Table[f[i],{i,n,m}]` creates the list

```
{f[n],f[n+1],...,f[m-1],f[m]};
```

4. `Table[f[i],{i,imin,imax,istep}]` creates the list

```
{f[imin],f[imin+istep],f[imin+2*istep],...,f[imax]};
```

   and
5. `Array[f,n]` creates the list `{f[1],...,f[n]}`.

   In particular,

```
Table[f[x],{x,a,b,(b-a)/(n-1)}]
```

returns a list of $f(x)$ values for n equally spaced values of x between a and b;

```
Table[{x,f[x]},{x,a,b,(b-a)/(n-1)}]
```

returns a list of points $(x, f(x))$ for n equally spaced values of x between a and b.

Mathematica by Example. http://dx.doi.org/10.1016/B978-0-12-812481-9.00004-1

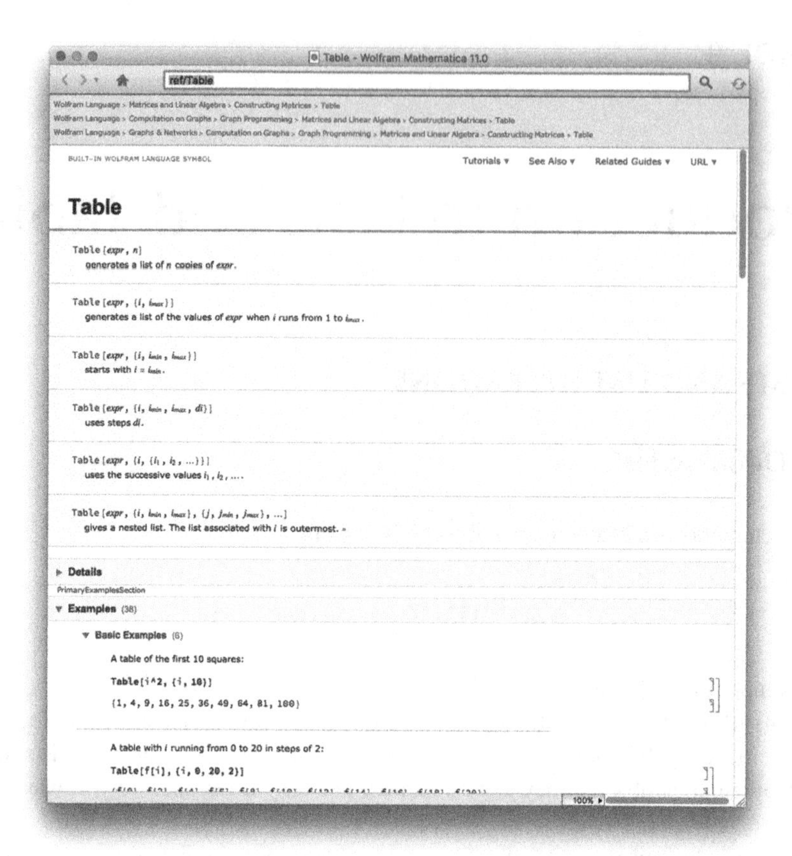

In addition to using `Table`, lists of numbers can be calculated using `Range`.

1. `Range[n]` generates the list `{1,2, ... , n}`;
2. `Range[n1,n2]` generates the list `{n1, n1+1, ... , n2-1, n2}`; and
3. `Range[n1,n2,nstep]` generates the list

`{n1, n1+nstep,n1+2*nstep, ... , n2-nstep,n2}`.

Example 4.1

Use Mathematica to generate the list `{1,2,3,4,5,6,7,8,9,10}`.

Solution. Generally, a given list can be constructed in several ways. In fact, each of the following five commands generates the list `{1,2,3,4,5,6,7,8,9,10}`.

{1, 2, 3, 4, 5, 6, 7, 8, 9, 10}

{1, 2, 3, 4, 5, 6, 7, 8, 9, 10}

Table[*i*, {*i*, 10}]

{1, 2, 3, 4, 5, 6, 7, 8, 9, 10}

Table[*i*, {*i*, 1, 10}]

{1, 2, 3, 4, 5, 6, 7, 8, 9, 10}

Table$\left[\frac{i}{2}, \{i, 2, 20, 2\}\right]$

$\{1, 2, 3, 4, 5, 6, 7, 8, 9, 10\}$

Range[10]

$\{1, 2, 3, 4, 5, 6, 7, 8, 9, 10\}$ □

Example 4.2

Use Mathematica to define `listone` to be the list of numbers {1, 3/2, 2, 5/2, 3, 7/2, 4}.

Solution. In this case, we generate a list and name the result `listone`. As in Example 4.1, we illustrate that `listone` can be created in several ways.

listone $= \left\{1, \frac{3}{2}, 2, \frac{5}{2}, 3, \frac{7}{2}, 4\right\}$

$\left\{1, \frac{3}{2}, 2, \frac{5}{2}, 3, \frac{7}{2}, 4\right\}$

listone = Table$\left[i, \left\{i, 1, 4, \frac{1}{2}\right\}\right]$

$\left\{1, \frac{3}{2}, 2, \frac{5}{2}, 3, \frac{7}{2}, 4\right\}$

Last, we define $i(n) = \frac{1}{2}n + \frac{1}{2}$ and use `Array` to create the table `listone`.

i**[n_]** $= \frac{n}{2} + \frac{1}{2}$;

listone = Array[i**, 7]**

$\left\{1, \frac{3}{2}, 2, \frac{5}{2}, 3, \frac{7}{2}, 4\right\}$ □

Example 4.3

Create a list of the first 25 prime numbers. What is the fifteenth prime number?

Solution. The command `Prime[n]` yields the nth prime number. We use `Table` to generate a list of the ordered pairs `{n,Prime[n]}` for $n = 1, 2, 3, \ldots, 25$ and name the resulting list `list`. We then use verb+Short+ to obtain an abbreviated portion of `list`. Generally, `Short` returns the first and last few elements of a list. The number of omitted terms between the first few and last few is indicated with `<<n>>`. In this case, we see that 13 terms are omitted.

list = Table[{n**, Prime[**n**]}, {**n**, 1, 25}];**

Short[list]

{{1, 2}, {2, 3}, {3, 5}, {4, 7}, {5, 11}, {6, 13}, ⟨⟨13⟩⟩, {20, 71}, {21, 73}, {22, 79}, {23, 83},

{24, 89}, {25, 97}}

The ith element of a list `list` is extracted from `list` with `list[[i]]` or `Part[list,i]`. From the resulting output, we see that the fifteenth prime number is 47.

list[[15]]

{15, 47}

Part[list, 15]

{15, 47} □

Remark 4.1. You can use the `Manipulate` function in nearly the exact same way as the `Table` function. With `Manipulate`, the result is an interactive dynamic object that can be saved as an application that can be run outside of Mathematica.

In addition, we can use `Table` to generate lists consisting of the same or similar objects.

Example 4.4

(a) Generate a list consisting of five copies of the letter a. (b) Generate a list consisting of ten random integers between -10 and 10 and then a list of ten random real numbers between -10 and 10.

Solution. Entering

Clear[*a*]

Table[*a*, {5}]

$\{a, a, a, a, a\}$

generates a list consisting of five copies of the letter a. For (b), we use the command `RandomInteger` and `RandomReal` to generate the desired lists. Because we are using `RandomInteger` and `RandomReal`, your results will certainly differ from those obtained here.

RandomInteger[{−10, 10}, 10]

{3, −7, 9, 5, 4, 10, 7, −8, 7, 8}

RandomReal[{−10, 10}, 10]

{4.71284, 1.69439, 6.60498, 8.12249, 5.20831, 6.73414, −1.84369, 9.16472, 8.96906,

−6.04237} □

`Manipulate` works in much the same way as `Table` but allows you to interactively see how adjusting parameters affects a given situation.

Example 4.5

For example, in polar coordinates, the graphs of $r = \sin n\theta$ and $r = \cos n\theta$ are n-leaved roses if n is odd and $2n$-leaved roses if n is even. If n is even, the area of the graph enclosed by the $2n$

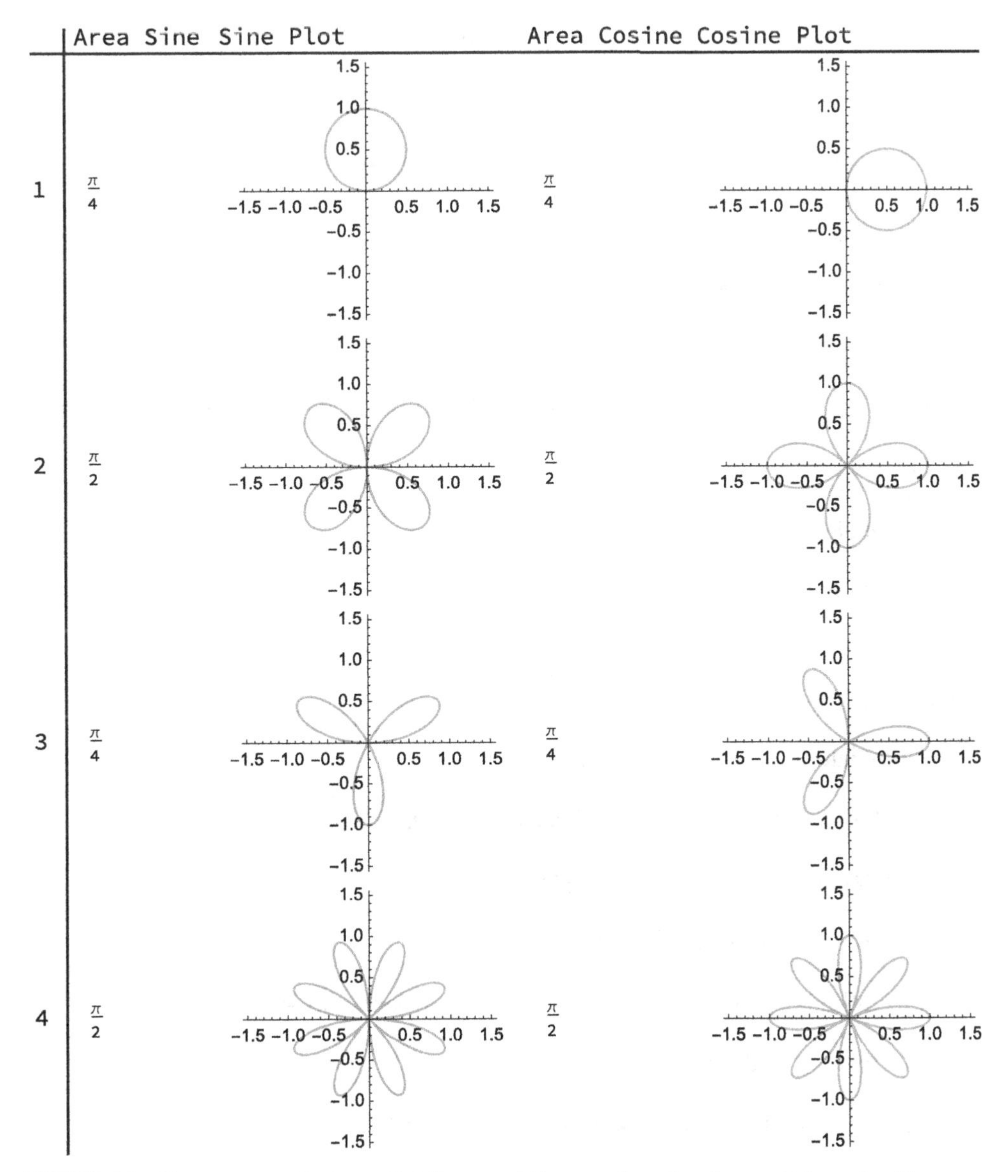

FIGURE 4.1 You can use `Table` to see that the area of the roses depends only on whether n is odd or even. (University of Colorado-Boulder colors)

roses is $A = \frac{1}{2}\int_0^{2\pi} r^2\,d\theta = \pi/2$. While if n is odd, the area of the graph enclosed by the n roses is $A = \frac{1}{2}\int_0^{2\pi} r^2\,d\theta = \pi/4$.

To see this with Mathematica, we can use `Table`. (See Figs. 4.1 and 4.2.) (Note that `If[condition,f,g]` returns `f` if `condition` is `True` and `g` if it is not.)

```
Clear[n, x];

t1 = Table[{

    If[Mod[n/2, 1]===0, Integrate[Sin[nx]^2, {x, 0, 2Pi}]/2,

      Integrate[Sin[nx]^2, {x, 0, Pi}]/2],

      PolarPlot[Sin[nx], {x, 0, 2Pi},
```

```
        PlotRange→ {{−3/2, 3/2}, {−3/2, 3/2}}, AspectRatio→ 1,
        PlotStyle→ {{CMYKColor[0, .1, .48, .22], Thickness[.01]}}],
      If[Mod[n/2, 1]===0, Integrate[Cos[nx]^2, {x, 0, 2Pi}]/2,
        Integrate[Cos[nx]^2, {x, 0, Pi}]/2],
        PolarPlot[Cos[nx], {x, 0, 2Pi},
        PlotRange→ {{−3/2, 3/2}, {−3/2, 3/2}}, AspectRatio→ 1,
        PlotStyle→ {{CMYKColor[0, .1, .48, .22], Thickness[.01]}}]}, {n, 1, 4}];
   TableForm[t1,
     TableHeadings→
     {Table[n, {n, 1, 4}], {"Area Sine", "Sine Plot", "Area Cosine", "Cosine Plot"}}]
```

Alternatively, you can use `Manipulate`.

```
Clear[n, x];
   Manipulate[{n,
      If[Mod[n/2, 1]===0, Integrate[Sin[nx]^2, {x, 0, 2Pi}]/2,
        Integrate[Sin[nx]^2, {x, 0, Pi}]/2],
        PolarPlot[Sin[nx], {x, 0, 2Pi},
        PlotRange→ {{−3/2, 3/2}, {−3/2, 3/2}}, AspectRatio→ 1,
        PlotStyle→ Black],
      If[Mod[n/2, 1]===0, Integrate[Cos[nx]^2, {x, 0, 2Pi}]/2,
        Integrate[Cos[nx]^2, {x, 0, Pi}]/2],
        PolarPlot[Cos[nx], {x, 0, 2Pi},
        PlotRange→ {{−3/2, 3/2}, {−3/2, 3/2}}, AspectRatio→ 1,
        PlotStyle→ Black]}, {{n, 5}, 1, 100, 1}]
```

4.1.2 Plotting Lists of Points

Lists are plotted with `ListPlot`.

1. `ListPlot[{{x1,y1},{x2,y2},...,{xn,yn}}]` plots the list of points $\{(x_1, y_1), (x_2, y_2), \ldots, (x_n, y_n)\}$. The size of the points in the resulting plot is controlled with the option `PlotStyle->PointSize[w]`, where w is the fraction of the total width of the graphic. For two-dimensional graphics, the default value is 0.008.

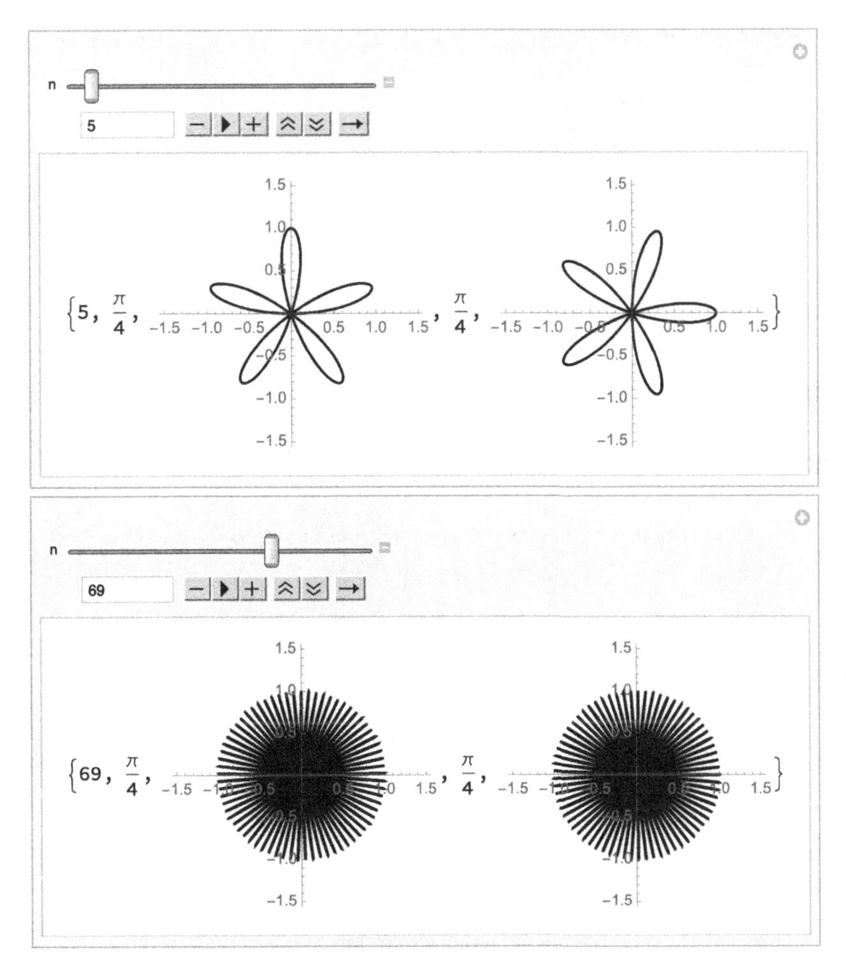

FIGURE 4.2 With `Manipulate` you can see that the area alternates from $\pi/2$ to $\pi/4$ as n alternates from even to odd.

2. `ListPlot[{y1,y2,...,yn}]` plots the list of points $\{(1, y_1), (2, y_2), \ldots, (n, y_n)\}$.
3. `ListLinePlot[{y1,y2,...,yn}]` plots the list of points $\{(1, y_1), (2, y_2), \ldots, (n, y_n)\}$ and connects consecutive points with line segments. Alternatively, you can use `ListPlot` together with the option `Joined->True` to connect consecutive points with line segments.

Example 4.6

Entering

```
t1 = Table[Sin[n], {n, 1, 1000}];

ListPlot[t1, PlotStyle → {CMYKColor[.09, 1, .64, .48], PointSize[.01]}]
```

creates a list consisting of $\sin n$ for $n = 1, 2, \ldots, 1000$ and then graphs the list of points $(n, \sin n)$ for $n = 1, 2, \ldots, 1000$. See Fig. 4.3.

When a semi-colon is included at the end of a command, the resulting output is suppressed.

Example 4.7: The Prime Difference Function and the Prime Number Theorem

In `t1`, we use `Prime` and `Table` to compute a list of the first 25,000 prime numbers.

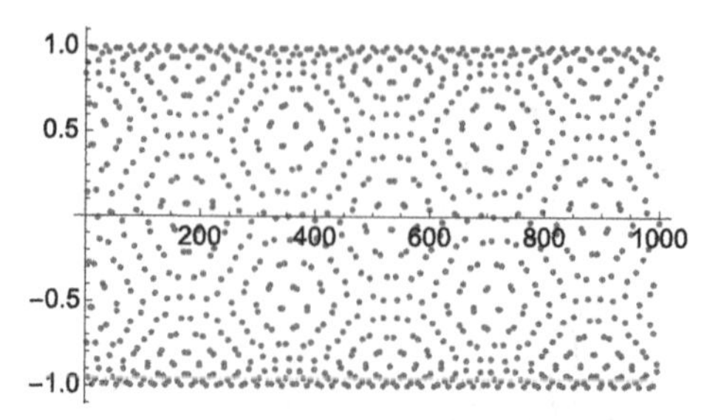

FIGURE 4.3 Plot of $(n, \sin n)$ for $n = 1, 2, \ldots, 1000$. (University of Massachusetts at Amherst colors)

```
t1 = Table[Prime[n], {n, 1, 25000}];
```

```
Length[t1]
```

We use `Length` to verify that `t1` has 25,000 elements and `Short` to see an abbreviated portion of `t1`.

25000

```
Short[t1]
```

{2, 3, 5, 7, 11, 13, 17, 19, ⟨⟨24984⟩⟩, 287047, 287057, 287059, 287087, 287093, 287099, 287107, 287117}

`First[list]` returns the first element of `list`; `Last[list]` returns the last element of `list`.

You can also use `Take` to extract elements of lists.

1. `Take[list,n]` returns the first n elements of `list`;
2. `Take[list,-n]` returns the last n elements of `list`; and
3. `Take[list,{n,m}]` returns the nth through mth elements of `list`.

```
Take[t1, 5]
```

{2, 3, 5, 7, 11}

```
Take[t1, -5]
```

{287087, 287093, 287099, 287107, 287117}

```
Take[t1, {12501, 12505}]
```

{134059, 134077, 134081, 134087, 134089}

`list[[i]]` returns the ith element of `list` so `list[[i + 1]] − list[[i]]` computes the difference between the $(i+1)$st and ith elements of `list`. `list[[i,j]]` returns the jth part of the ith part of `list`.

In `t2`, we compute the difference, d_n, between the successive prime numbers in `t1`. The result is plotted with `ListPlot` in Fig. 4.4.

```
t2 = Table[t1[[i + 1]] − t1[[i]], {i, 1, Length[t1] − 1}];
```

```
Short[t2]
```

{1, 2, 2, 4, 2, 4, 2, 4, 6, 2, 6, 4, 2, 4, 6, 6, 2, 6, ⟨⟨24964⟩⟩, 18, 28, 14, 54, 46, 8, 6, 12, 4, 44, 10, 2, 28, 6, 6, 8, 10}

```
ListPlot[t2, PlotRange → All,
  PlotStyle → {CMYKColor[.09, 1, .64, .48], PointSize[.01]}]
```

Let $\pi(n)$ denote the number of primes less than n and $Li(x)$ denote the **logarithmic integral:**

$$\texttt{LogIntegral[x]} = Li(x) = \int_0^x \frac{1}{\ln t}\,dt.$$

We use `Plot` to graph $Li(x)$ for $1 \le x \le 25,000$ in `p1`.

```
p1 = Plot[LogIntegral[x], {x, 1, 2500},
  PlotStyle → {{CMYKColor[.17, .36, .52, .38], Thickness[.01]}}]
```

Remember that `p1` is not displayed because a semi-colon is included at the end of the `Plot` command.

The **Prime Number Theorem** states that

$$\pi(n) \sim Li(n).$$

(See [17].) In the following, we use `Select` and `Length` to define $\pi(n)$. `Select[list,criteria]` returns the elements of `list` for which `criteria` is true. Note that `#<n` is called a `pure function`: given an argument `#`, `#<n` is true if `#<n` and false otherwise. The `&` symbol marks the end of a pure function. Thus, given n, `Select[t1,#<n&]` returns a list of the elements of `t1` less than n; `Select[t1,#<n&]//Length` returns the number of elements in the list.

```
smallpi[n_]:=Select[t1, # < n&]//Length
```

For example,

```
smallpi[100]

25
```

shows us that $\pi(100) = 25$. Note that because `t1` contains the first 25, 000 primes, `smallpi[n]` is valid for $1 \le n \le N$ where $\pi(N) = 25,000$. In `t3`, we compute $\pi(n)$ for $n = 1, 2, \ldots, 2500$

```
t3 = Table[smallpi[n], {n, 1, 2500}];

Short[t3]

{0, 0, 1, 2, 2, 3, 3, 4, 4, 4, 4, 5, 5, 6, <<2473>>, 367, 367, 367, 367, 367, 367, 367, 367, 367,
367, 367, 367, 367}
```

and plot the resulting list with `ListPlot`.

```
p2 = ListPlot[t3, PlotStyle → CMYKColor[.05, .71, 1, .23]]
```

`p1` and `p2` are displayed together with `Show` in Fig. 4.5.

```
Show[p1, p2]
```

Working in almost the same way as `Take`, `Span` (`;;`) selects elements of lists: `list[[n;;m]]` returns the n through mth elements of `list`.

`Span` is new Mathematica 11 but works in almost the same way as `Take`.

Example 4.8

Here are the first few terms of sequence A073184 (N.J.A. Sloane, 2007, *The On-Line Encyclopedia of Integer Sequences*, `www.research.att.com/njas/sequences/`), the number of cube free divisors of n.

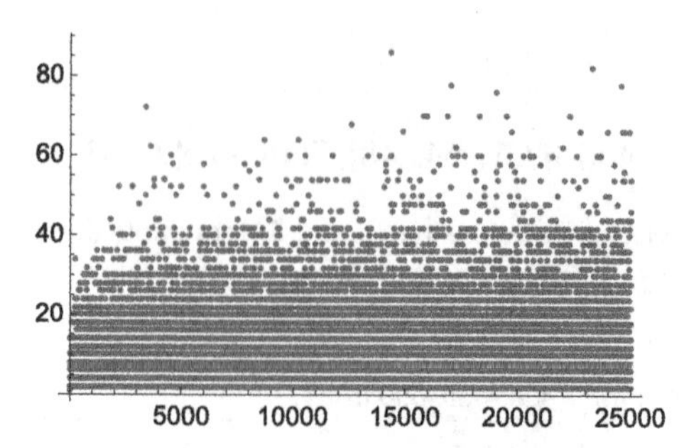

FIGURE 4.4 A plot of the difference, d_n, between successive prime numbers.

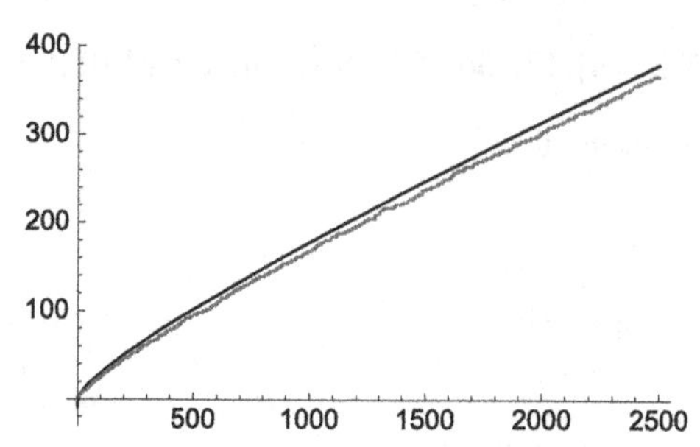

FIGURE 4.5 Graphs of $Li(x)$ and $\pi(n)$.

```
ashortlist = {1, 2, 2, 3, 2, 4, 2, 3, 3, 4, 2, 6,
      2, 4, 4, 3, 2, 6, 2, 6, 4, 4, 2, 6,
    3, 4, 3, 6, 2, 8, 2, 3, 4, 4, 4};
```

With `;;` (`Span`), we select the 2nd through 8th elements of `ashortlist`.

```
ashortlist[[2;;8]]
{2, 2, 3, 2, 4, 2, 3}
```

The same results are obtained with `Take`.

```
Take[ashortlist, {2, 8}]
{2, 2, 3, 2, 4, 2, 3}
```

You can count the number of elements of a list with `Length`.

```
Length[ashortlist]
35
```

With `Tally`, we count the number of occurrences of each digit in the list. Thus,

```
Tally[ashortlist]
{{1, 1}, {2, 11}, {3, 7}, {4, 10}, {6, 5}, {8, 1}}
```

shows us that there are 11 2's, 10 4's, and so on.

However, you can use `Table` together with `Part` (`[[...]]`) to obtain the same results as those obtained with `Take` or `Span`.

```
Table[t1[[i]], {i, 1, 5}]
```

```
Table[t1[[i]], {i, 24996, 25000}]
```

```
Table[t1[[i]], {i, 12501, 12505}]
```

{2, 3, 5, 7, 11}

{287087, 287093, 287099, 287107, 287117}

{134059, 134077, 134081, 134087, 134089}

You can iterate recursively with `Table`. Both

{{a[1, 2], a[2, 2], a[3, 2], a[4, 2], a[5, 2]}, {a[1, 4], a[2, 4], a[3, 4], a[4, 4], a[5, 4]},

{a[1, 6], a[2, 6], a[3, 6], a[4, 6], a[5, 6]}, {a[1, 8], a[2, 8], a[3, 8], a[4, 8], a[5, 8]},

{a[1, 10], a[2, 10], a[3, 10], a[4, 10], a[5, 10]}}

Length[t1]

5

and

t2 = Table[Table[a[i, j], {i, 1, 5}], {j, 2, 10, 2}]

{{a[1, 2], a[2, 2], a[3, 2], a[4, 2], a[5, 2]}, {a[1, 4], a[2, 4], a[3, 4], a[4, 4], a[5, 4]},

{a[1, 6], a[2, 6], a[3, 6], a[4, 6], a[5, 6]}, {a[1, 8], a[2, 8], a[3, 8], a[4, 8], a[5, 8]},

{a[1, 10], a[2, 10], a[3, 10], a[4, 10], a[5, 10]}}

compute tables of a_{ij}. The outermost iterator is evaluated first: in this case, i is followed by j as in `t1` and the result is a list of lists. To eliminate the inner lists (that is, the braces), use `Flatten`. Generally, `Flatten[list,n]` flattens `list` (removes braces) to level n.

Flatten[t1]

{a[1, 2], a[2, 2], a[3, 2], a[4, 2], a[5, 2], a[1, 4], a[2, 4], a[3, 4], a[4, 4], a[5, 4],

a[1, 6], a[2, 6], a[3, 6], a[4, 6], a[5, 6], a[1, 8], a[2, 8], a[3, 8], a[4, 8], a[5, 8], a[1, 10],

a[2, 10], a[3, 10], a[4, 10], a[5, 10]}

The observation is especially important when graphing lists of points obtained by iterating `Table`. For example,

`Length[list]` returns the number of elements in `list`.

t1 = Table[{Sin[x + y], Cos[x − y]}, {x, 1, 5}, {y, 1, 5}]

{{{Sin[2], 1}, {Sin[3], Cos[1]}, {Sin[4], Cos[2]}, {Sin[5], Cos[3]}, {Sin[6], Cos[4]}},

{{Sin[3], Cos[1]}, {Sin[4], 1}, {Sin[5], Cos[1]}, {Sin[6], Cos[2]}, {Sin[7], Cos[3]}},

{{Sin[4], Cos[2]}, {Sin[5], Cos[1]}, {Sin[6], 1}, {Sin[7], Cos[1]}, {Sin[8], Cos[2]}},

{{Sin[5], Cos[3]}, {Sin[6], Cos[2]}, {Sin[7], Cos[1]}, {Sin[8], 1}, {Sin[9], Cos[1]}},

{{Sin[6], Cos[4]}, {Sin[7], Cos[3]}, {Sin[8], Cos[2]}, {Sin[9], Cos[1]}, {Sin[10], 1}}}

Length[t1]

5

is not a list of 25 points: `t1` is a list of 5 lists each consisting of 5 points. `t1` has two levels. For example, the 3rd element of the second level is

t1[[3]]

{{Sin[4], Cos[2]}, {Sin[5], Cos[1]}, {Sin[6], 1}, {Sin[7], Cos[1]}, {Sin[8], Cos[2]}}

and the 2nd element of the third level (or the second part of the third part) is

t1[[3, 2]]

{Sin[5], Cos[1]}

To flatten `t2` to level 1, we use `Flatten`.

t2 = Flatten[t1, 1]

{{Sin[2], 1}, {Sin[3], Cos[1]}, {Sin[4], Cos[2]}, {Sin[5], Cos[3]}, {Sin[6], Cos[4]},

{Sin[3], Cos[1]}, {Sin[4], 1}, {Sin[5], Cos[1]}, {Sin[6], Cos[2]}, {Sin[7], Cos[3]},

{Sin[4], Cos[2]}, {Sin[5], Cos[1]}, {Sin[6], 1}, {Sin[7], Cos[1]}, {Sin[8], Cos[2]},

{Sin[5], Cos[3]}, {Sin[6], Cos[2]}, {Sin[7], Cos[1]}, {Sin[8], 1}, {Sin[9], Cos[1]},

{Sin[6], Cos[4]}, {Sin[7], Cos[3]}, {Sin[8], Cos[2]}, {Sin[9], Cos[1]}, {Sin[10], 1}}

The resulting list of ordered pairs (in Mathematica, `{x,y}` corresponds to (x, y)). This list of points are then plotted with `ListPlot` in Fig. 4.6 (a). We also illustrate the use of the `PlotStyle`, `PlotRange`, and `AspectRatio` options in the `ListPlot` command.

lp1 = ListPlot[t2, PlotStyle → {PointSize[.05], CMYKColor[.11, 0, 0, .64]},

PlotRange → {{−3/2, 3/2}, {−3/2, 3/2}}, AspectRatio → Automatic,

PlotLabel → "(a)"];

Increasing the number of points further illustrates the use of `Flatten`. Entering

t1 = Table[{Sin[*x* + *y*], Cos[*x* − *y*]}, {*x*, 1, 125}, {*y*, 1, 125}];

Length[t1]

125

results in a very long nested list. `t1` has 125 elements each of which has 125 elements.

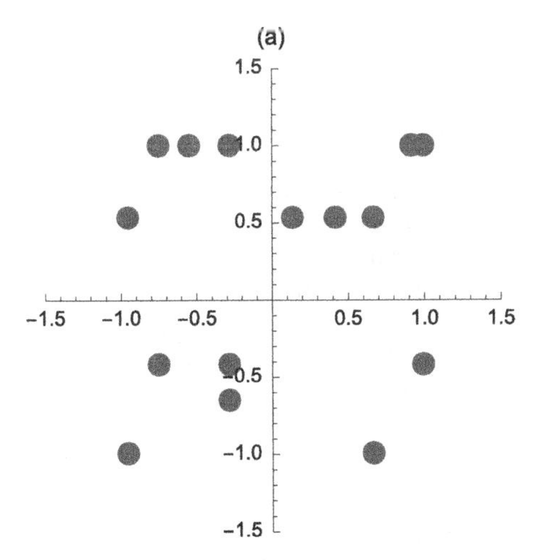

FIGURE 4.6 (a) and (b). (Rutgers University colors)

An abbreviated version is viewed with `Short`.

`Short[list]` yields an abbreviated version of `list`.

Short[t1]

{{{Sin[2], 1}, {Sin[3], Cos[1]}, ⟨⟨121⟩⟩, {Sin[125], Cos[123]}, {Sin[126], Cos[124]}},

⟨⟨123⟩⟩, {⟨⟨1⟩⟩}}

After using `Flatten`, we see with `Length` and `Short` that `t2` contains 15, 625 points,

t2 = Flatten[t1, 1];

Length[t2]

15625

Short[t2]

{{Sin[2], 1}, {Sin[3], Cos[1]}, ⟨⟨15621⟩⟩, {Sin[249], Cos[1]}, {Sin[250], 1}}

which are plotted with `ListPlot` in Fig. 4.6 (b).

lp2 = ListPlot[t2, PlotStyle → {{CMYKColor[0, 1, .81, .04]}}, AspectRatio → Automatic,

PlotLabel → "(b)"];

Show[GraphicsRow[{lp1, lp2}]]

Remark 4.2. Mathematica is very flexible and most calculations can be carried out in more than one way. Depending on how you think, some sequences of calculations may make more sense to you than others, even if they are less efficient than the most efficient way to perform the desired calculations. Often, the difference in time required for Mathematica to perform equivalent – but different – calculations is quite small. For the beginner, we think it is wisest to work with familiar calculations first and then efficiency.

Observe that $x_{n+1} = f(x_n)$ can also be computed with $x_{n+1} = f^n(x_0)$.

Example 4.9: Dynamical Systems

A sequence of the form $x_{n+1} = f(x_n)$ is called a **dynamical system**.

Sometimes, unusual behavior can be observed when working with dynamical systems. For example, consider the dynamical system with $f(x) = x + 2.5x(1 - x)$ and $x_0 = 1.2$. Note that we define x_n using the form `x[n_]:=x[n]=...` so that Mathematica "remembers" the functional values it computes and thus avoids recomputing functional values previously computed. This is particularly advantageous when we compute the value of x_n for large values of n.

```
Clear[f, x]

f[x_]:=x + 2.5x(1 − x)

x[n_]:=x[n] = f[x[n − 1]]

x[0] = 1.2;
```

In Fig. 4.7 (a), we see that the sequence x_n oscillates between the numbers 0.6 and 1.2. We say that the dynamical system has a **2-cycle** because the values of the sequence oscillate between two numbers.

```
tb = Table[x[n], {n, 1, 200}];

ListPlot[tb, PlotStyle → CMYKColor[1, .76, .12, .7],

PlotLabel → "(a)"]
```

In Fig. 4.7 (b), we see that changing x_0 from 1.2 to 1.201 results in a 4-cycle.

```
Clear[f, x]

f[x_]:=x + 2.5x(1 − x)

x[n_]:=x[n] = f[x[n − 1]]

x[0] = 1.201;

tb = Table[x[n], {n, 1, 200}];

Short[tb, 20]

{0.597497, 1.19873, 0.603163, 1.20156, 0.596102, 1.19801, 0.604957, 1.20242,

0.593943, 1.19688, 0.607777, 1.20374, 0.590622, 1.19509, 0.612212, 1.20573, 0.585585,

1.19227, 0.619168, 1.20867, 0.578149, 1.18788, 0.629931, 1.21273, 0.567781, 1.1813,

0.645888, 1.21768, 0.55502, 1.17245, 0.666974, 1.22227, 0.543077, 1.16344, 0.688063,
```

```
⟨⟨131⟩⟩, 0.701238, 1.225, 0.535948, 1.15772, 0.701238, 1.225, 0.535948, 1.15772,
0.701238, 1.225, 0.535948, 1.15772, 0.701238, 1.225, 0.535948, 1.15772, 0.701238,
1.225, 0.535948, 1.15772, 0.701238, 1.225, 0.535948, 1.15772, 0.701238, 1.225,
0.535948, 1.15772, 0.701238, 1.225, 0.535948, 1.15772, 0.701238, 1.225}
```

```
ListPlot[tb, PlotStyle → CMYKColor[.06, .27, 1, .12], PlotLabel → "(b)"]
```

The calculations indicate that the behavior of the system can change considerably for small changes in x_0. With the following, we adjust the definition of x so that x depends on $x_0 = c$: given c, $x_c(0) = c$.

```
Clear[f, x]

f[x_]:=x + 2.5x(1 − x)

x[c_][n_]:=x[c][n] = f[x[c][n − 1]]//N

x[c_][0]:=c//N;
```

In `tb`, we create a list of lists of the form $\{x_c(n)|n = 100, \ldots, 150\}$ for 150 equally spaced values of c between 0 and 1.5. Observe that Mathematica issues several error messages. When a Mathematica calculation is larger than the machine's precision, we obtain an `Overflow` warning. In numerical calculations, we interpret `Overflow` to correspond to ∞.

```
tb = Table[{c, x[c][n]}, {c, 0, 1.5, .01}, {n, 100, 150}];
```

```
General::ovfl
General::ovfl
General::ovfl
General::stop
```

We ignore the error messages and use `Short` to view an abbreviated form of `tb`.

`Short[expr]` prints an abbreviated form of `expr`.

```
Short[tb]
```

```
{{{0., 0.}, {0., 0.}, {0., 0.}, ⟨⟨45⟩⟩, {0., 0.}, {0., 0.}, {0., 0.}}, {⟨⟨1⟩⟩},
⟨⟨147⟩⟩, {⟨⟨1⟩⟩}, {⟨⟨1⟩⟩}}
```

We then use `Flatten` to convert `tb` to a list of points which are plotted with `ListPlot` in Fig. 4.8 (a).

```
tb2 = Flatten[tb, 1];

f1 = ListPlot[tb2, PlotStyle → CMYKColor[.31, .39, .89, .06],
  PlotLabel → "(a)"]
```

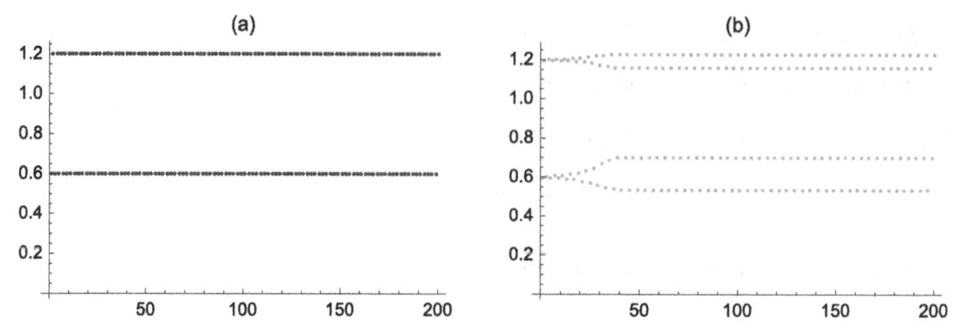

FIGURE 4.7 (a) A 2-cycle. (b) A 4-cycle. (University of Notre Dame colors)

Another interesting situation occurs if we fix x_0 and let c vary in $f(x) = x + cx(1-x)$. With the following we set $x_0 = 1.2$ and adjust the definition of f so that f depends on c: $f(x) = x + cx(1-x)$.

```
Clear[f, x]

f[c_][x_]:=x + c x(1 - x)//N

x[c_][n_]:=x[c][n] = f[c][x[c][n - 1]]//N

x[c_][0]:=1.2//N;
```

In `tb`, we create a list of lists of the form $\{x_c(n)|n = 200, \ldots, 300\}$ for 350 equally spaced values of c between 0 and 3.5. As before, Mathematica issues several error messages, which we ignore.

```
tb = Table[{c, x[c][n]}, {c, 0, 3.5, .01}, {n, 200, 300}];

General::ovfl
General::ovfl
General::ovfl
General::stop

Short[tb]

{{{0., 1.2}, {0., 1.2}, {0., 1.2}, ⟨⟨95⟩⟩, {0., 1.2}, {0., 1.2}, {0., 1.2}},
⟨⟨349⟩⟩, {⟨⟨1⟩⟩}}
```

`tb` is then converted to a list of points with `Flatten` and the resulting list is plotted in Fig. 4.8 (b) with `ListPlot`. This plot is called a **bifurcation diagram**.

```
tb2 = Flatten[tb, 1];

f2 = ListPlot[tb2, PlotStyle → CMYKColor[1, .76, .12, .7],
  PlotRange → {0, 2}, PlotLabel → "(b)"]

Show[GraphicsRow[{f1, f2}]]
```

A function f is **listable** if `f[list]` and `Map[f,list]` return the same results.

As indicated earlier, elements of lists can be numbers, ordered pairs, functions, and even other lists. You can also use Mathematica to manipulate lists in numerous ways. Most importantly, the `Map` function is used to apply a function to a list: `Map[f,{x1,x2,...,xn}]` returns the list $\{f(x_1), f(x_2), \ldots, f(x_n)\}$. We will discuss other operations that can be performed on lists in the following sections.

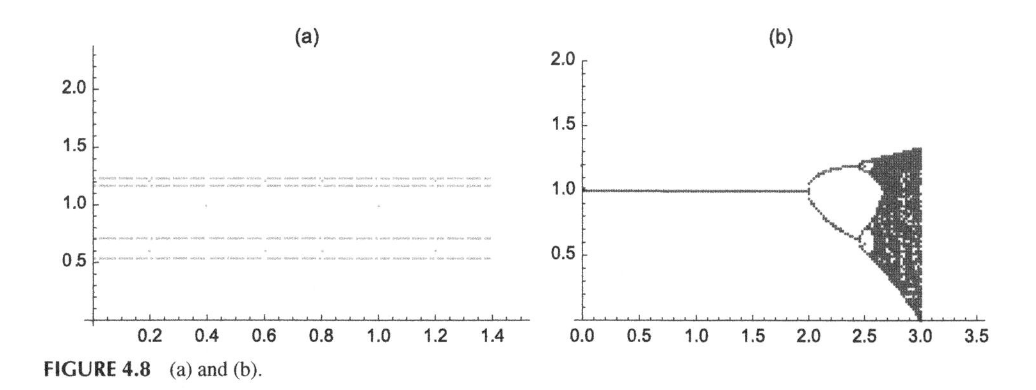

FIGURE 4.8 (a) and (b).

Example 4.10: Hermite Polynomials

The **Hermite polynomials,** $H_n(x)$, satisfy the differential equation $y'' - 2xy' + 2ny = 0$ and the orthogonality relation

$$\int_{-\infty}^{\infty} H_n(x)H_m(x)e^{-x^2}\,dx = \delta_{mn}2^n n!\sqrt{\pi}.$$

The Mathematica command `HermiteH[n,x]` yields the Hermite polynomial $H_n(x)$. (a) Create a table of the first five Hermite polynomials. (b) Evaluate each Hermite polynomial if $x = 1$. (c) Compute the derivative of each Hermite polynomial in the table. (d) Compute an antiderivative of each Hermite polynomial in the table. (e) Graph the five Hermite polynomials on the interval $[-1, 1]$. (f) Verify that $H_n(x)$ satisfies $y'' - 2xy' + 2ny = 0$ for $n = 1, 2 \ldots, 5$ ($'$ denotes d/dx).

Solution. We proceed by using `HermiteH` together with `Table` to define `hermitetable` to be the list consisting of the first five Hermite polynomials.

hermitetable = Table[HermiteH[*n*, *x*], {*n*, 1, 5}]

$\{2x, -2+4x^2, -12x+8x^3, 12-48x^2+16x^4, 120x-160x^3+32x^5\}$

We then use `ReplaceAll` (`->`) to evaluate each member of `hermitetable` if x is replaced by 1.

hermitetable/.*x* → 1

$\{2, 2, -4, -20, -8\}$

Functions like `D` and `Integrate` are *listable.* A function is *listable* if it automatically passes through a list. For example, since `D` is listable, the command `D[listoffunctions,x]` will return a list of the derivative of the list `listoffunctions` with respect to x. Thus, each of the following commands differentiate each element of `hermitetable` with respect to x. In the second case, we have used a *pure function*: given an argument `#`, `D[#,x]&` differentiates `#` with respect to x. Use the `&` symbol to indicate the end of a pure function.

***D*[hermitetable, *x*]**

$\{2, 8x, -12+24x^2, -96x+64x^3, 120-480x^2+160x^4\}$

Map[*D*[#, *x*]&, hermitetable]

$\{2, 8x, -12+24x^2, -96x+64x^3, 120-480x^2+160x^4\}$

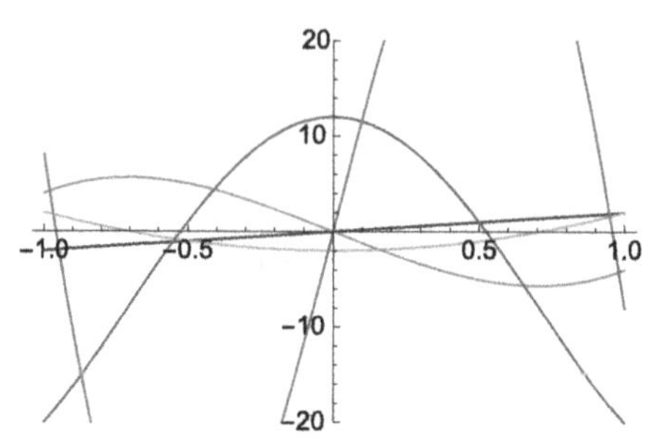

FIGURE 4.9 Graphs of $H_1(x)$, $H_2(x)$, $H_3(x)$, $H_4(x)$, and $H_5(x)$.

Similarly, we use `Integrate` to antidifferentiate each member of `hermitetable` with respect to x. Remember that Mathematica does not automatically include the "$+C$" that we include when we antidifferentiate.

Integrate[hermitetable, *x*]

$$\left\{x^2, -2x + \frac{4x^3}{3}, 4\left(-\frac{3x^2}{2} + \frac{x^4}{2}\right), 4\left(3x - 4x^3 + \frac{4x^5}{5}\right), 8\left(\frac{15x^2}{2} - 5x^4 + \frac{2x^6}{3}\right)\right\}$$

Map[Integrate[#, *x*]&, hermitetable]

$$\left\{x^2, -2x + \frac{4x^3}{3}, 4\left(-\frac{3x^2}{2} + \frac{x^4}{2}\right), 4\left(3x - 4x^3 + \frac{4x^5}{5}\right), 8\left(\frac{15x^2}{2} - 5x^4 + \frac{2x^6}{3}\right)\right\}$$

To graph the list `hermitetable`, we use `Plot` to plot each function in the set `hermitetable` on the interval $[-2, 2]$ in Fig. 4.9. In this case, we specify that the displayed y-values correspond to the interval $[-20, 20]$. Because we apply `Tooltip` to the set of functions being plotted, you can identify each curve by moving the cursor and placing it over each curve to see which function is being plotted.

Plot[Tooltip[Evaluate[hermitetable]], {*x*, −1, 1}, PlotRange → {−20, 20}]

`hermitetable[[n]]` returns the nth element of `hermitetable`, which corresponds to $H_n(x)$. Thus,

verifyde = Table[*D*[hermitetable[[*n*]], {*x*, 2}] − 2*x D*[hermitetable[[*n*]], *x*]+

2*n*hermitetable[[*n*]]//Simplify, {*n*, 1, 5}]

{0, 0, 0, 0, 0}

computes and simplifies $H_n'' - 2xH_n' + 2nH_n$ for $n = 1, 2, \ldots, 5$. We use `Table` and `Integrate` to compute $\int_{-\infty}^{\infty} H_n(x)H_m(x)e^{-x^2}\,dx$ for $n = 1, 2, \ldots, 5$ and $m = 1, 2, \ldots, 5$.

verifyortho = Table[Integrate[hermitetable[[*n*, 2]]hermitetable[[*m*, 2]]

Exp[−*x*^2], {*x*, −Infinity, Infinity}], {*n*, 1, 5}, {*m*, 1, 5}]

$\{\{\frac{\sqrt{\pi}}{2}, 0, 6\sqrt{\pi}, 0, -120\sqrt{\pi}\}, \{0, 12\sqrt{\pi}, 0, -144\sqrt{\pi}, 0\},$

$\{6\sqrt{\pi}, 0, 120\sqrt{\pi}, 0, -2400\sqrt{\pi}\}, \{0, -144\sqrt{\pi}, 0, 1728\sqrt{\pi}, 0\},$

$\{-120\sqrt{\pi}, 0, -2400\sqrt{\pi}, 0, 48000\sqrt{\pi}\}\}$

To view a table in traditional row-and-column form use `TableForm`, as we do here illustrating the use of the `TableHeadings` option.

TableForm[verifyortho,

TableHeadings → {{"m=1", "m=2", "m=3", "m=4", "m=5"},

{"n=1", "n=2", "n=3", "n=4", "n=5"}}]

	$n=1$	$n=2$	$n=3$	$n=4$	$n=5$
$m=1$	$\frac{\sqrt{\pi}}{2}$	0	$6\sqrt{\pi}$	0	$-120\sqrt{\pi}$
$m=2$	0	$12\sqrt{\pi}$	0	$-144\sqrt{\pi}$	0
$m=3$	$6\sqrt{\pi}$	0	$120\sqrt{\pi}$	0	$-2400\sqrt{\pi}$
$m=4$	0	$-144\sqrt{\pi}$	0	$1728\sqrt{\pi}$	0
$m=5$	$-120\sqrt{\pi}$	0	$-2400\sqrt{\pi}$	0	$48000\sqrt{\pi}$

Be careful when using `TableForm`: `TableForm[table]` is no longer a list and cannot be manipulated like a list. □

4.2 MANIPULATING LISTS: MORE ON PART AND MAP

Often, Mathematica's output is given to us as a list that we need to use in subsequent calculations. Elements of a list are extracted with `Part` (`[[...]]`): `list[[i]]` returns the ith element of `list`; `list[[i,j]]` (or `list[[i]][[j]]`) returns the jth element of the ith element of `list`, and so on.

Example 4.11

Let $f(x)=3x^4-8x^3-30x^2+72x$. Locate and classify the critical points of $y=f(x)$.

Solution. We begin by clearing all prior definitions of f and then defining $f(x)=3x^4-8x^3-30x^2+72x$. The critical numbers are found by solving the equation $f'(x)=0$. The resulting list is named `critnums`.

Clear[*f*]

$f[\text{x\_}]=3x^4-8x^3-30x^2+72x;$

critnums = Solve[*f*′[*x*]==0]

$\{\{x\to -2\},\{x\to 1\},\{x\to 3\}\}$

`critnums` is actually a list of lists. For example, the number -2 is the second part of the first part of the second part of `critnums`.

critnums[[1]]

$\{x \to -2\}$

critnums[[1, 1]]

$x \to -2$

critnums[[1, 1, 2]]

-2

Similarly, the numbers 1 and 3 are extracted with `critnums[[2,1,2]]` and `critnums[[3,1,2]]`, respectively.

critnums[[2, 1, 2]]

critnums[[3, 1, 2]]

1

3

We locate and classify the points by evaluating $f(x)$ and $f''(x)$ for each of the numbers in `critnums`. `f[x]/.x->a` replaces each occurrence of x in $f(x)$ by a, so entering

{x, f[x], f''[x]}/.critnums//TableForm

−2	−152	180
1	37	−72
3	−27	120

replaces each x in the list $\{x, f(x), f''(x)\}$ by each of the x-values in `critnums`.
By the Second Derivative Test, we conclude that $y = f(x)$ has relative minima at the points $(-2, -152)$ and $(3, -27)$ while $f(x)$ has a relative maximum at $(1, 37)$. In fact, because $\lim_{x \to \pm\infty} f(x) = \infty$, -152 is the absolute minimum value of $f(x)$. These results are confirmed by the graph of $y = f(x)$ in Fig. 4.10.

When you plot lists of functions and apply `Tooltip` to the list being plotted, you can identify each curve by sliding the cursor over the curve. When the cursor is on a curve, the definition of the curve being plotted is displayed.

Plot[Tooltip[{f[x], f'[x], f''[x]}], {x, −4, 4},

PlotStyle → {{CMYKColor[0, .1, 1, 0], Thickness[.01]},

{CMYKColor[1, .57, 0, .38], Thickness[.01]},

{CMYKColor[0, 0, 0, .17], Thickness[.01]}},

AspectRatio → 1, AxesLabel → {x, y}] □

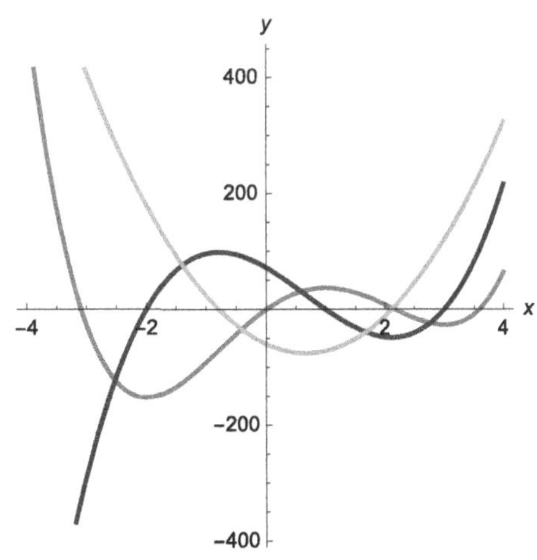

FIGURE 4.10 Graph of $f(x) = 3x^4 - 8x^3 - 30x^2 + 72x$, $f'(x)$, and $f''(x)$. (University of Rochester colors)

`Map` is a very powerful and useful function: `Map[f,list]` creates a list consisting of elements obtained by evaluating `f` for each element of `list`, provided that each member of `list` is an element of the domain of `f`. Note that if `f` is **listable**, `f[list]` produces the same result as `Map[f,list]`.

To determine if `f` is listable, enter `Attributes[f]`.

Example 4.12

Entering

```
t1 = Table[n, {n, 1, 100}];

t1b = Partition[t1, 10];

TableForm[t1b]
```

1	2	3	4	5	6	7	8	9	10
11	12	13	14	15	16	17	18	19	20
21	22	23	24	25	26	27	28	29	30
31	32	33	34	35	36	37	38	39	40
41	42	43	44	45	46	47	48	49	50
51	52	53	54	55	56	57	58	59	60
61	62	63	64	65	66	67	68	69	70
71	72	73	74	75	76	77	78	79	80
81	82	83	84	85	86	87	88	89	90
91	92	93	94	95	96	97	98	99	100

computes a list of the first 100 integers and names the result `t1`. To see `t1`, we use `Partition` to partition `t1` in 10 element subsets; the results are displayed in a standard row-and-column form with `TableForm`. We then define $f(x) = x^2$ and use `Map` to square each number in `t1`.

```
f[x_] = x^2;

t2 = Map[f, t1];

t2b = Partition[t2, 10];

TableForm[t2b]
```

1	4	9	16	25	36	49	64	81	100
121	144	169	196	225	256	289	324	361	400
441	484	529	576	625	676	729	784	841	900
961	1024	1089	1156	1225	1296	1369	1444	1521	1600
1681	1764	1849	1936	2025	2116	2209	2304	2401	2500
2601	2704	2809	2916	3025	3136	3249	3364	3481	3600
3721	3844	3969	4096	4225	4356	4489	4624	4761	4900
5041	5184	5329	5476	5625	5776	5929	6084	6241	6400
6561	6724	6889	7056	7225	7396	7569	7744	7921	8100
8281	8464	8649	8836	9025	9216	9409	9604	9801	10000

The same result is accomplished by the pure function that squares its argument. Note how # denotes the argument of the pure function; the & symbol marks the end of the pure function.

```
t3 = Map[#^2&, t1];

t3b = Partition[t3, 10];

TableForm[t3b]
```

1	4	9	16	25	36	49	64	81	100
121	144	169	196	225	256	289	324	361	400
441	484	529	576	625	676	729	784	841	900
961	1024	1089	1156	1225	1296	1369	1444	1521	1600
1681	1764	1849	1936	2025	2116	2209	2304	2401	2500
2601	2704	2809	2916	3025	3136	3249	3364	3481	3600
3721	3844	3969	4096	4225	4356	4489	4624	4761	4900
5041	5184	5329	5476	5625	5776	5929	6084	6241	6400
6561	6724	6889	7056	7225	7396	7569	7744	7921	8100
8281	8464	8649	8836	9025	9216	9409	9604	9801	10000

On the other hand, entering

```
t1 = Table[{a, b}, {a, 1, 5}, {b, 1, 5}];

Short[t1]
```

{{{1, 1}, {1, 2}, {1, 3}, {1, 4}, {1, 5}}, {⟨⟨1⟩⟩}, ⟨⟨1⟩⟩, {⟨⟨1⟩⟩}, {{5, 1}, {5, 2}, {5, 3}, {5, 4}, {5, 5}}}

is a list (of length 5) of lists (each of length 5). Use `Flatten` to obtain a list of 25 points, which we name `t2`.

```
t2 = Flatten[t1, 1];

Short[t2]
```

{{1, 1}, {1, 2}, {1, 3}, {1, 4}, {1, 5}, {2, 1}, ⟨⟨13⟩⟩, {4, 5}, {5, 1}, {5, 2}, {5, 3}, {5, 4}, {5, 5}}

We then use `Map` to apply f to `t2`.

```
f[{x_, y_}] = {{x, y}, x^2 + y^2};

t3 = Map[f, t2];

Short[t3]
```

{{{1, 1}, 2}, {{1, 2}, 5}, {{1, 3}, 10}, ⟨⟨19⟩⟩, {{5, 3}, 34}, {{5, 4}, 41}, {{5, 5}, 50}}

We accomplish the same result with a pure function. Observe how `#[[1]]` and `#[[2]]` are used to represent the first and second arguments: given a list of length 2, the pure function returns the list of ordered pairs consisting of the first element of the list, the second element of the list (as an ordered pair), and the sum of the squares of the first and second elements (of the first ordered pair).

```
t3b = Map[{{#[[1]], #[[2]]}, #[[1]]^2 + #[[2]]^2}&, t2];

Short[t3b]
```

{{{1, 1}, 2}, {{1, 2}, 5}, {{1, 3}, 10}, ⟨⟨19⟩⟩, {{5, 3}, 34}, {{5, 4}, 41}, {{5, 5}, 50}}

Example 4.13

Make a table of the values of the trigonometric functions $y = \sin x$, $y = \cos x$, and $y = \tan x$ for the principal angles.

Solution. We first construct a list of the principal angles which is accomplished by defining `t1` to be the list consisting of $n\pi/4$ for $n = 0, 1, \ldots, 8$ and `t2` to be the list consisting of $n\pi/6$ for $n = 0, 1, \ldots, 12$. The principal angles are obtained by taking the union of `t1` and `t2`. `Union[t1,t2]` joins the lists `t1` and `t2`, removes repeated elements, and sorts the results. If we did not wish to remove repeated elements and sort the result, the command `Join[t1,t2]` concatenates the lists `t1` and `t2`.

The **BasicMathInput** palette:

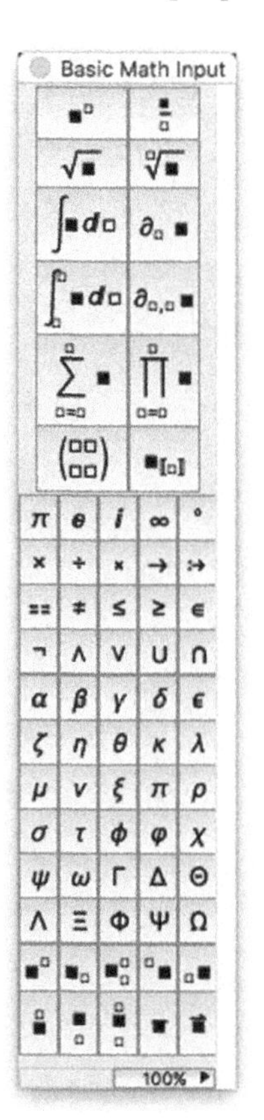

t1 = Table[$\frac{n\pi}{4}$, {n, 0, 8}];

t2 = Table[$\frac{n\pi}{6}$, {n, 0, 12}];

prinangles = Union[t1, t2]

$$\left\{0, \frac{\pi}{6}, \frac{\pi}{4}, \frac{\pi}{3}, \frac{\pi}{2}, \frac{2\pi}{3}, \frac{3\pi}{4}, \frac{5\pi}{6}, \pi, \frac{7\pi}{6}, \frac{5\pi}{4}, \frac{4\pi}{3}, \frac{3\pi}{2}, \frac{5\pi}{3}, \frac{7\pi}{4}, \frac{11\pi}{6}, 2\pi\right\}$$

We can also use the symbol ∪, which is obtained by clicking on the ∪ button on the **BasicMathInput** palette to represent `Union`.

prinangles = t1 ∪ t2

$$\left\{0, \frac{\pi}{6}, \frac{\pi}{4}, \frac{\pi}{3}, \frac{\pi}{2}, \frac{2\pi}{3}, \frac{3\pi}{4}, \frac{5\pi}{6}, \pi, \frac{7\pi}{6}, \frac{5\pi}{4}, \frac{4\pi}{3}, \frac{3\pi}{2}, \frac{5\pi}{3}, \frac{7\pi}{4}, \frac{11\pi}{6}, 2\pi\right\}$$

Next, we define $f(x)$ to be the function that returns the ordered quadruple

$$(x, \sin x, \cos x, \tan x)$$

and compute the value of $f(x)$ for each number in `prinangles` with `Map` naming the resulting table `prinvalues`. `prinvalues` is not displayed because a semi-colon is included at the end of the command.

Clear[*f*]

***f*[x_] = {*x*, Sin[*x*], Cos[*x*], Tan[*x*]};**

Remember that the result of using `TableForm` is not a list so cannot be manipulated like lists.

prinvalues = Map[*f*, prinangles];

Finally, we use `TableForm` illustrating the use of the `TableHeadings` option to display `prinvalues` in row-and-column form; the columns are labeled x, $\sin x$, $\cos x$, and $\tan x$.

TableForm[prinvalues, TableHeadings → {None, {x, "sin(x)", "cos(x)", "tan(x)"}}]

x	$\sin(x)$	$\cos(x)$	$\tan(x)$
0	0	1	0
$\frac{\pi}{6}$	$\frac{1}{2}$	$\frac{\sqrt{3}}{2}$	$\frac{1}{\sqrt{3}}$
$\frac{\pi}{4}$	$\frac{1}{\sqrt{2}}$	$\frac{1}{\sqrt{2}}$	1
$\frac{\pi}{3}$	$\frac{\sqrt{3}}{2}$	$\frac{1}{2}$	$\sqrt{3}$
$\frac{\pi}{2}$	1	0	ComplexInfinity
$\frac{2\pi}{3}$	$\frac{\sqrt{3}}{2}$	$-\frac{1}{2}$	$-\sqrt{3}$
$\frac{3\pi}{4}$	$\frac{1}{\sqrt{2}}$	$-\frac{1}{\sqrt{2}}$	-1
$\frac{5\pi}{6}$	$\frac{1}{2}$	$-\frac{\sqrt{3}}{2}$	$-\frac{1}{\sqrt{3}}$
π	0	-1	0
$\frac{7\pi}{6}$	$-\frac{1}{2}$	$-\frac{\sqrt{3}}{2}$	$\frac{1}{\sqrt{3}}$
$\frac{5\pi}{4}$	$-\frac{1}{\sqrt{2}}$	$-\frac{1}{\sqrt{2}}$	1
$\frac{4\pi}{3}$	$-\frac{\sqrt{3}}{2}$	$-\frac{1}{2}$	$\sqrt{3}$
$\frac{3\pi}{2}$	-1	0	ComplexInfinity
$\frac{5\pi}{3}$	$-\frac{\sqrt{3}}{2}$	$\frac{1}{2}$	$-\sqrt{3}$
$\frac{7\pi}{4}$	$-\frac{1}{\sqrt{2}}$	$\frac{1}{\sqrt{2}}$	-1
$\frac{11\pi}{6}$	$-\frac{1}{2}$	$\frac{\sqrt{3}}{2}$	$-\frac{1}{\sqrt{3}}$
2π	0	1	0

In the table, note that $y = \tan x$ is undefined at odd multiples of $\pi/2$ and Mathematica appropriately returns `ComplexInfinity` at those values of x for which $y = \tan x$ is undefined. □

Remark 4.3. The result of using `TableForm` is not a list (or table) and calculations on it using commands like `Map` cannot be performed. `TableForm` helps you see results in a more readable format. To avoid confusion, do not assign the results of using `TableForm` any name: adopting this convention avoids any possible attempted manipulation of `TableForm` objects.

`object=name` assigns the object `object` the name `name`.

We can use `Map` on any list, including lists of functions and/or other lists. Remember that if the function `f` is listable, `Map[f,list]` and `f[list]` produce the same result.

Lists of functions are graphed with `Plot`: `Plot[listoffunctions,{x,a,b}]` graphs the list of functions of x, `listoffunctions`, for $a \leq x \leq b$. If the command is entered as `Plot[Tooltip[listoffunctions],{x,a,b}]`, you can identify the curves in the plot by moving the cursor over the curves in the graphic.

Example 4.14: Bessel Functions

The **Bessel functions of the first kind**, $J_n(x)$, are nonsingular solutions of $x^2y'' + xy' + (x^2 - n^2)\,y = 0$. `BesselJ[n,x]` returns $J_n(x)$. Graph $J_n(x)$ for $n = 0, 1, 2, \ldots, 8$.

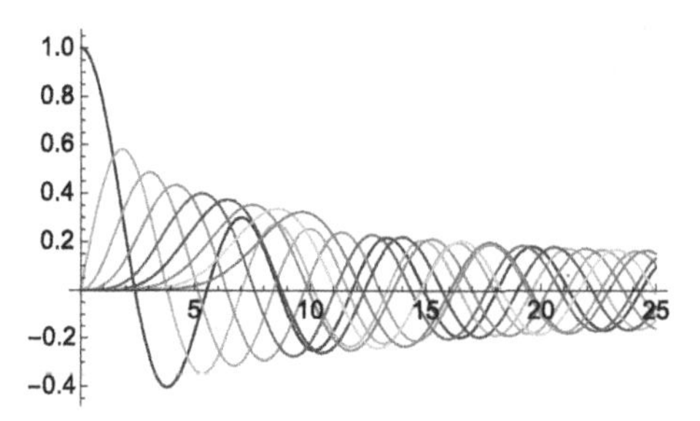

FIGURE 4.11 Graphs of $J_n(x)$ for $n = 0, 1, 2, \ldots, 8$.

Solution. In `t1`, we use `Table` and `BesselJ` to create a list of $J_n(x)$ for $n = 0, 1, 2, \ldots, 8$.

```
t1 = Table[BesselJ[n, x], {n, 0, 8}];
```

We then use `Plot` to graph each function in `t1` in Fig. 4.11. You can identify each curve by sliding the cursor over each.

```
Plot[Tooltip[Evaluate[t1]], {x, 0, 25}]
```

A different effect is achieved by graphing each function separately. To do so, we define the function `pfunc`. Given a function of x, `f`, `pfunc[f]` plots the function for $0 \leq x \leq 100$. The resulting graphic is not displayed because the option `DisplayFunction->Identity` is included in the `Plot` command. We then use `Map` to apply `pfunc` to each element of `t1`. The result is a list of 9 graphics objects, which we name `t2`. A nice way to display 9 graphics is as a 3×3 array so we use `Partition` to convert `t2` from a list of length 9 to a list of lists, each with length 3 – a 3×3 array. `Partition[list,n]` returns a list of lists obtained by partitioning `list` into n-element subsets.

Think of `Flatten` and `Partition` as inverse functions.

```
pfunc[f_]:=Plot[f, {x, 0, 100}];

t2 = Map[pfunc, t1];

t3 = Partition[t2, 3];
```

Instead of defining `pfunc`, you can use a pure function instead. The following accomplishes the same result. We display `t3` using `Show` together with `GraphicsGrid` in Fig. 4.12.

```
t2 = (Plot[#1, {x, 0, 100}, PlotStyle → CMYKColor[1, .47, .12, .62],

  AspectRatio → 1]&)/@t1;

t3 = Partition[t2, 3];

Show[GraphicsGrid[t3]]
```

□

Example 4.15: Dynamical Systems

Let $f_c(x) = x^2 + c$ and consider the dynamical system given by $x_0 = 0$ and $x_{n+1} = f_c(x_n)$. Generate a bifurcation diagram of f_c.

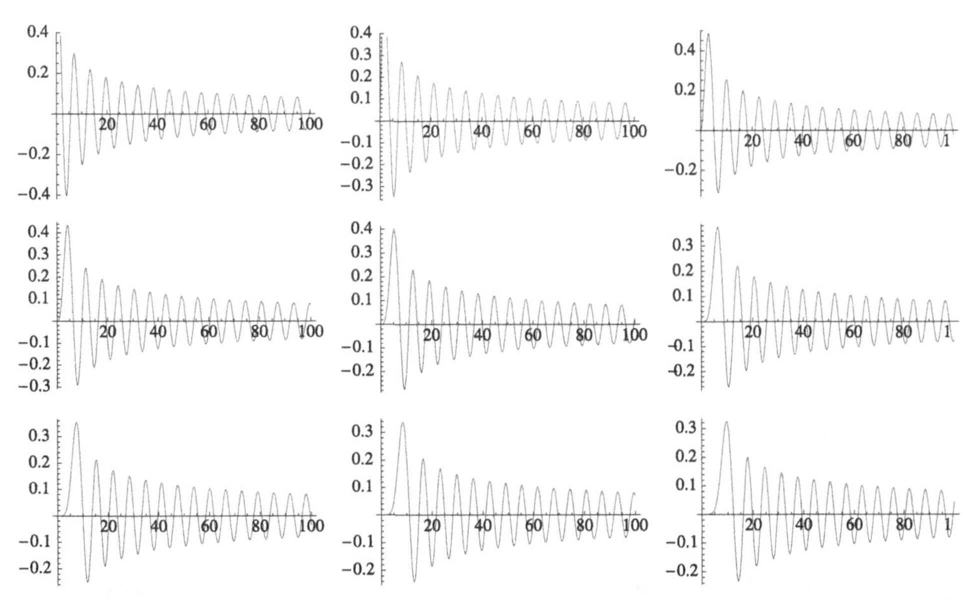

FIGURE 4.12 In the first row, from left to right, graphs of $J_0(x)$, $J_1(x)$, and $J_2(x)$; in the second row, from left to right, graphs of $J_3(x)$, $J_4(x)$, and $J_5(x)$; in the third row, from left to right, graphs of $J_6(x)$, $J_7(x)$, and $J_8(x)$. (Case Western Reserve University colors)

Solution. First, recall that `Nest[f,x,n]` computes the repeated composition $f^n(x)$. Then, in terms of a composition,

Compare the approach used here with the approach used in Example 4.9.

$$x_{n+1} = f_c(x_n) = f_c{}^n(0).$$

We will compute $f_c{}^n(0)$ for various values of c and "large" values of n so we begin by defining `cvals` to be a list of 300 equally spaced values of c between -2.5 and 1.

```
cvals = Table[c, {c, -2.5, 1., 3.5/299}];
```

We then define $f_c(x) = x^2 + c$. For a given value of c, `f[c]` is a function of one variable, x, while the form `f[c_][x_]:=...` results in a function of two variables that we think of as an indexed function that might represented using traditional mathematical notation as $f_c(x)$.

```
Clear[f]
```

```
f[c_][x_]:=x^2+c
```

To iterate f_c for various values of c, we define h. For a given value of c, $h(c)$ returns the list of points $\{(c, f_c{}^{100}(0)), (c, f_c{}^{101}(0)), \ldots, (c, f_c{}^{200}(0))\}$.

```
h[c_]:={Table[{c, Nest[f[c], 0, n]}, {n, 100, 500}]}
```

We then use `Map` to apply h to the list `cvals`. Observe that Mathematica generates several error messages when numerical precision is exceeded. We choose to disregard the error messages.

```
t1 = Map[h, cvals];
```

`t1` is a list (of length 300) of lists (each of length 101). To obtain a list of points (or, lists of length 2), we use `Flatten`. The resulting set of points is plotted with `ListPlot` in Fig. 4.13.

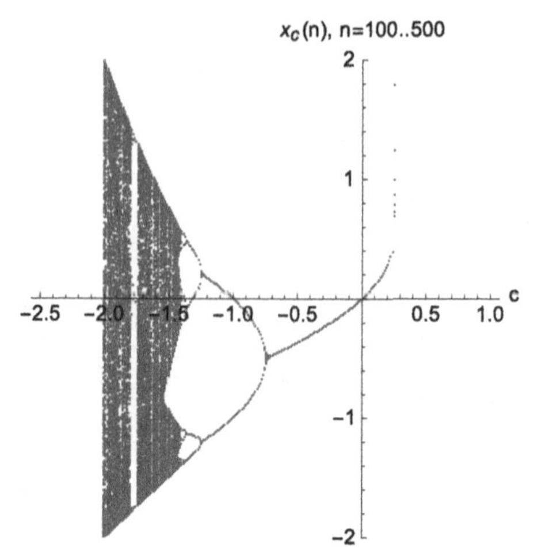

FIGURE 4.13 Bifurcation diagram of f_c. (Brown University Colors)

Observe that Mathematica again displays several error messages, which are not displayed here for length considerations, that we ignore: Mathematica only plots the points with real coordinates and ignores those containing `Overflow[]`.

```
t2 = Flatten[t1, 2];

ListPlot[t2, AxesLabel → {c, "xc(n), n=100..500"},

  PlotRange → {−2, 2}, AspectRatio → 1,

  PlotStyle → {CMYKColor[0, 1, 1, 0], PointSize[.005]}]
```

□

4.2.1 More on Graphing Lists; Graphing Lists of Points Using Graphics Primitives

Include the `Joined->True` option in a `ListPlot` command to connect successive points with line segments or use `ListLinePlot`.

Using *graphics primitives* like `Point` and `Line` gives you even more flexibility. `Point[{x,y}]` represents a point at (x, y).

```
Line[{{x1,y1},{x2,y2},...,{xn,yn}}]
```

represents a sequence of points (x_1, y_1), (x_2, y_2), ..., (x_n, y_n) connected with line segments. A graphics primitive is declared to be a graphics object with `Graphics`: `Show[Graphics[Point[{x,y}]]]` displays the point (x, y). The advantage of using primitives is that each primitive is affected by the options that directly precede it.

Example 4.16

Table 4.1 shows the percentage of the United States labor force that belonged to unions during certain years. Graph the data represented in the table.

Solution. We begin by entering the data represented in the table as `dataunion`:

```
dataunion = {{30, 11.6}, {35, 13.2}, {40, 26.9}, {45, 35.5},
```

TABLE 4.1 Union Membership as a Percentage of the Labor Force

Year	Union Membership as a Percentage of the Labor Force
1930	11.6
1935	13.2
1940	26.9
1945	35.5
1950	31.5
1955	33.2
1960	31.4
1965	28.4
1970	27.3
1975	25.5
1980	21.9
1985	18.0
1990	16.1

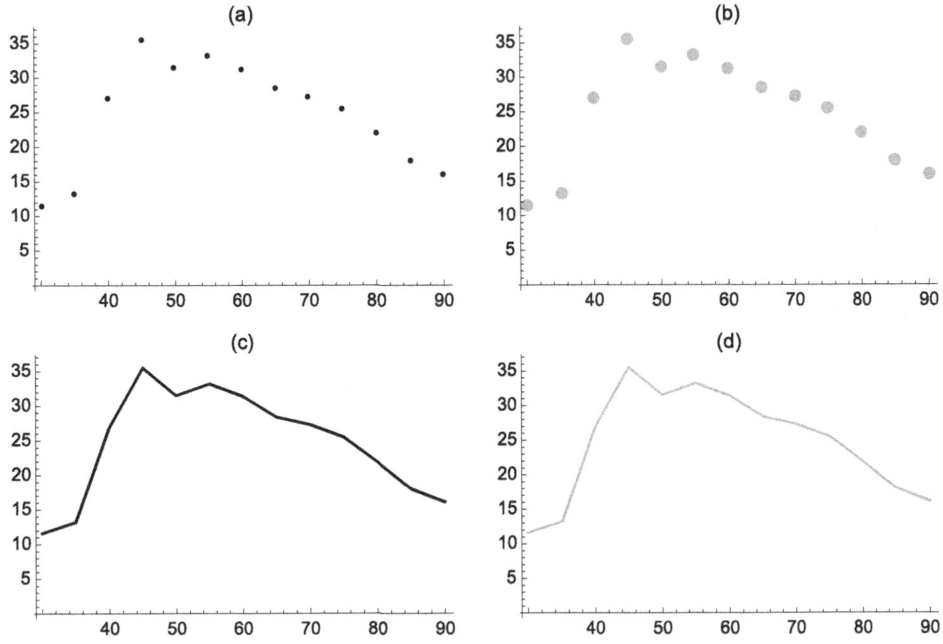

FIGURE 4.14 Union membership as a percentage of the labor force. (Vanderbilt University colors)

```
{50, 31.5}, {55, 33.2}, {60, 31.4}, {65, 28.4}, {70, 27.3},

{75, 25.5}, {80, 21.9}, {85, 18.0}, {90, 16.1}};
```

the x-coordinate of each point corresponds to the year, where x is the number of years past 1900, and the y-coordinate of each point corresponds to the percentage of the United States labor force that belonged to unions in the given year. We then use `ListPlot` to graph the set of points represented in `dataunion` in `lp1`, `lp2` (illustrating the `PlotStyle` option), and `lp3` (illustrating the `PlotJoined` option). All three plots are displayed side-by-side in Fig. 4.14 using `Show` together with `GraphicsRow`.

```
lp1 = ListPlot[dataunion, PlotLabel → "(a)",

  PlotStyle → Black];
```

```
lp2 = ListPlot[dataunion,
  PlotStyle → {{CMYKColor[.3, .4, .8, .15], PointSize[0.03]}},
  PlotLabel → "(b)"];
lp3 = ListPlot[dataunion, Joined → True,
  PlotLabel → "(c)", PlotStyle → Black];
lp4 = ListLinePlot[dataunion, PlotLabel → "(d)",
  PlotStyle->CMYKColor[.3, .4, .8, .15]];
Show[GraphicsGrid[{{lp1, lp2}, {lp3, lp4}}]]
```

An alternative to using `ListPlot` is to use `Show`, `Graphics`, and `Point` to view the data represented in `dataunion`. In the following command we use `Map` to apply the function `Point` to each pair of data in `dataunion`. The result is not a graphics object and cannot be displayed with `Show`.

```
datapts1 = Map[Point, dataunion];
Short[datapts1]
```

{Point[{30, 11.6}], Point[{35, 13.2}], ⟨⟨9⟩⟩, Point[{85, 18.}], Point[{90, 16.1}]}

Next, we use `Show` and `Graphics` to declare the set of points `Map[Point,dataunion]` as graphics objects and name the resulting graphics object `dp1`. The image is not displayed because a semi-colon is included at the end of the command. The `PointSize[.03]` command specifies that the points be displayed as filled circles of radius 0.03% of the displayed graphics object.

```
dp1 = Show[Graphics[{PointSize[0.03], datapts1},
  Axes → Automatic], PlotLabel → "(a)"];
```

The collection of all commands contained within a `Graphics` command is contained in braces `{...}`. Each graphics primitive is affected by the options like `PointSize`, `GrayLevel` (or `RGBColor`) directly preceding it. Thus,

```
datapts2 = ({GrayLevel[RandomReal[]], Point[#1]}&)/@dataunion;
Short[datapts2]
dp2 = Show[Graphics[{PointSize[0.03], datapts2},
  Axes → Automatic], PlotLabel → "(b)"];
```

displays the points in `dataunion` in various shades of gray in a graphic named `dp2` and

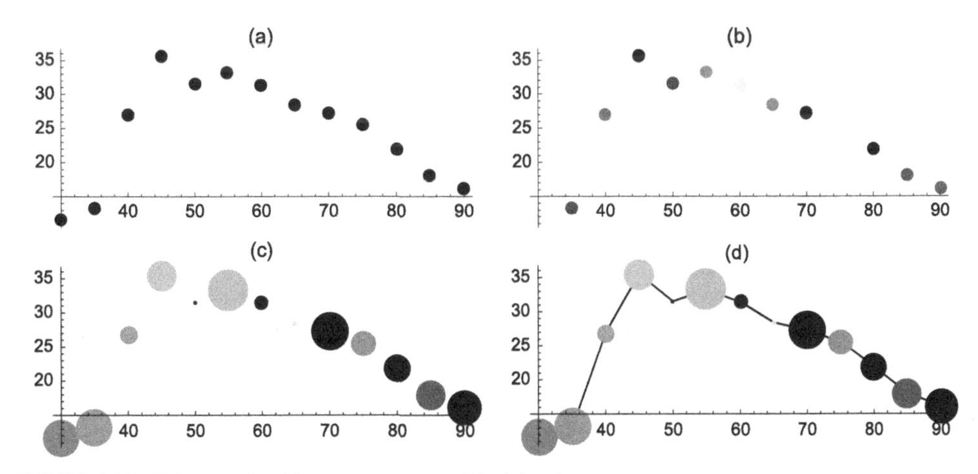

FIGURE 4.15 Union membership as a percentage of the labor force.

```
datapts3 = ({PointSize[RandomReal[{"0.008", "0.1"}]],
  GrayLevel[RandomReal[]], Point[#1]}&)/@dataunion;
dp3 = Show[Graphics[{datapts3}, Axes → Automatic],
  PlotLabel → "(c)"];
```

shows the points in `dataunion` in various sizes and in various shades of gray in a graphic named `dp3`. We connect successive points with line segments

```
connectpts = Graphics[Line[dataunion]];
dp4 = Show[connectpts, dp3, Axes → Automatic,
PlotLabel → "(d)"];
```

and show all four plots in Fig. 4.15 using `Show` and `GraphicsGrid`.

```
Show[GraphicsGrid[{{dp1, dp2}, {dp3, dp4}}]]
```
□

With the speed of today's computers and the power of Mathematica, it is relatively easy now to carry out many calculations that required supercomputers and sophisticated programming experience just a few years ago.

Example 4.17: Julia Sets

Plot Julia sets for $f(z) = \lambda \cos z$ if $\lambda = .66i$ and $\lambda = .665i$.

Solution. The sets are visualized by plotting the points (a, b) for which $|f^n(a + bi)|$ is *not* large in magnitude so we begin by forming our complex grid. Using `Table` and `Flatten`, we define `complexpts` to be a list of 62,500 points of the form $a + bi$ for 250 equally spaced real values of a between 0 and 8 and 300 equally spaced real values of b between -4 and 4 and then $f(z) = .66i \cos z$.

```
complexpts = Flatten[Table[a + bI, {a, 0., 8., 8/249}, {b, -4., 4., 6/249}], 1];

Clear[f]

f[z_] = .66ICos[z]
```

(0. + 0.66i)Cos[z]

For a given value of $c = a + bi$, $h(c)$ returns the ordered triple consisting of the real part of c, the imaginary part of c, and the value of $f^{200}(c)$.

```
h[c_]:={Re[c], Im[c], Nest[f, c, 200]}
```

We then use `Map` to apply h to `complexpts`. Observe that Mathematica generates several error messages. When machine precision is exceeded, we obtain an `Overflow[]` error message; numerical results smaller than machine precision results in an `Underflow[]` error message. Error messages can be machine specific, so if you don't get any, don't worry. For length considerations, we don't show any that we obtained here.

```
t1 = Map[h, complexpts]//Chop;
```

We use the error messages to our advantage. In `t2`, we select those elements of `t1` for which the third coordinate *is not* `Indeterminate`, which corresponds to the ordered triples $(a, b, f^n(a + bi))$ for which $|f^n(a + bi)|$ *is not* large in magnitude while in `t2b`, we select those elements of `t1` for which the third coordinate *is* `Indeterminate`, which corresponds to the ordered triples $(a, b, f^n(a + bi))$ for which $|f^n(a + bi)|$ *is* large in magnitude.

```
t2 = Select[t1, Not[#[[3]]===Indeterminate]&];

t2b = Select[t1, #[[3]]===Indeterminate&];

pt[{x_, y_, z_}]:={x, y}

t3 = Map[pt, t2];

t3b = Map[pt, t2b];
```

which are then graphed with `ListPlot` and shown side-by-side in Fig. 4.16 using `Show` and `GraphicsRow`. As expected, the images are inversions of each other.

```
lp1 = ListPlot[t3, PlotRange → {{0, 8}, {-4, 4}}, AspectRatio → Automatic,

  PlotStyle → CMYKColor[0, .09, .8, 0], PlotLabel → "(a)"];

lp2 = ListPlot[t3b, PlotRange → {{0, 8}, {-4, 4}}, AspectRatio → Automatic,

  PlotStyle → CMYKColor[0, .09, .8, 0], PlotLabel → "(b)"];
```

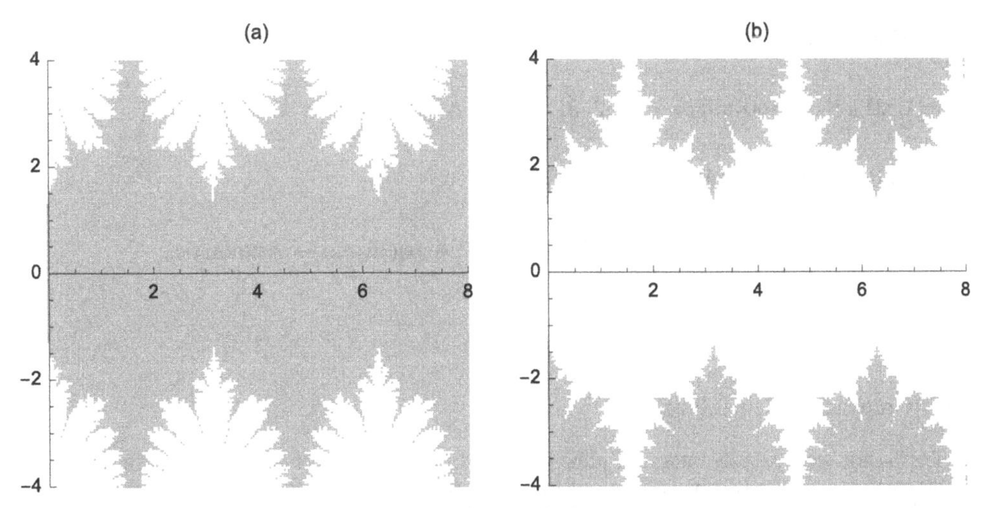

FIGURE 4.16 Julia set for $0.66i \cos z$. (University of Iowa colors)

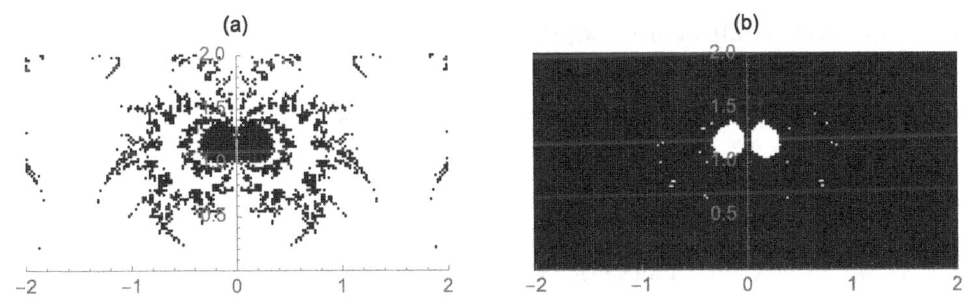

FIGURE 4.17 Julia set for $0.665i \cos z$.

We encountered similar error messages are as before but we have not included them due to length considerations.

```
Show[GraphicsRow[{lp1, lp2}]]
```

Changing λ from $0.66i$ to $0.665i$ results in a surprising difference in the plots. We proceed as before but increase the number of sample points to 120,000. See Fig. 4.17.

```
complexpts = Flatten[Table[a + bI, {a, −2., 2., 4/399}, {b, 0., 2., 2/299}], 1];

Clear[f]

f[z_] = .665ICos[z];

h[c_]:={Re[c], Im[c], Nest[f, c, 200]}

t1 = Map[h, complexpts]//Chop;

t2 = Select[t1, Not[#[[3]]===Indeterminate]&];

t2 = Select[t2, Not[#[[3]]===Overflow[]]&];

t2b = Select[t1, #[[3]]===Indeterminate&];

pt[{x_, y_, z_}]:={x, y}

t3 = Map[pt, t2];
```

```
t3b = Map[pt, t2b];

lp1 = ListPlot[t3, PlotRange → {{-2, 2}, {0, 2}}, AspectRatio → Automatic,

  PlotStyle → Black, PlotLabel → "(a)"];

lp2 = ListPlot[t3b, PlotRange → {{-2, 2}, {0, 2}}, AspectRatio → Automatic,

  PlotStyle → Black, PlotLabel → "(b)"];

Show[GraphicsRow[{lp1, lp2}]]
```

To see detail, we take advantage of pure functions, `Map`, and graphics primitives in three different ways. In Fig. 4.18, the shading of the point (a, b) is assigned according to the distance of $f^{200}(a + bi)$ from the origin. The color black indicates a distance of zero from the origin; as the distance increases, the shading of the point becomes lighter.

```
t2p = Map[{#[[1]], #[[2]], Min[Abs[#[[3]]], 3]}&, t2];

t2p2 = Map[{GrayLevel[#[[3]]/3], Point[{#[[1]], #[[2]]}]}&,

t2p];

jp1 = Show[Graphics[t2p2], PlotRange → {{-2, 2}, {0, 2}},

AspectRatio → 1, PlotLabel → "(a)"];

t2p = Map[{#[[1]], #[[2]], Min[Abs[Re[#[[3]]]], .25]}&, t2];

t2p2 = Map[{GrayLevel[#[[3]]/.25], Point[{#[[1]], #[[2]]}]}&,

t2p];

jp2 = Show[Graphics[t2p2], PlotRange → {{-2, 2}, {0, 2}}, AspectRatio → 1,

PlotLabel → "(b)"];

t2p = Map[{#[[1]], #[[2]], Min[Abs[Im[#[[3]]]], 2.5]}&, t2];

t2p2 = Map[{GrayLevel[#[[3]]/2.5], Point[{#[[1]], #[[2]]}]}&,

t2p];

jp3 = Show[Graphics[t2p2], PlotRange → {{-2, 2}, {0, 2}}, AspectRatio → 1,

PlotLabel → "(c)"];
```

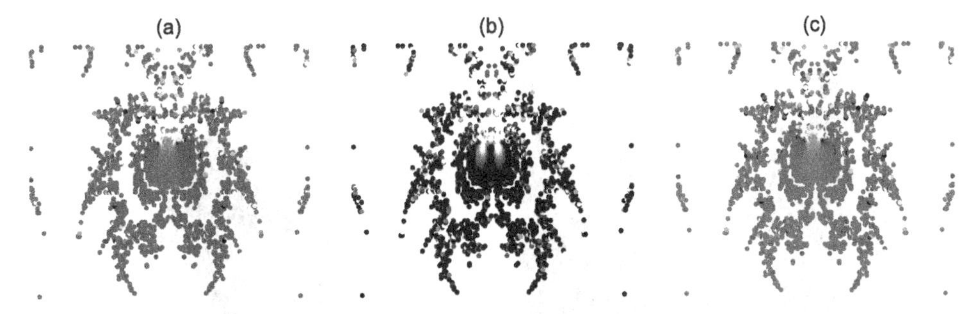

FIGURE 4.18 Shaded Julia sets for $0.665i \cos z$.

Show[GraphicsRow[{jp1, jp2, jp3}]]

Often times the *Julia set* is only defined for a rational function. With Mathematica, the command `JuliaSetPlot[f[z],z]` plots the Julia set of the rational function $y = f(z)$. To obtain good results, you will often want to take advantage of options such as `PlotRange` and `PlotStyle`. However, by the nature of the computations involved you should view `JuliaSetPlot` as a "fragile" commmand.
To illustrate approximating a Julia set for $f(z) = 0.66i \cos z$, we first use `Series` together with `Normal` to compute the Maclaurin series of degree 24 for $f(z)$. We then use `JuliaSetPlot` to graph the Julia set for this polynomial in `jp1`. See Fig. 4.19 (a). Observe that the approximation is fairly close to what we obtained previously.

p1 = Series[0.66*I*Cos[z], {z, 0, 24}]//Normal

$$(0. + 0.66i) - (0. + 0.33i)z^2 + (0. + 0.0275i)z^4 - (0. + 0.000916667i)z^6$$
$$+ (0. + 0.000016369i)z^8 - (0. + 1.8187830687830687*\wedge - 7i)z^{10}$$
$$+ (0. + 1.3778659611992948*\wedge - 9i)z^{12} - (0. + 7.570692094501619*\wedge - 12i)z^{14}$$
$$+ (0. + 3.1544550393756745*\wedge - 14i)z^{16} - (0. + 1.0308676599266909*\wedge - 16i)z^{18}$$
$$+ (0. + 2.712809631386029*\wedge - 19i)z^{20} - (0. + 5.871882319017379*\wedge - 22i)z^{22}$$
$$+ (0. + 1.0637467969234383*\wedge - 24i)z^{24}$$

jp1 = JuliaSetPlot[p1, z, PlotRange → {{0, 8}, {−4, 4}},

AspectRatio → 1, PlotLabel → "(a)"]

On the other hand, when we repeat the approximation process for $f(z) = 0.665i \cos z$, in Fig. 4.19 (b), we see that the approximation does not appear to agree as well as the approximation for $f(z) = 0.66i \cos z$.

p2 = Series[0.665*I*Cos[z], {z, 0, 24}]//Normal

$$(0. + 0.665i) - (0. + 0.3325i)z^2 + (0. + 0.0277083i)z^4 - (0. + 0.000923611i)z^6$$
$$+ (0. + 0.0000164931i)z^8 - (0. + 1.8325617283950616*\wedge - 7i)z^{10}$$
$$+ (0. + 1.3883043396932286*\wedge - 9i)z^{12} - (0. + 7.628045822490267*\wedge - 12i)z^{14}$$
$$+ (0. + 3.1783524260376114*\wedge - 14i)z^{16} - (0. + 1.038677263410984*\wedge - 16i)z^{18}$$

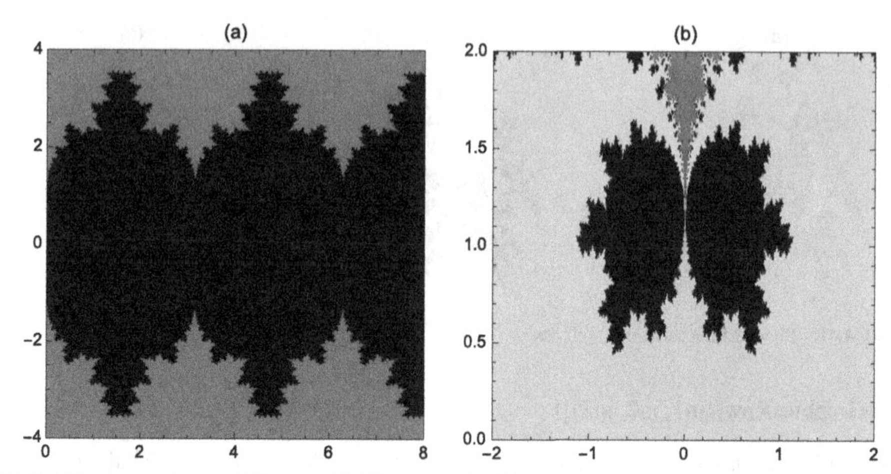

FIGURE 4.19 Approximating Julia sets of $0.66i \cos z$ and $0.665i \cos z$.

$+ (0. + 2.73336121950259*\wedge - 19i)z^{20} - (0. + 5.916366275979632*\wedge - 22i)z^{22}$

$+ (0. + 1.071854847789188*\wedge - 24i)z^{24}$

jp2 = JuliaSetPlot[p2, z, PlotRange → {{−2, 2}, {0, 2}},

AspectRatio → 1, PlotLabel → "(b)"]

Show[GraphicsRow[{jp1, jp2}]] □

4.2.2 Miscellaneous List Operations

4.2.2.1 Other List Operations

Some other Mathematica commands used with lists include:

1. `Append[list,element]`, which appends `element` to `list`;
2. `AppendTo[list,element]`, which appends `element` to `list` and names the result `list`;
3. `Drop[list,n]`, which returns the list obtained by dropping the first n elements from `list`;
4. `Drop[list,-n]`, which returns the list obtained by dropping the last n elements of `list`;
5. `Drop[list,{n,m}]`, which returns the list obtained by dropping the nth through mth elements of `list`;
6. `Drop[list,{n}]`, which returns the list obtained by dropping the nth element of `list`;
7. `Prepend[list,element]`, which prepends `element` to `list`; and
8. `PrependTo[list,element]`, which prepends `element` to `list` and names the result `list`.

4.2.2.2 Alternative Way to Evaluate Lists by Functions

Abbreviations of several of the commands discussed in this section are summarized in the following table.

@@ Apply	// (function application)	{...} List
/@ Map	[[...]] Part	

4.3 OTHER APPLICATIONS

We now present several other applications that we find interesting and require the manipulation of lists. The examples also illustrate (and combine) many of the techniques that were demonstrated in the earlier chapters.

4.3.1 Approximating Lists with Functions

Another interesting application of lists is that of curve-fitting. The commands

1. `Fit[data,functionset,variables]` fits the list of data points `data` using the functions in `functionset` by the method of least-squares. The functions in `functionset` are functions of the variables listed in `variables`; and
2. `InterpolatingPolynomial[data,x]` fits the list of n data points `data` with an $n-1$ degree polynomial in the variable x.

Example 4.18

Define `datalist` to be the list of numbers consisting of 1.14479, 1.5767, 2.68572, 2.5199, 3.58019, 3.84176, 4.09957, 5.09166, 5.98085, 6.49449, and 6.12113. (a) Find a quadratic approximation of the points in `datalist`. (b) Find a fourth degree polynomial approximation of the points in `datalist`.

Solution. The approximating function obtained via the least-squares method with `Fit` is plotted along with the data points in Fig. 4.20. Notice that many of the data points are not very close to the approximating function. A better approximation is obtained using a polynomial of higher degree (4).

```
Clear[datalist]

datalist = {1.14479, 1.5767, 2.68572, 2.5199, 3.58019, 3.84176,

    4.09957, 5.09166, 5.98085, 6.49449, 6.12113};

p1 = ListPlot[datalist, PlotStyle → Black];

Clear[y]

y[x_] = Fit[datalist, {1, x, x^2}, x]
```

$0.508266 + 0.608688x - 0.00519281x^2$

```
p2 = Plot[y[x], {x, −1, 11}, PlotStyle → {{Thickness[.02], CMYKColor[0, .09, .8, 0]}}];

pa = Show[p1, p2, PlotLabel → "(a)"];

Clear[y]

y[x_] = Fit[datalist, {1, x, x^2, x^3, x^4}, x]
```

$-0.54133 + 2.02744x - 0.532282x^2 + 0.0709201x^3 - 0.00310985x^4$

```
p3 = Plot[y[x], {x, −1, 11}, PlotStyle → {{Thickness[.02], CMYKColor[0, .09, .8, 0]}}];

pb = Show[p1, p3, PlotLabel → "(b)"];

Show[GraphicsRow[{pa, pb}]]
```

To check its accuracy, the second approximation is graphed simultaneously with the data points in Fig. 4.20 (b). □

Remember that when a semi-colon is placed at the end of the command the resulting output is *not* displayed by Mathematica.

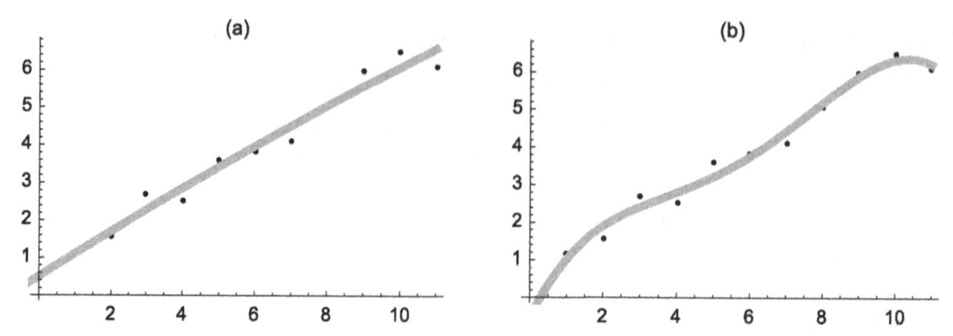

FIGURE 4.20 (a) The graph of a quadratic fit shown with the data points. (b) The graph of a quartic fit shown with the data points. (Appalachian State University colors)

TABLE 4.2 Petroleum Products Imported to the United States for Certain Years

Year	Percent	Year	Percent
1973	34.8105	1983	28.3107
1974	35.381	1984	29.9822
1975	35.8167	1985	27.2542
1976	40.6048	1986	33.407
1977	47.0132	1987	35.4875
1978	42.4577	1988	38.1126
1979	43.1319	1989	41.57
1980	37.3182	1990	42.1533
1981	33.6343	1991	39.5108
1982	28.0988		

Next, consider a list of data points made up of ordered pairs.

Example 4.19

Table 4.2 shows the average percentage of petroleum products imported to the United States for certain years. (a) Graph the points corresponding to the data in the table and connect the consecutive points with line segments. (b) Use `InterpolatingPolynomial` to find a function that approximates the data in the table. (c) Find a fourth degree polynomial approximation of the data in the table. (d) Find a trigonometric approximation of the data in the table.

Solution. We begin by defining `data` to be the set of ordered pairs represented in the table: the x-coordinate of each point represents the number of years past 1900 and the y-coordinate represents the percentage of petroleum products imported to the United States.

```
data = {{73., 34.8105}, {74., 35.381}, {75., 35.8167},
  {76., 40.6048}, {77., 47.0132}, {78., 42.4577},
  {79., 43.1319}, {80., 37.3182}, {81., 33.6343},
  {82., 28.0988}, {83., 28.3107}, {84., 29.9822},
  {85., 27.2542}, {86., 33.407}, {87., 35.4875},
  {88., 38.1126}, {89., 41.57}, {90., 42.1533}, {91., 39.5108}};
```

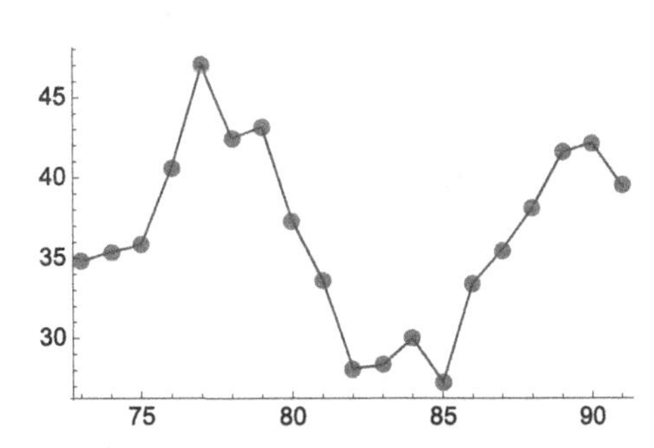

FIGURE 4.21 The points in Table 4.2 connected by line segments. (Northeastern University colors)

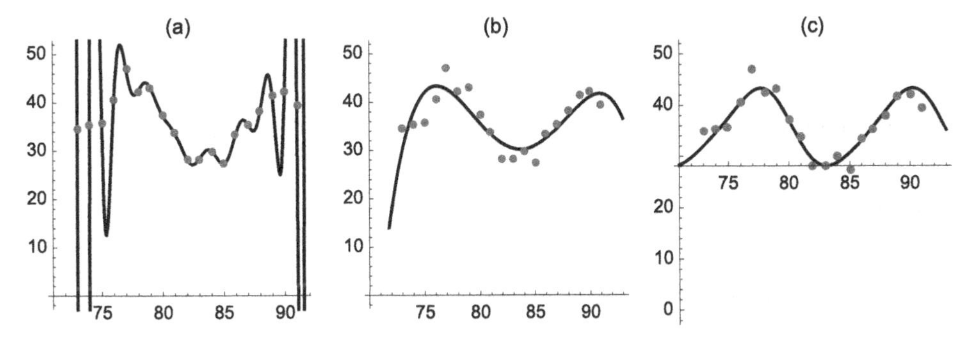

FIGURE 4.22 (a) Even though interpolating polynomials agree with the data exactly, they may have extreme oscillations, even for relatively small data sets. (b) Even though the fit does not agree with the data exactly, the oscillations seen in (a). (c) You can use `Fit` to approximate data by a variety of functions.

We use `ListPlot` to graph the ordered pairs in `data`. Note that because the option `PlotStyle->PointSize[.03]` is included within the `ListPlot` command, the points are larger than they would normally be. We also use `ListPlot` with the option `PlotJoined->True` to graph the set of points `data` and connect consecutive points with line segments. Then we use `Show` to display `lp1` and `lp2` together in Fig. 4.21. Note that in the result, the points are easy to distinguish because of their larger size.

```
lp1 = ListPlot[data, PlotStyle → {{CMYKColor[0, 1, .9, .05], PointSize[0.03]}}];

lp2 = ListPlot[data, Joined → True,

  PlotStyle → CMYKColor[0, .13, .3, .76]];

Show[lp1, lp2]
```

Next, we use `InterpolatingPolynomial` to find a polynomial approximation, p, of the data in the table. Note that the result is lengthy, so `Short` is used to display an abbreviated form of p. We then graph p and show the graph of p along with the data in the table for the years corresponding to 1971 to 1993 in Fig. 4.22 (a). Although the interpolating polynomial agrees with the data exactly, the interpolating polynomial oscillates wildly.

```
p = InterpolatingPolynomial[data, x];

Short[p, 3]
```

```
39.5108 + (0.261128 + (0.111875 + (0.0622256 + (−0.00714494 + (−0.00224511 +

(⟨⟨1⟩⟩)(−75. + x))(−88. + x))(−77. + x))(−82. + x))(−73. + x))(−91. + x)
```

```
plotp = Plot[p, {x, 71, 93}, PlotStyle → Black];

pa = Show[plotp, lp1, PlotRange → {0, 50},

  PlotLabel → "(a)", AspectRatio → 1];
```

To find a polynomial that approximates the data but does not oscillate wildly, we use `Fit`. Again, we graph the fit and display the graph of the fit and the data simultaneously. In this case, the fit does not identically agree with the data but does not oscillate wildly as illustrated in Fig. 4.22 (b).

```
Clear[p]

p = Fit[data, {1, x, x^2, x^3, x^4}, x]
```

$-198884. + 9597.83x - 173.196x^2 + 1.38539x^3 - 0.00414481x^4$

```
plotp = Plot[p, {x, 71, 93}, PlotStyle → Black];

pb = Show[plotp, lp1, PlotRange → {0, 50},

  AxesOrigin → {70, 0}, PlotLabel → "(b)",

  AspectRatio → 1]
```

See texts like Abell, Braselton, and Rafter's *Statistics with Mathematica* [15] for a more sophisticated discussion of curve-fitting and related statistical applications.

In addition to curve-fitting with polynomials, Mathematica can also fit the data with trigonometric functions. In this case, we use `Fit` to find an approximation of the data of the form $p = c_1 + c_2 \sin x + c_3 \sin(x/2) + c_4 \cos x + c_5 \cos(x/2)$. As in the previous two cases, we graph the fit and display the graph of the fit and the data simultaneously; the results are shown in Fig. 4.22 (c).

```
Clear[p]

p = Fit[data, {1, Sin[x], Sin[x/2], Cos[x], Cos[x/2]}, x]
```

$35.4237 + 4.25768\text{Cos}\left[\frac{x}{2}\right] - 0.941862\text{Cos}[x] + 6.06609\text{Sin}\left[\frac{x}{2}\right] + 0.0272062\text{Sin}[x]$

```
"35.4237" + "4.25768"Cos[x/2] - "0.941862"Cos[x]+

"6.06609"Sin[x/2] + "0.0272062"Sin[x]
```

$35.4237 + 4.25768\text{Cos}\left[\frac{x}{2}\right] - 0.941862\text{Cos}[x] + 6.06609\text{Sin}\left[\frac{x}{2}\right] + 0.0272062\text{Sin}[x]$

```
plotp = Plot[p, {x, 71, 93}, PlotStyle → Black];

pc = Show[plotp, lp1, PlotRange → {0, 50},
```

PlotLabel → "(c)", AspectRatio → 1];

Show[GraphicsRow[{pa, pb, pc}]] □

4.3.2 Introduction to Fourier Series

Many problems in applied mathematics are solved through the use of Fourier series. Mathematica assists in the computation of these series in several ways. Suppose that $y = f(x)$ is defined on $-p < x < p$. Then the Fourier series for $f(x)$ is

$$\frac{1}{2}a_0 + \sum_{n=1}^{\infty}\left(a_n \cos\frac{n\pi x}{p} + b_n \sin\frac{n\pi x}{p}\right) \tag{4.1}$$

where

$$\begin{aligned} a_0 &= \frac{1}{p}\int_{-p}^{p} f(x)\,dx \\ a_n &= \frac{1}{p}\int_{-p}^{p} f(x)\cos\left(\frac{n\pi x}{p}\right)dx \quad n = 1, 2\ldots \\ b_n &= \frac{1}{p}\int_{-p}^{p} f(x)\sin\left(\frac{n\pi x}{p}\right)dx \quad n = 1, 2\ldots \end{aligned} \tag{4.2}$$

The k**th term of the Fourier series** (4.1) is

$$a_n \cos\frac{n\pi x}{p} + b_n \sin\frac{n\pi x}{p}. \tag{4.3}$$

The k**th partial sum of the Fourier series** (4.1) is

$$\frac{1}{2}a_0 + \sum_{n=1}^{k}\left(a_n\left(\cos\left(\frac{n\pi x}{p}\right)\right) + b_n \sin\left(\frac{n\pi x}{p}\right)\right). \tag{4.4}$$

It is a well-known theorem that if $y = f(x)$ is a periodic function with period $2p$ and $f'(x)$ is continuous on $[-p, p]$ except at finitely many points, then at each point x the Fourier series for $f(x)$ converges and

$$\frac{1}{2}a_0 + \sum_{n=1}^{\infty}\left(a_n \cos\left(\frac{n\pi x}{p}\right) + b_n \sin\left(\frac{n\pi x}{p}\right)\right) = \frac{1}{2}\left(\lim_{z\to x^+} f(z) + \lim_{z\to x^-} f(z)\right).$$

In fact, if the series $\sum_{n=1}^{\infty}(|a_n| + |b_n|)$ converges, then the Fourier series converges uniformly on $(-\infty, \infty)$.

In the special case that $p = \pi$, `FourierSeries`, `FourierCosSeries`, and `FourierSinSeries` can be used to carry out these calculations. If $y = f(x)$ is defined on $[-\pi, \pi]$, `FourierSeries[f[x],x,n]` finds the first n terms of the Fourier series for $f(x)$, `FourierCosSeries[f[x],x,n]` finds the first n terms of the Fourier cosine series for $f(x)$ (even extension), and `FourierSinSeries[f[x],x,n]` finds the first n terms of the Fourier sin series for $f(x)$ (odd extension).

Example 4.20

Let $f(x) = \begin{cases} -x, & -1 \le x < 0 \\ 1, & 0 \le x < 1 \\ f(x-2), & x \ge 1 \end{cases}$. Compute and graph the first few partial sums of the Fourier series for $f(x)$.

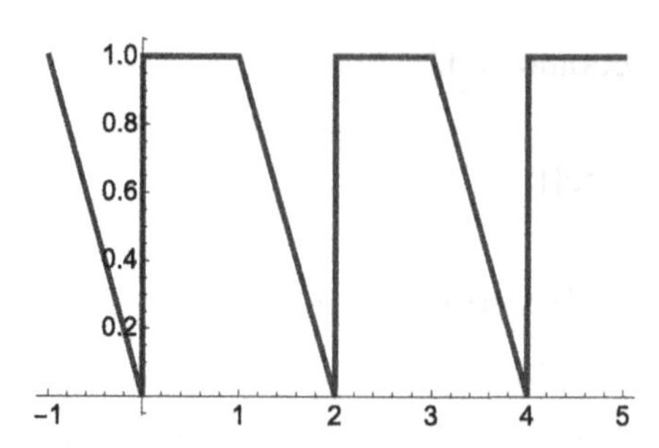

FIGURE 4.23 Plot of a few periods of $f(x)$. (University of Delaware colors)

We begin by clearing all prior definitions of f. We then define the piecewise function $f(x)$ and graph $f(x)$ on the interval $[-1, 5]$ in Fig. 4.23.

```
Clear[f]

f[x_]:=1/;0 ≤ x < 1

f[x_]:= − x/; − 1 ≤ x < 0

f[x_]:=f[x − 2]/;x ≥ 1

graphf = Plot[f[x], {x, −1, 5}, PlotStyle → {{CMYKColor[1, .38, 0, .15],

  Thickness[.01]}}]
```

The Fourier series coefficients are computed with the integral formulas in equation (4.2). Executing the following commands defines p to be 1, `a[0]` to be an approximation of the integral $a_0 = \frac{1}{p}\int_{-p}^{p} f(x)\,dx$, `a[n]` to be an approximation of the integral $a_n = \frac{1}{p}\int_{-p}^{p} f(x)\cos\left(\frac{n\pi x}{p}\right)dx$, and `b[n]` to be an approximation of the integral $b_n = \frac{1}{p}\int_{-p}^{p} f(x)\sin\left(\frac{n\pi x}{p}\right)dx$.

```
Clear[a, b, fs, L]

L = 1;

a[0] = NIntegrate[f[x],{x,−L,L}]/(2L)
```

0.75

```
a[n_]:= NIntegrate[f[x]Cos[nπx/L],{x,−L,L}]/L

b[n_]:= NIntegrate[f[x]Sin[nπx/L],{x,−L,L}]/L
```

A table of the coefficients `a[i]` and `b[i]` for $i = 1, 2, 3, \ldots, 10$ is generated with `Table` and named `coeffs`. Several error messages (which are not displayed here for length considerations) are generated because of the discontinuities but the resulting approximations are satisfactory for our purposes. The elements in the first column of the table represent the a_i's and the second column represents the b_i's. Notice how the elements of the table are extracted using double brackets with `coeffs`.

coeffs = Table[{a[i], b[i]}, {i, 1, 10}];

TableForm[coeffs]

−0.202642	0.31831
−2.7755575615628914*^ − 17	0.159155
−0.0225158	0.106103
−1.1796119636642288*^ − 16	0.0795775
−0.00810569	0.063662
−7.112366251504909*^ − 17	0.0530516
−0.00413556	0.0454728
−5.117434254131581*^ − 17	0.0397887
−0.00250176	0.0353678
−5.334274688628682*^ − 17	0.031831

The first element of the list is extracted with `coeffs[[1]]`.

coeffs[[1]]

{−0.202642, 0.31831}

The first element of the second element of `coeffs` and the second element of the third element of `coeffs` are extracted with `coeffs[[2,1]]` and `coeffs[[3,2]]`, respectively.

coeffs[[2, 1]]

−2.7755575615628914*^ − 17

coeffs[[3, 2]]

0.106103

After the coefficients are calculated, the nth partial sum of the Fourier series is obtained with `Sum`. The kth term of the Fourier series, $a_k \cos(k\pi x) + b_k \sin(k\pi x)$, is defined in `fs`. Hence, the nth partial sum of the series is given by

$$a_0 + \sum_{k=1}^{n} [a_k \cos(k\pi x) + b_k \sin(k\pi x)] = \texttt{a[0]} + \sum_{k=1}^{n} \texttt{fs[k, x]},$$

which is defined in `fourier` using `Sum`. We illustrate the use of `fourier` by finding `fourier[2,x]` and `fourier[3,x]`.

fs[k_, x_]:=coeffs[[k, 1]]Cos[$k\pi x$] + coeffs[[k, 2]]Sin[$k\pi x$]

fourier[n_, x_]:=a[0] + $\sum_{k=1}^{n}$ fs[k, x]

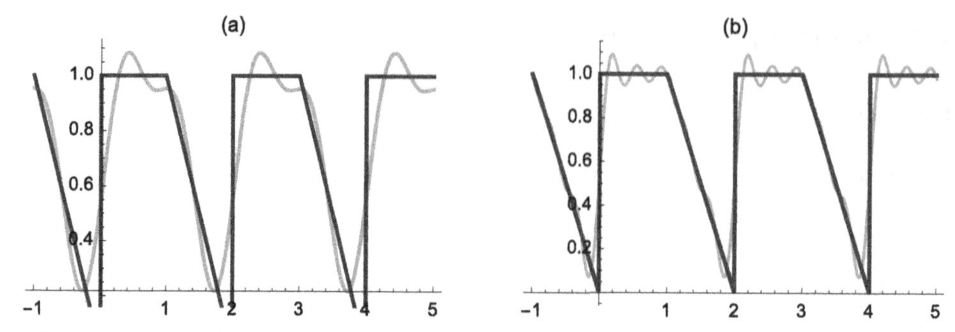

FIGURE 4.24 The first few terms of a Fourier series for a periodic function plotted with the function.

fourier[2, *x*]

0.75 − 0.202642Cos[πx] − 2.7755575615628914*^-17Cos[$2\pi x$] + 0.31831Sin[πx] + 0.159155Sin[$2\pi x$]

fourier[3, *x*]

0.75 − 0.202642Cos[πx] − 2.7755575615628914*^-17Cos[$2\pi x$] − 0.0225158Cos[$3\pi x$] + 0.31831Sin[πx] + 0.159155Sin[$2\pi x$] + 0.106103Sin[$3\pi x$]

To see how the Fourier series approximates the periodic function, we plot the function simultaneously with the Fourier approximation for $n = 2$ and $n = 5$. The results are displayed together using `GraphicsArray` in Fig. 4.24.

graphtwo = Plot[fourier[2, *x*], {*x*, −1, 5},

PlotStyle → {{Thickness[.01], CMYKColor[0, .09, .94, 0]}}];

bothtwo = Show[graphtwo, graphf, PlotLabel → "(a)"];

graphfive = Plot[fourier[5, *x*], {*x*, −1, 5},

PlotStyle → {{Thickness[.01], CMYKColor[0, .09, .94, 0]}}];

bothfive = Show[graphfive, graphf, PlotLabel → "(b)"];

Show[GraphicsRow[{bothtwo, bothfive}]] □

Example 4.21

For the first five terms of the Fourier series, Fourier cosine series, and Fourier sine series for $f(x) = x$ on $[-\pi, \pi]$.

Solution. Here we use `FourierSeries`, `FourierCosSeries`, and `FourierSinSeries`.

Clear[*f*]

***f*[x_] = *x*;**

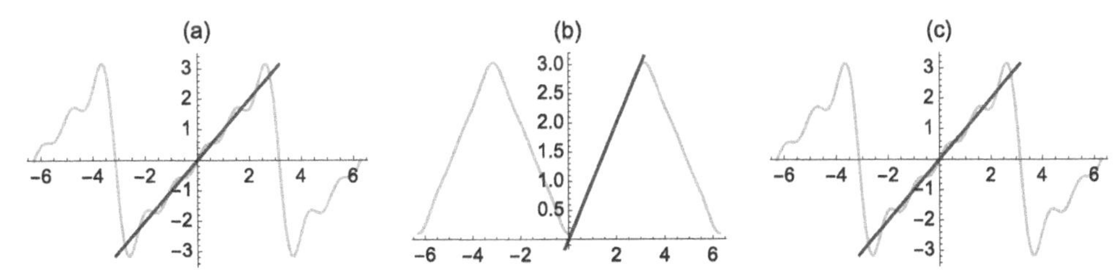

FIGURE 4.25 Approximating $f(x)$ with its Fourier series: the periodic extension in (a), the even extension in (b), and the odd extension in (c).

```
f1 = FourierSeries[x, x, 5]

f2 = FourierCosSeries[x, x, 5]

f3 = FourierSinSeries[x, x, 5]
```

$$ie^{-ix} - ie^{ix} - \frac{1}{2}ie^{-2ix} + \frac{1}{2}ie^{2ix} + \frac{1}{3}ie^{-3ix} - \frac{1}{3}ie^{3ix} - \frac{1}{4}ie^{-4ix} + \frac{1}{4}ie^{4ix} + \frac{1}{5}ie^{-5ix} - \frac{1}{5}ie^{5ix}$$

$$\frac{\pi}{2} - \frac{4\mathrm{Cos}[x]}{\pi} - \frac{4\mathrm{Cos}[3x]}{9\pi} - \frac{4\mathrm{Cos}[5x]}{25\pi}$$

$$-2\left(-\mathrm{Sin}[x] + \frac{1}{2}\mathrm{Sin}[2x] - \frac{1}{3}\mathrm{Sin}[3x] + \frac{1}{4}\mathrm{Sin}[4x] - \frac{1}{5}\mathrm{Sin}[5x]\right)$$

Next, we graph the Fourier approximations with $f(x) = x$ in Fig. 4.25. In the figure, observe that the Fourier series is the periodic extension of $f(x)$, the Fourier cosine series is the even periodic extension of $f(x)$, and the Fourier sine series is the odd periodic extension of $f(x)$.

```
q1 = Plot[x, {x, -Pi, Pi}, PlotStyle → {{CMYKColor[1, .38, 0, .15],
    Thickness[.01]}}]

p1 = Plot[f1, {x, -2Pi, 2Pi},
    PlotStyle → {{Thickness[.01], CMYKColor[0, .09, .94, 0]}}];

p2 = Plot[f2, {x, -2Pi, 2Pi},
    PlotStyle → {{Thickness[.01], CMYKColor[0, .09, .94, 0]}}];

p3 = Plot[f3, {x, -2Pi, 2Pi},
    PlotStyle → {{Thickness[.01], CMYKColor[0, .09, .94, 0]}}];

Show[GraphicsRow[{Show[p1, q1, PlotLabel → "(a)"], Show[p2, q1, PlotLabel → "(b)"],
Show[p3, q1, PlotLabel → "(c)"]}]]
```
□

Application: The One-Dimensional Heat Equation

A typical problem in applied mathematics that involves the use of Fourier series is that of the **one-dimensional heat equation**. The boundary value problem that describes the temperature

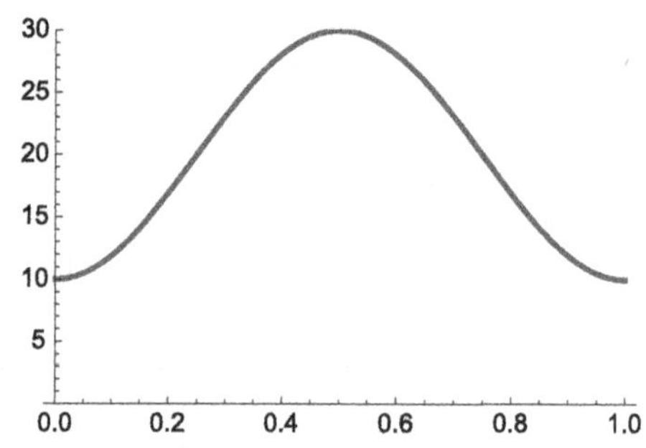

FIGURE 4.26 Graph of $f(x) = 10 + 20\sin^2 \pi x$. (University of Utah colors)

in a uniform rod with insulated surface is

$$\begin{aligned} &k\frac{\partial^2 u}{\partial x^2} = \frac{\partial u}{\partial t},\ 0 < x < a,\ t > 0, \\ &u(0,t) = T_0,\ t > 0, \\ &u(a,t) = T_a,\ t > 0, \text{ and} \\ &u(x,0) = f(x),\ 0 < x < a. \end{aligned} \tag{4.5}$$

In this case, the rod has "fixed end temperatures" at $x = 0$ and $x = a$ and $f(x)$ is the initial temperature distribution. The solution to the problem is

$$u(x,t) = \underbrace{T_0 + \frac{1}{a}(T_a - T_0)\,x}_{v(x)} + \sum_{n=1}^{\infty} b_n \sin(\lambda_n x)\, e^{-\lambda_n{}^2 kt}, \tag{4.6}$$

where

$$\lambda_n = n\pi/a \qquad \text{and} \qquad b_n = \frac{2}{a}\int_0^a (f(x) - v(x)) \sin\frac{n\pi x}{a}\,dx,$$

and is obtained through separation of variables techniques. The coefficient b_n in the solution equation (4.6) is the Fourier series coefficient b_n of the function $f(x) - v(x)$, where $v(x)$ is the **steady-state temperature**.

Example 4.22

Solve $\begin{cases} \dfrac{\partial^2 u}{\partial x^2} = \dfrac{\partial u}{\partial t},\ 0 < x < 1,\ t > 0, \\ u(0,t) = 10,\ u(1,t) = 10,\ t > 0, \\ u(x,0) = 10 + 20\sin^2 \pi x. \end{cases}$

Solution. In this case, $a = 1$ and $k = 1$. The fixed end temperatures are $T_0 = T_a = 10$, and the initial heat distribution is $f(x) = 10 + 20\sin^2 \pi x$. The steady-state temperature is $v(x) = 10$. The function $f(x)$ is defined and plotted in Fig. 4.26. Also, the steady-state temperature, $v(x)$, and the eigenvalue are defined. Finally, `Integrate` is used to define a function that will be used to calculate the coefficients of the solution.

```
Clear[f]

f[x_]:=10 + 20Sin[πx]^2

Plot[f[x], {x, 0, 1}, PlotRange → {0, 30},
  PlotStyle → {{Thickness[.01], CMYKColor[0, 1, .79, .2]}}]
```

v[x_]:=10

lambda[n_]:=$\frac{n\pi}{4}$

b[n_]:=b[n] = $\int_0^4 (f[x] - v[x])\text{Sin}\left[\frac{n\pi x}{4}\right]dx$

Notice that `b[n]` is defined using the form `b[n_]:=b[n]=...` so that Mathematica "remembers" the values of `b[n]` computed and thus avoids recomputing previously computed values. In the following table, we compute exact and approximate values of `b[1],...,b[10]`.

Table[{n, b[n], b[n]//N}, {n, 1, 10}]//TableForm

1	$\frac{5120}{63\pi}$	25.869
2	0	0.
3	$\frac{1024}{33\pi}$	9.87725
4	0	0.
5	$\frac{1024}{39\pi}$	8.35767
6	0	0.
7	$\frac{1024}{21\pi}$	15.5214
8	0	0.
9	$-\frac{5120}{153\pi}$	−10.6519
10	0	0.

Let $S_m = b_m \sin(\lambda_m x)\, e^{-\lambda_m{}^2 t}$. Then, the desired solution, $u(x,t)$, is given by

$$u(x,t) = v(x) + \sum_{m=1}^{\infty} S_m.$$

Let $u(x,t,n) = v(x) + \sum_{m=1}^{n} S_m$. Notice that $u(x,t,n) = u(x,t,n-1) + S_n$. Consequently, approximations of the solution to the heat equation are obtained recursively taking advantage of Mathematica's ability to compute recursively. The solution is first defined for $n = 1$ by `u[x,t,1]`. Subsequent partial sums, `u[x,t,n]`, are obtained by adding the nth term of the series, S_n, to `u[x,t,n-1]`.

u[x_, t_, 1]:=v[x] + b[1]Sin[lambda[1]x]Exp[−lambda[1]^2 t]

u[x_, t_, n_]:=u[x, t, n − 1] + b[n]Sin[lambda[n]x]Exp[−lambda[n]^2 t]

By defining the solution in this manner a table can be created that includes the partial sums of the solution. In the following table, we compute the first, fourth, and seventh partial sums of the solution to the problem. (See Fig. 4.27.)

Table[u[x, t, n], {n, 1, 7, 3}]

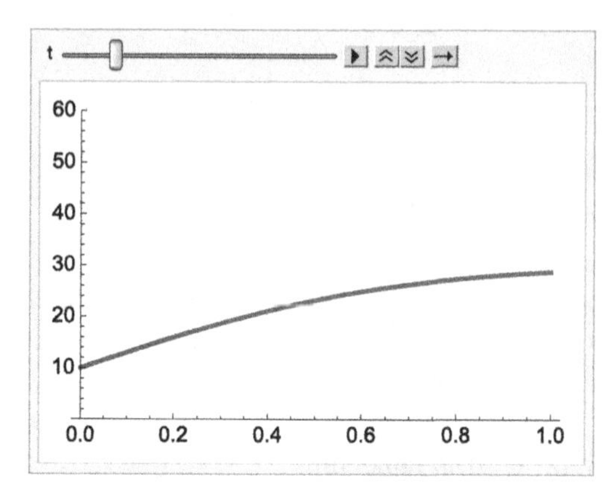

FIGURE 4.27 Animating the temperature distribution in a uniform rod with insulated surface. (University of Utah colors)

$$\{10+\frac{5120e^{-\frac{\pi^2 t}{16}}\mathrm{Sin}[\frac{\pi x}{4}]}{63\pi}, 10+\frac{5120e^{-\frac{\pi^2 t}{16}}\mathrm{Sin}[\frac{\pi x}{4}]}{63\pi}+\frac{1024e^{-\frac{9\pi^2 t}{16}}\mathrm{Sin}[\frac{3\pi x}{4}]}{33\pi}, 10+\frac{5120e^{-\frac{\pi^2 t}{16}}\mathrm{Sin}[\frac{\pi x}{4}]}{63\pi}+\frac{1024e^{-\frac{9\pi^2 t}{16}}\mathrm{Sin}[\frac{3\pi x}{4}]}{33\pi}+\frac{1024e^{-\frac{25\pi^2 t}{16}}\mathrm{Sin}[\frac{5\pi x}{4}]}{39\pi}+\frac{1024e^{-\frac{49\pi^2 t}{16}}\mathrm{Sin}[\frac{7\pi x}{4}]}{21\pi}\}$$

To generate graphics that can be animated, we use `Animate`. The 10th partial sum of the solution is plotted for $t=0$ to $t=1$.

```
Animate[Plot[u[x, t, 10], {x, 0, 1},

  PlotRange → {0, 60}, PlotStyle → {{Thickness[.01], CMYKColor[0, 1, .79, .2]}}],
  {t, 0, 1}]
```

Alternatively, we may generate several graphics and display the resulting set of graphics as a `GraphicsArray`. We plot the 10th partial sum of the solution for $t=0$ to $t=1$ using a step-size of $1/15$. The resulting 16 graphs are named `graphs` which are then partitioned into four element subsets with `Partition` and named `toshow`. We then use `Show` and `GraphicsGrid` to display `toshow` in Fig. 4.28.

```
graphs = Table[Plot[Evaluate[u[x, t, 10]], {x, 0, 1}, Ticks → None, PlotRange → {0, 60},

  PlotStyle → {{Thickness[.01], CMYKColor[0, 1, .79, .2]}}], {t, 0, 1, 1/15}];

toshow = Partition[graphs, 4];

Show[GraphicsGrid[toshow]]
```

□

Fourier series and generalized Fourier series arise in too many applications to list. Examples using them illustrate Mathematica's power to manipulate lists, symbolics, and graphics.

Application: The Wave Equation on a Circular Plate

For a classic approach to the subject see Graff's *Wave Motion in Elastic Solids*, [7].

The vibrations of a circular plate satisfy the equation

$$D\nabla^4 w(r,\theta,t)+\rho h\frac{\partial^2 w(r,\theta,t)}{\partial t^2}=q(r,\theta,t), \tag{4.7}$$

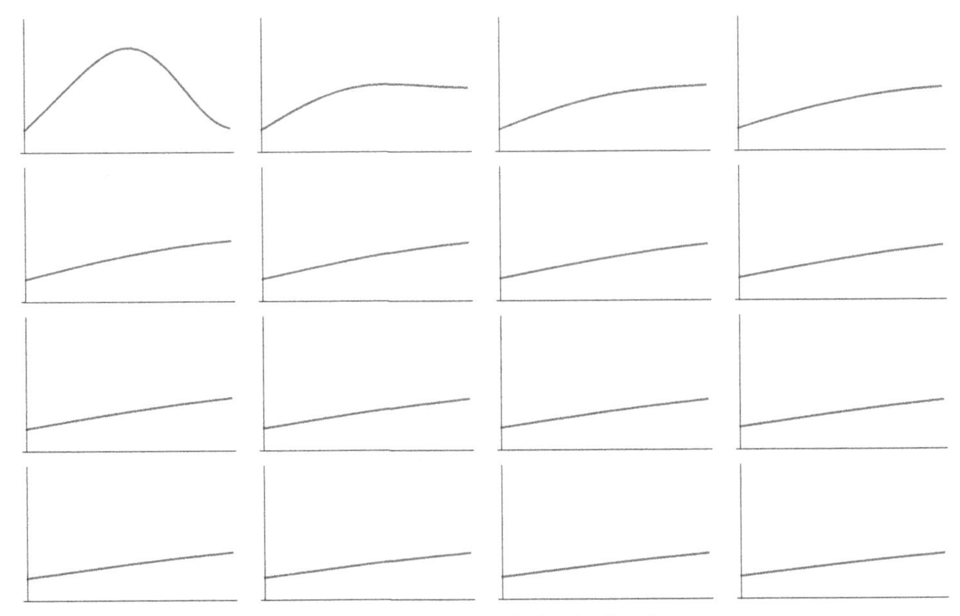

FIGURE 4.28 Temperature distribution in a uniform rod with insulated surface.

where $\nabla^4 w = \nabla^2 \nabla^2 w$ and ∇^2 is the **Laplacian in polar coordinates**, which is defined by

$$\nabla^2 = \frac{1}{r}\frac{\partial}{\partial r}\left(r\frac{\partial}{\partial r}\right) + \frac{1}{r^2}\frac{\partial^2}{\partial \theta^2} = \frac{\partial^2}{\partial r^2} + \frac{1}{r}\frac{\partial}{\partial r} + \frac{1}{r^2}\frac{\partial^2}{\partial \theta^2}.$$

Assuming no forcing so that $q(r,\theta,t)=0$ and $w(r,\theta,t)=W(r,\theta)e^{-i\omega t}$, equation (4.7) can be written as

$$\nabla^4 W(r,\theta) - \beta^4 W(r,\theta) = 0, \qquad \beta^4 = \omega^2 \rho h/D. \tag{4.8}$$

For a clamped plate, the boundary conditions are $W(a,\theta) = \partial W(a,\theta)/\partial r = 0$ and after *much work* (see [7]) the **normal modes** are found to be

$$W_{nm}(r,\theta) = \left[J_n(\beta_{nm} r) - \frac{J_n(\beta_{nm} a)}{I_n(\beta_{nm} a)} I_n(\beta_{nm} r)\right]\begin{pmatrix}\sin n\theta \\ \cos n\theta\end{pmatrix}. \tag{4.9}$$

In equation (4.9), $\beta_{nm} = \lambda_{nm}/a$ where λ_{nm} is the mth solution of

$$I_n(x)J_n{}'(x) - J_n(x)I_n{}'(x) = 0, \tag{4.10}$$

where $J_n(x)$ is the Bessel function of the first kind of order n and $I_n(x)$ is the **modified Bessel function of the first kind** of order n, related to $J_n(x)$ by $i^n I_n(x) = J_n(ix)$.

See Example 4.14.

The Mathematica command `BesselI[n,x]` returns $I_n(x)$.

Example 4.23

Graph the first few normal modes of the clamped circular plate.

Solution. We must determine the value of λ_{nm} for several values of n and m so we begin by defining `eqn[n][x]` to be $I_n(x)J_n{}'(x) - J_n(x)I_n{}'(x)$. The mth solution of equation (4.10) corresponds to the mth zero of the graph of `eqn[n][x]` so we graph `eqn[n][x]` for $n = 0$, 1, 2, and 3 with `Plot` in Fig. 4.29.

```
eqn[n_][x_]:=BesselI[n, x]D[BesselJ[n, x], x] − BesselJ[n, x]D[BesselI[n, x], x]
```

The result of the `Table` and `Plot` command is a list of length four, which is verified with `Length[p1]`.

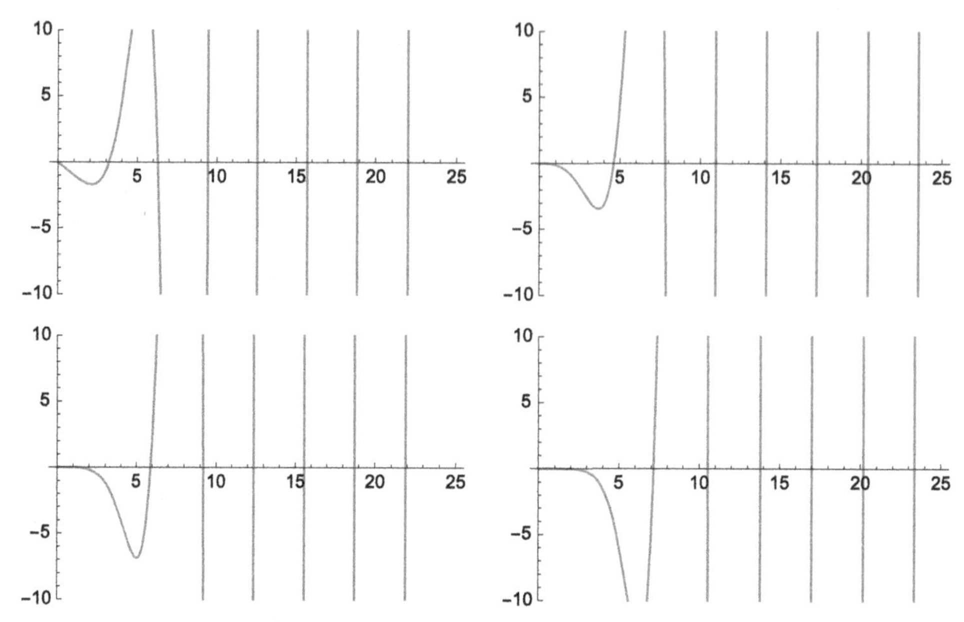

FIGURE 4.29 Plot of $I_n(x)J_n{}'(x) - J_n(x)I_n{}'(x)$ for $n = 0$ and 1 in the first row; $n = 2$ and 3 in the second row.

```
p1 = Table[Plot[Evaluate[eqn[n][x]], {x, 0, 25}, PlotRange → {−10, 10},
   PlotStyle → RGBColor[.3, 3.6, 7.7]], {n, 0, 3}];
```

so we use `Partition` to create a 2 × 2 array of graphics which is displayed using `Show` and `GraphicsGrid`.

```
p2 = Show[GraphicsGrid[Partition[p1, 2]]]
```

To determine λ_{nm} we use `FindRoot`. Recall that to use `FindRoot` to solve an equation an initial approximation of the solution must be given. For example,

```
l01 = FindRoot[eqn[0][x] == 0, {x, 3.04}]

{x → 3.19622}
```

approximates λ_{01}, the first solution of equation (4.10) if $n = 0$. However, the result of `FindRoot` is a list. The specific value of the solution is the second part of the first part of the list, `l01`, extracted from the list with `Part` (`[[...]]`).

```
l01[[1, 2]]

3.19622
```

Thus,

We use the graphs in Fig. 4.29 to obtain initial approximations of each solution.

```
λ0s = Map[FindRoot[eqn[0][x] == 0, {x, #}][[1, 2]]&, {3.04, 6.2, 9.36, 12.5, 15.7}]

{3.19622, 6.30644, 9.4395, 12.5771, 15.7164}
```

approximates the first five solutions of equation (4.10) if $n = 0$ and then returns the specific value of each solution. We use the same steps to approximate the first five solutions of equation (4.10) if $n = 1$, 2, and 3.

```
λ1s = Map[FindRoot[eqn[1][x] == 0, {x, #}][[1, 2]]&, {4.59, 7.75, 10.9, 14.1, 17.2}]
```

{4.6109, 7.79927, 10.9581, 14.1086, 17.2557}

```
λ2s = Map[FindRoot[eqn[2][x] == 0, {x, #}][[1, 2]]&, {5.78, 9.19, 12.4, 15.5, 18.7}]
```

{5.90568, 9.19688, 12.4022, 15.5795, 18.744}

```
λ3s = Map[FindRoot[eqn[3][x] == 0, {x, #}][[1, 2]]&, {7.14, 10.5, 13.8, 17, 20.2}]
```

{7.14353, 10.5367, 13.7951, 17.0053, 20.1923}

All four lists are combined together in `λs`.

```
λs = {λ0s, λ1s, λ2s, λ3s};
```

```
Short[λs]
```

⟨⟨1⟩⟩

For $n = 0$, 1, 2, and 3 and $m = 1$, 2, 3, 4, and 5, λ_{nm} is the mth part of the $(n+1)$st part of `λs`.

Observe that the value of a does not affect the shape of the graphs of the normal modes so we use $a = 1$ and then define β_{nm}.

```
a = 1;
```

```
β[n_, m_]:=λs[[n + 1, m]]/a
```

`ws` is defined to be the sine part of equation (4.9)

```
ws[n_, m_][r_, θ_]:=

  (BesselJ[n, β[n, m]r] − BesselJ[n, β[n, m]a]/BesselI[n, β[n, m]a]BesselI[n, β[n, m]r])
  Sin[nθ]
```

and `wc` to be the cosine part.

```
wc[n_, m_][r_, θ_]:=

  (BesselJ[n, β[n, m]r] − BesselJ[n, β[n, m]a]/BesselI[n, β[n, m]a]BesselI[n, β[n, m]r])
  Cos[nθ]
```

We use `ParametricPlot3D` to plot `ws` and `wc`. For example,

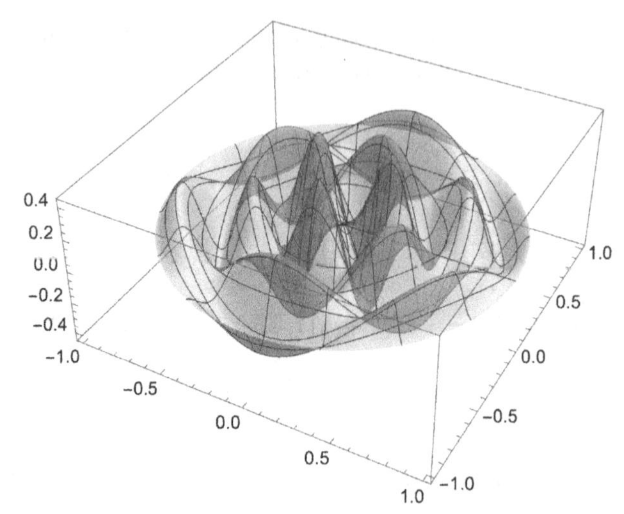

FIGURE 4.30 The sine part of $W_{34}(r, \theta)$. (Auburn University colors)

```
ParametricPlot3D[{rCos[θ], rSin[θ], ws[3, 4][r, θ]}, {r, 0, 1}, {θ, −Pi, Pi},

PlotPoints → 60, PlotStyle → Opacity[.5, RGBColor[2.21, .85, .12]]]
```

graphs the sine part of $W_{34}(r, \theta)$ shown in Fig. 4.30.

We use `Table` together with `ParametricPlot3D` followed by `Show` and `GraphicsGrid` to graph the sine part of $W_{nm}(r, \theta)$ for $n = 0$, 1, 2, and 3 and $m = 1$, 2, 3, and 4 shown in Fig. 4.31.

```
ms = Table[ParametricPlot3D[{rCos[θ], rSin[θ], ws[n, m][r, θ]}, {r, 0, 1},

  {θ, −Pi, Pi}, PlotStyle → Opacity[.5, RGBColor[3, 36, 77]],

  PlotPoints → 30, BoxRatios → {1, 1, 1}], {n, 0, 3}, {m, 1, 4}];

Show[GraphicsGrid[ms]]
```

Identical steps are followed to graph the cosine part shown in Fig. 4.32.

```
mc = Table[ParametricPlot3D[{rCos[θ], rSin[θ], wc[n, m][r, θ]}, {r, 0, 1},

  {θ, −Pi, Pi}, PlotStyle → Opacity[.5, RGBColor[3, 36, 77]],

  PlotPoints → 30, BoxRatios → {1, 1, 1}], {n, 0, 3}, {m, 1, 4}];

Show[GraphicsGrid[mc]]
```

□

See references like Barnsley's *Fractals Everywhere* [3] or Feldman's *Chaos and Fractals* [6] for detailed discussions regarding many of the topics briefly described in this section.

4.3.3 The Mandelbrot Set and Julia Sets

In Examples 4.9, 4.15, and 4.17 we illustrated several techniques for plotting bifurcation diagrams and Julia sets. For investigating Julia sets, try using built-in commands like `JuliaSetPlot` first. Similarly, for investigating the Mandelbrot set, try `MandelbrotSetPlot`

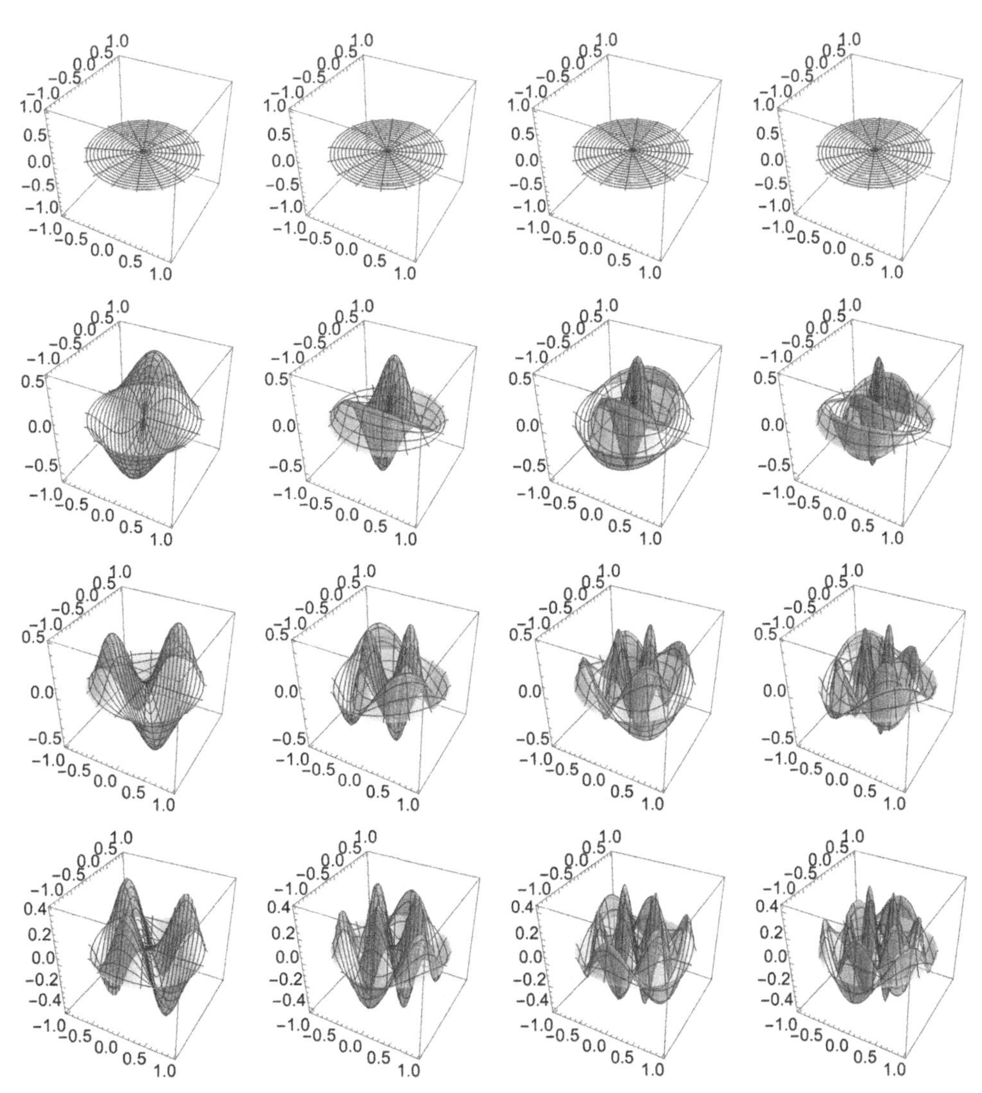

FIGURE 4.31 The sine part of $W_{nm}(r, \theta)$: $n = 0$ in row 1, $n = 1$ in row 2, $n = 2$ in row 3, and $n = 3$ in row 4 ($m = 1$ to 4 from left to right in each row).

along with its many options to see if you can obtain your desired results before spending a lot of time writing your own code.

Let $f_c(x) = x^2 + c$. In Example 4.15, we generated the c-values when plotting the bifurcation diagram of f_c. Depending upon how you think and approach various problems, some approaches may be easier to understand than others. With the exception of very serious calculations, the differences in the time needed to carry out the computations may be minimal so we encourage you to follow the approach that you understand.

$f_c(x) = x^2 + c$ is the special case of $p = 2$ for $f_{p,c}(x) = x^p + c$.

Example 4.24: Dynamical Systems

For example, entering

```
Clear[f, h]

f[c_][x_]:=x^2+c//N;
```

Compare the approach here with the approach used in Example 4.15.

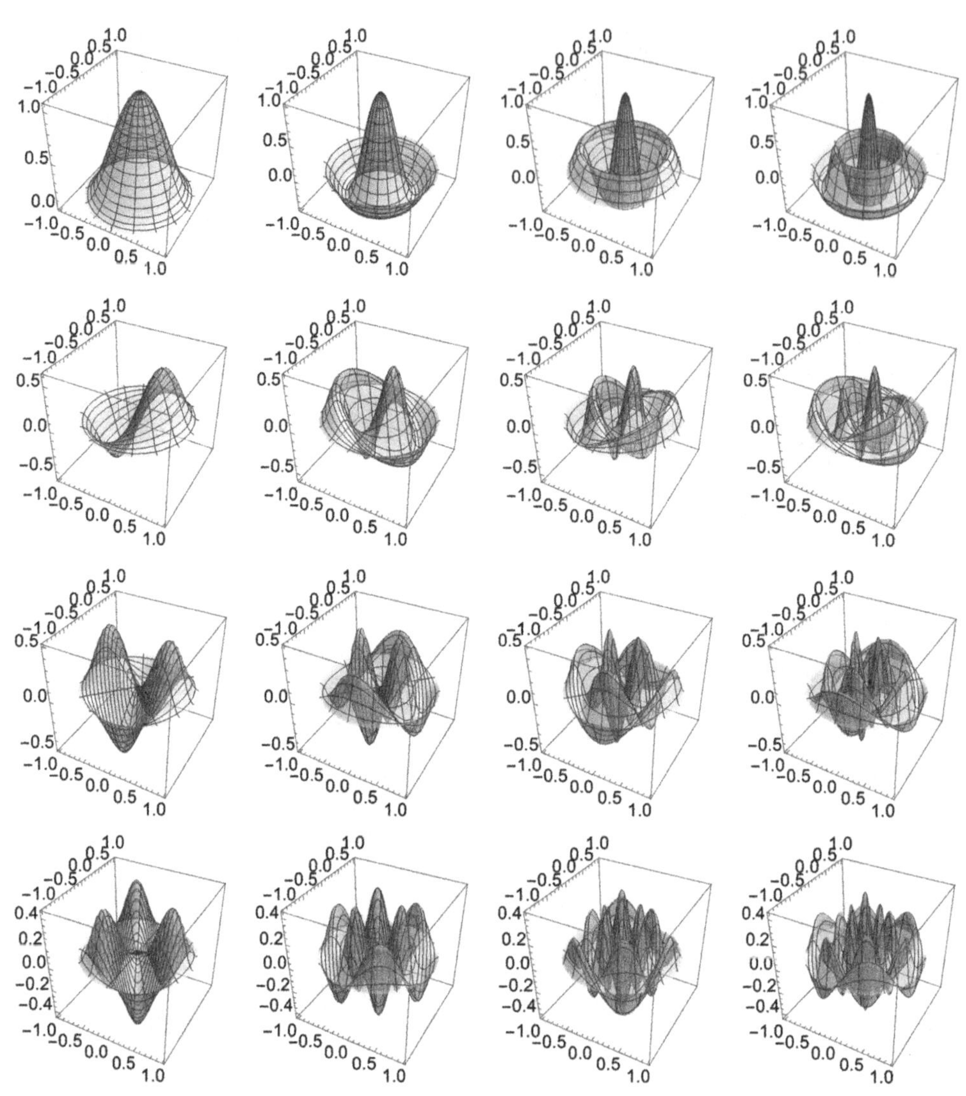

FIGURE 4.32 The cosine part of $W_{nm}(r, \theta)$: $n = 0$ in row 1, $n = 1$ in row 2, $n = 2$ in row 3, and $n = 3$ in row 4 ($m = 1$ to 4 from left to right in each row).

defines $f_c(x) = x^2 + c$ so

Nest[f[−1], x, 3]

$-1. + \left(-1. + \left(-1. + x^2\right)^2\right)^2$

computes $f_{-1}{}^3(x) = (f_{-1} \circ f_{-1} \circ f_{-1})(x)$ and

Table[Nest[f[1/4], 0, n], {n, 101, 200}]//Short

{0.490693, 0.490779, ⟨⟨96⟩⟩, 0.495148, 0.495171}

returns a list of $f_{1/4}{}^n(0)$ for $n = 101, 102, \ldots, 200$. Thus,

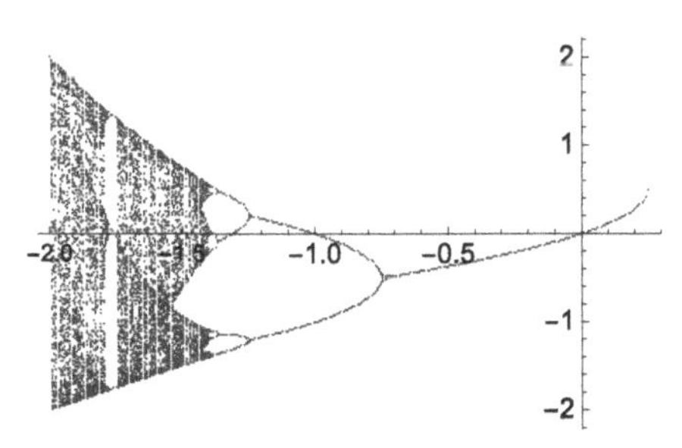

FIGURE 4.33 Another bifurcation diagram for f_c.

```
lgtable = Table[{c, Nest[f[c], 0, n]},
  {c, -2, 1/4, 9/(4 * 299)}, {n, 101, 200}];
Length[lgtable]
300
```

returns a list of lists of $f_c{}^n(0)$ for $n = 101, 102, \ldots, 200$ for 300 equally spaced values of c between −2 and 1. The list `lgtable` is converted to a list of points with `Flatten` and plotted with `ListPlot`. See Fig. 4.33 and compare this result to the result obtained in Example 4.15.

```
toplot = Flatten[lgtable, 1];
ListPlot[toplot,
  PlotStyle -> CMYKColor[1, .69, .07, .3]]
```

For a given complex number c the **Julia set**, J_c, of $f_c(x) = x^2 + c$ is the set of complex numbers, $z = a + bi$, a, b real, for which the sequence z, $f_c(z) = z^2 + c$, $f_c(f_c(z)) = (z^2+c)^2 + c, \ldots, f_c{}^n(z), \ldots$, does *not* tend to ∞ as $n \to \infty$:

We use the notation $f^n(x)$ to represent the composition $\underbrace{(f \circ f \circ \cdots \circ f)}_{n}(x)$.

$$J_c = \left\{ z \in \mathbf{C} | z,\ z^2 + c\ \left(z^2 + c\right)^2 + c,\ \cdots \not\to \infty \right\}.$$

Using a dynamical system, setting $z = z_0$ and computing $z_{n+1} = f_c(z_n)$ for large n can help us determine if z is an element of J_c. In terms of a composition, computing $f_c{}^n(z)$ for large n can help us determine if z is an element of J_c.

Example 4.25: Julia Sets

Plot the Julia set of $f_c(x) = x^2 + c$ if $c = -0.122561 + 0.744862i$.

As with previous examples, all error messages have been deleted.

Solution. After defining $f_c(x) = x^2 + c$, we use `Table` together with `Nest` to compute ordered triples of the form $\left(x, y, f_{-0.122561+0.744862i}{}^{200}(x+iy)\right)$ for 150 equally spaced values of x between $-3/2$ and $3/2$ and 150 equally spaced values of y between $-3/2$ and $3/2$.

You do not need to redefine $f_c(x)$ if you have already defined it during your current Mathematica session.

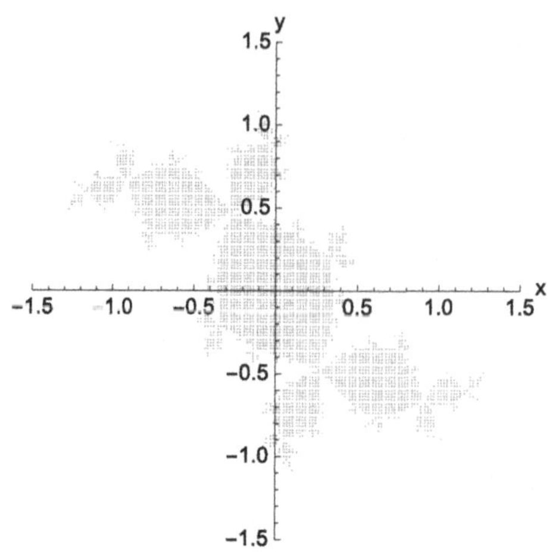

FIGURE 4.34 Filled Julia set for f_c.

```
Clear[f, h]

f[c_][x_]:=x^2 + c//N;

g1 = Table[{x, y, Nest[f[-0.12256117 + .74486177I], x + Iy, 200]},

  {x, -3/2, 3/2, 3/149}, {y, -3/2, 3/2, 3/149}];

g2 = Flatten[g1, 1];

Take[g2, 5]
```

$\{\{-\frac{3}{2}, -\frac{3}{2}, \text{Overflow}[]\}, \{-\frac{3}{2}, -\frac{441}{298}, \text{Overflow}[]\}, \{-\frac{3}{2}, -\frac{435}{298}, \text{Overflow}[]\},$
$\{-\frac{3}{2}, -\frac{429}{298}, \text{Overflow}[]\}, \{-\frac{3}{2}, -\frac{423}{298}, \text{Overflow}[]\}\}$

We remove those elements of `g2` for which the third coordinate is `Overflow[]` with `Select`,

```
g3 = Select[g2, Not[#[[3]]===Overflow[]]&];
```

extract a list of the first two coordinates, (x, y), from the elements of `g3`,

```
g4 = Map[{#[[1]], #[[2]]}&, g3];
```

and plot the resulting list of points in Fig. 4.34 using `ListPlot`.

```
lp1 = ListPlot[g4, PlotRange → {{-3/2, 3/2}, {-3/2, 3/2}},

  AxesLabel → {x, y}, AspectRatio → Automatic,

  PlotStyle → CMYKColor[0, .12, .98, 0]]
```

We can invert the image as well with the following commands. In the end result, we show the Julia set and its inverted image in Fig. 4.35

```
g3b = Select[g2, #[[3]]===Overflow[]&];

g4b = Map[{#[[1]], #[[2]]}&, g3b];
```

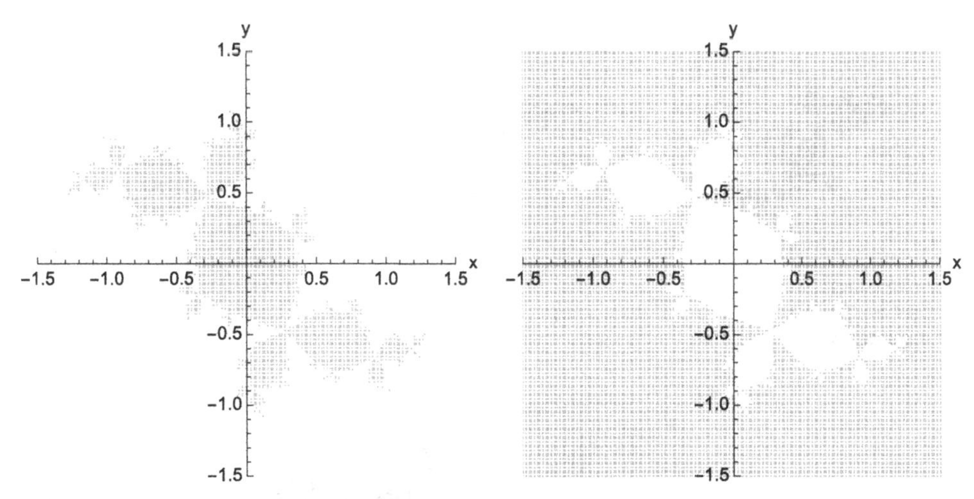

FIGURE 4.35 Filled Julia set for f_c on the left; the inverted set on the right.

```
lp2 = ListPlot[g4b, PlotRange → {{−3/2, 3/2}, {−3/2, 3/2}}, AxesLabel → {x, y},

AspectRatio → Automatic, PlotStyle → CMYKColor[0, .12, .98, 0]];

j1 = Show[GraphicsRow[{lp1, lp2}]]
```

For *rational functions* $f(z)$ you can use the command `JuliaSetPlot[f[z],z]` to plot the Julia set for $f(z)$. The default is $f(z) = z^2 + c$ so `JuliaSetPlot[c]` plots the Julia set for $f(z) = z^2 + c$. Use options like `PlotRange` to get the view of the plot that you expect.
Thus, in the following commands, all plot the Julia set for $f(z) = z^2 + c$ for $c = -0.12256117 + .74486177i$.

```
jsp1 = JuliaSetPlot[f[−0.12256117 + .74486177I][x], x, PlotRange →

  {{−3/2, 3/2}, {−3/2, 3/2}}]

jsp2 = JuliaSetPlot[f[−0.12256117 + .74486177I][x], x, PlotStyle →

  CMYKColor[0, .12, .98, 0],

PlotRange → {{−3/2, 3/2}, {−3/2, 3/2}}]

jsp3 = JuliaSetPlot[−0.12256117 + .74486177I, PlotStyle → CMYKColor[1, .69, .07, .3],

  PlotRange → {{−3/2, 3/2}, {−3/2, 3/2}}]
```

The plots are all shown together in Fig. 4.36.

```
Show[GraphicsRow[{jsp1, jsp2, jsp3}]]
```
□

Of course, one can consider functions other than $f_c(x) = x^2 + c$ as well as rearrange the order in which we carry out the computations.

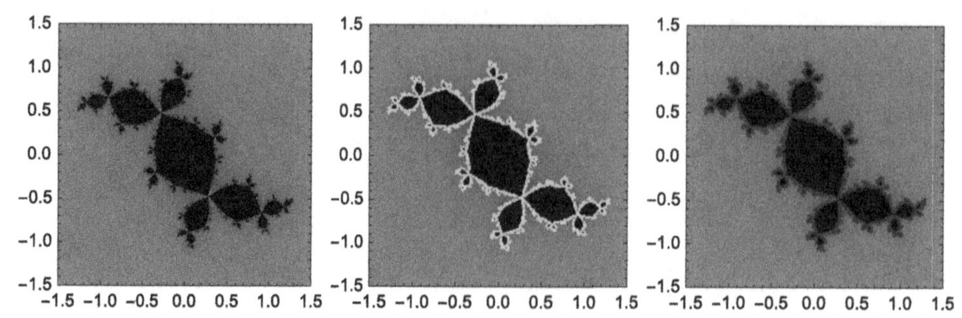

FIGURE 4.36 Various views of the Julia set for $f(z) = z^2 + c$ if $c = -0.12256117 + .74486177i$.

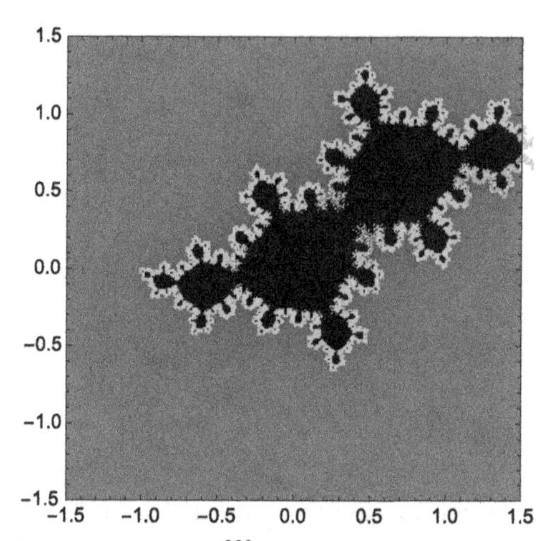

FIGURE 4.37 The Julia set for $f_{0.737369+0.67549i}{}^{200}(z)$.

Example 4.26: Julia Sets

Plot the Julia set for $f_c(z) = z^2 - cz$ if $c = 0.737369 + 0.67549i$.

As before, all error messages have been deleted.

Solution. Observe that $f_c(z) = z^2 - cz$ is a true rational function (it is a polynomial) so `JuliaSetPlot` will plot the Julia set.
After defining $f_c(z) = z^2 - cz$,

```
Clear[f, h]

f[c_][x_]:=x^2 − cx//N;
```

we use `JuliaSetPlot` as in the previous example. See Fig. 4.37.

```
JuliaSetPlot[f[0.737369 + 0.67549I][x], x, PlotRange → {{−3/2, 3/2}, {−3/2, 3/2}},

  PlotStyle → CMYKColor[0, .24, .94, 0]]
```

You can use commands like `Table` and `Manipulate` to investigate the Julia set for $J_c(x)$. For example,

```
t1 = Table[JuliaSetPlot[f[0.737369k + 0.67549I][x], x,
```

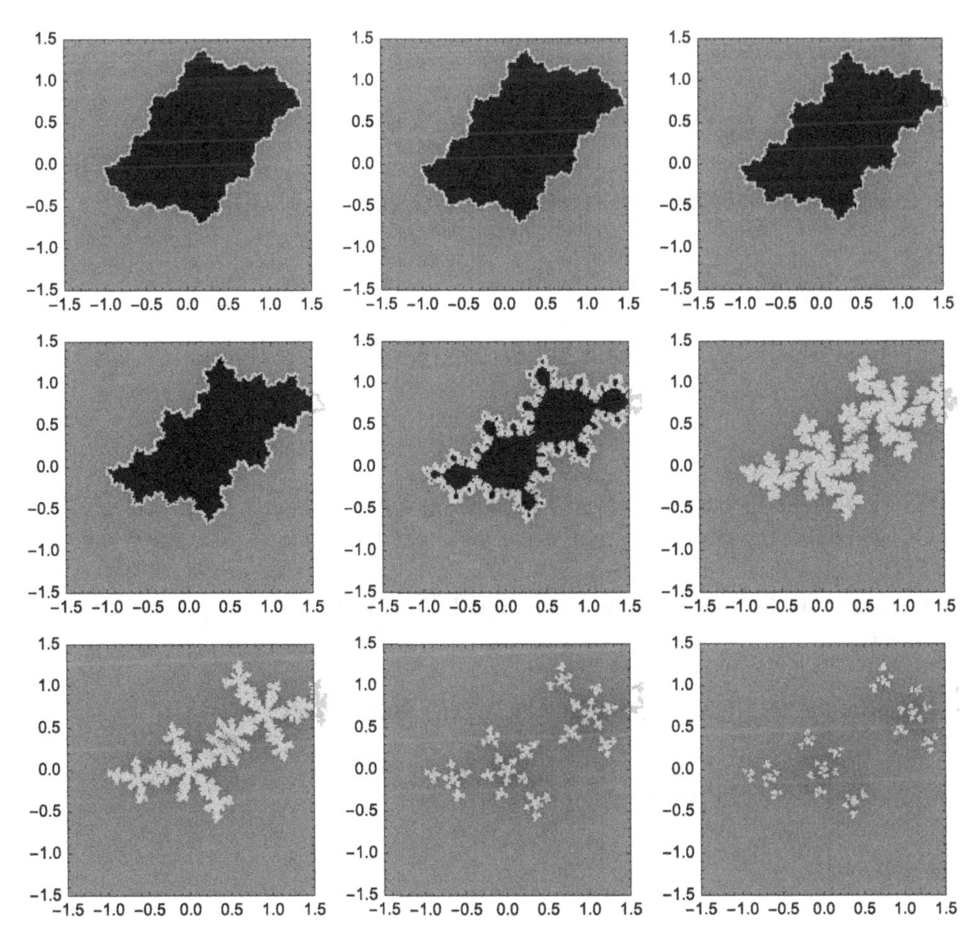

FIGURE 4.38 The Julia set for $f_{0.737369k+0.67549i}(x)$ for 9 equally spaced values of k between .5 and 1.5.

```
PlotRange → {{−3/2, 3/2}, {−3/2, 3/2}},

PlotStyle → CMYKColor[0, .24, .94, 0]], {k, .5, 1.5, 1/8}];
```

plots the Julia set for $f_{0.737369k+0.67549i}(x)$ for 9 equally spaced values of k between .5 and 1.5. The results are shown as an array with `GraphicsGrid` in Fig. 4.38. □

Example 4.27: The Ikeda Map

The **Ikeda map** is defined by

$$\mathbf{F}(x, y) = \langle \gamma + \beta (x \cos \tau - y \sin \tau), \beta (x \sin \tau + y \cos \tau) \rangle, \tag{4.11}$$

where $\tau = \mu - \alpha / (1 + x^2 + y^2)$. If $\beta = .9$, $\mu = .4$, and $\alpha = 4.0$, plot the *basins of attraction* for F if $\gamma = .92$ and $\gamma = 1.0$.

Solution. The *basins of attraction* for F are the set of points (x, y) for which $\|\mathbf{F}^n(x, y)\| \not\to \infty$ as $n \to \infty$.

After defining `f[γ][x, y]` to be equation (4.11) and then $\beta = .9$, $\mu = .4$, and $\alpha = 4.0$, we use `Table` followed by `Flatten` to define `pts` to be the list of 40,000 ordered pairs (x, y) for 200 equally spaced values of x between -2.3 and 1.3 and 200 equally spaced values of y between -2.8 and .8.

```
f[γ_][{x_, y_}]:=
  {γ + β(xCos[μ − α/(1 + x^2 + y^2)] − ySin[μ − α/(1 + x^2 + y^2)]),
  β(xSin[μ − α/(1 + x^2 + y^2)] + yCos[μ − α/(1 + x^2 + y^2)])}
β = .9; μ = .4; α = 4.0;
pts = Flatten[Table[{x, y}, {x, −2.3, 1.3, 3.6/199}, {y, −2.8, .8, 3.6/199}],
1];
```

In `l1`, we use `Map` to compute $(x, y, \mathbf{F}_{.92}{}^{200}(x, y))$ for each (x, y) in `pts`. In `pts2`, we use the graphics primitive `Point` and shade the points according to the maximum value of $\|\mathbf{F}^{200}(x, y)\|$ – those (x, y) for which $\mathbf{F}^{200}(x, y)$ is closest to the origin are darkest; the point (x, y) is shaded lighter as the distance of $\mathbf{F}^{200}(x, y)$ from the origin increases. (See Fig. 4.39 (a).)

```
l1 = Map[{#[[1]], #[[2]], Nest[f[.92], {#[[1]], #[[2]]}, 200]}&, pts];
g[{x_, y_, z_}]:={x, y, Sqrt[z[[1]]^2 + z[[2]]^2]};
l2 = Map[g, l1];
maxl2 = Table[l2[[i, 3]], {i, 1, Length[l2]}]//Max
```

4.33321

```
pts2 = Table[{GrayLevel[l2[[i, 3]]/(maxl2)],
  Point[{l2[[i, 1]], l2[[i, 2]]}]}, {i, 1, Length[l2]}];
ik1 = Show[Graphics[pts2], AspectRatio → 1]
```

For $\gamma = 1.0$, we proceed in the same way. The final results are shown in Fig. 4.39 (b).

```
l1 = Map[{#[[1]], #[[2]], Nest[f[1.0], {#[[1]], #[[2]]}, 200]}&, pts];
l2 = Map[g, l1];
maxl2 = Table[l2[[i, 3]], {i, 1, Length[l2]}]//Max
```

4.48421

```
pts2 = Table[{GrayLevel[l2[[i, 3]]/maxl2], Point[{l2[[i, 1]], l2[[i, 2]]}]},
```

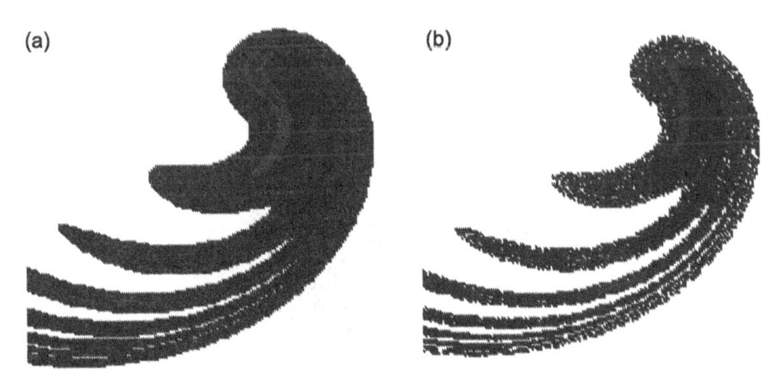

FIGURE 4.39 Basins of attraction for **F** if $\gamma = .92$ (a) and $\gamma = 1.0$ (b).

```
{i, 1, Length[l2]}];

ik2 = Show[Graphics[pts2], AspectRatio → 1]

Show[GraphicsRow[{ik1, ik2}]]
```

□

The **Mandelbrot set**, M, is the set of complex numbers, $z = a + bi$, a, b real, for which the sequence z, $f_z(z) = z^2 + z$, $f_z(f_z(z)) = \left(z^2+z\right)^2 + z, \ldots, f_z{}^n(z), \ldots$, does *not* tend to ∞ as $n \to \infty$:

$$M = \left\{z \in \mathbf{C}|z,\ z^2 + z\ \left(z^2 + z\right)^2 + z, \cdots \nrightarrow \infty\right\}.$$

Using a dynamical system, setting $z = z_0$ and computing $z_{n+1} = f_{z_0}(z_n)$ for large n can help us determine if z is an element of M. In terms of a composition, computing $f_z{}^n(z)$ for large n can help us determine if z is an element of M.
The command

```
MandelbrotSetPlot[{a+bi,c+di}]
```

plots the Mandelbrot set on the rectangle with lower left corner $a + bi$ and upper right corner $c + di$.

Example 4.28: Mandelbrot Set

Plot the Mandelbrot set.

Solution. We use `MandelbrotSetPlot`. The following gives us the image on the left in Fig. 4.40.

```
MandelbrotSetPlot[{−3/2 − I, 1 + I}]
```

□

The Mandelbrot set can be obtained (or more precisely, approximated) by repeatedly composing $f_z(z)$ for a grid of z-values and then deleting those for which the values exceed machine precision or other specified "large" value. Those values greater than `$MaxNumber` result in an `Overflow[]` message; computations with `Overflow[]` result in an `Indeterminate` message.

We can generalize by considering exponents other than 2 by letting $f_{p,c} = x^p + c$. The **generalized Mandelbrot set**, M_p, is the set of complex numbers, $z = a + bi$, a, b real, for which the sequence z, $f_{p,z}(z) = z^p + z$, $f_{p,z}\left(f_{p,z}(z)\right) = (z^p + z)^p + z, \ldots, f_{p,z}{}^n(z), \ldots$, does *not* tend to ∞ as $n \to \infty$:

$$M_p = \left\{z \in \mathbf{C}|z,\ z^p + z\ \left(z^p + z\right)^p + z, \cdots \nrightarrow \infty\right\}.$$

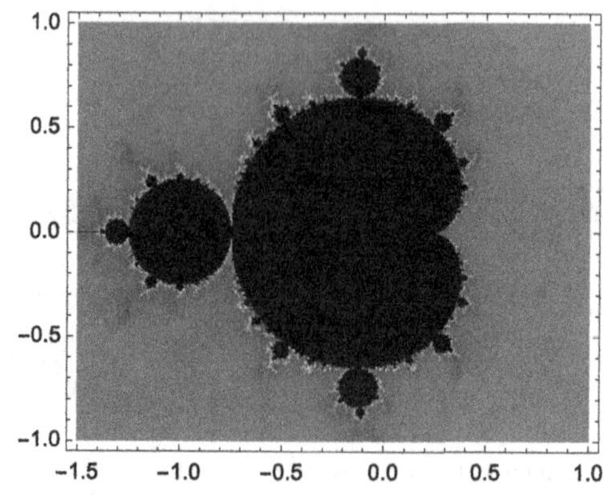

FIGURE 4.40 The Mandelbrot set.

Using a dynamical system, setting $z = z_0$ and computing $z_{n+1} = f_p(z_n)$ for large n can help us determine if z is an element of M_p. In terms of a composition, computing $f_p{}^n(z)$ for large n can help us determine if z is an element of M_p.

As with the previous examples, all error messages have been omitted.

Example 4.29: Generalized Mandelbrot Set

After defining $f_{p,c} = x^p + c$, we use `Table`, `Abs`, and `Nest` to compute a list of ordered triples of the form $(x, y, |f_{p,x+iy}{}^{100}(x+iy)|)$ for p-values from 1.625 to 2.625 spaced by equal values of 1/8 and 200 values of x (y) values equally spaced between −2 and 2, resulting in 40,000 sample points of the form $x + iy$.

```
Clear[f, p]

f[p_, c_][x_]:=x^p + c//N;

g1 =

Map[Table[{x, y, Abs[Nest[f[2, x + Iy], x + Iy, #]]}//N, {x, −1.5, 1., 5/(2 * 199)},

  {y, −1., 1., 2/199}]&, {5, 10, 15, 25, 50, 100}];

g2 = Map[Flatten[#, 1]&, g1];
```

Next, we extract those points for which the third coordinate is `Indeterminate` with `Select`, ordered pairs of the first two coordinates are obtained in `g4`. The resulting list of points is plotted with `ListPlot` in Fig. 4.41.

```
g3 = Table[Select[g2[[i]], Not[#[[3]]===Overflow[]]&], {i, 1, Length[g2]}];

h[{x_, y_, z_}]:={x, y};

g4 = Map[h, g3, {2}];
```

```
t1 = Table[ListPlot[g4[[i]], PlotRange -> {{-3/2, 1}, {-1, 1}},
  AspectRatio -> Automatic, DisplayFunction -> Identity], {i, 1, 6}];
  Show[GraphicsGrid[Partition[t1, 3]]]
```

More detail is observed if you use the graphics primitive `Point` as shown in Fig. 4.42. In this case, those points (x, y) for which $\left|f_{p,x+iy}{}^{100}(x+iy)\right|$ is small are shaded according to a darker `GrayLevel` than those points for which $\left|f_{p,x+iy}{}^{100}(x+iy)\right|$ is large.

```
h2[{x_, y_, z_}]:={GrayLevel[Min[{z, 1}]], Point[{x, y}]};

g5 = Map[h2, g3, {2}];

t1 = Table[Show[Graphics[g5[[i]]], PlotRange -> {{-3/2, 1}, {-1, 1}},
AspectRatio -> Automatic, DisplayFunction -> Identity], {i, 1, 6}];

Show[GraphicsGrid[Partition[t1, 3]]]
```

Throughout these examples, we have typically computed the iteration $f^n(z)$ for "large" n like values of n between 100 and 200. To indicate why we have selected those values of n, we revisit the Mandelbrot set plotted in Example 4.28.

Example 4.30: Mandelbrot Set

We proceed in essentially the same way as in the previous examples. After defining $f_{p,c}=x^p+c$,

As before, all error messages have been deleted.

```
Clear[f, p]

f[p_, c_][x_]:=x^p + c//N;
```

we use `Table` followed by `Map` to create a nested list. For each $n = 5$, 10, 15, 25, 50, and 100, a nested list is formed for 200 equally spaced values of y between -1 and 1 and then 200 equally spaced values of x between -1.5 and 1. between -1.5 and 1. At the bottom level of each nested list, the elements are of the form $\left(x, y, \left|f_{2,x+iy}{}^n(x+iy)\right|\right)$.

```
g1 = Table[{x, y, Abs[Nest[f[p, x + I y], x + I y, 100]]}//N,

{p, 1.625, 2.625, 1/8}, {x, -2., 2., 4/149}, {y, -2., 2., 4/149}];

pvals = Table[p, {p, 1.625, 2.625, 1/5}];

g1 = Map[Table[{x, y, Abs[Nest[f[#, x + I y], x + I y, 100]]}//N,
```

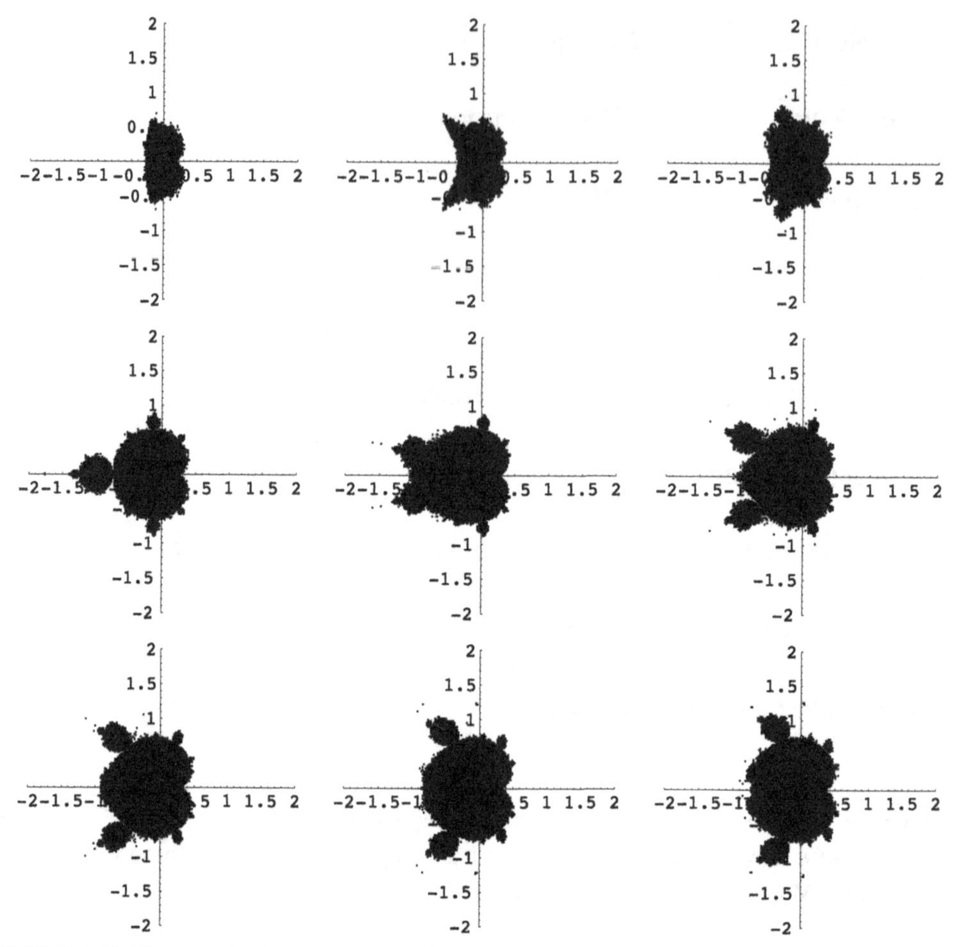

FIGURE 4.41 The generalized Mandelbrot set for 9 equally spaced values of p between 1.625 and 2.625.

```
{x, −1.5, 1., 5/(2 * 199)}, {y, −1., 1., 2/199}]&, pvals];
```

For each value of n, the corresponding list of ordered triples $(x, y, |f_{2,x+iy}{}^n(x+iy)|)$ is obtained using `Flatten`.

```
g2 = Map[Flatten[#, 1]&, g1];
```

We then remove those points for which the third coordinate, $|f_{2,x+iy}{}^n(x+iy)|$, is `Overflow[]` (corresponding to ∞),

```
g3 = Table[Select[g2[[i]], Not[#[[3]]===Indeterminate]&],
  {i, 1, Length[g2]}];
```

extract (x, y) from the remaining ordered triples,

```
h[{x_, y_, z_}]:={x, y};
```

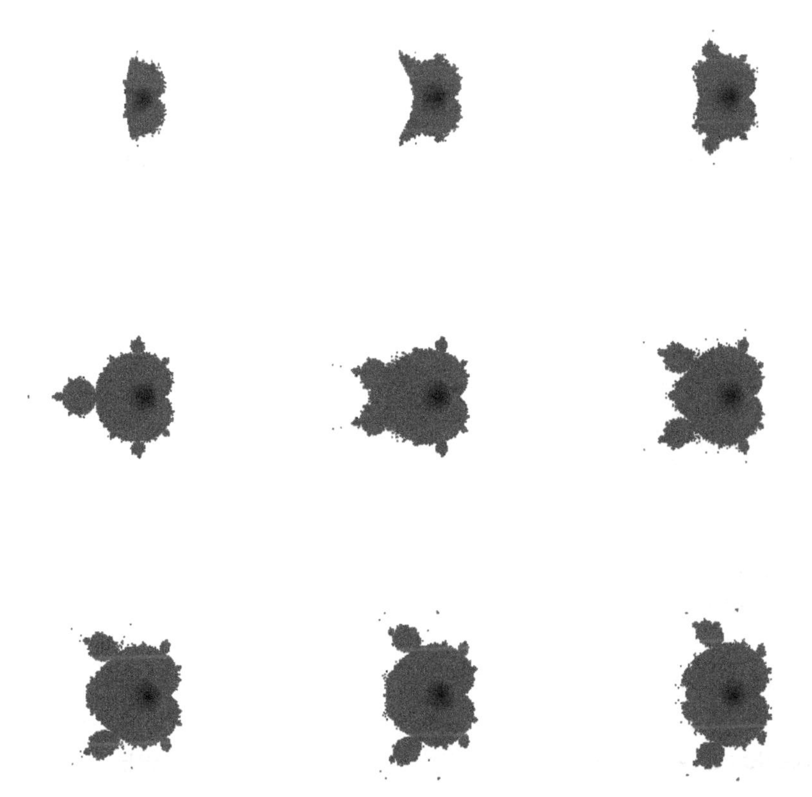

FIGURE 4.42 The generalized Mandelbrot set for 9 equally spaced values of p between 1.625 and 2.625 – the points (x, y) for which $\left|f_{p,x+iy}{}^{100}(x+iy)\right|$ is large are shaded lighter than those for which $\left|f_{p,x+iy}{}^{100}(x+iy)\right|$ is small.

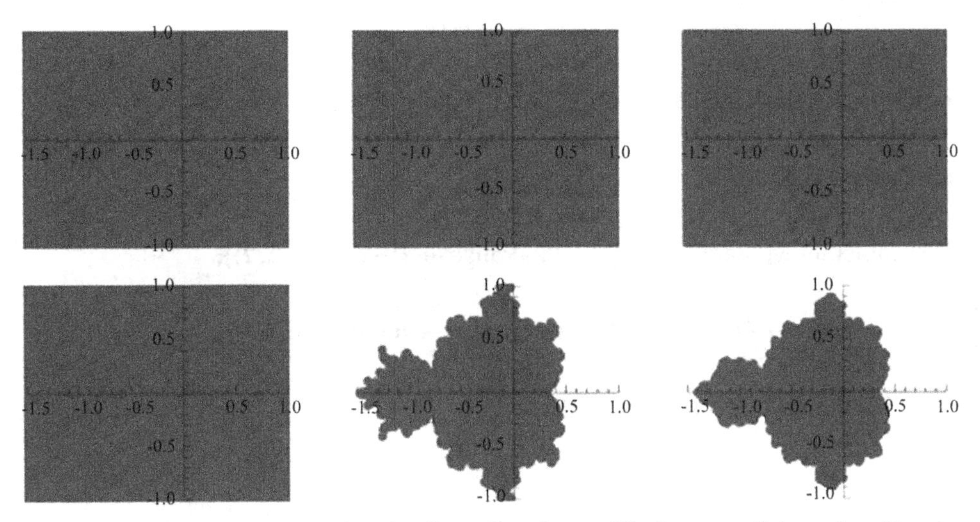

FIGURE 4.43 Without shading the points, the effects of iteration are difficult to see until the number of iterations is "large".

```
g4 = Map[h, g3, {2}];
```

and graph the resulting sets of points using `ListPlot` in Fig. 4.43. As shown in Fig. 4.43, we see that Mathematica's numerical precision (and consequently decent plots) are obtained when $n = 50$ or $n = 100$.

Fundamentally, we generated the previous plots by exceeding Mathematica's numerical precision.

FIGURE 4.44 Using graphics primitives and shading, we see that we can use a relatively small number of iterations to visualize the Mandelbrot set.

```
t1 = Table[ListPlot[g4[[i]], PlotRange → {{−2, 2}, {−2, 2}},
  AspectRatio → Automatic, DisplayFunction → Identity], {i, 1, 9}];
Show[GraphicsGrid[Partition[t1, 3]]]
```

If instead, we use graphics primitives like `Point` and then shade each point (x, y) according to $|f_{2,x+iy}{}^n(x+iy)|$ detail emerges quickly as shown in Fig. 4.44.

```
h2[{x_, y_, z_}]:={GrayLevel[Min[{z, .25}]], Point[{x, y}]};
g5 = Map[h2, g3, {2}];
t1 = Table[Show[Graphics[g5[[i]]], PlotRange → {{−2, 2}, {−2, 2}},
  AspectRatio → Automatic, DisplayFunction → Identity], {i, 1, 9}];
Show[GraphicsGrid[Partition[t1, 3]]]
```

The examples like the ones illustrated here indicate that similar results could have been accomplished using far smaller values of n than $n = 100$ or $n = 200$. With fast machines, the differences in the time needed to perform the calculations is minimal; $n = 100$ and $n = 200$ appear to be a "safe" large value of n for well-studied examples like these.

Chapter 5

Matrices and Vectors: Topics from Linear Algebra and Vector Calculus

5.1 NESTED LISTS: INTRODUCTION TO MATRICES, VECTORS, AND MATRIX OPERATIONS

5.1.1 Defining Nested Lists, Matrices, and Vectors

In Mathematica, a **matrix** is a list of lists where each list represents a row of the matrix. Therefore, the $m \times n$ matrix

$$\mathbf{A} = \begin{pmatrix} a_{11} & a_{12} & a_{13} & \cdots & a_{1n} \\ a_{21} & a_{22} & a_{23} & \cdots & a_{2n} \\ a_{31} & a_{32} & a_{33} & \cdots & a_{3n} \\ \vdots & \vdots & \vdots & & \vdots \\ a_{m1} & a_{m2} & a_{m3} & \cdots & a_{mn} \end{pmatrix}$$

is entered with

```
A={{a11,a12,...,a1n},{a21,a22,...,a2n},...,{am1,am2,...amn}}.
```

For example, to use Mathematica to define `m` to be the matrix $\mathbf{A} = \begin{pmatrix} a_{11} & a_{12} \\ a_{21} & a_{22} \end{pmatrix}$ enter the command

```
m={{a11,a12},{a21,a22}}.
```

The command `m=Array[a,{2,2}]` produces a result equivalent to this. Once a matrix `A` has been entered, it can be viewed in the traditional row-and-column form using the command `MatrixForm[A]`. You can quickly construct 2×2 matrices by clicking on the button from the **BasicMathInput** palette, which is accessed by going to **Palettes** followed by **BasicMathInput**.

Alternatively, you can construct matrices of any dimension by going to the Mathematica menu under **Insert** and selecting **Create Table/Matrix/Palette...**

Use `Part`, (`[[...]]`) to select elements of lists. Because of the construct of the matrix, `m[[i]]` returns the ith row of **M**. The **transpose** of **M**, $\mathbf{M}^t$, is the matrix obtained by interchanging the rows and columns of matrix **M**. Thus, to extract the ith column of **M**, use the commands `mt=Transpose[m]` followed by `mt[[i]]`.

As when using `TableForm`, the result of using `MatrixForm` is no longer a list that can be manipulated using Mathematica commands. Use `MatrixForm` to view a matrix in traditional row-and-column form. Do not attempt to perform matrix operations on a `MatrixForm` object.

Mathematica by Example. http://dx.doi.org/10.1016/B978-0-12-812481-9.00005-3

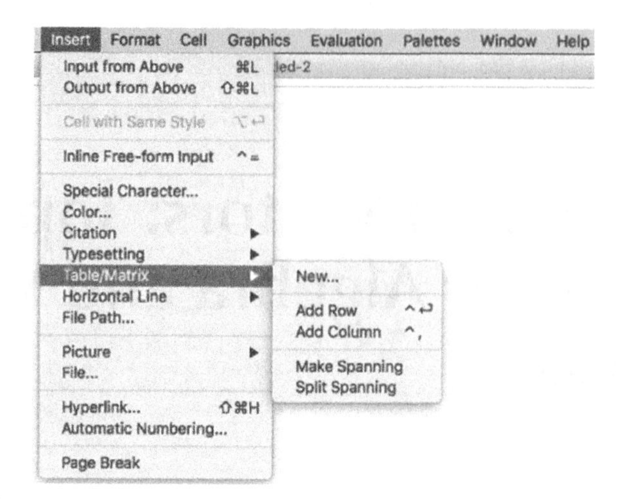

The resulting pop-up window allows you to create tables, matrices, and palettes. To create a matrix, select **Matrix**, enter the number of rows and columns of the matrix, and select any other options. Pressing the **OK** button places the desired matrix at the position of the cursor in the Mathematica notebook.

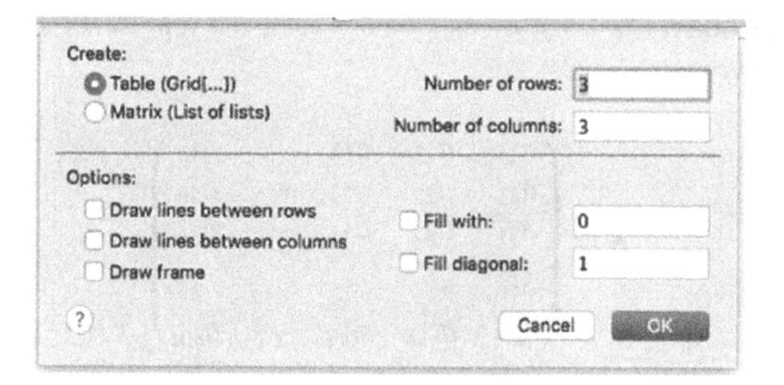

Example 5.1

Use Mathematica to define the matrices $\begin{pmatrix} a_{11} & a_{12} & a_{13} \\ a_{21} & a_{22} & a_{23} \\ a_{31} & a_{32} & a_{33} \end{pmatrix}$ and $\begin{pmatrix} b_{11} & b_{12} & b_{13} & b_{14} \\ b_{21} & b_{22} & b_{23} & b_{24} \end{pmatrix}$.

Solution. In this case, both `Table[a_{i,j}, {i, 1, 3}, {j, 1, 3}]` and `Array[a, {3,3}]` produce equivalent results when we define `matrixa` to be the matrix

$$\begin{pmatrix} a_{11} & a_{12} & a_{13} \\ a_{22} & a_{22} & a_{23} \\ a_{31} & a_{32} & a_{33} \end{pmatrix}.$$

The commands `MatrixForm` or `TableForm` are used to display the results in traditional matrix form.

Clear[*a*, *b*, matrixa, matrixb]

matrixa = Table$[a_{i,j}, \{i, 1, 3\}, \{j, 1, 3\}]$

$\{\{a_{1,1}, a_{1,2}, a_{1,3}\}, \{a_{2,1}, a_{2,2}, a_{2,3}\}, \{a_{3,1}, a_{3,2}, a_{3,3}\}\}$

MatrixForm[matrixa]

$$\begin{pmatrix} a_{1,1} & a_{1,2} & a_{1,3} \\ a_{2,1} & a_{2,2} & a_{2,3} \\ a_{3,1} & a_{3,2} & a_{3,3} \end{pmatrix}$$

matrixa = Array[*a*, {3, 3}]

{{*a*[1, 1], *a*[1, 2], *a*[1, 3]}, {*a*[2, 1], *a*[2, 2], *a*[2, 3]}, {*a*[3, 1], *a*[3, 2], *a*[3, 3]}}

{{*a*[1, 1], *a*[1, 2], *a*[1, 3]},

{*a*[2, 1], *a*[2, 2], *a*[2, 3]}, {*a*[3, 1], *a*[3, 2], *a*[3, 3]}}

MatrixForm[matrixa]

$$\begin{pmatrix} a[1,1] & a[1,2] & a[1,3] \\ a[2,1] & a[2,2] & a[2,3] \\ a[3,1] & a[3,2] & a[3,3] \end{pmatrix}$$

We may also use Mathematica to define non-square matrices.

matrixb = Array[*b*, {2, 4}]

{{*b*[1, 1], *b*[1, 2], *b*[1, 3], *b*[1, 4]}, {*b*[2, 1], *b*[2, 2], *b*[2, 3], *b*[2, 4]}}

MatrixForm[matrixb]

$$\begin{pmatrix} b[1,1] & b[1,2] & b[1,3] & b[1,4] \\ b[2,1] & b[2,2] & b[2,3] & b[2,4] \end{pmatrix}$$

Equivalent results would have been obtained by entering `Table[`$b_{i,j}$`, {i, 1, 2}, {j, 1, 4}]`. □

More generally the commands `Table[f[i,j],{i,imax},{j,jmax}]` and `Array[f,{imax,jmax}]` yield nested lists corresponding to the `imax` × `jmax` matrix

$$\begin{pmatrix} f(1,1) & f(1,2) & \cdots & f(1,\mathtt{jmax}) \\ f(2,1) & f(2,2) & \cdots & f(2,\mathtt{jmax}) \\ \vdots & \vdots & \vdots & \vdots \\ f(\mathtt{imax},1) & f(\mathtt{imax},2) & \cdots & f(\mathtt{imax},\mathtt{jmax}) \end{pmatrix}.$$

`Table[f[i,j],{i,imin,imax,istep},{j,jmin,jmax,jstep}]` returns the list of lists

```
{{f[imin,jmin],f[imin,jmin+jstep],...,f[imin,jmax]},
    {f[imin+istep,jmin],...,f[imin+istep,jmax]},
        ...,{f[imax,jmin],...,f[imax,jmax]}}
```

and the command

```
Table[f[i,j,k,...],{i,imin,imax,istep},{j,jmin,jmax,jstep},
    {k,kmin,kmax,kstep},...]
```

calculates a nested list; the list associated with i is outermost. If `istep` is omitted, the stepsize is one.

Example 5.2

Define **C** to be the 3×4 matrix (c_{ij}), where c_{ij}, the entry in the ith row and jth column of **C**, is the numerical value of $\cos(j^2 - i^2)\sin(i^2 - j^2)$.

Solution. After clearing all prior definitions of `c`, if any, we define `c[i,j]` to be the numerical value of $\cos(j^2 - i^2)\sin(i^2 - j^2)$ and then use `Array` to compute the 3×4 matrix `matrixc`.

Clear[*c*, matrixc]

c[i_, j_] = N[Cos[$j^2 - i^2$] Sin[$i^2 - j^2$]]

$\text{Cos}[i^2 - 1.j^2]\,\text{Sin}[i^2 - 1.j^2]$

matrixc = Array[*c*, {3, 4}]

{{0., 0.139708, 0.143952, 0.494016},

{−0.139708, 0., 0.272011, 0.452789},

{−0.143952, −0.272011, 0., −0.495304}}

MatrixForm[matrixc]

$$\begin{pmatrix} 0. & 0.139708 & 0.143952 & 0.494016 \\ -0.139708 & 0. & 0.272011 & 0.452789 \\ -0.143952 & -0.272011 & 0. & -0.495304 \end{pmatrix}$$

Mathematica provides several functions to help visualize a matrix. These commands include `MatrixPlot`, `ArrayPlot`, `ReliefPlot`, and `ListDensityPlot`. These commands have many of the same options as `Plot`.

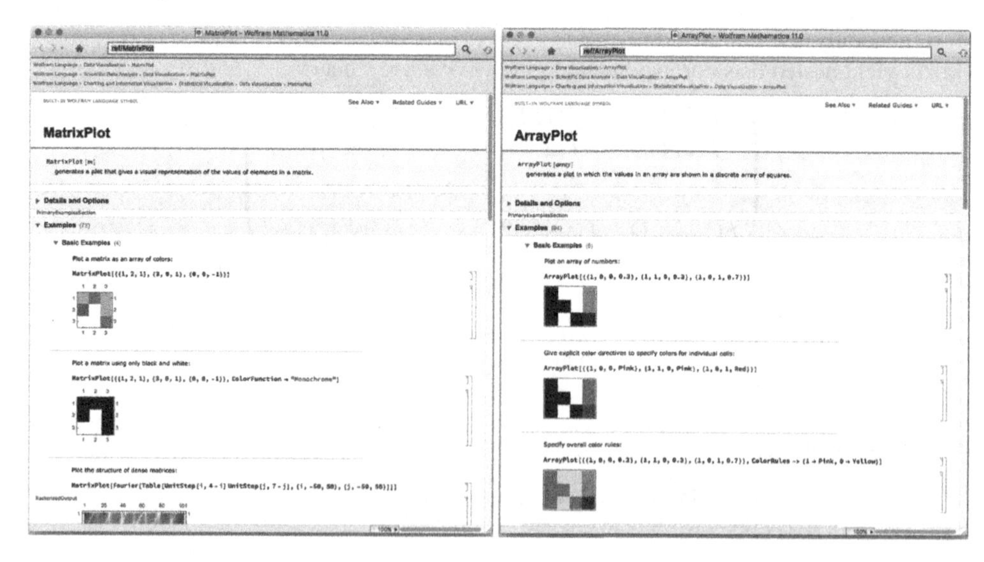

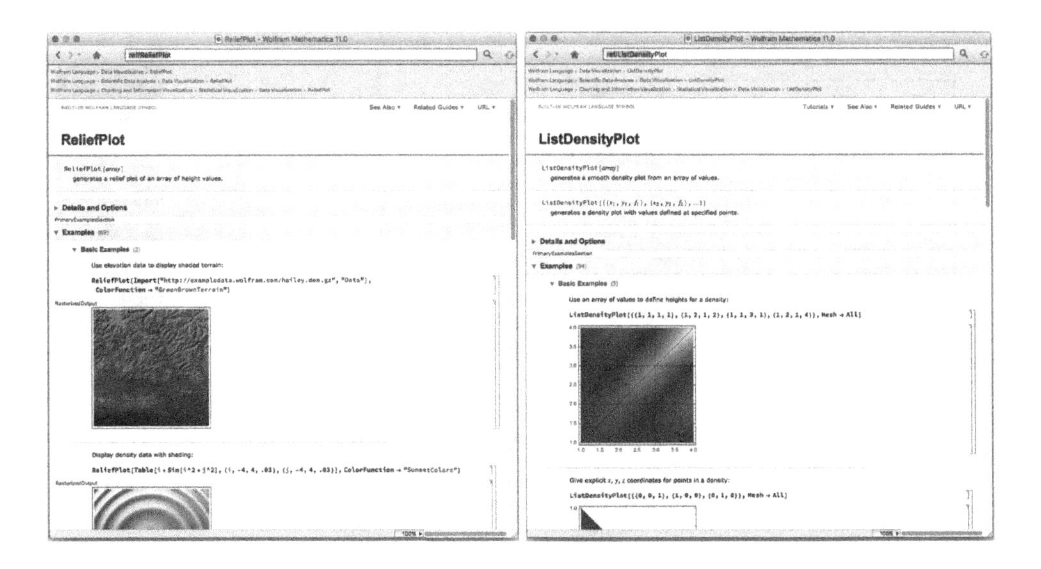

Each produces a slightly different result. To achieve the best result, we suggest that you try each and then adjust the options to finalize your result. In Fig. 5.1, we show the results of each command for `matrixc`.

```
m1 = MatrixPlot[matrixc,
PlotLabel → "(a)"]

a1 = ArrayPlot[matrixc, PlotLabel → "(b)"]

r1 = ReliefPlot[matrixc,
  PlotLabel → "(c)"]

lpd1 = ListDensityPlot[matrixc,
  PlotLabel → "(d)"]

Show[GraphicsGrid[{{m1, a1}, {r1, lpd1}}]]
```
□

Example 5.3

Define the matrix $\mathbf{I}_3 = \begin{pmatrix} 1 & 0 & 0 \\ 0 & 1 & 0 \\ 0 & 0 & 1 \end{pmatrix}$.

Solution. The matrix $\mathbf{I}_3$ is the 3×3 **identity matrix**. Generally, the $n \times n$ matrix with 1's on the diagonal and 0's elsewhere is the $n \times n$ identity matrix. The command `IdentityMatrix[n]` returns the $n \times n$ identity matrix.

```
IdentityMatrix[3]
```

{{1, 0, 0}, {0, 1, 0}, {0, 0, 1}}

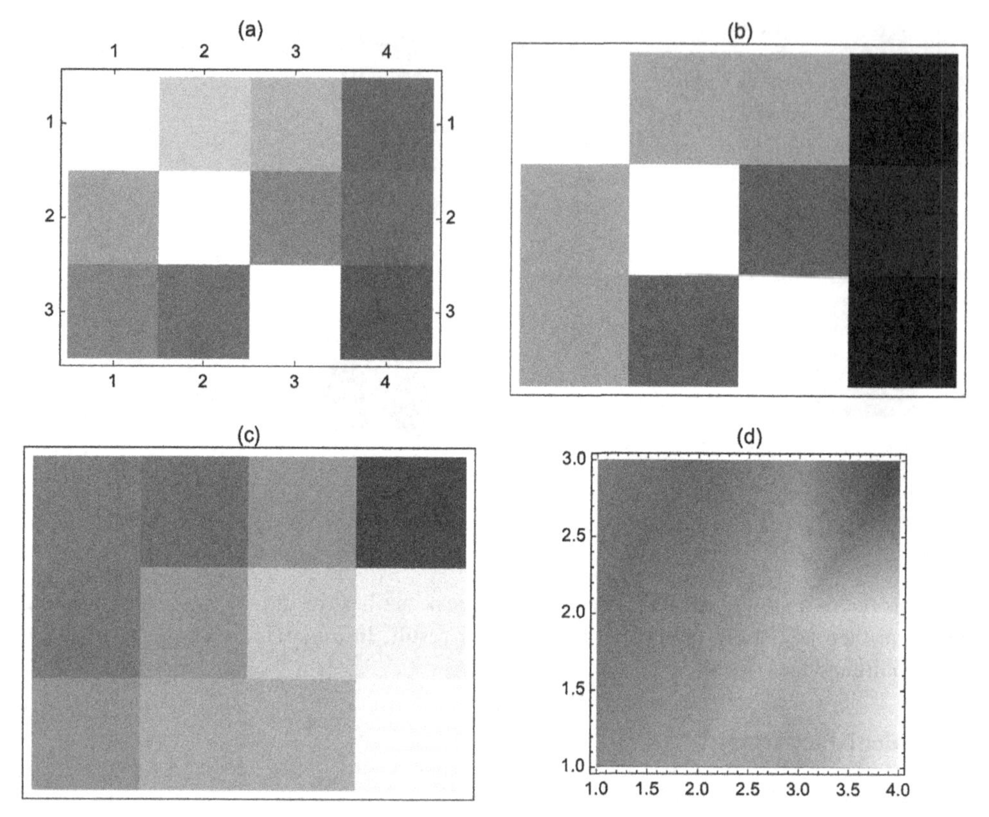

FIGURE 5.1 Illustrating different ways to visualize the entries of a matrix or array.

The same result is obtained by going to **Insert** under the Mathematica menu and selecting **Table/Matrix/** followed by **New....** We then check **Matrix**, **Fill with**: 0 and **Fill diagonal**: 1.

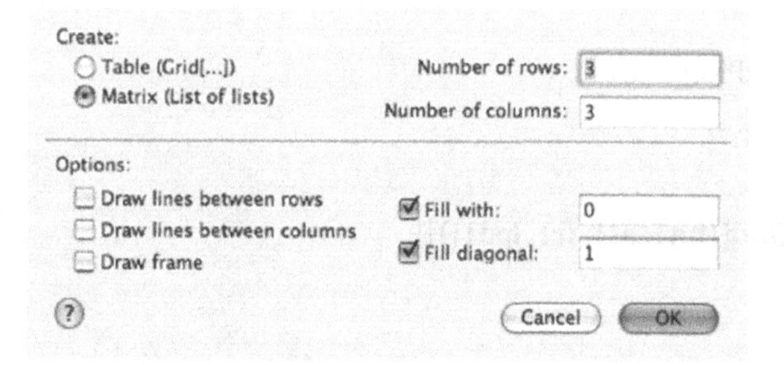

Pressing the **OK** button inserts the 3 × 3 identity matrix at the location of the cursor.

$$\begin{pmatrix} 1 & 0 & 0 \\ 0 & 1 & 0 \\ 0 & 0 & 1 \end{pmatrix}$$

{{1, 0, 0}, {0, 1, 0}, {0, 0, 1}} □

In Mathematica, a **vector** is a list of numbers and, thus, is entered in the same manner as lists. For example, to use Mathematica to define the row vector `vectorv` to be $\begin{pmatrix} v_1 & v_2 & v_3 \end{pmatrix}$

enter `vectorv={v1,v2,v3}`. Similarly, to define the column vector `vectorv` to be $\begin{pmatrix} v_1 \\ v_2 \\ v_3 \end{pmatrix}$ enter `vectorv={v1,v2,v3}` or `vectorv={{v1},{v2},{v3}}`.
Generally, with Mathematica you do not need to distinguish between row and column vectors: Mathematica usually performs computations with vectors and matrices correctly as long as the computations are well-defined.

With Mathematica, you do not need to distinguish between row and column vectors. Provided that computations are well-defined, Mathematica carries them out correctly. Mathematica warns of any ambiguities when they (rarely) occur.

Example 5.4

Define the vector $\mathbf{w} = \begin{pmatrix} -4 \\ -5 \\ 2 \end{pmatrix}$, `vectorv` to be the vector $\begin{pmatrix} v_1 & v_2 & v_3 & v_4 \end{pmatrix}$ and `zerovec` to be the vector $\begin{pmatrix} 0 & 0 & 0 & 0 & 0 \end{pmatrix}$.

Solution. To define **w**, we enter

w = {−4, −5, 2}

{−4, −5, 2}

or

w = {{−4}, {−5}, {2}};

MatrixForm[w]

$$\begin{pmatrix} -4 \\ -5 \\ 2 \end{pmatrix}$$

To define `vectorv`, we use `Array`.

vectorv = Array[v, 4]

{v[1], v[2], v[3], v[4]}

Equivalent results would have been obtained by entering `Table[v_i,{i,1,4}]`. To define `zerovec`, we use `Table`.

zerovec = Table[0, {5}]

{0, 0, 0, 0, 0}

The same result is obtained by going to **Insert** under the Mathematica menu and selecting **Table/Matrix** to create the zero vector.

$$\begin{pmatrix} 0 & 0 & 0 & 0 & 0 \end{pmatrix}$$

{{0, 0, 0, 0, 0}} □

5.1.2 Extracting Elements of Matrices

For the 2×2 matrix `m` $= \{\{a_{1,1}, a_{1,2}\}, \{a_{2,1}, a_{2,2}\}\}$ defined earlier, `m[[1]]` yields the first element of matrix `m` which is the list $\{a_{1,1}, a_{1,2}\}$ or the first row of `m`; `m[[2,1]]` yields the first element of the second element of matrix `m` which is $a_{2,1}$. In general, if **m** is an $i \times j$ matrix, `m[[i,j]]` or `Part[m,i,j]` returns the unique element in the ith row and jth column of **m**. More specifically, `m[[i,j]]` yields the jth part of the ith part of **m**; `list[[i]]` or `Part[list,i]` yields the ith part of `list`; `list[[i,j]]` or `Part[list,i,j]` yields the jth part of the ith part of list, and so on.

Example 5.5

Define `mb` to be the matrix $\begin{pmatrix} 10 & -6 & -9 \\ 6 & -5 & -7 \\ -10 & 9 & 12 \end{pmatrix}$. (a) Extract the third row of `mb`. (b) Extract the element in the first row and third column of `mb`. (c) Display `mb` in traditional matrix form.

Solution. We begin by defining `mb`. `mb[[i,j]]` yields the (unique) number in the ith row and jth column of `mb`. Observe how various components of `mb` (rows and elements) can be extracted and how `mb` is placed in `MatrixForm`.

```
mb = {{10, -6, -9}, {6, -5, -7}, {-10, 9, 12}};
```

```
MatrixForm[mb]
```

$$\begin{pmatrix} 10 & -6 & -9 \\ 6 & -5 & -7 \\ -10 & 9 & 12 \end{pmatrix}$$

```
mb[[3]]
```

{−10, 9, 12}

```
mb[[1, 3]]
```

−9

□

If `m` is a matrix, the ith row of `m` is extracted with `m[[i]]`. The command `Transpose[m]` yields the transpose of the matrix `m`, the matrix obtained by interchanging the rows and columns of `m`. We extract columns of `m` by computing `Transpose[m]` and then using `Part` to extract rows from the transpose. Namely, if `m` is a matrix, `Transpose[m][[i]]` extracts the ith row from the transpose of `m` which is the same as the ith column of `m`.

Alternatively, if **A** is $n \times m$ (rows $\times$ columns) the ith column of **A** is the vector that consists of the ith part of each row of the matrix so given an i-value `Table[A[[j,i]],{j,1,n}]` returns the ith column of **A**.

Example 5.6

Extract the second and third columns from $\mathbf{A} = \begin{pmatrix} 0 & -2 & 2 \\ -1 & 1 & -3 \\ 2 & -4 & 1 \end{pmatrix}$.

Solution. We first define `matrixa` and then use `Transpose` to compute the transpose of `matrixa`, naming the result `ta`, and then displaying `ta` in `MatrixForm`.

matrixa = {{0, −2, 2}, {−1, 1, −3}, {2, −4, 1}};

MatrixForm[matrixa]

$$\begin{pmatrix} 0 & -2 & 2 \\ -1 & 1 & -3 \\ 2 & -4 & 1 \end{pmatrix}$$

ta = Transpose[matrixa]

MatrixForm[ta]

{{0, −1, 2}, {−2, 1, −4}, {2, −3, 1}}

$$\begin{pmatrix} 0 & -1 & 2 \\ -2 & 1 & -4 \\ 2 & -3 & 1 \end{pmatrix}$$

Next, we extract the second column of `matrixa` using `Transpose` together with `Part` (`[[...]]`). Because we have already defined `ta` to be the transpose of `matrixa`, entering `ta[[2]]` would produce the same result.

Transpose[matrixa][[2]]

{−2, 1, −4}

To extract the third column, we take advantage of the fact that we have already defined `ta` to be the transpose of `matrixa`. Entering `Transpose[matrixa][[3]]` produces the same result.

ta[[3]]

{2, −3, 1}

You can also use `Take` to extract elements of lists and matrices. Entering

Take[matrixa, 2]

Take[matrixa, 2]//MatrixForm

{{0, −2, 2}, {−1, 1, −3}}

$$\begin{pmatrix} 0 & -2 & 2 \\ -1 & 1 & -3 \end{pmatrix}$$

returns the first two rows of `matrixa` because the first two parts of `matrixa` are the lists corresponding to those rows. Similarly,

Take[matrixa, {2}]

Take[matrixa, {2}]//MatrixForm

{{−1, 1, −3}}

$$\begin{pmatrix} -1 & 1 & -3 \end{pmatrix}$$

returns the second row while

Take[matrixa, {2, 3}]

Take[matrixa, {2, 3}]//MatrixForm

{{−1, 1, −3}, {2, −4, 1}}

$$\begin{pmatrix} -1 & 1 & -3 \\ 2 & -4 & 1 \end{pmatrix}$$

returns the second and third rows. □

The example illustrates that `Take[list,n]` returns the first n elements of `list`; `Take[list,{n}]` returns the nth element of `list`; `Take[list,{n1,n2,...}]` returns the n_1st, n_2nd, ... elements of `list`, and so on.

5.1.3 Basic Computations with Matrices

Mathematica performs all of the usual operations on matrices. Matrix addition ($\mathbf{A}+\mathbf{B}$), scalar multiplication ($k\mathbf{A}$), matrix multiplication (when defined) ($\mathbf{AB}$), and combinations of these operations are all possible. The **transpose** of $\mathbf{A}$, $\mathbf{A}^t$, is obtained by interchanging the rows and columns of $\mathbf{A}$ and is computed with the command `Transpose[A]`. If $\mathbf{A}$ is a square matrix, the determinant of $\mathbf{A}$ is obtained with `Det[A]`.

If $\mathbf{A}$ and $\mathbf{B}$ are $n \times n$ matrices satisfying $\mathbf{AB} = \mathbf{BA} = \mathbf{I}$, where $\mathbf{I}$ is the $n \times n$ matrix with 1's on the diagonal and 0's elsewhere (the $n \times n$ identity matrix), $\mathbf{B}$ is called the **inverse** of $\mathbf{A}$ and is denoted by $\mathbf{A}^{-1}$. If the inverse of a matrix $\mathbf{A}$ exists, the inverse is found with `Inverse[A]`. Thus, assuming that $\begin{pmatrix} a & b \\ c & d \end{pmatrix}$ has an inverse ($ad - bc \neq 0$), the inverse is

$$\mathbf{A}^{-1}\frac{1}{ad-bc}\begin{pmatrix} d & -b \\ -c & a \end{pmatrix}.$$

This easy-to-remember formula for finding the inverse of a 2 × 2 matrix is sometimes called **"the handy two-by-two inverse trick"** by instructors and students.

Inverse[{{*a*, *b*}, {*c*, *d*}}]

$$\left\{\left\{\frac{d}{-bc+ad}, -\frac{b}{-bc+ad}\right\}, \left\{-\frac{c}{-bc+ad}, \frac{a}{-bc+ad}\right\}\right\}$$

Example 5.7

Let $\mathbf{A} = \begin{pmatrix} 3 & -4 & 5 \\ 8 & 0 & -3 \\ 5 & 2 & 1 \end{pmatrix}$ and $\mathbf{B} = \begin{pmatrix} 10 & -6 & -9 \\ 6 & -5 & -7 \\ -10 & 9 & 12 \end{pmatrix}$. Compute (a) $\mathbf{A}+\mathbf{B}$; (b) $\mathbf{B}-4\mathbf{A}$; (c) the inverse of $\mathbf{AB}$; (d) the transpose of $(\mathbf{A}-2\mathbf{B})\,\mathbf{B}$; and (e) $\det\mathbf{A} = |\mathbf{A}|$.

Solution. We enter ma (corresponding to **A**) and mb (corresponding to **B**) as nested lists where each element corresponds to a row of the matrix. We suppress the output by ending each command with a semi-colon.

ma = {{3, −4, 5}, {8, 0, −3}, {5, 2, 1}};

mb = {{10, −6, −9}, {6, −5, −7}, {−10, 9, 12}};

Entering

ma + mb//MatrixForm

$$\begin{pmatrix} 13 & -10 & -4 \\ 14 & -5 & -10 \\ -5 & 11 & 13 \end{pmatrix}$$

adds matrix ma to mb and expresses the result in traditional matrix form. Entering

mb − 4ma//MatrixForm

$$\begin{pmatrix} -2 & 10 & -29 \\ -26 & -5 & 5 \\ -30 & 1 & 8 \end{pmatrix}$$

subtracts four times matrix ma from mb and expresses the result in traditional matrix form. Entering

Inverse[ma.mb]//MatrixForm

$$\begin{pmatrix} \frac{59}{380} & \frac{53}{190} & -\frac{167}{380} \\ -\frac{223}{570} & -\frac{92}{95} & \frac{979}{570} \\ \frac{49}{114} & \frac{18}{19} & -\frac{187}{114} \end{pmatrix}$$

computes the inverse of the matrix product **AB**. Similarly, entering

Matrix products, when defined, are computed by placing a period (.) between the matrices being multiplied. Note that a period is also used to compute the dot product of two vectors, when the dot product is defined.

Transpose[(ma − 2mb).mb]//MatrixForm

$$\begin{pmatrix} -352 & -90 & 384 \\ 269 & 73 & -277 \\ 373 & 98 & -389 \end{pmatrix}$$

computes the transpose of $(\mathbf{A} - 2\mathbf{B})\,\mathbf{B}$ and entering

Det[ma]

190

computes the determinant of **A**. □

Example 5.8

Compute **AB** and **BA** if $\mathbf{A} = \begin{pmatrix} -1 & -5 & -5 & -4 \\ -3 & 5 & 3 & -2 \\ -4 & 4 & 2 & -3 \end{pmatrix}$ and $\mathbf{B} = \begin{pmatrix} 1 & -2 \\ -4 & 3 \\ 4 & -4 \\ -5 & -3 \end{pmatrix}$.

Solution. Because **A** is a 3×4 matrix and **B** is a 4×2 matrix, **AB** is defined and is a 3×2 matrix. We define `matrixa` and `matrixb` with the following commands.

Remember that you can also define matrices by going to **Insert** under the Mathematica menu and selecting **Table/Matrix**. After entering the desired number of rows and columns and pressing the **OK** button, a matrix template is placed at the location of the cursor that you can fill in.

$$\textbf{matrixa} = \begin{pmatrix} -1 & -5 & -5 & -4 \\ -3 & 5 & 3 & -2 \\ -4 & 4 & 2 & -3 \end{pmatrix};$$

$$\textbf{matrixb} = \begin{pmatrix} 1 & -2 \\ -4 & 3 \\ 4 & -4 \\ -5 & -3 \end{pmatrix};$$

We then compute the product, naming the result `ab`, and display `ab` in `MatrixForm`.

ab = matrixa.matrixb;

MatrixForm[ab]

$$\begin{pmatrix} 19 & 19 \\ -1 & 15 \\ 3 & 21 \end{pmatrix}$$

However, the matrix product **BA** is not defined and Mathematica produces error messages, which are not displayed here, when we attempt to compute it.

matrixb.matrixa

{{1, −2}, {−4, 3}, {4, −4}, {−5, −3}}.{{−1, −5, −5, −4}, {−3, 5, 3, −2}, {−4, 4, 2, −3}} □

Special attention must be given to the notation that must be used in taking the product of a square matrix with itself. The following example illustrates how Mathematica interprets the expression `(matrixb)^n`. The command `(matrixb)^n` raises each element of the matrix `matrixb` to the nth power. The command `MatrixPower` is used to compute powers of matrices.

Example 5.9

Let $\mathbf{B} = \begin{pmatrix} -2 & 3 & 4 & 0 \\ -2 & 0 & 1 & 3 \\ -1 & 4 & -6 & 5 \\ 4 & 8 & 11 & -4 \end{pmatrix}$. (a) Compute $\mathbf{B}^2$ and $\mathbf{B}^3$. (b) Cube each entry of **B**.

Solution. After defining **B**, we compute $\mathbf{B}^2$. The same results would have been obtained by entering `MatrixPower[matrixb,2]`.

matrixb = {{−2, 3, 4, 0}, {−2, 0, 1, 3}, {−1, 4, −6, 5},

{4, 8, 11, −4}};

MatrixForm[matrixb.matrixb]

$$\begin{pmatrix} -6 & 10 & -29 & 29 \\ 15 & 22 & 19 & -7 \\ 20 & 13 & 91 & -38 \\ -51 & 24 & -86 & 95 \end{pmatrix}$$

Next, we use `MatrixPower` to compute $\mathbf{B}^3$. The same results would be obtained by entering `matrixb.matrixb.matrixb`.

MatrixForm[MatrixPower[matrixb, 3]]

$$\begin{pmatrix} 137 & 98 & 479 & -231 \\ -121 & 65 & -109 & 189 \\ -309 & 120 & -871 & 646 \\ 520 & 263 & 1381 & -738 \end{pmatrix}$$

Last, we cube each entry of **B** with ^.

MatrixForm[matrixb³]

$$\begin{pmatrix} -8 & 27 & 64 & 0 \\ -8 & 0 & 1 & 27 \\ -1 & 64 & -216 & 125 \\ 64 & 512 & 1331 & -64 \end{pmatrix}$$

□

If $|\mathbf{A}| \neq 0$, the inverse of **A** can be computed using the formula

$$\mathbf{A}^{-1} = \frac{1}{|\mathbf{A}|}\mathbf{A}^{a}, \tag{5.1}$$

where $\mathbf{A}^a$ is the *transpose of the cofactor matrix.*

If **A** has an inverse, reducing the matrix $(\mathbf{A}|\mathbf{I})$ to reduced row echelon form results in $(\mathbf{I}|\mathbf{A}^{-1})$. This method is often easier to implement than (5.1).

The **cofactor matrix**, $\mathbf{A}^c$, of **A** is the matrix obtained by replacing each element of **A** by its cofactor.

Example 5.10

Calculate $\mathbf{A}^{-1}$ if $\mathbf{A} = \begin{pmatrix} 2 & -2 & 1 \\ 0 & -2 & 2 \\ -2 & -1 & -1 \end{pmatrix}$.

Solution. After defining **A** and $\mathbf{I} = \begin{pmatrix} 1 & 0 & 0 \\ 0 & 1 & 0 \\ 0 & 0 & 1 \end{pmatrix}$, we compute $|\mathbf{A}| = 12$, so $\mathbf{A}^{-1}$ exists.

capa = {{2, −2, 1}, {0, −2, 2}, {−2, −1, −1}}

i3 = IdentityMatrix[3]

{{2, −2, 1}, {0, −2, 2}, {−2, −1, −1}}

{{1, 0, 0}, {0, 1, 0}, {0, 0, 1}}

Det[capa]
12

`Join[a,b,n]` concatenates lists `a` and `b` at level `n`. For matrices the level one objects (`capa[[i]]`) are the rows; the level two objects (`capa[[i,j]]`) are the entries. Thus, `Join[capa,i3]` returns the matrix $\begin{pmatrix} \mathbf{A} \\ \mathbf{I} \end{pmatrix}$ while `Join[capa,i3,2]` forms the matrix $(\mathbf{A}|\mathbf{I})$.

ai3 = Join[capa, i3, 2]

{{2, −2, 1, 1, 0, 0}, {0, −2, 2, 0, 1, 0}, {−2, −1, −1, 0, 0, 1}}

{{2, −2, 1, 1, 0, 0}, {0, −2, 2, 0, 1, 0},

{−2, −1, −1, 0, 0, 1}}

MatrixForm[ai3]

$$\begin{pmatrix} 2 & -2 & 1 & 1 & 0 & 0 \\ 0 & -2 & 2 & 0 & 1 & 0 \\ -2 & -1 & -1 & 0 & 0 & 1 \end{pmatrix}$$

We then use `RowReduce` to reduce $(\mathbf{A}|\mathbf{I})$ to row echelon form.

`RowReduce[A]` reduces **A** to **reduced row echelon form.**

rrai3 = RowReduce[ai3]

$\left\{\left\{1, 0, 0, \frac{1}{3}, -\frac{1}{4}, -\frac{1}{6}\right\}, \left\{0, 1, 0, -\frac{1}{3}, 0, -\frac{1}{3}\right\}, \left\{0, 0, 1, -\frac{1}{3}, \frac{1}{2}, -\frac{1}{3}\right\}\right\}$

$\left\{\left\{1, 0, 0, \frac{1}{3}, -\frac{1}{4}, -\frac{1}{6}\right\},\right.$

$\left.\left\{0, 1, 0, -\frac{1}{3}, 0, -\frac{1}{3}\right\}, \left\{0, 0, 1, -\frac{1}{3}, \frac{1}{2}, -\frac{1}{3}\right\}\right\}$

MatrixForm[rrai3]

$$\begin{pmatrix} 1 & 0 & 0 & \frac{1}{3} & -\frac{1}{4} & -\frac{1}{6} \\ 0 & 1 & 0 & -\frac{1}{3} & 0 & -\frac{1}{3} \\ 0 & 0 & 1 & -\frac{1}{3} & \frac{1}{2} & -\frac{1}{3} \end{pmatrix}$$

The result indicates that $\mathbf{A}^{-1} = \begin{pmatrix} 1/3 & -1/4 & -1/6 \\ -1/3 & 0 & -1/3 \\ -1/3 & 1/2 & -1/3 \end{pmatrix}$. □

5.1.4 Basic Computations with Vectors

5.1.4.1 Basic Operations on Vectors

Computations with vectors are performed in the same way as computations with matrices.

Example 5.11

Let $\mathbf{v} = \begin{pmatrix} 0 \\ 5 \\ 1 \\ 2 \end{pmatrix}$ and $\mathbf{w} = \begin{pmatrix} 3 \\ 0 \\ 4 \\ -2 \end{pmatrix}$. (a) Calculate $\mathbf{v} - 2\mathbf{w}$ and $\mathbf{v} \cdot \mathbf{w}$. (b) Find a unit vector with the same direction as $\mathbf{v}$ and a unit vector with the same direction as $\mathbf{w}$.

Solution. We begin by defining $\mathbf{v}$ and $\mathbf{w}$ and then compute $\mathbf{v} - 2\mathbf{w}$ and $\mathbf{v} \cdot \mathbf{w}$.

```
v = {0, 5, 1, 2};
w = {3, 0, 4, -2};
v - 2w
```

{−6, 5, −7, 6}

```
v.w
```

0

The **norm** of the vector $\mathbf{v} = \begin{pmatrix} v_1 \\ v_2 \\ \vdots \\ v_n \end{pmatrix}$ is

$$\|\mathbf{v}\| = \sqrt{{v_1}^2 + {v_2}^2 + \cdots + {v_n}^2} = \sqrt{\mathbf{v} \cdot \mathbf{v}}.$$

The command `Norm[v]` returns the norm of the vector $\mathbf{v}$.
If k is a scalar, the direction of $k\mathbf{v}$ is the same as the direction of $\mathbf{v}$. Thus, if $\mathbf{v}$ is a nonzero vector, the vector $\frac{1}{\|\mathbf{v}\|}\mathbf{v}$ has the same direction as $\mathbf{v}$ and because $\left\|\frac{1}{\|\mathbf{v}\|}\mathbf{v}\right\| = \frac{1}{\|\mathbf{v}\|}\|\mathbf{v}\| = 1$, $\frac{1}{\|\mathbf{v}\|}\mathbf{v}$ is a unit vector. First, we compute $\|\mathbf{v}\|$ with `Norm`. We then compute $\frac{1}{\|\mathbf{v}\|}\mathbf{v}$, calling the result `uv`, and $\frac{1}{\|\mathbf{w}\|}\mathbf{w}$. The results correspond to unit vectors with the same direction as $\mathbf{v}$ and $\mathbf{w}$, respectively.

```
Norm[v]
```

$\sqrt{30}$

```
uv = v/Norm[v]
```

$$\left\{0, \sqrt{\frac{5}{6}}, \frac{1}{\sqrt{30}}, \sqrt{\frac{2}{15}}\right\}$$

Norm[uv]

1

$\frac{w}{\text{Norm}[w]}$

$$\left\{\frac{3}{\sqrt{29}}, 0, \frac{4}{\sqrt{29}}, -\frac{2}{\sqrt{29}}\right\}$$

□

5.1.4.2 Basic Operations on Vectors in 3-Space

Vector calculus is discussed in Section 5.5.

We review the elementary properties of vectors in 3-space. Let

$$\mathbf{u} = \langle u_1, u_2, u_3 \rangle = u_1\mathbf{i} + u_2\mathbf{j} + u_3\mathbf{k}$$

and

$$\mathbf{v} = \langle v_1, v_2, v_3 \rangle = v_1\mathbf{i} + v_2\mathbf{j} + v_3\mathbf{k}$$

be vectors in space.

In space, the **standard unit vectors** are $\mathbf{i} = \langle 1, 0, 0 \rangle$, $\mathbf{j} = \langle 0, 1, 0 \rangle$, and $\mathbf{k} = \langle 0, 0, 1 \rangle$. With the exception of the cross product, the vector operations discussed here are performed in the same way for vectors in the plane as they are in space. In the plane, the **standard unit vectors** are $\mathbf{i} = \langle 1, 0 \rangle$ and $\mathbf{j} = \langle 0, 1 \rangle$.

1. **u** and **v** are **equal** if and only if their components are equal:
$$\mathbf{u} = \mathbf{v} \Leftrightarrow u_1 = v_1, u_2 = v_2, \text{ and } u_3 = v_3.$$
2. The **length** (or **norm**) of **u** is
$$\|\mathbf{u}\| = \sqrt{u_1^2 + u_2^2 + u_3^2}.$$
3. If c is a scalar (number),
$$c\mathbf{u} = \langle cu_1, cu_2, cu_3 \rangle.$$
4. The **sum** of **u** and **v** is defined to be the vector
$$\mathbf{u} + \mathbf{v} = \langle u_1 + v_1, u_2 + v_2, u_3 + v_3 \rangle.$$
5. If $\mathbf{u} \neq \mathbf{0}$, a unit vector with the same direction as **u** is
$$\frac{1}{\|\mathbf{u}\|}\mathbf{u} = \frac{1}{\sqrt{u_1^2 + u_2^2 + u_3^2}} \langle u_1, u_2, u_3 \rangle.$$
6. **u** and **v** are **parallel** if there is a scalar c so that $\mathbf{u} = c\mathbf{v}$.
7. The **dot product** of **u** and **v** is
$$\mathbf{u} \cdot \mathbf{v} = u_1v_1 + u_2v_2 + u_3v_3.$$
If θ is the angle between **u** and **v**,
$$\cos\theta = \frac{\mathbf{u} \cdot \mathbf{v}}{\|\mathbf{u}\| \, \|\mathbf{v}\|}.$$
Consequently, **u** and **v** are orthogonal if $\mathbf{u} \cdot \mathbf{v} = 0$.
8. The **cross product** of **u** and **v** is
$$\mathbf{u} \times \mathbf{v} = \begin{vmatrix} \mathbf{i} & \mathbf{j} & \mathbf{k} \\ u_1 & u_2 & u_3 \\ v_1 & v_2 & v_3 \end{vmatrix} = (u_2v_3 - u_3v_2)\mathbf{i} - (u_1v_3 - u_3v_1)\mathbf{j} + (u_1v_2 - u_2v_1)\mathbf{k}.$$
You should verify that $\mathbf{u} \cdot (\mathbf{u} \times \mathbf{v}) = 0$ and $\mathbf{v} \cdot (\mathbf{u} \times \mathbf{v}) = 0$. Hence, $\mathbf{u} \times \mathbf{v}$ is orthogonal to both **u** and **v**.

Topics from linear algebra (including determinants, which were mentioned previously) are discussed in more detail in the next sections. For now, we illustrate several of the basic operations listed above: `u.v` and `Dot[u,v]` compute $\mathbf{u} \cdot \mathbf{v}$; `Cross[u,v]` computes $\mathbf{u} \times \mathbf{v}$.

A **unit vector** is a vector with length 1.

Example 5.12

Let $\mathbf{u} = \langle 3, 4, 1 \rangle$ and $\mathbf{v} = \langle -4, 3, -2 \rangle$. Calculate (a) $\mathbf{u} \cdot \mathbf{v}$, (b) $\mathbf{u} \times \mathbf{v}$, (c) $\|\mathbf{u}\|$, and (d) $\|\mathbf{v}\|$. (e) Find the angle between $\mathbf{u}$ and $\mathbf{v}$. (f) Find unit vectors with the same direction as $\mathbf{u}$, $\mathbf{v}$, and $\mathbf{u} \times \mathbf{v}$.

Solution. We begin by defining $\mathbf{u} = \langle 3, 4, 1 \rangle$ and $\mathbf{v} = \langle -4, 3, -2 \rangle$. Notice that to define $\mathbf{u} = \langle u_1, u_2, u_3 \rangle$ with Mathematica, we use the form

```
u={u1,u2,u3}.
```

We illustrate the use of `Dot` and `Cross` to calculate (a)–(d).

Similarly, to define $\mathbf{u} = \langle u_1, u_2 \rangle$, we use the form `u={u1,u2}`.

```
u = {3, 4, 1};
v = {-4, 3, -2};
udv = Dot[u, v]
```

−2

```
u.v
```

−2

```
ucv = Cross[u, v]
```

{−11, 2, 25}

```
nu = Norm[u]
```

$\sqrt{26}$

```
nv = Sqrt[v.v]
```

$\sqrt{29}$

We use the formula $\theta = \cos^{-1}\left(\frac{\mathbf{u} \cdot \mathbf{v}}{\|\mathbf{u}\| \, \|\mathbf{v}\|}\right)$ to find the angle θ between $\mathbf{u}$ and $\mathbf{v}$.

```
ArcCos[u.v/(nu nv)]
N[%]
```

$\text{ArcCos}\left[-\sqrt{\frac{2}{377}}\right]$

1.6437

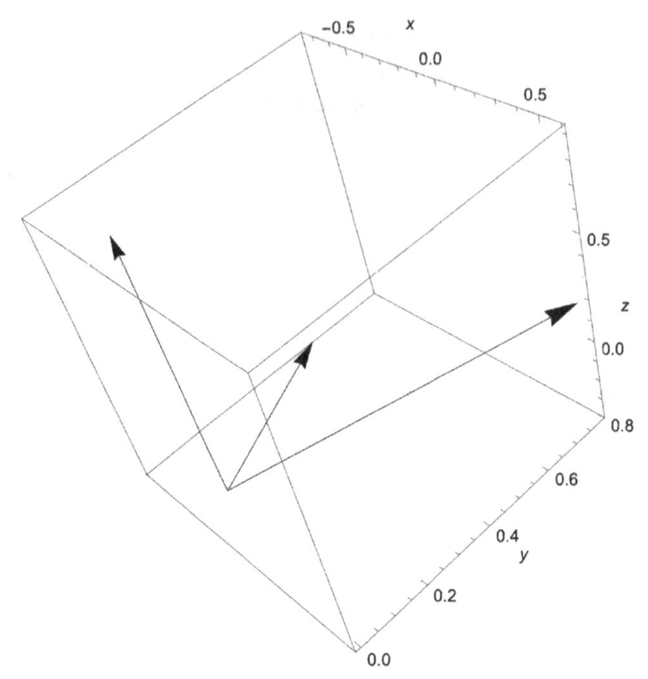

FIGURE 5.2 Orthogonal vectors.

Unit vectors with the same direction as **u**, **v**, and **u** × **v** are found next.

normu = *u*/nu

normv = *v*/nv

$$\left\{\frac{3}{\sqrt{26}}, 2\sqrt{\frac{2}{13}}, \frac{1}{\sqrt{26}}\right\}$$

$$\left\{-\frac{4}{\sqrt{29}}, \frac{3}{\sqrt{29}}, -\frac{2}{\sqrt{29}}\right\}$$

nucrossv = ucv/Norm[ucv]

$$\left\{-\frac{11}{5\sqrt{30}}, \frac{\sqrt{\frac{2}{15}}}{5}, \sqrt{\frac{5}{6}}\right\}$$

We can graphically confirm that these three vectors are orthogonal by graphing all three vectors with the graphics primitive `Arrow` together with `Graphics3D`. We show the vectors in Fig. 5.2.

p1 = Graphics3D[Arrow[{{{0, 0, 0}, normu}, {{0, 0, 0}, normv}, {{0, 0, 0}, nucrossv}}],

BoxRatios → {1, 1, 1}, AxesLabel → {*x*, *y*, *z*}, Axes → True]

In the plot, the vectors may not appear to be orthogonal (perpendicular) as expected because of the aspect ratio and viewing angles of the graphic. □

With the exception of the cross product, the calculations described above can also be performed on vectors in the plane.

Example 5.13

If **u** and **v** are nonzero vectors, the **projection** of **u** onto **v** is

$$\text{proj}_{\mathbf{v}}\mathbf{u} = \frac{\mathbf{u}\cdot\mathbf{v}}{\|\mathbf{v}\|^2}\mathbf{v}.$$

Find $\text{proj}_{\mathbf{v}}\mathbf{u}$ if $\mathbf{u} = \langle -1, 4\rangle$ and $\mathbf{v} = \langle 2, 6\rangle$.

Solution. First, we define $\mathbf{u} = \langle -1, 4\rangle$ and $\mathbf{v} = \langle 2, 6\rangle$ and then compute $\text{proj}_{\mathbf{v}}\mathbf{u}$.

```
u = {-1, 4};
v = {2, 6};
projvu = u.vv/v.v
```

$\left\{\frac{11}{10}, \frac{33}{10}\right\}$

Next, we graph **u**, **v**, and $\text{proj}_{\mathbf{v}}\mathbf{u}$ together using `Arrow`, `Show` and `GraphicsRow` in Fig. 5.3.

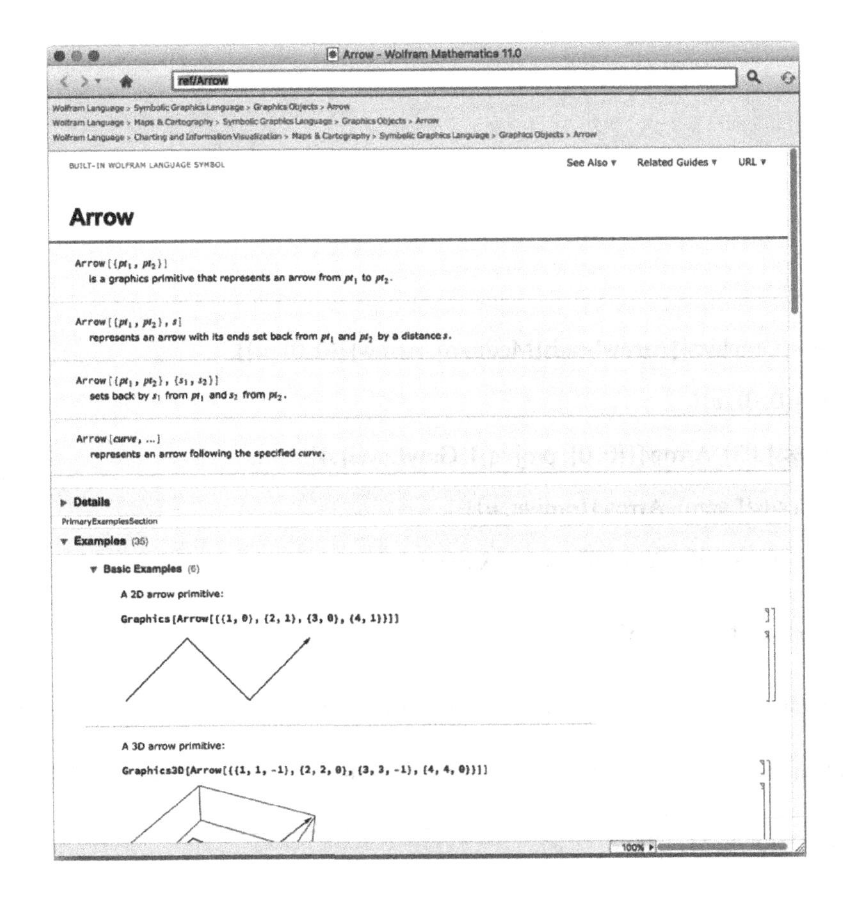


```
p1 = Show[Graphics[{Arrowheads[Medium], Arrow[{{0, 0}, u}], Arrow[{{0, 0}, v}],
Thickness[.05], Arrow[{{0, 0}, projvu}]}],
  Axes → Automatic, AspectRatio → Automatic];
```

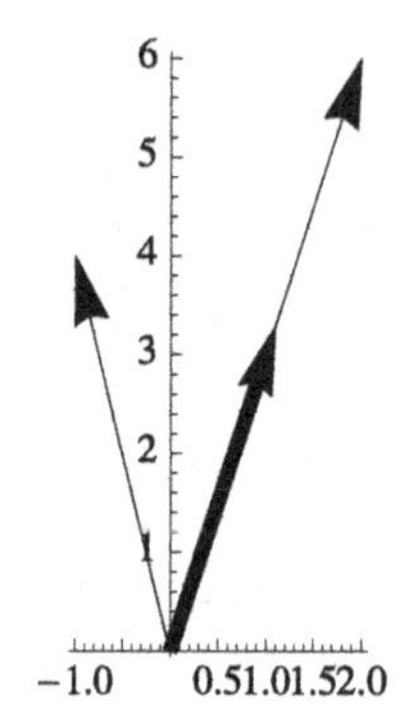

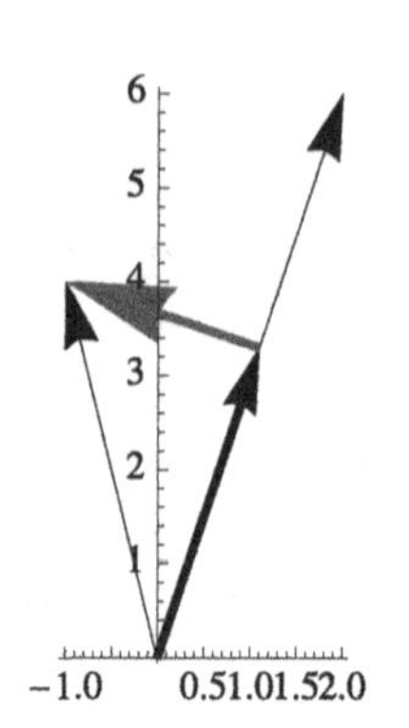

FIGURE 5.3 Projection of a vector.

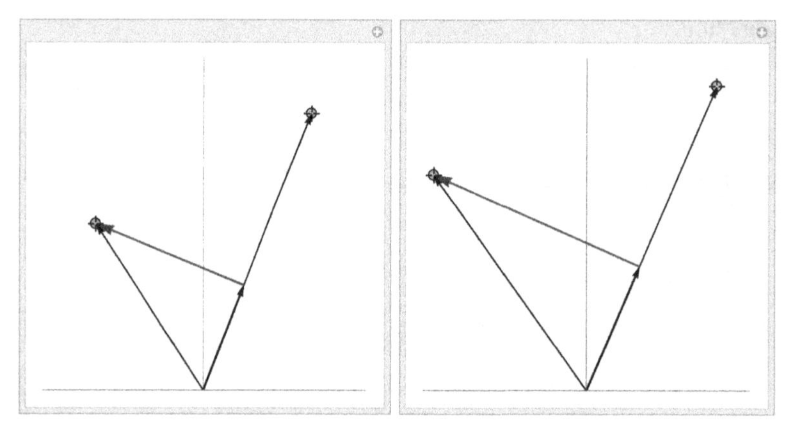

FIGURE 5.4 Using `Manipulate` to visualize the projection of one vector onto another.

```
p2 = Show[Graphics[{Arrowheads[Medium], Arrow[{{0, 0}, u}],
   Arrow[{{0, 0}, v}],
   Thickness[.03], Arrow[{{0, 0}, projvu}], GrayLevel[.4],
   Arrowheads[Large], Arrow[{projvu, u}]}],
   Axes → Automatic, AspectRatio → Automatic];

Show[GraphicsRow[{p1, p2}]]
```

In the graph, notice that $\mathbf{u} = \text{proj}_{\mathbf{v}}\mathbf{u} + (\mathbf{u} - \text{proj}_{\mathbf{v}}\mathbf{u})$ and the vector $\mathbf{u} - \text{proj}_{\mathbf{v}}\mathbf{u}$ is perpendicular to $\mathbf{v}$.

With the following, we use `Manipulate` to generalize the example. See Fig. 5.4.

```
Clear[u, v, projvu, p1, p2];

Manipulate[

projvu = u.vv/v.v;

Show[Graphics[{Arrowheads[Medium], Arrow[{{0, 0}, u}],
   Arrow[{{0, 0}, v}],
```

Thickness[.005], Arrow[{{0, 0}, projvu}], GrayLevel[.4],

Arrowheads[Large], Arrow[{projvu, *u*}]}],

Axes → Automatic, PlotRange → {{−3, 3}, {0, 6}},

AspectRatio → Automatic, Ticks → None], {{*u*, {−2, 3}}, Locator},

{{*v*, {2, 5}}, Locator}] □

If you only need to display a two-dimensional array in row-and-column form, it is easier to use `Grid` rather than `Table` together with `TableForm` or `MatrixForm`.

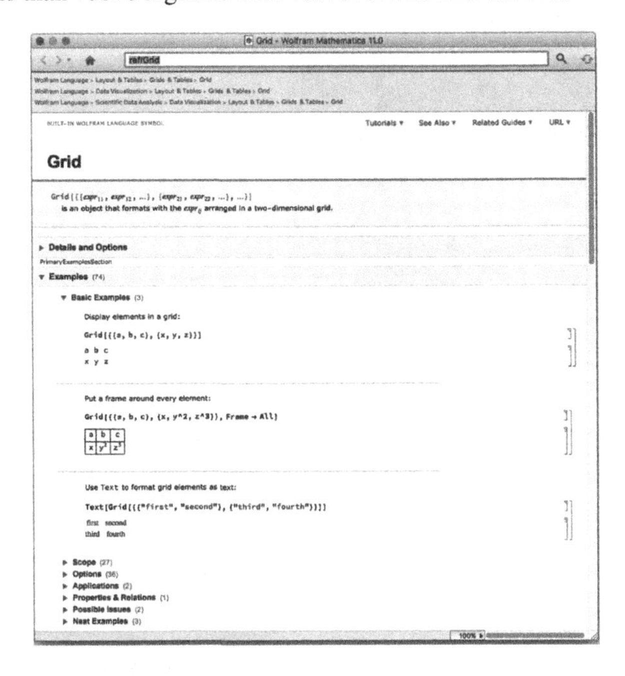

For a list of all the options associated with `Grid`, enter `Options[Grid]`.

```
Options[Grid]
{Alignment → {Center, Baseline}, AllowedDimensions → Automatic,
 AllowScriptLevelChange → True, AutoDelete → False, Background → None,
 BaselinePosition → Automatic, BaseStyle → {}, DefaultBaseStyle → Grid,
 DefaultElement → □, DeleteWithContents → True, Dividers → {},
 Editable → Automatic, Frame → None, FrameStyle → Automatic, ItemSize → Automatic,
 ItemStyle → None, Selectable → Automatic, Spacings → Automatic}
```

Thus,

p0 = Grid[{{*a*, *b*, *c*}, {*d*, *e*}, {*f*}}, Frame → All]

creates a basic grid. The first row consists of the entries a, b, and c; the second row d and e; and the third row f. See Fig. 5.5. Note that elements of grids can be any Mathematica object, including other grids.
You can create quite complex arrays with `Grid`. For example, elements of grids can be any Mathematica object, including grids.
In the following, we use `ExampleData` to generate several typical Mathematica objects.

`StringTake[string,n]` returns the first n characters of the string `string`.

p1 = ExampleData[{"AerialImage", "Earth"}];

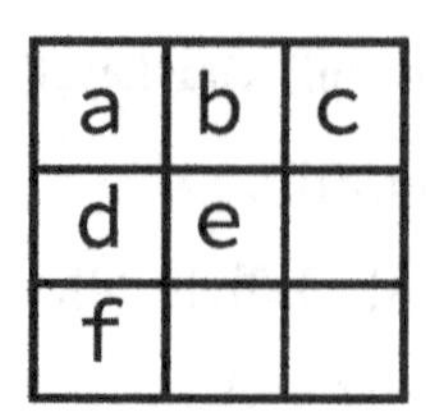

FIGURE 5.5 A basic grid.

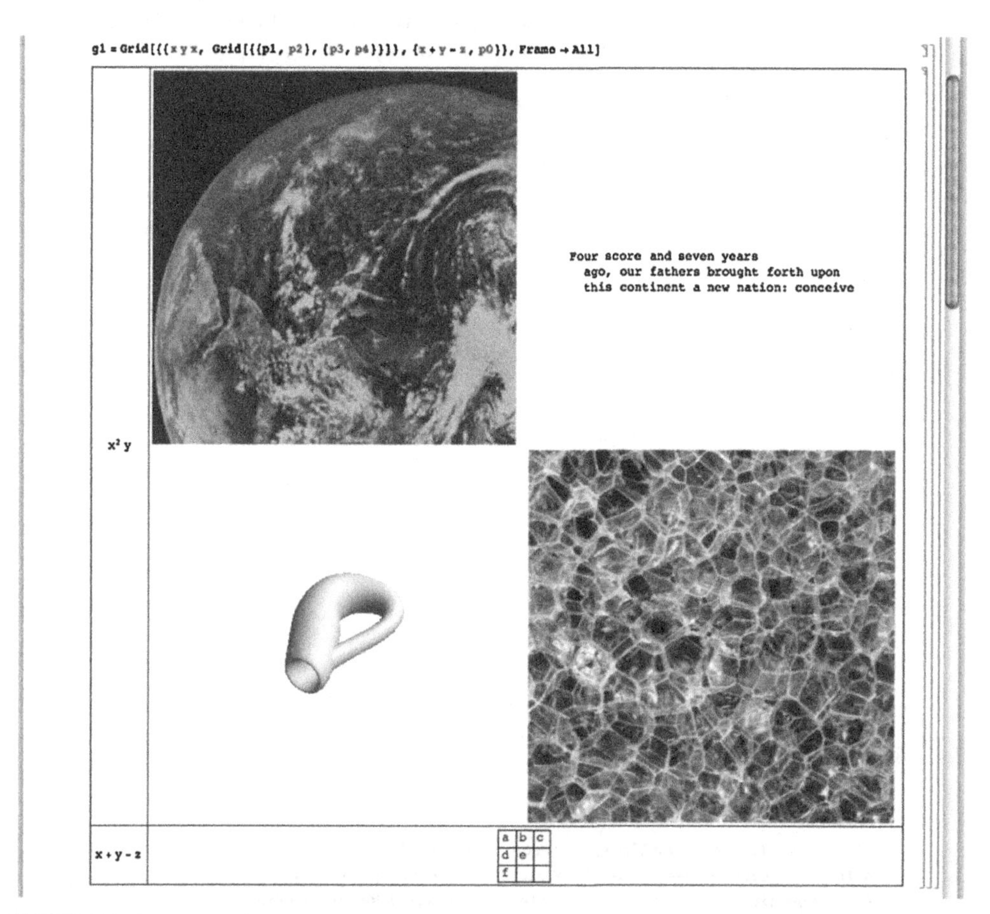

FIGURE 5.6 Very basic grids can appear to be quite complicated.

p2 = StringTake[ExampleData[{"Text", "GettysburgAddress"}], 100];

p3 = ExampleData[{"Geometry3D", "KleinBottle"}];

p4 = ExampleData[{"Texture", "Bubbles3"}];

Using our first grid, the above, and a few more strings we create a more sophisticated grid in Fig. 5.6.

g1 = Grid[{{xyx, Grid[{{p1, p2}, {p3, p4}}]}, {$x + y - z$, p0}}, Frame → All]

5.2 LINEAR SYSTEMS OF EQUATIONS

5.2.1 Calculating Solutions of Linear Systems of Equations

To solve the system of linear equations $\mathbf{Ax} = \mathbf{b}$, where $\mathbf{A}$ is the coefficient matrix, $\mathbf{b}$ is the known vector and $\mathbf{x}$ is the unknown vector, we often proceed as follows: if $\mathbf{A}^{-1}$ exists, then $\mathbf{AA}^{-1}\mathbf{x} = \mathbf{A}^{-1}\mathbf{b}$ so $\mathbf{x} = \mathbf{A}^{-1}\mathbf{b}$.
Mathematica offers several commands for solving systems of linear equations, however, that do not depend on the computation of the inverse of $\mathbf{A}$. The command

```
Solve[{eqn1,eqn2,...,eqnm},{var1,var2,...,varn}]
```

solves an $m \times n$ system of linear equations (m equations and n unknown variables). Note that both the equations as well as the variables are entered as lists. If one wishes to solve for all variables that appear in a system, the command `Solve[{eqn1,eqn2,...eqnn}]` attempts to solve `eqn1`, `eqn2`, ..., `eqnn` for all variables that appear in them. (Remember that a double equals sign (==) must be placed between the left and right-hand sides of each equation.)

Example 5.14

Solve the matrix equation $\begin{pmatrix} 3 & 0 & 2 \\ -3 & 2 & 2 \\ 2 & -3 & 3 \end{pmatrix} \begin{pmatrix} x \\ y \\ z \end{pmatrix} = \begin{pmatrix} 3 \\ -1 \\ 4 \end{pmatrix}$.

Solution. The solution is given by $\begin{pmatrix} x \\ y \\ z \end{pmatrix} = \begin{pmatrix} 3 & 0 & 2 \\ -3 & 2 & 2 \\ 2 & -3 & 3 \end{pmatrix}^{-1} \begin{pmatrix} 3 \\ -1 \\ 4 \end{pmatrix}$. We proceed by defining `matrixa` and `b` and then using `Inverse` to calculate `Inverse[matrixa].b` naming the resulting output `{x,y,z}`.

matrixa = {{3, 0, 2}, {−3, 2, 2}, {2, −3, 3}};

b = {3, −1, 4};

{x, y, z} = Inverse[matrixa].b

$\left\{\frac{13}{23}, -\frac{7}{23}, \frac{15}{23}\right\}$

We verify that the result is the desired solution by calculating `matrixa.{x,y,z}`. Because the result of this procedure is $\begin{pmatrix} 3 \\ -1 \\ 4 \end{pmatrix}$, we conclude that the solution to the system is $\begin{pmatrix} x \\ y \\ z \end{pmatrix} = \begin{pmatrix} 13/23 \\ -7/23 \\ 15/23 \end{pmatrix}$.

matrixa.{x, y, z}

{3, −1, 4}

We note that this matrix equation is equivalent to the system of equations

$$\begin{aligned} 3x + 2z &= 3 \\ -3x + 2y + 2z &= -1, \\ 2x - 3y + 3z &= 4 \end{aligned}$$

which we are able to solve with `Solve`. (Note that `Thread[{f1,f2,...}={g1,g2,...}]` returns the system of equations `{f1==g1,f2==g2,...}`.)

Clear[x, y, z]

sys = Thread[matrixa.{x, y, z}=={3, −1, 4}]

$\{3x + 2z == 3, -3x + 2y + 2z == -1, 2x - 3y + 3z == 4\}$

Solve[sys]

$\left\{\left\{x \to \frac{13}{23}, y \to -\frac{7}{23}, z \to \frac{15}{23}\right\}\right\}$

Shortly, we will discuss using row reduction to solve systems of equations. For now, we remark that given the augmented matrix for a system, $(\mathbf{A}|\mathbf{b})$, `RowReduce` reduces $(\mathbf{A}|\mathbf{b})$ to reduced row echelon form so that you can see the solution(s) to the linear system, if any.

RowReduce[{{3, 0, 2, 3}, {−3, 2, 2, −1}, {2, −3, 3, 4}}]

$\left\{\left\{1, 0, 0, \frac{13}{23}\right\}, \left\{0, 1, 0, -\frac{7}{23}\right\}, \left\{0, 0, 1, \frac{15}{23}\right\}\right\}$ □

In addition to using `Solve` to solve a system of linear equations, the command

```
LinearSolve[A,b]
```

calculates the solution vector $\mathbf{x}$ of the system $\mathbf{A}\mathbf{x} = \mathbf{b}$. `LinearSolve` generally solves a system more quickly than does `Solve` as we see from the comments in the **Documentation Center**.

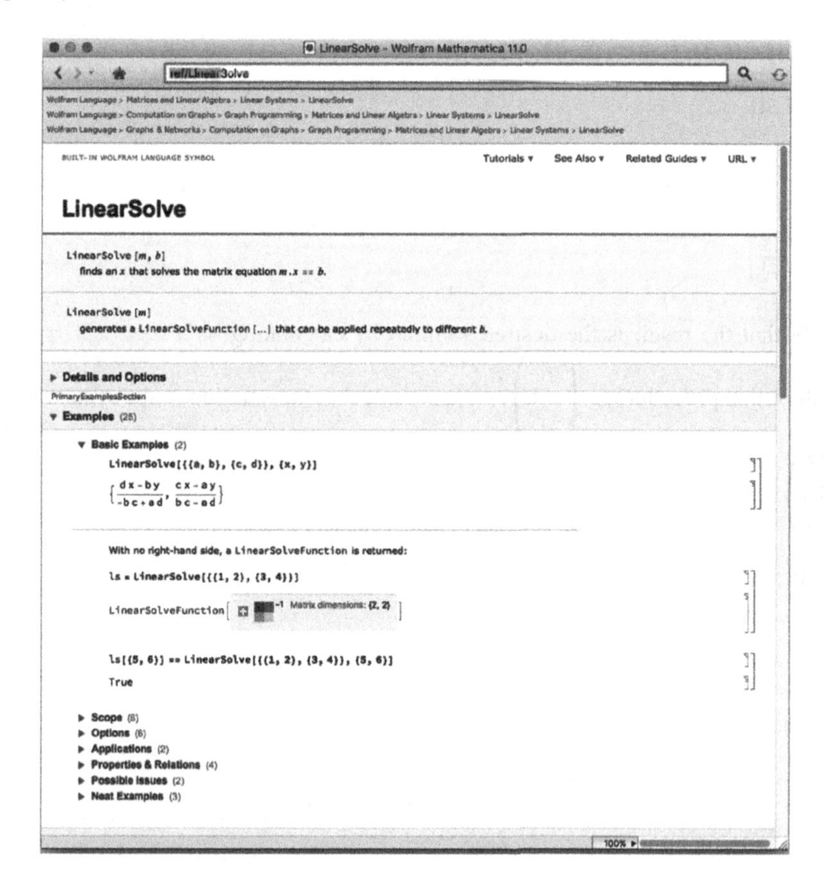

Example 5.15

Solve the system $\begin{cases} x-2y+z=-4 \\ 3x+2y-z=8 \\ -x+3y+5z=0 \end{cases}$ for x, y, and z.

Solution. In this case, entering either

```
Solve[{x-2y+z==-4,3x+2y-z==8,-x+3y+5z==0}]
```

or

```
Solve[{x-2y+z,3x+2y-z,-x+3y+5z}=={-4,8,0}]
```

give the same result.

Solve[{x − 2y + z== − 4, 3x + 2y − z==8, −x + 3y + 5z==0},

{x, y, z}]

$\{\{x \to 1, y \to 2, z \to -1\}\}$

Another way to solve systems of equations is based on the matrix form of the system of equations, $\mathbf{Ax} = \mathbf{b}$. This system of equations is equivalent to the matrix equation

$$\begin{pmatrix} 1 & -2 & 1 \\ 3 & 2 & -1 \\ -1 & 3 & 5 \end{pmatrix} \begin{pmatrix} x \\ y \\ z \end{pmatrix} = \begin{pmatrix} -4 \\ 8 \\ 0 \end{pmatrix}.$$

The matrix of coefficients in the previous example is entered as `matrixa` along with the vector of right-hand side values `vectorb`. After defining the vector of variables, `vectorx`, the system $\mathbf{Ax} = \mathbf{b}$ is solved explicitly with the command `Solve`.

matrixa = {{1, −2, 1}, {3, 2, −1}, {−1, 3, 5}};

vectorb = {−4, 8, 0};

vectorx = {x1, y1, z1};

Solve[matrixa.vectorx==vectorb, vectorx]

{{x1 → 1, y1 → 2, z1 → −1}}

LinearSolve[matrixa, vectorb]

{1, 2, −1} □

Example 5.16

Solve the system $\begin{cases} 2x-4y+z=-1 \\ 3x+y-2z=3 \\ -5x+y-2z=4 \end{cases}$. Verify that the result returned satisfies the system.

Solution. To solve the system using `Solve`, we define `eqs` to be the set of three equations to be solved and `vars` to be the variables x, y, and z and then use `Solve` to solve the set of equations `eqs` for the variables in `vars`. The resulting output is named `sols`.

eqs = {2x − 4y + z== − 1, 3x + y − 2z==3, −5x + y − 2z==4};

vars = {x, y, z};

sols = Solve[eqs, vars]

$\left\{\left\{x \to -\frac{1}{8}, y \to -\frac{15}{56}, z \to -\frac{51}{28}\right\}\right\}$

To verify that the result given in `sols` is the desired solution, we replace each occurrence of x, y, and z in `eqs` by the values found in `sols` using `ReplaceAll` (`/.`). Because the result indicates each of the three equations is satisfied, we conclude that the values given in `sols` are the components of the desired solution.

eqs/.sols

{{True, True, True}}

To solve the system using `LinearSolve`, we note that the system is equivalent to the matrix equation $\begin{pmatrix} 2 & -4 & 1 \\ 3 & 1 & -2 \\ -5 & 1 & -2 \end{pmatrix}\begin{pmatrix} x \\ y \\ z \end{pmatrix} = \begin{pmatrix} -1 \\ 3 \\ 4 \end{pmatrix}$, define `matrixa` and `vectorb`, and use `LinearSolve` to solve this matrix equation.

matrixa = {{2, −4, 1}, {3, 1, −2}, {−5, 1, −2}};

vectorb = {−1, 3, 4};

solvector = LinearSolve[matrixa, vectorb]

$\left\{-\frac{1}{8}, -\frac{15}{56}, -\frac{51}{28}\right\}$

To verify that the results are correct, we compute `matrixa.solvector`. Because the result is $\begin{pmatrix} -1 \\ 3 \\ 4 \end{pmatrix}$, we conclude that the solution to the system is $\begin{pmatrix} x \\ y \\ z \end{pmatrix} = \begin{pmatrix} -1/8 \\ -15/36 \\ -51/28 \end{pmatrix}$.

matrixa.solvector

{−1, 3, 4}

The command `LinearSolve[A]` returns a function that when given a vector **b** solves the equation **Ax** = **b**: `LinearSolve[A][b]` returns **x**.

LinearSolve[matrixa][{−1, 3, 4}]

$\left\{-\frac{1}{8}, -\frac{15}{56}, -\frac{51}{28}\right\}$ □

Enter indexed variables such $x_1, x_2, \ldots, x_n$ as `x[1]`, `x[2]`, ..., `x[n]`. If you need to include the entire list, `Table[x[i],{i,1,n}]` usually produces the desired result(s).

Example 5.17

Solve the system of equations $\begin{cases} 4x_1+5x_2-5x_3-8x_4-2x_5=5 \\ 7x_1+2x_2-10x_3-x_4-6x_5=-4 \\ 6x_1+2x_2+10x_3-10x_4+7x_5=-7 \\ -8x_1-x_2-4x_3+3x_5=5 \\ 8x_1-7x_2-3x_3+10x_4+5x_5=7 \end{cases}$.

Solution. We solve the system in two ways. First, we use `Solve` to solve the system. Note that in this case, we enter the equations in the form

```
set of left-hand sides==set of right-hand sides.
```

Solve[{4*x*[1] + 5*x*[2] − 5*x*[3] − 8*x*[4] − 2*x*[5],

7*x*[1] + 2*x*[2] − 10*x*[3] − *x*[4] − 6*x*[5],

6*x*[1] + 2*x*[2] + 10*x*[3] − 10*x*[4] + 7*x*[5],

−8*x*[1] − *x*[2] − 4*x*[3] + 3*x*[5],

8*x*[1] − 7*x*[2] − 3*x*[3] + 10*x*[4] + 5*x*[5]}=={5, −4, −7, 5, 7}]

$\left\{\left\{x[1]\to\frac{1245}{6626}, x[2]\to\frac{113174}{9939}, x[3]\to-\frac{7457}{9939}, x[4]\to\frac{38523}{6626}, x[5]\to\frac{49327}{9939}\right\}\right\}$

We also use `LinearSolve` after defining `matrixa` and `t2`. As expected, in each case, the results are the same.

Clear[matrixa]

matrixa = {{4, 5, −5, −8, −2}, {7, 2, −10, −1, −6},

{6, 2, 10, −10, 7},

{−8, −1, −4, 0, 3}, {8, −7, −3, 10, 5}};

t2 = {5, −4, −7, 5, 7};

LinearSolve[matrixa, t2]

$\left\{\frac{1245}{6626}, \frac{113174}{9939}, -\frac{7457}{9939}, \frac{38523}{6626}, \frac{49327}{9939}\right\}$ □

5.2.2 Gauss–Jordan Elimination

Given the matrix equation $\mathbf{Ax} = \mathbf{b}$, where

$$\mathbf{A} = \begin{pmatrix} a_{11} & a_{12} & \cdots & a_{1n} \\ a_{21} & a_{22} & \cdots & a_{2n} \\ \vdots & \vdots & \ddots & \vdots \\ a_{m1} & a_{m2} & \cdots & a_{mn} \end{pmatrix}, \quad \mathbf{x} = \begin{pmatrix} x_1 \\ x_2 \\ \vdots \\ x_n \end{pmatrix}, \quad \text{and} \quad \mathbf{b} = \begin{pmatrix} b_1 \\ b_2 \\ \vdots \\ b_m \end{pmatrix},$$

the $m \times n$ matrix $\mathbf{A}$ is called the **coefficient matrix** for the matrix equation $\mathbf{Ax} = \mathbf{b}$ and the $m \times (n+1)$ matrix

$$\begin{pmatrix} a_{11} & a_{12} & \cdots & a_{1n} & b_1 \\ a_{21} & a_{22} & \cdots & a_{2n} & b_2 \\ \vdots & \vdots & \ddots & \vdots & \vdots \\ a_{m1} & a_{m2} & \cdots & a_{mn} & b_m \end{pmatrix}$$

is called the **augmented** (or **associated**) **matrix** for the matrix equation. We may enter the augmented matrix associated with a linear system of equations directly or we can use commands like `Join` to help us construct the augmented matrix. For example, if $\mathbf{A}$ and $\mathbf{B}$ are rectangular matrices that have the same number of columns, `Join[A,B]` returns $\begin{pmatrix} \mathbf{A} \\ \mathbf{B} \end{pmatrix}$. On the other hand, if $\mathbf{A}$ and $\mathbf{B}$ are rectangular matrices that have the same number of rows, `Join[A,B,2]` returns the concatenated matrix $\begin{pmatrix} \mathbf{A} & \mathbf{B} \end{pmatrix}$.

Example 5.18

Solve the system $\begin{cases} -2x + y - 2x = 4 \\ 2x - 4y - 2z = -4 \\ x - 4y - 2z = 3 \end{cases}$ using Gauss–Jordan elimination.

Solution. The system is equivalent to the matrix equation

$$\begin{pmatrix} -2 & 1 & -2 \\ 2 & -4 & -2 \\ 1 & -4 & -2 \end{pmatrix} \begin{pmatrix} x \\ y \\ z \end{pmatrix} = \begin{pmatrix} 4 \\ -4 \\ 3 \end{pmatrix}.$$

The augmented matrix associated with this system is

$$\begin{pmatrix} -2 & 1 & -2 & 4 \\ 2 & -4 & -2 & -4 \\ 1 & -4 & -2 & 3 \end{pmatrix}$$

which we construct using the command `Join`.

```
matrixa = {{-2, 1, -2}, {2, -4, -2}, {1, -4, -2}};
b = {{4}, {-4}, {3}};
augm = Join[matrixa, b, 2];
MatrixForm[augm]
```

$$\begin{pmatrix} -2 & 1 & -2 & 4 \\ 2 & -4 & -2 & -4 \\ 1 & -4 & -2 & 3 \end{pmatrix}$$

We calculate the solution by row-reducing `augm` using `RowReduce`. Generally, `RowReduce[A]` reduces $\mathbf{A}$ to **reduced row echelon form**.

```
RowReduce[augm]//MatrixForm
```

$$\begin{pmatrix} 1 & 0 & 0 & -7 \\ 0 & 1 & 0 & -4 \\ 0 & 0 & 1 & 3 \end{pmatrix}$$

From this result, we see that the solution is

$$\begin{pmatrix} x \\ y \\ z \end{pmatrix} = \begin{pmatrix} -7 \\ -4 \\ 3 \end{pmatrix}.$$

We verify this by replacing each occurrence of x, y, and z on the left-hand side of the equations by -7, -4, and 3, respectively, and noting that the components of the result are equal to the right-hand side of each equation.

```
Clear[x, y, z]
{-2x + y - 2z, 2x - 4y - 2z, x - 4y - 2z}/.{x → -7, y → -4, z → 3}
```

{4, −4, 3} □

In the following example, we carry out the steps of the row reduction process.

Example 5.19

Solve

$$\begin{aligned} -3x + 2y - 2z &= -10 \\ 3x - y + 2z &= 7 \\ 2x - y + z &= 6. \end{aligned}$$

Solution. The augmented matrix is $\mathbf{A} = \begin{pmatrix} -3 & 2 & -2 & -10 \\ 3 & -1 & 2 & 7 \\ 2 & -1 & 1 & 6 \end{pmatrix}$, defined in `capa`, and then displayed in traditional row-and-column form with `MatrixForm`. Given the matrix **A**, the ith part of `A` corresponds to the ith row of **A**. Therefore, `A[[i]]` returns the ith row of **A**.

```
Clear[capa]
capa = {{-3, 2, -2, -10}, {3, -1, 2, 7}, {2, -1, 1, 6}};
MatrixForm[capa]
```

$$\begin{pmatrix} -3 & 2 & -2 & -10 \\ 3 & -1 & 2 & 7 \\ 2 & -1 & 1 & 6 \end{pmatrix}$$

We eliminate methodically. First, we multiply row 1 by $-1/3$ so that the first entry in the first column is 1.

```
capa = {-1/3capa[[1]], capa[[2]], capa[[3]]}
```

$\left\{\left\{1,-\frac{2}{3},\frac{2}{3},\frac{10}{3}\right\},\{3,-1,2,7\},\{2,-1,1,6\}\right\}$

We now eliminate below. First, we multiply row 1 by −3 and add it to row 2 and then we multiply row 1 by −2 and add it to row 3.

capa = {capa[[1]], −3capa[[1]] + capa[[2]],

−2capa[[1]] + capa[[3]]}

$\left\{\left\{1,-\frac{2}{3},\frac{2}{3},\frac{10}{3}\right\},\{0,1,0,-3\},\left\{0,\frac{1}{3},-\frac{1}{3},-\frac{2}{3}\right\}\right\}$

Observe that the first nonzero entry in the second row is 1. We eliminate below this entry by adding −1/3 times row 2 to row 3.

capa = {capa[[1]], capa[[2]], −1/3capa[[2]] + capa[[3]]}

$\left\{\left\{1,-\frac{2}{3},\frac{2}{3},\frac{10}{3}\right\},\{0,1,0,-3\},\left\{0,0,-\frac{1}{3},\frac{1}{3}\right\}\right\}$

We multiply the third row by −3 so that the first nonzero entry is 1.

capa = {capa[[1]], capa[[2]], −3capa[[3]]}

$\left\{\left\{1,-\frac{2}{3},\frac{2}{3},\frac{10}{3}\right\},\{0,1,0,-3\},\{0,0,1,-1\}\right\}$

This matrix is equivalent to the system

$$\begin{aligned} x-\frac{2}{3}y+\frac{2}{3}z&=\frac{10}{3}\\ y&=-3\\ z&=-1, \end{aligned}$$

which shows us that the solution is $x=2$, $y=-3$, $z=-1$.
Working backwards confirms this. Multiplying row 2 by 2/3 and adding to row 1 and then multiplying row 3 by −2/3 and adding to row 1 results in

capa = {2/3capa[[2]] + capa[[1]], capa[[2]], capa[[3]]}

$\left\{\left\{1,0,\frac{2}{3},\frac{4}{3}\right\},\{0,1,0,-3\},\{0,0,1,-1\}\right\}$

capa = {−2/3capa[[3]] + capa[[1]], capa[[2]], capa[[3]]}

$\{\{1,0,0,2\},\{0,1,0,-3\},\{0,0,1,-1\}\}$

which is equivalent to the system $x=2$, $y=-3$, $z=-1$.
Equivalent results are obtained with `RowReduce`.

Clear[capa]

capa = {{−3, 2, −2, −10}, {3, −1, 2, 7}, {2, −1, 1, 6}};

capa = RowReduce[capa];

MatrixForm[capa]

$$\begin{pmatrix} 1 & 0 & 0 & 2 \\ 0 & 1 & 0 & -3 \\ 0 & 0 & 1 & -1 \end{pmatrix}$$

Finally, we confirm the result directly with `Solve`.

Solve[{−3x + 2y − 2z == −10,

3x − y + 2z == 7, 2x − y + z == 6}]

{{x → 2, y → −3, z → −1}} □

It is important to remember that if you reduce the augmented matrix to reduced-row-echelon form, the results show you the solution to the problem. `RowReduce[A]` row reduces **A** to reduced-row-echelon form.

Example 5.20

Solve

$$\begin{aligned} -3x_1 + 2x_2 + 5x_3 &= -12 \\ 3x_1 - x_2 - 4x_3 &= 9 \\ 2x_1 - x_2 - 3x_3 &= 7. \end{aligned}$$

Solution. The augmented matrix is $\mathbf{A} = \begin{pmatrix} -3 & 2 & 5 & -12 \\ 3 & -1 & -4 & 9 \\ 2 & -1 & -3 & 7 \end{pmatrix}$, which is reduced to reduced row echelon form with `RowReduce`.

capa = {{−3, 2, 5, −12}, {3, −1, −4, 9}, {2, −1, −3, 7}};

rrcapa = RowReduce[capa];

MatrixForm[rrcapa]

$$\begin{pmatrix} 1 & 0 & -1 & 2 \\ 0 & 1 & 1 & -3 \\ 0 & 0 & 0 & 0 \end{pmatrix}$$

The result shows that the original system is equivalent to

$$\begin{aligned} x_1 - x_3 &= 2 \\ x_2 + x_3 &= -3 \end{aligned} \quad \text{or} \quad \begin{aligned} x_1 &= 2 + x_3 \\ x_2 &= -3 - x_3 \end{aligned}$$

so x_1, x_2, or x_3 is *free*. *Choosing* x_3 to be free, for any real number t, a solution to the system is

$$\begin{pmatrix} x_1 \\ x_2 \\ x_3 \end{pmatrix} = \begin{pmatrix} 2+t \\ -3-t \\ t \end{pmatrix} = \begin{pmatrix} 2 \\ -3 \\ 0 \end{pmatrix} + t \begin{pmatrix} 1 \\ -1 \\ 1 \end{pmatrix}.$$

The system has infinitely many solutions.
Equivalent results are obtained with `Solve`.

```
Solve[{−3x1 + 2x2 + 5x3 == −12, 3x1 − x2 − 4x3 == 9,
  2x1 − x2 − 3x3 == 7}]
```

```
{{x2 → −1 − x1, x3 → −2 + x1}}
```

□

Example 5.21

Solve

$$\begin{aligned} -3x_1 + 2x_2 + 5x_3 &= -14 \\ 3x_1 - x_2 - 4x_3 &= 11 \\ 2x_1 - x_2 - 3x_3 &= 8. \end{aligned}$$

Solution. The augmented matrix is $\mathbf{A} = \begin{pmatrix} -3 & 2 & 5 & -14 \\ 3 & -1 & -4 & 11 \\ 2 & -1 & -3 & 8 \end{pmatrix}$, which is reduced to reduced row echelon form with `RowReduce`.

```
Clear[x]
capa = {{−3, 2, 5, −14}, {3, −1, −4, 11}, {2, −1, −3, 8}};
rrcapa = RowReduce[capa];
MatrixForm[rrcapa]
```

$$\begin{pmatrix} 1 & 0 & -1 & 0 \\ 0 & 1 & 1 & 0 \\ 0 & 0 & 0 & 1 \end{pmatrix}$$

The result shows that the original system is equivalent to

$$\begin{aligned} x_1 - x_3 &= 0 \\ x_2 + x_3 &= 0 \\ 0 &= 1. \end{aligned}$$

Of course, 0 is not equal to 1: the last equation is false. The system has no solutions.
We check the calculation with `Solve`. In this case, the results indicate that `Solve` cannot find any solutions to the system.

```
Solve[{−3x[1] + 2x[2] + 5x[3] == −14,
3x[1] − x[2] − 4x[3] == 11, 2x[1] − x[2] − 3x[3] == 8}]
```

```
{}
```

Generally, if Mathematica returns nothing, the result means either that there is no solution or that Mathematica cannot solve the problem. Sometimes, Mathematica will return a warning message that solutions may exist but that it cannot find them. In these situations, we must always check using another method. □

Example 5.22

The **nullspace** of **A** is the set of solutions to the system of equations $\mathbf{Ax}=\mathbf{0}$. Find the nullspace of

$$\mathbf{A}=\begin{pmatrix} 3 & 2 & 1 & 1 & -2 \\ 3 & 3 & 1 & 2 & -1 \\ 2 & 2 & 1 & 1 & -1 \\ -1 & -1 & 0 & -1 & 0 \\ 5 & 4 & 2 & 2 & -3 \end{pmatrix}.$$

Solution. Observe that row reducing $(\mathbf{A}|\mathbf{0})$ is equivalent to row reducing **A**. After defining **A**, we use `RowReduce` to row reduce **A**.

```
capa = {{3, 2, 1, 1, −2}, {3, 3, 1, 2, −1},
  {2, 2, 1, 1, −1}, {−1, −1, 0, −1, 0},
  {5, 4, 2, 2, −3}};
RowReduce[capa]//MatrixForm
```

$$\begin{pmatrix} 1 & 0 & 0 & 0 & -1 \\ 0 & 1 & 0 & 1 & 1 \\ 0 & 0 & 1 & -1 & -1 \\ 0 & 0 & 0 & 0 & 0 \\ 0 & 0 & 0 & 0 & 0 \end{pmatrix}$$

The result indicates that the solutions of $\mathbf{Ax}=\mathbf{0}$ are

$$\mathbf{x}=\begin{pmatrix} x_1 \\ x_2 \\ x_3 \\ x_4 \\ x_5 \end{pmatrix}=\begin{pmatrix} t \\ -s-t \\ s+t \\ s \\ t \end{pmatrix}=s\begin{pmatrix} 0 \\ -1 \\ 1 \\ 1 \\ 0 \end{pmatrix}+t\begin{pmatrix} 1 \\ -1 \\ 1 \\ 0 \\ 1 \end{pmatrix},$$

where s and t are any real numbers. The dimension of the nullspace, the **nullity**, is 2; a basis for the nullspace is

$$\left\{\begin{pmatrix} 0 \\ -1 \\ 1 \\ 1 \\ 0 \end{pmatrix},\begin{pmatrix} 1 \\ -1 \\ 1 \\ 0 \\ 1 \end{pmatrix}\right\}.$$

You can use the command `NullSpace[A]` to find a basis of the nullspace of a matrix **A** directly.

```
NullSpace[capa]
```

{{1, −1, 1, 0, 1}, {0, −1, 1, 1, 0}}

A is **singular** because $|\mathbf{A}|=0$.

Det[capa]

0

Do *not* use `LinearSolve` on singular matrices:

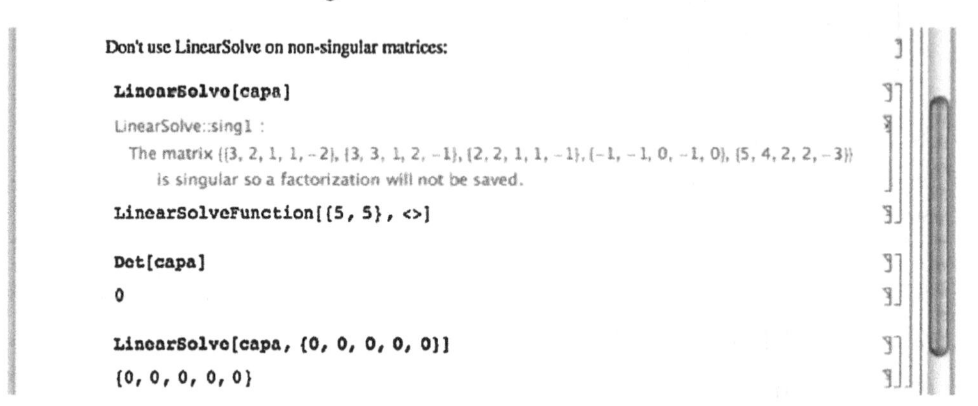

as the results returned may not be (completely) correct.

LinearSolve[capa]

LinearSolve[capa, {0, 0, 0, 0, 0}]

{0, 0, 0, 0, 0} □

5.3 SELECTED TOPICS FROM LINEAR ALGEBRA

5.3.1 Fundamental Subspaces Associated with Matrices

Let $\mathbf{A} = (a_{ij})$ be an $n \times m$ matrix with entry a_{ij} in the ith row and jth column. The **row space** of **A**, row(**A**), is the spanning set of the rows of **A**; the **column space** of **A**, col(**A**), is the spanning set of the columns of **A**. If **A** is any matrix, then the dimension of the column space of **A** is equal to the dimension of the row space of **A**. The dimension of the row space (column space) of a matrix **A** is called the **rank** of **A**. The **nullspace** of **A** is the set of solutions to the system of equations $\mathbf{Ax} = \mathbf{0}$. The nullspace of **A** is a subspace and its dimension is called the **nullity** of **A**. The rank of **A** is equal to the number of nonzero rows in the row echelon form of **A**, the nullity of **A** is equal to the number of zero rows in the row echelon form of **A**. Thus, if **A** is a square matrix, the sum of the rank of **A** and the nullity of **A** is equal to the number of rows (columns) of **A**.

1. `NullSpace[A]` returns a list of vectors which form a basis for the nullspace (or kernel) of the matrix **A**.
2. `RowReduce[A]` yields the reduced row echelon form of the matrix **A**.

Example 5.23

Place the matrix

$$\mathbf{A} = \begin{pmatrix} -1 & -1 & 2 & 0 & -1 \\ -2 & 2 & 0 & 0 & -2 \\ 2 & -1 & -1 & 0 & 1 \\ -1 & -1 & 1 & 2 & 2 \\ 1 & -2 & 2 & -2 & 0 \end{pmatrix}$$

in reduced row echelon form. What is the rank of **A**? Find a basis for the nullspace of **A**.

Solution. We begin by defining the matrix `matrixa`. Then, `RowReduce` is used to place `matrixa` in reduced row echelon form.

capa = {{−1, −1, 2, 0, −1}, {−2, 2, 0, 0, −2},
{2, −1, −1, 0, 1}, {−1, −1, 1, 2, 2},
{1, −2, 2, −2, 0}};
RowReduce[capa]//MatrixForm

$$\begin{pmatrix} 1 & 0 & 0 & -2 & 0 \\ 0 & 1 & 0 & -2 & 0 \\ 0 & 0 & 1 & -2 & 0 \\ 0 & 0 & 0 & 0 & 1 \\ 0 & 0 & 0 & 0 & 0 \end{pmatrix}$$

Because the row-reduced form of `matrixa` contains four nonzero rows, the rank of **A** is 4 and thus the nullity is 1. We obtain a basis for the nullspace with `NullSpace`.

NullSpace[capa]

{{2, 2, 2, 1, 0}}

As expected, because the nullity is 1, a basis for the nullspace contains one vector. □

Example 5.24

Find a basis for the column space of

$$\mathbf{B} = \begin{pmatrix} 1 & -2 & 2 & 1 & -2 \\ 1 & 1 & 2 & -2 & -2 \\ 1 & 0 & 0 & 2 & -1 \\ 0 & 0 & 0 & -2 & 0 \\ -2 & 1 & 0 & 1 & 2 \end{pmatrix}.$$

Solution. A basis for the column space of **B** is the same as a basis for the row space of the transpose of **B**. We begin by defining `matrixb` and then using `Transpose` to compute the transpose of `matrixb`, naming the resulting output `tb`.

matrixb = {{1, −2, 2, 1, −2}, {1, 1, 2, −2, −2},
{1, 0, 0, 2, −1}, {0, 0, 0, −2, 0},
{−2, 1, 0, 1, 2}};
tb = Transpose[matrixb]

{{1, 1, 1, 0, −2}, {−2, 1, 0, 0, 1}, {2, 2, 0, 0, 0}, {1, −2, 2, −2, 1}, {−2, −2, −1, 0, 2}}

Next, we use `RowReduce` to row reduce `tb` and name the result `rrtb`. A basis for the column space consists of the first four elements of `rrtb`. We also use `Transpose` to show that the first four elements of `rrtb` are the same as the first four columns of the transpose of `rrtb`. Thus, the jth column of a matrix **A** can be extracted from **A** with `Transpose [A][[j]]`.

```
rrtb = RowReduce[tb];
Transpose[rrtb]//MatrixForm
```

$$\begin{pmatrix} 1 & 0 & 0 & 0 & 0 \\ 0 & 1 & 0 & 0 & 0 \\ 0 & 0 & 1 & 0 & 0 \\ 0 & 0 & 0 & 1 & 0 \\ -\frac{1}{3} & \frac{1}{3} & -2 & -3 & 0 \end{pmatrix}$$

We extract the first four elements of `rrtb` with `Take`. The results correspond to a basis for the column space of **B**.

```
Take[rrtb, 4]
```

$$\left\{\left\{1, 0, 0, 0, -\frac{1}{3}\right\}, \left\{0, 1, 0, 0, \frac{1}{3}\right\}, \{0, 0, 1, 0, -2\}, \{0, 0, 0, 1, -3\}\right\}$$ □

5.3.2 The Gram–Schmidt Process

A set of vectors $\{\mathbf{v}_1, \mathbf{v}_2, \ldots, \mathbf{v}_n\}$ is **orthonormal** means that $\|\mathbf{v}_i\| = 1$ for all values of i and $\mathbf{v}_i \cdot \mathbf{v}_j = 0$ for $i \neq j$. Given a set of linearly independent vectors $S = \{\mathbf{v}_1, \mathbf{v}_2, \ldots, \mathbf{v}_n\}$, the set of all linear combinations of the elements of S, $V = \text{span}\, S$, is a vector space. Note that if S is an orthonormal set and $\mathbf{u} \in \text{span}\, S$, then $\mathbf{u} = (\mathbf{u} \cdot \mathbf{v}_1)\mathbf{v}_1 + (\mathbf{u} \cdot \mathbf{v}_2)\mathbf{v}_2 + \cdots + (\mathbf{u} \cdot \mathbf{v}_n)\mathbf{v}_n$. Thus, we may easily express $\mathbf{u}$ as a linear combination of the vectors in S. Consequently, if we are given any vector space, V, it is frequently convenient to be able to find an orthonormal basis of V. We may use the **Gram–Schmidt process** to find an orthonormal basis of the vector space $V = \text{span}\,\{\mathbf{v}_1, \mathbf{v}_2, \ldots, \mathbf{v}_n\}$.

We summarize the algorithm of the Gram–Schmidt process so that given a set of n linearly independent vectors $S = \{\mathbf{v}_1, \mathbf{v}_2, \ldots, \mathbf{v}_n\}$, where $V = \text{span}\,\{\mathbf{v}_1, \mathbf{v}_2, \ldots, \mathbf{v}_n\}$, we can construct a set of orthonormal vectors $\{\mathbf{u}_1, \mathbf{u}_2, \ldots, \mathbf{u}_n\}$ so that $V = \text{span}\,\{\mathbf{u}_1, \mathbf{u}_2, \ldots, \mathbf{u}_n\}$.

1. Let $\mathbf{u}_1 = \dfrac{1}{\|\mathbf{v}\|}\mathbf{v}$;
2. Compute $\text{proj}_{\{\mathbf{u}_1\}}\mathbf{v}_2 = (\mathbf{u}_1 \cdot \mathbf{v}_2)\mathbf{u}_1$, $\mathbf{v}_2 - \text{proj}_{\{\mathbf{u}_1\}}\mathbf{v}_2$, and let
$$\mathbf{u}_2 = \frac{1}{\left\|\mathbf{v}_2 - \text{proj}_{\{\mathbf{u}_1\}}\mathbf{v}_2\right\|}\left(\mathbf{v}_2 - \text{proj}_{\{\mathbf{u}_1\}}\mathbf{v}_2\right).$$
Then, $\text{span}\,\{\mathbf{u}_1, \mathbf{u}_2\} = \text{span}\,\{\mathbf{v}_1, \mathbf{v}_2\}$ and $\text{span}\,\{\mathbf{u}_1, \mathbf{u}_2, \mathbf{v}_3, \ldots, \mathbf{v}_n\} = \text{span}\,\{\mathbf{v}_1, \mathbf{v}_1, \ldots, \mathbf{v}_n\}$;
3. Generally, for $3 \le i \le n$, compute
$$\text{proj}_{\{\mathbf{u}_1, \mathbf{u}_2, \ldots, \mathbf{u}_n\}}\mathbf{v}_i = (\mathbf{u}_1 \cdot \mathbf{v}_i)\mathbf{u}_1 + (\mathbf{u}_2 \cdot \mathbf{v}_i)\mathbf{u}_2 + \cdots + (\mathbf{u}_{i-1} \cdot \mathbf{v}_i)\mathbf{u}_{i-1},$$
$\mathbf{v}_i - \text{proj}_{\{\mathbf{u}_1, \mathbf{u}_2, \ldots, \mathbf{u}_n\}}\mathbf{v}_i$, and let
$$\mathbf{u}_1 = \frac{1}{\left\|\text{proj}_{\{\mathbf{u}_1, \mathbf{u}_2, \ldots, \mathbf{u}_n\}}\mathbf{v}_i\right|}\left(\text{proj}_{\{\mathbf{u}_1, \mathbf{u}_2, \ldots, \mathbf{u}_n\}}\mathbf{v}_i\right).$$

Then, span $\{\mathbf{u}_1, \mathbf{u}_2, \ldots, \mathbf{u}_i\} = \text{span}\ \{\mathbf{v}_1, \mathbf{v}_2, \ldots, \mathbf{v}_i\}$ and

$$\text{span}\ \{\mathbf{u}_1, \mathbf{u}_2, \ldots, \mathbf{u}_i, \mathbf{v}_{i+1}, \ldots, \mathbf{v}_n\} = \text{span}\ \{\mathbf{v}_1, \mathbf{v}_2, \mathbf{v}_3, \ldots, \mathbf{v}_n\};$$

and

4. Because span $\{\mathbf{u}_1, \mathbf{u}_2, \ldots, \mathbf{u}_n\} = \text{span}\ \{\mathbf{v}_1, \mathbf{v}_2, \ldots, \mathbf{v}_n\}$ and $\{\mathbf{u}_1, \mathbf{u}_2, \ldots, \mathbf{u}_n\}$ is an orthonormal set, $\{\mathbf{u}_1, \mathbf{u}_2, \ldots, \mathbf{u}_n\}$ is an orthonormal basis of V.

The Gram–Schmidt procedure is well-suited to computer arithmetic. The following code performs each step of the Gram–Schmidt process on a set of n linearly independent vectors $\{\mathbf{v}_1, \mathbf{v}_1, \ldots, \mathbf{v}_n\}$. At the completion of each step of the procedure, `gramschmidt[vecs]` prints the list of vectors corresponding to $\{\mathbf{u}_1, \mathbf{u}_2, \ldots, \mathbf{u}_i, \mathbf{v}_{i+1}, \ldots, \mathbf{v}_n\}$ and returns the list of vectors $\{\mathbf{u}_1, \mathbf{u}_2, \ldots, \mathbf{u}_n\}$. Note how comments are inserted into the code using (*...*).

```
gramschmidt[vecs_]:=Module[{n, proj, u, capw},
    (*n represents the number of vectors in the list vecs*)
    n = Length[vecs];
    (*proj[v, capw]computestheprojectionofvontocapw*)
    proj[v_, capw_]:=
Sum[capw[[i]].v capw[[i]], {i, 1, Length[capw]}];
    u[1] = vecs[[1]]/Sqrt[vecs[[1]].vecs[[1]]];
    capw = {};
    u[i_]:=u[i] = Module[{stepone},
        stepone = vecs[[i]] - proj[vecs[[i]], capw];
        Together[stepone/Sqrt[stepone.stepone]]];
    Do[
      u[i];
      AppendTo[capw, u[i]];
Print[Join[capw, Drop[vecs, i]]], {i, 1, n - 1}];
    u[n];
    AppendTo[capw, u[n]]]
```

Example 5.25

Use the Gram-Schmidt process to transform the basis $S = \left\{ \begin{pmatrix} -2 \\ -1 \\ -2 \end{pmatrix}, \begin{pmatrix} 0 \\ -1 \\ 2 \end{pmatrix}, \begin{pmatrix} 1 \\ 3 \\ -2 \end{pmatrix} \right\}$ of $\mathbf{R}^3$ into an orthonormal basis.

Solution. We proceed by defining `v1`, `v2`, and `v3` to be the vectors in the basis S and using `gramschmidt[{v1,v2,v3}]` to find an orthonormal basis.

v1 = {−2, −1, −2};

v2 = {0, −1, 2};

v3 = {1, 3, −2};

gramschmidt[{v1, v2, v3}]

$\left\{\left\{-\frac{2}{3}, -\frac{1}{3}, -\frac{2}{3}\right\}, \{0, -1, 2\}, \{1, 3, -2\}\right\}$

$\left\{\left\{-\frac{2}{3}, -\frac{1}{3}, -\frac{2}{3}\right\}, \left\{-\frac{1}{3}, -\frac{2}{3}, \frac{2}{3}\right\}, \{1, 3, -2\}\right\}$

$\left\{\left\{-\frac{2}{3}, -\frac{1}{3}, -\frac{2}{3}\right\}, \left\{-\frac{1}{3}, -\frac{2}{3}, \frac{2}{3}\right\}, \left\{-\frac{2}{3}, \frac{2}{3}, \frac{1}{3}\right\}\right\}$

On the first line of output, the result $\{\mathbf{u}_1, \mathbf{v}_2, \mathbf{v}_3\}$ is given; $\{\mathbf{u}_1, \mathbf{u}_2, \mathbf{v}_3\}$ appears on the second line; $\{\mathbf{u}_1, \mathbf{u}_2, \mathbf{u}_3\}$ follows on the third. □

Example 5.26

Compute an orthonormal basis for the subspace of $\mathbf{R}^4$ spanned by the vectors $\begin{pmatrix} 2 \\ 4 \\ 4 \\ 1 \end{pmatrix}$, $\begin{pmatrix} -4 \\ 1 \\ -3 \\ 2 \end{pmatrix}$, and $\begin{pmatrix} 1 \\ 4 \\ 4 \\ -1 \end{pmatrix}$. Also, verify that the basis vectors are orthogonal and have norm 1.

Solution. With `gramschmidt`, we compute the orthonormal basis vectors. Note that Mathematica names `oset` the last result returned by `gramschmidt`. The orthogonality of these vectors is then verified. Notice that `Together` is used to simplify the result in the case of `oset[[2]].oset[[3]]`. The norm of each vector is then found to be 1.

oset = gramschmidt[{{2, 4, 4, 1}, {−4, 1, −3, 2}, {1, 4, 4, −1}}]

$\left\{\left\{\frac{2}{\sqrt{37}}, \frac{4}{\sqrt{37}}, \frac{4}{\sqrt{37}}, \frac{1}{\sqrt{37}}\right\}, \{-4, 1, -3, 2\}, \{1, 4, 4, -1\}\right\}$

$\left\{\left\{\frac{2}{\sqrt{37}}, \frac{4}{\sqrt{37}}, \frac{4}{\sqrt{37}}, \frac{1}{\sqrt{37}}\right\}, \left\{-60\sqrt{\frac{2}{16909}}, \frac{93}{\sqrt{33818}}, -\frac{55}{\sqrt{33818}}, 44\sqrt{\frac{2}{16909}}\right\}, \{1, 4, 4, -1\}\right\}$

$\left\{\left\{\frac{2}{\sqrt{37}}, \frac{4}{\sqrt{37}}, \frac{4}{\sqrt{37}}, \frac{1}{\sqrt{37}}\right\}, \left\{-60\sqrt{\frac{2}{16909}}, \frac{93}{\sqrt{33818}}, -\frac{55}{\sqrt{33818}}, 44\sqrt{\frac{2}{16909}}\right\},\right.$
$\left.\left\{-\frac{449}{\sqrt{934565}}, \frac{268}{\sqrt{934565}}, \frac{156}{\sqrt{934565}}, -\frac{798}{\sqrt{934565}}\right\}\right\}$

The three vectors are extracted with `oset` using `oset[[1]]`, `oset[[2]]`, and `oset[[3]]`.

oset[[1]].oset[[2]]

oset[[1]].oset[[3]]

oset[[2]].oset[[3]]

0

0

0

Sqrt[oset[[1]].oset[[1]]]

Sqrt[oset[[2]].oset[[2]]]

Sqrt[oset[[3]].oset[[3]]]

1

1

1 □

Mathematica contains functions that perform most of the operations discussed here.

1. `Orthogonalize[{v1,v2,...},Method->GramSchmidt]` returns an orthonormal set of vectors given the set of vectors $\{\mathbf{v}_1, \mathbf{v}_2, \ldots, \mathbf{v}_n\}$. Note that this command does not illustrate each step of the Gram–Schmidt procedure as the `gramschmidt` function defined above.
2. `Normalize[v]` returns $\frac{1}{\|\mathbf{v}\|}\mathbf{v}$ given the nonzero vector $\mathbf{v}$.
3. `Projection[v1,v2]` returns the projection of $\mathbf{v}_1$ onto $\mathbf{v}_2$: $\text{proj}_{\mathbf{v}_2}\mathbf{v}_1 = \frac{\mathbf{v}_1 \cdot \mathbf{v}_2}{\|\mathbf{v}_2\|^2}\mathbf{v}_2$.

Thus,

Orthogonalize[{{2, 4, 4, 1}, {−4, 1, −3, 2}, {1, 4, 4, −1}}, Method → "GramSchmidt"]

$$\left\{\left\{\frac{2}{\sqrt{37}}, \frac{4}{\sqrt{37}}, \frac{4}{\sqrt{37}}, \frac{1}{\sqrt{37}}\right\}, \left\{-60\sqrt{\frac{2}{16909}}, \frac{93}{\sqrt{33818}}, -\frac{55}{\sqrt{33818}}, 44\sqrt{\frac{2}{16909}}\right\}, \left\{-\frac{449}{\sqrt{934565}}, \frac{268}{\sqrt{934565}}, \frac{156}{\sqrt{934565}}, -\frac{798}{\sqrt{934565}}\right\}\right\}$$

returns an orthonormal basis for the subspace of $\mathbf{R}^4$ spanned by the vectors $\begin{pmatrix} 2 \\ 4 \\ 4 \\ 1 \end{pmatrix}$, $\begin{pmatrix} -4 \\ 1 \\ -3 \\ 2 \end{pmatrix}$, and $\begin{pmatrix} 1 \\ 4 \\ 4 \\ -1 \end{pmatrix}$. The command

Normalize[{2, 4, 4, 1}]

$$\left\{\frac{2}{\sqrt{37}}, \frac{4}{\sqrt{37}}, \frac{4}{\sqrt{37}}, \frac{1}{\sqrt{37}}\right\}$$

finds a unit vector with the same direction as the vector $\mathbf{v} = \begin{pmatrix} 2 \\ 4 \\ 4 \\ 1 \end{pmatrix}$. Entering

Projection[{2, 4, 4, 1}, {−4, 1, −3, 2}]

$$\left\{\frac{28}{15}, -\frac{7}{15}, \frac{7}{5}, -\frac{14}{15}\right\}$$

finds the projection of $\mathbf{v} = \begin{pmatrix} 2 \\ 4 \\ 4 \\ 1 \end{pmatrix}$ onto $\mathbf{w} = \begin{pmatrix} -4 \\ 1 \\ -3 \\ 2 \end{pmatrix}$.

5.3.3 Linear Transformations

A function $T : \mathbf{R}^n \longrightarrow \mathbf{R}^m$ is a **linear transformation** means that T satisfies the properties $T(\mathbf{u} + \mathbf{v}) = T(\mathbf{u}) + T(\mathbf{v})$ and $T(c\mathbf{u}) = cT(\mathbf{u})$ for all vectors $\mathbf{u}$ and $\mathbf{v}$ in $\mathbf{R}^n$ and all real numbers c. Let $T : \mathbf{R}^n \longrightarrow \mathbf{R}^m$ be a linear transformation and suppose $T(\mathbf{e}_1) = \mathbf{v}_1$, $T(\mathbf{e}_2) = \mathbf{v}_2$, ..., $T(\mathbf{e}_n) = \mathbf{v}_n$ where $\{\mathbf{e}_1, \mathbf{e}_2, \ldots, \mathbf{e}_n\}$ represents the standard basis of $\mathbf{R}^n$ and $\mathbf{v}_1$, $\mathbf{v}_2$, ..., $\mathbf{v}_n$ are (column) vectors in $\mathbf{R}^m$. The **associated matrix** of T is the $m \times n$ matrix $\mathbf{A} = \begin{pmatrix} \mathbf{v}_1 & \mathbf{v}_2 & \cdots & \mathbf{v}_n \end{pmatrix}$:

$$\text{if } \mathbf{x} = \begin{pmatrix} x_1 \\ x_2 \\ \vdots \\ x_n \end{pmatrix}, \quad T(\mathbf{x}) = T\left(\begin{pmatrix} x_1 \\ x_2 \\ \vdots \\ x_n \end{pmatrix}\right) = \mathbf{A}\mathbf{x} = \begin{pmatrix} \mathbf{v}_1 & \mathbf{v}_2 & \cdots & \mathbf{v}_n \end{pmatrix} \begin{pmatrix} x_1 \\ x_2 \\ \vdots \\ x_n \end{pmatrix}$$

Moreover, if $\mathbf{A}$ is any $m \times n$ matrix, then $\mathbf{A}$ is the associated matrix of the linear transformation defined by $T(\mathbf{x}) = \mathbf{A}\mathbf{x}$. In fact, a linear transformation T is completely determined by its action on any basis.

The **kernel** of the linear transformation T, $\ker(T)$, is the set of all vectors $\mathbf{x}$ in $\mathbf{R}^n$ such that $T(\mathbf{x}) = \mathbf{0}$: $\ker(T) = \{x \in \mathbf{R}^n | T(\mathbf{x}) = \mathbf{0}\}$. The kernel of T is a subspace of $\mathbf{R}^n$. Because $T(\mathbf{x}) = \mathbf{A}\mathbf{x}$ for all $\mathbf{x}$ in $\mathbf{R}^n$, $\ker(T) = \{x \in \mathbf{R}^n | T(\mathbf{x}) = \mathbf{0}\} = \{x \in \mathbf{R}^n | \mathbf{A}\mathbf{x} = \mathbf{0}\}$ so the kernel of T is the same as the nullspace of $\mathbf{A}$.

Example 5.27

Let $T : \mathbf{R}^5 \longrightarrow \mathbf{R}^3$ be the linear transformation defined by $T(\mathbf{x}) = \begin{pmatrix} 0 & -3 & -1 & -3 & -1 \\ -3 & 3 & -3 & -3 & -1 \\ 2 & 2 & -1 & 1 & 2 \end{pmatrix} \mathbf{x}$.

(a) Calculate a basis for the kernel of the linear transformation. (b) Determine which of the vectors $\begin{pmatrix} 4 \\ 2 \\ 0 \\ 0 \\ -6 \end{pmatrix}$ and $\begin{pmatrix} 1 \\ 2 \\ -1 \\ -2 \\ 3 \end{pmatrix}$ is in the kernel of T.

Solution. We begin by defining `matrixa` to be the matrix $\mathbf{A} = \begin{pmatrix} 0 & -3 & -1 & -3 & -1 \\ -3 & 3 & -3 & -3 & -1 \\ 2 & 2 & -1 & 1 & 2 \end{pmatrix}$ and then defining `t`. A basis for the kernel of T is the same as a basis for the nullspace of $\mathbf{A}$ found with `NullSpace`.

```
Clear[t, x, matrixa]

matrixa = {{0, -3, -1, -3, -1}, {-3, 3, -3, -3, -1},

{2, 2, -1, 1, 2}};

t[x_] = matrixa.x;

NullSpace[matrixa]

{{-2, -1, 0, 0, 3}, {-6, -8, -15, 13, 0}}
```

Because $\begin{pmatrix} 4 \\ 2 \\ 0 \\ 0 \\ -6 \end{pmatrix}$ is a linear combination of the vectors that form a basis for the kernel, $\begin{pmatrix} 4 \\ 2 \\ 0 \\ 0 \\ -6 \end{pmatrix}$ is in the kernel while $\begin{pmatrix} 1 \\ 2 \\ -1 \\ -2 \\ 3 \end{pmatrix}$ is not. These results are verified by evaluating `t` for each vector.

t[{4, 2, 0, 0, −6}]

{0, 0, 0}

t[{1, 2, −1, −2, 3}]

{−2, 9, 11} □

Application: Rotations

Let $\mathbf{x} = \begin{pmatrix} x_1 \\ x_2 \end{pmatrix}$ be a vector in $\mathbf{R}^2$ and θ an angle. Then, there are numbers r and ϕ given by $r = \sqrt{x_1{}^2 + x_2{}^2}$ and $\phi = \tan^{-1}(x_2/x_1)$ so that $x_1 = r\cos\phi$ and $x_2 = r\sin\phi$. When we rotate $\mathbf{x} = \begin{pmatrix} x_1 \\ x_2 \end{pmatrix} = \begin{pmatrix} r\cos\phi \\ r\sin\phi \end{pmatrix}$ through the angle θ, we obtain the vector $\mathbf{x}' = \begin{pmatrix} r\cos(\theta+\phi) \\ r\sin(\theta+\phi) \end{pmatrix}$. Using the trigonometric identities $\sin(\theta \pm \phi) = \sin\theta\cos\phi \pm \sin\phi\cos\theta$ and $\cos(\theta \pm \phi) = \cos\theta\cos\phi \mp \sin\theta\sin\phi$ we rewrite

$$\mathbf{x}' = \begin{pmatrix} r\cos(\theta+\phi) \\ r\sin(\theta+\phi) \end{pmatrix} = \begin{pmatrix} r\cos\theta\cos\phi - r\sin\theta\sin\phi \\ r\sin\theta\cos\phi + r\sin\phi\cos\theta \end{pmatrix} = \begin{pmatrix} \cos\theta & -\sin\theta \\ \sin\theta & \cos\theta \end{pmatrix} \begin{pmatrix} r\cos\phi \\ r\sin\phi \end{pmatrix}$$
$$= \begin{pmatrix} \cos\theta & -\sin\theta \\ \sin\theta & \cos\theta \end{pmatrix} \begin{pmatrix} x_1 \\ x_2 \end{pmatrix}.$$

Thus, the vector $\mathbf{x}'$ is obtained from $\mathbf{x}$ by computing $\begin{pmatrix} \cos\theta & -\sin\theta \\ \sin\theta & \cos\theta \end{pmatrix} \mathbf{x}$. Generally, if θ represents an angle, the linear transformation $T : \mathbf{R}^2 \longrightarrow \mathbf{R}^2$ defined by $T(\mathbf{x}) = \begin{pmatrix} \cos\theta & -\sin\theta \\ \sin\theta & \cos\theta \end{pmatrix} \mathbf{x}$ is called the **rotation of $\mathbf{R}^2$ through the angle** θ. We write code to rotate a polygon through an angle θ. The procedure `rotate` uses a list of n points and the rotation matrix defined in `r` to produce a new list of points that are joined using the `Line` graphics directive. Entering

```
Line[{{x1,y1},{x2,y2},...,{xn,yn}}]
```

represents the graphics primitive for a line in two dimensions that connects the points listed in `{{x1,y1},{x2,y2},...,{xn,yn}}`. Entering

```
Show[Graphics[Line[{{x1,y1},{x2,y2},...,{xn,yn}}]]]
```

displays the line. This rotation can be determined for one value of θ. However, a more interesting result is obtained by creating a list of rotations for a sequence of angles and then displaying the graphics objects. This is done for $\theta = 0$ to $\theta = \pi/2$ using increments of $\pi/16$. Hence, a list of nine graphs is given for the square with vertices $(-1, 1)$, $(1, 1)$, $(1, -1)$, and $(-1, -1)$ and displayed in Fig. 5.7.

FIGURE 5.7 A rotated square.

$$r[\theta\_]=\begin{pmatrix} \text{Cos}[\theta] & -\text{Sin}[\theta] \\ \text{Sin}[\theta] & \text{Cos}[\theta] \end{pmatrix};$$

$$r[\theta\_]=\begin{pmatrix} \text{Cos}[\theta] & -\text{Sin}[\theta] \\ \text{Sin}[\theta] & \text{Cos}[\theta] \end{pmatrix};$$

```
rotate[pts_, angle_]:=Module[{newpts},
    newpts = Table[r[angle].pts[[i]], {i, 1, Length[pts]}];
newpts = AppendTo[newpts, newpts[[1]]];
    figure = Line[newpts];
    Show[Graphics[figure], AspectRatio → 1,
  PlotRange → {{−1.5, 1.5}, {−1.5, 1.5}},
  ]]
```

```
graphs = Table[rotate[{{−1, 1}, {1, 1}, {1, −1}, {−1, −1}}, t], {t, 0, π/2, π/16}];
array = Partition[graphs, 3];
Show[GraphicsGrid[array]]
```

5.3.4 Eigenvalues and Eigenvectors

Let **A** be an $n \times n$ matrix. λ is an **eigenvalue** of **A** if there is a *nonzero* vector, **v**, called an **eigenvector**, satisfying $\mathbf{Av} = \lambda\mathbf{v}$. Because $(\mathbf{A} - \lambda\mathbf{I})\mathbf{v} = \mathbf{0}$ has a unique solution of $\mathbf{v} = \mathbf{0}$ if $|\mathbf{A} - \lambda\mathbf{I}| \neq 0$, to find non-zero solutions, **v**, of $(\mathbf{A} - \lambda\mathbf{I})\mathbf{v} = \mathbf{0}$, we begin by solving $|\mathbf{A} - \lambda\mathbf{I}| = 0$. That is, we find the eigenvalues of **A** by solving the **characteristic polynomial** $|\mathbf{A} - \lambda\mathbf{I}| = 0$ for λ. Once we find the eigenvalues, the corresponding eigenvectors are found by solving $(\mathbf{A} - \lambda\mathbf{I})\,\mathbf{v} = \mathbf{0}$ for **v**.

If $\mathbf{A}$ is $n \times n$, `Eigenvalues[A]` finds the eigenvalues of $\mathbf{A}$, `Eigenvectors[A]` finds the eigenvectors, and `Eigensystem[A]` finds the eigenvalues and corresponding eigenvectors. `CharacteristicPolynomial[A,lambda]` finds the characteristic polynomial of $\mathbf{A}$ as a function of λ.

Example 5.28

Find the eigenvalues and corresponding eigenvectors for each of the following matrices. (a) $\mathbf{A} = \begin{pmatrix} -3 & 2 \\ 2 & -3 \end{pmatrix}$, (b) $\mathbf{A} = \begin{pmatrix} 1 & -1 \\ 1 & 3 \end{pmatrix}$, (c) $\mathbf{A} = \begin{pmatrix} 0 & 1 & 1 \\ 1 & 0 & 1 \\ 1 & 1 & 0 \end{pmatrix}$, and (d) $\mathbf{A} = \begin{pmatrix} -1/4 & 2 \\ -8 & -1/4 \end{pmatrix}$.

Solution. (a) We begin by finding the eigenvalues. Solving

$$|\mathbf{A} - \lambda \mathbf{I}| = \begin{vmatrix} -3-\lambda & 2 \\ 2 & -3-\lambda \end{vmatrix} = \lambda^2 + 6\lambda + 5 = 0$$

gives us $\lambda_1 = -5$ and $\lambda_2 = -1$.

Observe that the same results are obtained using `CharacteristicPolynomial` and `Eigenvalues`.

```
capa = {{-3, 2}, {2, -3}};

cp1 = CharacteristicPolynomial[capa, λ]

5 + 6λ + λ^2

Solve[cp1 == 0]

{{λ → -5}, {λ → -1}}

e1 = Eigenvalues[capa]

{-5, -1}
```

We now find the corresponding eigenvectors. Let $\mathbf{v}_1 = \begin{pmatrix} x_1 \\ y_1 \end{pmatrix}$ be an eigenvector corresponding to λ_1, then

$$(\mathbf{A} - \lambda_1 \mathbf{I})\,\mathbf{v}_1 = \mathbf{0}$$

$$\left[\begin{pmatrix} -3 & 2 \\ 2 & -3 \end{pmatrix} - (-5)\begin{pmatrix} 1 & 0 \\ 0 & 1 \end{pmatrix}\right]\begin{pmatrix} x_1 \\ y_1 \end{pmatrix} = \begin{pmatrix} 0 \\ 0 \end{pmatrix}$$

$$\begin{pmatrix} 2 & 2 \\ 2 & 2 \end{pmatrix}\begin{pmatrix} x_1 \\ y_1 \end{pmatrix} = \begin{pmatrix} 0 \\ 0 \end{pmatrix},$$

which row reduces to

$$\begin{pmatrix} 1 & 1 \\ 0 & 0 \end{pmatrix}\begin{pmatrix} x_1 \\ y_1 \end{pmatrix} = \begin{pmatrix} 0 \\ 0 \end{pmatrix}.$$

That is, $x_1 + y_1 = 0$ or $x_1 = -y_1$. Hence, for any value of $y_1 \neq 0$,

$$\mathbf{v}_1 = \begin{pmatrix} x_1 \\ y_1 \end{pmatrix} = \begin{pmatrix} -y_1 \\ y_1 \end{pmatrix} = y_1 \begin{pmatrix} -1 \\ 1 \end{pmatrix}$$

is an eigenvector corresponding to λ_1. Of course, this represents infinitely many vectors. But, they are all linearly dependent. Choosing $y_1 = 1$ yields $\mathbf{v}_1 = \begin{pmatrix} -1 \\ 1 \end{pmatrix}$. Note that you might have chosen $y_1 = -1$ and obtained $\mathbf{v}_1 = \begin{pmatrix} 1 \\ -1 \end{pmatrix}$. However, both of our results are "correct" because these vectors are linearly dependent.

Similarly, letting $\mathbf{v}_2 = \begin{pmatrix} x_2 \\ y_2 \end{pmatrix}$ be an eigenvector corresponding to λ_2 we solve $(\mathbf{A} - \lambda_2 \mathbf{I}) \times \mathbf{v}_1 = \mathbf{0}$:

$$\begin{pmatrix} -2 & 2 \\ 2 & -2 \end{pmatrix} \begin{pmatrix} x_2 \\ y_2 \end{pmatrix} = \begin{pmatrix} 0 \\ 0 \end{pmatrix} \quad \text{or} \quad \begin{pmatrix} 1 & -1 \\ 0 & 0 \end{pmatrix} \begin{pmatrix} x_2 \\ y_2 \end{pmatrix} = \begin{pmatrix} 0 \\ 0 \end{pmatrix}.$$

Thus, $x_2 - y_2 = 0$ or $x_2 = y_2$. Hence, for any value of $y_2 \neq 0$,

$$\mathbf{v}_2 = \begin{pmatrix} x_2 \\ y_2 \end{pmatrix} = \begin{pmatrix} y_2 \\ y_2 \end{pmatrix} = y_2 \begin{pmatrix} 1 \\ 1 \end{pmatrix}$$

is an eigenvector corresponding to λ_2. Choosing $y_2 = 1$ yields $\mathbf{v}_2 = \begin{pmatrix} 1 \\ 1 \end{pmatrix}$. We confirm these results using `RowReduce`.

```
i2 = IdentityMatrix[2];
ev1 = capa − e1[[1]]i2
```

{{2, 2}, {2, 2}}

```
RowReduce[ev1]
```

{{1, 1}, {0, 0}}

We obtain the same results using `Eigenvectors` and `Eigensystem`.

```
Eigenvectors[capa]
Eigensystem[capa]
```

{{−1, 1}, {1, 1}}

{{−5, −1}, {{−1, 1}, {1, 1}}}

(b) In this case, we see that $\lambda = 2$ has multiplicity 2. There is only one linearly independent eigenvector, $\mathbf{v} = \begin{pmatrix} -1 \\ 1 \end{pmatrix}$, corresponding to λ.

```
capa = {{1, −1}, {1, 3}};
Factor[CharacteristicPolynomial[capa, λ]]
```

$(-2 + \lambda)^2$

```
Eigenvectors[capa]
```

{{−1, 1}, {0, 0}}

Eigensystem[capa]

{{2, 2}, {{−1, 1}, {0, 0}}}

(c) The eigenvalue $\lambda_1 = 2$ has corresponding eigenvector $\mathbf{v}_1 = \begin{pmatrix} 1 \\ 1 \\ 1 \end{pmatrix}$. The eigenvalue $\lambda_{2,3} = -1$ has multiplicity 2. In this case, there are two linearly independent eigenvectors corresponding to this eigenvalue: $\mathbf{v}_2 = \begin{pmatrix} -1 \\ 0 \\ 1 \end{pmatrix}$ and $\mathbf{v}_3 = \begin{pmatrix} -1 \\ 1 \\ 0 \end{pmatrix}$.

capa = {{0, 1, 1}, {1, 0, 1}, {1, 1, 0}};

Factor[CharacteristicPolynomial[capa, λ]]

$-(-2+\lambda)(1+\lambda)^2$

Eigenvectors[capa]

{{1, 1, 1}, {−1, 0, 1}, {−1, 1, 0}}

Eigensystem[capa]

{{2, −1, −1}, {{1, 1, 1}, {−1, 0, 1}, {−1, 1, 0}}}

(d) In this case, the eigenvalues $\lambda_{1,2} = -\frac{1}{4} \pm 4i$ are complex conjugates. We see that the eigenvectors $\mathbf{v}_{1,2} = \begin{pmatrix} 0 \\ 2 \end{pmatrix} \pm \begin{pmatrix} 1 \\ 0 \end{pmatrix} i$ are complex conjugates as well.

capa = {{−1/4, 2}, {−8, −1/4}};

Factor[CharacteristicPolynomial[capa, λ],

GaussianIntegers → True]

$\frac{1}{16}((1-16i)+4\lambda)((1+16i)+4\lambda)$

Eigenvalues[capa]

$\left\{-\frac{1}{4}+4i, -\frac{1}{4}-4i\right\}$

Eigenvectors[capa]

{{−i, 2}, {i, 2}}

Eigensystem[capa]

$\left\{\left\{-\frac{1}{4}+4i, -\frac{1}{4}-4i\right\}, \{\{-i, 2\}, \{i, 2\}\}\right\}$ □

5.3.5 Jordan Canonical Form

Let $\mathbf{N}_k = (n_{ij}) = \begin{cases} 1, & j = i+1 \\ 0, & \text{otherwise} \end{cases}$ represent a $k \times k$ matrix with the indicated elements. The $k \times k$ **Jordan block matrix** is given by $\mathbf{B}(\lambda) = \lambda\mathbf{I} + \mathbf{N}_k$ where λ is a constant:

$$\mathbf{N}_k = \begin{pmatrix} 0 & 1 & 0 & \cdots & 0 \\ 0 & 0 & 1 & \cdots & 0 \\ \vdots & \vdots & \vdots & & \vdots \\ 0 & 0 & 0 & \cdots & 1 \\ 0 & 0 & 0 & \cdots & 0 \end{pmatrix} \quad \text{and} \quad \mathbf{B}(\lambda) = \lambda\mathbf{I} + \mathbf{N}_k = \begin{pmatrix} \lambda & 1 & 0 & \cdots & 0 \\ 0 & \lambda & 1 & \cdots & 0 \\ \vdots & \vdots & \vdots & & \vdots \\ 0 & 0 & 0 & \cdots & 1 \\ 0 & 0 & 0 & \cdots & \lambda \end{pmatrix}.$$

Hence, $\mathbf{B}(\lambda)$ can be defined as $\mathbf{B}(\lambda) = (b_{ij}) = \begin{cases} \lambda, & i = j \\ 1, & j = i+1 \\ 0, & \text{otherwise} \end{cases}$. A **Jordan matrix** has the form

$$\mathbf{J} = \begin{pmatrix} \mathbf{B}_1(\lambda) & 0 & \cdots & 0 \\ 0 & \mathbf{B}_2(\lambda) & \cdots & 0 \\ \vdots & \vdots & & \vdots \\ 0 & 0 & \cdots & \mathbf{B}_n(\lambda) \end{pmatrix}$$

where the entries $\mathbf{B}_j(\lambda)$, $j = 1, 2, \ldots, n$ represent Jordan block matrices.

Suppose that $\mathbf{A}$ is an $n \times n$ matrix. Then there is an invertible $n \times n$ matrix $\mathbf{C}$ such that $\mathbf{C}^{-1}\mathbf{AC} = \mathbf{J}$ where $\mathbf{J}$ is a Jordan matrix with the eigenvalues of $\mathbf{A}$ as diagonal elements. The matrix $\mathbf{J}$ is called the **Jordan canonical form** of $\mathbf{A}$. The command

```
JordanDecomposition[m]
```

yields a list of matrices `{s,j}` such that `m==s.j.Inverse[s]` and `j` is the Jordan canonical form of the matrix `m`.

For a given matrix $\mathbf{A}$, the unique monic polynomial q of least degree satisfying $q(\mathbf{A}) = 0$ is called the **minimal polynomial of A**. Let p denote the characteristic polynomial of $\mathbf{A}$. Because $p(\mathbf{A}) = 0$, it follows that q divides p. We can use the Jordan canonical form of a matrix to determine its minimal polynomial.

Example 5.29

Find the Jordan canonical form, $\mathbf{J_A}$, of $\mathbf{A} = \begin{pmatrix} 2 & 9 & -9 \\ 0 & 8 & -6 \\ 0 & 9 & -7 \end{pmatrix}$.

Solution. After defining `matrixa`, we use `JordanDecomposition` to find the Jordan canonical form of `a` and name the resulting output `ja`.

```
matrixa = {{2, 9, -9}, {0, 8, -6}, {0, 9, -7}};
ja = JordanDecomposition[matrixa]
```

{{{3, 0, 1}, {2, 1, 0}, {3, 1, 0}}, {{−1, 0, 0}, {0, 2, 0}, {0, 0, 2}}}

The Jordan matrix corresponds to the second element of `ja` extracted with `ja[[2]]` and displayed in `MatrixForm`.

ja[[2]]//MatrixForm

$$\begin{pmatrix} -1 & 0 & 0 \\ 0 & 2 & 0 \\ 0 & 0 & 2 \end{pmatrix}$$

We also verify that the matrices `ja[[1]]` and `ja[[2]]` satisfy

```
matrixa=ja[[1]].ja[[2]].Inverse[ja[[1]]].
```

ja[[1]].ja[[2]].Inverse[ja[[1]]]

$\{\{2, 9, -9\}, \{0, 8, -6\}, \{0, 9, -7\}\}$

Next, we use `CharacteristicPolynomial` to find the characteristic polynomial of `matrixa` and then verify that `matrixa` satisfies its characteristic polynomial.

p = CharacteristicPolynomial[matrixa, x]

$-4 + 3x^2 - x^3$

−4IdentityMatrix[3] + 3MatrixPower[matrixa, 2] − MatrixPower[matrixa, 3]

$\{\{0, 0, 0\}, \{0, 0, 0\}, \{0, 0, 0\}\}$

From the Jordan form, we see that the minimal polynomial of **A** is $(x + 1)(x - 2)$. We define the minimal polynomial to be `q` and then verify that `matrixa` satisfies its minimal polynomial.

q = Expand[(x + 1)(x − 2)]

$-2 - x + x^2$

−2IdentityMatrix[3] − matrixa + MatrixPower[matrixa, 2]

$\{\{0, 0, 0\}, \{0, 0, 0\}, \{0, 0, 0\}\}$

As expected, `q` divides `p`.

Cancel[p/q]

$2 - x$ □

Example 5.30

If $\mathbf{A} = \begin{pmatrix} 3 & 8 & 6 & -1 \\ -3 & 2 & 0 & 3 \\ 3 & -3 & -1 & -3 \\ 4 & 8 & 6 & -2 \end{pmatrix}$, find the characteristic and minimal polynomials of **A**.

Solution. As in the previous example, we first define `matrixa` and then use `JordanDecomposition` to find the Jordan canonical form of **A**.

```
matrixa = {{3, 8, 6, -1}, {-3, 2, 0, 3}, {3, -3, -1, -3},
{4, 8, 6, -2}};
ja = JordanDecomposition[matrixa]
```

$$\left\{\left\{\{3,-1,1,0\},\left\{-1,-1,0,\tfrac{1}{2}\right\},\left\{0,2,0,-\tfrac{1}{2}\right\},\{4,0,1,0\}\right\},\right.$$
$$\left.\{\{-1,0,0,0\},\{0,-1,0,0\},\{0,0,2,1\},\{0,0,0,2\}\}\right\}$$

The Jordan canonical form of **A** is the second element of `ja`, extracted with `ja[[2]]` and displayed in `MatrixForm`.

```
ja[[2]]//MatrixForm
```

$$\begin{pmatrix} -1 & 0 & 0 & 0 \\ 0 & -1 & 0 & 0 \\ 0 & 0 & 2 & 1 \\ 0 & 0 & 0 & 2 \end{pmatrix}$$

From this result, we see that the minimal polynomial of **A** is $(x+1)(x-2)^2$. We define `q` to be the minimal polynomial of **A** and then verify that `matrixa` satisfies `q`.

```
q = Expand[(x - 2)^2(x + 1)]
```

$4 - 3x^2 + x^3$

```
4IdentityMatrix[4] - 3MatrixPower[matrixa, 2] + MatrixPower[matrixa, 3]
```

$\{\{0,0,0,0\},\{0,0,0,0\},\{0,0,0,0\},\{0,0,0,0\}\}$

The characteristic polynomial is obtained next and named `p`. As expected, `q` divides `p`, verified with `Cancel`.

```
p = CharacteristicPolynomial[matrixa, x]
```

$4 + 4x - 3x^2 - 2x^3 + x^4$

```
Cancel[p/q]
```

$1 + x$ □

5.3.6 The QR Method

The **conjugate transpose** (or **Hermitian adjoint matrix**) of the $m \times n$ complex matrix $\mathbf{A}$ which is denoted by $\mathbf{A}^*$ is the transpose of the complex conjugate of $\mathbf{A}$. Symbolically, we have $\mathbf{A}^* = (\bar{\mathbf{A}})^t$. A complex matrix $\mathbf{A}$ is **unitary** if $\mathbf{A}^* = \mathbf{A}^{-1}$. Given a matrix $\mathbf{A}$, there is a unitary matrix $\mathbf{Q}$ and an upper triangular matrix $\mathbf{R}$ such that $\mathbf{A} = \mathbf{QR}$. The product matrix $\mathbf{QR}$ is called the **QR factorization of A**. The command

```
QRDecomposition[N[m]]
```

determines the QR decomposition of the matrix `m` by returning the list `{q,r}`, where `q` is an orthogonal matrix, `r` is an upper triangular matrix and `m=Transpose[q].r`.

Example 5.31

Find the QR factorization of the matrix $\mathbf{A} = \begin{pmatrix} 4 & -1 & 1 \\ -1 & 4 & 1 \\ 1 & 1 & 4 \end{pmatrix}$.

Solution. We define `matrixa` and then use `QRDecomposition` to find the QR decomposition of `matrixa`, naming the resulting output `qrm`.

matrixa = {{4, −1, 1}, {−1, 4, 1}, {1, 1, 4}};

qrm = QRDecomposition[*N*[matrixa]]

{{{−0.942809, 0.235702, −0.235702}, {−0.142134, −0.92387, −0.355335},

{−0.301511, −0.301511, 0.904534}}, {{−4.24264, 1.64992, −1.64992},

{0., −3.90868, −2.48734}, {0., 0., 3.01511}}}

The first matrix in `qrm` is extracted with `qrm[[1]]` and the second with `qrm[[2]]`.

qrm[[1]]//MatrixForm

$$\begin{pmatrix} -0.942809 & 0.235702 & -0.235702 \\ -0.142134 & -0.92387 & -0.355335 \\ -0.301511 & -0.301511 & 0.904534 \end{pmatrix}$$

qrm[[2]]//MatrixForm

$$\begin{pmatrix} -4.24264 & 1.64992 & -1.64992 \\ 0. & -3.90868 & -2.48734 \\ 0. & 0. & 3.01511 \end{pmatrix}$$

We verify that the results returned are the QR decomposition of **A**.

Transpose[qrm[[1]]].qrm[[2]]//MatrixForm

$$\begin{pmatrix} 4. & -1. & 1. \\ -1. & 4. & 1. \\ 1. & 1. & 4. \end{pmatrix}$$

□

One of the most efficient and most widely used methods for numerically calculating the eigenvalues of a matrix is the QR Method. Given a matrix **A**, then there is a Hermitian matrix **Q** and an upper triangular matrix **R** such that $\mathbf{A} = \mathbf{QR}$. If we define a sequence of matrices $\mathbf{A}_1 = \mathbf{A}$, factored as $\mathbf{A}_1 = \mathbf{Q}_1\mathbf{R}_1$; $\mathbf{A}_2 = \mathbf{R}_1\mathbf{Q}_1$, factored as $\mathbf{A}_2 = \mathbf{R}_2\mathbf{Q}_2$; $\mathbf{A}_3 = \mathbf{R}_2\mathbf{Q}_2$, factored as $\mathbf{A}_2 = \mathbf{R}_3\mathbf{Q}_3$; and in general, $\mathbf{A}_k = \mathbf{R}_{k+1}\mathbf{Q}_{k+1}$, $k = 1, 2, \ldots$ then the sequence $\{\mathbf{A}_n\}$ converges to a triangular matrix with the eigenvalues of **A** along the diagonal or to a nearly triangular matrix from which the eigenvalues of **A** can be calculated rather easily.

Example 5.32

Consider the 3×3 matrix $\mathbf{A} = \begin{pmatrix} 4 & -1 & 1 \\ -1 & 4 & 1 \\ 1 & 1 & 4 \end{pmatrix}$. Approximate the eigenvalues of **A** with the QR Method.

Solution. We define the sequence `a` and `qr` recursively. We define `a` using the form `a[n_]:=a[n]=...` and `qr` using the form `qr[n_]:=qr[n]=...` so that Mathematica "remembers" the values of `a` and `qr` computed, and thus Mathematica avoids recomputing values previously computed. This is of particular advantage when computing `a[n]` and `qr[n]` for large values of n.

```
matrixa = {{4, −1, 1}, {−1, 4, 1}, {1, 1, 4}};
a[1] = N[matrixa];
qr[1] = QRDecomposition[a[1]];
```

```
a[n_]:=a[n] = qr[n − 1][[2]].Transpose[qr[n − 1][[1]]];
qr[n_]:=qr[n] = QRDecomposition[a[n]];
```

We illustrate `a[n]` and `qr[n]` by computing `qr[9]` and `a[10]`. Note that computing `a[10]` requires the computation of `qr[9]`. From the results, we suspect that the eigenvalues of **A** are 5 and 2.

```
qr[9]
```

```
{{{−1., 2.2317292355411738*^−7, −0.000278046},
{−8.926920162652915*^−8, −1., −0.000481589}, {−0.000278046, −0.000481589, 1.}},
{{−5., 1.5622104650343012*^−6, −0.00194632}, {0., −5., −0.00337112}, {0., 0., 2.}}}
```

a[10]//MatrixForm

$$\begin{pmatrix} 5. & -1.7853841951182494*\wedge-7 & -0.000556091 \\ -1.7853841934911076*\wedge-7 & 5. & -0.000963178 \\ -0.000556091 & -0.000963178 & 2. \end{pmatrix}$$

Next, we compute `a[n]` for $n = 5$, 10, and 15, displaying the result in `TableForm`. We obtain further evidence that the eigenvalues of **A** are 5 and 2.

Table[a[n]//MatrixForm, {n, 5, 15, 5}]//TableForm

$$\begin{pmatrix} 4.99902 & -0.001701 & 0.0542614 \\ -0.001701 & 4.99706 & 0.0939219 \\ 0.0542614 & 0.0939219 & 2.00393 \end{pmatrix}$$

$$\begin{pmatrix} 5. & -1.7853841951182494*\wedge-7 & -0.000556091 \\ -1.7853841934911076*\wedge-7 & 5. & -0.000963178 \\ -0.000556091 & -0.000963178 & 2. \end{pmatrix}$$

$$\begin{pmatrix} 5. & -1.872126315868421*\wedge-11 & 5.694375937113059*\wedge-6 \\ -1.872110026712737*\wedge-11 & 5. & 9.862948441373512*\wedge-6 \\ 5.694375937403863*\wedge-6 & 9.862948440910027*\wedge-6 & 2. \end{pmatrix}$$

We verify that the eigenvalues of **A** are indeed 5 and 2 with `Eigenvalues`.

Eigenvalues[matrixa]

{5, 5, 2} □

5.4 MAXIMA AND MINIMA USING LINEAR PROGRAMMING

5.4.1 The Standard Form of a Linear Programming Problem

We call the linear programming problem of the following form the **standard form** of the linear programming problem:

$$\text{Minimize } Z = \underbrace{c_1x_1 + c_2x_2 + \cdots + c_nx_n}_{\text{function}}, \text{ subject to the restrictions}$$

$$\begin{cases} a_{11}x_1 + a_{12}x_2 + \cdots + a_{1n}x_n \le b_1 \\ a_{21}x_1 + a_{22}x_2 + \cdots + a_{2n}x_n \le b_2 \\ \vdots \\ a_{m1}x_1 + a_{m2}x_2 + \cdots + a_{mn}x_n \le b_m \end{cases}$$

and $x_1 \ge 0, x_2 \ge 0, \ldots, x_n \ge 0$. (5.2)

The command

```
Minimize[{function,inequalities},{variables}]
```

solves the standard form of the linear programming problem. Similarly, the command

```
Maximize[{function,inequalities},{variables}]
```

solves the linear programming problem: Maximize $Z = \underbrace{c_1x_1 + c_2x_2 + \cdots + c_nx_n}_{\text{function}}$, subject to the restrictions

$$\begin{cases} a_{11}x_1 + a_{12}x_2 + \cdots + a_{1n}x_n \le b_1 \\ a_{21}x_1 + a_{22}x_2 + \cdots + a_{2n}x_n \le b_2 \\ \vdots \\ a_{m1}x_1 + a_{m2}x_2 + \cdots + a_{mn}x_n \le b_m \end{cases}$$

and $x_1 \ge 0, x_2 \ge 0, \ldots, x_n \ge 0$.

Example 5.33

Maximize $Z(x_1, x_2, x_3) = 4x_1 - 3x_2 + 2x_3$ subject to the constraints $3x_1 - 5x_2 + 2x_3 \le 60$, $x_1 - x_2 + 2x_3 \le 10$, $x_1 + x_2 - x_3 \le 20$, and x_1, x_2, x_3 all nonnegative.

Solution. In order to solve a linear programming problem with Mathematica, the variables `{x1,x2,x3}` and objective function `z[x1,x2,x3]` are first defined. In an effort to limit the amount of typing required to complete the problem, the set of inequalities is assigned the name `ineqs` while the set of variables is called `vars`. The symbol "<=", obtained by typing the "<" key and then the "=" key, represents "less than or equal to" and is used in `ineqs`. Hence, the maximization problem is solved with the command

```
Maximize[{z[x1,x2,x3],ineqs},vars].
```

```
Clear[x1, x2, x3, z, ineqs, vars]

vars = {x1, x2, x3};

z[x1_, x2_, x3_] = 4x1 - 3x2 + 2x3;

ineqs = {3x1 - 5x2 + x3 ≤ 60, x1 - x2 + 2x3 ≤ 10, x1 + x2 - x3 ≤ 20,
  x1 ≥ 0, x2 ≥ 0, x3 ≥ 0};

Maximize[{z[x1, x2, x3], ineqs}, vars]

{45, {x1 → 15, x2 → 5, x3 → 0}}
```

The solution gives the maximum value of `z` subject to the given constraints as well as the values of `x1`, `x2`, and `x3` that maximize `z`. Thus, we see that the maximum value of Z is 45 if $x_1 = 15$, $x_2 = 5$, and $x_3 = 0$. □

We demonstrate the use of `Minimize` in the following example.

Example 5.34

Minimize $Z(x, y, z) = 4x - 3y + 2z$ subject to the constraints $3x - 5y + z \le 60$, $x - y + 2z \le 10$, $x + y - z \le 20$, and x, y, z all nonnegative.

Solution. After clearing all previously used names of functions and variable values, the variables, objective function, and set of constraints for this problem are defined and entered as they were in the first example. By using `Minimize`, the minimum value of the objective function is obtained as well as the variable values that give this minimum.

Clear[x1, x2, x3, z, ineqs, vars]

vars = {x1, x2, x3};

z[x1_, x2_, x3_] = 4x1 − 3x2 + 2x3;

ineqs = {3x1 − 5x2 + x3 ≤ 60, x1 − x2 + 2x3 ≤ 10, x1 + x2 − x3 ≤ 20,

x1 ≥ 0, x2 ≥ 0, x3 ≥ 0};

Minimize[{z[x1, x2, x3], ineqs}, vars]

{−90, {x1 → 0, x2 → 50, x3 → 30}}

We conclude that the minimum value is -90 and occurs if $x_1 = 0$, $x_2 = 50$, and $x_3 = 30$. □

5.4.2 The Dual Problem

Given the standard form of the linear programming problem in equations (5.2), the **dual problem** is as follows: "Maximize $Y = \sum_{i=1}^{m} b_i y_y$ subject to the constraints $\sum_{i=1}^{m} a_{ij} y_i \le c_{ij}$ for $j = 1, 2, \ldots, n$ and $y_i \ge 0$ for $i = 1, 2, \ldots, m$." Similarly, for the problem: "Maximize $Z = \sum_{j=1}^{n} c_j x_j$ subject to the constraints $\sum_{j=1}^{n} a_{ij} x_j \le b_j$ for $i = 1, 2, \ldots, m$ and $x_j \ge 0$ for $j = 1, 2, \ldots, n$," the dual problem is as follows: "Minimize $Y = \sum_{i=1}^{m} b_i y_i$ subject to the constraints $\sum_{i=1}^{m} a_{ij} y_i \ge c_j$ for $j = 1, 2, \ldots, n$ and $y_i \ge 0$ for $i = 1, 2, \ldots, m$."

Example 5.35

Maximize $Z = 6x + 8y$ subject to the constraints $5x + 2y \le 20$, $x + 2y \le 10$, $x \ge 0$, and $y \ge 0$. State the dual problem and find its solution.

Solution. First, the original (or *primal*) problem is solved. The objective function for this problem is represented by `zx`. Finally, the set of inequalities for the primal is defined to be `ineqsx`. Using the command

```
Maximize[{zx,ineqsx},{x[1],x[2]}],
```

the maximum value of `zx` is found to be 45.

Clear[zx, zy, *x*, *y*, valsx, valsy, ineqsx, ineqsy]

zx = 6*x*[1] + 8*x*[2]; ineqsx = {5*x*[1] + 2*x*[2] ≤ 20, *x*[1] + 2*x*[2] ≤ 10, *x*[1] ≥ 0,

***x*[2] ≥ 0};**

Maximize[{zx, ineqsx}, {*x*[1], *x*[2]}]

$\left\{45, \left\{x[1] \to \frac{5}{2}, x[2] \to \frac{15}{4}\right\}\right\}$

Because in this problem we have $c_1 = 6$, $c_2 = 8$, $b_1 = 20$, and $b_2 = 10$, the dual problem is as follows: Minimize $Z = 20y_1 + 10y_2$ subject to the constraints $5y_1 + y_2 \geq 6$, $2y_1 + 2y_2 \geq 8$, $y_1 \geq 0$, and $y_2 \geq 0$. The dual is solved in a similar fashion by defining the objective function `zy` and the collection of inequalities `ineqsy`. The minimum value obtained by `zy` subject to the constraints `ineqsy` is 45, which agrees with the result of the primal and is found with

```
Minimize[{zy,ineqsy},{y[1],y[2]}].
```

```
zy = 20y[1] + 10y[2]; ineqsy = {5y[1] + y[2] ≥ 6, 2y[1] + 2y[2] ≥ 8,
  y[1] ≥ 0, y[2] ≥ 0};
```

```
Minimize[{zy, ineqsy}, {y[1], y[2]}]
```

$\left\{45, \left\{y[1] \to \frac{1}{2}, y[2] \to \frac{7}{2}\right\}\right\}$ □

Of course, linear programming models can involve numerous variables. Consider the following: given the standard form linear programming problem in equations (5.2), let $\mathbf{x} = \begin{pmatrix} x_1 \\ x_2 \\ \vdots \\ x_n \end{pmatrix}$, $\mathbf{b} = \begin{pmatrix} b_1 \\ b_2 \\ \vdots \\ b_m \end{pmatrix}$, $\mathbf{c} = \begin{pmatrix} c_1 & c_2 & \cdots & c_n \end{pmatrix}$, and $\mathbf{A}$ denote the $m \times n$ matrix $\mathbf{A} = \begin{pmatrix} a_{11} & a_{12} & \cdots & a_{1n} \\ a_{21} & a_{22} & \cdots & a_{2n} \\ \vdots & \vdots & & \vdots \\ a_{m1} & a_{m2} & \cdots & a_{mn} \end{pmatrix}$. Then the standard form of the linear programming problem is equivalent to finding the vector $\mathbf{x}$ that maximizes $Z = \mathbf{c} \cdot \mathbf{x}$ subject to the restrictions $\mathbf{Ax} \geq \mathbf{b}$ and $x_1 \geq 0$, $x_2 \geq 0$, ..., $x_n \geq 0$. The dual problem is: "Minimize $Y = \mathbf{y} \cdot \mathbf{b}$ where $\mathbf{y} = \begin{pmatrix} y_1 & y_2 & \cdots & y_m \end{pmatrix}$ subject to the restrictions $\mathbf{yA} \leq \mathbf{c}$ (componentwise) and $y_1 \geq 0$, $y_2 \geq 0$, ..., $y_m \geq 0$."
The command

```
LinearProgramming[c,A,b]
```

finds the vector $\mathbf{x}$ that minimizes the quantity `Z=c.x` subject to the restrictions `A.x>=b` and `x>=0`. `LinearProgramming` does not yield the minimum value of `Z` as did `Minimize` and `Maximize` and the value must be determined from the resulting vector.

Example 5.36

Maximize $Z = 5x_1 - 7x_2 + 7x_3 + 5x_4 + 6x_5$ subject to the constraints $2x_1 + 3x_2 + 3x_3 + 2x_4 + 2x_5 \geq 10$, $6x_1 + 5x_2 + 4x_3 + x_4 + 4x_5 \geq 30$, $-3x_1 - 2x_2 - 3x_3 - 4x_4 \geq -5$, $-x_1 - x_2 - x_4 \geq -10$, and $x_i \geq 0$ for $i = 1, 2, 3, 4$, and 5. State the dual problem. What is its solution?

Solution. For this problem, $\mathbf{x} = \begin{pmatrix} x_1 \\ x_2 \\ x_3 \\ x_4 \\ x_5 \end{pmatrix}$, $\mathbf{b} = \begin{pmatrix} 10 \\ 30 \\ -5 \\ -10 \end{pmatrix}$, $\mathbf{c} = \begin{pmatrix} 5 & -7 & 7 & 5 & 6 \end{pmatrix}$, and $\mathbf{A} = \begin{pmatrix} 2 & 3 & 3 & 2 & 2 \\ 6 & 5 & 4 & 1 & 4 \\ -3 & -2 & -3 & -4 & 0 \\ -1 & -1 & 0 & -1 & 0 \end{pmatrix}$. First, the vectors $\mathbf{c}$ and $\mathbf{b}$ are entered and then matrix $\mathbf{A}$ is entered and named `matrixa`.

```
Clear[matrixa, x, y, c, b]
c = {5, −7, 7, 5, 6}; b = {10, 30, −5, −10};
matrixa = {{2, 3, 3, 2, 2}, {6, 5, 4, 1, 4},
   {−3, −2, −3, −4, 0}, {−1, −1, 0, −1, 0}};
```

Next, we use `Array[x,5]` to create the list of five elements `{x[1],x[2],...,x[5]}` named `xvec`. The command `Table[x[i], {i,1,5}]` returns the same list. These variables must be defined before attempting to solve this linear programming problem.

```
xvec = Array[x, 5]
```

$\{x[1], x[2], x[3], x[4], x[5]\}$

After entering the objective function coefficients with the vector **c**, the matrix of coefficients from the inequalities with `matrixa`, and the right-hand side values found in **b**; the problem is solved with

`LinearProgramming[c,matrixa,b].`

The solution is called `xvec`. Hence, the maximum value of the objective function is obtained by evaluating the objective function at the variable values that yield a maximum. Because these values are found in `xvec`, the maximum is determined with the dot product of the vector **c** and the vector `xvec`. (Recall that this product is entered as `c.xvec`.) This value is found to be 35/4.

```
xvec = LinearProgramming[c, matrixa, b]
```

$\left\{0, \frac{5}{2}, 0, 0, \frac{35}{8}\right\}$

```
c.xvec
```

$\frac{35}{4}$

Because the dual of the problem is "Minimize the number `Y=y.b` subject to the restrictions `y.A<c` and `y>0`," we use Mathematica to calculate `y.b` and `y.A`. A list of the dual variables `{y[1],y[2],y[3],y[4]}` is created with `Array[y,4]`. This list includes four elements because there are four constraints in the original problem. The objective function of the dual problem is, therefore, found with `yvec.b`, and the left-hand sides of the set of inequalities are given with `yvec.matrixa`.

```
yvec = Array[y, 4]
```

$\{y[1], y[2], y[3], y[4]\}$

```
yvec.b
```

$10y[1] + 30y[2] - 5y[3] - 10y[4]$

yvec.matrixa

{2y[1]+6y[2]−3y[3]−y[4], 3y[1]+5y[2]−2y[3]−y[4], 3y[1]+4y[2]−3y[3], 2y[1]+y[2]−4y[3]−y[4], 2y[1]+4y[2]}

Hence, we may state the dual problem as:

Minimize $Y = 10y_1 + 30y_2 - 5y_3 - 10y_4$ subject to the constraints

$$\begin{cases} 2y_1 + 6y_2 - 3y_3 - y_4 \le 5 \\ 3y_1 + 5y_2 - 2y_3 - y_4 \le -7 \\ 3y_1 + 4y_2 - 3y_3 \le 7 \\ 2y_1 + y_2 - 4y_3 - y_4 \le 5 \\ 2y_1 + 4y_2 \le 6 \end{cases}$$

and $y_i \ge 0$ for $i = 1, 2, 3,$ and 4. □

Application: A Transportation Problem

A certain company has two factories, F1 and F2, each producing two products, P1 and P2, that are to be shipped to three distribution centers, D1, D2, and D3. The following table illustrates the cost associated with shipping each product from the factory to the distribution center, the minimum number of each product each distribution center needs, and the maximum output of each factory. How much of each product should be shipped from each plant to each distribution center to minimize the total shipping costs?

	F1/P1	**F1/P2**	**F2/P1**	**F2/P2**	**Minimum**
D1/P1	\$0.75		\$0.80		500
D1/P2		\$0.50		\$0.40	400
D2/P1	\$1.00		\$0.90		300
D2/P2		\$0.75		\$1.20	500
D3/P1	\$0.90		\$0.85		700
D3/P2		\$0.80		\$0.95	300
Maximum Output	1000	400	800	900	

Solution. Let x_1 denote the number of units of P1 shipped from F1 to D1; x_2 the number of units of P2 shipped from F1 to D1; x_3 the number of units of P1 shipped from F1 to D2; x_4 the number of units of P2 shipped from F1 to D2; x_5 the number of units of P1 shipped from F1 to D3; x_6 the number of units of P2 shipped from F1 to D3; x_7 the number of units of P1 shipped from F2 to D1; x_8 the number of units of P2 shipped from F2 to D1; x_9 the number of units of P1 shipped from F2 to D2; x_{10} the number of units of P2 shipped from F2 to D2; x_{11} the number of units of P1 shipped from F2 to D3; and x_{12} the number of units of P2 shipped from F2 to D3.

Then, it is necessary to minimize the number

$$Z = .75x_1 + .5x_2 + x_3 + .75x_4 + .9x_5 + .8x_6 + .8x_7 + .4x_8 + .9x_9 + 1.2x_{10} + .85x_{11} + .95x_{12}$$

subject to the constraints $x_1 + x_3 + x_5 \le 1000$, $x_2 + x_4 + x_6 \le 400$, $x_7 + x_9 + x_{11} \le 800$, $x_8 + x_{10} + x_{12} \le 900$, $x_1 + x_7 \ge 500$, $x_3 + x_9 \ge 500$, $x_5 + x_{11} \ge 700$, $x_2 + x_8 \ge 400$, $x_4 + x_{10} \ge 500$, $x_6 + x_{12} \ge 300$, and x_i nonnegative for $i = 1, 2, \ldots, 12$. In order to solve this linear programming problem, the objective function which computes the total cost, the 12 variables, and the set of inequalities must be entered. The coefficients of the objective function are given in the vector `c`. Using the command `Array[x,12]` illustrated in the previous example to define the list of 12 variables `{x[1],x[2],...,x[12]}`, the objective function is given by the product `z=xvec.c`, where `xvec` is the name assigned to the list of variables.

```
Clear[xvec, z, constraints, vars, c]

c = {0.75, 0.5, 1, 0.75, 0.9, 0.8, 0.8, 0.4, 0.9, 1.2, 0.85, 0.95};

xvec = Array[x, 12]
```

```
{x[1], x[2], x[3], x[4], x[5], x[6], x[7], x[8], x[9], x[10], x[11], x[12]}
```


```
{x[1], x[2], x[3], x[4], x[5], x[6],
x[7], x[8], x[9], x[10], x[11], x[12]}
```


```
z = xvec.c
```

```
0.75x[1] + 0.5x[2] + x[3] + 0.75x[4] + 0.9x[5] + 0.8x[6] + 0.8x[7] + 0.4x[8] + 0.9x[9] +
1.2x[10] + 0.85x[11] + 0.95x[12]
```


```
0.75x[1] + 0.5x[2] + x[3] + 0.75x[4]+
0.9x[5] + 0.8x[6] + 0.8x[7] + 0.4x[8]+
0.9x[9] + 1.2x[10] + 0.85x[11] + 0.95x[12]
```


The set of constraints are then entered and named `constraints` for easier use. Therefore, the minimum cost and the value of each variable which yields this minimum cost are found with the command

```
Minimize[{z,constraints},xvec].
```

```
constraints = {x[1] + x[3] + x[5] ≤ 1000, x[2] + x[4] + x[6] ≤ 400,
  x[7] + x[9] + x[11] ≤ 800, x[8] + x[10] + x[12] ≤ 900,
  x[1] + x[7] ≥ 500, x[3] + x[9] ≥ 300, x[5] + x[11] ≥ 700,
  x[2] + x[8] ≥ 400, x[4] + x[10] > 500, x[6] + x[12] > 300,
  x[1] ≥ 0, x[2] ≥ 0, x[3] ≥ 0, x[4] ≥ 0,
  x[5] ≥ 0, x[6] ≥ 0, x[7] ≥ 0, x[8] ≥ 0,
  x[9] ≥ 0, x[10] ≥ 0, x[11] ≥ 0, x[12] ≥ 0};

values = Minimize[{z, constraints}, xvec]
```

```
{2115., {x[1] → 500., x[2] → 0., x[3] → 0., x[4] → 400., x[5] → 200., x[6] → 0., x[7] →
0., x[8] → 400., x[9] → 300., x[10] → 100., x[11] → 500., x[12] → 300.}}
```

```
{2115., {x[1] → 500., x[2] → 0., x[3] → 0., x[4] → 400.,
```

$x[5] \to 200., x[6] \to 0., x[7] \to 0., x[8] \to 400.,$

$x[9] \to 300., x[10] \to 100., x[11] \to 500., x[12] \to 300.\}\}$

Notice that `values` is a list consisting of two elements: the minimum value of the cost function, 2115, and the list of the variable values `{x[1]->500,x[2]->0, ...}`. Hence, the minimum cost is obtained with the command `values[[1]]` and the list of variable values that yield the minimum cost is extracted with `values[[2]]`.

values[[1]]

2115.

values[[2]]

$\{x[1] \to 500., x[2] \to 0., x[3] \to 0., x[4] \to 400., x[5] \to 200., x[6] \to 0., x[7] \to 0., x[8]$

$\to 400., x[9] \to 300., x[10] \to 100., x[11] \to 500., x[12] \to 300.\}$

Using these extraction techniques, the number of units produced by each factory can be computed. Because x_1 denotes the number of units of P1 shipped from F1 to D1, x_3 the number of units of P1 shipped from F1 to D2, and x_5 the number of units of P1 shipped from F1 to D3, the total number of units of Product 1 produced by Factory 1 is given by the command `x[1]+x[3]+x[5] /. values[[2]]` which evaluates this sum at the values of `x[1]`, `x[3]`, and `x[5]` given in the list `values[[2]]`.

x[1] + x[3] + x[5]/.values[[2]]

700.

Also, the number of units of Products 1 and 2 received by each distribution center can be computed. The command `x[3]+x[9] //values[[2]]` gives the total amount of P1 received at D2 because `x[3]`=amount of P1 received by D2 from F1 and `x[9]`= amount of P1 received by D2 from F2. Notice that this amount is the minimum number of units (300) of P1 requested by D2.

x[3] + x[9]/.values[[2]]

300.

The number of units of each product that each factory produces can be calculated and the amount of P1 and P2 received at each distribution center is calculated in a similar manner.

{x[1] + x[3] + x[5], x[2] + x[4] + x[6], x[7] + x[9] + x[11],

x[8] + x[10] + x[12], x[1] + x[7], x[3] + x[9],

x[5] + x[11], x[2] + x[8],

x[4] + x[10], x[6] + x[12]}/.values[[2]]//

TableForm

700.

400.

800.

800.

500.

300.

700.

400.

500.

300.

From these results, we see that F1 produces 700 units of P1, F1 produces 400 units of P2, F2 produces 800 units of P1, F2 produces 800 units of P2, and each distribution center receives exactly the minimum number of each product it requests. □

5.5 SELECTED TOPICS FROM VECTOR CALCULUS

5.5.1 Vector-Valued Functions

Basic operations on two and three-dimensional vectors are discussed in Section 5.1.4.2.

We now turn our attention to vector-valued functions. In particular, we consider vector-valued functions of the following forms.

$$\begin{aligned}
\text{Plane curves:}\quad & \mathbf{r}(t) = x(t)\mathbf{i} + y(t)\mathbf{j} && (5.3)\\
\text{Space curves:}\quad & \mathbf{r}(t) = x(t)\mathbf{i} + y(t)\mathbf{j} + z(t)\mathbf{k} && (5.4)\\
\text{Parametric surfaces:}\quad & \mathbf{r}(s,t) = x(s,t)\mathbf{i} + y(s,t)\mathbf{j} + z(s,t)\mathbf{k} && (5.5)\\
\text{Vector fields in the plane:}\quad & \mathbf{F}(x,y) = P(x,y)\mathbf{i} + Q(x,y)\mathbf{j} && (5.6)\\
\text{Vector fields in space:}\quad & \mathbf{F}(x,y,z) = P(x,y,z)\mathbf{i} + Q(x,y,z)\mathbf{j} + R(x,y,z)\mathbf{k} && (5.7)
\end{aligned}$$

For the vector-valued functions (5.3) and (5.4), differentiation and integration are carried out term-by-term, provided that all the terms are differentiable and integrable. Suppose that C is a smooth curve defined by $\mathbf{r}(t)$, $a \le t \le b$.

In 2-space, $\mathbf{i} =< 1, 0 >$ and $\mathbf{j} =< 0, 1 >$. In 3-space $\mathbf{i} =< 1, 0, 0 >$, $\mathbf{j} =< 0, 1, 0 >$, and $\mathbf{k} =< 0, 0, 1 >$.

1. If $\mathbf{r}'(t) \neq \mathbf{0}$, the **unit tangent vector**, $\mathbf{T}(t)$, is $\mathbf{T}(t) = \dfrac{\mathbf{r}'(t)}{\|\mathbf{r}'(t)\|}$.
2. If $\mathbf{T}'(t) \neq \mathbf{0}$, the **principal unit normal vector**, $\mathbf{N}(t)$, is $\mathbf{N}(t) = \dfrac{\mathbf{T}'(t)}{\|\mathbf{T}'(t)\|}$.
3. The **arc length function**, $s(t)$, is $s(t) = \int_a^t \|\mathbf{r}'(u)\|\, du$. In particular, the length of C on the interval $[a, b]$ is $\int_a^b \|\mathbf{r}'(t)\|\, dt$.
4. The **curvature**, κ, of C is

$$\kappa = \frac{\|\mathbf{T}'(t)\|}{\|\mathbf{r}'(t)\|} = \frac{\mathbf{a}(t)\cdot\mathbf{N}(t)}{\|\mathbf{v}(t)\|^2} = \frac{\|\mathbf{r}'(t)\times\mathbf{r}''(t)\|}{\|\mathbf{r}'(t)\|^3},$$

where $\mathbf{v}(t) = \mathbf{r}'(t)$ and $\mathbf{a}(t) = \mathbf{r}''(t)$.

It is a good exercise to show that the curvature of a circle of radius r is $1/r$.

Example 5.37: Folium of Descartes

The **Folium of Descartes** is defined by

$$\mathbf{r}(t) = \frac{3at}{1+t^3}\mathbf{i} + \frac{3at^2}{1+t^3}\mathbf{j}$$

for $t \neq -1$, if $a = 1$. (a) Find $\mathbf{r}'(t)$, $\mathbf{r}''(t)$ and $\int \mathbf{r}(t)\,dt$. (b) Find $\mathbf{T}(t)$ and $\mathbf{N}(t)$. (c) Find the curvature, κ. (d) Find the length of the loop of the folium.

Solution. (a) After defining $\mathbf{r}(t)$,

r[t_] = {3at/(1 + t^3), 3at^2/(1 + t^3)};

a = 1;

we compute $\mathbf{r}'(t)$ and $\int \mathbf{r}(t)\,dt$ with $'$, $''$ and `Integrate`, respectively. We name $\mathbf{r}'(t)$ `dr`, $\mathbf{r}''(t)$ `dr2`, and $\int \mathbf{r}(t)\,dt$ `ir`.

dr = Simplify[r'[t]]

dr2 = Simplify[r''[t]]

ir = Integrate[r[t], t]

$$\left\{\frac{3-6t^3}{(1+t^3)^2}, -\frac{3t(-2+t^3)}{(1+t^3)^2}\right\}$$

$$\left\{\frac{18t^2(-2+t^3)}{(1+t^3)^3}, \frac{6(1-7t^3+t^6)}{(1+t^3)^3}\right\}$$

$$\left\{3\left(\frac{\text{ArcTan}\left[\frac{-1+2t}{\sqrt{3}}\right]}{\sqrt{3}} - \frac{1}{3}\text{Log}[1+t] + \frac{1}{6}\text{Log}\left[1-t+t^2\right]\right), \text{Log}\left[1+t^3\right]\right\}$$

(b) Mathematica does not automatically make assumptions regarding the value of t, so it does not algebraically simplify $\|\mathbf{r}'(t)\|$ as we might typically do unless we use `PowerExpand`

`PowerExpand[Sqrt[x^2]]` returns x rather than $|x|$.

nr = PowerExpand[Sqrt[dr.dr]//Simplify]

$$\frac{3\sqrt{1+4t^2-4t^3-4t^5+4t^6+t^8}}{(1+t^3)^2}$$

The unit tangent vector, $\mathbf{T}(t)$ is formed in `ut`.

ut = dr/nr//Simplify

$$\left\{\frac{1-2t^3}{\sqrt{1+4t^2-4t^3-4t^5+4t^6+t^8}}, -\frac{t(-2+t^3)}{\sqrt{1+4t^2-4t^3-4t^5+4t^6+t^8}}\right\}$$

We perform the same steps to compute the unit normal vector, $\mathbf{N}(t)$. In particular, note that `dutb` $= \|\mathbf{T}'(t)\|$.

dut = D[ut, t]//Simplify

$$\left\{\frac{2t(-2-3t^3+t^9)}{(1+4t^2-4t^3-4t^5+4t^6+t^8)^{3/2}}, \frac{2-6t^6-4t^9}{(1+4t^2-4t^3-4t^5+4t^6+t^8)^{3/2}}\right\}$$

duta = dut.dut//Simplify

$$\frac{4(1+t^3)^4}{(1+4t^2-4t^3-4t^5+4t^6+t^8)^2}$$

dutb = PowerExpand[Sqrt[duta]]

$$\frac{2(1+t^3)^2}{1+4t^2-4t^3-4t^5+4t^6+t^8}$$

nt = dut/dutb//Simplify

$$\{\frac{t(-2+t^3)}{\sqrt{1+4t^2-4t^3-4t^5+4t^6+t^8}}, \frac{1-2t^3}{\sqrt{1+4t^2-4t^3-4t^5+4t^6+t^8}}\}$$

(c) We use the formula $\kappa = \dfrac{\|\mathbf{T}'(t)\|}{\|\mathbf{r}'(t)\|}$ to determine the curvature in `curvature`.

curvature = Simplify[dutb/nr]

$$\frac{2(1+t^3)^4}{3(1+4t^2-4t^3-4t^5+4t^6+t^8)^{3/2}}$$

We graphically illustrate the unit tangent and normal vectors at $\mathbf{r}(1) = \langle 3/2, 3/2\rangle$. First, we compute the unit tangent and normal vectors if $t = 1$ using `/.` (`ReplaceAll`).

ut1 = ut/.*t* → 1

nt1 = nt/.*t* → 1

$$\{-\frac{1}{\sqrt{2}}, \frac{1}{\sqrt{2}}\}$$

$$\{-\frac{1}{\sqrt{2}}, -\frac{1}{\sqrt{2}}\}$$

We then compute the curvature if $t = 1$ in `smallk`. The center of the osculating circle at $\mathbf{r}(1)$ is found in `x0` and `y0`.

The radius of the osculating circle is $1/\kappa$; the position vector of the center is $\mathbf{r} + \frac{1}{\kappa}\mathbf{N}$.

smallk = curvature/.*t* → 1

***N*[smallk]**

x0 = (*r*[*t*] + 1/curvaturent/.*t* → 1)[[1]]

y0 = (*r*[*t*] + 1/curvaturent/.*t* → 1)[[2]]

$$\frac{8\sqrt{2}}{3}$$

3.77124

$$\frac{21}{16}$$

$$\frac{21}{16}$$

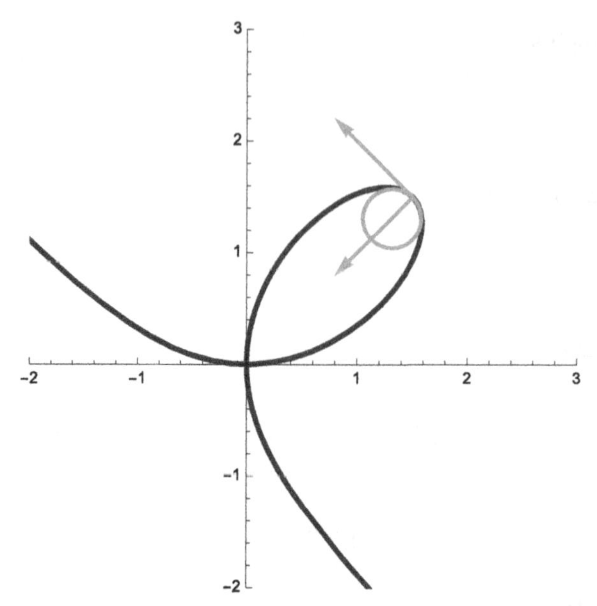

FIGURE 5.8 The folium with an osculating circle (Brigham Young University colors).

We now graph $\mathbf{r}(t)$ with `ParametricPlot`. The unit tangent and normal vectors at $\mathbf{r}(1)$ are graphed with `Arrow` in `a1` and `a2`. The osculating circle at $\mathbf{r}(1)$ is graphed with `Circle` in `c1`. All four graphs are displayed together with `Show` in Fig. 5.8.

`Graphics[Circle[{x0, y0}, r]]` is a two-dimensional graphics object that represents a circle of radius r centered at the point (x_0, y_0). Use `Show` to display the graph.

```
p1 = ParametricPlot[r[t], {t, −100, 100},
  PlotRange → {{−2, 3}, {−2, 3}}, AspectRatio → 1,
  PlotStyle → {{Thickness[.01], CMYKColor[1, .62, 0, .52]}}];
p2 = Graphics[{Thickness[.0075], CMYKColor[.18, .31, .56, 0],
  Circle[{x0, y0}, 1/smallk],
  Arrow[{r[1], r[1] + ut1}], Arrow[{r[1], r[1] + nt1}]}];
Show[p1, p2, AspectRatio → Automatic]
```

(d) The loop is formed by graphing $\mathbf{r}(t)$ for $t \geq 0$. Hence, the length of the loop is given by the improper integral $\int_0^\infty \|\mathbf{r}(t)\| \, dt$, which we compute with `NIntegrate`.

```
NIntegrate[nr, {t, 0, Infinity}]
```

4.91749 □

In the example, we computed the curvature at $t = 1$. Of course, we could choose other t values. With `Manipulate`,

```
Manipulate[
r[t_] = {3t/(1 + t^3), 3t^2/(1 + t^3)};
dr = Simplify[r'[t]];
```

```
dr2 = Simplify[r''[t]];
ir = Integrate[r[t], t];
nr = PowerExpand[Sqrt[dr.dr]//Simplify];
ut = dr/nr//Simplify;
dut = D[ut, t]//Simplify;
duta = dut.dut//Simplify;
dutb = PowerExpand[Sqrt[duta]];
nt = dut/dutb//Simplify;
curvature = Simplify[dutb/nr];
ut1 = ut/.t -> t0;
nt1 = nt/.t -> t0;
smallk = curvature/.t -> t0;
  x0 = (r[t] + 1/curvaturent/.t -> t0)[[1]];
  y0 = (r[t] + 1/curvaturent/.t -> t0)[[2]];
p1 = ParametricPlot[r[t], {t, -10, 10},
  PlotStyle -> {{Thickness[.01], CMYKColor[1, .62, 0, .52]}},
  PlotRange -> {{-2, 3}, {-2, 3}}, AspectRatio -> 1, PlotPoints -> 200];
p2 = Graphics[{Circle[{x0, y0}, 1/smallk],
  Arrow[{r[t0], r[t0] + ut1}], Arrow[{r[t0], r[t0] + nt1}]}];
.Show[p1, p2], {{t0, 1}, -5, 10}]
```

we can see the osculating circle at various values of $y_0 \neq -1$. See Fig. 5.9.

Of course, this particular choice of using the folium to illustrate the procedure could be modified as well. With

```
Manipulate[
folium[t_] = {3t/(1 + t^3), 3t^2/(1 + t^3)};
cycloid[t_] = {1/(2Pi)(t - Sin[t]), (1 - Cos[t])/(2Pi)};
rose[t_] = {3/2Cos[2t]Cos[t], 3/2Cos[2t]Sin[t]};
squiggle[t_] = {Cos[t] - Sin[2t], Sin[2t] + Cos[5t]};
cornu[t_] = {2.5FresnelC[t], 2.5FresnelS[t]};
lissajous[t_] = {2Cos[t], Sin[2t]};
evolute[t_] = {Cos[t]^3, 2Sin[t]^3};
dr = Simplify[r'[t]];
```

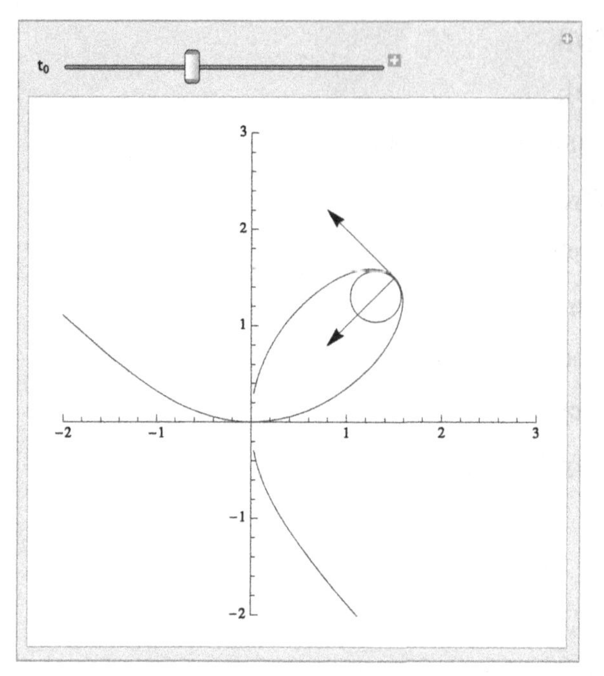

FIGURE 5.9 Using Manipulate to see the osculating circle at various values of t_0.

```
dr2 = Simplify[r''[t]];
ir = Integrate[r[t], t];
nr = PowerExpand[Sqrt[dr.dr]//Simplify];
ut = dr/nr//Simplify;
dut = D[ut, t]//Simplify;
duta = dut.dut//Simplify;
dutb = PowerExpand[Sqrt[duta]];
nt = dut/dutb//Simplify;
curvature = Simplify[dutb/nr];
ut1 = ut/.t → t0;
nt1 = nt/.t → t0;
smallk = curvature/.t → t0;
  x0 = (r[t] + 1/curvaturent/.t → t0)[[1]];
  y0 = (r[t] + 1/curvaturent/.t → t0)[[2]];
p1 = ParametricPlot[r[t], {t, −10, 10},
  PlotRange → {{−3, 3}, {−3, 3}}, AspectRatio → 1, PlotPoints → 200];
p2 = Graphics[{Thickness[.0075], CMYKColor[.18, .31, .56, 0],
  Circle[{x0, y0}, 1/smallk],
```

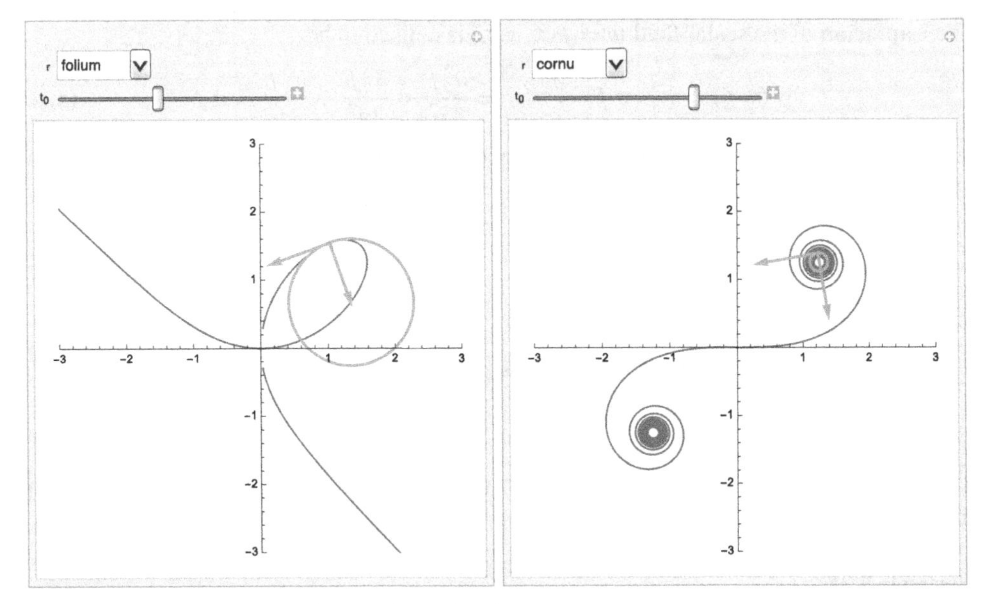

FIGURE 5.10 The osculating circle for various $\mathbf{r}(t)$ and t_0.

```
Arrow[{r[t0], r[t0] + ut1}], Arrow[{r[t0], r[t0] + nt1}]}];
Show[p1, p2], {{r, folium},
  {folium, cycloid, rose, squiggle, cornu, lissajous, evolute}}, {{t0, 3/2}, -5, 10}]
```

we not only allow t_0 to vary but also $\mathbf{r}(t)$. Note that the resulting `Manipulate` object is quite slow on all except the fastest computers. See Fig. 5.10 (b).

Recall that the **gradient** of $z = f(x, y)$ is the vector-valued function $\nabla f(x, y) = \langle f_x(x, y), f_y(x, y)\rangle$. Similarly, we define the **gradient** of $w = f(x, y, z)$ to be

$$\nabla f(x, y, z) = \langle f_x(x, y, z), f_y(x, y, z), f_z(x, y, z)\rangle = \frac{\partial f}{\partial x}\mathbf{i} + \frac{\partial f}{\partial y}\mathbf{j} + \frac{\partial f}{\partial z}\mathbf{k}. \tag{5.8}$$

A vector field $\mathbf{F}$ is **conservative** if there is a function f, called a **potential function**, satisfying $\nabla f = \mathbf{F}$. In the special case that $\mathbf{F}(x, y) = P(x, y)\mathbf{i} + Q(x, y)\mathbf{j}$, $\mathbf{F}$ is conservative if and only if

$$\frac{\partial P}{\partial y} = \frac{\partial Q}{\partial x}.$$

The **divergence** of the vector field $\mathbf{F}(x, y, z) = P(x, y, z)\mathbf{i} + Q(x, y, z)\mathbf{j} + R(x, y, z)\mathbf{k}$ is the scalar field

$$\operatorname{div}\mathbf{F} = \nabla \cdot \mathbf{F} = \frac{\partial P}{\partial x} + \frac{\partial Q}{\partial y} + \frac{\partial R}{\partial z}. \tag{5.9}$$

The `Div` command can be used to find the divergence of a vector field:

```
Div[{P(x,y,z),Q(x,y,z),R(x,y,z)},"Cartesian"]
```

computes the divergence of $\mathbf{F}(x, y, z) = P(x, y, z)\mathbf{i} + Q(x, y, z)\mathbf{j} + R(x, y, z)\mathbf{k}$ in the Cartesian coordinate system. If you omit "Cartesian," the default coordinates are the Cartesian coordinate system. However, if you are using a non-Cartesian coordinates system such as cylindrical or spherical coordinates, be sure to replace "Cartesian" with "Cylindrical," "Spherical" or the name of the coordinate system you are using. Note that Mathematica supports nearly all coordinate systems used by scientists.

The **Laplacian** of the scalar field $w = f(x, y, z)$ is defined to be

$$\operatorname{div}(\nabla f) = \nabla \cdot (\nabla f) = \nabla^2 f = \frac{\partial^2 f}{\partial x^2} + \frac{\partial^2 f}{\partial y^2} + \frac{\partial^2 f}{\partial z^2} = \Delta f. \tag{5.10}$$

In the same way that `Div` computes the divergence of a vector field `Laplacian` computes the Laplacian of a scalar field.

The **curl** of the vector field $\mathbf{F}(x, y, z) = P(x, y, z)\mathbf{i} + Q(x, y, z)\mathbf{j} + R(x, y, z)\mathbf{k}$ is

$$\begin{aligned} \operatorname{curl}\mathbf{F}(x, y, z) &= \nabla \times \mathbf{F}(x, y, z) \\ &= \begin{vmatrix} \mathbf{i} & \mathbf{j} & \mathbf{k} \\ \frac{\partial}{\partial x} & \frac{\partial}{\partial y} & \frac{\partial}{\partial z} \\ P(x, y, z) & Q(x, y, z) & R(x, y, z) \end{vmatrix} \\ &= \left(\frac{\partial R}{\partial y} - \frac{\partial Q}{\partial z}\right)\mathbf{i} - \left(\frac{\partial R}{\partial x} - \frac{\partial P}{\partial z}\right)\mathbf{j} + \left(\frac{\partial Q}{\partial x} - \frac{\partial P}{\partial y}\right)\mathbf{k}. \end{aligned} \tag{5.11}$$

If $\mathbf{F}(x, y, z) = P(x, y, z)\mathbf{i} + Q(x, y, z)\mathbf{j} + R(x, y, z)\mathbf{k}$, $\mathbf{F}$ is conservative if and only if $\operatorname{curl}\mathbf{F}(x, y, z) = \mathbf{0}$, in which case $\mathbf{F}$ is said to be **irrotational**.

Example 5.38

Determine if

$$\mathbf{F}(x, y) = \left(1 - 2x^2\right) y e^{-x^2-y^2}\mathbf{i} + \left(1 - 2y^2\right) x e^{-x^2-y^2}\mathbf{j}$$

is conservative. If $\mathbf{F}$ is conservative find a potential function for $\mathbf{F}$.

Solution. We define $P(x, y) = \left(1 - 2x^2\right) y e^{-x^2-y^2}$ and $Q(x, y) = \left(1 - 2y^2\right) x e^{-x^2-y^2}$. Then we use `D` and `Simplify` to see that $P_y(x, y) = Q_x(x, y)$. Hence, $\mathbf{F}$ is conservative.

```
p[x_, y_] = (1 - 2x^2)yExp[-x^2 - y^2];
q[x_, y_] = (1 - 2y^2)xExp[-x^2 - y^2];
```

```
Simplify[D[p[x, y], y]]
Simplify[D[q[x, y], x]]
```

$e^{-x^2-y^2}(-1 + 2x^2)(-1 + 2y^2)$

$e^{-x^2-y^2}(-1 + 2x^2)(-1 + 2y^2)$

We use `Integrate` to find f satisfying $\nabla f = \mathbf{F}$.

```
i1 = Integrate[p[x, y], x] + g[y]
```

$e^{-x^2-y^2} xy + g[y]$

```
Solve[D[i1, y] == q[x, y], g'[y]]
```

$\{\{g'[y] \to 0\}\}$

Therefore, $g(y) = C$, where C is an arbitrary constant. Letting $C = 0$ gives us the following potential function.

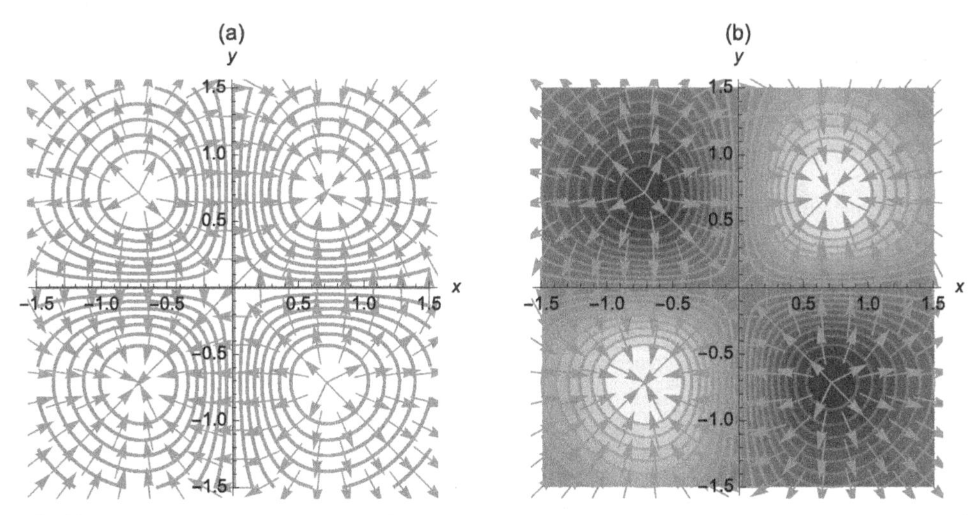

FIGURE 5.11 Two different views illustrating that the vectors **F** are perpendicular to the level curves of f (University of Vermont colors).

```
f = i1/.g[y]->0
```

$e^{-x^2-y^2}xy$

Remember that the vectors **F** are perpendicular to the level curves of f. To see this, we normalize **F** in `uv`.

```
uv = {p[x, y], q[x, y]}/
Sqrt[{p[x, y], q[x, y]}.{p[x, y], q[x, y]}]//Simplify
```

$$\left\{\frac{e^{-x^2-y^2}(y-2x^2y)}{\sqrt{e^{-2(x^2+y^2)}(y^2+4x^4y^2+x^2(1-8y^2+4y^4))}}, \frac{e^{-x^2-y^2}(x-2xy^2)}{\sqrt{e^{-2(x^2+y^2)}(y^2+4x^4y^2+x^2(1-8y^2+4y^4))}}\right\}$$

We then graph several level curves of f in `cp1` and `cp2` with `ContourPlot` and several vectors of `uv` with `VectorPlot`. We show the graphs together with `Show` in Fig. 5.11.

```
cp1 = ContourPlot[f, {x, -3/2, 3/2}, {y, -3/2, 3/2}, Contours → 15,.
  ContourShading → False, PlotPoints → 60, Frame → False,
  Axes → Automatic, AxesOrigin → {0, 0}, AxesLabel → {x, y},
  ContourStyle → {{Thickness[.0075], CMYKColor[.85, 0, .64, .45]}}];
cp2 = ContourPlot[f, {x, -3/2, 3/2}, {y, -3/2, 3/2}, Contours → 20, Frame → False,.
  Axes → Automatic, AxesOrigin → {0, 0}, AxesLabel → {x, y},
  PlotPoints → 60, ContourStyle → CMYKColor[.6, .01, .93, .1]];

fp = VectorPlot[uv, {x, -3/2, 3/2}, {y, -3/2, 3/2},
  VectorStyle->CMYKColor[.35, 0, .95, 0]];

Show[GraphicsRow[{Show[cp1, fp, PlotLabel → "(a)"],
Show[cp2, fp, PlotLabel → "(b)"]}]]
```

□

Example 5.39

(a) Show that

$$\mathbf{F}(x, y, z) = -10xy^2\mathbf{i} + \left(3z^3 - 10x^2y\right)\mathbf{j} + 9yz^2\mathbf{k}$$

is irrotational. (b) Find a function $y = f(x, y)$ satisfying $\nabla f = \mathbf{F}$. (c) Compute div $\mathbf{F}$ and $\nabla^2 f$.

Solution. (a) After defining $\mathbf{F}(x, y, z)$, we use `Curl`, to see that curl $\mathbf{F}(x, y, z) = \mathbf{0}$.

```
Clear[f, x, y, z]
f[x_, y_, z_] = {-10xy^2, 3z^3 - 10x^2y, 9yz^2}
```

$\{-10xy^2, -10x^2y + 3z^3, 9yz^2\}$

```
Curl[f[x, y, z], {x, y, z}]
```

$\{0, 0, 0\}$

(b) We then use `Integrate` to find $w = f(x, y, z)$ satisfying $\nabla f = \mathbf{F}$.

```
i1 = Integrate[f[x, y, z][[1]], x] + g[y, z]
```

$-5x^2y^2 + g[y, z]$

```
i2 = D[i1, y]
```

$-10x^2y + g^{(1,0)}[y, z]$

```
Solve[i2 == f[x, y, z][[2]], g^(1,0)[y, z]]
```

$\{\{g^{(1,0)}[y, z] \to 3z^3\}\}$

```
i3 = Integrate[3z^3, y] + h[z]
```

$3yz^3 + h[z]$

```
i4 = i1/.g[y, z]->i3
```

$-5x^2y^2 + 3yz^3 + h[z]$

```
Solve[D[i4, z] == f[x, y, z][[3]]]
```

$\{\{z \to \text{InverseFunction}[h', 1, 1][0]\}\}$

With $h(z) = C$ and $C = 0$ we have $f(x, y, z) = -5x^2y^2 + 3yz^3$.

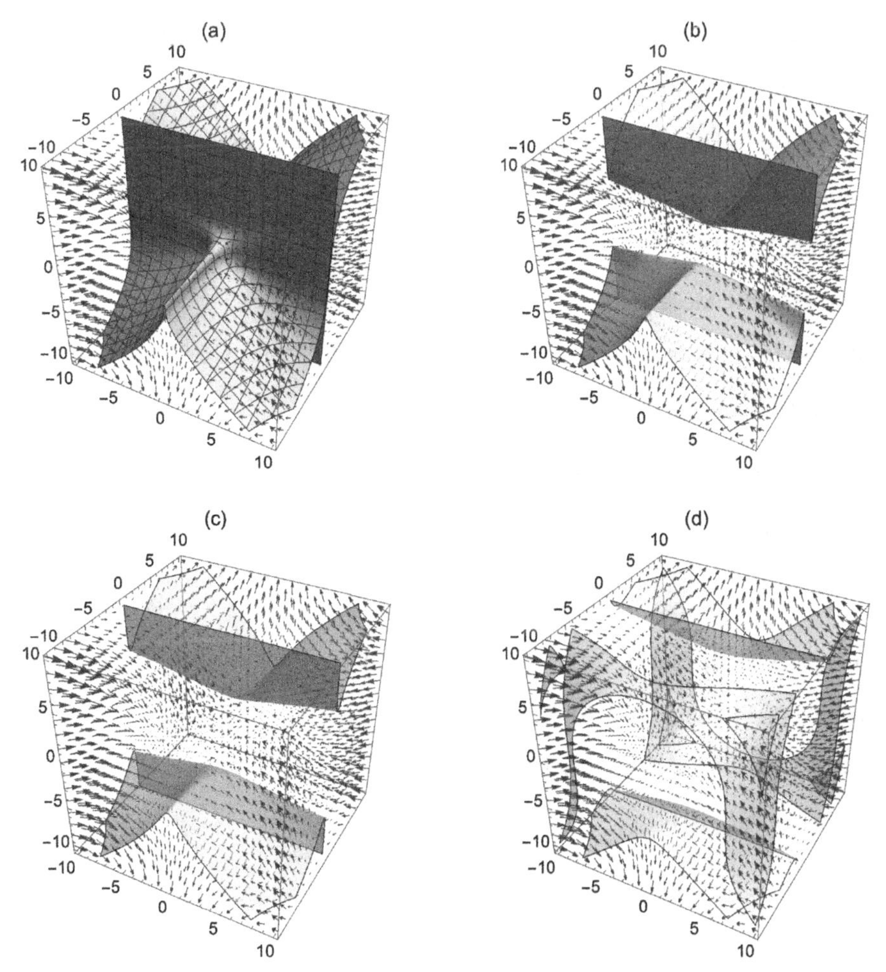

FIGURE 5.12 ∇f is orthogonal to the level surfaces of f (University of Oregon colors).

```
lf = -5x^2y^2 + 3yz^3;
```

∇f is orthogonal to the level surfaces of f. To illustrate this, we use `ContourPlot3D` to graph several level surfaces of $w = f(x, y, z)$ for $-10 \le x \le 10$, $-10 \le y \le 10$, and $-10 \le z \le 10$ in `pf`. We then use `GradientFieldPlot3D`, which is contained in the `VectorFieldPlots` package, to graph several vectors in the gradient field of f over the same domain in `gradf`. The two plots are shown together with `Show` in Fig. 5.12. In the plot, notice that the vectors appear to be perpendicular to the surface.

```
pf1 = ContourPlot3D[lf == -5, {x, -10, 10}, {y, -10, 10},
  {z, -10, 10}, PlotPoints -> 40,
  ContourStyle -> {{Thickness[.01], CMYKColor[.02, .02, .95, 0]}}];
pf2 = ContourPlot3D[lf == 10, {x, -10, 10}, {y, -10, 10},
  {z, -10, 10}, PlotPoints -> 40, Mesh -> None,
  ContourStyle -> Directive[CMYKColor[.02, .02, .95, 0], Opacity[0.8],
  Specularity[White, 10]]];
```

```
pf3 = ContourPlot3D[lf == 100, {x, -10, 10}, {y, -10, 10},
  {z, -10, 10}, Mesh -> None,
  ContourStyle -> Directive[CMYKColor[.02, .02, .95, 0], Opacity[0.5]],
  PlotPoints -> 40];
pf4 = ContourPlot3D[lf, {x, -10, 10}, {y, -10, 10}, {z, -10, 10},
  PlotPoints -> 50, Mesh -> None,
  ContourStyle -> Directive[CMYKColor[.02, .02, .95, 0], Opacity[0.3],
  Specularity[White, 30]]];

gf = VectorPlot3D[Evaluate[{D[lf, x], D[lf, y], D[lf, z]}], {x, -10, 10},
{y, -10, 10}, {z, -10, 10}, VectorPoints -> 15,
VectorStyle -> CMYKColor[.87, .45, .78, .49]]

Show[GraphicsGrid[{{Show[pf1, gf, PlotLabel -> "(a)"],
  Show[pf2, gf, PlotLabel -> "(b)"]},
  {Show[pf3, gf, PlotLabel -> "(c)"], Show[pf4, gf,
  PlotLabel -> "(d)"]}}]]
```

For (c), we take advantage of `Div` and `Laplacian`. As expected, the results are the same. □

5.5.2 Line Integrals

If **F** is continuous on the smooth curve C with parametrization $\mathbf{r}(t)$, $a \leq t \leq b$, the **line integral** of **F** on C is

$$\int_C \mathbf{F} \cdot d\mathbf{r} = \int_a^b \mathbf{F} \cdot \mathbf{r}'(t)\, dt \tag{5.12}$$

If **F** is conservative and C is piecewise smooth, line integrals can be evaluated using the *Fundamental Theorem of Line Integrals*.

Theorem 5.1 (Fundamental Theorem of Line Integrals). *If* **F** *is conservative and the curve C defined by $\mathbf{r}(t)$, $a \leq t \leq b$, is piecewise smooth,*

$$\int_C \mathbf{F} \cdot d\mathbf{r} = f(\mathbf{r}(b)) - f(\mathbf{r}(a)) \tag{5.13}$$

where $\mathbf{F} = \nabla f$.

Example 5.40

Find $\int_C \mathbf{F} \cdot d\mathbf{r}$ where $\mathbf{F}(x, y) = (e^{-y} - ye^{-x})\mathbf{i} + (e^{-x} - xe^{-y})\mathbf{j}$ and C is defined by $\mathbf{r}(t) = \cos t\,\mathbf{i} + \ln(2t/\pi)\,\mathbf{j}$, $\pi/2 \leq t \leq 4\pi$.

Solution. We see that **F** is conservative with `D` and find that $f(x, y) = xe^{-y} + ye^{-x}$ satisfies $\nabla f = \mathbf{F}$ with `Integrate`.

```
f[x_, y_] = {Exp[-y] - yExp[-x], Exp[-x] - xExp[-y]};
r[t_] = {Cos[t], Log[2t/Pi]};

D[f[x, y][[1]], y]//Simplify
D[f[x, y][[2]], x]//Simplify
```

$-e^{-x}-e^{-y}$

$-e^{-x}-e^{-y}$

```
lf = Integrate[f[x, y][[1]], x]
```

$e^{-y}x+e^{-x}y$

Hence, using (5.13),

$$\int_C \mathbf{F}\cdot d\mathbf{r} = \left(xe^{-y}+ye^{-x}\right)\Big]_{x=0,y=0}^{x=1,y=\ln 8} = \frac{3\ln 2}{e}+\frac{1}{8}\approx 0.890.$$

```
xr[t_] = Cos[t];
yr[t_] = Log[2t/Pi];
{xr[Pi/2], yr[Pi/2]}
{xr[4Pi], yr[4Pi]}
```

{0, 0}

{1, Log[8]}

```
Simplify[lf/.{x->1, y->Log[8]}]
N[%]
```

$\frac{1}{8}+\frac{\text{Log}[8]}{e}$

0.889984 □

If C is a piecewise smooth simple closed curve and $P(x,y)$ and $Q(x,y)$ have continuous partial derivatives, *Green's Theorem* relates the line integral $\oint_C (P(x,y)\,dx+Q(x,y)\,dy)$ to a double integral.

Theorem 5.2 (Green's Theorem). *Let C be a piecewise smooth simple closed curve in the plane and R the region bounded by C. If $P(x,y)$ and $Q(x,y)$ have continuous partial derivatives on R,*

$$\oint_C (P(x,y)\,dx+Q(x,y)\,dy) = \iint_R \left(\frac{\partial Q}{\partial x}-\frac{\partial P}{\partial y}\right) dA. \tag{5.14}$$

We assume that the symbol $\oint$ means to evaluate the integral in the positive (or counter-clockwise) direction.

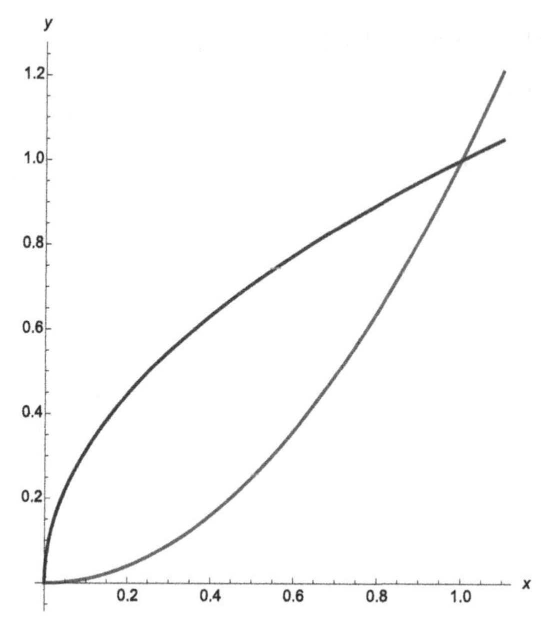

FIGURE 5.13 $y=x^2$ and $y=\sqrt{x}$, $0\le x\le 1$ (University of Arizona colors).

Example 5.41

Evaluate

$$\oint_C \left(e^{-x}-\sin y\right)\,dx+\left(\cos x-e^{-y}\right)\,dy$$

where C is the boundary of the region between $y=x^2$ and $x=y^2$.

Solution. After defining $P(x,y)=e^{-x}-\sin y$ and $Q(x,y)=\cos x-e^{-y}$, we use `Plot` to determine the region R bounded by C in Fig. 5.13.

```
p[x_, y_] = Exp[-x] - Sin[y];
q[x_, y_] = Cos[x] - Exp[-y];
Plot[{x^2, Sqrt[x]}, {x, 0, 1.1},
  PlotStyle->{{Thickness[.01], CMYKColor[0, 1, .65, .15]},
  {Thickness[.01], CMYKColor[1, .72, 0, .38]}},
  AspectRatio->Automatic, AxesLabel → {x, y}]
```

Using equation (5.14),

$$\begin{aligned}\oint_C \left(e^{-x}-\sin y\right)\,dx+\left(\cos x-e^{-y}\right)\,dy &= \iint_R \left(\frac{\partial Q}{\partial x}-\frac{\partial P}{\partial y}\right)dA\\ &= \iint_R (\cos y-\sin x)\,dA\\ &= \int_0^1\int_{x^2}^{\sqrt{x}} (\cos y-\sin x)\,dy\,dx,\end{aligned}$$

```
dqdp = Simplify[D[q[x, y], x] - D[p[x, y], y]]
```

Cos[y] − Sin[x]

which we evaluate with `Integrate`.

Integrate[dqdp, {x, 0, 1}, {y, x^2, Sqrt[x]}]

***N*[%]**

$\frac{1}{2}(-4-\sqrt{2\pi}(\text{FresnelC}[\sqrt{\frac{2}{\pi}}]+\text{FresnelS}[\sqrt{\frac{2}{\pi}}])+8\text{Sin}[1])$

0.151091

Notice that the result is given in terms of the `FresnelS` and `FresnelC` functions, which are defined by

$$\texttt{FresnelS[x]} = \int_0^x \sin\left(\frac{\pi}{2}t^2\right) dt \quad \text{and} \quad \texttt{FresnelC[x]} = \int_0^x \cos\left(\frac{\pi}{2}t^2\right) dt.$$

A more meaningful approximation is obtained with `N`. We conclude that

$$\int_0^1 \int_{x^2}^{\sqrt{x}} (\cos y - \sin x)\, dy\, dx \approx 0.151. \qquad \square$$

5.5.3 Surface Integrals

Let S be the graph of $z = f(x,y)$ $(y = h(x,z),\ x = k(y,z))$ and let R_{xy} (R_{xz}, R_{yz}) be the projection of S onto the xy (xz, yz) plane. Then,

$$\iint_S g(x,y,z)\,dS = \iint_{R_{xy}} g(x,y,f(x,y))\sqrt{[f_x(x,y)]^2+[f_y(x,y)]^2+1}\,dA \tag{5.15}$$

$$= \iint_{R_{xz}} g(x,h(x,z),z)\sqrt{[h_x(x,z)]^2+[h_z(x,z)]^2+1}\,dA \tag{5.16}$$

$$= \iint_{R_{yz}} g(k(y,z),y,z)\sqrt{[k_y(y,z)]^2+[k_z(y,z)]^2+1}\,dA. \tag{5.17}$$

If S is defined parametrically by

$$\mathbf{r}(s,t) = x(s,t)\mathbf{i} + y(s,t)\mathbf{j} + z(s,t)\mathbf{k}, \quad (s,t) \in R$$

the formula

$$\iint_S g(x,y,z)\,dS = \iint_R g(\mathbf{r}(s,t))\,\|\mathbf{r}_s \times \mathbf{r}_t\|\,dA, \tag{5.18}$$

where

$$\mathbf{r}_s = \frac{\partial x}{\partial s}\mathbf{i} + \frac{\partial y}{\partial s}\mathbf{j} + \frac{\partial z}{\partial s}\mathbf{k} \quad \text{and} \quad \mathbf{r}_t = \frac{\partial x}{\partial t}\mathbf{i} + \frac{\partial y}{\partial t}\mathbf{j} + \frac{\partial z}{\partial t}\mathbf{k},$$

is also useful.

Theorem 5.3 (The Divergence Theorem). *Let Q be any domain with the property that each line through any interior point of the domain cuts the boundary in exactly two points, and such that the boundary S is a piecewise smooth closed, oriented surface with unit normal $\mathbf{n}$. If $\mathbf{F}$ is a vector field that has continuous partial derivatives on Q, then*

$$\iiint_Q \nabla \cdot \mathbf{F}\,dV = \iiint_Q \mathit{div}\mathbf{F}\,dV = \iint_S \mathbf{F}\cdot\mathbf{n}\,dS. \tag{5.19}$$

For our purposes, a surface is **oriented** if it has two distinct sides.

In (5.19), $\iint_S \mathbf{F}\cdot\mathbf{n}\,dS$ is called the **outward flux** of the vector field **F** across the surface S. If S is a portion of the level curve $g(x, y) = C$ for some g, then a unit normal vector **n** may be taken to be either

$$\mathbf{n} = \frac{\nabla g}{\|\nabla g\|} \quad \text{or} \quad \mathbf{n} = -\frac{\nabla g}{\|\nabla g\|}.$$

If S is defined parametrically by

$$\mathbf{r}(s,t) = x(s,t)\mathbf{i} + y(s,t)\mathbf{j} + z(s,t)\mathbf{k}, \quad (s,t) \in R,$$

a unit normal vector to the surface is $\mathbf{n} = \dfrac{\mathbf{r}_s \times \mathbf{r}_t}{\|\mathbf{r}_s \times \mathbf{r}_t\|}$ and (5.19) becomes $\iint_S \mathbf{F}\cdot\mathbf{n}\,dS = \iint_R \mathbf{F}\cdot(\mathbf{r}_s \times \mathbf{r}_t)\,dA$.

Example 5.42

Find the outward flux of the vector field

$$\mathbf{F}(x,y,z) = \left(xz + xyz^2\right)\mathbf{i} + \left(xy + x^2yz\right)\mathbf{j} + \left(yz + xy^2z\right)\mathbf{k}$$

through the surface of the cube cut from the first octant by the planes $x = 1$, $y = 1$, and $z = 1$.

Solution. By the Divergence theorem,

$$\iint_{\text{cube surface}} \mathbf{F}\cdot\mathbf{n}\,dA = \iiint_{\text{cube interior}} \nabla\cdot\mathbf{F}\,dV.$$

Hence, without the Divergence theorem, calculating the outward flux would require six separate integrals, corresponding to the six faces of the cube. After defining **F**, we compute $\nabla\cdot\mathbf{F}$ with `Div`.

```
f[x_, y_, z_] = {xz + xyz^2, xy + x^2yz, yz + xy^2z};
```

```
divf = Div[f[x, y, z], {x, y, z}]
```

$$x + y + xy^2 + z + x^2z + yz^2$$

Remember, "Cartesian" is the default coordinate system so the same result as that obtained above is obtained with the following commands. Also keep in mind that Mathematica understands a wide number of coordinate systems. Thus, if you need results in your `coordinatesystem` be sure to include the explicit system in your command.

```
divf = Div[f[x, y, z], {x, y, z}, "Cartesian"]
```

$$x + y + xy^2 + z + x^2z + yz^2$$

The outward flux is then given by

$$\iiint_{\text{cube interior}} \nabla\cdot\mathbf{F}\,dV = \int_0^1\int_0^1\int_0^1 \nabla\cdot\mathbf{F}\,dz\,dy\,dx = 2,$$

which we compute with `Integrate`.

```
Integrate[divf, {z, 0, 1}, {y, 0, 1}, {x, 0, 1}]
```

2

□

Theorem 5.4 (Stokes' Theorem). *Let S be an oriented surface with finite surface area, unit normal $\mathbf{n}$, and boundary C. Let $\mathbf{F}$ be a continuous vector field defined on S such that the components of $\mathbf{F}$ have continuous partial derivatives at each nonboundary point of S. Then,*

$$\oint_C \mathbf{F} \cdot d\mathbf{r} = \iint_S \text{curl } \mathbf{F} \cdot \mathbf{n}\, dS. \tag{5.20}$$

In other words, the surface integral of the normal component of the curl of $\mathbf{F}$ taken over S equals the line integral of the tangential component of the field taken over C. In particular, if $\mathbf{F} = P(x, y, z)\mathbf{i} + Q(x, y, z)\mathbf{j} + R(x, y, z)\mathbf{k}$, then

$$\int_C (P(x, y, z)dx + Q(x, y, z)dy + R(x, y, z)dz) = \iint_S \text{curl } \mathbf{F} \cdot \mathbf{n}\, dS.$$

Example 5.43

Verify Stokes' theorem for the vector field

$$\mathbf{F}(x, y, z) = \left(x^2 - y\right)\mathbf{i} + \left(y^2 - z\right)\mathbf{j} + \left(x + z^2\right)\mathbf{k}$$

and S the portion of the paraboloid $z = f(x, y) = 9 - (x^2 + y^2)$, $z \geq 0$.

Solution. The curl of $\mathbf{F}$ is computed with `Curl` in `curlcapf`.

```
capf[x_, y_, z_] = {x^2 - y, y^2 - z, x + z^2};
f[x_, y_] = 9 - (x^2 + y^2);
curlcapf = Curl[capf[x, y, z], {x, y, z}]
```

$\{1, -1, 1\}$

Next, we define the function $h(x, y, z) = z - f(x, y)$. A normal vector to the surface is given by ∇h. A unit normal vector, $\mathbf{n}$, is then given by $\mathbf{n} = \dfrac{\nabla h}{\| \nabla h\|}$, which is computed in `un`.

```
h[x_, y_, z_] = z - f[x, y]
normtosurf = Grad[h[x, y, z], {x, y, z}]
```

$-9 + x^2 + y^2 + z$

$\{2x, 2y, 1\}$

```
un = Simplify[normtosurf/Sqrt[normtosurf.normtosurf]]
```

$\left\{\frac{2x}{\sqrt{1+4x^2+4y^2}}, \frac{2y}{\sqrt{1+4x^2+4y^2}}, \frac{1}{\sqrt{1+4x^2+4y^2}}\right\}$

The dot product curl $\mathbf{F} \cdot \mathbf{n}$ is computed in `g`.

```
g = Simplify[curlcapf.un]
```

$\frac{1+2x-2y}{\sqrt{1+4x^2+4y^2}}$

Using the surface integral evaluation formula (5.15),

$$\iint_S \text{curl } \mathbf{F} \cdot \mathbf{n}\, dS = \iint_R g(x, y, f(x, y)) \sqrt{[f_x(x, y)]^2 + [f_y(x, y)]^2 + 1}\, dA$$
$$= \int_{-3}^{3} \int_{-\sqrt{9-x^2}}^{\sqrt{9-x^2}} g(x, y, f(x, y)) \sqrt{[f_x(x, y)]^2 + [f_y(x, y)]^2 + 1}\, dy\, dx$$
$$= 9\pi,$$

which we compute with `Integrate`.

In this example, R, the projection of $f(x, y)$ onto the xy-plane, is the region bounded by the graph of the circle $x^2 + y^2 = 9$.

```
tointegrate = Simplify[(g/.z → f[x, y])]*
  Sqrt[D[f[x, y], x]^2 + D[f[x, y], y]^2 + 1]
```

$1 + 2x - 2y$

```
i1 = Integrate[tointegrate, {x, −3, 3},
  {y, −Sqrt[9 − x^2], Sqrt[9 − x^2]}]
```

9π

To verify Stokes' theorem, we must compute the associated line integral. Notice that the boundary of $z = f(x, y) = 9 - (x^2 + y^2)$, $z = 0$, is the circle $x^2 + y^2 = 9$ with parametrization $x = 3\cos t$, $y = 3\sin t$, $z = 0$, $0 \le t \le 2\pi$. This parametrization is substituted into $\mathbf{F}(x, y, z)$ and named `pvf`.

```
pvf = capf[3Cos[t], 3Sin[t], 0]
```

$\{9\text{Cos}[t]^2 - 3\text{Sin}[t], 9\text{Sin}[t]^2, 3\text{Cos}[t]\}$

To evaluate the line integral along the circle, we next define the parametrization of the circle and calculate $d\mathbf{r}$. The dot product of `pvf` and `dr` represents the integrand of the line integral.

```
r[t_] = {3Cos[t], 3Sin[t], 0};
dr = r'[t]
tointegrate = pvf.dr;
```

$\{-3\text{Sin}[t], 3\text{Cos}[t], 0\}$

As before with x and y, we instruct Mathematica to assume that t is real, compute the dot product of `pvf` and `dr`, and evaluate the line integral with `Integrate`.

```
Integrate[tointegrate, {t, 0, 2Pi}]
```

9π

As expected, the result is 9π. □

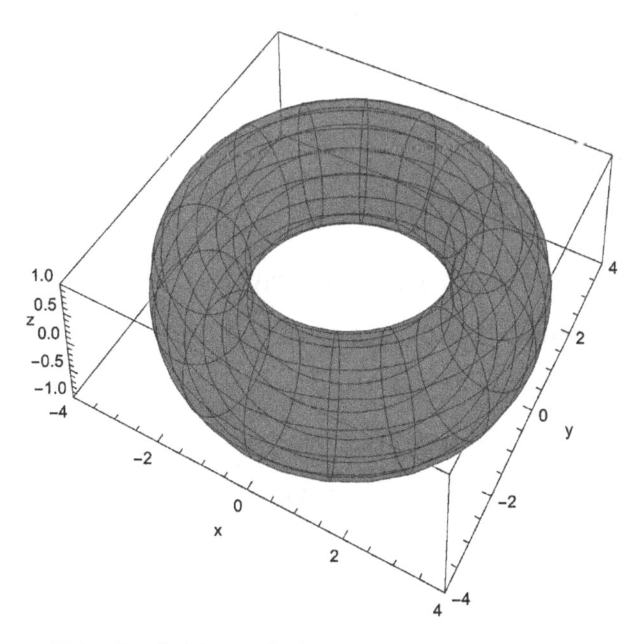

FIGURE 5.14 A torus (University of Alabama colors).

5.5.4 A Note on Nonorientability

Suppose that S is the surface determined by

$$\mathbf{r}(s,t) = x(s,t)\mathbf{i} + y(s,t)\mathbf{j} + z(s,t)\mathbf{k}, \quad (s,t) \in R$$

and let

$$\mathbf{n} = \frac{\mathbf{r}_s \times \mathbf{r}_t}{\|\mathbf{r}_s \times \mathbf{r}_t\|} \quad \text{or} \quad \mathbf{n} = -\frac{\mathbf{r}_s \times \mathbf{r}_t}{\|\mathbf{r}_s \times \mathbf{r}_t\|}, \tag{5.21}$$

where

$$\mathbf{r}_s = \frac{\partial x}{\partial s}\mathbf{i} + \frac{\partial y}{\partial s}\mathbf{j} + \frac{\partial z}{\partial s}\mathbf{k} \quad \text{and} \quad \mathbf{r}_t = \frac{\partial x}{\partial t}\mathbf{i} + \frac{\partial y}{\partial t}\mathbf{j} + \frac{\partial z}{\partial t}\mathbf{k},$$

if $\|\mathbf{r}_s \times \mathbf{r}_t\| \neq 0$. If $\mathbf{n}$ is defined, $\mathbf{n}$ is orthogonal (or perpendicular) to S. We state three familiar definitions of *orientable*.

See "When is a surface *not* orientable?" by Braselton, Abell, and Braselton [4] for a detailed discussion regarding the examples in this section.

- S is **orientable** if S has a unit normal vector field, $\mathbf{n}$, that varies continuously between any two points (x_0, y_0, z_0) and (x_1, y_1, z_1) on S. (See [5].)
- S is **orientable** if S has a continuous unit normal vector field, $\mathbf{n}$. (See [5] and [16].)
- S is **orientable** if a unit vector $\mathbf{n}$ can be defined at every nonboundary point of S in such a way that the normal vectors vary continuously over the surface S. (See [11].)

A path is **order preserving** if our chosen orientation is preserved as we move along the path. Thus, a surface like a torus is orientable.

Example 5.44: The Torus

Using the standard parametrization of the torus, we use `ParametricPlot3D` to plot the torus if $c = 3$ and $a = 1$ in Fig. 5.14.

Also see Example 2.33.

```
Clear[r]
c = 3;
a = 1;
```

```
x[s_, t_] = (c + aCos[s])Cos[t];
y[s_, t_] = (c + aCos[s])Sin[t];
z[s_, t_] = aSin[s];
r[s_, t_] = {x[s, t], y[s, t], z[s, t]};

threedp1t = ParametricPlot3D[r[s, t], {s, -Pi, Pi},
  {t, -Pi, Pi}, PlotPoints->{30, 30}, AspectRatio->1,
  PlotRange->{{-4, 4}, {-4, 4}, {-1, 1}},
  BoxRatios->{4, 4, 1}, AxesLabel->{x, y, z},
  PlotStyle → {{CMYKColor[.25, 1, .79, .2], Opacity[.3]}}]
```

To plot a normal vector field on the torus, we compute $\frac{\partial}{\partial s}\mathbf{r}(s,t)$,

```
rs = D[r[s, t], s]

{-Cos[t]Sin[s], -Sin[s]Sin[t], Cos[s]}

rt = D[r[s, t], t]

{-(3 + Cos[s])Sin[t], (3 + Cos[s])Cos[t], 0}
```

The cross product $\frac{\partial}{\partial s}\mathbf{r}(s,t) \times \frac{\partial}{\partial t}$ is formed in `rscrossrt`.

```
rscrossrt = Cross[rs, rt]//Simplify

{-Cos[s](3 + Cos[s])Cos[t], -Cos[s](3 + Cos[s])Sin[t], -(3 + Cos[s])Sin[s]}

Sqrt[rscrossrt.rscrossrt]//FullSimplify
```

$\sqrt{(3+\text{Cos}[s])^2}$

Using equation (5.24), we define un: given s and t, `un[s,t]` returns a unit normal to the torus.

```
Clear[un]
  un[s_, t_] =
  -rscrossrt/Sqrt[rscrossrt.
  rscrossrt]//PowerExpand//FullSimplify
```

$\left\{\frac{\text{Cos}[s](3+\text{Cos}[s])\text{Cos}[t]}{\sqrt{(3+\text{Cos}[s])^2}}, \frac{\text{Cos}[s](3+\text{Cos}[s])\text{Sin}[t]}{\sqrt{(3+\text{Cos}[s])^2}}, \frac{(3+\text{Cos}[s])\text{Sin}[s]}{\sqrt{(3+\text{Cos}[s])^2}}\right\}$

```
Map[PowerExpand, un[s, t]]
```

{Cos[s]Cos[t], Cos[s]Sin[t], Sin[s]}

```
r[s, t]
```

{(3 + Cos[s])Cos[t], (3 + Cos[s])Sin[t], Sin[s]}

```
un[s, t]
```

$\left\{\frac{\text{Cos}[s](3+\text{Cos}[s])\text{Cos}[t]}{\sqrt{(3+\text{Cos}[s])^2}}, \frac{\text{Cos}[s](3+\text{Cos}[s])\text{Sin}[t]}{\sqrt{(3+\text{Cos}[s])^2}}, \frac{(3+\text{Cos}[s])\text{Sin}[s]}{\sqrt{(3+\text{Cos}[s])^2}}\right\}$

To plot the normal vector field on the torus, we take advantage of the command `Arrow`. The command

```
Arrow[{{u1,u2,u3},{v1,v2,v3}}]
```

generates an arrow from (u_1, u_2, u_3) to (v_1, v_2, v_3), which we interpret to be the vector going from (u_1, u_2, u_3) to (v_1, v_2, v_3). To display the `Arrow` object, use `Show` together with `Graphics` (for two-dimensional arrows) or `Graphics3D` (for three dimensional arrows). See Fig. 5.15.

```
Clear[vecs]
vecs = Flatten[Table[{r[s, t], r[s, t] + un[s, t]},
   {s, −Pi, Pi, 2Pi/14}, {t, −Pi, Pi, 2Pi/29}], 1];
pp2 = Show[Graphics3D[Arrow[vecs]]]
```

We use `Show` (illustrating the use of the `ViewPoint` option) together with `GraphicsGrid` to see the vector field on the torus together from various angles in Fig. 5.16. Regardless of the viewing angle, the figure looks the same; the torus is orientable.

```
Show[threedp1t, pp2, AspectRatio->1,
   PlotRange->{{−5, 5}, {−5, 5}, {−2, 2}},
   BoxRatios->{4, 4, 1}, AxesLabel->{x, y, z}]
g1 = Show[threedp1t, pp2, AspectRatio->1,
   PlotRange->{{−5, 5}, {−5, 5}, {−2, 2}},
   BoxRatios->{4, 4, 1}, AxesLabel->{x, y, z},
   ViewPoint->{2.729, −0.000, 2.000}];
g2 = Show[threedp1t, pp2, AspectRatio->1,
   PlotRange->{{−5, 5}, {−5, 5}, {−2, 2}},
   BoxRatios->{4, 4, 1}, AxesLabel->{x, y, z},
   ViewPoint->{1.365, −2.364, 2.000}];
```

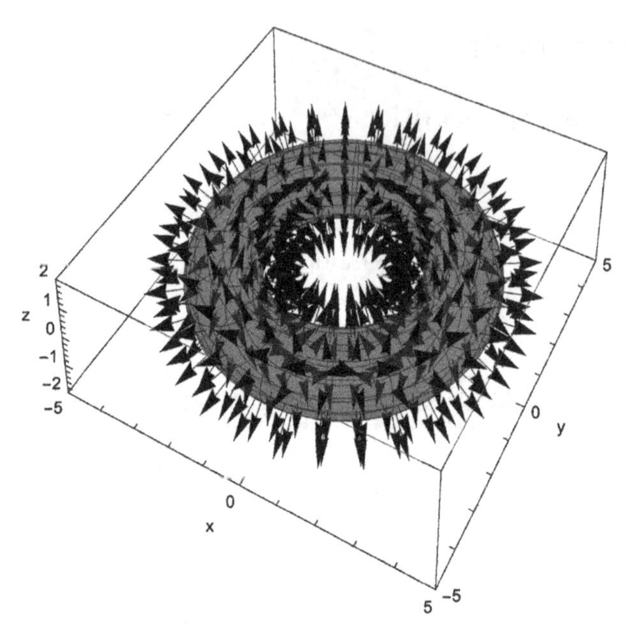

FIGURE 5.15 Unit normal vector field on a torus.

```
g3 = Show[threedp1t, pp2, AspectRatio->1,
  PlotRange->{{-5, 5}, {-5, 5}, {-2, 2}},
  BoxRatios->{4, 4, 1}, AxesLabel->{x, y, z},
  ViewPoint->{-1.365, -2.364, 2.000}];

g4 = Show[threedp1t, pp2, AspectRatio->1,
  PlotRange->{{-5, 5}, {-5, 5}, {-2, 2}},
  BoxRatios->{4, 4, 1}, AxesLabel->{x, y, z},
  ViewPoint->{-2.729, 0.000, 2.000}];

g5 = Show[threedp1t, pp2, AspectRatio->1,
  PlotRange->{{-5, 5}, {-5, 5}, {-2, 2}},
  BoxRatios->{4, 4, 1}, AxesLabel->{x, y, z},
  ViewPoint->{-1.365, 2.364, 2.000}];

g6 = Show[threedp1t, pp2, AspectRatio->1,
  PlotRange->{{-5, 5}, {-5, 5}, {-2, 2}},
  BoxRatios->{4, 4, 1}, AxesLabel->{x, y, z},
  ViewPoint->{1.365, 2.364, 2.000}];

Show[GraphicsGrid[{{g1, g2}, {g3, g4}, {g5, g6}}]]
```

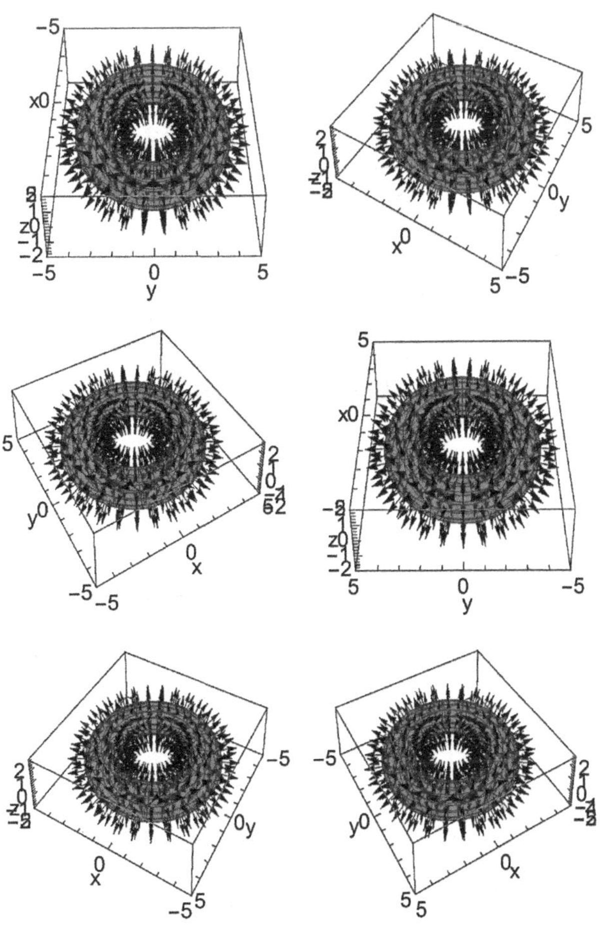

FIGURE 5.16 The torus is orientable.

If a 2-manifold, S, has an **order reversing path** (or **not order preserving path**), S is **nonorientable** (or **not orientable**).

Determining whether a given surface S is orientable or not may be a difficult problem.

Example 5.45: The Möbius Strip

The *Möbius strip* is frequently cited as an example of a nonorientable surface with boundary: it has one side and is physically easy to construct by hand by half twisting and taping (or pasting) together the ends of a piece of paper (for example, see [5], [11], and [16]). A parametrization of the Möbius strip is $\mathbf{r}(s,t) = x(s,t)\mathbf{i} + y(s,t)\mathbf{j} + z(s,t)\mathbf{k}$, $-1 \le s \le 1$, $-\pi \le t \le \pi$, where

$$x = \left[c + s\cos\left(\frac{1}{2}t\right)\right]\cos t, \quad y = \left[c + s\cos\left(\frac{1}{2}t\right)\right]\sin t, \text{ and}$$
$$z = s\sin\left(\frac{1}{2}t\right), \tag{5.22}$$

and we assume that $c > 1$. In Fig. 5.17, we graph the Möbius strip using $c = 3$.

```
c = 3;
x[s_, t_] = (c + sCos[t/2])Cos[t];
y[s_, t_] = (c + sCos[t/2])Sin[t];
```

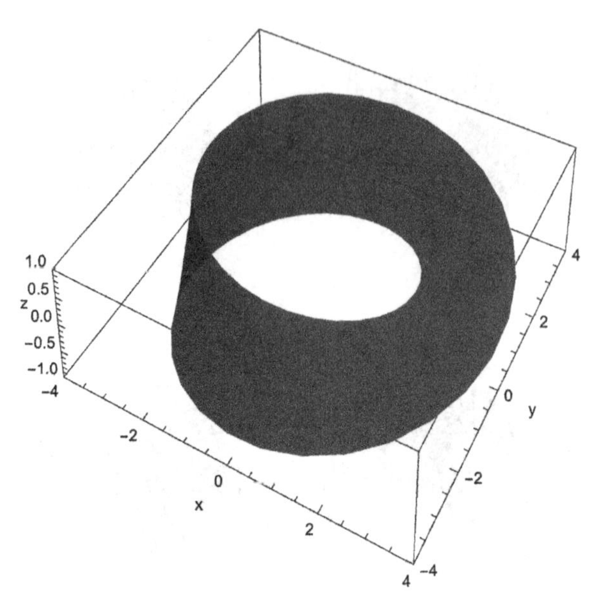

FIGURE 5.17 Parametric plot of equations (5.22) if $c = 3$ (Mississippi State University colors).

```
z[s_, t_] = sSin[t/2];
r[s_, t_] = {x[s, t], y[s, t], z[s, t]};

threedp1 = ParametricPlot3D[r[s, t], {s, -1, 1},
  {t, -Pi, Pi}, PlotPoints->{30, 30},
  AspectRatio->1, PlotRange ->
  {{-4, 4}, {-4, 4}, {-1, 1}}, BoxRatios->{4, 4, 1},
  AxesLabel->{x, y, z},
  Mesh -> False, PlotStyle -> {{CMYKColor[.5, 1, 1.25], Opacity[.8]}}]
```

Although it is relatively easy to see in the plot that the Möbius strip has only one side, the fact that a unit vector, $\mathbf{n}$, normal to the Möbius strip at a point P reverses its direction as $\mathbf{n}$ moves around the strip to P is not obvious to the novice.

With Mathematica, we compute $\|\mathbf{r}_s \times \mathbf{r}_t\|$ and $\mathbf{n} = \dfrac{\mathbf{r}_s \times \mathbf{r}_t}{\|\mathbf{r}_s \times \mathbf{r}_t\|}$.

```
rs = D[r[s, t], s]
```

$\{\text{Cos}[\tfrac{t}{2}]\text{Cos}[t], \text{Cos}[\tfrac{t}{2}]\text{Sin}[t], \text{Sin}[\tfrac{t}{2}]\}$

```
rt = D[r[s, t], t]
```

$\{-\tfrac{1}{2}s\text{Cos}[t]\text{Sin}[\tfrac{t}{2}] - (3 + s\text{Cos}[\tfrac{t}{2}])\text{Sin}[t], (3 + s\text{Cos}[\tfrac{t}{2}])\text{Cos}[t] - \tfrac{1}{2}s\text{Sin}[\tfrac{t}{2}]\text{Sin}[t],$
$\tfrac{1}{2}s\text{Cos}[\tfrac{t}{2}]\}$

```
rscrossrt = Cross[rs, rt]//Simplify
```

$\{-\frac{1}{2}(-s\text{Cos}[\frac{t}{2}]+6\text{Cos}[t]+s\text{Cos}[\frac{3t}{2}])\text{Sin}[\frac{t}{2}], \frac{1}{4}(-s-6\text{Cos}[\frac{t}{2}]-2s\text{Cos}[t]+6\text{Cos}[\frac{3t}{2}]+ s\text{Cos}[2t]), \text{Cos}[\frac{t}{2}](3+s\text{Cos}[\frac{t}{2}])\}$

```
Sqrt[rscrossrt.rscrossrt]//FullSimplify
```

$\sqrt{9+\frac{3s^2}{4}+6s\text{Cos}[\frac{t}{2}]+\frac{1}{2}s^2\text{Cos}[t]}$

```
Clear[un]
  un[s_, t_] =
rscrossrt/Sqrt[rscrossrt.rscrossrt]//FullSimplify
```

$\{\frac{2\text{Sin}[\frac{t}{2}](-3\text{Cos}[t]+s\text{Sin}[\frac{t}{2}]\text{Sin}[t])}{\sqrt{36+3s^2+24s\text{Cos}[\frac{t}{2}]+2s^2\text{Cos}[t]}}, -\frac{3\text{Cos}[\frac{t}{2}]-3\text{Cos}[\frac{3t}{2}]+s(\text{Cos}[t]+\text{Sin}[t]^2)}{\sqrt{36+3s^2+24s\text{Cos}[\frac{t}{2}]+2s^2\text{Cos}[t]}}, \frac{s+6\text{Cos}[\frac{t}{2}]+s\text{Cos}[t]}{\sqrt{36+3s^2+24s\text{Cos}[\frac{t}{2}]+2s^2\text{Cos}[t]}}\}$

Consider the path C given by $\mathbf{r}(0,t)$, $-\pi \le t \le \pi$ that begins and ends at $\langle -3,0,0\rangle$. On C, $\mathbf{n}(0,t)$ is given by

```
un[0, t]
```

$\{-\text{Cos}[t]\text{Sin}[\frac{t}{2}], \frac{1}{6}(-3\text{Cos}[\frac{t}{2}]+3\text{Cos}[\frac{3t}{2}]), \text{Cos}[\frac{t}{2}]\}$

At $t=-\pi$, $\mathbf{n}(0,-\pi)=\langle 1,0,0\rangle$, while at $t=\pi$, $\mathbf{n}(0,\pi)=\langle -1,0,0\rangle$.

```
r[0, -Pi]
r[0, Pi]
```

$\{-3,0,0\}$

$\{-3,0,0\}$

As $\mathbf{n}$ moves along C from $\mathbf{r}(0,-\pi)$ to $\mathbf{r}(0,\pi)$, the orientation of $\mathbf{n}$ reverses, as shown in Fig. 5.18.

```
l1 = Table[r[0, t], {t, -Pi, Pi, 2Pi/179}];

threedp2 = Show[Graphics3D[{Thickness[.02],
   CMYKColor[.3, .2, .17, .57], Line[l1]}], Axes->Automatic,
   PlotRange->{{-4, 4}, {-4, 4}, {-1, 1}},
   BoxRatios->{4, 4, 1}, AspectRatio->1];
```

```
vecs = Table[Arrow[{r[0, t], r[0, t] + un[0, t]}], {t, -π, π, 2π/59}];
pp2 = Show[Graphics3D[vecs]];

Show[threedp2, pp2, ViewPoint →
  {-2.093, 2.124, 1.600}, AxesLabel->{x, y, z},
  Boxed->False]
```

Several different views of Fig. 5.18 on the Möbius strip shown in Fig. 5.17 are shown in Fig. 5.19. C is an orientation reversing path and we can conclude that the Möbius strip is not orientable.

An animation is particularly striking.

```
g1 = Show[threedp1, threedp2, pp2,
  ViewPoint->{2.729, -0.000, 2.000},
  AxesLabel->{x, y, z}, Boxed->False];

g2 = Show[threedp1, threedp2, pp2,
  ViewPoint->{1.365, -2.364, 2.000},
  AxesLabel->{x, y, z}, Boxed->False];

g3 = Show[threedp1, threedp2, pp2,
  ViewPoint->{-1.365, -2.364, 2.000},
  AxesLabel->{x, y, z}, Boxed->False];

g4 = Show[threedp1, threedp2, pp2,
  ViewPoint->{-2.729, 0.000, 2.000},
  AxesLabel->{x, y, z}, Boxed->False];

g5 = Show[threedp1, threedp2, pp2,
  ViewPoint->{-1.365, 2.364, 2.000},
  AxesLabel->{x, y, z}, Boxed->False];

g6 = Show[threedp1, threedp2, pp2,
  ViewPoint->{1.365, 2.364, 2.000},
  AxesLabel->{x, y, z}, Boxed->False];

Show[GraphicsGrid[{{g1, g2}, {g3, g4}, {g5, g6}}]]
```

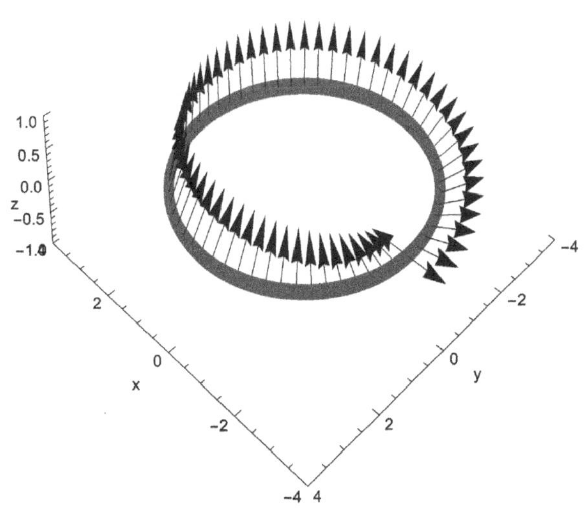

FIGURE 5.18 Parametric plot of equations (5.22) if $c = 3$.

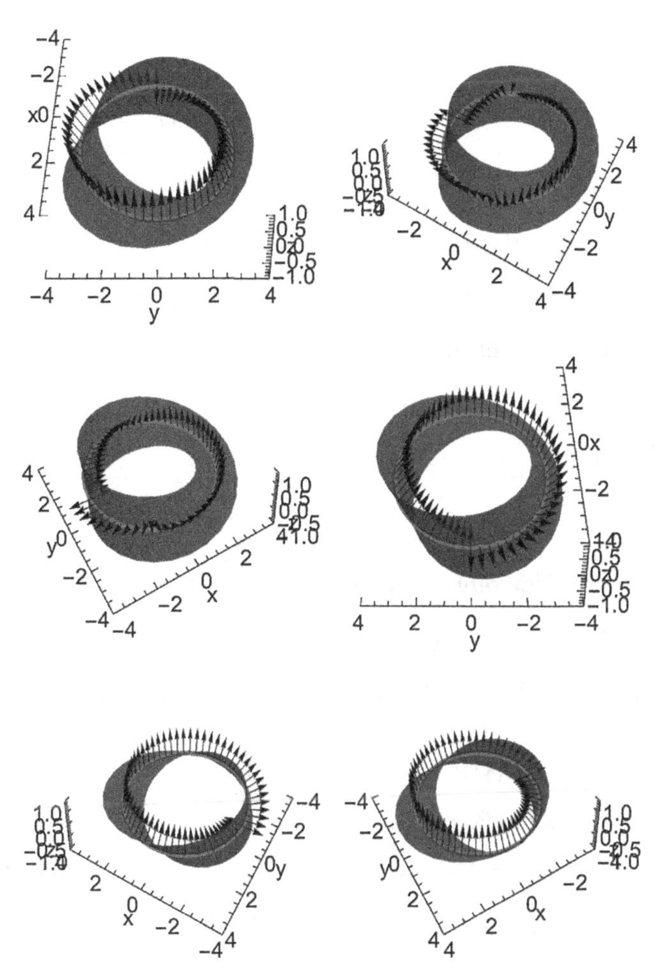

FIGURE 5.19 Different views of a Möbius strip with an orientation reversing path.

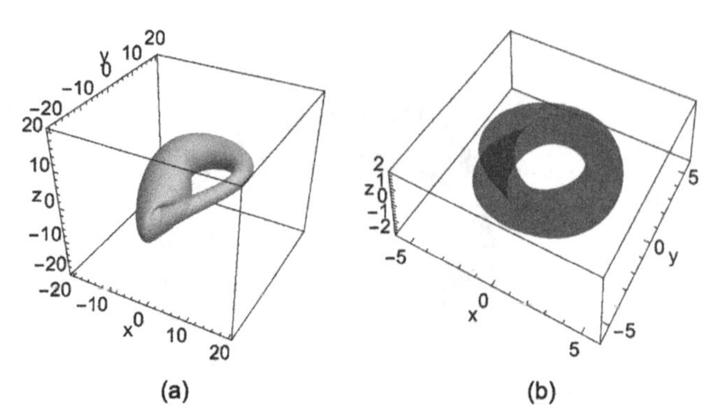

FIGURE 5.20 Two different immersions of the Klein bottle: (a) the "usual" immersion; (b) the Figure-8 immersion (Louisiana State University colors).

Example 5.46

The *Klein bottle* is an interesting surface with neither an inside nor an outside, which indicates to us that it is not orientable. In Fig. 5.20 (a) we show the "usual" *immersion* of the Klein bottle. Although the Klein bottle does not intersect itself, it is not possible to visualize it in Euclidean 3-space without it doing so. Visualizations of 2-manifolds like the Klein bottle's "usual" rendering in Euclidean 3-space are called *immersions*. (See [14] for a nontechnical discussion of immersions.)

```
r = 4(1 - 1/2Cos[u]);
x1[u_, v_] = 6(1 + Sin[u])Cos[u] + rCos[u]Cos[v];
x2[u_, v_] = 6(1 + Sin[u])Cos[u] + rCos[v + Pi];
y1[u_, v_] = 16Sin[u] + rSin[u]Cos[v];
y2[u_, v_] = 16Sin[u];
z[u_, v_] = rSin[v];

kb1a = ParametricPlot3D[{x1[s, t], y1[s, t], z[s, t]},
  {s, 0, Pi}, {t, 0, 2Pi}, PlotPoints->{30, 30},
  AspectRatio->1, AxesLabel->{x, y, z},
  Mesh → False, PlotStyle → {{CMYKColor[0, .19, .89, 0], Opacity[.8]}}];

kb1b = ParametricPlot3D[{x1[s, t], y1[s, t], z[s, t]},
  {s, Pi, 2Pi}, {t, 0, 2Pi}, PlotPoints->{30, 30},
  AspectRatio->1, AxesLabel->{x, y, z},
  Mesh → False, PlotStyle → {{CMYKColor[0, .19, .89, 0], Opacity[.8]}}]

kb1 = Show[kb1a, kb1b, PlotRange → {{-20, 20}, {-20, 20}, {-20, 20}}]
```

Fig. 5.20 (b) shows the *Figure-8* immersion of the Klein bottle. Notice that it is not easy to see that the Klein bottle has neither an inside nor an outside in Fig. 5.14.

```
a = 3;
x[u_, v_] = (a + Cos[u/2]Sin[v] - Sin[u/2]Sin[2v])Cos[u];
y[u_, v_] = (a + Cos[u/2]Sin[v] - Sin[u/2]Sin[2v])Sin[u];
z[u_, v_] = Sin[u/2]Sin[v] + Cos[u/2]Sin[2v];
r[u_, v_] = {x[u, v], y[u, v], z[u, v]};

ParametricPlot3D[r[t, t], {t, 0, 2Pi}]

kb2 = ParametricPlot3D[r[s, t], {s, -Pi, Pi}, {t, -Pi, Pi},
  PlotPoints->{30, 30}, AspectRatio->1,
  AxesLabel->{x, y, z},
  PlotRange->{{-6, 6}, {-6, 6}, {-2, 2}}, BoxRatios->{4, 4, 1}, Mesh → False,
  PlotStyle → {{CMYKColor[.82, .98, 0, .12], Opacity[.4]}}]

Show[GraphicsRow[{kb1, kb2}]]
```

In fact, to many readers it may not be clear whether the Klein bottle is orientable or nonorientable, especially when we compare the graph to the graphs of the Möbius strip and torus in the previous examples.

A parametrization of the Figure-8 immersion of the Klein bottle (see [17]) is $\mathbf{r}(s,t) = x(s,t)\mathbf{i} + y(s,t)\mathbf{j} + z(s,t)\mathbf{k}$, $-\pi \le s \le \pi$, $-\pi \le t \le \pi$, where

$$x = \left[c + \cos\left(\frac{1}{2}s\right)\sin t - \sin\left(\frac{1}{2}s\right)\sin 2t\right]\cos s,$$

$$y = \left[c + \cos\left(\frac{1}{2}s\right)\sin t - \sin\left(\frac{1}{2}s\right)\sin 2t\right]\sin s, \tag{5.23}$$

and

$$z = \sin\left(\frac{1}{2}s\right)\sin t + \cos\left(\frac{1}{2}s\right)\sin 2t.$$

The plot in Fig. 5.20 (b) uses equation (5.23) if $c = 3$.

Using (5.21), let

$$\mathbf{n} = \frac{\mathbf{r}_s \times \mathbf{r}_t}{\|\mathbf{r}_s \times \mathbf{r}_t\|}.$$

Let C be the path given by

$$\mathbf{r}(t,t) = x(t,t)\mathbf{i} + y(t,t)\mathbf{j} + z(t,t)\mathbf{k}, \quad -\pi \le t \le \pi \tag{5.24}$$

that begins and ends at $\mathbf{r}(-\pi,-\pi) = \mathbf{r}(\pi,\pi) = \langle -3, 0, 0\rangle$ and where the components are given by (5.23). The components of $\mathbf{r}$ and $\mathbf{n}$ are computed with Mathematica. The final calculations are quite lengthy so we suppress the output of the last few by placing a semicolon (;) at the end of those commands.

```
rs = D[r[s, t], s]//Simplify
```

$\{-\frac{1}{2}\text{Cos}[s](\text{Sin}[\frac{s}{2}]\text{Sin}[t]+\text{Cos}[\frac{s}{2}]\text{Sin}[2t])+\text{Sin}[s](-3-\text{Cos}[\frac{s}{2}]\text{Sin}[t]+\text{Sin}[\frac{s}{2}]\text{Sin}[2t]),$

$-\frac{1}{2}\text{Sin}[s](\text{Sin}[\frac{s}{2}]\text{Sin}[t]+\text{Cos}[\frac{s}{2}]\text{Sin}[2t])+\text{Cos}[s](3+\text{Cos}[\frac{s}{2}]\text{Sin}[t]-\text{Sin}[\frac{s}{2}]\text{Sin}[2t]),$

$\frac{1}{2}(\text{Cos}[\frac{s}{2}]-2\text{Cos}[t]\text{Sin}[\frac{s}{2}])\text{Sin}[t]\}$

```
rt = D[r[s, t], t]//Simplify
```

$\{\text{Cos}[s](\text{Cos}[\frac{s}{2}]\text{Cos}[t]-2\text{Cos}[2t]\text{Sin}[\frac{s}{2}]),(\text{Cos}[\frac{s}{2}]\text{Cos}[t]-2\text{Cos}[2t]\text{Sin}[\frac{s}{2}])\text{Sin}[s],$

$2\text{Cos}[\frac{s}{2}]\text{Cos}[2t]+\text{Cos}[t]\text{Sin}[\frac{s}{2}]\}$

```
rscrossrt = Cross[rs, rt];

normcross = Sqrt[rscrossrt.rscrossrt];

Clear[un]
  un[s_, t_] = -rscrossrt/Sqrt[rscrossrt.rscrossrt];
```

At $t=-\pi$, $\mathbf{n}(-\pi,-\pi)=\left\langle\frac{1}{\sqrt{5}},0,\frac{2}{\sqrt{5}}\right\rangle$, while at $t=\pi$, $\mathbf{n}(\pi,\pi)=\left\langle-\frac{1}{\sqrt{5}},0,-\frac{2}{\sqrt{5}}\right\rangle$ so as $\mathbf{n}$ moves along C from $\mathbf{r}(-\pi,-\pi)$ to $\mathbf{r}(\pi,\pi)$, the orientation of $\mathbf{n}$ reverses. Several different views of the orientation reversing path on the Klein bottle shown in Fig. 5.20 (b) are shown in Fig. 5.21.

```
l1 = Table[r[s, s], {s, -Pi, Pi, 2Pi/179}];

threedp2 = Show[Graphics3D[{Thickness[.02],
  GrayLevel[.6], Line[l1]}], Axes->Automatic,
  PlotRange->{{-4, 4}, {-4, 4}, {-4, 4}},
  BoxRatios->{4, 4, 1}, AspectRatio->1];

vecs = Table[Arrow[{{r[s, s], r[s, s] + un[s, s]}}], {s, -π, π, 2π/59}];
pp2 = Show[Graphics3D[vecs]];

pp3 = Show[threedp2, pp2,
  AxesLabel->{x, y, z},
  Boxed->False, PlotRange → {{-5, 5},
  {-5, 5}, {-5, 5}}]

g1 = Show[kb2, threedp2, pp2, AspectRatio->1,
  PlotRange->{{-6, 6}, {-6, 6}, {-2, 2}},
  BoxRatios->{4, 4, 1}, AxesLabel->{x, y, z},
  ViewPoint->{2.729, -0.000, 2.000}]
```

```
g2 = Show[kb2, threedp2, pp2,
  AspectRatio->1,
  PlotRange->{{-6, 6}, {-6, 6}, {-2, 2}},
  BoxRatios->{4, 4, 1},
  AxesLabel->{x, y, z},
  ViewPoint->{1.365, -2.364, 2.000}]

g3 = Show[kb2, threedp2, pp2, AspectRatio->1,
  PlotRange->{{-6, 6}, {-6, 6}, {-6, 6}},
  BoxRatios->{4, 4, 1}, AxesLabel->{x, y, z},
  ViewPoint->{-1.365, -2.364, 2.000}]

g4 = Show[kb2, threedp2, pp2, AspectRatio->1,
  PlotRange->{{-6, 6}, {-6, 6}, {-6, 6}},
  BoxRatios->{4, 4, 1}, AxesLabel->{x, y, z},
  ViewPoint->{-2.729, 0.000, 2.000}]

g5 = Show[kb2, pp2, AspectRatio->1,
  PlotRange->{{-6, 6}, {-6, 6}, {-6, 6}},
  BoxRatios->{4, 4, 1}, AxesLabel->{x, y, z},
  ViewPoint->{-1.365, 2.364, 2.000}]

g6 = Show[kb2, pp3, AspectRatio->1,
  PlotRange->{{-6, 6}, {-6, 6}, {-2, 2}},
  BoxRatios->{4, 4, 1}, AxesLabel->{x, y, z},
  ViewPoint->{1.365, 2.364, 2.000}]

Show[GraphicsGrid[{{g1, g2}, {g3, g4}, {g5, g6}}]]
```

C is an orientation reversing path and we can conclude that the Klein bottle is not orientable.

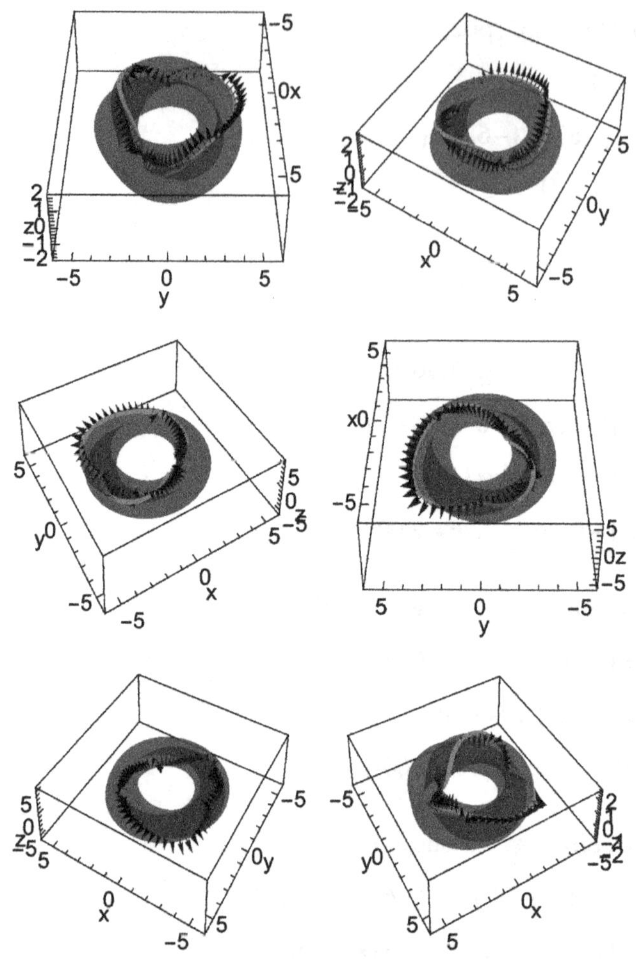

FIGURE 5.21 Different views of the Figure-8 immersion of the Klein bottle with an orientation reversing path.

5.5.5 More on Tangents, Normals, and Curvature in $\mathcal{R}^3$

Earlier, we discussed the unit tangent and normal vectors and curvature for a vector-valued function $\gamma : (a, b) \to \mathcal{R}^2$. These concepts can be extended to curves and surfaces in space.

These concepts are presented *beautifully* and extensively for the Mathematica user in *Modern Differential Geometry of Curves and Surfaces with Mathematica*, Third Edition, CRC Press, 2006, by Alfred Gray, Simon Salamon, and Elsa Abbena. Our treatment just touches on a few of the topics discussed by Gray et al. and updates some of their wonderful and elegant work to Mathematica 11.
For many good reasons, sometimes the "Frenet formulas" are also called the "Frenet-Serret formulas."

For $\gamma : (a, b) \to \mathcal{R}^3$, the **Frenet frame field** is the ordered triple $\{\mathbf{T}, \mathbf{N}, \mathbf{B}\}$, where $\mathbf{T}$ is the **unit tangent vector field**, $\mathbf{N}$ is the **unit normal vector field**, and $\mathbf{B}$ is the **unit binormal vector field**. Each of these vectors has norm 1 and each is orthogonal to the other (the dot product of one with another is 0) and the ***Frenet formulas*** are satisfied: $\mathbf{T}' = \kappa\mathbf{N}$, $\mathbf{N}' = -\kappa\mathbf{T} + \tau\mathbf{B}$, $\mathbf{B}' = -\tau\mathbf{N}$. τ is the **torsion** of the curve γ; κ is the curvature. For the curve $\gamma : (a, b) \to \mathcal{R}^3$, formulas for these quantities are given by:

$$\mathbf{T} = \frac{\gamma'}{\|\gamma'\|}, \qquad \mathbf{N} = \mathbf{B} \times \mathbf{T}, \qquad \mathbf{B} = \frac{\gamma' \times \gamma''}{\|\gamma' \times \gamma''\|},$$
$$\kappa = \frac{\|\gamma' \times \gamma''\|}{\|\gamma'\|^3}, \qquad \tau = \frac{\gamma' \times \gamma'' \cdot \gamma'''}{\|\gamma' \times \gamma''\|^2}. \tag{5.25}$$

We adjust Gray's routines slightly for Mathematica 11. Here is the unit tangent vector,

```
tangent[α_][t_]:=D[α[tt], tt]/FullSimplify[Norm[D[α[tt], tt]],
  Assumptions → tt ∈ Reals]/.tt → t
```

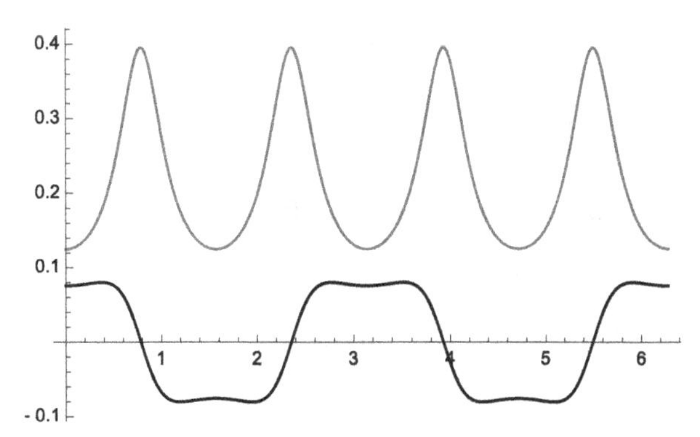

FIGURE 5.22 The curvature and torsion for a spherical spiral (University of Tennessee colors).

Similarly, the binormal is defined with

```
binormal[α_][t_]:=FullSimplify[
  Cross[D[α[tt], tt], D[α[tt], {tt, 2}]]]/
  FullSimplify[Norm[Cross[D[α[tt], tt], D[α[tt], {tt, 2}]]],
  Assumptions → tt ∈ Reals]/.tt → t
```

so the unit normal is defined with

```
normal[α_][t_]:=Cross[binormal[α][t], tangent[α][t]];
```

Notice how we use `Assumptions` to instruct Mathematica to assume that the domain of γ consists of real numbers. In the same manner, we define the curvature and torsion.

```
curve2[α_][t_]:=Simplify[Norm[Cross[D[α[tt], tt],
  D[α[tt], {tt, 2}]]]/
  Norm[D[α[tt], tt]]^3,
  Assumptions → tt ∈ Reals]/.tt → t;

torsion2[α_][t_]:=Simplify[Cross[D[α[tt], tt],
  D[α[tt], {tt, 2}]].D[α[tt], {tt, 3}]/
  Norm[Cross[D[α[tt], tt], D[α[tt], {tt, 2}]]]^2,
  Assumptions → tt ∈ Reals]/.tt → t;
```

In even the simplest situations, these calculations are quite complicated. Graphically seeing the results may be more meaningful that the explicit formulas.

Example 5.47

Consider the spherical spiral given by $\gamma(t) = \langle 8\cos 3t\, \cos 2t, 8\sin 3t\, \cos 2t, 8\sin 2t\rangle$. The curvature and torsion for the curve are graphed with `Plot` and shown in Fig. 5.22.

```
γ[t_] = {8Cos[3t]Cos[2t], 8Sin[3t]Cos[2t], 8Sin[2t]}
```

{8Cos[2t]Cos[3t], 8Cos[2t]Sin[3t], 8Sin[2t]}

```
Plot[Tooltip[{curve2[γ][t], torsion2[γ][t]}], {t, 0, 2Pi},
  PlotStyle → {{Thickness[.01], CMYKColor[0, .5, 1, 0]},
  {Thickness[.01], CMYKColor[0, 0, 0, .6]}}]
```

We now compute **T**, **B**, and **N**. For length considerations, we display an abbreviated portion of **B** with `Short`.

```
tangent[γ][t]
binormal[γ][t]
normal[γ][t]//Short
```

$$\left\{\frac{-16\text{Cos}[3t]\text{Sin}[2t]-24\text{Cos}[2t]\text{Sin}[3t]}{4\sqrt{34+18\text{Cos}[4t]}}, \frac{24\text{Cos}[2t]\text{Cos}[3t]-16\text{Sin}[2t]\text{Sin}[3t]}{4\sqrt{34+18\text{Cos}[4t]}}, \frac{4\text{Cos}[2t]}{\sqrt{34+18\text{Cos}[4t]}}\right\}$$

$$\left\{\frac{2(-3\text{Sin}[t]+34\text{Sin}[3t]+15\text{Sin}[7t])}{\sqrt{21886+12456\text{Cos}[4t]+810\text{Cos}[8t]}}, -\frac{2(3\text{Cos}[t]+34\text{Cos}[3t]+15\text{Cos}[7t])}{\sqrt{21886+12456\text{Cos}[4t]+810\text{Cos}[8t]}}, \frac{6(21+5\text{Cos}[4t])}{\sqrt{21886+12456\text{Cos}[4t]+810\text{Cos}[8t]}}\right\}$$

$$\left\{\langle\langle 8\rangle\rangle + \frac{60\langle\langle 2\rangle\rangle\text{Sin}[3t]}{\sqrt{17+9\langle\langle 1\rangle\rangle}\sqrt{\langle\langle 1\rangle\rangle}}, \langle\langle 1\rangle\rangle, \langle\langle 1\rangle\rangle\right\}$$

It is difficult to see how these complicated formulas relate to this spherical spiral. To help us understand what they mean, we first plot the spiral with `ParametricPlot3D`. See Fig. 5.23 (a).

```
p1 = ParametricPlot3D[γ[t], {t, 0, 2Pi},
  PlotRange → {{-8.5, 8.5}, {-8.5, 8.5}, {-8.5, 8.5}},
  PlotStyle → {{CMYKColor[0, .5, 1, 0], Thick}}]
```

Next, we use `Table` to compute lists of two ordered triples. For each list, the first ordered triple consists of $\gamma(t)$ and the second the value of the sum of $\gamma(t)$ and $\mathbf{T}(\gamma(t))$ ($\mathbf{B}(\gamma(t))$, $\mathbf{N}(\gamma(t))$). These ordered triples that correspond to vectors are plotted with `Arrow` together with `Show` and `Graphics3D`.

```
ts = Table[Arrow[{{γ[t], γ[t] + tangent[γ][t]}}//N], {t, 0, 2Pi, 2Pi/99}];
bs = Table[Arrow[{{γ[t], γ[t] + binormal[γ][t]}}//N], {t, 0, 2Pi, 2Pi/99}];
ns = Table[Arrow[{{γ[t], γ[t] + normal[γ][t]}}//N], {t, 0, 2Pi, 2Pi/99}];
ysplot = Show[Graphics3D[ts]];
bsplot = Show[Graphics3D[bs]];
nsplot = Show[Graphics3D[ns]];
p2 = Show[ysplot, bsplot, nsplot]
```

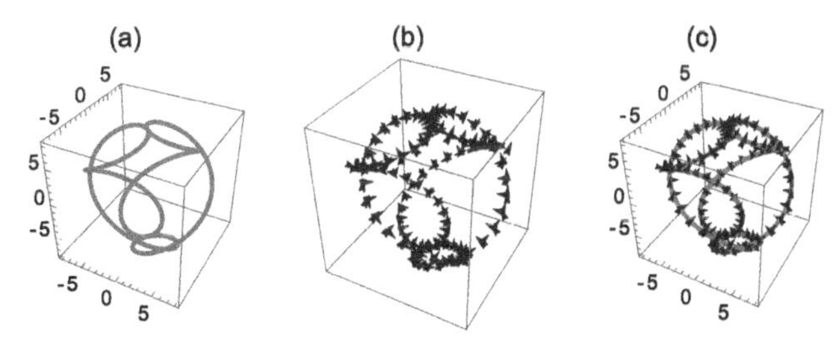

FIGURE 5.23 (a) The spherical spiral. (b) Various **T**, **N**, and **B** for the spherical spiral. (c) The spherical spiral with various **T**, **N**, and **B** shown together.

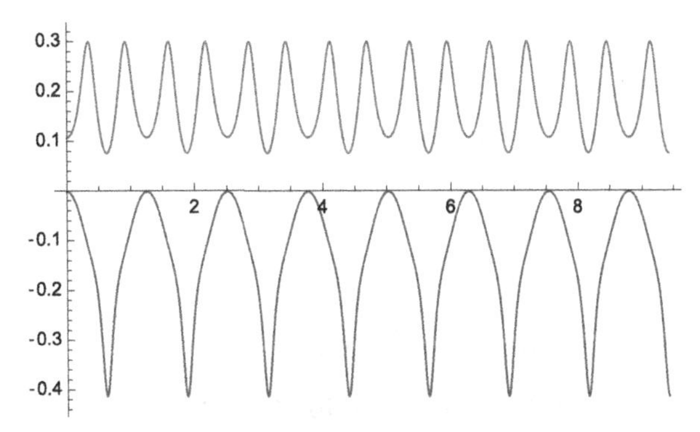

FIGURE 5.24 The curvature and torsion for the torus knot, `torusknot[8,3,5][2,5]`.

For a nice view of `p1` and `p2`, display them together with `Show`. See Fig. 5.23 (c).

```
Show[p1, p2]

Show[GraphicsRow[{p1, p2, Show[p1, p2]}]]
```

The previous example illustrates that capturing the depth of three-dimensional curves by projections into two dimensions can be difficult. Sometimes taking advantage of three-dimensional surface plots can help. For a basic space curve, `tubecurve` places a "tube" of radius r around the space curve.

```
Clear[tubecurve, γ, r]

tubecurve[γ_][r_][t_, θ_] = γ[t]+
  r(Cos[θ]normal[γ][t] + Sin[θ]binormal[γ][t]);
```

To illustrate the utility, we redefine `torusknot` that was presented in Chapter 2.

The results displayed in the text are in black-and-white and do not reflect the stunning color images generated by these commands.

```
torusknot[a_, b_, c_][p_, q_][t_]:=
  {(a + bCos[qt])Cos[pt], (a + bCos[qt])Sin[pt],
  cSin[qt]}
```

Example 5.48

For the knot `torusknot}[8,3,5][2,5]` we plot the curvature and torsion with `Plot` in Fig. 5.24.

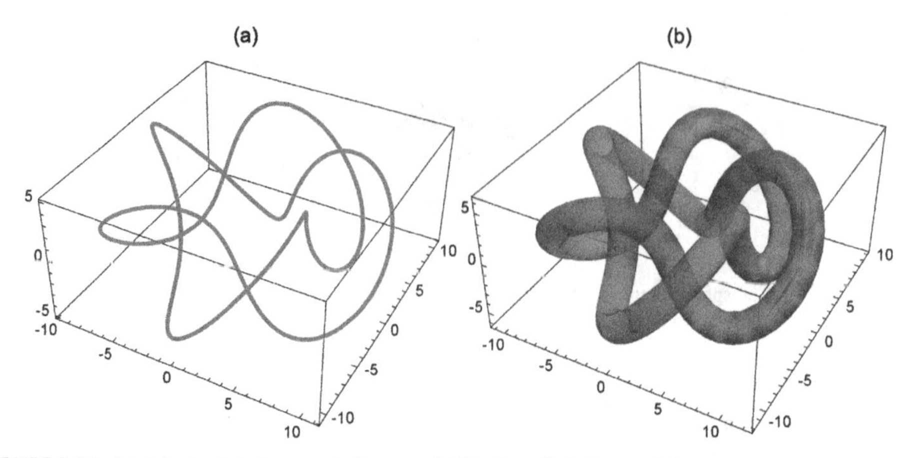

FIGURE 5.25 (a) A basic plot of a curve in 3-space. (b) Placing a "tube" around the curve.

```
Plot[Tooltip[{curve2[torusknot[8, 3, 5][2, 5]][t],
  torsion2[torusknot[8, 3, 5][2, 5]][t]}], {t, 0, 3Pi},
  PlotStyle → {{Thickness[.01], CMYKColor[0, .5, 1, 0]},
  {Thickness[.01], CMYKColor[0, 0, 0, .6]}}]
```

We generate a basic plot of this torus knot in 3-space with `ParametricPlot3D`. See Fig. 5.25 (a).

```
ParametricPlot3D[torusknot[8, 3, 5][2, 5][t], {t, 0, 3Pi},
  PlotStyle → {{Thickness[.01], CMYKColor[0, .5, 1, 0]}}]
```

Using `tubecurve`, we place a "tube" around the knot. See Fig. 5.25 (b).

```
p1 = ParametricPlot3D[tubecurve[torusknot[8, 3, 5][2, 5]][1.3][t, θ],
  {t, 0, 2Pi}, {θ, 0, 2Pi}, Mesh → False,
  PlotStyle → {{CMYKColor[0, .5, 1, 0], Opacity[.5]}},
  PlotPoints → {40, 40}]
```

A more interesting graphic is obtained by placing a transparent tube around the curve

```
p1b = ParametricPlot3D[tubecurve[torusknot[8, 3, 5][2, 5]][1][t, θ],
  {t, 0, 2Pi}, {θ, 0, 2Pi}, Mesh → False,
  PlotStyle → {{CMYKColor[0, .5, 1, 0], Opacity[.3]}},
  PlotPoints → {40, 40}]
```

and then creating a thicker version of the curve.

```
p2 = ParametricPlot3D[torusknot[8, 3, 5][2, 5][t], {t, 0, 3Pi},
  PlotStyle → {{Thickness[.01], CMYKColor[0, .5, 1, 0]}}]
```

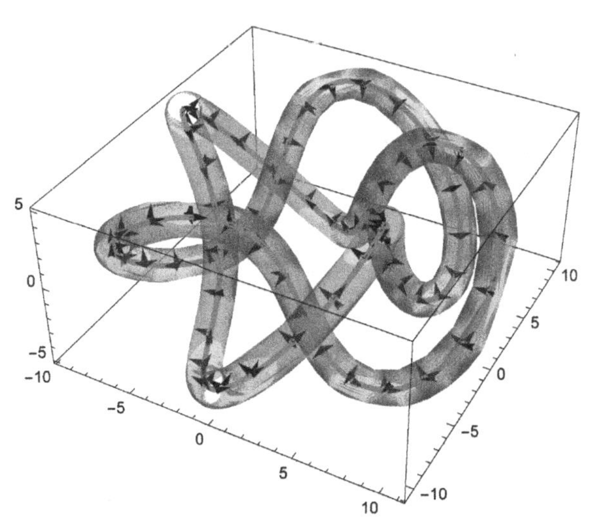

FIGURE 5.26 The torus knot, `torusknot[8,3,5][2,5]`, with its Frenet field.

As before, we use `tangent`, `normal`, and `binormal` to create a vector field on the curve.

```
ts = Table[Arrow[{{torusknot[8, 3, 5][2, 5][t],
   torusknot[8, 3, 5][2, 5][t] + tangent[torusknot[8, 3, 5][2, 5]][t]}}//N],
   {t, 0, 3Pi, 3Pi/99}];
bs = Table[Arrow[{{torusknot[8, 3, 5][2, 5][t],
   torusknot[8, 3, 5][2, 5][t] + binormal[torusknot[8, 3, 5][2, 5]][t]}}//N],
   {t, 0, 3Pi, 3Pi/99}];
ns = Table[Arrow[{{torusknot[8, 3, 5][2, 5][t],
   torusknot[8, 3, 5][2, 5][t] + normal[torusknot[8, 3, 5][2, 5]][t]}}//N],
   {t, 0, 3Pi, 3Pi/99}];
```

A striking graph is generated by showing the three graphs together. See Fig. 5.26

```
ysplot = Show[Graphics3D[{Arrowheads[.025], ts}]];
bsplot = Show[Graphics3D[{Arrowheads[.025], bs}]];
nsplot = Show[Graphics3D[{Arrowheads[.025], ns}]];
p3 = Show[ysplot, bsplot, nsplot]
```

```
Show[p2, p1b, p3]
```

Alternatively, display the results as an array with `GraphicsGrid`. See Fig. 5.27.

```
Show[GraphicsGrid[{{p1, p2}, {Show[p2, p1b], Show[p2, p1b, p3]}}]]
```

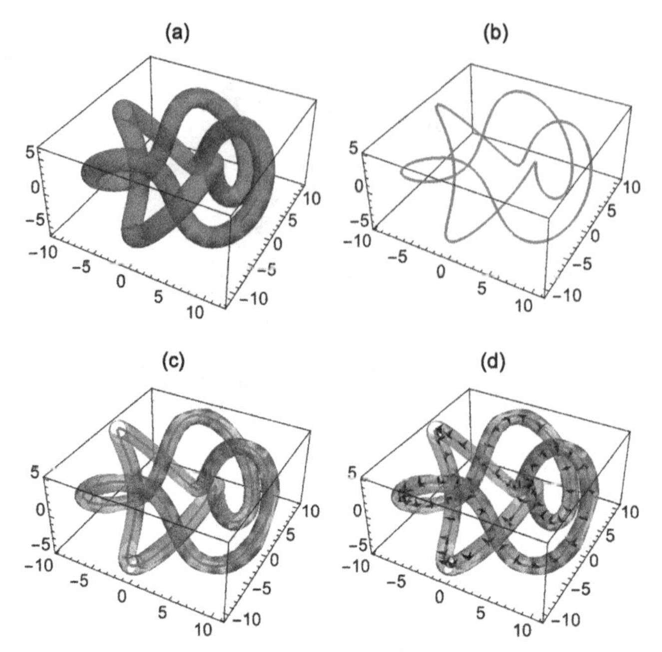

FIGURE 5.27 (a) A "tubed" knot. (b) A thick knot. (c) A knot within a tube around it. (d) A knot within a tube illustrating the Frenet field.

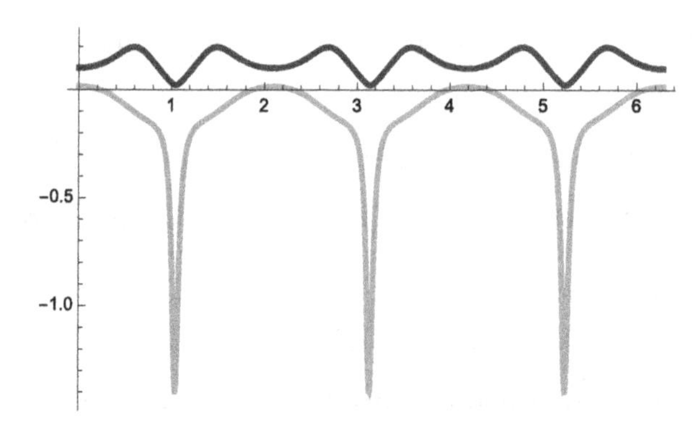

FIGURE 5.28 The curvature and torsion for the Trefoil knot (University of Alaska colors).

Example 5.49

The **Trefoil knot** is the special case of `torusknot[8,3,5][2,3]][t]`. We use `Plot` to graph its curvature and torsion in Fig. 5.28. Because we have used `Tooltip`, you can identify each plot by moving the cursor over the curve in Fig. 5.28.

```
Plot[Tooltip[{curve2[torusknot[8, 3, 5][2, 3]][t],
  torsion2[torusknot[8, 3, 5][2, 3]][t]}], {t, 0, 2Pi}, PlotRange → All,
    PlotStyle → {{Thickness[.01], CMYKColor[.98, 0, .72, .61]},
  {Thickness[.01], CMYKColor[0, .24, .94, 0]}}]
```

Next, we generate a thickened version of the Trefoil knot.

```
p1 = ParametricPlot3D[torusknot[8, 3, 5][2, 3][t], {t, 0, 2Pi},
  PlotStyle → {Thickness[.01], CMYKColor[.98, 0, .72, .61]}]
```

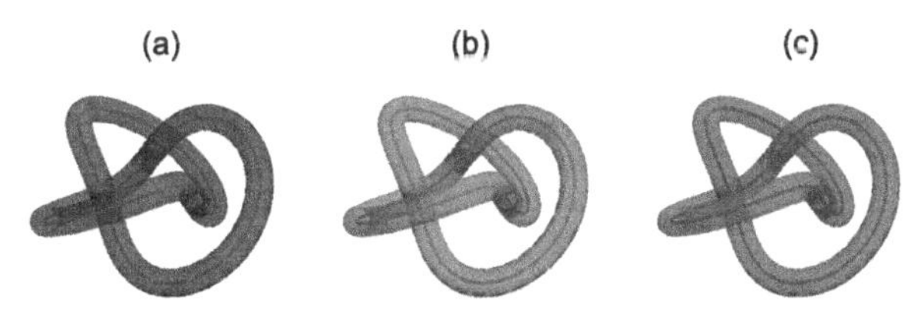

FIGURE 5.29 (a) The Trefoil knot with a tube around it; (b) and (c) Changing the color of the tube.

Three different tube plots of the Trefoil knot are generated. In p2, the result is a basic plot. In p2b, The plot is shaded according to the Rainbow color gradient. In p2c, the plot is shaded according to the knots curvature. The knot together with the three surfaces are shown together in Fig. 5.29.

```
p2 = ParametricPlot3D[tubecurve[torusknot[8, 3, 5][2, 3]][1.3][t, θ],
  {t, 0, 2Pi}, {θ, 0, 2Pi}, Mesh → False,
  PlotStyle → {{CMYKColor[.98, 0, .72, .61], Opacity[.5]}},
  PlotPoints → {40, 40}]

p2b = ParametricPlot3D[tubecurve[torusknot[8, 3, 5][2, 3]][1.3][t, θ],
  {t, 0, 2Pi}, {θ, 0, 2Pi}, Mesh → False,
  PlotPoints → {40, 40},
  PlotStyle->{{CMYKColor[.70, .14, .59, .01], Opacity[.5]}}]

p2c = ParametricPlot3D[tubecurve[torusknot[8, 3, 5][2, 3]][1.3][t, θ],
  {t, 0, 2Pi}, {θ, 0, 2Pi}, Mesh → False,
  PlotStyle->{{CMYKColor[0, .6, 1, 0], Opacity[.5]}},
  PlotPoints → {40, 40}]

ba1 = Show[p2, p1, Boxed → False, Axes → None]

ba2 = Show[p2b, p1, Boxed → False, Axes → None]

ba3 = Show[p2c, p1, Boxed → False, Axes → None]

Show[GraphicsRow[{ba1, ba2, ba3}]]
```

For surfaces in $\mathcal{R}^3$, extending and stating these definitions precisely becomes even more complicated. First, define the **vector triple product** **(xyz)**, where $\mathbf{x} = \begin{pmatrix} x_1 \\ x_2 \\ x_3 \end{pmatrix}$, $\mathbf{y} = \begin{pmatrix} y_1 \\ y_2 \\ y_3 \end{pmatrix}$, and $\mathbf{z} = \begin{pmatrix} z_1 \\ z_2 \\ z_3 \end{pmatrix}$, by $(\mathbf{xyz}) = \begin{vmatrix} x_1 & x_2 & x_3 \\ y_1 & y_2 & y_3 \\ z_1 & z_2 & z_3 \end{vmatrix}$. We assume that $\gamma = \gamma(u, v)$ is a vector-valued function with domain contained in a "nice" region $U \subset \mathcal{R}^2$ and range in $\mathcal{R}^3$. The **Gaussian curvature**, $\mathcal{K}$, and the **mean curvature**, $\mathcal{H}$, under reasonable conditions, are given by the formulas

$$\mathcal{K}=\frac{(\gamma_{uu}\gamma_u\gamma_v)(\gamma_{vv}\gamma_u\gamma_v)-(\gamma_{uv}\gamma_u\gamma_v)^2}{\left(\|\gamma_u\|^2\|\gamma_v\|^2-(\gamma_u\cdot\gamma_v)^2\right)^2}$$

and (5.26)

$$\mathcal{H}=\frac{(\gamma_{uu}\gamma_u\gamma_v)\|\gamma_v\|^2-2(\gamma_{uv}\gamma_u\gamma_v)(\gamma_u\cdot\gamma_v)+(\gamma_{vv}\gamma_u\gamma_v)\|\gamma_u\|^2}{2\left(\|\gamma_u\|^2\|\gamma_v\|^2-(\gamma_u\cdot\gamma_v)^2\right)^{3/2}}.$$

For the parametrically defined surface $\gamma=\gamma(u,v)$, the **unit normal field**, **U**, is $\mathbf{U}=\frac{\gamma_u\times\gamma_v}{\|\gamma_u\times\gamma_v\|}$. Observe that the expressions that result from explicitly computing **U**, $\mathcal{K}$, and $\mathcal{H}$ are almost always so complicated that they are impossible to understand.

After defining `vtp` to return the vector triple product of three vectors, we define `gaussianc` and `meanc` to compute $\mathcal{K}$ and $\mathcal{H}$ for a parametrically defined surface $\gamma(u,v)=\langle x(u,v),y(u,v),z(u,v)\rangle$.

```
vtp[x_, y_, z_]:=Det[{{x[[1]], x[[2]], x[[3]]},
{y[[1]], y[[2]], y[[3]]}, {z[[1]], z[[2]], z[[3]]}}]
```

```
gaussianc[γ_][u_, v_]:=
Module[{lu, lv, vtp},
  vtp[x_, y_, z_]:=Det[{{x[[1]], x[[2]], x[[3]]},
  {y[[1]], y[[2]], y[[3]]}, {z[[1]], z[[2]], z[[3]]}}];
  (vtp[D[γ[lu, lv], {lu, 2}], D[γ[lu, lv], lu], D[γ[lu, lv], lv]]
  vtp[D[γ[lu, lv], {lv, 2}], D[γ[lu, lv], lu], D[γ[lu, lv], lv]]−
  vtp[D[γ[lu, lv], lu, lv], D[γ[lu, lv], lu], D[γ[lu, lv], lv]]^2)/
  (Norm[D[γ[lu, lv], lu]]^2Norm[D[γ[lu, lv], lv]]^2−
  (D[γ[lu, lv], lu].D[γ[lu, lv], lv])^2)^2/.
  {lu → u, lv → v}//PowerExpand//Simplify
  ]
```

```
meanc[γ_][u_, v_]:=
Module[{lu, lv, vtp},
  vtp[x_, y_, z_]:=Det[{{x[[1]], x[[2]], x[[3]]},
  {y[[1]], y[[2]], y[[3]]}, {z[[1]], z[[2]], z[[3]]}}];
  (vtp[D[γ[lu, lv], {lu, 2}], D[γ[lu, lv], lu], D[γ[lu, lv], lv]]
  Norm[D[γ[lu, lv], lv]]^2−
  2vtp[D[γ[lu, lv], lu, lv], D[γ[lu, lv], lu], D[γ[lu, lv], lv]]
  (D[γ[lu, lv], lu].D[γ[lu, lv], lv])+
  vtp[D[γ[lu, lv], {lv, 2}], D[γ[lu, lv], lu], D[γ[lu, lv], lv]]
```

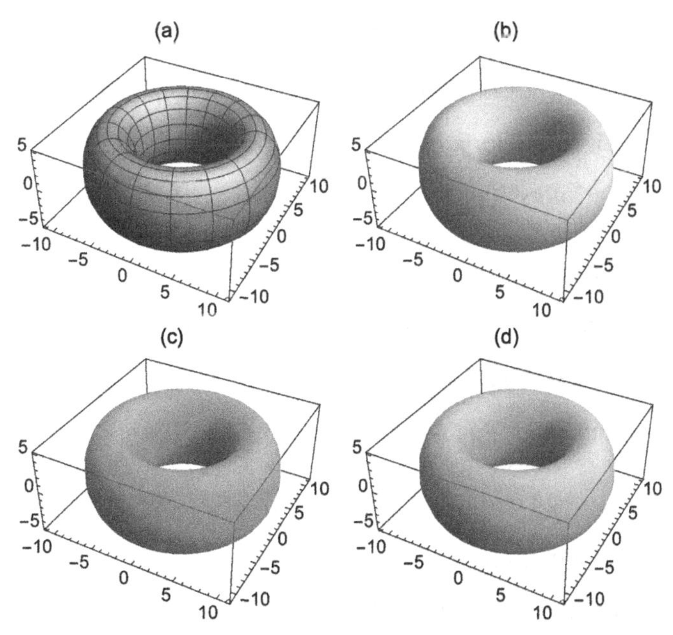

FIGURE 5.30 (a) A basic torus. (b) Changing the coloring of the torus. (c) Shading according to Gaussian curvature. (d) Shading according to mean curvature.

```
Norm[D[γ[lu, lv], lu]]^2)/
(2(Norm[D[γ[lu, lv], lu]]^2Norm[D[γ[lu, lv], lv]]^2−
(D[γ[lu, lv], lu].D[γ[lu, lv], lv])^2)^(3/2))/.
{lu → u, lv → v}//PowerExpand//Simplify
]
```

Example 5.50

We illustrate the commands with the torus, first discussed in Chapter 2, and `ParametricPlot3D`. For convenience, we redefine `torus`.

```
torus[a_, b_, c_][p_, q_][u_, v_]:={(a + bCos[u])
Cos[v], (a + bCos[u])Sin[v], cSin[u]}
```

In `pp1`, we generate a basic plot of the torus. The shading is changed in `pp2`. In `pp3` the surface is shaded according to its Gaussian curvature while in `pp4` the surface is shaded according to its mean curvature. All four plots are shown together in Fig. 5.30.

```
pp1 = ParametricPlot3D[torus[8, 3, 5][2, 5][u, v], {u, 0, 2Pi},
  {v, 0, 2Pi}]
```

```
pp1 = ParametricPlot3D[Evaluate[torus[8, 3, 5][2, 5][u, v]], {u, 0, 2Pi},
  {v, 0, 2Pi}, PlotPoints → 60]
```

```
pp2 = ParametricPlot3D[torus[8, 3, 5][2, 5][u, v],
  {u, 0, 2Pi}, {v, 0, 2Pi}, Mesh → False, PlotStyle → Opacity[.75],
  PlotPoints → {25, 25}, ColorFunction →
  ColorData["MintColors"]]

pp3 = ParametricPlot3D[torus[8, 3, 5][2, 5][u, v],
  {u, 0, 2Pi}, {v, 0, 2Pi}, Mesh → False, PlotStyle → Opacity[.5],
  PlotPoints → {25, 25}, ColorFunction →
  (ColorData["MintColors"][gaussianc[torus[8, 3, 5][2, 5]][#1, #2]//N//
  Chop]&)]

pp4 = ParametricPlot3D[torus[8, 3, 5][2, 5][u, v],
  {u, 0, 2Pi}, {v, 0, 2Pi}, Mesh → False, PlotStyle → Opacity[.5],
  PlotPoints → {25, 25}, ColorFunction →
  (ColorData["MintColors"][meanc[torus[8, 3, 5][2, 5]][#1, #2]//N//Chop]&)]

Show[GraphicsGrid[{{pp1, pp2}, {pp3, pp4}}]]
```

5.6 MATRICES AND GRAPHICS

5.6.1 Manipulating Photographs with Built-In Functions

As introduced in Chapter 2, Mathematica contains a wide range of functions that allow you to manipulate images such as photographs quickly to achieve a variety of photographic effects. To import a photograph or other image into Mathematica, use `Import`. Generally, the object that is to be imported should be in your root directory. However, Mathematica supports clicking and dragging: you can simply click on the file and drag it to the desired location in your Mathematica notebook.

For "standard" effects and manipulation of photographs, try using `ImageEffect` first. `ImageEffect` is fast and has a wide range of options. We illustrate just a few here. First, we import a graphic,

```
p1 = Import["1000182.jpg"];
Show[p1, ImageSize → Small]
```

Next, we use `ImageEffect` to apply a variety of commonly used enhancements to the image, `p1`. The results are shown in Fig. 5.31.

FIGURE 5.31 Using `ImageEffect` with the `Charcoal`, `Comics`, `SaltPepperNoise`, and `Embossing` options.

```
p1a = ImageEffect[p1, {"Charcoal", 2}];
p1b = ImageEffect[p1, {"Comics", {.25, .5}}];
p1c = ImageEffect[p1, {"SaltPepperNoise", .2}];
p1d = ImageEffect[p1, {"Embossing", 4, 0}];
Show[GraphicsGrid[{{p1a, p1b}, {p1c, p1d}}]]
```

Once you have imported an image either using `Import` or by selecting and dragging the image into the desired location in your Mathematica notebook, Mathematica gives you a wide range of functions to obtain information about the image.

```
p1 = Import["IMG5209.jpg"];
Show[p1, ImageSize → Small]
```

To determine the size of your image, number of pixel columns by number of pixel rows, use `ImageDimensions`.

```
ImageDimensions[p1]
```

{640, 640}

FIGURE 5.32 Using ImageAdjust with the LaplacianGaussianFilter and ColorQuantize options.

Here, we illustrate ImageAdjust with a few of its options (see Fig. 5.32).

```
p2a = ImageAdjust[p1, {1, 1}];

p2b = ImageAdjust[LaplacianGaussianFilter[p1, 5]];

p2c = ImageAdjust[p1, {0., 3.7, 2.7}];

p2d = ImageAdjust[ColorQuantize[p1, 5], {.15, .75}];

Show[GraphicsGrid[{{p2a, p2b}, {p2c, p2d}}]]
```

ColorQuantize[image,n] approximates image with *n* colors. To see the resulting colors, use DominantColors. Observe that the colors are displayed in small squares.

```
dcp2a = DominantColors[p2a]
```

{□, ■, ■, □, ■, ■, ■}

To see the RGB code, click on the color or use InputForm to see the actual color codes displayed as a list.

FIGURE 5.33 Illustrating elementary uses of `ImageApply`.

```
InputForm[dcp2a]
```

```
{RGBColor[0.9993332719608886, 0.9983786009836767, 0.9976452825923762],
RGBColor[0.010466076648957105, 0.004068849634045346, 0.004172871274537774],
RGBColor[0.4284221221669545, 0.531302194561179, 0.06417935218806642],
RGBColor[0.9885278855002001, 0.9959965192142667, 0.13330881577073309],
RGBColor[0.6554690223969046, 0.47494092958610595, 0.34335869307359934],
RGBColor[1., 0.4073167704732596, 0.8661078573314603],
RGBColor[0.9994877730131592, 0.17160548964165695, 0.4709315807968833]}
```

`ImageApply` may give the most flexibility for dealing with a graphic as it allows you to manipulate the graphic pixel-by-pixel. Closely related commands to those discussed include `ImageFilter`, `ImageConvolve`, and `ImageCorrelate`. Use `?<command>` to obtain detailed help regarding the capabilities of each (see Fig. 5.33).

```
p1 = Import["IMG3065.jpg"];
Show[p1, ImageSize → Small]
```

```
p2a = ImageApply[(#[[1]] + #[[2]] + #[[3]])/3&, p1];
p2b = ImageApply[(#[[2]]^2 + #[[3]])/2&, p1];
p2c = ImageApply[#^(1/3)&, p1];
```

```
Show[GraphicsRow[{p2a, p2b, p2c}]]
```

5.6.2 Manipulating Photographs by Viewing Them as a Matrix or Array

With Mathematica, virtually every object is a list or list of lists, including images. In situations where you want to manipulate your image or photograph with more detail than that provided

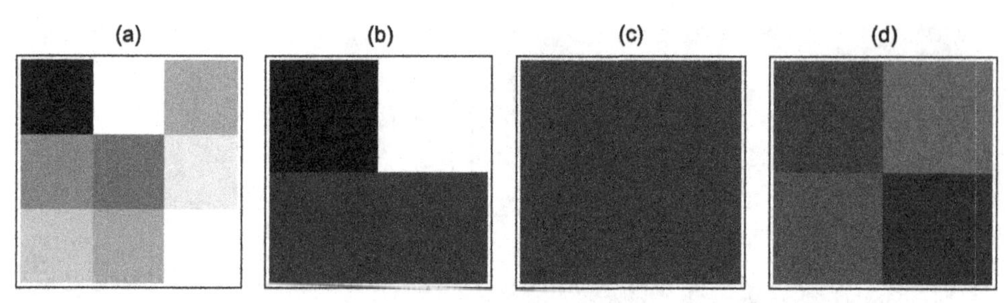

FIGURE 5.34 (a) Mathematica shades all cells according to their heights. (b) Mathematica does not know how to shade the cells in the second row. (c) Mathematica cannot shade any of the cells. (d) Mathematica shades all four cells using `RGBColor`.

by the built-in Mathematica functions for manipulating `Image` objects, you may want to manipulate the data that determines the image directly. To do so, it is important to understand that the underlying data of an `Image` object is generally a matrix. Typically, the entries of the matrix for the image are single digits for black-and-white images or ordered triples of the form `{r,g,b}` for color jpegs. The easiest way to determine the size of the image is to use `ImageDimensions[image]`. The resulting list `{n,m}` indicates that the dimensions of image are n pixels of rows by m pixels of columns.

Mathematica contains several functions that allow you to represent matrices graphically. These commands are analogous to the corresponding ones for dealing with lists (like `ListPlot`) or functions (such as `Plot`, `Plot3D`, and `ContourPlot`).

1. `MatrixPlot[A]` generates a grid with the same dimensions as **A**. The cells are shaded according to the entries of **A**. The default is in color.
2. `ArrayPlot[A]` generates a grid with the same dimensions as **A**. The cells are shaded according to the entries of **A**. The default is in black and white.
3. `ListContourPlot[A]` generates a contour plot using the entries of **A** as the height values.
4. `ReliefPlot[A]` generates a relief plot using the entries of **A** as the height values.

As the figures in the text are in black and white, refer to the on-line version of the text to see the images in color.

Observe that `ArrayPlot` and `MatrixPlot` are virtually interchangeable. However, the entries of `ArrayPlot` need not be numbers. If Mathematica cannot determine how to shade a cell, the default is to shade it in a dark maroon color. Although these functions generate graphics that depend on the entries of the matrix, loosely speaking we will use phrases like "we use `MatrixPlot` to plot **A**" and "we use `ArrayPlot` to graph **A**" to describe the graphic that results from applying one of these functions to an array.

For example, consider the arrays $\mathbf{A} = \begin{pmatrix} 1 & 0 & .3 \\ .4 & .5 & .1 \\ .2 & .3 & 0 \end{pmatrix}$, $\mathbf{B} = \begin{pmatrix} 1 & 0 & \\ 0 & 1 & 0 \\ .1 & .2 & .3 \end{pmatrix}$, and $\mathbf{C} = \begin{pmatrix} \begin{pmatrix} 1 & 0 & 0 \\ .3 & .4 & .5 \end{pmatrix} & \begin{pmatrix} 0 & 1 & 0 \\ .1 & .2 & .3 \end{pmatrix} \end{pmatrix}$.

In the first command, Mathematica shades all the cells according to its `GrayLevel` value. However, in the second and third commands, Mathematica cannot shade the cells in the second row and all the cells, respectively, because ordered triples cannot be evaluated by `GrayLevel`. However, `RGBColor` evaluates ordered triples so Mathematica shades the cells in Fig. 5.34 (c) according to their `RGBColor` value. See Fig. 5.34.

```
ap1 = ArrayPlot[{{1, 0, .3}, {.4, .5, .1}, {.2, .3, 0}}];

ap2 = ArrayPlot[{{1, 0}, {{.3, .4, .5}, {.1, .2, .3}}}];

ap3 = ArrayPlot[{{{1, 0, 0}, {0, 1, 0}}, {{.3, .4, .5},
  {.1, .2, .3}}}];
```

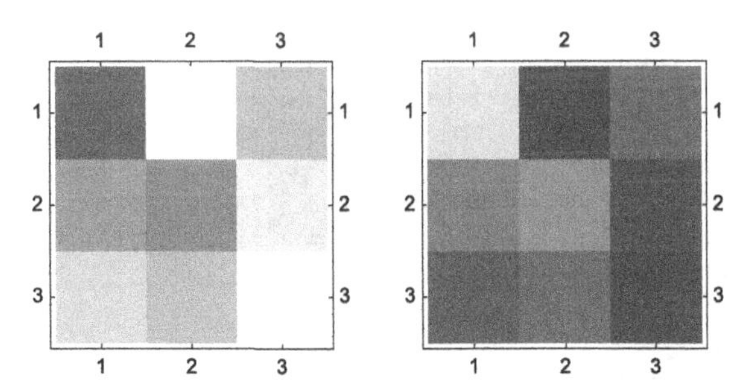

FIGURE 5.35 By default, `MatrixPlot` uses a color scheme. Use `ColorFunction` to change the colors.

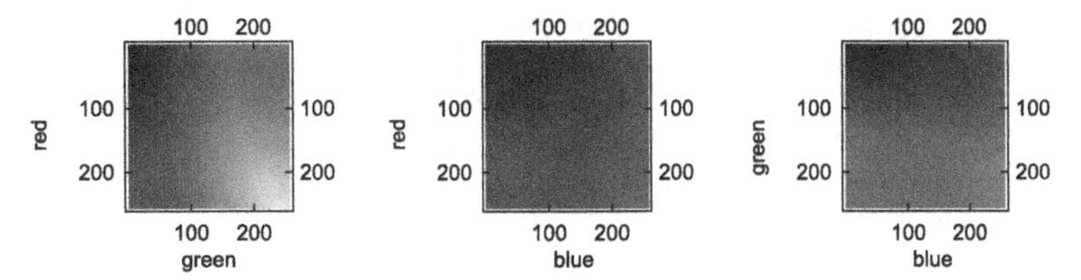

FIGURE 5.36 A comparison of how red, green, and blue affect `RGBColor[r,g,b]`.

```
ap4 = ArrayPlot[{{{1, 0, 0}, {0, 1, 0}}, {{.3, .4, .5},
  {.1, .2, .3}}},
  ColorFunction → RGBColor];
Show[GraphicsRow[{ap1, ap2, ap3, ap4}]]
```

`MatrixPlot` is unable to graphically represent **B** or **C**. However, coloring is automatic with `MatrixPlot`. See Fig. 5.35.

```
mp1 = MatrixPlot[{{1, 0, .3}, {.4, .5, .1}, {.2, .3, 0}}];
mp2 = MatrixPlot[{{1, 0, .3}, {.4, .5, .1}, {.2, .3, 0}},
  ColorFunction → "PlumColors"];
Show[GraphicsRow[{mp1, mp2}]]
```

If you need to adjust the color of a graphic, usually you can use the **ColorSchemes** palette to select an appropriate gradient or color function. In other situations, you might wish to create your own using `Blend`. To use `Blend`, you might need to know how various `RGBColor`s or `CMYKColor`s vary as the variables affecting the color change.

`ArrayPlot` can help us see the variability in the colors. With the following, we see how `RGBColor[r,g,b]` affects color for $b = 0$, $g = 0$, and then $r = 0$. The results are shown together in Fig. 5.36. The figure can help us select appropriate values to generate our own color blending function using `Blend` rather than relying on Mathematica's built-in color schemes and gradients.

In these calculations, `t1` is a 256×256 array for which each entry is an ordered triple. In the first `t1`, the ordered triple has the form $(r, g, 0)$, in the second the form $(r, 0, b)$, and so on.

```
t1 = Table[{r, g, 0}//N, {r, 0, 255}, {g, 0, 255}];

redgreen = ArrayPlot[t1, Axes → Automatic, AxesOrigin → {0, 0},
  FrameTicks → Automatic, FrameLabel → {red, green},
  LabelStyle → Medium, ColorFunction → RGBColor];
```

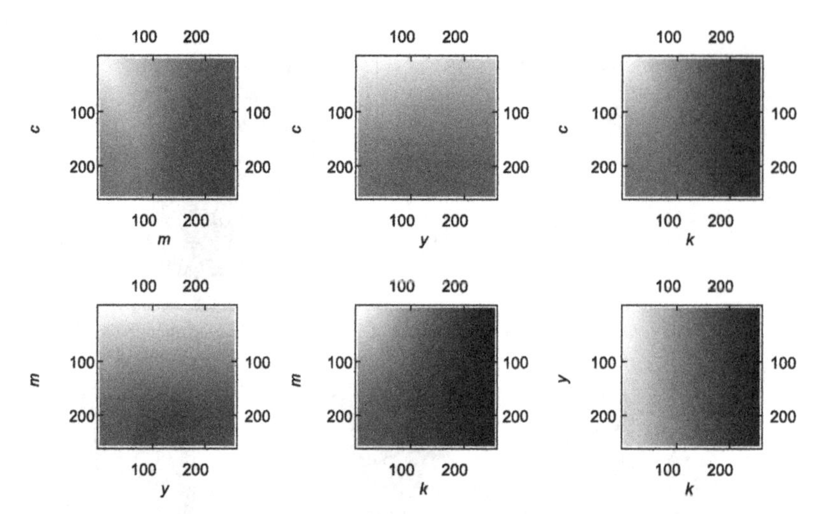

FIGURE 5.37 A comparison of how c, m, y, and k affect `CMYKColor[c,m,y,k]`.

```
t2 = Table[{r, 0, b}//N, {r, 0, 255}, {b, 0, 255}];

redblue = ArrayPlot[t2, Axes → Automatic, AxesOrigin → {0, 0},
  FrameTicks → Automatic, FrameLabel → {red, blue},
  LabelStyle → Medium, ColorFunction → RGBColor];

t2 = Table[{0, g, b}//N, {g, 0, 255}, {b, 0, 255}];

greenblue = ArrayPlot[t2, Axes → Automatic, AxesOrigin → {0, 0},
  FrameTicks → Automatic, FrameLabel → {green, blue},
  LabelStyle → Medium, ColorFunction → RGBColor];

s1 = Show[GraphicsRow[{redgreen, redblue, greenblue}]]
```

We modify the calculation slightly to see how `CMYKColor` varies as we adjust two parameters. Keep in mind that each `t2` is a 256 × 256 array. Each entry of `t2` is an ordered quadruple, which is illustrated in the first calculation where we use `Part` to take the 5th element of the 8th part of `t2` (see Fig. 5.37).

```
t2 = Table[{c, m, 0, 0}//N, {c, 0, 255}, {m, 0, 255}];

t2[[8, 5]]

{7., 4., 0., 0.}

cmplot = ArrayPlot[t2, Axes → Automatic, AxesOrigin → {0, 0},
  FrameTicks → Automatic, FrameLabel → {c, m}, ColorFunction → CMYKColor];

t2 = Table[{c, 0, y, 0}//N, {c, 0, 255}, {y, 0, 255}];

cyplot = ArrayPlot[t2, Axes → Automatic, AxesOrigin → {0, 0},
  FrameTicks → Automatic, FrameLabel → {c, y}, ColorFunction → CMYKColor];
```

```
t2 = Table[{c, 0, 0, k}//N, {c, 0, 255}, {k, 0, 255}];

ckplot = ArrayPlot[t2, Axes → Automatic, AxesOrigin → {0, 0},
  FrameTicks → Automatic, FrameLabel → {c, k}, ColorFunction → CMYKColor];

t2 = Table[{0, m, y, 0}//N, {m, 0, 255}, {y, 0, 255}];

myplot = ArrayPlot[t2, Axes → Automatic, AxesOrigin → {0, 0},
  FrameTicks → Automatic, FrameLabel → {m, y}, ColorFunction → CMYKColor];

t2 = Table[{0, m, 0, k}//N, {m, 0, 255}, {k, 0, 255}];

mkplot = ArrayPlot[t2, Axes → Automatic, AxesOrigin → {0, 0},
  FrameTicks → Automatic, FrameLabel → {m, k}, ColorFunction → CMYKColor];

t2 = Table[{0, 0, y, k}//N, {y, 0, 255}, {k, 0, 255}];

ykplot = ArrayPlot[t2, Axes → Automatic, AxesOrigin → {0, 0},
  FrameTicks → Automatic, FrameLabel → {y, k}, ColorFunction → CMYKColor];

Show[GraphicsGrid[{{cmplot, cyplot, ckplot}, {myplot, mkplot, ykplot}}]]
```

Keep in mind that you can load files into Mathematica with `Import` or by clicking and dragging the image to the desired location in your Mathematica notebook. Generally, the underlying structure of the loaded file is relatively easy to understand. Be careful when you import data into Mathematica. We recommend that you use `ExampleData` to investigate your routines before finalizing them. Although importing external files into Mathematica is easy, understanding the underlying structure of the imported data may take some time but be necessary to produce the results you desire.
We illustrate a few of the subtle differences that can be encountered with several images.
Using `Import`, we import a graphic into Mathematica. The result is shown in Fig. 5.38 (a).

```
p1 = Import["9306COL.jpg"];

Show[p1, ImageSize → Small]
```

We use `ImageDimensions` to determine the size of the image.

```
ImageDimensions[p1]

{640, 439}
```

Alternatively, using `Length` we see that `p2` has 439 rows (pixels)

```
Length[p2]

439
```

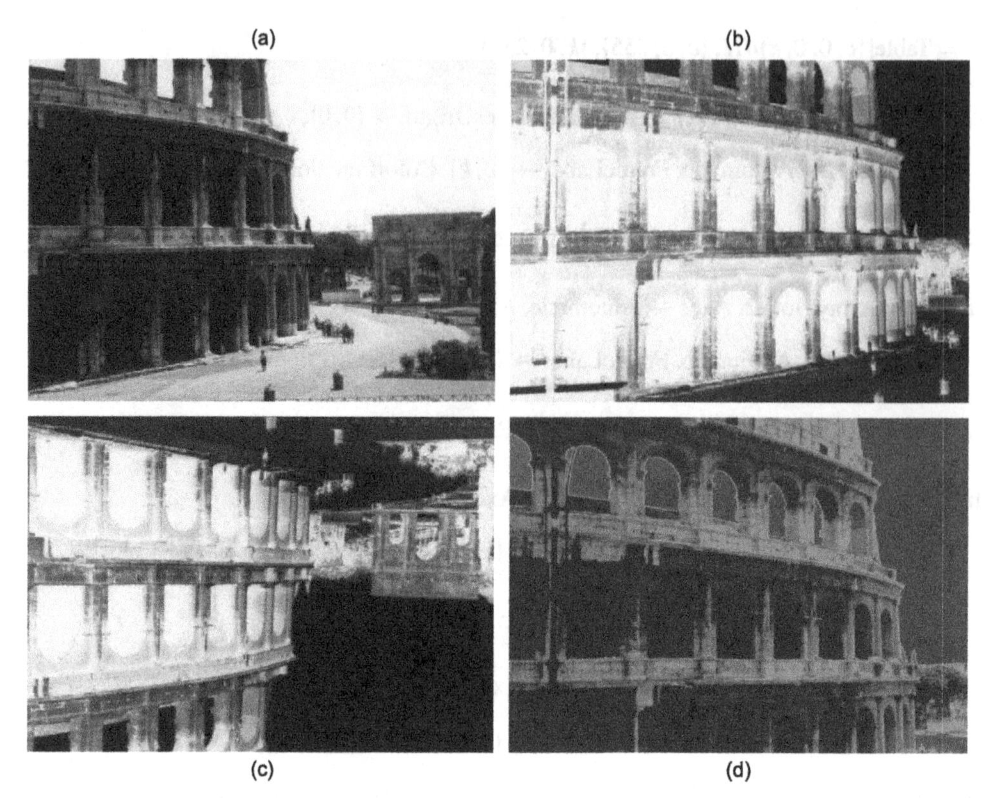

FIGURE 5.38 (a) The original. (b) Applying `ArrayPlot` to the original data points. (c) Reorienting the image. (d) Applying a color function to the data points.

and then counting the number of entries in the first row of `p2`, we see that `p2` has 640 columns (pixels).

Length[p2[[1]]]

640

To convert the image to an array (matrix) that we can manipulate, we use `ImageData`.

p2 = ImageData[p1];

After we have obtained the image data from `p1` in `p2`, we see that it is a matrix where each entry is a list of the form (r, g, b), corresponding to the RGB color code for that pixel.

Short[p2]

{{{0.137255, ⟨⟨20⟩⟩, 0.160784}, ⟨⟨638⟩⟩, ⟨⟨1⟩⟩}, ⟨⟨437⟩⟩, {⟨⟨1⟩⟩}}

`ArrayPlot` produces the negative of an image. See Fig. 5.38 (a).

g1a = ArrayPlot[p2]

To manipulate `p2`, it is important to understand that `p2` is a 2×2 array where each entry is a 3×1 list, corresponding to the RGB color codes for that particular pixel.

```
p2[[1]][[2]]
```

{0.133333, 0.0901961, 0.156863}

```
Length[p2[[1]]]
```

640

To convert the image to a different color, we first convert the matrix to a list of ordered triples, `p3`, define $f(\{x, y, z\}) = (x + y + z)/3$, apply f to `p3`, and then apply a color function to the result. In this example, we chose to use the `Rainbow` color function.

```
f[{x_, y_, z_}]:=(x + y + z)/3;
p3 = Flatten[p2, 1];
Take[p3, {1, 5}]
p4 = Flatten[Map[f, p3]];
p5 = Partition[p4, 640];
```

{{0.137255, 0.0941176, 0.160784}, {0.133333, 0.0901961, 0.156863},

{0.121569, 0.0784314, 0.145098}, {0.113725, 0.0705882, 0.137255},

{0.113725, 0.0705882, 0.137255}}

```
g2a = ArrayPlot[p5, ColorFunction → "Rainbow"]
```

We show the results in Fig. 5.38. On a color printer, the results are amazing.

```
Show[GraphicsGrid[{{p1, g1a}, {g1b, g2a}}]]
```

```
g2a = ArrayPlot[Reverse[p1[[1, 1]]], ColorFunction → "Pastel"]
```

The results are shown side-by-side in Fig. 5.38. Printed on a color printer, the results are amazing.

```
Show[GraphicsRow[{p1, g1a, g1b, g2a}]]
```

Now that we understand how to manipulate an image, we can be creative. In the following, the image is scaled so that the width of the image is 70 pixels (because of `ImageSize->70`). We then display the small image with another graphic. Using `Inset`, we put the Colliseum next to a sine graph that is plotted using the same coloring gradient. See Fig. 5.39.

```
g1 = ArrayPlot[p5, ColorFunction → "BrightBands", ImageSize → 70];

p2 = Plot[Sin[x], {x, 0, 2Pi}, Epilog->Inset[g1, {3Pi/2, 1/2}],
  ColorFunction → "BrightBands", PlotStyle → Thickness[.05]]
```

An alternative way to visualize the data is to use `ListContourPlot`. To assure that the aspect ratio of the original image is preserved, include the `AspectRatio->Automatic` option in the `ListContourPlot` command (see Fig. 5.40).

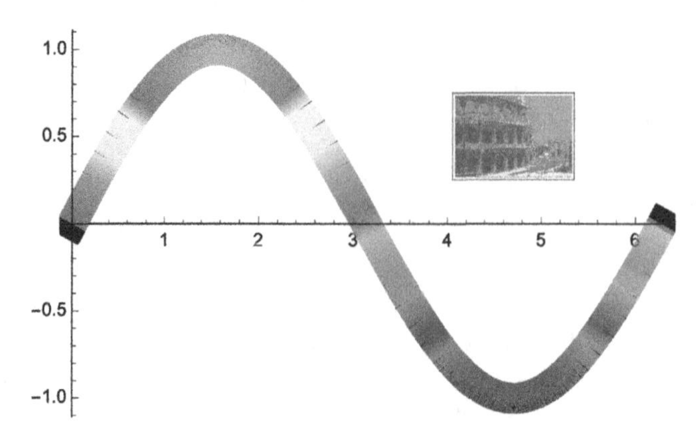

FIGURE 5.39 Use Inset to place one graphic within another.

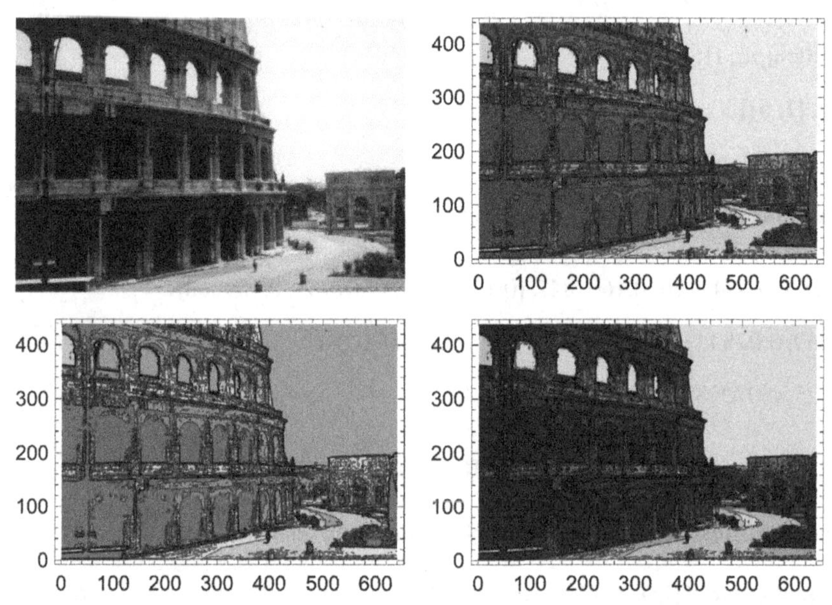

FIGURE 5.40 Using ListContourPlot rather than ArrayPlot.

```
glp1 = ListContourPlot[p5//Reverse, AspectRatio → Automatic];

glp2 = ListContourPlot[p5//Reverse, ColorFunction → "Pastel",
  AspectRatio → Automatic]

glp3 = ListContourPlot[p5//Reverse, ColorFunction → "GrayTones",
  AspectRatio → Automatic]

Show[GraphicsGrid[{{p1, glp1}, {glp2, glp3}}]]
```

The structure of a black and white jpeg differs from that of a color one. To see so, we import a *very old* picture of the second author of this text,

```
p1 = Import["littlejim.jpg"];
Show[p1, ImageSize → Small]
```

and name the result `p1`. With `ImageDimensions`, we see that `p1` is 428 pixels wide by 600 pixels tall.

```
ImageDimensions[p1]
```

{428, 600}

We obtain the data for the image with `ImageData`. With `Length`, we see that the resulting array has 600 rows and 428 columns, confirming the result obtained with `ImageDimensions`. Using `Short`, we see the form of each entry. For the black-and-white image, each entry is a number that corresponds to a `GrayLevel`.

```
p2 = ImageData[p1];
Length[p2]
```

600

```
Length[p2[[1]]]
```

428

```
Short[p2]
```

{{0.529412, ⟨⟨426⟩⟩, 0.419608}, ⟨⟨598⟩⟩, {⟨⟨1⟩⟩}}

In Fig. 5.41, we illustrate the use of `ListContourPlot` and `ReliefPlot` along with various options.

```
g2 = ReliefPlot[p2, AspectRatio → Automatic,
  ColorFunction → "GrayTones", FrameTicks → None];
g3 = ListContourPlot[p2//Reverse, AspectRatio → Automatic,
  ColorFunction → "GrayTones", FrameTicks → None];
g4 = ListContourPlot[p2//Reverse, AspectRatio → Automatic,
  ContourStyle → Black, ContourShading → False,
  FrameTicks → None];
Show[GraphicsRow[{g2, g3, g4}]]
```

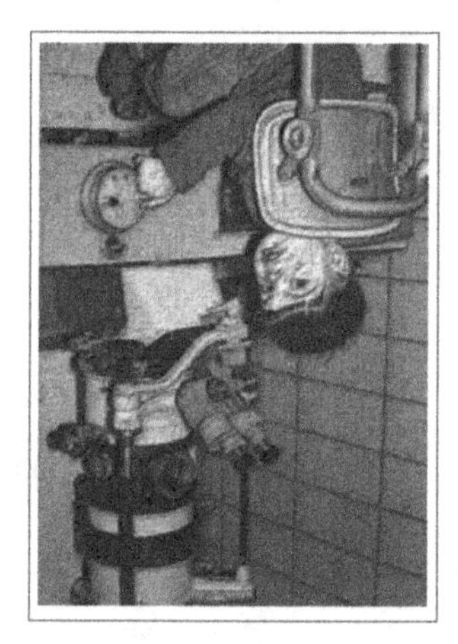  

FIGURE 5.41 Using `ListContourPlot` and `ReliefPlot` along with various options to graphically represent a matrix.

`ReliefPlot` can help add insight to images, especially when they have geographical or biological meaning. For example, this jpeg

p1 = Import["071105fertx2c.jpg"];

Show[p1, ImageSize → Small]

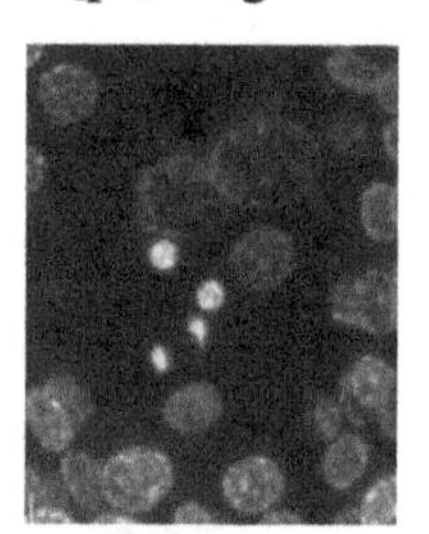

shows the beginning of a biological process of a cell.
With `ImageDimensions`, we see that `p1` is a 400 pixels wide by 500 pixels tall. After obtaining the image data with `ImageData`, these calculations are confirmed with length.

ImageDimensions[p1]

{400, 500}

p2 = ImageData[p1];

Length[p2]

500

Length[p1[[1, 1]]]

500

Length[p1[[1, 1, 1]]]

400

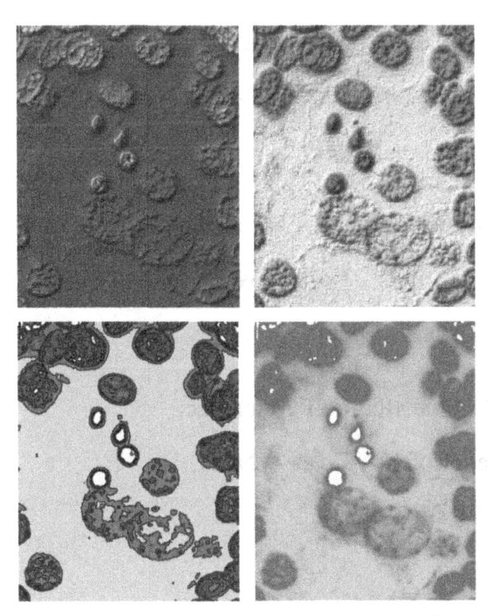

FIGURE 5.42 Using ReliefPlot, ListContourPlot, and ListDensityPlot along with various options to graphically represent a matrix.

Viewing p2 as a 500×400 array, each entry is 1×3 array/vector. To easily apply a function, f, that assigns a number to each ordered triple, we use Flatten to convert the nested list/array p2 to a list of ordered triples in p3.

```
p3 = Flatten[p2, 1];
Short[p3]
Length[p3]
```

200000

{{0.384314, ⟨⟨20⟩⟩, 0.12549}, ⟨⟨199998⟩⟩, {⟨⟨1⟩⟩}}

To apply our own color function to this data set, we convert the ordered triples to some other form. For illustrative purposes, we convert each ordered triple (x, y, z) in p3 to the number $x + y^2$. The result is converted back to a 500×400 array, with Partition in p4.

```
f[y_]:=y[[1]] + y[[2]]^2
p4 = Partition[Map[f, p3], 400];
Length[p4]
```

500

We then use ReliefPlot, ListContourPlot, and ListDensityPlot along with various options to graph the result in Fig. 5.42.

```
g1 = ReliefPlot[p4, AspectRatio → Automatic,
  ColorFunction → "DarkRainbow"]
```

```
g2 = ReliefPlot[p4, AspectRatio → Automatic,
   ColorFunction → "NeonColors", Ticks → None,
   Axes → None, FrameTicks → None]

g3 = ListContourPlot[p4, AspectRatio → Automatic,
   ColorFunction → "NeonColors", Ticks → None,
   Axes → None, FrameTicks → None]

g4 = ListDensityPlot[p4, AspectRatio → Automatic,
   ColorFunction → "NeonColors", Ticks → None,
   Axes → None, FrameTicks → None]
```

Chapter 6

Applications Related to Ordinary and Partial Differential Equations

6.1 FIRST-ORDER DIFFERENTIAL EQUATIONS

6.1.1 Separable Equations

Because they are solved by integrating, separable differential equations are usually the first introduced in the introductory differential equations course.

Definition 6.1 (Separable Differential Equation). A differential equation of the form

$$f(y)\,dy = g(t)\,dt \tag{6.1}$$

is called a first-order **separable differential equation**.

We solve separable differential equations by first separating variables and then integrating.

Remark 6.1. The command

```
DSolve[y'[t]==f[t,y[t]],y[t],t]
```

attempts to solve $y' = dy/dt = f(t, y)$ for y.

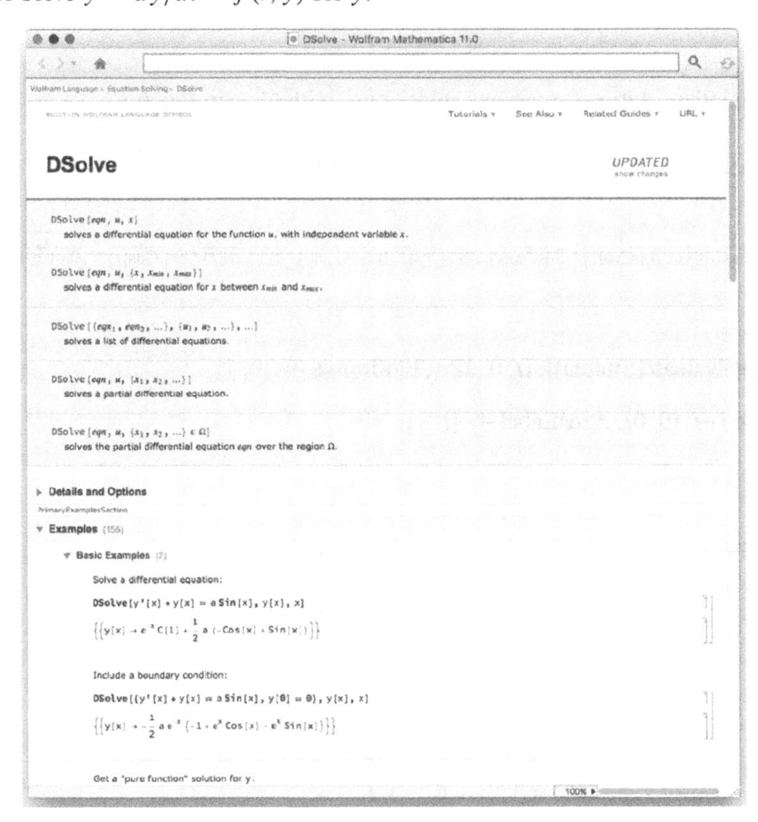

Mathematica by Example. http://dx.doi.org/10.1016/B978-0-12-812481-9.00006-5

Example 6.1

Solve each of the following equations: (a) $\frac{dy}{dt} - y^2 \sin t = 0$; (b) $\frac{dy}{dt} = \alpha y\left(1 - \frac{1}{K}y\right)$, K, $\alpha > 0$ constant.

Solution. (a) The equation is separable so we separate and then integrate:

$$\frac{1}{y^2}dy = \sin t\, dt$$
$$\int \frac{1}{y^2}dy = \int \sin t\, dt$$
$$-\frac{1}{y} = -\cos t + C$$
$$y = \frac{1}{\cos t + C}.$$

We check our result with `DSolve`.

sola = DSolve[y′[t] − y[t]^2Sin[t] == 0, y[t], t]

$\left\{\left\{y[t] \to \frac{1}{-C[1]+\mathrm{Cos}[t]}\right\}\right\}$

Observe that the result is given as a list. The formula for the solution is the second part of the first part of the first part of `sola`.

sola[[1, 1, 2]]

$\frac{1}{-C[1]+\mathrm{Cos}[t]}$

We then graph the solution for various values of C with `Plot` in Fig. 6.1.

`expression /. x->y` replaces all occurrences of x in `expression` by y. `Table[a[k],{k,n,m}]` generates the list $a_n, a_{n+1}, \ldots, a_{m-1}, a_m$.

To graph the list of functions `{list}` for $a \le x \le b$, enter `Plot[list,{x,a,b}]`.

toplota = Table[sola[[1, 1, 2]]/.C[1] → −i, {i, 2, 10}]

$\left\{\frac{1}{2+\mathrm{Cos}[t]}, \frac{1}{3+\mathrm{Cos}[t]}, \frac{1}{4+\mathrm{Cos}[t]}, \frac{1}{5+\mathrm{Cos}[t]}, \frac{1}{6+\mathrm{Cos}[t]}, \frac{1}{7+\mathrm{Cos}[t]}, \frac{1}{8+\mathrm{Cos}[t]}, \frac{1}{9+\mathrm{Cos}[t]}, \frac{1}{10+\mathrm{Cos}[t]}\right\}$

Plot[Tooltip[Evaluate[toplota]], {t, 0, 2Pi}, PlotRange → {0, 1},

AxesOrigin → {0, 0}, AxesLabel → {t, y},

AspectRatio → 1]

(b) After separating variables, we use partial fractions to integrate:

$$y' = \alpha y\left(1 - \frac{1}{K}y\right)$$
$$\frac{1}{\alpha y\left(1 - \frac{1}{K}y\right)}dy = dt$$
$$\frac{1}{\alpha}\left(\frac{1}{y} + \frac{1}{K-y}\right) = dt$$

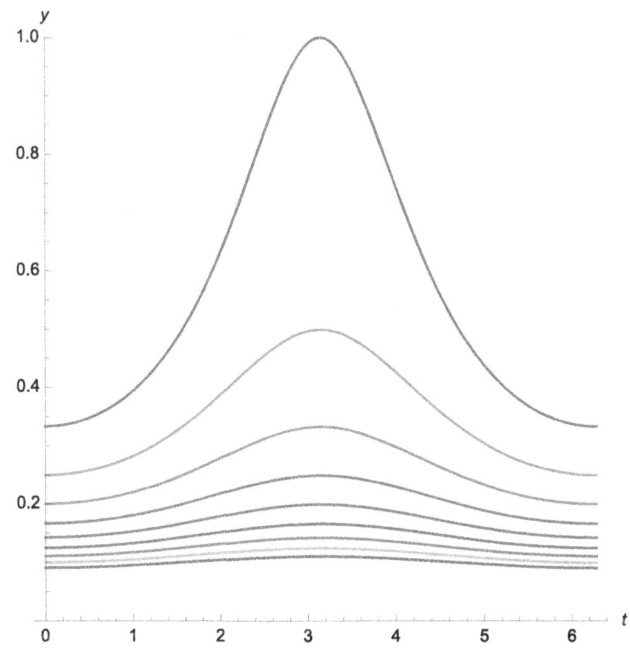

FIGURE 6.1 Several solutions of $y' - y^2 \sin t = 0$.

$$\frac{1}{\alpha}(\ln|y| - \ln|K - y|) = C_1 + t$$
$$\frac{y}{K - y} = Ce^{\alpha t}$$
$$y = \frac{CKe^{\alpha t}}{Ce^{\alpha t} - 1}.$$

We check the calculations with Mathematica. First, we use `Apart` to find the partial fraction decomposition of $\dfrac{1}{\alpha y\left(1 - \frac{1}{K}y\right)}$.

s1 = Apart[1/(αy(1 − 1/ky)), y]

$\frac{1}{y\alpha} - \frac{1}{(-k+y)\alpha}$

Then, we use `Integrate` to check the integration.

s2 = Integrate[s1, y]

$k\left(\frac{\text{Log}[y]}{k\alpha} - \frac{\text{Log}[-k+y]}{k\alpha}\right)$

Last, we use `Solve` to solve $\frac{1}{\alpha}(\ln|y| - \ln|K - y|) = ct$ for y.

Solve[s2 == c + t, y]

$\{\{y \to \frac{e^{c\alpha+t\alpha}k}{-1+e^{c\alpha+t\alpha}}\}\}$

We can use `DSolve` to find a general solution of the equation

```
solb = DSolve[y'[t] == αy[t](1 − 1/ky[t]), y[t], t]
```

$$\{\{y[t] \to \frac{e^{t\alpha+kC[1]}k}{-1+e^{t\alpha+kC[1]}}\}\}$$

as well as find the solution that satisfies the initial condition $y(0) = y_0$, although Mathematica generates several error messages because inverse functions are being used so the resulting solution set may not be complete.

```
solc = DSolve[{y'[t] == αy[t](1 − 1/ky[t]), y[0] == y0}, y[t], t]
```

$$\{\{y[t] \to \frac{e^{t\alpha}ky0}{k-y0+e^{t\alpha}y0}\}\}$$

The equation $\frac{dy}{dt} = \alpha y \left(1 - \frac{1}{K}y\right)$ is called the **logistic equation** (or **Verhulst equation**) and is used to model the size of a population that is not allowed to grow in an unbounded manner. Assuming that $y(0) > 0$, then all solutions of the equation have the property that $\lim_{t\to\infty} y(t) = K$.

To see this, we set $\alpha = K = 1$ and use `StreamPlot` or `VectorPlot` to graph the direction field associated with the equation in Fig. 6.2.

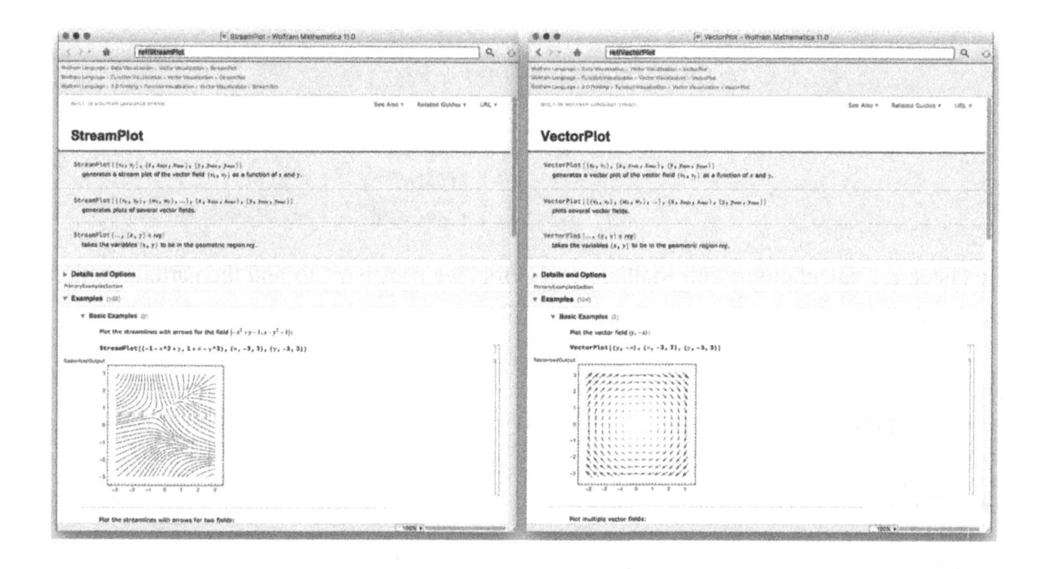


```
pvf1 = StreamPlot[{1, y(1 − y)}, {t, 0, 5}, {y, 0, 5/2}, Axes → Automatic,
  AxesOrigin → {0, 0}, StreamStyle → CMYKColor[.16, 1, .87, .06],
  Frame → False, Axes → Automatic, AxesOrigin → {0, 0},
  AxesLabel → {t, y}, PlotLabel → "(a)"];

pvf2 = VectorPlot[{1, y(1 − y)}, {t, 0, 5}, {y, 0, 5/2}, Axes → Automatic,
  AxesOrigin → {0, 0}, VectorStyle → CMYKColor[.16, 1, .87, .06],
  Frame → False, Axes → Automatic, AxesOrigin → {0, 0},
```

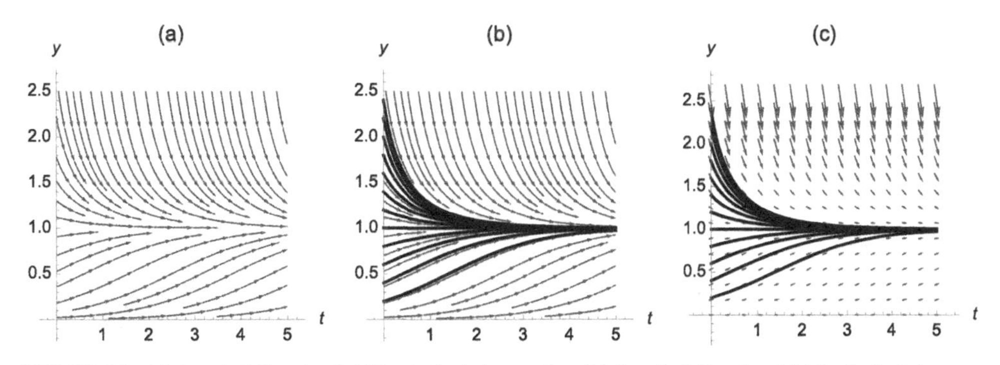

FIGURE 6.2 (a) A typical direction field for the logistic equation. (b) A typical direction field for the logistic equation along with several solutions. (c) Using `VectorPlot` rather than `StreamPlot` to see the direction field. (University of Richmond colors)

```
    AxesLabel → {t, y}]
```

The property is more easily seen when we graph various solutions along with the direction field as done next in Fig. 6.2.

```
toplot = Table[solc[[1, 1, 2]]/.{α → 1, k → 1, y0 → i/5}, {i, 1, 12}];

sols = Plot[toplot, {t, 0, 5}, PlotStyle → GrayLevel[0], PlotRange → All,

    PlotStyle → CMYKColor[1, .91, .32, .34],

    Frame → False, Axes → Automatic, AxesOrigin → {0, 0},

    AxesLabel → {t, y}];

plot2 = Show[pvf1, sols, PlotLabel → "(b)"];

plot3 = Show[pvf2, sols, PlotLabel → "(c)"];

Show[GraphicsRow[{pvf1, plot2, plot3}]]
```

□

When Mathematica encounters inverse functions, it sometimes chooses the incorrect *branch* to form a continuous solution to an initial-value problem.

Example 6.2

Solve $dy/dt = \sin t \cos y$, $y(1) = 3$.

Solution. When we use `DSolve` to solve the equation and the initial-value problem, Mathematica warns us that inverse functions are being used. For length considerations, Mathematica's warning messages are not displayed here.

```
sol = DSolve[y'[t] == Sin[t]Cos[y[t]], y[t], t]
```

$\{\{y[t] \to 2\text{ArcTan}[\text{Tanh}[\frac{1}{4}(C[1] - 2\text{Cos}[t])]]\}\}$

From the direction field, we see that the solution satisfying $y(1) = 3$ is continuous for (at least) $0 \le t \le 4\pi$. However, the explicit solution returned by `DSolve` is not the solution that is continuous on $[0, 4\pi]$. See Fig. 6.3 (a).

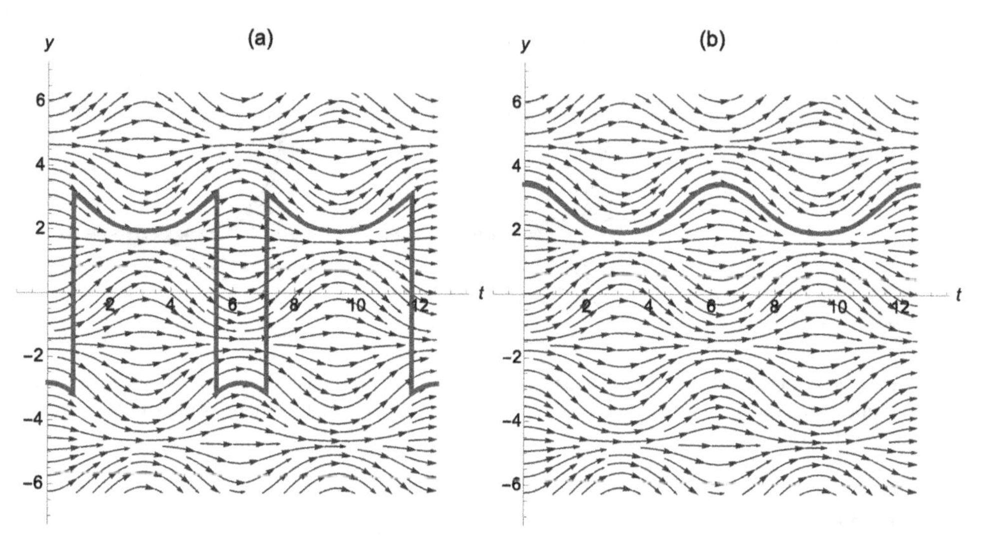

FIGURE 6.3 (a) The solution returned by `DSolve` is discontinuous on $[0, 4\pi]$. (b) We use `NDSolve` to find the continuous solution of the initial-value problem. (College of Charleston colors)

```
pvfl = StreamPlot[{1, Sin[t]Cos[y]}, {t, 0, 4Pi}, {y, −2Pi, 2Pi},
  StreamPoints → Fine, Axes → Automatic, AxesOrigin → {0, 0},
  StreamStyle → CMYKColor[0, .97, 1, .5, 1.02]];

psol1 = Plot[y[t]/.sol1, {t, 0, 4Pi},
  PlotStyle → {{CMYKColor[.38, .88, 0, 0], Thickness[.01]}}];
discont = Show[pvfl, psol1, Frame → False, Axes → Automatic,
  AxesOrigin → {0, 0}, AxesLabel → {t, y}]
```

To see the continuous solution, we use `NDSolve` to generate a numerical solution to the initial value problem. If possible, `NDSolve[{y'[t]==f[t,y[t]],y[t0]==y0},y[t],{t,a,b}]` attempts to numerically solve $y' = f(t, y)$, $y(t_0) = y_0$ for $a \le t \le b$.

`NDSolve` is discussed in more detail later in the section.

```
sol2 = NDSolve[{y'[t] == Sin[t]Cos[y[t]], y[1] == 3}, y[t], {t, 0, 4Pi}]

{{y[t] → InterpolatingFunction[][t]}}
```

In Fig. 6.3 (b), we see that the result returned by `NDSolve` is continuous on $[0, 4\pi]$.

```
psol2 = Plot[y[t]/.sol2, {t, 0, 4Pi},
  PlotStyle → {{CMYKColor[.38, .88, 0, 0], Thickness[.01]}}];
cont = Show[pvfl, psol2, Frame → False, Axes → Automatic,
  AxesOrigin → {0, 0}, AxesLabel → {t, y}]
```

With `Manipulate`, you can see how varying the initial conditions affects the solution. See Fig. 6.4. When you drag the locator points, the solution changes accordingly.

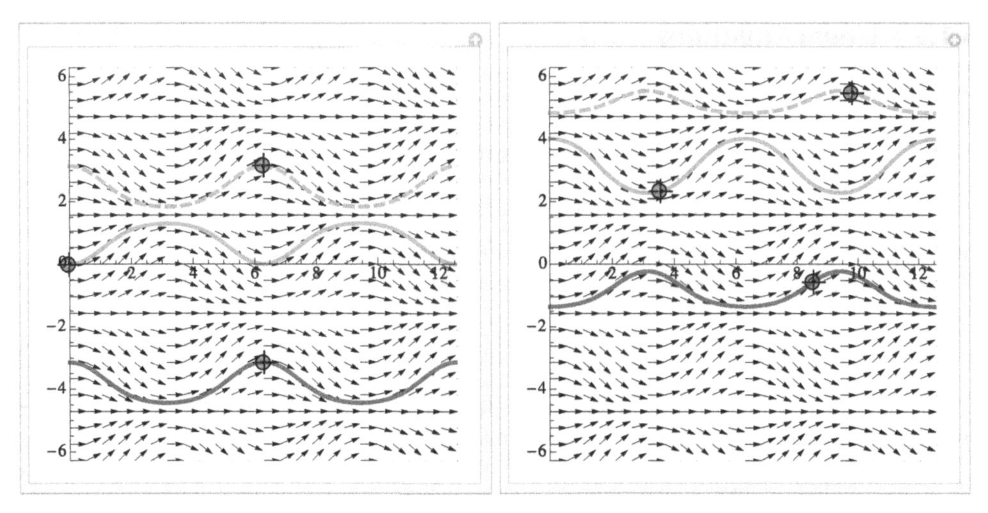

FIGURE 6.4 Visualizing how changing t_0 and y_0 affect the solution that satisfies $y(t_0) = y_0$.

```
Manipulate[
sol1 = NDSolve[{y'[t] == Sin[t]Cos[y[t]], y[pt[[1, 1]]] == pt[[1, 2]]}, y[t],
{t, 0, 4Pi}];
psol1 = Plot[y[t]/.sol1, {t, 0, 4Pi},
  PlotStyle → {{CMYKColor[.38, .88, 0, 0], Thickness[.01]}}];
sol2 = NDSolve[{y'[t] == Sin[t]Cos[y[t]], y[pt[[2, 1]]] == pt[[2, 2]]},
  y[t], {t, 0, 4Pi}];
psol2 = Plot[y[t]/.sol2, {t, 0, 4Pi},
  PlotStyle → {{CMYKColor[0, .9, .86, 0], Dashing[{0.02}], Thickness[.01]}}];
sol3 = NDSolve[{y'[t] == Sin[t]Cos[y[t]], y[pt[[3, 1]]] == pt[[3, 2]]}, y[t],
  {t, 0, 4Pi}];
psol3 = Plot[y[t]/.sol3, {t, 0, 4Pi},
  PlotStyle → {{CMYKColor[0, .18, 1, .27], Thickness[.01]}}];
initialpt = Graphics[Point[{pt}], PlotRange → {{0, 4Pi}, {−2Pi, 2Pi}}];
Show[pvf1, psol1, psol2, psol3, initialpt, Axes → Automatic,
  PlotRange → {{0, 4Pi}, {−2Pi, 2Pi}}, AspectRatio → Automatic,
  Frame → False, Axes → Automatic, AxesOrigin → {0, 0}, AxesLabel → {t, y}],
  {{pt, {{0, 0}, {2Pi, Pi}, {2Pi, −Pi}}}, Locator}
  ]
```
□

6.1.2 Linear Equations

Definition 6.2 (First-Order Linear Equation). A differential equation of the form

$$a_1(t)\frac{dy}{dt} + a_0(t)y = f(t), \tag{6.2}$$

where $a_1(t)$ is not identically the zero function, is a first-order **linear differential equation**.

Assuming that $a_1(t)$ is not identically the zero function, dividing equation (6.2) by $a_1(t)$ gives us the **standard form** of the first-order linear equation:

$$\frac{dy}{dt} + p(t)y = q(t). \tag{6.3}$$

If $q(t)$ is identically the zero function, we say that the equation is **homogeneous**. The **corresponding homogeneous equation** of equation (6.3) is

$$\frac{dy}{dt} + p(t)y = 0. \tag{6.4}$$

Observe that equation (6.4) is separable:

$$\begin{aligned}\frac{dy}{dt} + p(t)y &= 0\\ \frac{1}{y}dy &= -p(t)\,dt\\ \ln|y| &= -\int p(t)\,dt + C\\ y &= Ce^{-\int p(t)\,dt}.\end{aligned}$$

Notice that any constant multiple of a solution to a linear homogeneous equation is also a solution. Now suppose that y is any solution of equation (6.3) and y_p is a particular solution of equation (6.3). Then,

A **particular solution** is a specific solution to the equation that does not contain any arbitrary constants.

$$\begin{aligned}\left(y - y_p\right)' + p(t)\left(y - y_p\right) &= y' + p(t)y - \left(y_p{}' + p(t)y_p\right)\\ &= q(t) - q(t) = 0.\end{aligned}$$

Thus, $y - y_p$ is a solution to the corresponding homogeneous equation of equation (6.3). Hence,

$$\begin{aligned}y - y_p &= Ce^{-\int p(t)\,dt}\\ y &= Ce^{-\int p(t)\,dt} + y_p\\ y &= y_h + y_p,\end{aligned}$$

where $y_h = Ce^{-\int p(t)\,dt}$. That is, a general solution of equation (6.3) is $y = y_h + y_p$, where y_p is a particular solution to the nonhomogeneous equation and y_h is a general solution to the corresponding homogeneous equation. Thus, to solve equation (6.3), we need to first find a general solution to the corresponding homogeneous equation, y_h, which we can accomplish through separation of variables, and then find a particular solution, y_p, to the nonhomogeneous equation.

A fundamental concept is that a general solution of a nonhomogeneous linear equation is the sum of the solution to the corresponding homogeneous equation, y_h, and a particular solution, y_p to the nonhomogeneous equation: $y = y_h + y_p$.

If y_h is a solution to the corresponding homogeneous equation of equation (6.3) then for any constant C, Cy_h is also a solution to the corresponding homogeneous equation. Therefore, it is impossible to find a particular solution to equation (6.3) of this form. Instead, we search for a particular solution of the form $y_p = u(t)y_h$, where $u(t)$ is *not* a constant function. Assuming that a particular solution, y_p, to equation (6.3) has the form $y_p = u(t)y_h$, differentiating gives us $y_p{}' = u'y_h + uy_h{}'$ and substituting into equation (6.3) results in

$$y_p{}' + p(t)y_p = u'y_h + uy_h{}' + p(t)uy_h = q(t).$$

Because $u y_h' + p(t) u y_h = u\left[y_h' + p(t) y_h\right] = u \cdot 0 = 0$, we obtain

y_h is a solution to the corresponding homogeneous equation so $y_h' + p(t) y_h = 0$.

$$\begin{aligned} u' y_h &= q(t) \\ u' &= \frac{1}{y_h} q(t) \\ u' &= e^{\int p(t)\,dt} q(t) \\ u &= \int e^{\int p(t)\,dt} q(t)\,dt \end{aligned}$$

so

$$y_p = u(t)\, y_h = C e^{-\int p(t)\,dt} \int e^{\int p(t)\,dt} q(t)\,dt.$$

Because we can include an arbitrary constant of integration when evaluating $\int e^{\int p(t)\,dt} q(t)\,dt$, it follows that we can write a general solution of equation (6.3) as

$$y = e^{-\int p(t)\,dt} \int e^{\int p(t)\,dt} q(t)\,dt. \tag{6.5}$$

Alternatively, multiplying equation (6.3) by the **integrating factor** $\mu(t) = e^{\int p(t)\,dt}$ gives us the same result:

$$\begin{aligned} e^{\int p(t)\,dt} \frac{dy}{dt} + p(t) e^{\int p(t)\,dt} y &= q(t) e^{\int p(t)\,dt} \\ \frac{d}{dt}\left(e^{\int p(t)\,dt} y\right) &= q(t) e^{\int p(t)\,dt} \\ e^{\int p(t)\,dt} y &= \int q(t) e^{\int p(t)\,dt}\,dt \\ y &= e^{-\int p(t)\,dt} \int q(t) e^{\int p(t)\,dt}\,dt. \end{aligned}$$

Thus, first-order linear equations can always be solved, although the resulting integrals may be difficult or impossible to evaluate exactly.

Mathematica is able to solve the general form of the first-order equation, the initial-value problem $y' + p(t) y = q(t)$, $y(0) = y_0$,

DSolve[y′[*t*] + *p*[*t*]y[*t*] == *q*[*t*], y[*t*], *t*]

$$\{\{y[t] \to e^{\int_1^t -p[K[1]]\,dK[1]} C[1] + e^{\int_1^t -p[K[1]]\,dK[1]} \int_1^t e^{-\int_1^{K[2]} -p[K[1]]\,dK[1]} q[K[2]]\,dK[2]\}\}$$

DSolve[{y′[*t*] + *p*[*t*]y[*t*] == *q*[*t*], y[0] == y0}, y[*t*], *t*]

$$\{\{y[t] \to -e^{-\int_1^0 -p[K[1]]\,dK[1] + \int_1^t -p[K[1]]\,dK[1]} (-\text{y0}$$
$$+ e^{\int_1^0 -p[K[1]]\,dK[1]} \int_1^0 e^{-\int_1^{K[2]} -p[K[1]]\,dK[1]} q[K[2]]\,dK[2]$$
$$- e^{\int_1^0 -p[K[1]]\,dK[1]} \int_1^t e^{-\int_1^{K[2]} -p[K[1]]\,dK[1]} q[K[2]]\,dK[2])\}\}$$

as well as the corresponding homogeneous equation,

DSolve[y′[*t*] + *p*[*t*]y[*t*] == 0, y[*t*], *t*]

$$\{\{y[t] \to e^{\int_1^t -p[K[1]]\,dK[1]} C[1]\}\}$$

DSolve[{y′[*t*] + *p*[*t*]y[*t*] == 0, y[0] == y0}, y[*t*], *t*]

$\{\{y[t] \to e^{-\int_1^0 -p[K[1]]dK[1]+\int_1^t -p[K[1]]dK[1]}y0\}\}$

although the results contain unevaluated integrals.

Example 6.3: Exponential Growth

Let $y = y(t)$ denote the size of a population at time t. If y grows at a rate proportional to the amount present, y satisfies

$$\frac{dy}{dt} = \alpha y, \tag{6.6}$$

where α is the **growth constant**. If $y(0) = y_0$, using equation (6.5) results in $y = y_0 e^{\alpha t}$. We use `DSolve` to confirm this result.

```
DSolve[{y'[t] == αy[t], y[0] == y0}, y[t], t]
```

$\{\{y[t] \to e^{t\alpha}y0\}\}$

Example 6.4

Solve each of the following equations: (a) $dy/dt = k(y - y_s)$, $y(0) = y_0$, k and y_s constant (b) $y' - 2ty = t$ (c) $ty' - y = 4t\cos 4t - \sin 4t$.

$dy/dt = k(y - y_s)$ models *Newton's Law of Cooling*: the rate at which the temperature, $y(t)$, changes in a heating/cooling body is proportional to the difference between the temperature of the body and the constant temperature, y_s, of the surroundings.

This will turn out to be a lucky guess. If there is not a solution of this form, we would not find one of this form.

Solution. (a) By hand, we rewrite the equation and obtain $y' - ky = -ky_s$. A general solution of the corresponding homogeneous equation $y' - ky = 0$ is $y_h = e^{kt}$. Because k and $-ky_s$ are constants, we suppose that a particular solution of the nonhomogeneous equation, y_p, has the form $y_p = A$, where A is a constant.

Assuming that $y_p = A$, we have $y_p' = 0$ and substitution into the nonhomogeneous equation gives us

$$y_p{}' - ky_p = -KA = -ky_s \quad \text{so} \quad A = y_s.$$

Thus, a general solution is $y = y_h + y_p = Ce^{kt} + y_s$. Applying the initial condition $y(0) = y_0$ results in $y = y_s + (y_0 - y_s)e^{kt}$.

We obtain the same result with `DSolve`. We graph the solution satisfing $y(0) = 75$ assuming that $k = -1/2$ and $y_s = 300$ in Fig. 6.5. Notice that $y(t) \to y_s$ as $t \to \infty$.

```
sola = DSolve[{y'[t] == k(y[t] - ys), y[0] == y0}, y[t], t]
```

$\{\{y[t] \to e^{kt}y0 + ys - e^{kt}ys\}\}$

```
Plot[y[t]/.sola/.{k → −1/2, ys → 300, y0 → 75}, {t, 0, 10},
  PlotStyle → {{Thickness[.01], CMYKColor[1, 0, .79, .6]}}]
```

(b) The equation is in standard form and we identify $p(t) = -2t$. Then, the integrating factor is $\mu(t) = e^{\int p(t)\,dt} = e^{-t^2}$. Multiplying the equation by the integrating factor, $\mu(t)$, results in

$$e^{-t^2}(y' - 2ty) = te^{-t^2} \quad \text{or} \quad \frac{d}{dt}\left(ye^{-t^2}\right) = te^{-t^2}.$$

Integrating gives us $ye^{-t^2} = -\frac{1}{2}e^{-t^2} + C$ or $y = -\frac{1}{2} + Ce^{t^2}$. We confirm the result with `DSolve`.

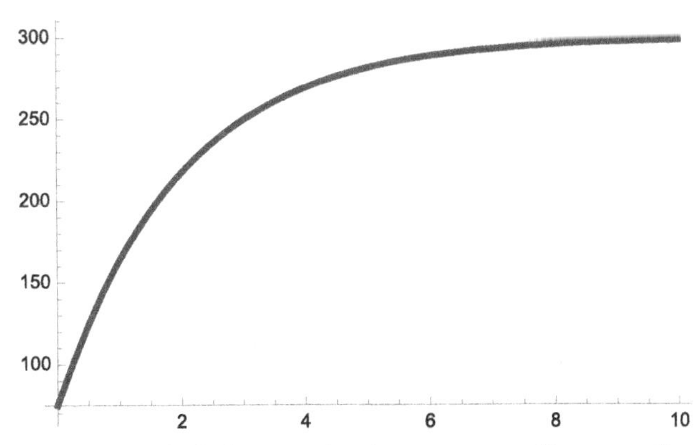

FIGURE 6.5 The temperature of the body approaches the temperature of its surroundings. (Tulane University colors)

```
DSolve[y'[t] - 2ty[t] == t, y[t], t]
```

$\{\{y[t] \to -\frac{1}{2} + e^{t^2} C[1]\}\}$

(c) In standard form, the equation is $y' - y/t = (4t \cos 4t - \sin 4t)/t$ so $p(t) = -1/t$. The integrating factor is $\mu(t) = e^{\int p(t)\,dt} = e^{-\ln t} = 1/t$ and multiplying the equation by the integrating factor and then integrating gives us

$$\begin{aligned} \frac{1}{t}\frac{dy}{dt} - \frac{1}{t^2}y &= \frac{1}{t^2}(4t\cos 4t - \sin 4t) \\ \frac{d}{dt}\left(\frac{1}{t}y\right) &= \frac{1}{t^2}(4t\cos 4t - \sin 4t) \\ \frac{1}{t}y &= \frac{\sin 4t}{t} + C \\ y &= \sin 4t + Ct, \end{aligned}$$

where we use the `Integrate` function to evaluate $\int \frac{1}{t^2}(4t\cos 4t - \sin 4t)\,dt = \frac{\sin 4t}{t} + C$.

```
Integrate[(4tCos[4t] - Sin[4t])/t^2, t]
```

$\frac{\text{Sin}[4t]}{t}$

We confirm this result with `DSolve`.

```
sol = DSolve[y'[t] - y[t]/t ==
  (4tCos[4t] - Sin[4t])/t, y[t], t]
```

$\{\{y[t] \to tC[1] + \text{Sin}[4t]\}\}$

In the general solution, observe that *every* solution satisfies $y(0) = 0$. That is, the initial-value problem

$$\frac{dy}{dt} - \frac{1}{t}y = \frac{1}{t^2}(4t\cos 4t - \sin 4t), \qquad y(0) = 0$$

has infinitely many solutions. We see this in the plot of several solutions that is generated with `Plot` in Fig. 6.6.

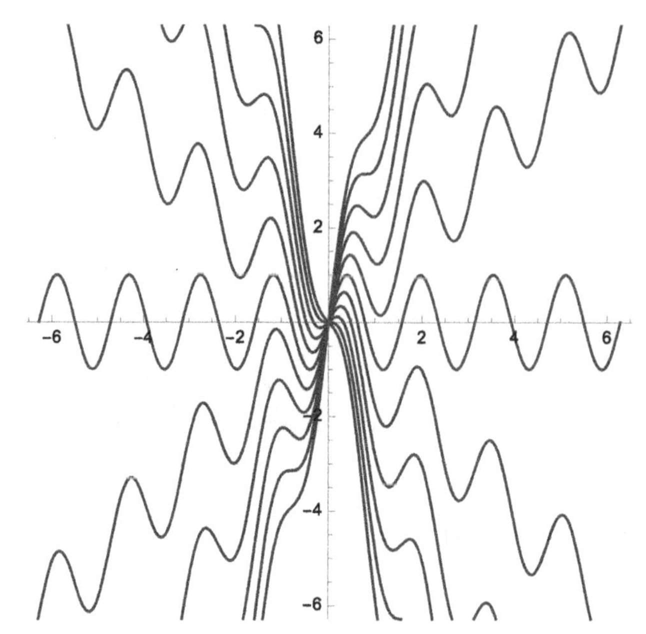

FIGURE 6.6 Every solution satisfies $y(0) = 0$.

```
toplot = Table[sol/.C[1] → i, {i, −5, 5}];
  Plot[y[t]/.toplot, {t, −2Pi, 2Pi}, PlotRange → {−2Pi, 2Pi},
  PlotStyle → CMYKColor[1, 0, .79, .6],
  AspectRatio → Automatic]
```

□

6.1.2.1 Application: Free-Falling Bodies

The motion of objects can be determined through the solution of first-order initial-value problems. We begin by explaining some of the theory that is needed to set up the differential equation that models the situation.

> ***Newton's Second Law of Motion:*** *The rate at which the momentum of a body changes with respect to time is equal to the resultant force acting on the body.*

Because the body's momentum is defined as the product of its mass and velocity, this statement is modeled as

$$\frac{d}{dt}(mv) = F,$$

where m and v represent the body's mass and velocity, respectively, and F is the sum of the forces (the resultant force) acting on the body. Because m is constant, differentiation leads to the well-known equation

$$m\frac{dv}{dt} = F.$$

If the body is subjected only to the force due to gravity, then its velocity is determined by solving the differential equation

$$m\frac{dv}{dt} = mg \qquad \text{or} \qquad \frac{dv}{dt} = g,$$

where $g = 32\ \text{ft}/\text{s}^2$ (English system) and $g = 9.8\ \text{m}/\text{s}^2$ (metric system). This differential equation is applicable only when the resistive force due to the medium (such as air resistance) is ignored. If this offsetting resistance is considered, we must discuss all of the forces acting on the object. Mathematically, we write the equation as

$$m\frac{dv}{dt} = \sum (\text{forces acting on the object})$$

where the direction of motion is taken to be the positive direction. Because air resistance acts against the object as it falls and g acts in the same direction of the motion, we state the differential equation in the form

$$m\frac{dv}{dt} = mg + (-F_R) \qquad \text{or} \qquad m\frac{dv}{dt} = mg - F_R,$$

where F_R represents this resistive force. Note that down is assumed to be the positive direction. The resistive force is typically proportional to the body's velocity, v, or the square of its velocity, v^2. Hence, the differential equation is linear or nonlinear based on the resistance of the medium taken into account.

Example 6.5

An object of mass $m = 1$ is dropped from a height of 50 feet above the surface of a small pond. While the object is in the air, the force due to air resistance is v. However, when the object is in the pond, it is subjected to a buoyancy force equivalent to $6v$. Determine how much time is required for the object to reach a depth of 25 feet in the pond.

Solution. This problem must be broken into two parts: an initial-value problem for the object above the pond, and an initial-value problem for the object below the surface of the pond. The initial-value problem above the pond's surface is found to be

$$\begin{cases} dv/dt = 32 - v \\ v(0) = 0. \end{cases}$$

However, to define the initial-value problem to find the velocity of the object beneath the pond's surface, the velocity of the object when it reaches the surface must be known. Hence, the velocity of the object above the surface must be determined by solving the initial-value problem above. The equation $dv/dt = 32 - v$ is separable and solved with `DSolve` in `d1`.

Clear[v, y]

d1 = DSolve[{v'[t] == 32 − v[t], v[0] == 0}, v[t], t]

$\{\{v[t] \to 32e^{-t}(-1 + e^t)\}\}$

In order to find the velocity when the object hits the pond's surface we must know the time at which the distance traveled by the object (or the displacement of the object) is 50. Thus, we must find the displacement function, which is done by integrating the velocity function obtaining $s(t) = 32e^{-t} + 32t - 32$.

p1 = DSolve[{y'[t] == v[t]/.d1, y[0] == 0}, y[t], t]

$\{\{y[t] \to 32e^{-t}(1 - e^t + e^t t)\}\}$

The displacement function is graphed with `Plot` in Fig. 6.7 (a). The value of t at which the object has traveled 50 feet is needed. This time appears to be approximately 2.5 seconds.

Plot[{y[t]/.p1, 50}, {t, 0, 5},

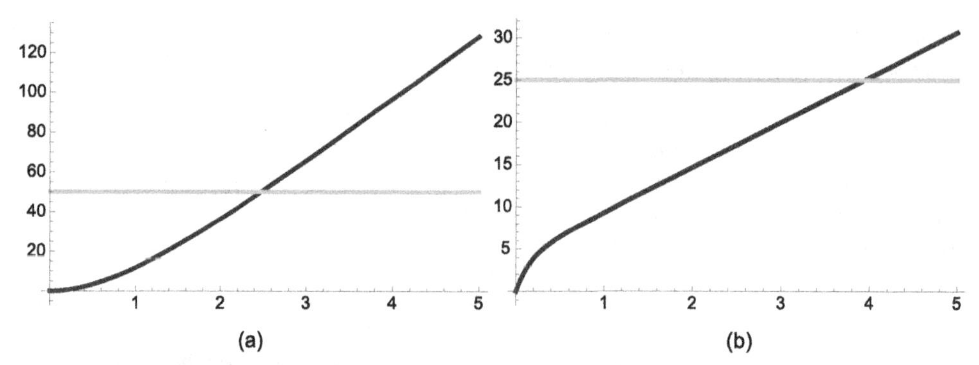

FIGURE 6.7 (a) The object has traveled 50 feet when $t \approx 2.5$. (b) After approximately 4 seconds, the object is 25 feet below the surface of the pond. (Louisiana State University colors)

PlotStyle → {{Thickness[.01], CMYKColor[.82, .98, 0, .12]},

{Thickness[.01], CMYKColor[0, .19, .89, 0]}}]

A more accurate value of the time at which the object hits the surface is found using `FindRoot`. In this case, we obtain $t \approx 2.47864$. The velocity at this time is then determined by substitution into the velocity function resulting in $v(2.47864) \approx 29.3166$. Note that this value is the initial velocity of the object when it hits the surface of the pond.

t1 = FindRoot[Evaluate[y[*t*]/.p1] == 50, {*t*, 2.5}]

$\{t \to 2.47864\}$

v1 = d1/.t1

$\{\{v[2.47864] \to 29.3166\}\}$

Thus, the initial-value problem that determines the velocity of the object beneath the surface of the pond is given by

$$\begin{cases} dv/dt = 32 - 6v \\ v(0) = 29.3166. \end{cases}$$

The solution of this initial-value problem is $v(t) = \frac{16}{3} + 23.9833e^{-t}$ and integrating to obtain the displacement function (the initial displacement is 0) we obtain $s(t) = 3.99722 - 3.99722e^{-6t} + \frac{16}{3}t$. These steps are carried out in `d2` and `p2`.

d2 = DSolve[{*v*′[*t*] == 32 − 6*v*[*t*], *v*[0] == v1[[1, 1, 2]]}, *v*[*t*], *t*]

$\{\{v[t] \to 5.33333e^{-6t}(4.49686 + e^{6t})\}\}$

p2 = DSolve[{*y*′[*t*] == *v*[*t*]/.d2, *y*[0] == 0}, *y*[*t*], *t*]

$\{\{y[t] \to 5.33333e^{-6.t}(-0.749476 + 0.749476e^{6.t} + 1.e^{6.t}t)\}\}$

This displacement function is then plotted in Fig. 6.7 (b) to determine when the object is 25 feet beneath the surface of the pond. This time appears to be near 4 seconds.

```
Plot[{y[t]/.p2, 25}, {t, 0, 5},
  PlotStyle → {{Thickness[.01], CMYKColor[.82, .98, 0, .12]},
  {Thickness[.01], CMYKColor[0, .19, .89, 0]}}]
```

A more accurate approximation of the time at which the object is 25 feet beneath the pond's surface is obtained with `FindRoot`. In this case, we obtain $t \approx 3.93802$. Finally, the time required for the object to reach the pond's surface is added to the time needed for it to travel 25 feet beneath the surface to see that approximately 6.41667 seconds are required for the object to travel from a height of 50 feet above the pond to a depth of 25 feet below the surface.

```
t2 = FindRoot[Evaluate[y[t]/.p2] == 25, {t, 4}]
```

$\{t \to 3.93802\}$

```
t1[[1, 2]] + t2[[1, 2]]
```

6.41667 □

6.1.3 Nonlinear Equations

Mathematica can solve or help solve a variety of nonlinear first-order equations that are typically encountered in the introductory differential equations course.

Example 6.6

Solve each: (a) $(\cos x + 2xe^y)\,dx + (\sin y + x^2e^y - 1)\,dy = 0$; (b) $(y^2 + 2xy)\,dx - x^2dy = 0$.

Solution. (a) Notice that $(\cos x + 2xe^y)\,dx + (\sin y + x^2e^y - 1)\,dy = 0$ can be written as $dy/dx = -(\cos x + 2xe^y)/(\sin x + x^2e^y - 1)$. The equation is an example of an *exact equation*. A theorem tells us that the equation

$$M(x, y)dx + N(x, y)dy = 0$$

is **exact** if and only if $\partial M/\partial y = \partial N/\partial x$.

```
m = Cos[x] + 2xExp[y];
n = Sin[y] + x^2Exp[y] - 1;
D[m, y]
D[n, x]
```

$2e^y x$

$2e^y x$

We solve exact equations by integrating. Let $F(x, y) = C$ satisfy $(y\cos x + 2xe^y)dx + (\sin y + x^2e^y - 1)\,dy = 0$. Then,

$$F(x, y) = \int (\cos x + 2xe^y)\,dx = \sin x + x^2e^y + g(y),$$

The function f satisfying $df = f_x(x, y)\,dx + f_y(x, y)\,dy$ is called the **potential function**.

where $g(y)$ is a function of y.

```
f1 = Integrate[m, x]
```

$e^y x^2 + \text{Sin}[x]$

We next find that $g'(y) = \sin y - 1$ so $g(y) = -\cos y - y$. Hence, a general solution of the equation is

$$\sin x + x^2 e^y - \cos y - y = C.$$

```
f2 = D[f1, y]
```

$e^y x^2$

```
f3 = Solve[f2 + c == n, c]
```

$\{\{c \to -1 + \text{Sin}[y]\}\}$

```
Integrate[f3[[1, 1, 2]], y]
```

$-y - \text{Cos}[y]$

We confirm this result with `DSolve`. Notice that Mathematica warns us that it cannot solve for y explicitly and returns the same implicit solution obtained by us.

```
mf = m/.y → y[x];
nf = n/.y → y[x];
sol = DSolve[mf + nfy'[x] == 0, y[x], x]
```

$\text{Solve}[e^{y[x]}x^2 - \text{Cos}[y[x]] + \text{Sin}[x] - y[x] == C[1], y[x]]$

Graphs of several solutions using the values of C generated in `cvals` are graphed with `ContourPlot` in Fig. 6.8.

```
sol2 = sol[[1, 1]]/.y[x] → y
```

$e^y x^2 - y - \text{Cos}[y] + \text{Sin}[x]$

```
cvals = Table[sol2/.{x → −3Pi/2, y → i}, {i, 0, 6Pi, 6Pi/24}];
```

```
ContourPlot[sol2, {x, −3Pi, 3Pi}, {y, 0, 6Pi}, Contours → cvals,
  ContourShading → False, Axes → Automatic, Frame → False,
  AxesOrigin → {0, 0}, ContourStyle → CMYKColor[.07, 1, .68, .32]]
```

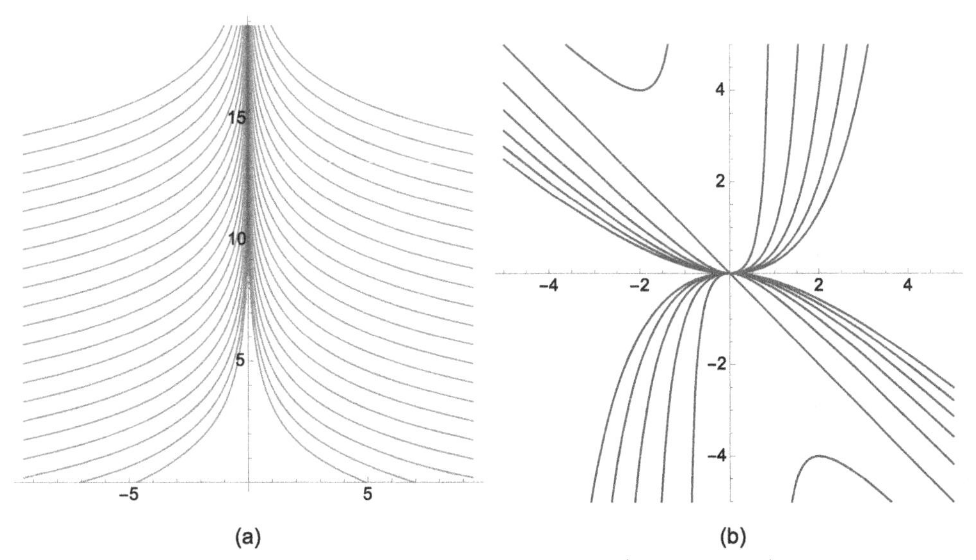

FIGURE 6.8 (a) Graphs of several solutions of $(\cos x + 2xe^y)\,dx + \left(\sin y + x^2e^y - 1\right)dy = 0$. (b) Graphs of several solutions of $\left(y^2 + 2xy\right)dx - x^2dy = 0$. (University of Arkansas colors)

(b) We can write $(y^2 + 2xy)\,dx - x^2dy = 0$ as $dy/dx = (y^2 + 2xy)/x^2$. A first-order equation is **homogeneous** if it can be written in the form $dy/dx = F(y/x)$. Homogeneous equations are reduced to separable equations with either the substitution $y = ux$ or $x = vy$. In this case, we have that $dy/dx = (y/x)^2 + 2(y/x)$ so the equation is homogeneous.

Let $y = ux$. Then, $dy = u\,dx + x\,du$. Substituting into $(y^2 + 2xy)\,dx - x^2dy = 0$ and separating gives us

$$\begin{aligned}\left(y^2 + 2xy\right)dx - x^2dy &= 0\\ \left(u^2x^2 + 2ux^2\right)dx - x^2(u\,dx + x\,du) &= 0\\ \left(u^2 + 2u\right)dx - (u\,dx + x\,du) &= 0\\ \left(u^2 + u\right)dx &= x\,du\\ \frac{1}{u\,(u+1)}du &= \frac{1}{x}dx.\end{aligned}$$

Integrating the left and right-hand sides of this equation with `Integrate`,

Integrate[1/(u(u + 1)), u]

Log[u] − Log[1 + u]

Integrate[1/x, x]

Log[x]

exponentiating, resubstituting $u = y/x$, and solving for y gives us

$$\begin{aligned}\ln|u| - \ln|u+1| &= \ln|x| + C\\ \frac{u}{u+1} &= Cx\end{aligned}$$

$$\frac{\frac{y}{x}}{\frac{y}{x}+1} = Cx$$

$$y = \frac{Cx^2}{1 - Cx}.$$

sol1 = Solve[(y/x)/(y/x + 1)==cx, y]

$\{\{y \to -\frac{cx^2}{-1+cx}\}\}$

We confirm this result with `DSolve` and then graph several solutions with `Plot` in Fig. 6.8 (b).

sol2 = DSolve[y[x]^2 + 2xy[x] − x^2y′[x] == 0, y[x], x]

$\{\{y[x] \to -\frac{x^2}{x-C[1]}\}\}$

toplot = Table[sol2[[1, 1, 2]]/.C[1] → i, {i, −5, 5}];

Plot[Tooltip[toplot], {x, −5, 5}, PlotRange → {−5, 5},

AspectRatio → Automatic, PlotStyle->CMYKColor[.07, 1, .68, .32]] □

6.1.4 Numerical Methods

If numerical results are desired, use `NDSolve`:

```
NDSolve[{y'[t]==f[t,y[t]],y[t0]==y0},y[t],{t,a,b}]
```

attempts to generate a numerical solution of $dy/dt = f(t, y)$, $y(t_0) = y_0$, valid for $a \le t \le b$.

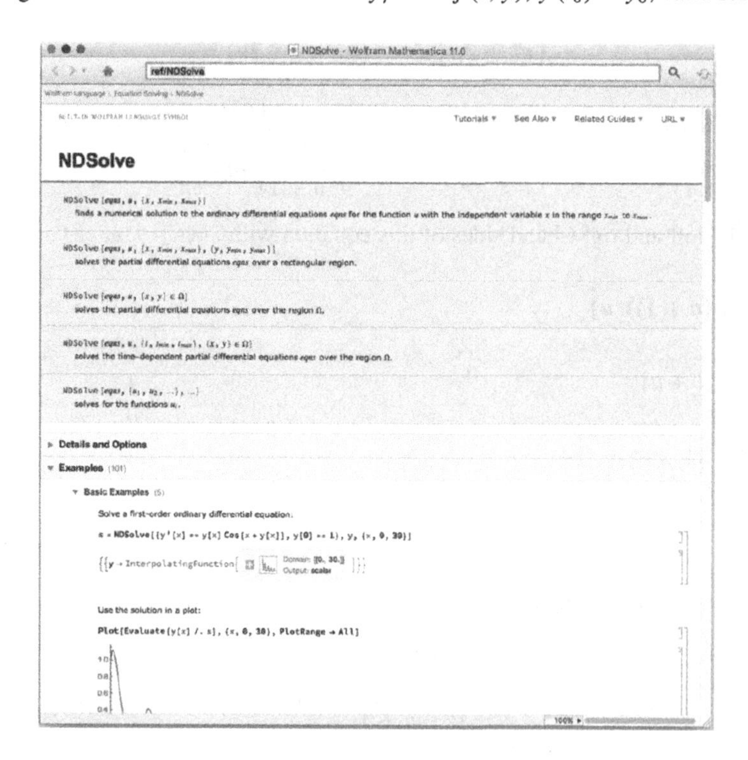

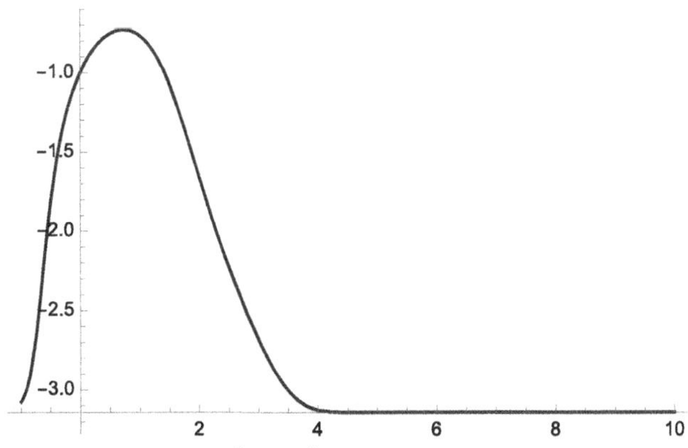

FIGURE 6.9 Graph of the solution to $y' = \left(t^2 - y^2\right) \sin y$, $y(0) = -1$. (University of Connecticut colors)

Example 6.7

Consider $dy/dt = (t^2 - y^2) \sin y$, $(0) = -1$. (a) Determine $y(1)$. (b) Graph $y(t)$, $-1 \le t \le 10$.

Solution. We first remark that `DSolve` can neither exactly solve the differential equation $y' = \left(t^2 - y^2\right) \sin y$ nor find the solution that satisfies $y(0) = -1$.

When Mathematica returns a command unevaluated, interpret the result to mean that Mathematica cannot evaluate the command.

```
sol = DSolve[y'[t] == (t^2 - y[t]^2)Sin[y[t]], y[t], t]
```

DSolve[$y'[t]$ == Sin[$y[t]$]$(t^2 - y[t]^2)$, $y[t]$, t]

```
sol = DSolve[{y'[t] == (t^2 - y[t]^2)Sin[y[t]], y[0] == -1}, y[t], t]
```

DSolve[{$y'[t]$ == Sin[$y[t]$]$(t^2 - y[t]^2)$, $y[0]$ == -1}, $y[t]$, t]

However, we obtain a numerical solution valid for $-1 \le t \le 10$ using the `NDSolve` function.

```
sol = NDSolve[{y'[t] == (t^2 - y[t]^2)Sin[y[t]], y[0] == -1}, y[t],
  {t, -1, 10}]
```

{{$y[t]$ → InterpolatingFunction[][t]}}

Entering `sol /.t->1` evaluates the numerical solution if $t = 1$.

```
sol/.t -> 1
```

{{y[1] → −0.766013}}

The result means that $y(1) \approx -.766$. We use the `Plot` command to graph the solution for $0 \le t \le 10$ in Fig. 6.9.

```
Plot[y[t]/.sol, {t, -1, 10},
  PlotStyle -> CMYKColor[1, .76, .12, .7]]
```

□

Example 6.8: Logistic Equation with Predation

Incorporating predation into the **logistic equation**, $y' = \alpha y\left(1 - \frac{1}{K}y\right)$, results in $\frac{dy}{dt} = \alpha y\left(1 - \frac{1}{K}y\right) - P(y)$, where $P(y)$ is a function of y describing the rate of predation. A typical choice for P is $P(y) = ay^2/(b^2 + y^2)$ because $P(0) = 0$ and P is bounded above: $\lim_{t\to\infty} P(y) < \infty$.

Remark 6.2. Of course, if $\lim_{t\to\infty} y(t) = Y$, then $\lim_{t\to\infty} P(y) = aY^2/(b^2 + Y^2)$. Generally, however, $\lim_{t\to\infty} P(y) \neq a$ because $\lim_{t\to\infty} y(t) \leq K \neq \infty$, for some $K \geq 0$, in the predation situation.

If $\alpha = 1$, $a = 5$, and $b = 2$, graph the direction field associated with the equation as well as various solutions if (a) $K = 19$ and (b) $K = 20$.

Solution. (a) We define `eqn[k]` to be $\frac{dy}{dt} = y\left(1 - \frac{1}{K}y\right) - \frac{5y^2}{4 + y^2}$.

```
eqn[k_] = y'[t] == y[t](1 - y[t]/k) - 5y[t]^2/(4 + y[t]^2);
```

We use `StreamPlot` to graph the direction field in Fig. 6.10 (a) and then the direction field along with the solutions that satisfy $y(0) = .5$, $y(0) = .2$, and $y(0) = 4$ in Fig. 6.10 (b).

```
pvf19 = StreamPlot[{1, y(1 - 1/19y) - 5y^2/(4 + y^2)}, {t, 0, 10},
  {y, 0, 6}, StreamStyle → {Fine, CMYKColor[0, .23, .53, .35]},
  Axes → Automatic, AxesOrigin → {0, 0},
  Frame → False]
```

```
numsols = Map[NDSolve[{eqn[19], y[0] == #}, y[t], {t, 0, 10}]&,
  Table[i, {i, 0.5, 6, 5.5/9}]];
  solplot = Plot[y[t]/.numsols, {t, 0, 10}, PlotRange → All,
  PlotStyle → {{CMYKColor[0, 0, 0, .5], Thickness[.01]}}]
```

```
Show[GraphicsRow[{pvf19, Show[pvf19, solplot]}]]
```

In the plot, notice that all nontrivial solutions appear to approach an equilibrium solution. We determine the equilibrium solution by solving $y' = 0$

```
eqn[19][[2]]
```

$\left(1 - \frac{y[t]}{19}\right)y[t] - \frac{5y[t]^2}{4+y[t]^2}$

```
Solve[eqn[19.][[2]]==0, y[t]]
```

$\{\{y[t] \to 0.\}, \{y[t] \to 0.923351\}, \{y[t] \to 9.03832 - 0.785875i\},$

$\{y[t] \to 9.03832 + 0.785875i\}\}$

to see that the equilibrium solution $(dy/dt = 0)$ is $y \approx 0.923$.

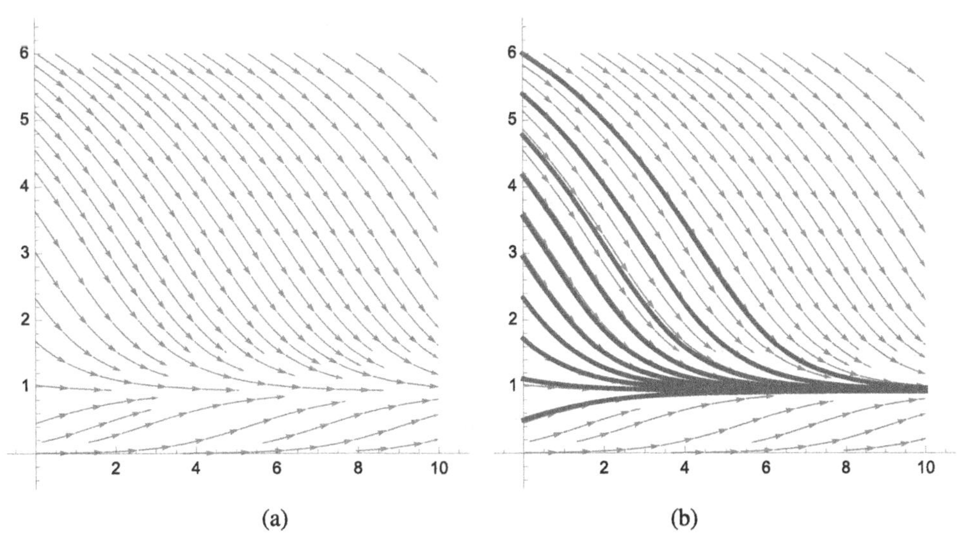

FIGURE 6.10 (a) Direction field and (b) direction field with three solutions. (University of Idaho colors)

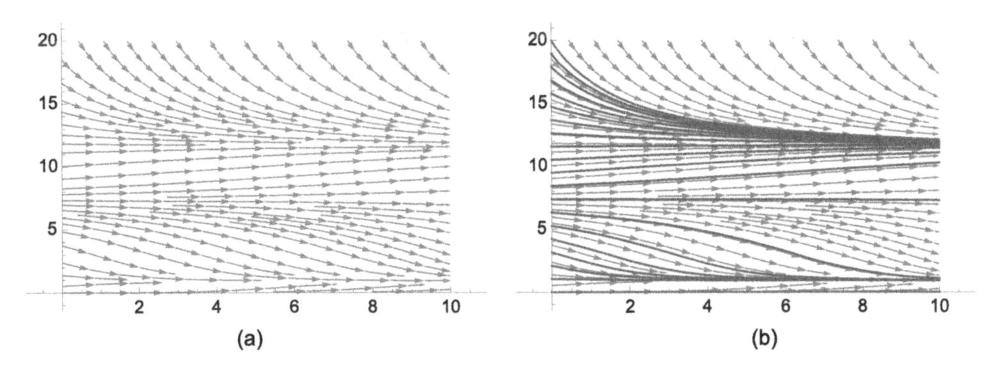

FIGURE 6.11 (a) Direction field. (b) Direction field with several solutions.

(b) We carry out similar steps for (b). First, we graph the direction field with `StreamPlot` in Fig. 6.11 (a).

```
pvf20 = StreamPlot[{1, y(1 − 1/20y) − 5y^2/(4 + y^2)}, {t, 0, 10},
  {y, 0, 20}, Axes → Automatic, AxesOrigin → {0, 0},
  AspectRatio → 1/GoldenRatio,
  StreamStyle → {Fine, CMYKColor[0, .23, .53, .35]},
  Frame → False];
```

We then use `Map` together with `NDSolve` to numerically find the solution satisfying $y(0) = i$, for 20 equally spaced values of i between 0 and 20 and name the resulting list `numsols`. The functions contained in `numsols` are graphed with `Plot` in `solplot`.

```
numsols = Map[NDSolve[{eqn[20], y[0] == #}, y[t], {t, 0, 10}]&,
  Table[i, {i, 0, 20, 20/19}]];
solplot = Plot[y[t]/.numsols, {t, 0, 10}, PlotRange → All,
```

```
PlotStyle → {{CMYKColor[0, 0, 0, .5], Thickness[.005]}}];
```

Last, we display the direction field along with the solution graphs in `solplot` using `Show` in Fig. 6.11 (b).

```
Show[GraphicsRow[{pvf20, Show[pvf20, solplot, Frame → False]}]]
```

Notice that there are three nontrivial equilibrium solutions that are found by solving $y' = 0$.

```
Solve[eqn[20.][[2]]==0, y[t], t]
```

```
{{y[t] → 0.}, {y[t] → 0.926741}, {y[t] → 7.38645}, {y[t] → 11.6868}}
```

In this case, $y \approx .926$ and $y \approx 11.687$ are stable while $y \approx 7.386$ is unstable. □

6.2 SECOND-ORDER LINEAR EQUATIONS

We now present a concise discussion of second-order linear equations, which are extensively discussed in the introductory differential equations course.

6.2.1 Basic Theory

The **general form** of the **second-order linear equation** is

$$a_2(t)\frac{d^2y}{dt^2} + a_1(t)\frac{dy}{dt} + a_0(t)y = f(t), \tag{6.7}$$

where $a_2(t)$ is not identically the zero function.

The **standard form** of the second-order linear equation (6.7) is

$$\frac{d^2y}{dt^2} + p(t)\frac{dy}{dt} + q(t)y = f(t). \tag{6.8}$$

The **corresponding homogeneous equation** of equation (6.8) is

$$\frac{d^2y}{dt^2} + p(t)\frac{dy}{dt} + q(t)y = 0. \tag{6.9}$$

A **general solution** of equation (6.9) is $y = c_1y_1 + c_2y_2$ where

1. y_1 and y_2 are solutions of equation (6.9), and
2. y_1 and y_2 are *linearly independent*.

A particular solution, y_p, is a solution that does not contain any arbitrary constants.

If y_1 and y_2 are solutions of equation (6.9), then y_1 and y_2 are **linearly independent** if and only if the **Wronskian**,

$$W(\{y_1, y_2\}) = \begin{vmatrix} y_1 & y_2 \\ y_1' & y_2' \end{vmatrix} = y_1y_2' - y_1'y_2, \tag{6.10}$$

is not the zero function. If y_1 and y_2 are linearly independent solutions of equation (6.9), we call the set $S = \{y_1, y_2\}$ a **fundamental set of solutions** for equation (6.9).

Key concept: A general solution of the nonhomogeneous linear equation is $y = y_h + y_p$, where y_h is a general solution of the corresponding homogeneous equation and y_p is a particular solution of the nonhomogeneous equation.

Let y be a general solution of equation (6.8) and y_p be a particular solution of equation (6.8). It follows that $y - y_p$ is a solution of equation (6.9) so $y - y_p = y_h$ where y_h is a general solution of equation (6.9). Hence, $y = y_h + y_p$. That is, to solve the nonhomogeneous equation, we need a general solution, y_h, of the corresponding homogeneous equation and a particular solution, y_p, of the nonhomogeneous equation.

6.2.2 Constant Coefficients

Suppose that the coefficient functions of equation (6.7) are constants: $a_2(t) = a$, $a_1(t) = b$, and $a_0(t) = c$ and that $f(t)$ is identically the zero function. In this case, equation (6.7) becomes

$$ay'' + by' + cy = 0. \tag{6.11}$$

Now suppose that $y = e^{kt}$, k constant, is a solution of equation (6.11). Then, $y' = ke^{kt}$ and $y'' = k^2e^{kt}$. Substitution into equation (6.11) then gives us

$$\begin{aligned} ay'' + by' + cy &= ak^2e^{kt} + bke^{kt} + ce^{kt} \\ &= e^{kt}\left(ak^2 + bk + c\right) = 0. \end{aligned}$$

Because $e^{kt} \neq 0$, the solutions of equation (6.11) are determined by the solutions of

$$ak^2 + bk + c = 0, \tag{6.12}$$

called the **characteristic equation** of equation (6.11).

Theorem 6.1. *Let k_1 and k_2 be the solutions of equation* (6.12).

1. *If $k_1 \neq k_2$ are real and distinct, two linearly independent solutions of equation* (6.11) *are $y_1 = e^{k_1t}$ and $y_2 = e^{k_2t}$; a general solution of equation* (6.11) *is*
$$y = c_1e^{k_1t} + c_2e^{k_2t}.$$
2. *If $k_1 = k_2$, two linearly independent solutions of equation* (6.11) *are $y_1 = e^{k_1t}$ and $y_2 = te^{k_1t}$; a general solution of equation* (6.11) *is*
$$y = c_1e^{k_1t} + c_2te^{k_1t}.$$
3. *If $k_{1,2} = \alpha \pm \beta i$, $\beta \neq 0$, two linearly independent solutions of equation* (6.11) *are $y_1 = e^{\alpha t}\cos\beta t$ and $y_2 = e^{\alpha t}\sin\beta t$; a general solution of equation* (6.11) *is*
$$y = e^{\alpha t}\left(c_1\cos\beta t + c_2\sin\beta t\right).$$

Example 6.9

Solve each of the following equations: (a) $6y'' + y' - 2y = 0$; (b) $y'' + 2y' + y = 0$; (c) $16y'' + 8y' + 145y = 0$.

Solution. (a) The characteristic equation is $6k^2 + k - 2 = (3k + 2)(2k - 1) = 0$ with solutions $k = -2/3$ and $k = 1/2$. We check with either `Factor` or `Solve`.

```
Factor[6k^2 + k - 2]
```

$(-1 + 2k)(2 + 3k)$

```
Solve[6k^2 + k - 2 == 0]
```

$\{\{k \to -\frac{2}{3}\}, \{k \to \frac{1}{2}\}\}$

Then, a fundamental set of solutions is $\left\{e^{-2t/3}, e^{t/2}\right\}$ and a general solution is

$$y = c_1e^{-2t/3} + c_2e^{t/2}.$$

Of course, we obtain the same result with `DSolve`.

Clear[y]

DSolve[6y''[t] + y'[t] − 2y[t] == 0, y[t], t]

$\{\{y[t] \to e^{-2t/3}C[1] + e^{t/2}C[2]\}\}$

(b) The characteristic equation (polynomial) is $k^2 + 2k + 1 = (k+1)^2 = 0$ with solution $k = -1$, which has multiplicity two, so a fundamental set of solutions is $\{e^{-t}, te^{-t}\}$ and a general solution is

$$y = c_1 e^{-t} + c_2 t e^{-t}.$$

We check the calculation in the same way as in (a).

Factor[k^2 + 2k + 1]

Solve[k^2 + 2k + 1 == 0]

DSolve[y''[t] + 2y'[t] + y[t] == 0, y[t], t]

$(1+k)^2$

$\{\{k \to -1\}, \{k \to -1\}\}$

(c) The characteristic equation is $16k^2 + 8k + 145 = 0$ with solutions $k_{1,2} = -\frac{1}{4} \pm 3i$ so a fundamental set of solutions is $\{e^{-t/4}\cos 3t, e^{-t/4}\sin 3t\}$ and a general solution is

$$y = e^{-t/4}\left(c_1 \cos 3t + c_2 \sin 3t\right).$$

The calculation is verified in the same way as in (a) and (b).

$\{\{y[t] \to e^{-t}C[1] + e^{-t}tC[2]\}\}$

Factor[16k^2 + 8k + 145]

$145 + 8k + 16k^2$

Factor[16k^2 + 8k + 145, GaussianIntegers → True]

$((1-12i)+4k)((1+12i)+4k)$

Solve[16k^2 + 8k + 145 == 0]

$\{\{k \to -\frac{1}{4} - 3i\}, \{k \to -\frac{1}{4} + 3i\}\}$

DSolve[16y''[t] + 8y'[t] + 145y[t] == 0, y[t], t]

$\{\{y[t] \to e^{-t/4}C[2]\text{Cos}[3t] + e^{-t/4}C[1]\text{Sin}[3t]\}\}$ □

Example 6.10

Solve $64\frac{d^2y}{dt^2}+16\frac{dy}{dt}+1025y=0$, $y(0)=1$, $\frac{dy}{dt}(0)=3$.

Solution. A general solution of $64y''+16y'+1025y=0$ is $y=e^{-t/8}(c_1\sin 4t+c_2\cos 4t)$.

gensol = DSolve[64y"[*t*] + 16y'[*t*] + 1025y[*t*] == 0, y[*t*], *t*]

$\{\{y[t]\to e^{-t/8}C[2]\mathrm{Cos}[4t]+e^{-t/8}C[1]\mathrm{Sin}[4t]\}\}$

Applying $y(0)=1$ shows us that $c_2=1$.

e1 = gensol[[1, 1, 2]]/.*t* → 0

$C[2]$

Computing y'

_D_[y[*t*]/.gensol[[1]], *t*]

$4e^{-t/8}C[1]\mathrm{Cos}[4t]-\frac{1}{8}e^{-t/8}C[2]\mathrm{Cos}[4t]-\frac{1}{8}e^{-t/8}C[1]\mathrm{Sin}[4t]-4e^{-t/8}C[2]\mathrm{Sin}[4t]$

and then $y'(0)$, shows us that $-4c_1-\frac{1}{8}c_2=3$.

e2 = *D*[y[*t*]/.gensol[[1]], *t*]/.*t* → 0

$4C[1]-\frac{C[2]}{8}$

Solving for c_1 and c_2 with `Solve` shows us that $c_1=-25/32$ and $c_1=1$.

cvals = Solve[{e1 == 1, e2 == 3}]

$\{\{C[1]\to\frac{25}{32},C[2]\to 1\}\}$

Thus, $y=e^{-t/8}\left(\frac{-25}{32}\sin 4t+\cos 4t\right)$, which we graph with `Plot` in Fig. 6.12.

sol = y[*t*]/.gensol[[1]]/.cvals[[1]]

$e^{-t/8}\mathrm{Cos}[4t]+\frac{25}{32}e^{-t/8}\mathrm{Sin}[4t]$

Plot[sol, {*t*, 0, 8Pi}, PlotStyle → {{Thickness[.005],

CMYKColor[0, 1, .65, .34]}}]

We verify the calculation with `DSolve`.

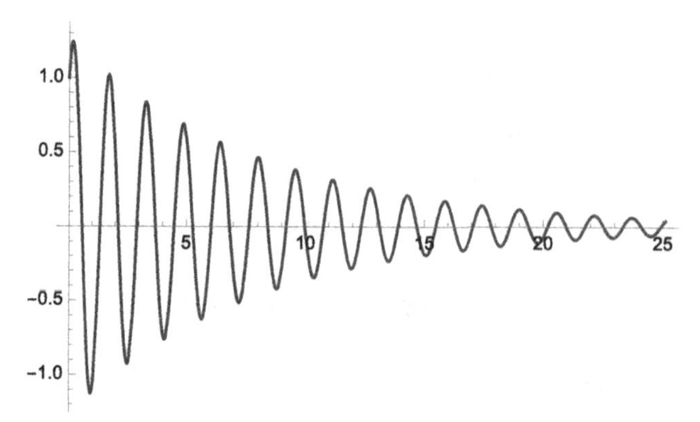

FIGURE 6.12 The solution to the initial-value problem tends to 0 as $t \to \infty$. (Indiana University colors)

```
DSolve[{64y''[t] + 16y'[t] + 1025y[t] == 0, y[0] == 1,
  y'[0] == 3}, y[t], t]
```

$\{\{y[t] \to \frac{1}{32}e^{-t/8}(32\text{Cos}[4t] + 25\text{Sin}[4t])\}\}$ □

Application: Harmonic Motion

Suppose that a mass is attached to an elastic spring that is suspended from a rigid support such as a ceiling. According to Hooke's law, the spring exerts a restoring force in the upward direction that is proportional to the displacement of the spring.

> ***Hooke's Law:*** *$F = ks$, where $k > 0$ is the constant of proportionality or spring constant, and s is the displacement of the spring.*

Using Hooke's law and assuming that $x(t)$ represents the displacement of the mass from the equilibrium position at time t, we obtain the initial-value problem

$$m\frac{d^2x}{dt^2} + kx = 0, \quad x(0) = \alpha, \quad \frac{dx}{dt}(0) = \beta.$$

Note that the initial conditions give the initial displacement and velocity, respectively. This differential equation disregards all retarding forces acting on the motion of the mass and a more realistic model which takes these forces into account is needed. Studies in mechanics reveal that resistive forces due to damping are proportional to a power of the velocity of the motion. Hence, $F_R = a\,dx/dt$ or $F_R = a\,(dx/dt)^3$, where $a > 0$, are typically used to represent the damping force. Then, we have the following initial-value problem assuming that $F_R = a\,dx/dt$:

$$m\frac{d^2x}{dt^2} + a\frac{dx}{dt} + kx = 0, \quad x(0) = \alpha, \quad \frac{dx}{dt}(0) = \beta.$$

Problems of this type are characterized by the value of $a^2 - 4mk$ as follows.

1. $a^2 - 4mk > 0$. This situation is said to be **overdamped** because the damping coefficient a is large in comparison to the spring constant k.
2. $a^2 - 4mk = 0$. This situation is described as **critically damped** because the resulting motion is oscillatory with a slight decrease in the damping coefficient a.
3. $a^2 - 4mk < 0$. This situation is called **underdamped** because the damping coefficient a is small in comparison with the spring constant k.

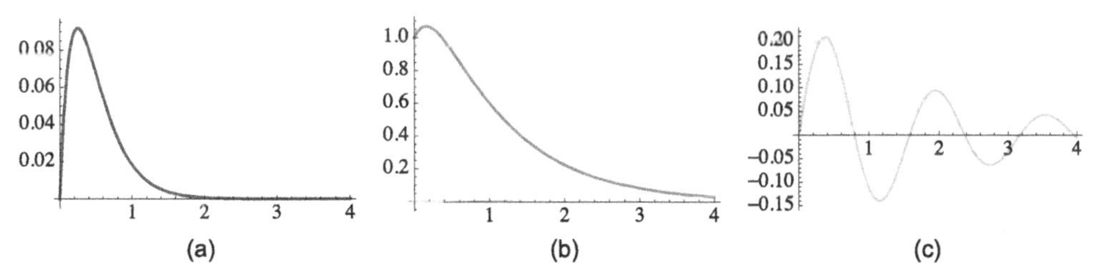

FIGURE 6.13 (a) Critically damped motion. (b) Overdamped motion. (c) Underdamped motion. (Iowa State University colors)

Example 6.11

Classify the following differential equations as overdamped, underdamped, or critically damped. Also, solve the corresponding initial-value problem using the given initial conditions and investigate the behavior of the solutions.

(a) $\frac{d^2x}{dt^2}+8\frac{dx}{dt}+16x=0$ subject to $x(0)=0$ and $\frac{dx}{dt}(0)=1$;

(b) $\frac{d^2x}{dt^2}+5\frac{dx}{dt}+4x=0$ subject to $x(0)=1$ and $\frac{dx}{dt}(0)=1$; and

(c) $\frac{d^2x}{dt^2}+\frac{dx}{dt}+16x=0$ subject to $x(0)=0$ and $\frac{dx}{dt}(0)=1$.

Solution. For (a), we identify $m=1$, $a=8$, and $k=16$ so that $a^2-4mk=0$, which means that the differential equation $x''+8x'+16x=0$ is critically damped. After defining `de1`, we solve the equation subject to the initial conditions and name the resulting output `sol1`. We then graph the solution shown in Fig. 6.13 (a).

```
Clear[de1, x, t]

de1 = x''[t] + 8x'[t] + 16x[t]==0;

sol1 = DSolve[{de1, x[0]==0, x'[0]==1}, x[t], t]
```

$\{\{x[t]\to e^{-4t}t\}\}$

```
p1 = Plot[sol1[[1, 1, 2]], {t, 0, 4},
   PlotStyle → {{Thickness[.01], CMYKColor[.02, 1, .85, .06]}}]
```

For (b), we proceed in the same manner. We identify $m=1$, $a=5$, and $k=4$ so that $a^2-4mk=9$ and the equation $x''+5x'+4x=0$ is overdamped. We then define `de2` to be the equation and the solution to the initial-value problem obtained with `DSolve`, `sol2`, and then graph $x(t)$ on the interval $[0,4]$ in Fig. 6.13 (b).

```
Clear[de2, x, t]

de2 = x''[t] + 5x'[t] + 4x[t]==0;

sol2 = DSolve[{de2, x[0]==1, x'[0]==1}, x[t], t]
```

$\{\{x[t]\to \frac{1}{3}e^{-4t}(-2+5e^{3t})\}\}$

```
p2 = Plot[sol2[[1, 1, 2]], {t, 0, 4},
```

```
PlotStyle → {{Thickness[.01], CMYKColor[.21, .15, .54, .31]}}]
```

For (c), we proceed in the same manner as in (a) and (b) to show that the equation is under-damped because the value of $a^2 - 4mk$ is -63. See Fig. 6.13 (c).

```
Clear[de3, x, t]
de3 = x''[t] + x'[t] + 16x[t]==0;
sol3 = DSolve[{de3, x[0]==0, x'[0]==1}, x[t], t]
```

$\{\{x[t] \to \frac{2e^{-t/2}\mathrm{Sin}[\frac{3\sqrt{7}t}{2}]}{3\sqrt{7}}\}\}$

```
p3 = Plot[sol3[[1, 1, 2]], {t, 0, 4},
  PlotStyle → {{Thickness[.01], CMYKColor[0, .24, .78, 0]}}]
Show[GraphicsRow[{p1, p2, p3}]]
```

□

You can also use `Manipulate` to help you visualize harmonic motion. With

```
Manipulate[
sol = DSolve[{mx''[t] + ax'[t] + kx[t] == 0, x[0] == 0, x'[0] == 1}, x[t], t];
  Plot[x[t]/.sol, {t, 0, 5}, PlotRange → {−1/2, 1/2}, AspectRatio → 1,
  PlotStyle → {{Thickness[.01], CMYKColor[.02, 1, .85, .06]}}], {{m, 1}, 0, 5},
  {{a, 8}, 0, 15, 1}, {{k, 16}, 0, 20, 1}]
```

we generate a `Manipulate` object that lets us investigate harmonic motion for various values of m, a, and k if the initial position is zero ($x(0) = 0$) and the initial velocity is one ($x'(0) = 1$). See Fig. 6.14. (Note that m is centered at 1, a at 8, and k at 16.)

6.2.3 Undetermined Coefficients

If equation (6.7) has constant coefficients and $f(t)$ is a product of terms t^n, $e^{\alpha t}$, α constant, $\cos\beta t$, and/or $\sin\beta t$, β constant, *undetermined coefficients* can often be used to find a particular solution of equation (6.7). The key to implementing the method is to *judiciously* choose the correct form of y_p.

Assume that a general solution, y_h, of the corresponding homogeneous equation has been found and that each term of $f(t)$ has the form

$$t^n e^{\alpha t}\cos\beta t \qquad \text{or} \qquad t^n e^{\alpha t}\sin\beta t.$$

For *each* term of $f(t)$, write down the *associated set*

$$F = \left\{t^n e^{\alpha t}\cos\beta t, t^n e^{\alpha t}\sin\beta t, t^{n-1}e^{\alpha t}\cos\beta t, t^{n-1}e^{\alpha t}\sin\beta t, \ldots, e^{\alpha t}\cos\beta t, e^{\alpha t}\sin\beta t, \right\}.$$

If any element of F is a solution to the corresponding homogeneous equation, multiply each element of F by t^m, where m is the smallest positive integer so that none of the elements of $t^m F$ are solutions to the corresponding homogeneous equation. A particular solution will be a linear combination of the functions in all the F's.

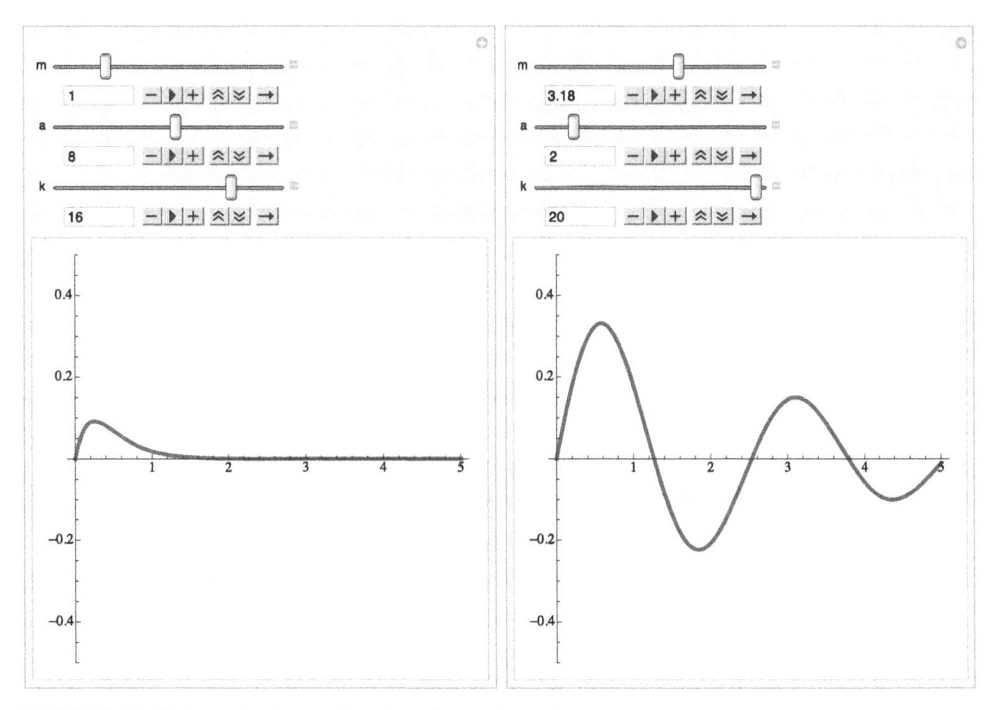

FIGURE 6.14 Using `Manipulate` to investigate harmonic motion.

Example 6.12

Solve $4\frac{d^2y}{dt^2} - y = t - 2 - 5\cos t - e^{-t/2}$.

Solution. The corresponding homogeneous equation is $4y'' - y = 0$ with general solution $y_h = c_1 e^{-t/2} + c_2 e^{t/2}$.

Clear[y, e1, e2, e3, e4, e5, eqn]

DSolve[4y''[t] − y[t] == 0, y[t], t]

$\{\{y[t] \to e^{t/2}C[1] + e^{-t/2}C[2]\}\}$

A fundamental set of solutions for the corresponding homogeneous equation is $S = \{e^{-t/2}, e^{t/2}\}$. The associated set of functions for $t - 2$ is $F_1 = \{1, t\}$, the associated set of functions for $-5\cos t$ is $F_2 = \{\cos t, \sin t\}$, and the associated set of functions for $-e^{-t/2}$ is $F_3 = \{e^{-t/2}\}$. Note that $e^{-t/2}$ is an element of S so we multiply F_3 by t resulting in $tF_3 = \{te^{-t/2}\}$. Then, we search for a particular solution of the form

$$y_p = A + Bt + C\cos t + D\sin t + Ete^{-t/2},$$

where A, B, C, D, and E are constants to be determined.

No element of F_1 is contained in S and no element of F_2 is contained in S.

We don't use capital letters to avoid any confusion with built-in Mathematica commands.

yp[t_] = *a* + *bt* + *a*Cos[*t*] + *d*Sin[*t*] + *et*Exp[−*t*/2]

$a + bt + ee^{-t/2}t + a\text{Cos}[t] + d\text{Sin}[t]$

Computing y_p' and y_p''

dyp = yp'[t]

$b + ee^{-t/2} - \frac{1}{2}ee^{-t/2}t + d\text{Cos}[t] - a\text{Sin}[t]$

d2yp = yp''[t]

$-ee^{-t/2} + \frac{1}{4}ee^{-t/2}t - a\text{Cos}[t] - d\text{Sin}[t]$

and substituting into the nonhomogeneous equation results in

$$-A - Bt - 5C\cos t - 5D\sin t - 4Ee^{-t/2} = t - 2 - 5\cos t - e^{-t/2}.$$

eqn = 4yp''[t] − yp[t] == t − 2 − 5Cos[t] − Exp[−t/2]

$-a - bt - ee^{-t/2}t - a\text{Cos}[t] - d\text{Sin}[t] + 4(-ee^{-t/2} + \frac{1}{4}ee^{-t/2}t - a\text{Cos}[t] - d\text{Sin}[t]) ==$
$-2 - e^{-t/2} + t - 5\text{Cos}[t]$

Equating coefficients results in

$$-A = -2 \qquad -B = 1 \qquad -5C = -5 \qquad -5D = 0 \qquad -4E = -1$$

so $A = 2$, $B = -1$, $C = 1$, $D = 0$, and $E = 1/4$.

cvals = Solve[{−a == −2, −b == 1, −5c == −5, −5d == 0, −4e == −1}]

$\{\{a \to 2, b \to -1, c \to 1, d \to 0, e \to \frac{1}{4}\}\}$

y_p is then given by $y_p = 2 - t + \cos t + \frac{1}{4}te^{-t/2}$

yp[t]/.cvals[[1]]

$2 - t + \frac{1}{4}e^{-t/2}t + 2\text{Cos}[t]$

and a general solution is given by

$$y = y_h + y_p = c_1e^{-t/2} + c_2e^{t/2} + 2 - t + \cos t + \frac{1}{4}te^{-t/2}.$$

We check our result with `DSolve` and `Simplify`.

DSolve[4y''[t] − y[t]==t − 2 − 5Cos[t] − Exp[−t/2],

y[t], t]//Simplify

$\{\{y[t] \to \frac{1}{4}e^{-t/2}(1 - 4e^{t/2}(-2 + t) + t + 4e^{t}C[1] + 4C[2] + 4e^{t/2}\text{Cos}[t])\}\}$ □

Example 6.13

Solve $y'' + 4y = \cos 2t$, $y(0) = 0$, $y'(0) = 0$.

Solution. A general solution of the corresponding homogeneous equation is $y_h = c_1 \cos 2t + c_2 \sin 2t$. For this equation, $F = \{\cos 2t, \sin 2t\}$. Because elements of F are solutions to the corresponding homogeneous equation, we multiply each element of F by t resulting in $tF = \{t \cos 2t, t \sin 2t\}$. Therefore, we assume that a particular solution has the form

$$y_p = At \cos 2t + Bt \sin 2t,$$

where A and B are constants to be determined. Proceeding in the same manner as before, we compute y_p' and y_p''

```
yp[t_] = atCos[2t] + btSin[2t]
yp'[t]
yp''[t]
```

$at\text{Cos}[2t] + bt\text{Sin}[2t]$

$a\text{Cos}[2t] + 2bt\text{Cos}[2t] + b\text{Sin}[2t] - 2at\text{Sin}[2t]$

$4b\text{Cos}[2t] - 4at\text{Cos}[2t] - 4a\text{Sin}[2t] - 4bt\text{Sin}[2t]$

and then substitute into the nonhomogeneous equation.

```
eqn = yp''[t] + 4yp[t] == Cos[2t]
```

$4b\text{Cos}[2t] - 4at\text{Cos}[2t] - 4a\text{Sin}[2t] - 4bt\text{Sin}[2t] + 4(at\text{Cos}[2t] + bt\text{Sin}[2t]) == \text{Cos}[2t]$

Equating coefficients readily yields $A = 0$ and $B = 1/4$. Alternatively, remember that $-4A \sin 2t + 4B \cos 2t = \cos 2t$ is true for *all* values of t. Evaluating for two values of t and then solving for A and B gives the same result.

```
e1 = eqn/.t -> 0
e2 = eqn/.t -> 1
cvals = Solve[{e1, e2}]
```

$4b == 1$

$-4a\text{Cos}[2] + 4b\text{Cos}[2] - 4a\text{Sin}[2] - 4b\text{Sin}[2] + 4(a\text{Cos}[2] + b\text{Sin}[2]) == \text{Cos}[2]$

$\{\{a \to 0, b \to \frac{1}{4}\}\}$

It follows that $y_p = \frac{1}{4}t \sin 2t$ and $y = c_1 \cos 2t + c_2 \sin 2t + \frac{1}{4}t \sin 2t$.

```
yp[t]/.cvals[[1]]
```

$\frac{1}{4}t\text{Sin}[2t]$

y[t_] = c1Cos[2*t*] + c2Sin[2*t*] + 1/4*t*Sin[2*t*]

$\text{c1Cos}[2t] + \text{c2Sin}[2t] + \frac{1}{4}t\text{Sin}[2t]$

Applying the initial conditions after finding y'

y′[*t*]

$2\text{c2Cos}[2t] + \frac{1}{2}t\text{Cos}[2t] + \frac{1}{4}\text{Sin}[2t] - 2\text{c1Sin}[2t]$

cvals = Solve[{y[0] == 0, y′[0] == 0}]

$\{\{\text{c1} \to 0, \text{c2} \to 0\}\}$

results in $y = \frac{1}{4}t \sin 2t$, which we graph with `Plot` in Fig. 6.15.

y[*t*]/.cvals[[1]]

$\frac{1}{4}t\text{Sin}[2t]$

Plot[y[*t*]/.cvals, {*t*, 0, 16Pi},

PlotStyle → {{Thickness[.0075], CMYKColor[.82, 1, 0, .12]}}]

We verify the calculation with `DSolve`.

Clear[y]

DSolve[{y"[*t*] + 4y[*t*] == Cos[2*t*], y[0] == 0, y′[0] == 0},

y[*t*], *t*]//Simplify

$\{\{y[t] \to \frac{1}{4}t\text{Sin}[2t]\}\}$ □

Use `Manipulate` to help you see how changing parameter values and equations affect a system. With

Manipulate[

sol1 = DSolve[{*mx*"[*t*] + *ax*′[*t*] + *kx*[*t*] == ΓCos[*ωt*], *x*[0] == 0, *x*′[0] == 0}, *x*[*t*], *t*];

sol2 = NDSolve[{*mx*"[*t*] + *ax*′[*t*] + *k*Sin[*x*[*t*]] == ΓCos[*ωt*], *x*[0] == 0, *x*′[0] == 0},

***x*[*t*], {*t*, 0, 50}];**

p1 = Plot[*x*[*t*]/.sol1, {*t*, 0, 50}, PlotRange → {−5, 5}, AspectRatio → 1,

PlotStyle → {{Thickness[.0075], CMYKColor[.82, 1, 0, .12]}}];

p2 = Plot[*x*[*t*]/.sol2, {*t*, 0, 50}, PlotRange → {−5, 5}, AspectRatio → 1,

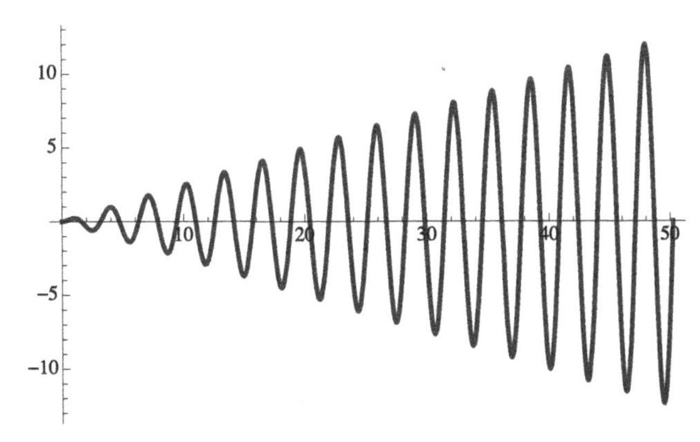

FIGURE 6.15 The forcing function causes the solution to become unbounded as $t \to \infty$. (Kansas State University colors)

```
PlotStyle → {{Thickness[.0075], CMYKColor[.82, 1, 0, .12]}}];
Show[GraphicsRow[{p1, p2}]], {{m, 1}, 0, 5}, {{a, 0}, 0, 15, 1}, {{k, 4}, 0, 20, 1},
  {{ω, 2}, 0, 20, 1}, {{Γ, 1}, 0, 10, 1}]
```

we can compare the solution of $mx'' + ax' + kx = \Gamma \cos \omega t$, $x(0) = 0$, $x'(0) = 0$ to the solution of $mx'' + ax' + k \sin x = \Gamma \cos \omega t$, $x(0) = 0$, $x'(0) = 0$ for various values of m, a, k, ω, and Γ. See Fig. 6.16.

Example 6.14: Hearing Beats and Resonance

In order to *hear* beats and resonance, we solve the initial-value problem

$$x'' + \omega^2 x = F \cos \beta t, \quad x(0) = \alpha, \quad x'(0) = \beta, \tag{6.13}$$

for each of the following parameter values: (a) $\omega^2 = 6000^2$, $\beta = 5991.62$, $F = 2$; and (a) $\omega^2 = 6000^2$, $\beta = 6000$, $F = 2$.

First, we define the function `sol` which when given the parameters, solves the initial-value problem (6.13).

```
Clear[x, t, f, sol]

sol[ω_, β_, f_]:=
DSolve[{x''[t] + ω^2x[t] == fCos[βt], x[0] == 0,
x'[0] == 0}, x[t], t][[1, 1, 2]]
```

Thus, our solution for (a) is obtained by entering

```
a = sol[6000, 5991.62, 2]

0.0000198886(−1.0007Cos[6000.t] + 1.Cos[8.38t]Cos[6000.t]
+ 0.000698821Cos[6000.t]Cos[11991.6t] + 1.Sin[8.38t]Sin[6000.t]
+ 0.000698821Sin[6000.t]Sin[11991.6t])
```

To *hear* the function we use `Play` in the same way that we use `Plot` to *see* functions.

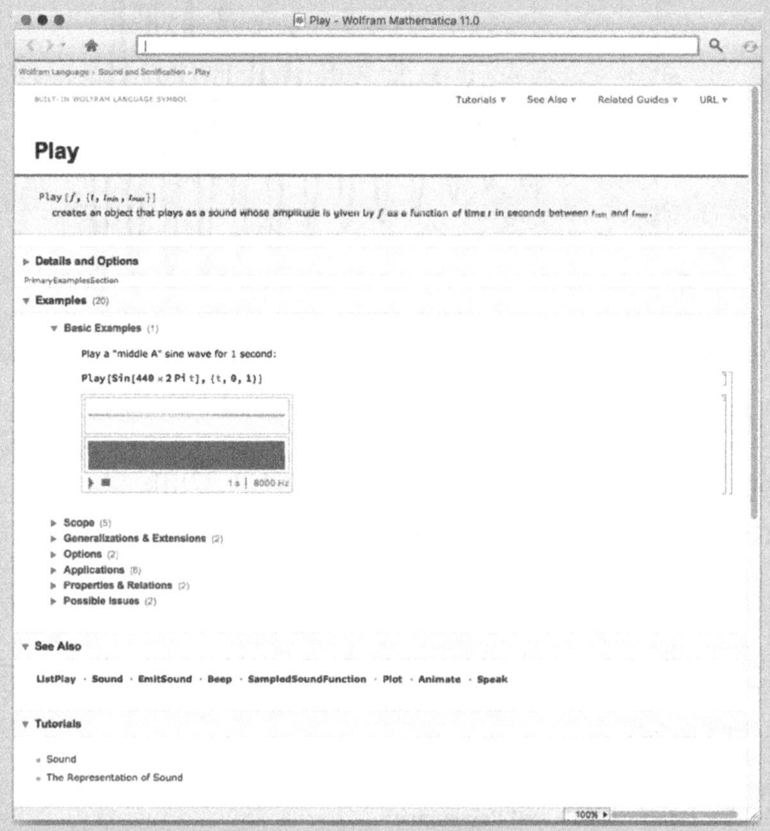

Play - Wolfram Mathematica 11.0

Play

Play[*f*, {*t*, t_{min}, t_{max}}]
creates an object that plays as a sound whose amplitude is given by *f* as a function of time *t* in seconds between t_{min} and t_{max}.

▸ Details and Options
▾ Examples (20)
▾ Basic Examples (1)
Play a "middle A" sine wave for 1 second:
Play[Sin[440 × 2 Pi t], {t, 0, 1}]
▸ Scope (5)
▸ Generalizations & Extensions (2)
▸ Options (2)
▸ Applications (6)
▸ Properties & Relations (2)
▸ Possible Issues (2)
▾ See Also
ListPlay · Sound · EmitSound · Beep · SampledSoundFunction · Plot · Animate · Speak
▾ Tutorials
Sound
The Representation of Sound

The values of a correspond to the amplitude of the sound as a function of time. See Fig. 6.17 (a).

Play[*a*, {*t*, 0, 6}]

Similarly, the solution for (b) is obtained by entering

b **= sol[6000, 6000, 2]**

$$\frac{-\text{Cos}[6000t]+\text{Cos}[6000t]\text{Cos}[12000t]+12000t\,\text{Sin}[6000t]+\text{Sin}[6000t]\text{Sin}[12000t]}{72000000}$$

We hear resonance with `Play`. See Fig. 6.17 (b).

Play[*b*, {*t*, 0, 6}]

6.2.4 Variation of Parameters

A particular solution, y_p, is a solution that does not contain any arbitrary constants.

Let $S = \{y_1, y_2\}$ be a fundamental set of solutions for equation (6.9). To solve the nonhomogeneous equation (6.8), we need to find a particular solution, y_p of equation (6.8). We search

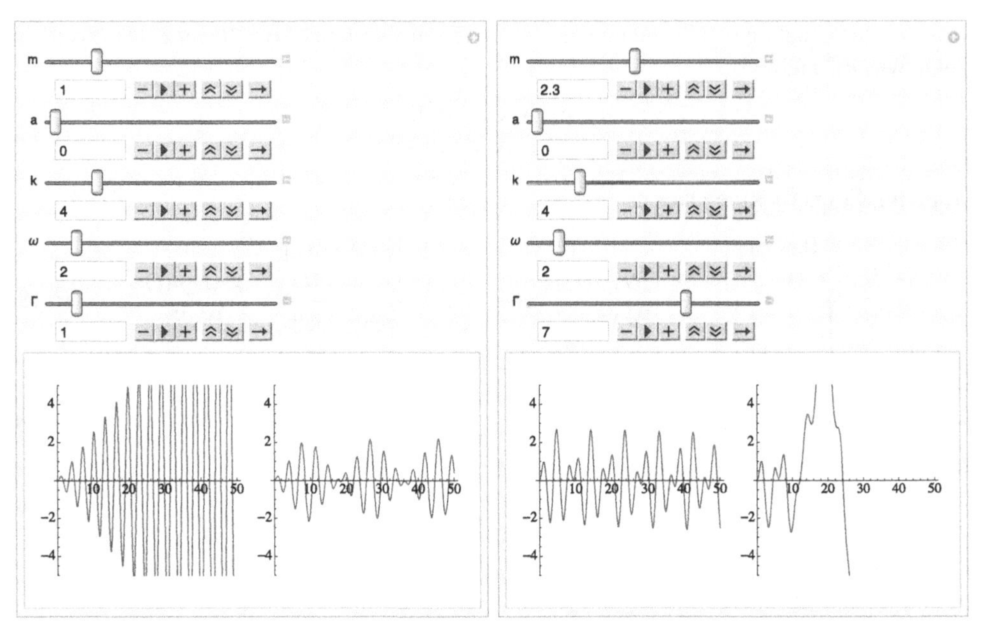

FIGURE 6.16 Comparing solutions of nonlinear initial-value problems to their corresponding linear approximations.

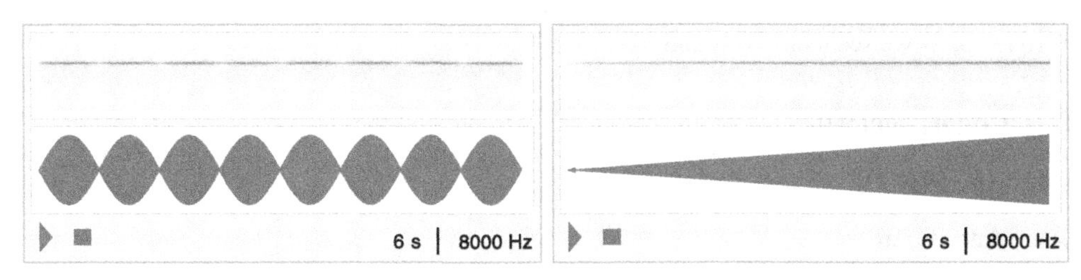

FIGURE 6.17 Hearing and seeing beats and resonance: (a) Beats. (b) Resonance.

for a particular solution of the form

$$y_p = u_1(t)y_1(t) + u_2(t)y_2(t), \tag{6.14}$$

where u_1 and u_2 are functions of t. Differentiating equation (6.14) gives us

$$y_p{}' = u_1{}'y_1 + u_1 y_1{}' + u_2{}'y_2 + u_2 y_2{}'.$$

Assuming that

$$y_1 u_1{}' + y_2 u_2{}' = 0 \tag{6.15}$$

results in $y_p{}' = u_1 y_1{}' + u_2 y_2{}'$. Computing the second derivative then yields

$$y_p{}'' = u_1{}'y_1{}' + u_1 y_1{}'' + u_2{}'y_2{}' + u_2 y_2{}''.$$

Substituting y_p, $y_p{}'$, and $y_p{}''$ into equation (6.8) and using the facts that

$$u_1\left(y_1{}'' + p\,y_1{}' + q\,y_1\right) = 0 \quad \text{and} \quad u_2\left(y_2{}'' + p\,y_2{}' + q\,y_2\right) = 0$$

(because y_1 and y_2 are solutions to the corresponding homogeneous equation) results in

$$\begin{aligned}\frac{d^2y_p}{dt^2} + p(t)\frac{dy_p}{dt} + q(t)y_p &= u_1{}'y_1{}' + u_1 y_1{}'' + u_2{}'y_2{}' + u_2 y_2{}'' + p(t)\left(u_1 y_1{}' + u_2 y_2{}'\right) \\ &\quad + q(t)\left(u_1 y_1 + u_2 y_2\right) \\ &= y_1{}'u_1{}' + y_2{}'u_2{}' = f(t).\end{aligned} \tag{6.16}$$

Key concept: A general solution of the nonhomogeneous linear equation is $y = y_h + y_p$, where y_h is a general solution of the corresponding homogeneous equation and y_p is a particular solution of the nonhomogeneous equation.

Observe that it is pointless to search for solutions of the form $y_p = c_1 y_1 + c_2 y_2$ where c_1 and c_2 are constants because for every choice of c_1 and c_2, $c_1 y_1 + c_2 y_2$ is a solution to the corresponding homogeneous equation.

Observe that equation (6.15) and equation (6.16) form a system of two linear equations in the unknowns u_1' and u_2':

$$\begin{aligned} y_1 u_1' + y_2 u_2' &= 0 \\ y_1' u_1' + y_2' u_2' &= f(t). \end{aligned} \tag{6.17}$$

Applying Cramer's Rule gives us

$$u_1' = \frac{\begin{vmatrix} 0 & y_2 \\ f(t) & y_2' \end{vmatrix}}{\begin{vmatrix} y_1 & y_2 \\ y_1' & y_2' \end{vmatrix}} = -\frac{y_2(t)f(t)}{W(S)} \quad \text{and} \quad u_2' = \frac{\begin{vmatrix} y_1 & 0 \\ y_1' & f(t) \end{vmatrix}}{\begin{vmatrix} y_1 & y_2 \\ y_1' & y_2' \end{vmatrix}} = \frac{y_1(t)f(t)}{W(S)}, \tag{6.18}$$

where $W(S)$ is the Wronskian, $W(S) = \begin{vmatrix} y_1 & y_2 \\ y_1' & y_2' \end{vmatrix}$. After integrating to obtain u_1 and u_2, we form y_p and then a general solution, $y = y_h + y_p$.

Example 6.15

Solve $y'' + 9y = \sec 3t$, $y(0) = 0$, $y'(0) = 0$, $0 \le t < \pi/6$.

Solution. The corresponding homogeneous equation is $y'' + 9y = 0$ with general solution $y_h = c_1 \cos 3t + c_2 \sin 3t$. Then, a fundamental set of solutions is $S = \{\cos 3t, \sin 3t\}$ and $W(S) = 3$, as we see using `Det`, and `Simplify`.

```
fs = {Cos[3t], Sin[3t]};
wm = {fs, D[fs, t]};
wm//MatrixForm
wd = Det[wm]//Simplify
```

$$\begin{pmatrix} \text{Cos}[3t] & \text{Sin}[3t] \\ -3\text{Sin}[3t] & 3\text{Cos}[3t] \end{pmatrix}$$

3

We use equation (6.18) to find $u_1 = \frac{1}{9} \ln \cos 3t$ and $u_2 = \frac{1}{3}t$.

```
u1 = Integrate[-Sin[3t]Sec[3t]/3, t]
u2 = Integrate[Cos[3t]Sec[3t]/3, t]
```

$\frac{1}{9}$Log[Cos[3t]]

$\frac{t}{3}$

Absolute value is not needed in the antiderivatives because we are restricting the domain to $0 \le t < \pi/6$ and $\cos t > 0$ on this interval.

It follows that a particular solution of the nonhomogeneous equation is $y_p = \frac{1}{9} \cos 3t \times \ln \cos 3t + \frac{1}{3} t \sin 3t$ and a general solution is $y = y_h + y_p = c_1 \cos 3t + c_2 \sin 3t + \frac{1}{9} \cos 3t \ln \cos 3t + \frac{1}{3} t \sin 3t$.

```
yp = u1Cos[3t] + u2Sin[3t]
```

$\frac{1}{9}$Cos[3t]Log[Cos[3t]] + $\frac{1}{3}t$Sin[3t]

Identical results are obtained using `DSolve`.

The negative sign in the output does not affect the result because `C[1]` is arbitrary.

DSolve[y''[t] + 9y[t] == Sec[3t], y[t], t]

{{y[t] → C[1]Cos[3t] + C[2]Sin[3t] + $\frac{1}{9}$(Cos[3t]Log[Cos[3t]] + 3tSin[3t])}}

Applying the initial conditions gives us $c_1 = c_2 = 0$ so we conclude that the solution to the initial-value problem is $y = \frac{1}{9}\cos 3t\ \ln\cos 3t + \frac{1}{3}t\sin 3t$.

sol = DSolve[{y''[t] + 9y[t] == Sec[3t], y[0] == 0, y'[0] == 0}, y[t], t]

{{y[t] → $\frac{1}{9}$(Cos[3t]Log[Cos[3t]] + 3tSin[3t])}}

We graph the solution with `Plot` in Fig. 6.18.

Plot[y[t]/.sol, {t, 0, Pi/6},

PlotStyle → {{Thickness[.01], CMYKColor[1, .75, 0, 0]}}] □

6.3 HIGHER-ORDER LINEAR EQUATIONS

6.3.1 Basic Theory

The **standard form of the nth-order linear equation** is

$$\frac{d^n y}{dt^n} + a_{n-1}(t)\frac{d^{n-1}y}{dt^{n-1}} + \cdots + a_1(t)\frac{dy}{dt} + a_0(t)y = f(t). \tag{6.19}$$

The **corresponding homogeneous equation** of equation (6.19) is

$$\frac{d^n y}{dt^n} + a_{n-1}(t)\frac{d^{n-1}y}{dt^{n-1}} + \cdots + a_1(t)\frac{dy}{dt} + a_0(t)y = 0. \tag{6.20}$$

Let $y_1, y_2, \ldots, y_n$ be n solutions of equation (6.20). The set $S = \{y_1, y_2, \ldots, y_n\}$ is **linearly independent** if and only if the **Wronskian**,

$$W(S) = \begin{vmatrix} y_1 & y_2 & y_3 & \cdots & y_n \\ y_1{}' & y_2{}' & y_3{}' & \cdots & y_n{}' \\ y_1{}'' & y_2{}'' & y_3{}'' & \cdots & y_n{}'' \\ y_1{}^{(3)} & y_2{}^{(3)} & y_3{}^{(3)} & \cdots & y_n{}^{(3)} \\ \vdots & \vdots & \vdots & \cdots & \vdots \\ y_1{}^{(n-1)} & y_2{}^{(n-1)} & y_3{}^{(n-1)} & \cdots & y_n{}^{(n-1)} \end{vmatrix} \tag{6.21}$$

is not identically the zero function. S is **linearly dependent** if S is not linearly independent.

If $y_1, y_2, \ldots, y_n$ are n linearly independent solutions of equation (6.20), we say that $S = \{y_1, y_2, \ldots, y_n\}$ is a **fundamental set** for equation (6.20) and a **general solution** of equation (6.20) is $y = c_1y_1 + c_2y_2 + c_3y_3 + \cdots + c_ny_n$.

A **general solution** of equation (6.19) is $y = y_h + y_p$ where y_h is a general solution of the corresponding homogeneous equation and y_p is a particular solution of equation (6.19).

Key concept: A general solution of the nonhomogeneous linear equation is $y = y_h + y_p$, where y_h is a general solution of the corresponding homogeneous equation and y_p is a particular solution of the nonhomogeneous equation.

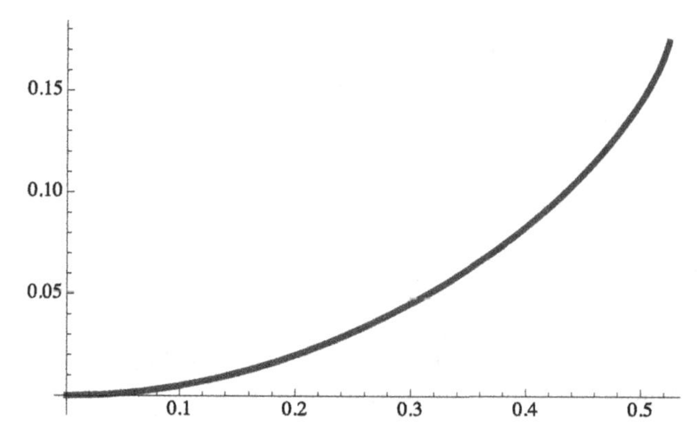

FIGURE 6.18 The domain of the solution is $0 \le t < \pi/6$. (University of Kentucky colors)

6.3.2 Constant Coefficients

If

$$\frac{d^n y}{dt^n} + a_{n-1}\frac{d^{n-1} y}{dt^{n-1}} + \cdots + a_1 \frac{dy}{dt} + a_0 y = 0$$

has real constant coefficients, we assume that $y = e^{kt}$ and find that k satisfies the **characteristic equation**

$$k^n + a_{n-1}k^{n-1} + \cdots + a_1 k + a_0 = 0. \tag{6.22}$$

If a solution k of equation (6.22) has multiplicity m, m linearly independent solutions corresponding to k are

$$e^{kt}, te^{kt}, \ldots, t^{m-1}e^{kt}.$$

If a solution $k = \alpha + \beta i$, $\beta \ne 0$, of equation (6.22) has multiplicity m, $2m$ linearly independent solutions corresponding to $k = \alpha + \beta i$ (and $k = \alpha - \beta i$) are

$$e^{\alpha t}\cos\beta t, e^{\alpha t}\sin\beta t, te^{\alpha t}\cos\beta t, te^{\alpha t}\sin\beta t, \ldots, t^{m-1}e^{\alpha t}\cos\beta t, t^{m-1}e^{\alpha t}\sin\beta t.$$

Example 6.16

Solve $12y''' - 5y'' - 6y' - y = 0$.

Solution. The characteristic equation is

$$12k^3 - 5k^2 - 6k - 1 = (k-1)(3k+1)(4k+1) = 0$$

with solutions $k_1 = -1/3$, $k_2 = -1/4$, and $k_3 = 1$.

`Factor[expression]` attempts to factor `expression`.

Factor[12*k*^3 − 5*k*^2 − 6*k* − 1]

$(-1+k)(1+3k)(1+4k)$

Thus, three linearly independent solutions of the equation are $y_1 = e^{-t/3}$, $y_2 = e^{-t/4}$, and $y_3 = e^t$; a general solution is $y = c_1 e^{-t/3} + c_2 e^{-t/4} + c_3 e^t$. We check with `DSolve`.

Clear[y]

DSolve[12y'''[*t*] − 5y''[*t*] − 6y'[*t*] − y[*t*] == 0, y[*t*], *t*]

$\{\{y[t] \to e^{-t/4}C[1] + e^{-t/3}C[2] + e^t C[3]\}\}$ □

Example 6.17

Solve $y''' + 4y' = 0$, $y(0) = 0$, $y'(0) = 1$, $y''(0) = -1$.

Solution. The characteristic equation is $k^3 + 4k = k(k^2 + 4) = 0$ with solutions $k_1 = 0$ and $k_{2,3} = \pm 2i$ that are found with `Solve`.

Enter `?Solve` to obtain basic help regarding the `Solve` function.

Solve[k^3 + 4k == 0]

$\{\{k \to 0\}, \{k \to -2i\}, \{k \to 2i\}\}$

Three linearly independent solutions of the equation are $y_1 = 1$, $y_2 = \cos 2t$, and $y_3 = \sin 2t$. A general solution is $y = c_1 + c_2 \sin 2t + c_3 \cos 2t$.

gensol = DSolve[y'''[t] + 4y'[t] == 0, y[t], t]

$\{\{y[t] \to C[3] - \frac{1}{2}C[2]\text{Cos}[2t] + \frac{1}{2}C[1]\text{Sin}[2t]\}\}$

Application of the initial conditions shows us that $c_1 = -1/4$, $c_2 = 1/2$, and $c_3 = 1/4$ so the solution to the initial-value problem is $y = -\frac{1}{4} + \frac{1}{2}\sin 2t + \frac{1}{4}\cos 2t$. We verify the computation with `DSolve` and graph the result with `Plot` in Fig. 6.19.

e1 = y[t]/.gensol[[1]]/.t → 0

$-\frac{C[2]}{2} + C[3]$

e2 = D[y[t]/.gensol[[1]], t]/.t → 0

e3 = D[y[t]/.gensol[[1]], {t, 2}]/.t → 0

$C[1]$

$2C[2]$

cvals = Solve[{e1 == 0, e2 == 1, e3 == −1}]

$\{\{C[1] \to 1, C[2] \to -\frac{1}{2}, C[3] \to -\frac{1}{4}\}\}$

Clear[y]

partsol = DSolve[{y'''[t] + 4y'[t] == 0, y[0]==0,

y'[0] == 1, y''[0] == −1}, y[t], t]

$\{\{y[t] \to \frac{1}{4}(-1 + \text{Cos}[2t] + 2\text{Sin}[2t])\}\}$

Plot[y[t]/.partsol, {t, 0, 2Pi}, AspectRatio → Automatic,

PlotStyle → {{Thickness[.01], CMYKColor[0, 1, 1, .15]}}]

□

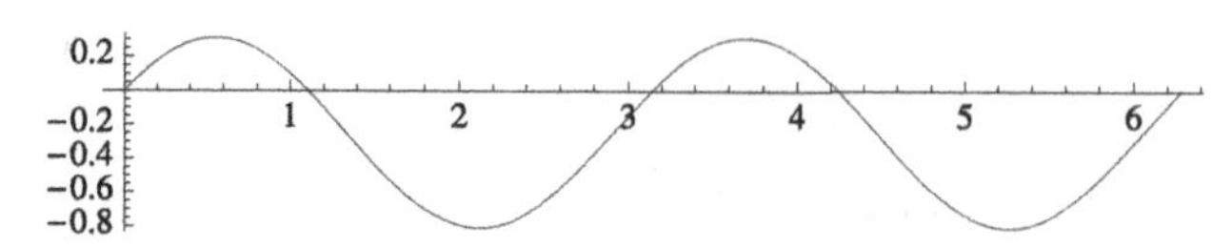

FIGURE 6.19 Graph of $y=-\frac{1}{4}+\frac{1}{2}\sin 2t+\frac{1}{4}\cos 2t$. (University of Louisiana colors)

Example 6.18

Find a linear homogeneous differential equation with constant coefficients that has general solution $y=c_1e^{-2t/3}+c_2te^{-2t/3}+c_3t^2e^{-2t/3}+c_4\cos t+c_5\sin t+c_6t\cos t+c_7t\sin t+c_8t^2\cos t+c_9t^2\sin t$.

Solution. A linear homogeneous differential equation with constant coefficients that has this general solution has fundamental set of solutions

$$S=\left\{e^{-2t/3},te^{-2t/3},t^2e^{-2t/3},\cos t,\sin t,t\cos t,t\sin t,t^2\cos t,t^2\sin t\right\}.$$

Hence, in the characteristic equation $k=-2/3$ has multiplicity 3 while $k=\pm i$ has multiplicity 3. The characteristic equation is

$$27\left(k+\frac{2}{3}\right)^3(k-i)^3(k+i)^3=k^9+2k^8+\frac{13}{3}k^7+\frac{170}{27}k^6+7k^5+\frac{62}{9}k^4 +5k^3+\frac{26}{9}k^2+\frac{4}{3}k+\frac{8}{27},$$

where we use Mathematica to compute the multiplication with `Expand`.

```
Expand[27(k + 2/3)^3(k^2 + 1)^3]
```

$8+36k+78k^2+135k^3+186k^4+189k^5+170k^6+117k^7+54k^8+27k^9$

Thus, a linear homogeneous differential equation with constant coefficients obtained after dividing by 27 with the indicated general solution is

$$\frac{d^9y}{dt^9}+2\frac{d^8y}{dt^8}+\frac{13}{3}\frac{d^7y}{dt^7}+\frac{170}{27}\frac{d^6y}{dt^6}+7\frac{d^5y}{dt^5}+\frac{62}{9}\frac{d^4y}{dt^4} +5\frac{d^3y}{dt^3}+\frac{26}{9}\frac{d^2y}{dt^2}+\frac{4}{3}\frac{dy}{dt}+\frac{8}{27}y=0.$$ □

6.3.3 Undetermined Coefficients

For higher-order linear equations with constant coefficients, the method of undetermined coefficients is the same as for second-order equations discussed in Section 6.2.3, provided that the forcing function involves the terms discussed in Section 6.2.3.

Example 6.19

Solve $\frac{d^3y}{dt^3}+\frac{2}{3}\frac{d^2y}{dt^2}+\frac{145}{9}\frac{dy}{dt}=e^{-t}$, $y(0)=1$, $\frac{dy}{dt}(0)=2$, $\frac{d^2y}{dt^2}(0)=-1$.

Solution. The corresponding homogeneous equation, $y'''+\frac{2}{3}y''+\frac{145}{9}y'=0$, has general solution $y_h=c_1+(c_2\sin 4t+c_3\cos 4t)e^{-t/3}$ and a fundamental set of solutions for the corresponding homogeneous equation is $S=\left\{1,e^{-t/3}\cos 4t,e^{-t/3}\sin 4t\right\}$.

```
DSolve[y'''[t] + 2/3y''[t] + 145/9y'[t] == 0,
```

y[t], t]

$\{\{y[t] \to C[3] - \frac{3}{145}e^{-t/3}((12C[1] + C[2])\text{Cos}[4t] + (C[1] - 12C[2])\text{Sin}[4t])\}\}$

For e^{-t}, the associated set of functions is $F = \{e^{-t}\}$. Because no element of F is an element of S, we assume that $y_p = Ae^{-t}$, where A is a constant to be determined. After defining y_p, we compute the necessary derivatives

Clear[yp, a]

yp[t_] = aExp[−t];

yp′[t]

yp″[t]

yp‴[t]

$-ae^{-t}$

ae^{-t}

$-ae^{-t}$

and substitute into the nonhomogeneous equation.

eqn = yp‴[t] + 2/3yp″[t] + 145/9yp′[t] == Exp[−t]

$-\frac{148}{9}ae^{-t} == e^{-t}$

Equating coefficients and solving for A gives us $A = -9/148$ so $y_p = -\frac{9}{148}e^{-t}$ and a general solution is $y = y_h + y_p$.

Remark 6.3. `SolveAlways[equation,variable]` attempts to solve `equation` so that it is true for all values of `variable`.

SolveAlways[eqn, t]

$\{\{a \to -\frac{9}{148}\}\}$

We verify the result with `DSolve`.

gensol = DSolve[y‴[t] + 2/3y″[t]+

145/9y′[t] == Exp[−t], y[t], t]//FullSimplify

$\{\{y[t] \to -\frac{9e^{-t}}{148} + C[3] + \frac{1}{145}e^{-t/3}(-3(12C[1] + C[2])\text{Cos}[4t] - 3(C[1] - 12C[2])\text{Sin}[4t])\}\}$

To obtain a real-valued solution, we use `ComplexExpand`. If you are using a version of Mathematica older than version 11, you might receive a complex valued function rather than the

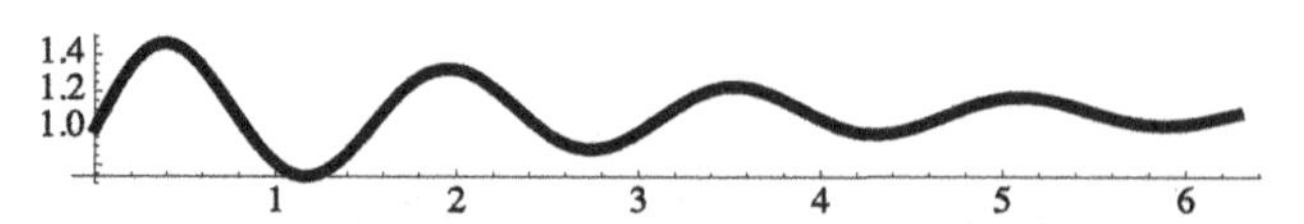

FIGURE 6.20 The solution of the equation that satisfies $y(0) = 1$, $y'(0) = 2$, and $y''(0) = -1$.

real-valued function that we obtained. In those cases, `ComplexExpand` can help you rewrite your complex solution as a real-valued solution.
The result indicates that the form returned by `DSolve`. To apply the initial conditions, we compute $y(0) = 1$, $y'(0) = 2$, and $y''(0) = -1$ and solve for c_1, c_2, and c_3. The solution of the initial-value problem is obtained by substituting these values into the general solution and we graph the result with `Plot` in Fig. 6.20.

```
initsol = DSolve[{y'''[t] + 2/3y''[t]+
  145/9y'[t] == Exp[-t], y[0] == 1, y'[0] == 2, y''[0] == -1},
  y[t], t]//FullSimplify
```

$$\left\{\left\{y[t] \to \frac{e^{-t}(-2610+46472e^{t}+e^{2t/3}(-942\text{Cos}[4t]+20729\text{Sin}[4t]))}{42920}\right\}\right\}$$

```
Plot[y[t]/.initsol, {t, 0, 2Pi}, AspectRatio -> Automatic,
  PlotStyle -> {{Thickness[.01], Black}}]
```

□

Example 6.20

Solve

$$\frac{d^8y}{dt^8} + \frac{7}{2}\frac{d^7y}{dt^7} + \frac{73}{2}\frac{d^6y}{dt^6} + \frac{229}{2}\frac{d^5y}{dt^5} + \frac{801}{2}\frac{d^4y}{dt^4} + 976\frac{d^3y}{dt^3} + 1168\frac{d^2y}{dt^2} + 640\frac{dy}{dt} + 128y = te^{-t} + \sin 4t + t.$$

Solution. Solving the characteristic equation

```
Solve[k^8 + 7/2k^7 + 73/2k^6 + 229/2k^5+
  801/2k^4 + 976k^3 + 1168k^2 + 640k + 128 == 0]
```

$$\{\{k \to -1\}, \{k \to -1\}, \{k \to -1\}, \{k \to -\tfrac{1}{2}\}, \{k \to -4i\}, \{k \to -4i\}, \{k \to 4i\}, \{k \to 4i\}\}$$

shows us that the solutions are $k_1 = -1/2$, $k_2 = -1$ with multiplicity 3, and $k_{3,4} = \pm 4i$, each with multiplicity 2. A fundamental set of solutions for the corresponding homogeneous equation is

$$S = \left\{e^{-t/2}, e^{-t}, te^{-t}, t^2e^{-t}, \cos 4t, t\cos 4t, \sin 4t, t\sin 4t\right\}.$$

A general solution of the corresponding homogeneous equation is

$$y_h = c_1e^{-t/2} + \left(c_2 + c_3t + c_4t^2\right)e^{-t} + (c_5 + c_7t)\sin 4t + (c_6 + c_8t)\cos 4t.$$

```
gensol = DSolve[D[y[t], {t, 8}] + 7/2D[y[t], {t, 7}]+
```

```
73/2D[y[t], {t, 6}] + 229/2D[y[t], {t, 5}]+
801/2D[y[t], {t, 4}] + 976y'''[t] + 1168y''[t]+
640y'[t] + 128y[t] == 0, y[t], t]
```

```
{{y[t] → e^(-t/2)C[5] + e^(-t)C[6] + e^(-t)tC[7] + e^(-t)t^2C[8] + C[1]Cos[4t] + tC[2]Cos[4t] +
C[3]Sin[4t] + tC[4]Sin[4t]}}
```

The associated set of functions for te^{-t} is $F_1 = \{e^{-t}, te^{-t}\}$. We multiply F_1 by t^n, where n is the smallest nonnegative integer so that no element of $t^n F_1$ is an element of S: $t^3 F_1 = \{t^3 e^{-t}, t^4 e^{-t}\}$. The associated set of functions for $\sin 4t$ is $F_2 = \{\cos 4t, \sin 4t\}$. We multiply F_2 by t^n, where n is the smallest nonnegative integer so that no element of $t^n F_2$ is an element of S: $t^2 F_2 = \{t^2 \cos 4t, t^2 \sin 4t\}$. The associated set of functions for t is $F_3 = \{1, t\}$. No element of F_3 is an element of S.

Thus, we search for a particular solution of the form

$$y_p = A_1 t^3 e^{-t} + A_2 t^4 e^{-t} + A_3 t^2 \cos 4t + A_4 t^2 \sin 4t + A_5 + A_6 t,$$

where the A_i are constants to be determined.

After defining y_p,

```
yp[t_] = a[1]t^3Exp[-t] + a[2]t^4Exp[-t]+
  a[3]t^2Cos[4t] + a[4]t^2Sin[4t] + a[5] + a[6]t;
```

we substitute into the nonhomogeneous equation, naming the result eqn. At this point we can either equate coefficients and solve for A_i or use the fact that eqn is true for *all* values of t.

```
eqn = D[yp[t], {t, 8}] + 7/2D[yp[t], {t, 7}]+
  73/2D[yp[t], {t, 6}] + 229/2D[yp[t], {t, 5}]+
  801/2D[yp[t], {t, 4}] + 976yp'''[t] + 1168yp''[t]+
  640yp'[t] + 128yp[t] == tExp[-t] + Sin[4t] + t//Simplify
```

```
e^(-t)(-867a[1]+7752a[2]-3468ta[2]+128e^t a[5]+640e^t a[6]+128e^t ta[6]-64e^t(369a[3]
- 428a[4])Cos[4t] - 64e^t(428a[3] + 369a[4])Sin[4t]) == t + e^(-t)t + Sin[4t]
```

We substitute in six values of t

```
sysofeqs = Table[eqn/.t → n//N, {n, 0, 5}];
```

and then solve for A_i.

```
coeffs = Solve[sysofeqs, {a[1.], a[2.], a[3.], a[4.], a[5.], a[6.]}]
```

```
{{a[1.] → -0.00257819, a[2.] → -0.000288351, a[3.] → -0.0000209413, a[4.] →
- 0.0000180545, a[5.] → -0.0390625, a[6.] → 0.0078125}}
```

y_p is obtained by substituting the values for A_i into y_p and a general solution is $y = y_h + y_p$. `DSolve` is able to find an exact solution.

```
gensol = DSolve[D[y[t], {t, 8}] + 7/2D[y[t], {t, 7}]+
  73/2D[y[t], {t, 6}] + 229/2D[y[t], {t, 5}]+
  801/2D[y[t], {t, 4}] + 976y'''[t] + 1168y''[t]+
  640y'[t] + 128y[t] == tExp[-t] + Sin[4t] + t, y[t], t]//Simplify//Expand
```

$\{\{y[t] \to -\frac{5}{128} - \frac{2924806e^{-t}}{24137569} + \frac{t}{128} - \frac{86016e^{-t}t}{1419857} - \frac{1270e^{-t}t^2}{83521} - \frac{38e^{-t}t^3}{14739} - \frac{e^{-t}t^4}{3468} + e^{-t/2}C[5] +$
$e^{-t}C[6] + e^{-t}tC[7] + e^{-t}t^2C[8] + \frac{9041976373\text{Cos}[4t]}{199643253056000} - \frac{1568449t\text{Cos}[4t]}{45168156800} - \frac{107t^2\text{Cos}[4t]}{5109520} +$
$C[1]\text{Cos}[4t] + tC[2]\text{Cos}[4t] + \frac{13794625331\text{Sin}[4t]}{798573012224000} + \frac{20406t\text{Sin}[4t]}{352876225} - \frac{369t^2\text{Sin}[4t]}{20438080} + C[3]\text{Sin}[4t] +$
$tC[4]\text{Sin}[4t]\}\}$ □

Variation of Parameters

In the same way as with second-order equations, we assume that a particular solution of the nth-order linear equation (6.19) has the form $y_p = u_1(t)y_1 + u_2(t)y_2 + \cdots + u_n(t)y_n$, where $S = \{y_1, y_2, \ldots, y_n\}$ is a fundamental set of solutions to the corresponding homogeneous equation (6.20). With the assumptions

$$\begin{aligned} y_p' &= y_1u_1' + y_2u_2' + \cdots + y_nu_n' = 0 \\ y_p'' &= y_1'u_1' + y_2'u_2' + \cdots + y_n'u_n' = 0 \\ &\vdots \\ y_p^{(n-1)} &= y_1^{(n-2)}u_1' + y_2^{(n-2)}u_2' + \cdots + y_n^{(n-2)}u_n' = 0 \end{aligned} \tag{6.23}$$

we obtain the equation

$$y_1^{(n-1)}u_1' + y_2^{(n-1)}u_2' + \cdots + y_n^{(n-1)}u_n' = f(t). \tag{6.24}$$

Equations (6.23) and (6.24) form a system of n linear equations in the unknowns u_1', u_2', $\ldots$, u_n'. Applying Cramer's Rule,

$$u_i' = \frac{W_i(S)}{W(S)}, \tag{6.25}$$

where $W(S)$ is given by equation (6.21) and $W_i(S)$ is the determinant of the matrix obtained by replacing the ith column of

$$\begin{pmatrix} y_1 & y_2 & \cdots & y_n \\ y_1' & y_2' & \cdots & y_n' \\ \vdots & \vdots & \cdots & \vdots \\ y_1^{(n-1)} & y_2^{(n-1)} & \cdots & y_n^{(n-1)} \end{pmatrix} \quad \text{by} \quad \begin{pmatrix} 0 \\ 0 \\ \vdots \\ f(t) \end{pmatrix}.$$

Example 6.21

Solve $y^{(3)} + 4y' = \sec 2t$.

Solution. A general solution of the corresponding homogeneous equation is $y_h = c_1 + c_2 \cos 2t + c_3 \sin 2t$; a fundamental set is $S = \{1, \cos 2t, \sin 2t\}$ with Wronskian $W(S) = 8$.

```
yh = DSolve[y'''[t] + 4y'[t] == 0, y[t], t]
```

$\{\{y[t] \to C[3] - \frac{1}{2}C[2]\text{Cos}[2t] + \frac{1}{2}C[1]\text{Sin}[2t]\}\}$

s = {1, Cos[2t], Sin[2t]};

ws = {s, D[s, t], D[s, {t, 2}]};

MatrixForm[ws]

$$\begin{pmatrix} 1 & \text{Cos}[2t] & \text{Sin}[2t] \\ 0 & -2\text{Sin}[2t] & 2\text{Cos}[2t] \\ 0 & -4\text{Cos}[2t] & -4\text{Sin}[2t] \end{pmatrix}$$

dws = Simplify[Det[ws]]

8

Using variation of parameters to find a particular solution of the nonhomogeneous equation, we let $y_1 = 1$, $y_2 = \cos 2t$, and $y_3 = \sin 2t$ and assume that a particular solution has the form $y_p = u_1 y_1 + u_2 y_2 + u_3 y_3$. Using the variation of parameters formula, we obtain

$$u_1' = \frac{1}{8}\begin{vmatrix} 0 & \cos 2t & \sin 2t \\ 0 & -2\sin 2t & 2\cos 2t \\ \sec 2t & -4\cos 2t & -4\sin 2t \end{vmatrix} = \frac{1}{4}\sec 2t \quad \text{so} \quad u_1 = \frac{1}{8}\ln|\sec 2t + \tan 2t|,$$

$$u_2' = \frac{1}{8}\begin{vmatrix} 1 & 0 & \sin 2t \\ 0 & 0 & 2\cos 2t \\ 0 & \sec 2t & -4\sin 2t \end{vmatrix} = -\frac{1}{4} \quad \text{so} \quad u_2 = -\frac{1}{4}t,$$

and

$$u_3' = \frac{1}{8}\begin{vmatrix} 1 & \cos 2t & 0 \\ 0 & -2\sin 2t & 0 \\ 0 & -4\cos 2t & \sec 2t \end{vmatrix} = -\frac{1}{2}\tan 2t \quad \text{so} \quad u_3 = \frac{1}{8}\ln|\cos 2t|,$$

where we use `Det` and `Integrate` to evaluate the determinants and integrals. In the case of u_1, the output given by Mathematica looks different than the result we obtained by hand but differentiating the difference between the two results yields 0 so the results obtained by hand and with Mathematica are the same.

u1p = 1/8Det[{{0, Cos[2t], Sin[2t]}, {0, −2Sin[2t], 2Cos[2t]},

{Sec[2t], −4Cos[2t], −4Sin[2t]}}]//Simplify

$\frac{1}{4}\text{Sec}[2t]$

u1 = Integrate[u1p, t]

$\frac{1}{4}(-\frac{1}{2}\text{Log}[\text{Cos}[t] - \text{Sin}[t]] + \frac{1}{2}\text{Log}[\text{Cos}[t] + \text{Sin}[t]])$

s1 = D[u1 − 1/8Log[Sec[2t] + Tan[2t]], t]

$\frac{1}{4}\left(-\frac{-\text{Cos}[t]-\text{Sin}[t]}{2(\text{Cos}[t]-\text{Sin}[t])} + \frac{\text{Cos}[t]-\text{Sin}[t]}{2(\text{Cos}[t]+\text{Sin}[t])}\right) - \frac{2\text{Sec}[2t]^2+2\text{Sec}[2t]\text{Tan}[2t]}{8(\text{Sec}[2t]+\text{Tan}[2t])}$

Simplify[s1]

0

u2p = 1/8Det[{{1, 0, Sin[2*t*]}, {0, 0, 2Cos[2*t*]},

{0, Sec[2*t*], −4Sin[2*t*]}}]//Simplify

$-\frac{1}{4}$

u2 = Integrate[u2p, *t*]

$-\frac{t}{4}$

u3p = 1/8Det[{{1, Cos[2*t*], 0}, {0, −2Sin[2*t*], 0},

{0, −4Cos[2*t*], Sec[2*t*]}}]//Simplify

$-\frac{1}{4}$Tan[2*t*]

u3 = Integrate[u3p, *t*]

$\frac{1}{8}$Log[Cos[2*t*]]

Thus, a particular solution of the nonhomogeneous equation is

$$y_p = \frac{1}{8}\ln|\sec 2t + \tan 2t| - \frac{1}{4}t\cos 2t + \frac{1}{8}\ln|\cos 2t|\sin 2t$$

and a general solution is $y = y_h + y_p$. We verify that the calculations using `DSolve` return an equivalent solution.

gensol = DSolve[*y*‴[*t*] + 4*y*′[*t*] == Sec[2*t*], *y*[*t*], *t*]//Simplify

{{*y*[*t*] → $\frac{1}{8}$(8*C*[3] − 8*C*[2]Cos[*t*]^2 − 2*t*Cos[2*t*] − Log[Cos[*t*] − Sin[*t*]] + Log[Cos[*t*] + Sin[*t*]] + 4*C*[1]Sin[2*t*] + Log[Cos[2*t*]]Sin[2*t*])}} □

6.3.4 Laplace Transform Methods

The *method of Laplace transforms* can be useful when the forcing function is piecewise-defined or periodic.

Definition 6.3 (Laplace Transform and Inverse Laplace Transform). Let $y = f(t)$ be a function defined on the interval $[0, \infty)$. The **Laplace transform** is the function (of s)

$$F(s) = \mathcal{L}\{f(t)\} = \int_0^\infty e^{-st} f(t)\,dt, \tag{6.26}$$

provided the improper integral exists. $f(t)$ is the **inverse Laplace transform** of $F(s)$ means that $\mathcal{L}\{f(t)\} = F(s)$ and we write $\mathcal{L}^{-1}\{F(s)\} = f(t)$.

1. `LaplaceTransform[f[t],t,s]` computes $\mathcal{L}\{f(t)\} = F(s)$.
2. `InverseLaplaceTransform[F[s],t,s]` computes $\mathcal{L}^{-1}\{F(s)\} = f(t)$.

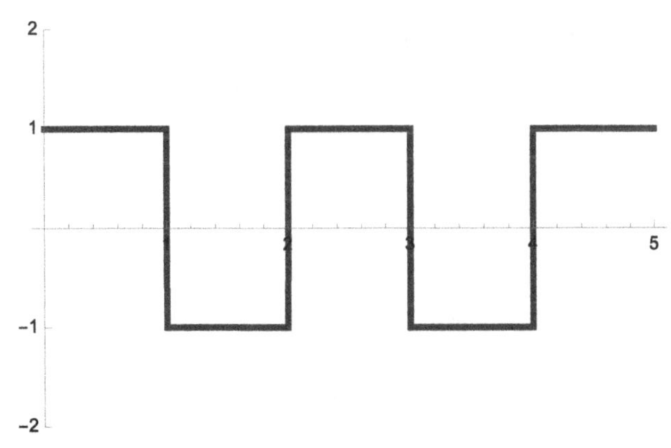

FIGURE 6.21 Plot of $f(t)$ for $0 \le t \le 5$. (University of Maine colors)

3. `UnitStep[t]` returns $\mathcal{U}(t) = \begin{cases} 0, & t < 0 \\ 1, & t \ge 0. \end{cases}$

Typically, when we use Laplace transforms to solve a differential equation for a function $y(t)$, we will compute the Laplace transform of each term of the differential equation, solve the resulting algebraic equation for the Laplace transform of $y(t)$, $\mathcal{L}\{y(t)\} = Y(s)$, and finally determine $y(t)$ by computing the inverse Laplace transform of $Y(s)$, $\mathcal{L}^{-1}\{Y(s)\} = y(t)$.

Example 6.22

Let $y = f(t)$ be defined recursively by $f(t) = \begin{cases} 1, & 0 \le t < 1 \\ -1, & 1 \le t < 2 \end{cases}$ and $f(t) = f(t-2)$ if $t \ge 2$. Solve $y'' + 4y' + 20y = f(t)$.

Solution. We begin by defining and graphing $y = f(t)$ for $0 \le t \le 5$ in Fig. 6.21.

```
Clear[f, g, u, y1, y2, sol]
f[t_]:=1/;0 ≤ t < 1
f[t_]:= − 1/;1 ≤ t ≤ 2
f[t_]:=f[t − 2]/;t > 2
Plot[f[t], {t, 0, 5}, Ticks → {Automatic, {−2, −1, 0, 1, 2}},
  PlotRange → {−2, 2}, PlotStyle → {{Thickness[.01],
  CMYKColor[1, .64, 0, .6]}}]
```

We then define `lhs` to be the left-hand side of the equation $y'' + 4y' + 20y = f(t)$,

```
Clear[y, x, lhs, stepone, steptwo]
lhs = y''[t] + 4y'[t] + 20y[t];
```

and compute the Laplace transform of `lhs` with `LaplaceTransform`, naming the result `stepone`.

```
stepone = LaplaceTransform[lhs, t, s]
```

$20\text{LaplaceTransform}[y[t],t,s]+s^2\text{LaplaceTransform}[y[t],t,s]+$

$4(s\text{LaplaceTransform}[y[t],t,s]-y[0])-sy[0]-y'[0]$

Let `lr` denote the Laplace transform of the right-hand side of the equation, $f(t)$. We now solve the equation $20\text{ly}+4s\text{ly}+s^2\text{ly}-4y(0)-sy(0)-y'(0)=\text{lr}$ for `ly` and name the resulting output `steptwo`.

steptwo = Solve[stepone==lr, LaplaceTransform[y[t], t, s]]

$\left\{\left\{\text{LaplaceTransform}[y[t],t,s]\to\frac{\text{lr}+4y[0]+sy[0]+y'[0]}{20+4s+s^2}\right\}\right\}$

stepthree = ExpandNumerator[steptwo[[1, 1, 2]], 1r]

$\frac{\text{lr}+4y[0]+sy[0]+y'[0]}{20+4s+s^2}$

To find $y(t)$, we must compute the inverse Laplace transform of $\mathcal{L}\{y(t)\}$; the formula for which is explicitly obtained from `steptwo` with `steptwo[[1,1,2]]`. First, we rewrite: $\mathcal{L}\{y(t)\}$. Then,

$$\begin{aligned}y(t)&=\mathcal{L}^{-1}\left\{\frac{\mathcal{L}\{f(t)\}}{s^2+4s+20}+\frac{4y(0)+sy(0)+y'(0)}{s^2+4s+20}\right\}\\&=\mathcal{L}^{-1}\left\{\frac{\mathcal{L}\{f(t)\}}{s^2+4s+20}\right\}+\mathcal{L}^{-1}\left\{\frac{4y(0)+sy(0)+y'(0)}{s^2+4s+20}\right\}.\end{aligned}$$

Completing the square yields $s^2+4s+20=(s+2)^2+16$. Because

$$\mathcal{L}^{-1}\left\{\frac{b}{(s-a)^2+b^2}\right\}=e^{at}\sin bt\quad\text{and}\quad\mathcal{L}^{-1}\left\{\frac{s-a}{(s-a)^2+b^2}\right\}=e^{at}\cos bt,$$

the inverse Laplace transform of

$$\frac{4y(0)+sy(0)+y'(0)}{s^2+4s+20}=y(0)\frac{s+2}{(s+2)^2+4^2}+\frac{y'(0)+2y(0)}{4}\frac{4}{(s+2)^2+4^2}$$

is

$$y(0)e^{-2t}\cos 4t+\frac{y'(0)+2y(0)}{4}e^{-2t}\sin 4t,$$

which is defined as $y_1(t)$. We perform these steps with Mathematica by first using `InverseLaplaceTransform` of the expression in `stepthree` after solving for lr to calculate $\mathcal{L}^{-1}\left\{\frac{4y(0)+sy(0)+y'(0)}{s^2+4s+20}\right\}$, naming the result `stepfour`.

stepfour = InverseLaplaceTransform$\left[-\frac{-4y[0]-sy[0]-y'[0]}{20+4s+s^2},s,t\right]$

$-\frac{1}{8}ie^{(-2-4i)t}\left((-2+4i)y[0]+(2+4i)e^{8it}y[0]-y'[0]+e^{8it}y'[0]\right)$

To see that this is a real-valued function, we use `ComplexExpand` together with `Simplify`.

stepfive = ComplexExpand[stepfour]//Simplify

$\frac{1}{4}e^{-2t}\left(4\text{Cos}[4t]y[0]+\text{Sin}[4t]\left(2y[0]+y'[0]\right)\right)$

If the result in `stepfive` is given in terms of real and imaginary parts of $y(0)$ and $y'(0)$, because $y'(0)$ is assumed to be a real number, the imaginary part of $y'(0)$ is 0; the real part of $y'(0)$ is $y'(0)$.

y1[t_] = stepfive/. {Im[y'[0]] → 0, Re[y'[0]] → y'[0]} //Simplify

$\frac{1}{4}e^{-2t}\left(4\text{Cos}[4t]y[0] + \text{Sin}[4t]\left(2y[0] + y'[0]\right)\right)$

To compute the inverse Laplace transform of $\dfrac{\mathcal{L}\{f(t)\}}{s^2+4s+20}$, we begin by computing `lr` $=$ $\mathcal{L}\{f(t)\}$. Let $\mathcal{U}_a(t) = \begin{cases} 1, & t \ge a \\ 0, & t < a \end{cases}$. Then, $\mathcal{U}_a(t) = \mathcal{U}(t-a) =$ `UnitStep[t - a]`.

The periodic function $f(t) = \begin{cases} 1, & 0 \le t < 1 \\ -1, & 1 \le t < 2 \end{cases}$ and $f(t) = f(t-2)$ if $t \ge 2$ can be written in terms of step functions as

$$\begin{aligned} f(t) &= \mathcal{U}_0(t) - 2\mathcal{U}_1(t) + 2\mathcal{U}_2(t) - 2\mathcal{U}_3(t) + 2\mathcal{U}_4(t) - \dots \\ &= \mathcal{U}(t) - 2\mathcal{U}(t-1) + 2\mathcal{U}(t-2) - 2\mathcal{U}(t-3) + 2\mathcal{U}(t-4) - \dots \\ &= \mathcal{U}(t) + 2\sum_{n=1}^{\infty}(-1)^n\mathcal{U}(t-n). \end{aligned}$$

The Laplace transform of $\mathcal{U}_a(t) = \mathcal{U}(t-a)$ is $\dfrac{1}{s}e^{-as}$ and the Laplace transform of $f(t)\mathcal{U}_a(t) = f(t)\mathcal{U}(t-a)$ is $e^{-as}F(s)$, where $F(s)$ is the Laplace transform of $f(t)$. Then,

$$\begin{aligned} \texttt{lr} &= \frac{1}{s} - \frac{2}{s}e^{-s} + \frac{2}{s}e^{-2s} - \frac{2}{s}e^{-3s} + \dots \\ &= \frac{1}{s}\left(1 - 2e^{-s} + 2e^{-2s} - 2e^{-3s} + \dots\right) \end{aligned}$$

and

$$\begin{aligned} \frac{\texttt{lr}}{s^2+4s+20} &= \frac{1}{s\left(s^2+4s+20\right)}\left(1 - 2e^{-s} + 2e^{-2s} - 2e^{-3s} + \dots\right) \\ &= \frac{1}{s\left(s^2+4s+20\right)} + 2\sum_{n=1}^{\infty}(-1)^n\frac{e^{-ns}}{s\left(s^2+4s+20\right)}. \end{aligned}$$

Because $\dfrac{1}{s^2+4s+20} = \dfrac{1}{4}\dfrac{1}{(s+2)^2+4^2}$,

$$\mathcal{L}^{-1}\left\{\frac{1}{s\left(s^2+4s+20\right)}\right\} = \int_0^t \frac{1}{4}e^{-2\alpha}\sin 4\alpha\, d\alpha,$$

computed and defined to be the function $g(t)$.

g[t_] = $\int_0^t \frac{1}{4}$Exp[−2α]Sin[4α] dα

$\frac{1}{40}\left(2 - e^{-2t}(2\text{Cos}[4t] + \text{Sin}[4t])\right)$

Alternatively, we can use `InverseLaplaceTransform` to obtain the same result.

g[t_] =

InverseLaplaceTransform$\left[\frac{1}{s(s^2+4s+20)}, s, t\right]$ //ExpToTrig//

Simplify

$\frac{1}{80}(4+(2\text{Cos}[4t]+\text{Sin}[4t])(-2\text{Cosh}[2t]+2\text{Sinh}[2t]))$

Then, $\mathcal{L}^{-1}\left\{2(-1)^n\frac{e^{-ns}}{s\left(s^2+4s+20\right)}\right\}=2(-1)^n g(t-n)\mathcal{U}(t-n)$ and the inverse Laplace transform of

$$\frac{1}{s\left(s^2+4s+20\right)}+2\sum_{n=1}^{\infty}(-1)^n\frac{e^{-ns}}{s\left(s^2+4s+20\right)}$$

is

$$y_2(t)=g(t)+2\sum_{n=1}^{\infty}(-1)^n g(t-n)\mathcal{U}(t-n).$$

It then follows that

$$\begin{aligned} y(t) &= y_1(t)+y_2(t) \\ &= y(0)e^{-2t}\cos 4t+\frac{y'(0)+2y(0)}{4}e^{-2t}\sin 4t+2\sum_{n=1}^{\infty}(-1)^n g(t-n)\mathcal{U}(t-n), \end{aligned}$$

where $g(t)=\frac{1}{20}-\frac{1}{20}e^{-2t}\cos 4t-\frac{1}{40}e^{-2t}\sin 4t$.

To graph the solution for various initial conditions on the interval $[0,5]$, we define $y_2(t)=g(t)+2\sum_{n=1}^{5}(-1)^n g(t-n)\mathcal{U}(t-n)$, `sol`, and `inits`. (Note that we can graph the solution for various initial conditions on the interval $[0,m]$ by defining $y_2(t)=g(t)+2\sum_{n=1}^{m}(-1)^n g(t-n)\mathcal{U}(t-n)$.)

y2[t_]:=g[t] + 2$\sum_{n=1}^{5}(-1)^n$g[t − n]UnitStep[t − n]

Clear[sol]

sol[t_]:=y1[t] + y2[t]

inits = {−1/2, 0, 1/2};

We then create a table of graphs of `sol[t]` on the interval $[0,5]$ corresponding to replacing $y(0)$ and $y'(0)$ by the values $-1/2$, 0, and $1/2$ and then displaying the resulting graphics array in Fig. 6.22.

```
graphs =
  Table[Plot[sol[t]/.{y[0] → inits[[i]], y'[0] → inits[[j]]},
   {t, 0, 5}, PlotStyle → {{Thickness[.01],
   CMYKColor[1, .64, 0, .6]}}], {i, 1, 3}, {j, 1, 3}];
Show[GraphicsGrid[graphs]]
```

□

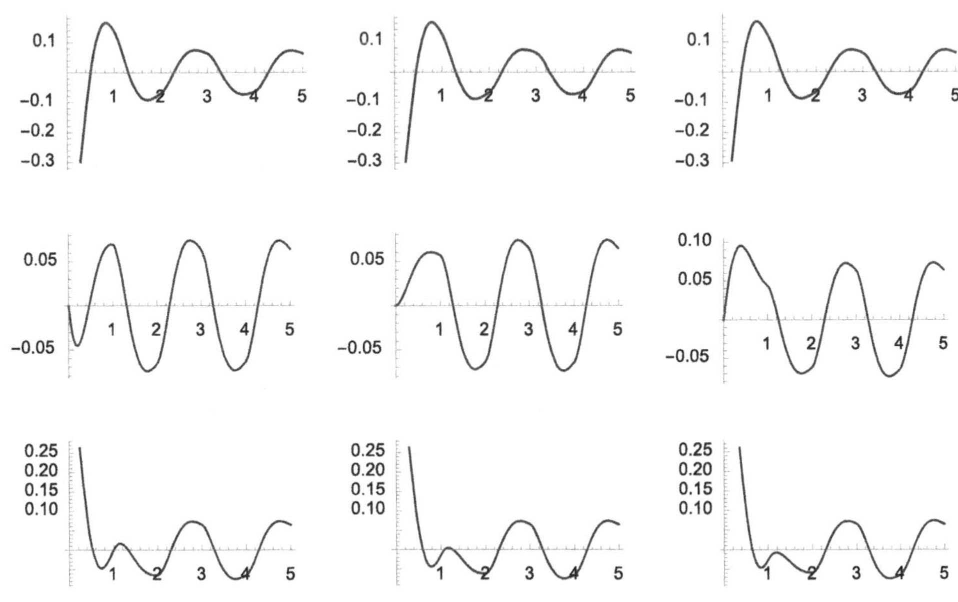

FIGURE 6.22 Solutions to a differential equation with a piecewise-defined periodic forcing function.

Application: The Convolution Theorem

Sometimes we are required to determine the inverse Laplace transform of a product of two functions. Just as in differential and integral calculus when the derivative and integral of a product of two functions did not produce the product of the derivatives and integrals, respectively, neither does the inverse Laplace transform of the product yield the product of the inverse Laplace transforms. *The Convolution Theorem* tells us how to compute the inverse Laplace transform of a product of two functions.

Theorem 6.2 (The Convolution Theorem)**.** *Suppose that $f(t)$ and $g(t)$ are piecewise continuous on $[0, \infty)$ and both are of exponential order. Further, suppose that the Laplace transform of $f(t)$ is $F(s)$ and that of $g(t)$ is $G(s)$. Then,*

$$\mathcal{L}^{-1}\{F(s)G(s)\} = \mathcal{L}^{-1}\{\mathcal{L}\{(f*g)(t)\}\} = \int_0^t f(t-v)g(v)\,dv. \tag{6.27}$$

Note that $(f*g)(t) = \displaystyle\int_0^t f(t-v)g(v)\,dv$ is called the **convolution integral**.

Example 6.23: L–R–C Circuits

The initial-value problem used to determine the charge $q(t)$ on the capacitor in an L–R–C circuit is

$$L\frac{d^2Q}{dt^2} + R\frac{dQ}{dt} + \frac{1}{C}Q = f(t), \quad Q(0)=0, \quad \frac{dQ}{dt}(0)=0,$$

where L denotes inductance, $dQ/dt = I$, $I(t)$ current, R resistance, C capacitance, and $E(t)$ voltage supply. Because $dQ/dt = I$, this differential equation can be represented as

$$L\frac{dI}{dt} + RI + \frac{1}{C}\int_0^t I(u)\,du = E(t).$$

Note also that the initial condition $Q(0)=0$ is satisfied because $Q(0) = \frac{1}{C}\int_0^0 I(u)\,du = 0$. The condition $dQ/dt(0) = 0$ is replaced by $I(0) = 0$. (a) Solve this *integrodifferential equation*, an equation that involves a derivative as well as an integral of the unknown function, by using the Convolution theorem. (b) Consider this example with constant values $L = C = R = 1$ and $E(t) = \begin{cases} \sin t, & 0 \le t < \pi/2 \\ 0, & t \ge \pi/2 \end{cases}$. Determine $I(t)$ and graph the solution.

Solution. We proceed as in the case of a differential equation by taking the Laplace transform of both sides of the equation. The Convolution theorem, equation (6.27), is used in determining the Laplace transform of the integral with

$$\mathcal{L}\left\{\int_0^t I(u)\,du\right\} = \mathcal{L}\{1 * I(t)\} = \mathcal{L}\{1\}\mathcal{L}\{I(t)\} = \frac{1}{s}\mathcal{L}\{I(t)\}.$$

Therefore, application of the Laplace transform yields

$$Ls\mathcal{L}\{I(t)\} - LI(0) + R\mathcal{L}\{I(t)\} + \frac{1}{C}\frac{1}{s}\mathcal{L}\{I(t)\} = \mathcal{L}\{E(t)\}.$$

Because $I(0) = 0$, we have $Ls\mathcal{L}\{I(t)\} + R\mathcal{L}\{I(t)\} + \frac{1}{C}\frac{1}{s}\mathcal{L}\{I(t)\} = \mathcal{L}\{E(t)\}$. Simplifying and solving for $\mathcal{L}\{I(t)\}$ results in $\mathcal{L}\{I(t)\} = \dfrac{Cs\mathcal{L}\{E(t)\}}{LCs^2 + RCs + 1}$.

```
Clear[i]

LaplaceTransform[li'[t] + ri[t], t, s]
```

rLaplaceTransform[i[t], t, s] + l(−i[0] + sLaplaceTransform[i[t], t, s])

```
Solve[lslapi + rlapi + lapi/(cs) == lape, lapi]
```

$\left\{\left\{\text{lapi} \to \frac{\text{clapes}}{1+crs+cls^2}\right\}\right\}$

so that $I(t) = \mathcal{L}^{-1}\left\{\dfrac{Cs\mathcal{L}\{E(t)\}}{LCs^2 + RCs + 1}\right\}$. In the `Solve` command we use `lapi` to denote $\mathcal{L}\{I(t)\}$ and `lape` to denote $\mathcal{L}\{E(t)\}$. For (b), we note that $E(t) = \begin{cases} \sin t, & 0 \le t < \pi/2 \\ 0, & t \ge \pi/2 \end{cases}$ can be written as $E(t) = \sin t\,(\mathcal{U}(t) - \mathcal{U}(t - \pi/2))$. We define and plot the forcing function $E(t)$ on the interval $[0, \pi]$ in Fig. 6.23 (a).

We use lowercase letters to avoid any possible ambiguity with built-in Mathematica functions, like `E` and `I`.

```
e[t_]:=Sin[t] (UnitStep[t] − UnitStep[t − π/2])

p1 = Plot[e[t], {t, 0, π}, PlotStyle →
    {{Thickness[.01], CMYKColor[0, .25, .9, .05]}}]
```

Next, we compute the Laplace transform of $\mathcal{L}\{E(t)\}$ with `LaplaceTransform`. We call this result `lcape`.

```
lcape = LaplaceTransform[e[t], t, s]
```

$\frac{1}{1+s^2} - \frac{e^{-\frac{\pi s}{2}}s}{1+s^2}$

Using the general formula obtained for the Laplace transform of $I(t)$, we note that the denominator of this expression is given by $s^2 + s + 1$ which is entered as `denom`. Hence, the Laplace transform of $I(t)$, called `lcapi`, is given by the ratio `s lcape/denom`.

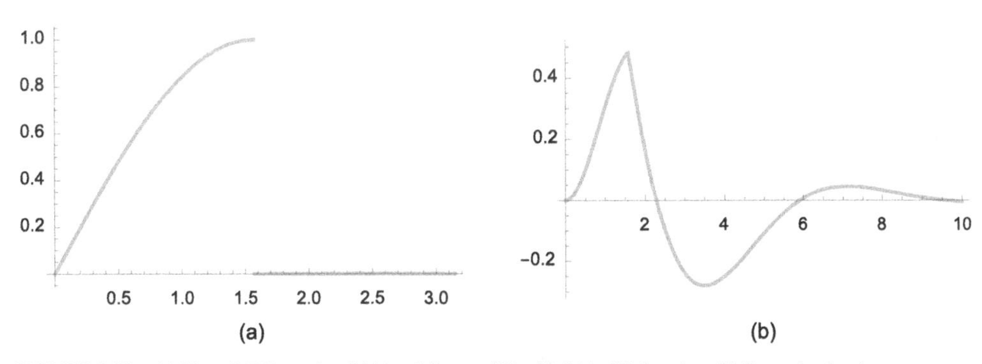

FIGURE 6.23 (a) Plot of $E(t) = \sin t\,(\mathcal{U}(t) - \mathcal{U}(t - \pi/2))$. (b) $I(t)$. (University of Missouri colors)

denom = $s^2 + s + 1$;

lcapi = $s$lcape/denom;

lcapi = Simplify[lcapi]

$\frac{s-e^{-\frac{\pi s}{2}}s^2}{1+s+2s^2+s^3+s^4}$

We determine $I(t)$ with `InverseLaplaceTransform`. Note that `HeavisideTheta[x]` is defined by $\theta(x) = \begin{cases} 0, & \text{if } x < 0 \\ 1, & \text{if } x > 0 \end{cases}$.

i[t_] = InverseLaplaceTransform[lcapi, s, t]

Sin[t] − HeavisideTheta[$-\frac{\pi}{2}+t$]($-\frac{e^{\frac{1}{4}(\pi-2t)}(\sqrt{3}\text{Cos}[\frac{1}{4}\sqrt{3}(\pi-2t)]+\text{Sin}[\frac{1}{4}\sqrt{3}(\pi-2t)])}{\sqrt{3}}$ + Sin[t]) − $\frac{2e^{-t/2}\text{Sin}[\frac{\sqrt{3}t}{2}]}{\sqrt{3}}$

This solution is plotted in `p2` (in black) and displayed with the forcing function (in gray) in Fig. 6.23 (b). Notice the effect that the forcing function has on the solution to the differential equation.

p2 = Plot[i[t], {t, 0, 10}, PlotStyle →

{{Thickness[.01], CMYKColor[0, .25, .9, .05]}}];

Show[p1, p2, PlotRange → All]

Show[GraphicsRow[{p1, p2}]]

In this case, we see that we can use `DSolve` to solve the initial-value problem

$$Q'' + Q' + Q = E(t), \quad Q(0) = 0, \quad Q'(0) = 0$$

as well. However, the unsimplified result is very lengthy so we use `FullSimplify` to attempt to simplify the result as much as possible.

Clear[q]

sol = DSolve[{q''[t] + q'[t] + q[t]==e[t], q[0]==0, q'[0]==0}, .

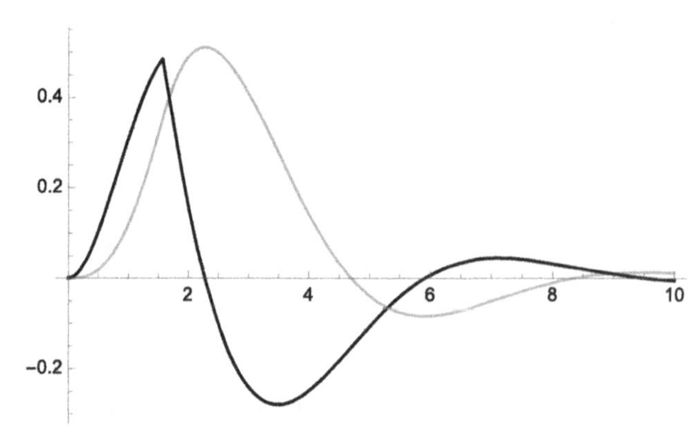

FIGURE 6.24 $Q(t)$ and $I(t) = Q'(t)$.

q[t], t]//FullSimplify

$$\{\{q[t] \to \begin{cases} -\text{Cos}[t] + \frac{1}{3}e^{-t/2}(3\text{Cos}[\frac{\sqrt{3}t}{2}] + \sqrt{3}\text{Sin}[\frac{\sqrt{3}t}{2}]) & t \geq 0\&\&2t \leq \pi \\ \frac{1}{3}e^{-t/2}(3\text{Cos}[\frac{\sqrt{3}t}{2}] + \sqrt{3}(-2e^{\pi/4}\text{Sin}[\frac{1}{4}\sqrt{3}(\pi - 2t)] + \text{Sin}[\frac{\sqrt{3}t}{2}])) & 2t > \pi \\ 0 & \text{True} \end{cases}\}\}$$

We see that this result is a real-valued function using `ComplexExpand` followed by `Simplify`.

q[t_] = ComplexExpand[sol[[1, 1, 2]]]//Simplify

$$\begin{cases} -\text{Cos}[t] + \frac{3\text{Cos}[\frac{\sqrt{3}t}{2}]+\sqrt{3}\text{Sin}[\frac{\sqrt{3}t}{2}]}{3\sqrt{e^t}} & t \geq 0\&\&2t \leq \pi \\ \frac{3\text{Cos}[\frac{\sqrt{3}t}{2}]+\sqrt{3}(-2e^{\pi/4}\text{Sin}[\frac{1}{4}\sqrt{3}(\pi-2t)]+\text{Sin}[\frac{\sqrt{3}t}{2}])}{3\sqrt{e^t}} & 2t > \pi \\ 0 & \text{True} \end{cases}$$

We use this result to graph $Q(t)$ and $I(t) = Q'(t)$ in Fig. 6.24.

```
Plot[{q[t], q'[t]}, {t, 0, 10},
  PlotStyle → {{CMYKColor[0, .25, .9, .05]},
  {CMYKColor[.6, .5, .4, 1]}}]
```

□

Application: The Dirac Delta Function

Let $\delta(t - t_0)$ denote the (generalized) function with the two properties

1. $\delta(t - t_0) = 0$ if $t \neq t_0$ and
2. $\int_{-\infty}^{\infty} \delta(t - t_0)\, dt = 1$

which is called the **Dirac delta function** and is quite useful in the definition of impulse forcing functions that arise in some differential equations. The Laplace transform of $\delta(t - t_0)$ is $\mathcal{L}\{\delta(t - t_0)\} = e^{-st_0}$. The Mathematica function `DiracDelta` represents the δ distribution.

LaplaceTransform[DiracDelta[t − t0], t, s]

e^{-st0}HeavisideTheta[t0]

Example 6.24

Solve $\begin{cases} x''+x'+x=\delta(t)+\mathcal{U}(t-2\pi) \\ x(0)=0,\ x'(0)=0 \end{cases}$.

Solution. We define `eq` to be the equation $x''+x'+x=\delta(t)+\mathcal{U}(t-2\pi)$ and then use `LaplaceTransform` to compute the Laplace transform of `eq`, naming the resulting output `leq`. The symbol `LaplaceTransform [x[t],t,s]` represents the Laplace transform of `x[t]`. We then apply the initial conditions $x(0)=0$ and $x'(0)=0$ to `leq` and name the resulting output `ics`.

Clear[x, eq]

eq = x''[t] + x'[t] + x[t]==DiracDelta[t] + UnitStep[t − 2π];

leq = LaplaceTransform[eq, t, s]

$\text{LaplaceTransform}[x[t],t,s]+s\text{LaplaceTransform}[x[t],t,s]$
$+s^2\text{LaplaceTransform}[x[t],t,s]-x[0]-sx[0]-x'[0]==1+\frac{e^{-2\pi s}}{s}$

ics = leq/.{x[0] → 0, x'[0] → 0}

$\text{LaplaceTransform}[x[t],t,s]+s\text{LaplaceTransform}[x[t],t,s]$
$+s^2\text{LaplaceTransform}[x[t],t,s]==1+\frac{e^{-2\pi s}}{s}$

Next, we use `Solve` to solve the equation `ics` for the Laplace transform of $x(t)$. The expression for the Laplace transform is extracted from `lapx` with `lapx[[1,1,2]]`.

lapx = Solve[ics, LaplaceTransform[x[t], t, s]]

$\{\{\text{LaplaceTransform}[x[t],t,s]\to\frac{e^{-2\pi s}(1+e^{2\pi s}s)}{s(1+s+s^2)}\}\}$

To find $x(t)$, we must compute the inverse Laplace transform of the Laplace transform of $\mathcal{L}\{x(t)\}$ obtained in `lapx`. We use `InverseLaplaceTransform` to compute the inverse Laplace transform of `lapx[[1,1,2]]` and name the resulting function `x[t]`.

x[t_] = InverseLaplaceTransform[lapx[[1, 1, 2]], s, t]

$$\frac{2e^{-t/2}\text{Sin}\left[\frac{\sqrt{3}t}{2}\right]}{\sqrt{3}}+\frac{1}{3}\text{HeavisideTheta}[-2\pi+t]$$
$$\left(3-e^{\pi-\frac{t}{2}}\left(3\text{Cos}\left[\tfrac{1}{2}\sqrt{3}(-2\pi+t)\right]+\sqrt{3}\text{Sin}\left[\tfrac{1}{2}\sqrt{3}(-2\pi+t)\right]\right)\right)$$

With some earlier versions of Mathematica, the above result may include imaginary components. If needed, we see that the result is a real-valued function using `ComplexExpand` followed by `Simplify`.

x[t_] = ComplexExpand[x[t]]//Simplify

$$\frac{1}{3\sqrt{e^t}}(2\sqrt{3}\text{Sin}[\tfrac{\sqrt{3}t}{2}]+(3\sqrt{e^t}-3e^{\pi}\text{Cos}[\tfrac{1}{2}\sqrt{3}(-2\pi+t)]$$
$$-\sqrt{3}e^{\pi}\text{Sin}[\tfrac{1}{2}\sqrt{3}(-2\pi+t)])\text{UnitStep}[-2\pi+t])$$

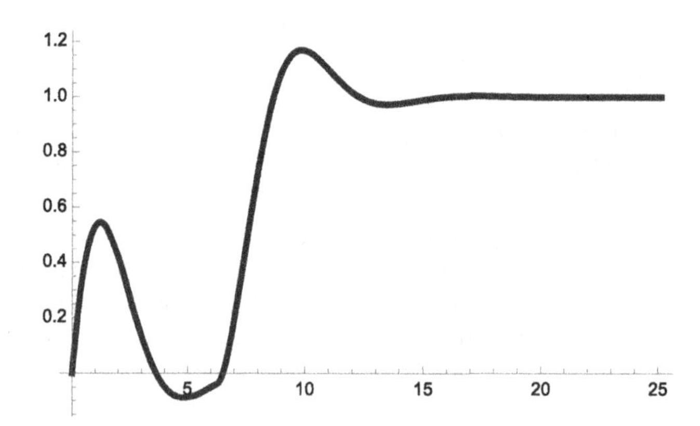

FIGURE 6.25 Plot of $x(t)$ on the interval $[0, 8\pi]$. (University of Montana colors)

We use `Plot` to graph the solution on the interval $[0, 8\pi]$ in Fig. 6.25.

```
Plot[x[t], {t, 0, 8π}, PlotStyle →
  {{Thickness[.01], CMYKColor[.1, 1, .4, .6]}}]
```

Finally, we note that `DSolve` is able to solve the initial-value problem directly as well.

```
Clear[x]
sol =
DSolve[{x''[t] + x'[t] + x[t]==DiracDelta[t] + UnitStep[t − 2π],
  x[0]==0, x'[0]==0}, x[t], t]//FullSimplify
```

$\{\{x[t] \to \frac{1}{3}e^{-t/2}(2\sqrt{3}(-\text{HeavisideTheta}[0] + \text{HeavisideTheta}[t])\text{Sin}[\frac{\sqrt{3}t}{2}] - (3e^{t/2}$
$+ e^{\pi}(-3\text{Cos}[\frac{1}{2}\sqrt{3}(2\pi - t)] + \sqrt{3}\text{Sin}[\frac{1}{2}\sqrt{3}(2\pi - t)]))(-1 + \text{UnitStep}[2\pi - t]))\}\}$

```
ComplexExpand[sol[[1, 1, 2]]]//Simplify
```

$$\begin{cases} \frac{3\sqrt{e^t} - 3e^{\pi}\text{Cos}[\frac{1}{2}\sqrt{3}(2\pi - t)] + \sqrt{3}e^{\pi}\text{Sin}[\frac{1}{2}\sqrt{3}(2\pi - t)]}{3\sqrt{e^t}} & t > 2\pi \\ -\frac{2\text{Sin}[\frac{\sqrt{3}t}{2}]}{\sqrt{3}\sqrt{e^t}} & t < 0 \\ 0 & \text{True} \end{cases}$$

□

6.3.5 Nonlinear Higher-Order Equations

Generally, rigorous results regarding nonlinear equations are very difficult to obtain. In some cases, analysis is best carried out numerically and/or graphically. In other situations, rewriting the equation as a system can be of benefit, which is discussed in the next section. (See Examples 6.30, 6.29, and 6.32.)

6.4 SYSTEMS OF EQUATIONS

6.4.1 Linear Systems

We now consider first-order linear systems of differential equations:

$$\mathbf{X}' = \mathbf{A}(t)\mathbf{X} + \mathbf{F}(t), \tag{6.28}$$

where

$$\mathbf{X}(t) = \begin{pmatrix} x_1(t) \\ x_2(t) \\ \vdots \\ x_n(t) \end{pmatrix}, \quad \mathbf{A}(t) = \begin{pmatrix} a_{11}(t) & a_{12}(t) & \dots & a_{1n}(t) \\ a_{21}(t) & a_{22}(t) & \dots & a_{2n}(t) \\ \vdots & \vdots & \dots & \vdots \\ a_{n1}(t) & a_{n2}(t) & \dots & a_{nn}(t) \end{pmatrix}, \quad \text{and} \quad \mathbf{F}(t) = \begin{pmatrix} f_1(t) \\ f_2(t) \\ \vdots \\ f_n(t) \end{pmatrix}.$$

6.4.1.1 *Homogeneous Linear Systems*

The corresponding homogeneous system of equation (6.28) is

$$\mathbf{X}' = \mathbf{A}\mathbf{X}. \tag{6.29}$$

In the same way as with the previously discussed linear equations, a **general solution** of equation (6.28) is $\mathbf{X} = \mathbf{X}_h + \mathbf{X}_p$ where $\mathbf{X}_h$ is a *general solution* of equation (6.29) and $\mathbf{X}_p$ is a *particular solution* of the nonhomogeneous system equation (6.28).

A **particular solution** to a system of ordinary differential equations is a set of functions that satisfy the system but do not contain any arbitrary constants. That is, a particular solution to a system is a set of specific functions, *containing no arbitrary constants*, that satisfy the system.

If $\mathbf{\Phi}_1 = \begin{pmatrix} \phi_{11} \\ \phi_{21} \\ \vdots \\ \phi_{n1} \end{pmatrix}$, $\mathbf{\Phi}_2 = \begin{pmatrix} \phi_{12} \\ \phi_{22} \\ \vdots \\ \phi_{n2} \end{pmatrix}$, ..., $\mathbf{\Phi}_n = \begin{pmatrix} \phi_{1n} \\ \phi_{2n} \\ \vdots \\ \phi_{nn} \end{pmatrix}$ are n linearly independent solutions of equation (6.29), a **general solution** of equation (6.29) is

$$\mathbf{X} = c_1\mathbf{\Phi}_1 + c_2\mathbf{\Phi}_2 + \dots + c_n\mathbf{\Phi}_n = \begin{pmatrix} \mathbf{\Phi}_1 & \mathbf{\Phi}_2 & \dots & \mathbf{\Phi}_n \end{pmatrix} \begin{pmatrix} c_1 \\ c_2 \\ \vdots \\ c_n \end{pmatrix} = \Phi\mathbf{C},$$

where

$$\mathbf{\Phi} = \begin{pmatrix} \mathbf{\Phi}_1 & \mathbf{\Phi}_2 & \dots & \mathbf{\Phi}_n \end{pmatrix} \quad \text{and} \quad \mathbf{C} = \begin{pmatrix} c_1 \\ c_2 \\ \vdots \\ c_n \end{pmatrix}.$$

$\mathbf{\Phi} = \begin{pmatrix} \phi_{11} & \phi_{12} & \dots & \phi_{1n} \\ \phi_{21} & \phi_{22} & \dots & \phi_{2n} \\ \vdots & \vdots & \dots & \vdots \\ \phi_{n1} & \phi_{n2} & \dots & \phi_{nn} \end{pmatrix}$ is called a **fundamental matrix** for equation (6.29). If $\mathbf{\Phi}$ is a fundamental matrix for equation (6.29), $\mathbf{\Phi}' = \mathbf{A}\Phi$ or $\mathbf{\Phi}' - \mathbf{A}\Phi = \mathbf{0}$.

$\mathbf{A}(t)$ *Constant*

Suppose that $\mathbf{A}(t) = \mathbf{A}$ has constant real entries. Let λ be an eigenvalue of $\mathbf{A}$ with corresponding eigenvector $\mathbf{v}$. Then, $\mathbf{\Phi} = \mathbf{v}e^{\lambda t}$ is a solution of $\mathbf{X}' = \mathbf{A}\mathbf{X}$.
If $\lambda = \alpha + \beta i$, $\beta \neq 0$, is an eigenvalue of $\mathbf{A}$ and has corresponding eigenvector $\mathbf{v} = \mathbf{a} + \mathbf{b}i$, two linearly independent solutions of $\mathbf{X}' = \mathbf{A}\mathbf{X}$ are

$$\mathbf{\Phi}_1 = e^{\alpha t}\left(\mathbf{a}\cos\beta t - \mathbf{b}\sin\beta t\right) \quad \text{and} \quad \mathbf{\Phi}_2 = e^{\alpha t}\left(\mathbf{a}\sin\beta t + \mathbf{b}\cos\beta t\right). \tag{6.30}$$

Example 6.25

Solve each of the following systems. (a) $\mathbf{X}' = \begin{pmatrix} -3 & 2 \\ -1 & 0 \end{pmatrix} \mathbf{X}$; (b) $\begin{cases} x' = -y \\ y' = -2x - y \end{cases}$; (c) $\begin{cases} dx/dt = -2y \\ dy/dt = x + 3y \end{cases}$; (d) $\begin{cases} dx/dt = 3x - 5y \\ dy/dt = 5x - 3y \end{cases}$; (e) $\begin{cases} dx/dt = 2x - 8y \\ dy/dt = 4x - 6y \end{cases}$.

Solution. (a) With `Eigensystem`, we see that the eigenvalues and eigenvectors of $\mathbf{A} = \begin{pmatrix} -3 & 2 \\ -1 & 0 \end{pmatrix}$ are $\lambda_1 = -2$ and $\lambda_2 = -1$ and $\mathbf{v}_1 = \begin{pmatrix} 2 \\ 1 \end{pmatrix}$ and $\mathbf{v}_2 = \begin{pmatrix} 1 \\ 1 \end{pmatrix}$, respectively.

```
capa = {{-3, 2}, {-1, 0}};
Eigensystem[capa]
```

$\{\{-2, -1\}, \{\{2, 1\}, \{1, 1\}\}\}$

Then $\mathbf{X}_1 = \begin{pmatrix} 2 \\ 1 \end{pmatrix} e^{-2t}$ and $\mathbf{X}_2 = \begin{pmatrix} 1 \\ 1 \end{pmatrix} e^{-t}$ are two linearly independent solutions of the system so a general solution is $\mathbf{X} = \begin{pmatrix} 2e^{-2t} & e^{-t} \\ e^{-2t} & e^{-t} \end{pmatrix} \begin{pmatrix} c_1 \\ c_2 \end{pmatrix}$; a fundamental matrix is $\mathbf{\Phi} = \begin{pmatrix} 2e^{-2t} & e^{-t} \\ e^{-2t} & e^{-t} \end{pmatrix}$.
We use `DSolve` to find a general solution of the system by entering

```
Clear[x, y]
gensol = DSolve[{x'[t] == -3x[t] + 2y[t],
  y'[t] == -x[t]}, {x[t], y[t]}, t]
```

$\{\{x[t] \to -e^{-2t}(-2 + e^t)C[1] + 2e^{-2t}(-1 + e^t)C[2], y[t] \to -e^{-2t}(-1 + e^t)C[1] + e^{-2t}(-1 + 2e^t)C[2]\}\}$

We graph the direction field with `StreamPlot` in Fig. 6.26.

Remark 6.4. `StreamPlot[{f[x,y],g[x,y]},{x,a,b},{y,c,d}]`
generates a basic direction field for the system $\{x' = f(x, y), y' = g(x, y)\}$ for $a \leq x \leq b$ and $c \leq y \leq d$. You can also use `VectorPlot` but we find that `StreamPlot` generally produces better results than `VectorPlot`.

For this example, we draw the eigenlines, $y = x/2$ and $y = x$. Because both eigenvalues of the coefficient matrix are negative, all solutions tend to the origin, $(0, 0)$, as $t \to \infty$.

```
sp1 = StreamPlot[{-3x + 2y, -x}, {x, -1, 1}, {y, -1, 1},
  StreamStyle -> CMYKColor[.02, 1, .85, .06], StreamPoints -> Fine, Frame -> False,
  Axes -> Automatic, AxesOrigin -> {0, 0}, AxesLabel -> {x, y}];
p1 = Plot[{1/2x, x}, {x, -1, 1},
  PlotStyle -> {{CMYKColor[0, 0, 0, .25], Thickness[.01]},
  {CMYKColor[0, 0, 0, .25], Thickness[.01]}}];
```

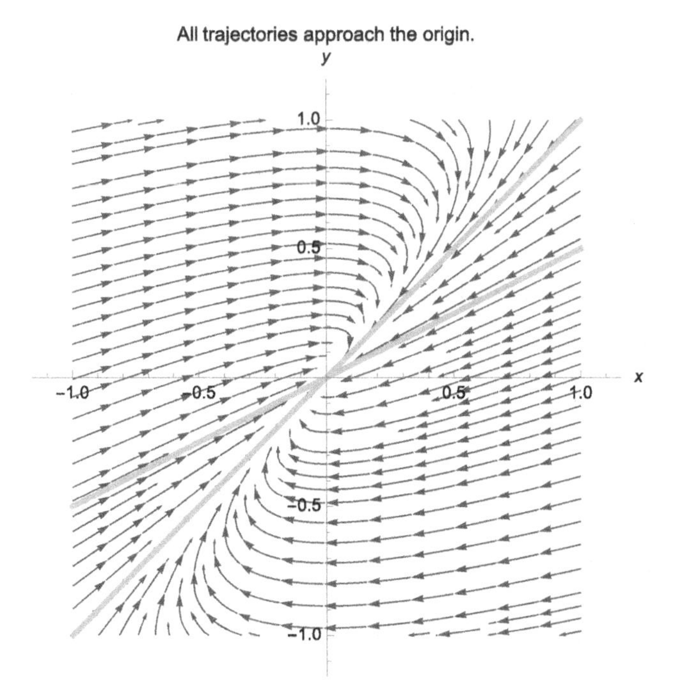

FIGURE 6.26 Direction field for $\mathbf{X}' = \mathbf{AX}$. (University of Nebraska colors)

Show[sp1, p1, PlotLabel → "All trajectories approach the origin."]

We proceed in much the same way for (b). First, we write the system in matrix form, $\mathbf{X}' = \begin{pmatrix} 0 & -1 \\ -2 & -1 \end{pmatrix}\mathbf{X}$. Using `Eigensystem`, we find that the eigenvalues are $\lambda_1 = -2$ and $\lambda_2 = 1$ with corresponding eigenvectors $\mathbf{v}_1 = \begin{pmatrix} 1 \\ 2 \end{pmatrix}$ and $\mathbf{v}_2 = \begin{pmatrix} -1 \\ 1 \end{pmatrix}$.

capa = {{0, −1}, {−2, −1}};

Eigensystem[capa]

{{−2, 1}, {{1, 2}, {−1, 1}}}

Several solutions are also graphed with `ParametricPlot` and shown together with the direction field in Fig. 6.27. To do so, we first solve the system if $x(0) = x_0$ and $y(0) = y_0$.

initsol = DSolve[{x′[t] == −y[t],

y′[t] == −2x[t] − y[t], x[0] == x0, y[0] == y0}, {x[t], y[t]}, t]

$\{\{x[t] \to \frac{1}{3}e^{-2t}(\text{x0} + 2e^{3t}\text{x0} + \text{y0} - e^{3t}\text{y0}), y[t] \to -\frac{1}{3}e^{-2t}(-2\text{x0} + 2e^{3t}\text{x0} - 2\text{y0} - e^{3t}\text{y0})\}\}$

Given an ordered pair, `solplot` parametrically graphs the solution satisfying $x(0) = x_0$ and $y(0) = y_0$ for $0 \le t \le 15$.

solplot[pair_]:=

ParametricPlot[

```
Evaluate[{x[t], y[t]}/.initsol/.{x0 → pair[[1]], y0 → pair[[2]]}],
{t, 0, 15}, PlotStyle → {{CMYKColor[0, 0, 0, .25], Thickness[.005]}}]
```

We then define a list of ordered pairs with `Table` followed by `Flatten`

```
Clear[i, j]
orderedpairs = Flatten[Table[{i, j}, {i, −1, 1, 1/4}, {j, −1, 1, 1/4}], 1];
Short[orderedpairs]
```

$\{\{-1,-1\},\{-1,-\frac{3}{4}\},\{-1,-\frac{1}{2}\},\{-1,-\frac{1}{4}\},\langle\langle 73\rangle\rangle,\{1,\frac{1}{4}\},\{1,\frac{1}{2}\},\{1,\frac{3}{4}\},\{1,1\}\}$

and use `Map` to apply `solplot` to `orderedpairs`.

```
toshow = Map[solplot, orderedpairs];
```

The resulting list of graphics objects is displayed together with `Show`. See Fig. 6.27.

```
pvf = StreamPlot[{−y, −2x − y}, {x, −1, 1}, {y, −1, 1},
  StreamStyle → CMYKColor[.02, 1, .85, .06], StreamPoints → Fine, Frame → False,
  Axes → Automatic, AxesOrigin → {0, 0}, AxesLabel → {x, y}];

Show[toshow, pvf, PlotRange → {{−1, 1}, {−1, 1}},
  PlotLabel → "The origin is a saddle."]
```

(c) In matrix form the system is equivalent to the system $\mathbf{X}' = \begin{pmatrix} 0 & -2 \\ 1 & 3 \end{pmatrix}\mathbf{X}$. As in (a), we use `Eigensystem` to see that the eigenvalues and eigenvectors of $\mathbf{A} = \begin{pmatrix} 0 & -2 \\ 1 & 3 \end{pmatrix}$ are $\lambda_1 = 2$ and $\lambda_1 = 1$ with corresponding eigenvectors $\mathbf{v}_1 = \begin{pmatrix} -1 \\ 1 \end{pmatrix}$ and $\mathbf{v}_2 = \begin{pmatrix} -2 \\ 1 \end{pmatrix}$.

```
capa = {{0, −2}, {1, 3}};
Eigensystem[capa]
```

{{2, 1}, {{−1, 1}, {−2, 1}}}

Two linearly independent solutions are then $\mathbf{X}_1 = \begin{pmatrix} -1 \\ 1 \end{pmatrix} e^{2t}$ and $\mathbf{X}_2 = \begin{pmatrix} -2 \\ 1 \end{pmatrix} e^{t}$. A general solution is $\mathbf{X} = c_1\mathbf{X}_1 + c_2\mathbf{X}_2 = \begin{pmatrix} -e^{2t} & -2e^{t} \\ e^{2t} & e^{t} \end{pmatrix}\begin{pmatrix} c_1 \\ c_2 \end{pmatrix}$ or $x = -c_1e^{2t} - 2c_2e^{t}$ and $y = c_1e^{2t} + c_2e^{t}$.

As before, we use `DSolve` to find a general solution and confirm the result obtained above.

```
Clear[x, y]
gensol = DSolve[{x'[t] == −2y[t],
```

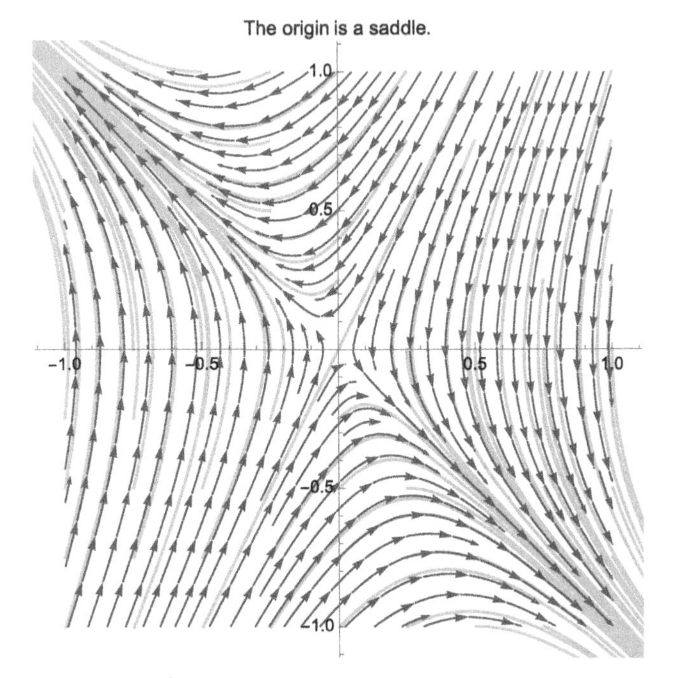

FIGURE 6.27 Direction field for $\mathbf{X}' = \mathbf{AX}$ along with various solution curves.

```
y'[t] == x[t] + 3y[t]}, {x[t], y[t]}, t]
```

$\{\{x[t] \to -e^t(-2 + e^t)C[1] - 2e^t(-1 + e^t)C[2], y[t] \to e^t(-1 + e^t)C[1] + e^t(-1 + 2e^t)C[2]\}\}$

We graph the direction field using `StreamPlot` and the eigenlines in Fig. 6.28. Observe that all nontrivial solutions move away from the origin following the eigenlines.

```
sp1 = StreamPlot[{−2y, x + 3y}, {x, −1, 1}, {y, −1, 1},
  StreamStyle → CMYKColor[.02, 1, .85, .06], StreamPoints → Fine, Frame → False,
  Axes → Automatic, AxesOrigin → {0, 0}, AxesLabel → {x, y}];
p1 = Plot[{−x, −1/2x}, {x, −1, 1},
  PlotStyle → {{CMYKColor[0, 0, 0, .25], Thickness[.01]},
  {CMYKColor[0, 0, 0, .25], Thickness[.01]}}];
Show[sp1, p1,
  PlotLabel → "All nontrivial trajectories move away from the origin.",
  PlotRange → {{−1, 1}, {−1, 1}}]
```

(d) In matrix form, the system is $\mathbf{X}' = \begin{pmatrix} 3 & -5 \\ 5 & -3 \end{pmatrix} \mathbf{X}$. The coefficient matrix $\mathbf{A} = \begin{pmatrix} 3 & -5 \\ 5 & -3 \end{pmatrix}$ has eigenvalues $\lambda_{1,2} = \pm 4i$ with corresponding eigenvectors $\mathbf{v}_{1,2} = \begin{pmatrix} 3 \pm 4i \\ 5 \end{pmatrix}$.

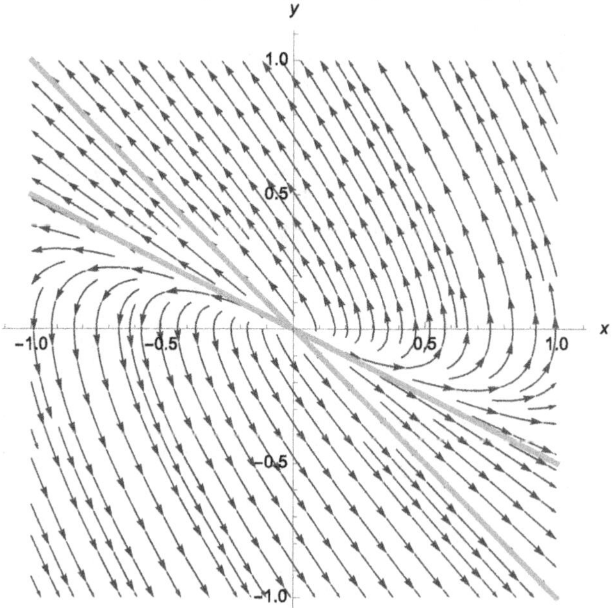

FIGURE 6.28 Direction field for $\mathbf{X}' = \mathbf{A}\mathbf{X}$ with eigenlines.

```
capa = {{3, -5}, {5, -3}};
Eigensystem[capa]
```

$\{\{4i, -4i\}, \{\{3+4i, 5\}, \{3-4i, 5\}\}\}$

As we did previously, we solve for the solutions satisfying $x(0) = x_0$ and $y(0) = y_0$ with `DSolve`.

```
Clear[x, y]
gensol = DSolve[{x'[t] == 3x[t] - 5y[t],
  y'[t] == 5x[t] - 3y[t], x[0] == x0, y[0] == y0}, {x[t], y[t]}, t]
```

$\{\{x[t] \to \frac{1}{4}(4\text{x0Cos}[4t] + 3\text{x0Sin}[4t] - 5\text{y0Sin}[4t]), y[t] \to \frac{1}{4}(4\text{y0Cos}[4t] + 5\text{x0Sin}[4t] - 3\text{y0Sin}[4t])\}\}$

Given an ordered pair, `solplot` parametrically graphs the solution satisfying $x(0) = x_0$ and $y(0) = y_0$ for $0 \le t \le 15$.

```
solplot[pair_]:=
  ParametricPlot[
  Evaluate[{x[t], y[t]}/.gensol/.{x0 → pair[[1]], y0 → pair[[2]]}],
  {t, 0, 15}, PlotStyle → {{CMYKColor[0, 0, 0, .25], Thickness[.005]}}]
```

We then define a list of ordered pairs with `Table` followed by `Flatten`

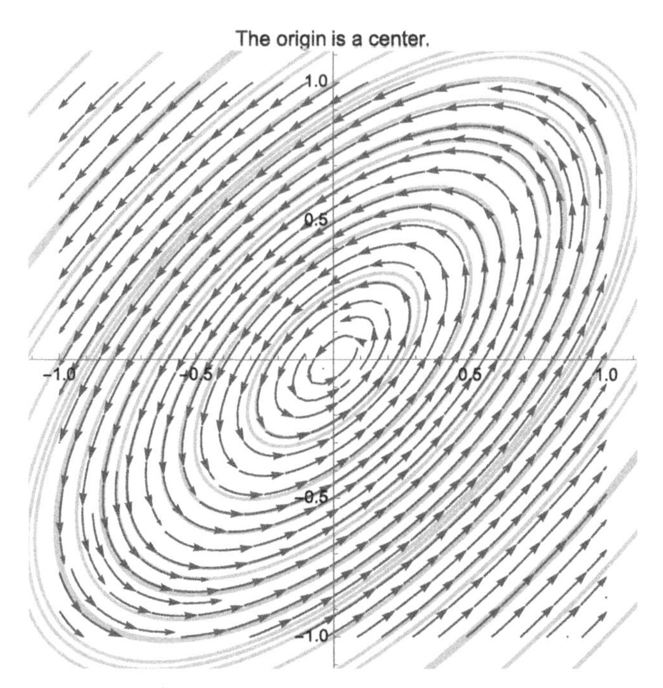

FIGURE 6.29 Direction field for $\mathbf{X}' = \mathbf{AX}$ with solution curves.

```
Clear[i, j]
orderedpairs = Flatten[Table[{i, j}, {i, −1, 1, 1/4}, {j, −1, 1, 1/4}], 1];
Short[orderedpairs]
```

and use `Map` to apply `solplot` to `orderedpairs`.

```
toshow = Map[solplot, orderedpairs];
```

```
sp1 = StreamPlot[{3x − 5y, 5x − 3y}, {x, −1, 1}, {y, −1, 1},
  StreamStyle → CMYKColor[.02, 1, .85, .06], StreamPoints → Fine, Frame → False,
  Axes → Automatic, AxesOrigin → {0, 0}, AxesLabel → {x, y},
  PlotLabel → "The origin is a center."]
```

We use `StreamPlot` to graph the direction field and show the various solutions obtained in `toshow` together with the direction field in Fig. 6.29.

```
Show[toshow, sp1, PlotRange → {{−1, 1}, {−1, 1}},
  PlotLabel → "The origin is a center."]
```

For (e), we proceed in much the same way as with the previous examples. In matrix form, observe that the system can be written as $\mathbf{X}' = \begin{pmatrix} 2 & -8 \\ 4 & -6 \end{pmatrix} \mathbf{X}$. We find the eigenvalues and corresponding eigenvectors of $\mathbf{A} = \begin{pmatrix} 2 & -8 \\ 4 & -6 \end{pmatrix}$ with `Eigensystem`.

```
capa = {{2, -8}, {4, -6}};
Eigensystem[capa]
```

{{−2 + 4i, −2 − 4i}, {{1 + i, 1}, {1 − i, 1}}}

Similarly, we use DSolve to solve $\mathbf{X}' = \begin{pmatrix} 2 & -8 \\ 4 & -6 \end{pmatrix}\mathbf{X}$, $\mathbf{X}(0) = \mathbf{X}_0$.

```
Clear[x, y]
gensol = DSolve[{x'[t] == 2x[t] - 8y[t],
  y'[t] == 4x[t] - 6y[t], x[0] == x0, y[0] == y0}, {x[t], y[t]}, t]
```

{{x[t] → −e^{−2t}(−x0Cos[4t] − x0Sin[4t] + 2y0Sin[4t]), y[t] → e^{−2t}(y0Cos[4t] + x0Sin[4t] − y0Sin[4t])}}

We use StreamPlot to generate a direction field in sp1.

```
sp1 = StreamPlot[{2x - 8y, 4x - 6y}, {x, -1, 1}, {y, -1, 1},
  StreamStyle → CMYKColor[.02, 1, .85, .06], StreamPoints → Fine, Frame → False,
  Axes → Automatic, AxesOrigin → {0, 0}, AxesLabel → {x, y},
  PlotLabel → "The origin is a stable spiral."]
```

Next, we define solplot to graph the solution of the system that satisfies $x(0) = x_0$ and $y(0) = y_0$.

```
solplot[pair_]:=
ParametricPlot[
Evaluate[{x[t], y[t]}/.gensol/.{x0 → pair[[1]], y0 → pair[[2]]}],
{t, 0, 15}, PlotStyle → {{CMYKColor[0, 0, 0, .25], Thickness[.005]}}]

Clear[i, j]
orderedpairs = Flatten[Table[{i, j}, {i, -1, 1, 1/4}, {j, -1, 1, 1/4}], 1];
Short[orderedpairs]
```

$\{\{-1, -1\}, \{-1, -\frac{3}{4}\}, \{-1, -\frac{1}{2}\}, \{-1, -\frac{1}{4}\}, \langle\langle 73\rangle\rangle, \{1, \frac{1}{4}\}, \{1, \frac{1}{2}\}, \{1, \frac{3}{4}\}, \{1, 1\}\}$

We use Map to apply solplot to orderedpairs in toshow.

```
toshow = Map[solplot, orderedpairs];
```

Finally, we use Show to display the direction field together with various solutions in Fig. 6.30.

```
Show[toshow, sp1, PlotRange → {{-1, 1}, {-1, 1}},
  PlotLabel → "The origin is a stable spiral."]
```

□

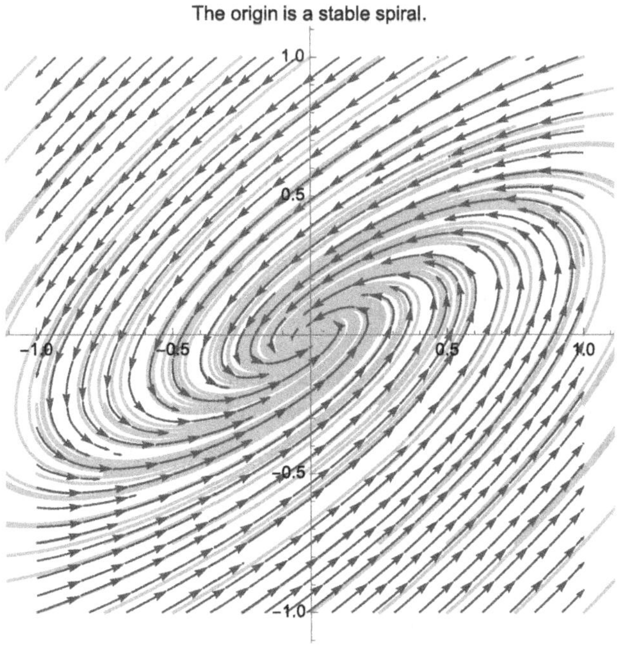

The origin is a stable spiral.

FIGURE 6.30 Direction field for $\mathbf{X}' = \mathbf{AX}$ with solution curves. Observe that all nontrivial solutions spiral into the origin.

Application: The Double Pendulum

The motion of a double pendulum is modeled by the system of differential equations

$$\begin{cases} (m_1+m_2)l_1{}^2\dfrac{d^2\theta_1}{dt^2}+m_2l_1l_2\dfrac{d^2\theta_2}{dt^2}+(m_1+m_2)l_1g\theta_1=0 \\ m_2l_2{}^2\dfrac{d^2\theta_2}{dt^2}+m_2l_1l_2\dfrac{d^2\theta_1}{dt^2}+m_2l_2g\theta_2=0 \end{cases}$$

using the approximation $\sin\theta \approx \theta$ for small displacements. θ_1 represents the displacement of the upper pendulum and θ_2 that of the lower pendulum. Also, m_1 and m_2 represent the mass attached to the upper and lower pendulums, respectively, while the length of each is given by l_1 and l_2.

Example 6.26

Suppose that $m_1 = 3$, $m_2 = 1$, and each pendulum has length 16. If $\theta_1(0) = 1$, $\theta_1{}'(0) = 0$, $\theta_2(0) = -1$, and $\theta_2{}'(0) = 0$, solve the double pendulum problem using $g = 32$. Plot the solution.

Solution. In this case, the system to be solved is

$$\begin{cases} 4\cdot 16^2\dfrac{d^2\theta_1}{dt^2}+16^2\dfrac{d^2\theta_2}{dt^2}+4\cdot 16\cdot 32\theta_1=0 \\ 16^2\dfrac{d^2\theta_2}{dt^2}+16^2\dfrac{d^2\theta_1}{dt^2}+16\cdot 32\theta_2=0, \end{cases}$$

which we simplify to obtain

$$\begin{cases} 4\dfrac{d^2\theta_1}{dt^2}+\dfrac{d^2\theta_2}{dt^2}+8\theta_1=0 \\ \dfrac{d^2\theta_2}{dt^2}+\dfrac{d^2\theta_1}{dt^2}+2\theta_2=0. \end{cases}$$

In the following code, we let $x(t)$ and $y(t)$ represent $\theta_1(t)$ and $\theta_2(t)$, respectively. First, we use `DSolve` to solve the initial-value problem.

```
sol = DSolve[{4x''[t] + y''[t] + 8x[t]==0, x''[t] + y''[t] + 2y[t]==0,
  x[0]==1, x'[0]==1, y[0]==0, y'[0]== - 1}, {x[t], y[t]}, t]
```

$$\{\{x[t] \to \tfrac{1}{8}(4\text{Cos}[2t] + 4\text{Cos}[\tfrac{2t}{\sqrt{3}}] + 3\text{Sin}[2t] + \sqrt{3}\text{Sin}[\tfrac{2t}{\sqrt{3}}]), y[t] \to \tfrac{1}{4}(-4\text{Cos}[2t]$$
$$+ 4\text{Cos}[\tfrac{2t}{\sqrt{3}}] - 3\text{Sin}[2t] + \sqrt{3}\text{Sin}[\tfrac{2t}{\sqrt{3}}])\}\}$$

The **Laplace transform** of $y = f(t)$ is $F(s) = \mathcal{L}\{f(t)\} = \int_0^\infty e^{-st} f(t)\,dt$.

To solve the initial-value problem using traditional methods, we use the *method of Laplace transforms*. To do so, we define `sys` to be the system of equations and use `LaplaceTransform` to compute the Laplace transform of each equation.

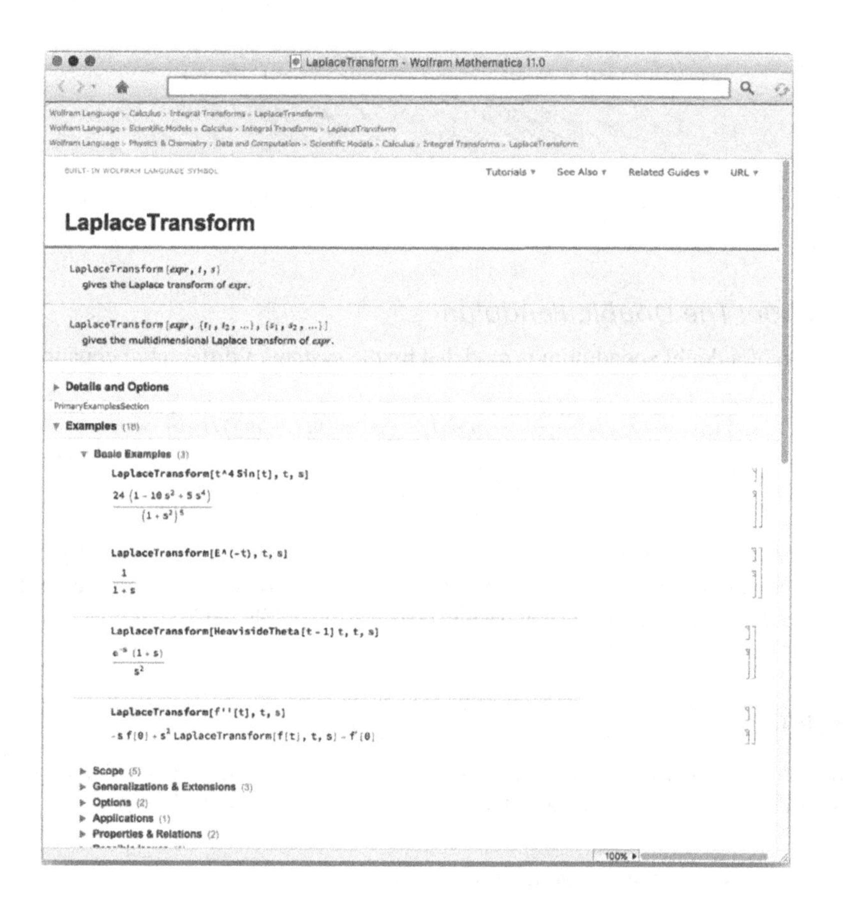


```
step1 = LaplaceTransform[sys, t, s]
```

$$\{8\text{LaplaceTransform}[x[t], t, s] + s^2\text{LaplaceTransform}[y[t], t, s] - sy[0]$$
$$+ 4(s^2\text{LaplaceTransform}[x[t], t, s] - sx[0] - x'[0]) - y'[0] == 0,$$
$$s^2\text{LaplaceTransform}[x[t], t, s] + 2\text{LaplaceTransform}[y[t], t, s]$$
$$+ s^2\text{LaplaceTransform}[y[t], t, s] - sx[0] - sy[0] - x'[0] - y'[0] == 0\}$$

Next, we apply the initial conditions and solve the resulting system of equations for $\mathcal{L}\{\theta_1(t)\} = X(s)$ and $\mathcal{L}\{\theta_2(t)\} = Y(s)$.

step2 = step1/.{x[0]->1, x'[0]->1, y[0]->0, y'[0]-> − 1}

{1 + 8LaplaceTransform[x[t], t, s] + 4(−1 − s + s²LaplaceTransform[x[t], t, s])

+ s²LaplaceTransform[y[t], t, s] == 0, −s + s²LaplaceTransform[x[t], t, s]

+ 2LaplaceTransform[y[t], t, s] + s²LaplaceTransform[y[t], t, s] == 0}

step3 =

Solve[step2, {LaplaceTransform[x[t], t, s],

LaplaceTransform[y[t], t, s]}]

{{LaplaceTransform[x[t], t, s] → $-\frac{-6-8s-3s^2-3s^3}{16+16s^2+3s^4}$, LaplaceTransform[y[t], t, s]

→ $-\frac{s(-8+3s)}{16+16s^2+3s^4}$}}

`InverseLaplaceTransform` is then used to find $\theta_1(t)$ and $\theta_2(t)$.

$f(t)$ is the **inverse Laplace transform** of $F(s)$ if $\mathcal{L}\{f(t)\} = F(s)$; we write $\mathcal{L}^{-1}\{F(s)\} = f(t)$.

x[t_] = InverseLaplaceTransform[$-\frac{-6-8s-3s^2-3s^3}{16+16s^2+3s^4}$, s, t]

$\frac{1}{8}$(4Cos[2t] + 4Cos[$\frac{2t}{\sqrt{3}}$] + 3Sin[2t] + $\sqrt{3}$Sin[$\frac{2t}{\sqrt{3}}$])

y[t_] = InverseLaplaceTransform[$-\frac{-8s+3s^2}{16+16s^2+3s^4}$, s, t]

$\frac{1}{4}$(−4Cos[2t] + 4Cos[$\frac{2t}{\sqrt{3}}$] − 3Sin[2t] + $\sqrt{3}$Sin[$\frac{2t}{\sqrt{3}}$])

These two functions are graphed together in Fig. 6.31 (a) and parametrically in Fig. 6.31 (b).

p1 = Plot[{x[t], y[t]}, {t, 0, 20},

PlotStyle → {{CMYKColor[.93, 0, 1, 0]}, {CMYKColor[0, .7, 1, 0]}}]

p2 = ParametricPlot[{x[t], y[t]}, {t, 0, 20},

PlotRange->{{−5/2, 5/2}, {−5/2, 5/2}}, AspectRatio->1,

PlotStyle->CMYKColor[0, .39, .1, 0]]

Show[GraphicsRow[{p1, p2}]]

We can illustrate the motion of the pendulum as follows. First, we define the function `pen2`.

Clear[pen2]

pen2[t_, len1_, len2_]:=Module[{pt1, pt2},

pt1 = {len1Cos[$\frac{3\pi}{2}$ + x[t]], len1Sin[$\frac{3\pi}{2}$ + x[t]]};

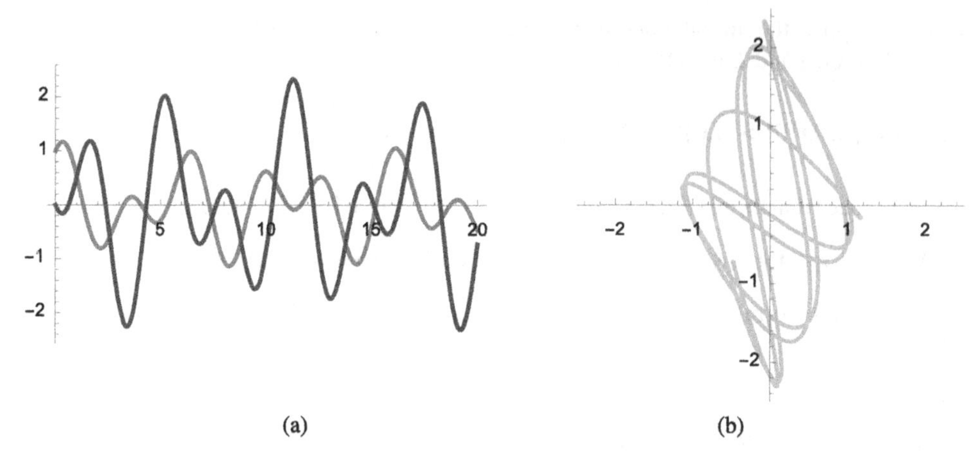

FIGURE 6.31 (a) $\theta_1(t)$ and $\theta_2(t)$ as functions of t. (b) Parametric plot of $\theta_1(t)$ versus $\theta_2(t)$. (University of North Dakota colors)

```
    pt2 = {len1Cos[3π/2 + x[t]] + len2Cos[3π/2 + y[t]],
  len1Sin[3π/2 + x[t]] + len2Sin[3π/2 + y[t]]};


Show[Graphics[{CMYKColor[.93, 0, 1, 0], Line[{{0, 0}, pt1}],
  PointSize[.05], Point[pt1], Line[{pt1, pt2}], PointSize[.05],
  Point[pt2]}], Axes → Automatic, Ticks → None, AxesStyle → GrayLevel[.5],
  PlotRange → {{−32, 32}, {−34, 0}}]];
```

Next, we define `tvals` to be a list of sixteen evenly spaced numbers between 0 and 10. `Map` is then used to apply `pen2` to the list of numbers in `tvals`. The resulting set of graphics is partitioned into four element subsets and displayed using `Show` and `GraphicsGrid` in Fig. 6.32.

```
tvals = Table[t, {t, 0, 10, 10/15}];

graphs = Map[pen2[#, 16, 16]&, tvals];

toshow = Partition[graphs, 4];

Show[GraphicsGrid[toshow]]
```

To animate the double pendulum, use `Animate` (see Fig. 6.33).

```
Clear[pen2]
pen2[t_, len1_, len2_]:=Module[{pt1, pt2},
    pt1 = {len1Cos[3π/2 + x[t]], len1 * Sin[3π/2 + x[t]]};


pt2 = {len1Cos[3π/2 + x[t]] + len2Cos[3π/2 + y[t]],
```

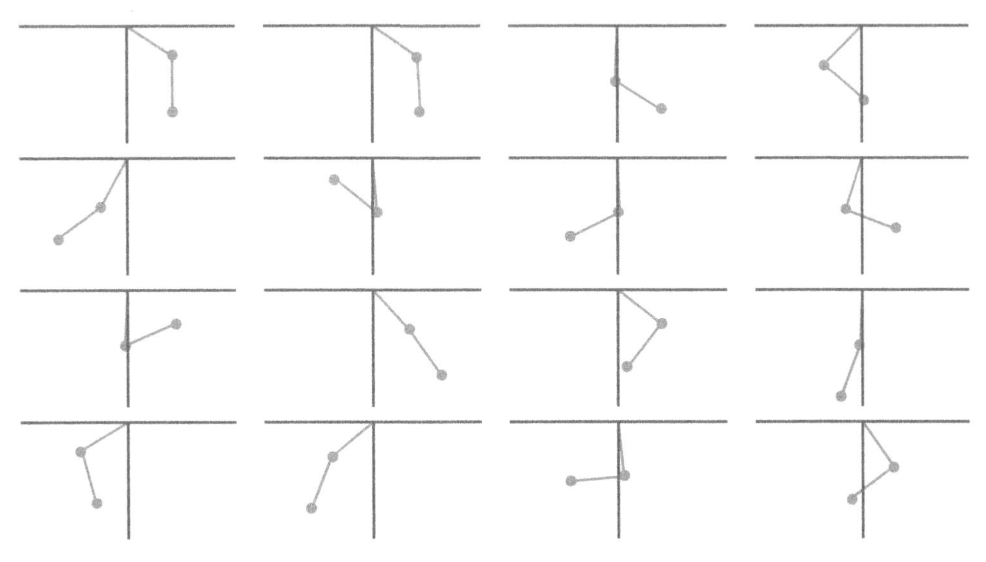

FIGURE 6.32 The double pendulum for 16 equally spaced values of t between 0 and 10.

$\text{len1} * \text{Sin}[\frac{3\pi}{2} + x[t]] + \text{len2} * \text{Sin}[\frac{3\pi}{2} + y[t]]\}$;

```
Show[Graphics[{CMYKColor[0, .7, 1, 0], Line[{{0, 0}, pt1}], PointSize[.05],
  Point[pt1], Line[{pt1, pt2}], PointSize[.05], Point[pt2]}],
  Axes → Automatic, Ticks → None, AxesStyle → GrayLevel[.5],
  PlotRange → {{−32, 32}, {−34, 0}}]]
```

We show a frame from the animation that results from the `Animate` command.

```
Animate[pen2[t, 16, 16], {t, 0, 10}]
```

Alternatively, you can use `Manipulate`

```
Clear[pen2]
Manipulate[
```

$x[\text{t\_}] = \frac{1}{8}(4\text{Cos}[2t] + 4\text{Cos}[\frac{2t}{\sqrt{3}}] + 3\text{Sin}[2t] + \sqrt{3}\text{Sin}[\frac{2t}{\sqrt{3}}]);$

$y[\text{t\_}] = \frac{1}{4}(-4\text{Cos}[2t] + 4\text{Cos}[\frac{2t}{\sqrt{3}}] - 3\text{Sin}[2t] + \sqrt{3}\text{Sin}[\frac{2t}{\sqrt{3}}]);$

pen2[t_, len1_, len2_]:=Module[{pt1, pt2},

$\text{pt1} = \{\text{len1Cos}[\frac{3\pi}{2} + x[t]], \text{len1Sin}[\frac{3\pi}{2} + x[t]]\};$

$\text{pt2} = \{\text{len1Cos}[\frac{3\pi}{2} + x[t]] + \text{len2Cos}[\frac{3\pi}{2} + y[t]],$

$\text{len1Sin}[\frac{3\pi}{2} + x[t]] + \text{len2Sin}[\frac{3\pi}{2} + y[t]]\};$

```
Show[Graphics[{CMYKColor[.93, 0, 1, 0], Line[{{0, 0}, pt1}],
```

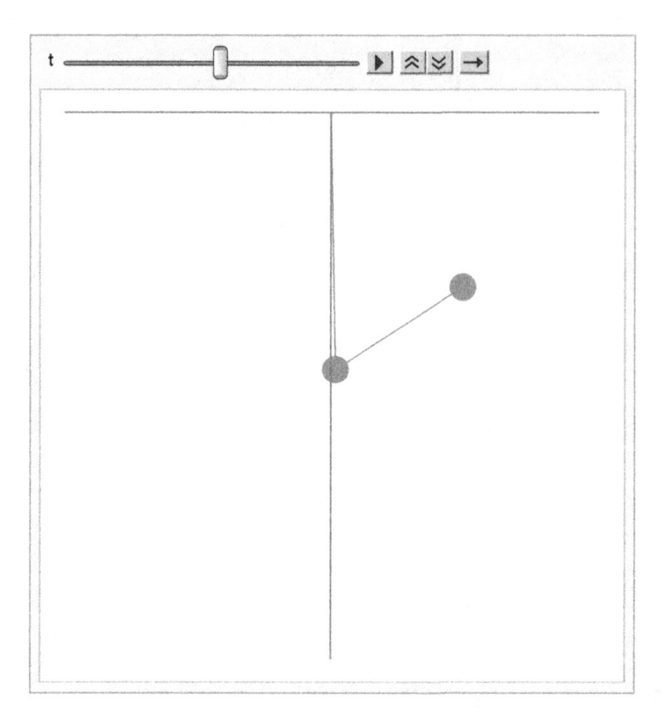

FIGURE 6.33 Animating the double pendulum with `Animate`.

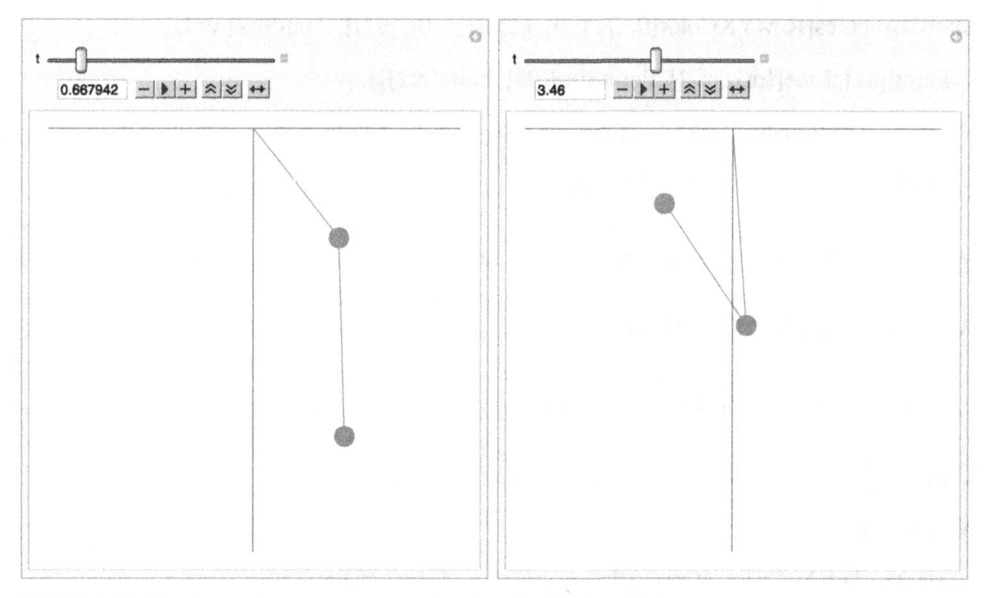

FIGURE 6.34 Two frames from a `Manipulate` animation of a double pendulum.

```
PointSize[.05], Point[pt1], Line[{pt1, pt2}], PointSize[.05],
Point[pt2]}], Axes → Automatic, Ticks → None, AxesStyle → GrayLevel[.5],
AspectRatio → 1, PlotRange → {{−32, 32}, {−34, 0}}]];
pen2[t, 16, 16], {{t, 0}, 0, 6}]
```

to generate a nearly identical animation as shown in Fig. 6.34. □

6.4.2 Nonhomogeneous Linear Systems

Generally, the method of undetermined coefficients is difficult to implement for nonhomogeneous linear systems as the choice for the particular solution must be very carefully made. Variation of parameters is implemented in much the same way as for first-order linear equations.

Let $\mathbf{X}_h$ be a general solution to the corresponding homogeneous system of equation (6.28), $\mathbf{X}$ a general solution of equation (6.28), and $\mathbf{X}_p$ a particular solution of equation (6.28). It then follows that $\mathbf{X}-\mathbf{X}_p$ is a solution to the corresponding homogeneous system so $\mathbf{X}-\mathbf{X}_p = \mathbf{X}_h$ and, consequently, $\mathbf{X}=\mathbf{X}_h+\mathbf{X}_p$. A particular solution of equation (6.28) is found in much the same way as with first-order linear equations. Let $\mathbf{\Phi}$ be a fundamental matrix for the corresponding homogeneous system. We assume that a particular solution has the form $\mathbf{X}_p=\mathbf{\Phi}\mathbf{U}(t)$. Differentiating $\mathbf{X}_p$ gives us

$$\mathbf{X}_p{}' = \mathbf{\Phi}'\mathbf{U}+\mathbf{\Phi}\mathbf{U}'.$$

Substituting into equation (6.28) results in

Because the fundamental matrix $\mathbf{\Phi}$ satisfies the homogeneous system $\mathbf{X}'=\mathbf{A}(t)\mathbf{X}$, $\mathbf{\Phi}'=\mathbf{A}(t)\mathbf{\Phi}$.

$$\begin{aligned}\mathbf{\Phi}'\mathbf{U}+\mathbf{\Phi}\mathbf{U}' &= \mathbf{A}\mathbf{\Phi}\mathbf{U}+\mathbf{F}(t)\\ \mathbf{\Phi}\mathbf{U}' &= \mathbf{F}(t)\\ \mathbf{U}' &= \mathbf{\Phi}^{-1}\mathbf{F}(t)\\ \mathbf{U} &= \int \mathbf{\Phi}^{-1}\mathbf{F}(t)\,dt,\end{aligned}$$

where we have used the fact that $\mathbf{\Phi}'\mathbf{U}-\mathbf{A}\mathbf{\Phi}\mathbf{U}=\left(\mathbf{\Phi}'-\mathbf{A}\mathbf{\Phi}\right)\mathbf{U}=\mathbf{0}$. It follows that

$$\mathbf{X}_p=\mathbf{\Phi}\int \mathbf{\Phi}^{-1}\mathbf{F}(t)\,dt. \tag{6.31}$$

A general solution is then

$$\begin{aligned}\mathbf{X} &= \mathbf{X}_h+\mathbf{X}_p\\ &= \mathbf{\Phi}\mathbf{C}+\mathbf{\Phi}\int \mathbf{\Phi}^{-1}\mathbf{F}(t)\,dt\\ &= \mathbf{\Phi}\left(\mathbf{C}+\int \mathbf{\Phi}^{-1}\mathbf{F}(t)\,dt\right)=\mathbf{\Phi}\int \mathbf{\Phi}^{-1}\mathbf{F}(t)\,dt,\end{aligned}$$

where we have incorporated the constant vector $\mathbf{C}$ into the indefinite integral $\int \mathbf{\Phi}^{-1}\mathbf{F}(t)\,dt$.

Example 6.27

Solve the initial-value problem

$$\mathbf{X}'=\begin{pmatrix}1 & -1\\ 10 & -1\end{pmatrix}\mathbf{X}-\begin{pmatrix}t\cos 3t\\ t\sin t+t\cos 3t\end{pmatrix},\quad \mathbf{X}(0)=\begin{pmatrix}1\\ -1\end{pmatrix}.$$

Remark 6.5. In traditional form, the system is equivalent to

$$\begin{cases}x'=x-y-t\cos 3t\\ y'=10x-y-t\sin t-t\cos 3t,\end{cases}\qquad x(0)=1,\ y(0)=-1.$$

Solution. The corresponding homogeneous system is $\mathbf{X}_h'=\begin{pmatrix}1 & -1\\ 10 & -1\end{pmatrix}\mathbf{X}_h$. The eigenvalues and corresponding eigenvectors of $\mathbf{A}=\begin{pmatrix}1 & -1\\ 10 & -1\end{pmatrix}$ are $\lambda_{1,2}=\pm 3i$ and $\mathbf{v}_{1,2}=\begin{pmatrix}1\\ 10\end{pmatrix}\pm\begin{pmatrix}-3\\ 0\end{pmatrix}i$, respectively.

```
capa = {{1, −1}, {10, −1}};
```

Eigensystem[capa]

$\{\{3i, -3i\}, \{\{1+3i, 10\}, \{1-3i, 10\}\}\}$

A fundamental matrix is $\mathbf{\Phi} = \begin{pmatrix} \sin 3t & \cos 3t \\ \sin 3t - 3\cos 3t & \cos 3t + 3\sin 3t \end{pmatrix}$ with inverse $\mathbf{\Phi}^{-1} = \begin{pmatrix} \frac{1}{3}\cos 3t + \sin 3t & -\frac{1}{3}\cos 3t \\ -\frac{1}{3}\sin 3t + \cos 3t & \frac{1}{3}\sin 3t \end{pmatrix}$.

fm = {{Sin[3t], Sin[3t] − 3Cos[3t]}, {Cos[3t], Cos[3t] + 3Sin[3t]}};

fminv = Inverse[fm]//Simplify

$\left\{\left\{\frac{1}{3}\text{Cos}[3t] + \text{Sin}[3t], \text{Cos}[3t] - \frac{1}{3}\text{Sin}[3t]\right\}, \left\{-\frac{1}{3}\text{Cos}[3t], \frac{1}{3}\text{Sin}[3t]\right\}\right\}$

We now compute $\mathbf{\Phi}^{-1}\mathbf{F}(t)$

ft = {−tCos[3t], −tSin[t] − tCos[3t]};

step1 = fminv.ft

$\{(-t\text{Cos}[3t] - t\text{Sin}[t])(\text{Cos}[3t] - \frac{1}{3}\text{Sin}[3t]) - t\text{Cos}[3t](\frac{1}{3}\text{Cos}[3t] + \text{Sin}[3t]), \frac{1}{3}t\text{Cos}[3t]^2 + \frac{1}{3}(-t\text{Cos}[3t] - t\text{Sin}[t])\text{Sin}[3t]\}$

and $\int \mathbf{\Phi}^{-1}\mathbf{F}(t)\,dt$.

step2 = Integrate[step1, t]

$\{-\frac{t^2}{3} + \frac{1}{24}\text{Cos}[2t] - \frac{1}{4}t\text{Cos}[2t] - \frac{1}{96}\text{Cos}[4t] + \frac{1}{8}t\text{Cos}[4t] - \frac{1}{54}\text{Cos}[6t] + \frac{1}{18}t\text{Cos}[6t] + \frac{1}{8}\text{Sin}[2t] + \frac{1}{12}t\text{Sin}[2t] - \frac{1}{32}\text{Sin}[4t] - \frac{1}{24}t\text{Sin}[4t] - \frac{1}{108}\text{Sin}[6t] - \frac{1}{9}t\text{Sin}[6t], \frac{t^2}{12} - \frac{1}{24}\text{Cos}[2t] + \frac{1}{96}\text{Cos}[4t] + \frac{1}{216}\text{Cos}[6t] + \frac{1}{36}t\text{Cos}[6t] - \frac{1}{12}t\text{Sin}[2t] + \frac{1}{24}t\text{Sin}[4t] - \frac{1}{216}\text{Sin}[6t] + \frac{1}{36}t\text{Sin}[6t]\}$

A general solution of the nonhomogeneous system is then $\mathbf{\Phi}\left(\int \mathbf{\Phi}^{-1}\mathbf{F}(t)\,dt + \mathbf{C}\right)$.

Simplify[fm.step2]

$\{\frac{1}{288}(27\text{Cos}[t] - 4((1 + 6t + 18t^2)\text{Cos}[3t] + 27t\text{Sin}[t] + (-1 + 6t + 18t^2)\text{Sin}[3t])), \frac{1}{288}(-36t\text{Cos}[t] - 4(1 - 6t + 18t^2)\text{Cos}[3t] - 45\text{Sin}[t] - 4\text{Sin}[3t] - 24t\text{Sin}[3t] + 72t^2\text{Sin}[3t])\}$

It is easiest to use `DSolve` to solve the initial-value problem directly as we do next.

Clear[x, y]

check = DSolve[{x′[t]==x[t] − y[t] − tCos[3t], y′[t] ==

10x[t] − y[t] − tSin[t] − tCos[3t], x[0]==1, y[0]== − 1},

{x[t], y[t]}, t]

$\{\{x[t] \to \frac{1}{288}(301\text{Cos}[3t] - 72t^2\text{Cos}[3t] - 12\text{Cos}[2t]\text{Cos}[3t] + 3\text{Cos}[3t]\text{Cos}[4t] -$

$4\text{Cos}[3t]\text{Cos}[6t] - 24t\text{Cos}[3t]\text{Sin}[2t] + 192\text{Sin}[3t] + 24t\text{Cos}[2t]\text{Sin}[3t] -$

$12t\text{Cos}[4t]\text{Sin}[3t] + 24t\text{Cos}[6t]\text{Sin}[3t] - 12\text{Sin}[2t]\text{Sin}[3t] + 12t\text{Cos}[3t]\text{Sin}[4t] +$

$3\text{Sin}[3t]\text{Sin}[4t] - 24t\text{Cos}[3t]\text{Sin}[6t] - 4\text{Sin}[3t]\text{Sin}[6t]), y[t] \to \frac{1}{288}(-275\text{Cos}[3t] -$

$72t^2\text{Cos}[3t] - 12\text{Cos}[2t]\text{Cos}[3t] - 72t\text{Cos}[2t]\text{Cos}[3t] + 3\text{Cos}[3t]\text{Cos}[4t] +$

$36t\text{Cos}[3t]\text{Cos}[4t] - 4\text{Cos}[3t]\text{Cos}[6t] - 72t\text{Cos}[3t]\text{Cos}[6t] + 36\text{Cos}[3t]\text{Sin}[2t] -$

$24t\text{Cos}[3t]\text{Sin}[2t] + 1095\text{Sin}[3t] - 216t^2\text{Sin}[3t] - 36\text{Cos}[2t]\text{Sin}[3t] + 24t\text{Cos}[2t]\text{Sin}[3t] +$

$9\text{Cos}[4t]\text{Sin}[3t] - 12t\text{Cos}[4t]\text{Sin}[3t] - 12\text{Cos}[6t]\text{Sin}[3t] + 24t\text{Cos}[6t]\text{Sin}[3t] -$

$12\text{Sin}[2t]\text{Sin}[3t] - 72t\text{Sin}[2t]\text{Sin}[3t] - 9\text{Cos}[3t]\text{Sin}[4t] + 12t\text{Cos}[3t]\text{Sin}[4t] +$

$3\text{Sin}[3t]\text{Sin}[4t] + 36t\text{Sin}[3t]\text{Sin}[4t] + 12\text{Cos}[3t]\text{Sin}[6t] - 24t\text{Cos}[3t]\text{Sin}[6t] -$

$4\text{Sin}[3t]\text{Sin}[6t] - 72t\text{Sin}[3t]\text{Sin}[6t])\}\}$

The solutions are graphed with `Plot` and `ParametricPlot` in Fig. 6.35.

```
p1 = Plot[Evaluate[{x[t], y[t]}/.check], {t, 0, 8π}, PlotRange → All,
  PlotStyle → {{CMYKColor[0, 1, .65, .34]}, {CMYKColor[.06, .08, .23, 0]}},
  AxesOrigin → {0, 0}];
p2 = ParametricPlot[Evaluate[{x[t], y[t]}/.check], {t, 0, 8π},
  PlotStyle → CMYKColor[0, 1, .65, .34],
  AspectRatio → Automatic];
Show[GraphicsRow[{p1, p2}]]
```

□

In the case that **A** is constant, $\mathbf{X}' = \mathbf{AX}$ is called an *autonomous system* and the only *equilibrium* (*rest point*) solution is the zero solution: $\mathbf{X} = \mathbf{0}$. The stability of the solution is determined by the eigenvalues of **A**. If all the eigenvalues of **A** have negative real part, then $\mathbf{X} = \mathbf{0}$ is **globally asymptotically stable** because $\lim_{t\to 0}\mathbf{X}(t) = \mathbf{0}$ for *all* solutions. In the case when the real parts of all the eigenvalues of **A** are 0, the origin, $\mathbf{X} = \mathbf{0}$ is a **center** and classified as stable. If not all the eigenvalues of **A** have negative real part, unless they are all 0, then $\mathbf{X} = \mathbf{0}$ is unstable.

For the 2×2 system, $\mathbf{X}' = \begin{pmatrix} a & b \\ c & d \end{pmatrix}\mathbf{X}$ or, equivalently, $x' = ax + by$, $y' = cx + dy$, the stability of $(0, 0)$ is easily seen by examining the direction field for the system. If all vectors lead to the origin, the origin, $(0, 0)$, is stable. With the exception of a center, if not all solutions tend to the origin, then the origin is not stable; it is unstable.

Example 6.28

The eigenvalues of $\begin{pmatrix} -\alpha & \beta \\ -\beta & 0 \end{pmatrix}$ are $\lambda_{1,2} = \frac{1}{2}\left(-a \pm \sqrt{a^2 - 4b^2}\right)$. Thus, $(0, 0)$ is globally asymptotically stable for the system $\mathbf{X}' = \begin{pmatrix} -\alpha & \beta \\ -\beta & 0 \end{pmatrix}\mathbf{X}$.

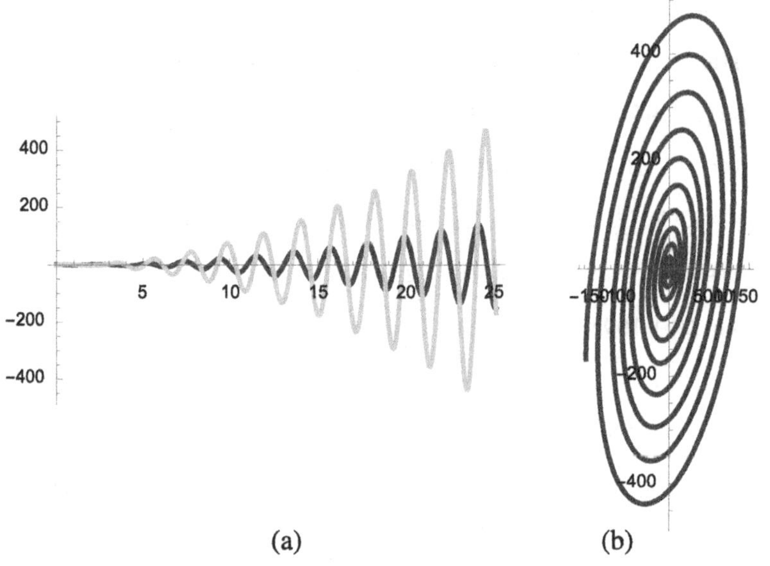

FIGURE 6.35 (a) Graph of $x(t)$ and $y(t)$; (b) Parametric plot of $x(t)$ versus $y(t)$. (University of Oklahoma colors)

With `Manipulate`, you can investigate the various situations. In the following, we can vary α and β and then plot the solution passing through each locator point. Several results are shown in Fig. 6.36.

```
Manipulate[
sol1 = DSolve[{x'[t] == -αx[t] + βy[t], y'[t]== - βx[t],
  x[0] == init1[[1]], y[0] == init1[[2]]}, {x[t], y[t]}, t];
sol2 = DSolve[{x'[t] == -αx[t] + βy[t], y'[t]== - βx[t],
  x[0] == init2[[1]], y[0] == init2[[2]]}, {x[t], y[t]}, t];
psol1 = ParametricPlot[{x[t], y[t]}/.sol1, {t, -20, 20},
  PlotStyle → {{GrayLevel[.3], Thickness[.01]}}, PlotPoints → 200];
psol2 = ParametricPlot[{x[t], y[t]}/.sol2, {t, -20, 20},
  PlotStyle → {{GrayLevel[.6], Thickness[.01]}}, PlotPoints → 200];
p1 = Show[{StreamPlot[{-αx + βy, -βx}, {x, -1, 1}, {y, -1, 1},
  StreamStyle → CMYKColor[0, .63, 1, 0]], psol1, psol2},
  Axes → Automatic, AxesOrigin → {0, 0}, PlotRange → {{-1, 1}, {-1, 1}},
  AspectRatio → Automatic],
{{α, 1}, -2.5, 5}, {{β, 2}, -2.5, 5}, {{init1, {.5, .5}}, Locator},
{{init2, {-.5, -.5}}, Locator}]
```

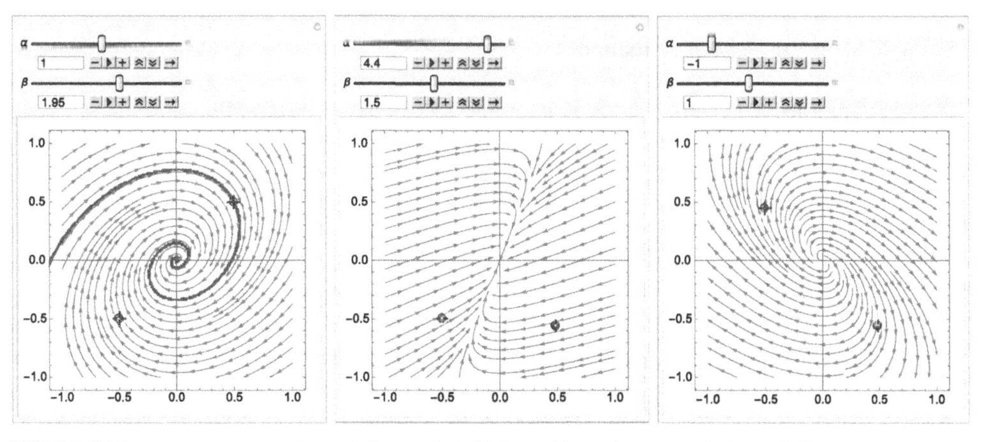

FIGURE 6.36 As we vary α and β and change the initial conditions, the system behaves differently. (a) A stable spiral; (b) A node; (c) An unstable spiral. (Oklahoma State University colors)

6.4.3 Nonlinear Systems

Nonlinear systems of differential equations arise in numerous situations. Rigorous analysis of the behavior of solutions to nonlinear systems is usually very difficult, if not impossible.

To generate numerical solutions of equations (when possible), use `NDSolve`.

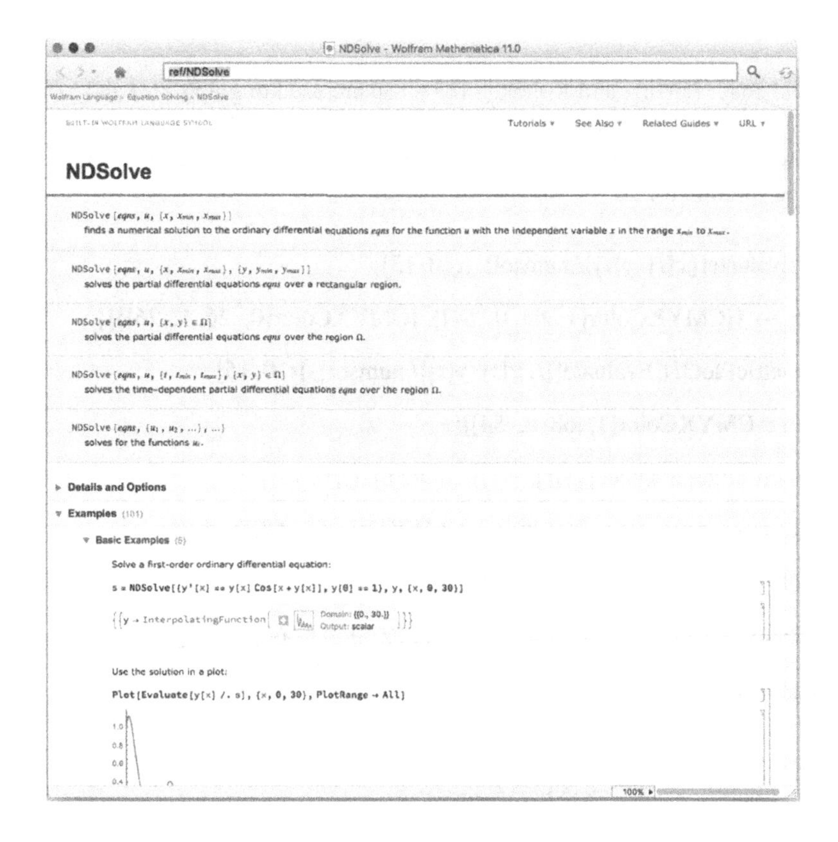

Example 6.29: Van-der-Pol's equation

Van-der-Pol's Equation $x'' + \mu(x^2 - 1)x' + x = 0$ can be written as the system

$$x' = y$$
$$y' = -x - \mu\left(x^2 - 1\right)y. \tag{6.32}$$

If $\mu = 2/3$, $x(0) = 1$, and $y(0) = 0$, (a) find $x(1)$ and $y(1)$. (b) Graph the solution that satisfies these initial conditions.

Also see Example 6.32.

Solution. We use `NDSolve` together to solve equation (6.32) with $\mu = 2/3$ subject to $x(0) = 1$ and $y(0) = 0$. We name the resulting numerical solution `numsol`.

```
numsol =
NDSolve[{x'[t]==y[t], y'[t]== − x[t] − 2/3(x[t]^2 − 1)y[t],
  x[0]==1, y[0]==0}, {x[t], y[t]}, {t, 0, 30}]
```

{{$x[t]$ → InterpolatingFunction[][t], $y[t]$ → InterpolatingFunction[][t]}}

We evaluate `numsol` if $t = 1$ to see that $x(1) \approx .5128$ and $y(1) \approx -.9692$.

```
{x[t], y[t]}/.numsol/.t->1
```

{{0.512848, −0.969204}}

`Plot`, `ParametricPlot`, and `ParametricPlot3D` are used to graph $x(t)$ and $y(t)$ together in Fig. 6.37 (a); a three-dimensional plot, $(t, x(t), y(t))$ is shown in Fig. 6.37 (b); a parametric plot is shown in Fig. 6.37 (c); and the limit cycle is shown more clearly in Fig. 6.37 (d) by graphing the solution for $20 \le t \le 30$.

```
p1 = Plot[Evaluate[{x[t], y[t]}/.numsol], {t, 0, 15},
  PlotStyle → {{CMYKColor[1, .68, 0, .54]}, {CMYKColor[0, .26, 1, .26]}}];
p2 = ParametricPlot3D[Evaluate[{t, x[t], y[t]}/.numsol], {t, 0, 15},
  PlotStyle->CMYKColor[1, .68, 0, .54]];
p3 = ParametricPlot[Evaluate[{x[t], y[t]}/.numsol], {t, 0, 15},
  AspectRatio → Automatic, PlotStyle->CMYKColor[1, .68, 0, .54]];
p4 = ParametricPlot[Evaluate[{x[t], y[t]}/.numsol], {t, 20, 30},
  AspectRatio → Automatic, PlotStyle->CMYKColor[1, .68, 0, .54]];
Show[GraphicsGrid[{{p1, p2}, {p3, p4}}]]
```

To avoid conflicts with the variables in the `Manipulate` consider quitting Mathematica, restarting, and then entering the `Manipulate` command in a new notebook.

To consider other μ values, decide on a μ range, combine the above commands, replace 2/3, with μ, and use `Manipulate`. See Fig. 6.38.

```
Clear[x, y]
```

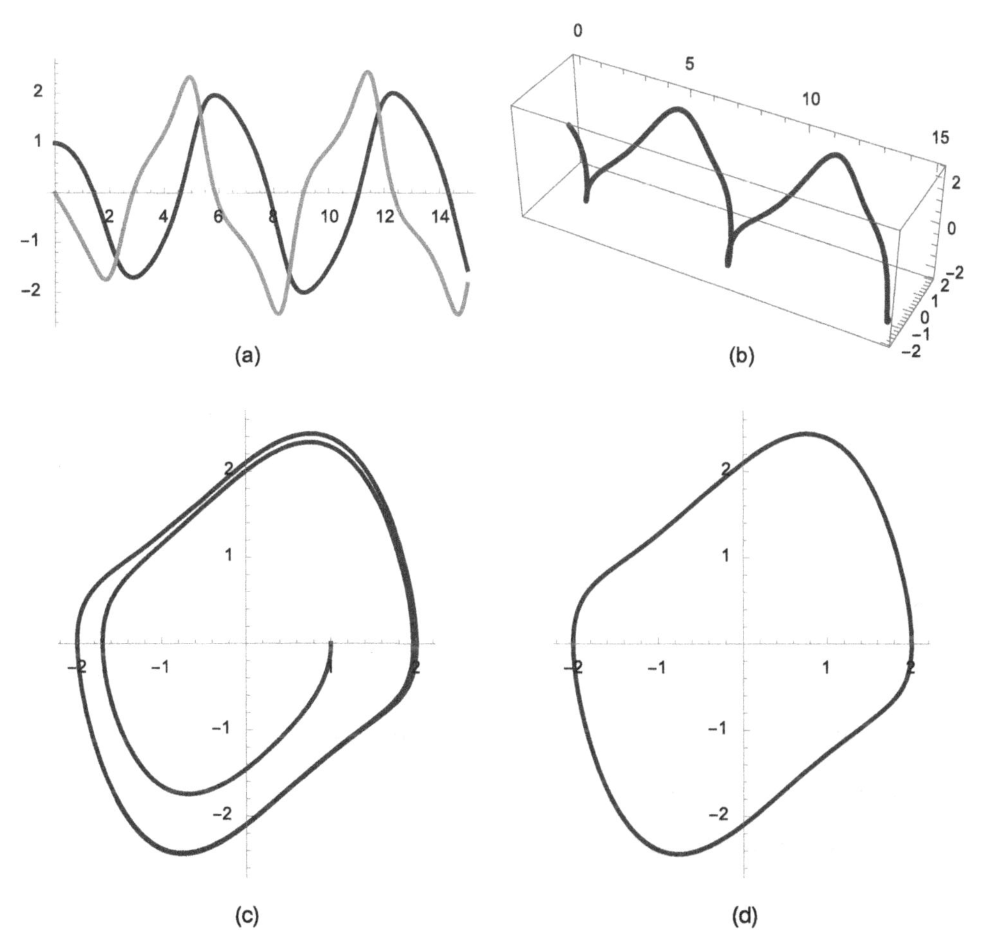

FIGURE 6.37 (a) $x(t)$ and $y(t)$. (b) A three-dimensional plot. (c) $x(t)$ versus $y(t)$. (d) $x(t)$ versus $y(t)$ for $20 \le t \le 30$. (University of Rhode Island colors)

```
Manipulate[
numsol = NDSolve[{x'[t]==y[t], y'[t]== − x[t] − μ(x[t]^2 − 1)y[t],
   x[0]==1, y[0]==0}, {x[t], y[t]}, {t, 0, 30}];
p1 = Plot[Evaluate[{x[t], y[t]}/.numsol], {t, 0, 15},
   PlotRange → All, AspectRatio → 1,
   PlotStyle → {{CMYKColor[1, .68, 0, .54]},
   {CMYKColor[0, .26, 1, .26]}}];
p2 = ParametricPlot3D[Evaluate[{t, x[t], y[t]}/.numsol], {t, 0, 15},
   BoxRatios → {4, 1, 1}, PlotStyle->CMYKColor[1, .68, 0, .54]];
p3 = ParametricPlot[Evaluate[{x[t], y[t]}/.numsol], {t, 0, 15},
   AspectRatio → 1, PlotRange → All,
   PlotStyle->CMYKColor[1, .68, 0, .54]];
p4 = ParametricPlot[Evaluate[{x[t], y[t]}/.numsol], {t, 20, 30},
```

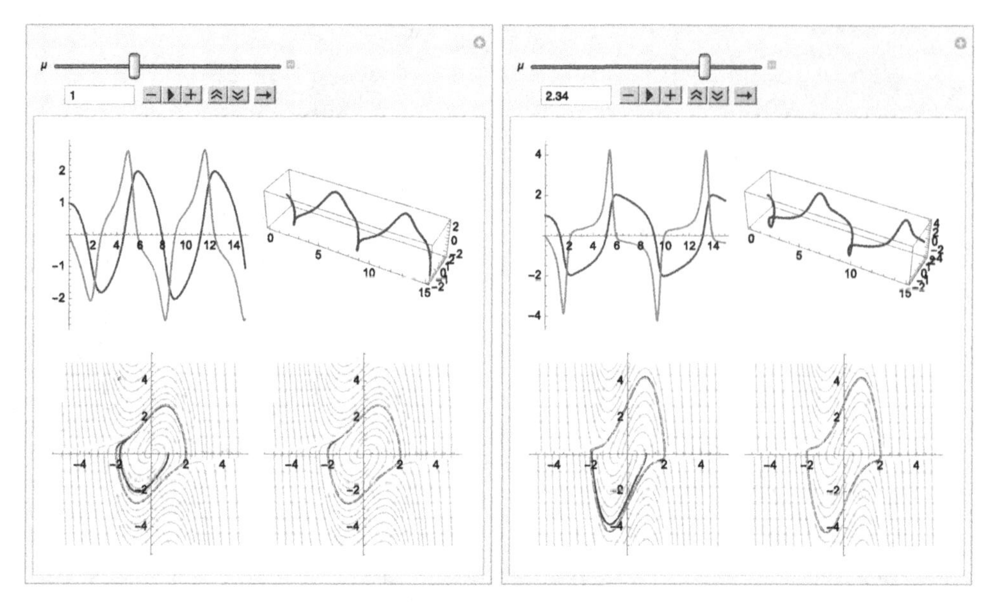

FIGURE 6.38 Plots of solutions of Van der Pol's equation for various values of μ.

```
    AspectRatio → 1, PlotRange → All,
    AspectRatio → Automatic, PlotStyle->CMYKColor[1, .68, 0, .54]];
p5 = StreamPlot[{y, −x − μ(x^2 − 1)y}, {x, −5, 5}, {y, −5, 5},
    StreamStyle → CMYKColor[0, .12, 1, .07]];
Show[GraphicsGrid[{{p1, p2}, {Show[p3, p5], Show[p4, p5]}}]],
{{μ, 1}, 0, 3}]
```

□

Linearization

Consider the autonomous system of the form

An **autonomous system** does not explicitly depend on the independent variable, t. That is, if you write the system omitting all arguments, the independent variable (typically t) does not appear.

$$\begin{aligned} x_1{}' &= f_1(x_1, x_2, \ldots, x_n) \\ x_2{}' &= f_2(x_1, x_2, \ldots, x_n) \\ &\vdots \\ x_n{}' &= f_n(x_1, x_2, \ldots, x_n). \end{aligned} \tag{6.33}$$

An **equilibrium** (or **rest**) **point**, $E = (x_1{}^*, x_2{}^*, \ldots, x_n{}^*)$, of equation (6.33) is a solution of the system

$$\begin{aligned} f_1(x_1, x_2, \ldots, x_n) &= 0 \\ f_2(x_1, x_2, \ldots, x_n) &= 0 \\ &\vdots \\ f_n(x_1, x_2, \ldots, x_n) &= 0. \end{aligned} \tag{6.34}$$

The **Jacobian** of equation (6.33) is

$$\mathbf{J}(x_1, x_2, \ldots, x_n) = \begin{pmatrix} \frac{\partial f_1}{\partial x_1} & \frac{\partial f_1}{\partial x_2} & \cdots & \frac{\partial f_1}{\partial x_n} \\ \frac{\partial f_2}{\partial x_1} & \frac{\partial f_2}{\partial x_2} & \cdots & \frac{\partial f_2}{\partial x_n} \\ \vdots & \vdots & \cdots & \vdots \\ \frac{\partial f_n}{\partial x_1} & \frac{\partial f_n}{\partial x_2} & \cdots & \frac{\partial f_n}{\partial x_n} \end{pmatrix}.$$

The rest point, E, is **locally stable** if and only if all the eigenvalues of $\mathbf{J}(E)$ have negative real part. If E is not locally stable, E is **unstable**.

Establishing global stability of an equilibrium point for a nonlinear system is *significantly* more difficult than establishing global stability of an equilibrium point ($E = (0, 0)$) for a linear autonomous system.

Example 6.30: Duffing's Equation

Consider the forced **pendulum equation** with damping,

$$x'' + kx' + \omega \sin x = F(t). \tag{6.35}$$

Recall the Maclaurin series for $\sin x$: $\sin x = x - \frac{1}{3!}x^3 + \frac{1}{5!}x^5 - \frac{1}{7!}x^7 + \ldots$. Using $\sin x \approx x$, equation (6.35) reduces to the linear equation $x'' + kx' + \omega x = F(t)$.

On the other hand, using the approximation $\sin x \approx x - \frac{1}{6}x^3$, we obtain $x'' + kx' + \omega\left(x - \frac{1}{6}x^3\right) = F(t)$. Adjusting the coefficients of x and x^3 and assuming that $F(t) = F\cos\omega t$ gives us **Duffing's equation**:

$$x'' + kx' + cx + \epsilon x^3 = F\cos\omega t, \tag{6.36}$$

where k and c are positive constants.

Let $y = x'$. Then, $y' = x'' = F\cos\omega t - kx' - cx - \epsilon x^3 = F\cos\omega t - ky - cx - \epsilon x^3$ and we can write equation (6.36) as the system

$$\begin{aligned} x' &= y \\ y' &= F\cos\omega t - ky - cx - \epsilon x^3. \end{aligned} \tag{6.37}$$

Assuming that $F = 0$ results in the autonomous system

$$\begin{aligned} x' &= y \\ y' &= -cx - \epsilon x^3 - ky. \end{aligned} \tag{6.38}$$

The rest points of system equation (6.38) are found by solving

$$\begin{aligned} x' &= 0 \\ y' &= -cx - \epsilon x^3 - ky = 0 \end{aligned}$$

resulting in $E_0 = (0, 0)$.

```
Solve[{y==0, −cx − εx^3 − ky==0}, {x, y}]
```

$\{\{x \to 0, y \to 0\}, \{x \to -\frac{i\sqrt{c}}{\sqrt{\epsilon}}, y \to 0\}, \{x \to \frac{i\sqrt{c}}{\sqrt{\epsilon}}, y \to 0\}\}$

We find the Jacobian of equation (6.38) in `s1` and then evaluate the Jacobian at E_0,

```
s1 = {{0, 1}, {−c − 3εx^2, −k}};
s2 = s1/.x->0
```

$\{\{0, 1\}, \{-c, -k\}\}$

and then compute the eigenvalues with `Eigenvalues`.

```
s3 = Eigenvalues[s2]
```

$\{\frac{1}{2}(-k - \sqrt{-4c + k^2}), \frac{1}{2}(-k + \sqrt{-4c + k^2})\}$

Because k and c are positive, $k^2 - 4c < k^2$ so the real part of each eigenvalue is always negative if $k^2 - 4c \neq 0$. Thus, E_0 is locally stable.
For the autonomous system

$$\begin{aligned} x' &= f(x, y) \\ y' &= g(x, y), \end{aligned} \tag{6.39}$$

Bendixson's theorem states that if $f_x(x, y) + g_y(x, y)$ is a continuous function that is either always positive or always negative in a particular region R of the plane, then system (6.39) has no limit cycles in R. For equation (6.38) we have

$$\frac{d}{dx}(y) + \frac{d}{dy}\left(-cx - \epsilon x^3 - ky\right) = -k,$$

which is always negative. Hence, equation (6.38) has no limit cycles and it follows that E_0 is globally, asymptotically stable.

```
D[y, x] + D[-cx - ϵx^3 - ky, y]
```

$-k$

We use `StreamPlot` and `ParametricPlot` to illustrate two situations that occur. In Fig. 6.39 (a), we use $c = 1$, $\epsilon = 1/2$, and $k = 3$. In this case, E_0 is a *stable node*. On the other hand, in Fig. 6.39 (b), we use $c = 10$, $\epsilon = 1/2$, and $k = 3$. In this case, E_0 is a *stable spiral*.

```
pvf1 = StreamPlot[{y, -x - x^3/2 - 3y}, {x, -2.5, 2.5},
  {y, -2.5, 2.5}, StreamPoints → Fine, StreamStyle →
  CMYKColor[0, 1, .61, .43]];

numgraph[init_, c_, opts___]:=Module[{numsol},
numsol = NDSolve[{x'[t]==y[t], y'[t]== - cx[t] - 1/2x[t]^3 - 3y[t],
  x[0]==init[[1]], y[0]==init[[2]]}, {x[t], y[t]}, {t, 0, 10}];
  ParametricPlot[Evaluate[{x[t], y[t]}/.numsol], {t, 0, 10},
  PlotStyle → Black, opts]];

i1 = Table[numgraph[{2.5, i}, 1], {i, -2.5, 2.5, 1/2}];
i2 = Table[numgraph[{-2.5, i}, 1], {i, -2.5, 2.5, 1/2}];
i3 = Table[numgraph[{i, 2.5}, 1], {i, -2.5, 2.5, 1/2}];
i4 = Table[numgraph[{i, -2.5}, 1], {i, -2.5, 2.5, 1/2}];

c1 = Show[i1, i2, i3, i4, pvf1, PlotRange → {{-2.5, 2.5}, {-2.5, 2.5}},
  AspectRatio → Automatic]

pvf2 = StreamPlot[{y, -10x - x^3/2 - 3y}, {x, -2.5, 2.5},
  {y, -2.5, 2.5}, StreamPoints → Fine, StreamStyle →
```

```
   CMYKColor[0, 1, .61, .43]];

i1 = Table[numgraph[{2.5, i}, 10], {i, −2.5, 2.5, 1/2}];
i2 = Table[numgraph[{−2.5, i}, 10], {i, −2.5, 2.5, 1/2}];
i3 = Table[numgraph[{i, 2.5}, 10], {i, −2.5, 2.5, 1/2}];
i4 = Table[numgraph[{i, −2.5}, 10], {i, −2.5, 2.5, 1/2}];

c2 = Show[i1, i2, i3, i4, pvf2, PlotRange → {{−2.5, 2.5}, {−2.5, 2.5}},
   AspectRatio → Automatic]

Show[GraphicsRow[{c1, c2}]]
```

To experiment with different parameter values, use `Manipulate`. In the following, we investigate how varying c from 0 to 10 affects the solutions of Duffing's equation (see Fig. 6.40).

```
Clear[pvf, i1, i2, i3, i4];
Manipulate[
   numgraph[init_, c_, opts___]:=Module[{numsol},
numsol = NDSolve[{x'[t]==y[t], y'[t]== − cx[t] − 1/2x[t]^3 − 3y[t],
   x[0]==init[[1]], y[0]==init[[2]]}, {x[t], y[t]}, {t, 0, 10}];
   ParametricPlot[Evaluate[{x[t], y[t]}/.numsol], {t, 0, 10},
   PlotStyle → Black,
   opts]];
pvf = StreamPlot[{y, −cx − x^3/2 − 3y}, {x, −2.5, 2.5},
   {y, −2.5, 2.5}, StreamPoints → Fine, StreamStyle →
   CMYKColor[0, 1, .61, .43]];
i1 = Table[numgraph[{2.5, i}, c], {i, −2.5, 2.5, 1/2}];
i2 = Table[numgraph[{−2.5, i}, c], {i, −2.5, 2.5, 1/2}];
i3 = Table[numgraph[{i, 2.5}, c], {i, −2.5, 2.5, 1/2}];
i4 = Table[numgraph[{i, −2.5}, c], {i, −2.5, 2.5, 1/2}];
Show[i1, i2, i3, i4, pvf, PlotRange → {{−2.5, 2.5}, {−2.5, 2.5}},
   AspectRatio → Automatic], {{c, 1}, 0, 10}]
```

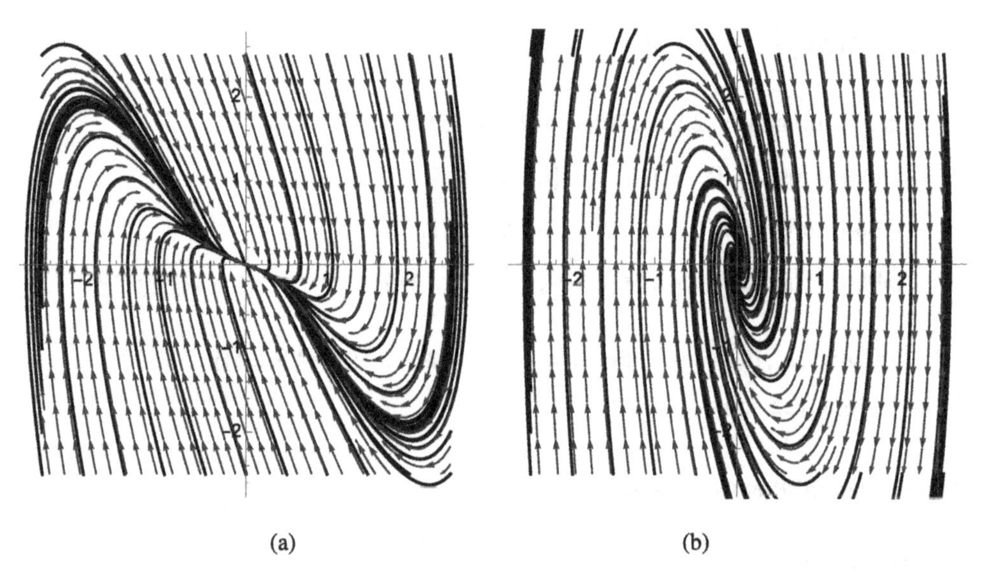

FIGURE 6.39 (a) The origin is a stable node. (b) The origin is a stable spiral. (University of South Carolina colors)

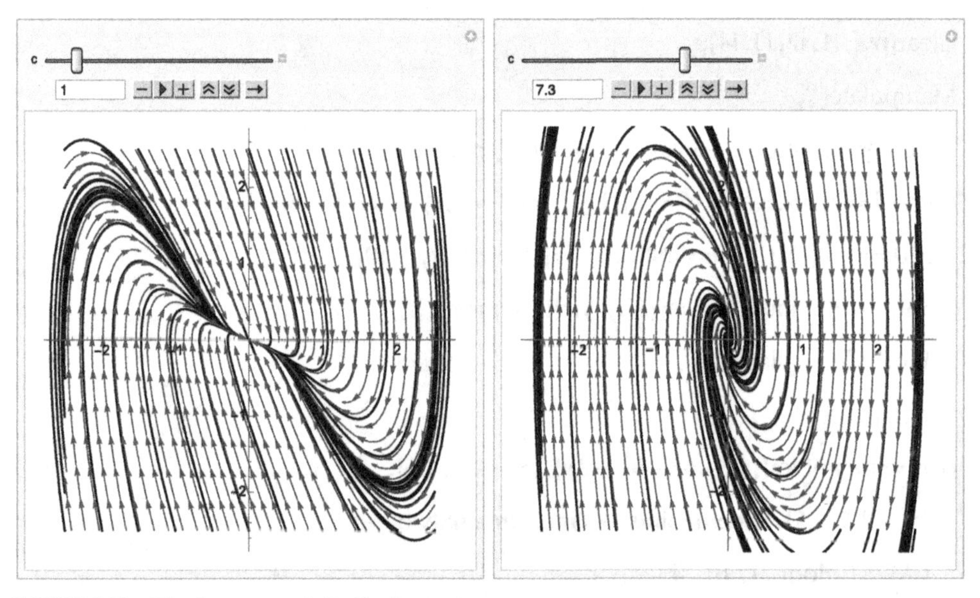

FIGURE 6.40 Allowing c to vary in Duffing's equation.

There are *many* other predator–prey models.

Example 6.31: Predator–Prey

One form of the **predator–prey** equations is

$$\frac{dx}{dt} = ax - bxy$$
$$\frac{dy}{dt} = dxy - cy$$

where a, b, c, and d are positive constants. x represents the size of the prey population at time t while y represents the size of the predator population at time t.

We use `Solve` to calculate the rest points. In this case, there is one boundary rest point, $E_0 = (0, 0)$ and one interior rest point, $E_1 = (c/d, a/b)$.

```
rps = Solve[{ax - bxy==0, dxy - cy==0}, {x, y}]
```

$\{\{x \to 0, y \to 0\}, \{x \to \frac{c}{d}, y \to \frac{a}{b}\}\}$

The Jacobian is then found using `D`.

```
jac = {{D[ax - bxy, x], D[ax - bxy, y]},
{D[dxy - cy, x], D[dxy - cy, y]}};
MatrixForm[jac]
```

$$\begin{pmatrix} a - by & -bx \\ dy & -c + dx \end{pmatrix}$$

E_0 is unstable because one eigenvalue of $\mathbf{J}(E_0)$ is positive. For the linearized system, E_1 is a center because the eigenvalues of $\mathbf{J}(E_1)$ are complex conjugates.

```
Eigenvalues[jac/.rps[[2]]]
```

$\{-i\sqrt{a}\sqrt{c}, i\sqrt{a}\sqrt{c}\}$

In fact, E_1 is a center for the nonlinear system as illustrated in Fig. 6.41, where we have used $a = 1$, $b = 2$, $c = 2$, and $d = 1$. Notice that there are multiple limit cycles around $E_1 = (1/2, 1/2)$.

```
pvf = StreamPlot[{x - 2xy, 2xy - y}, {x, 0, 2},
  {y, 0, 2}, StreamPoints → Fine, StreamStyle → CMYKColor[0, 0, 0, .34]]

numgraph[init_, opts___]:=Module[{numsol},
numsol = NDSolve[{x'[t]==x[t] - 2x[t]y[t], y'[t]==2x[t]y[t] - y[t],
  x[0]==init[[1]], y[0]==init[[2]]}, {x[t], y[t]}, {t, 0, 50}];
  ParametricPlot[Evaluate[{x[t], y[t]}/.numsol], {t, 0, 10},
  PlotStyle → {{Thickness[.0075], CMYKColor[0, 1, .63, .12]}},
  opts]]

i1 = Table[numgraph[{i, i}], {i, 3/20, 1/2, 1/20}];
Show[i1, pvf, PlotRange->{{0, 2}, {0, 2}}, AspectRatio->Automatic]
```

As illustrated previously, if you want to play around with the system, use `Manipulate`. In this case, we allow a, b, c, and d to vary. The solution plotted is the one that passes through the locator point. See Fig. 6.42.

```
Manipulate[
```

```
pvf = StreamPlot[{ax − bxy, dxy − cy}, {x, 0, 5},
  {y, 0, 5}, StreamPoints → Fine,
  StreamStyle → CMYKColor[0, 0, 0, .34]];
numsol = NDSolve[{x'[t] == ax[t] − bx[t]y[t], y'[t] == dx[t]y[t] − cy[t],
  x[0]==init[[1]], y[0]==init[[2]]}, {x[t], y[t]}, {t, 0, 25}];
p1 = ParametricPlot[Evaluate[{x[t], y[t]}/.numsol], {t, 0, 25},
  PlotStyle → {{Thickness[.0075], CMYKColor[0, 1, .63, .12]}}];
Show[p1, pvf, PlotRange->{{0, 5}, {0, 5}}, AspectRatio → 1,
  AxesOrigin → {0, 0}],
  {{a, 1}, 0, 5}, {{b, 2}, 0, 5}, {{c, 1}, 0, 5}, {{d, 2}, 0, 5},
    {{init, {1, 1}}, Locator}]
```

In this model, a stable interior rest state is not possible.
The complexity of the behavior of solutions to the system increases based on the assumptions made. Typical assumptions include adding satiation terms for the predator (y) and/or limiting the growth of the prey (x). The **standard predator–prey equations of the Kolmogorov type,**

$$\begin{aligned} x' &= \alpha x\left(1 - \frac{1}{K}x\right) - \frac{mxy}{a+x} \\ y' &= y\left(\frac{mx}{a+x} - s\right), \end{aligned} \tag{6.40}$$

incorporate both of these assumptions.
We use `Solve` to find the three rest points of system (6.40). Let $E_0 = (0, 0)$ and $E_1 = (k, 0)$ denote the two boundary rest points, and let E_2 represent the interior rest point.

```
Clear[α, x, k, m, a, s]
rps = Solve[{αx(1 − 1/kx) − mxy/(a + x)==0, y(mx/(a + x) − s)==0},
  {x, y}]
```

$\{\{x \to k, y \to 0\}, \{x \to \frac{as}{m-s}, y \to \frac{akm\alpha - a^2 s\alpha - aks\alpha}{k(m-s)^2}\}, \{x \to 0, y \to 0\}\}$

The Jacobian, **J**, is calculated next in `s1`.

```
s1 = {{D[αx(1 − 1/kx) − mxy/(a + x), x],
  D[αx(1 − 1/kx) − mxy/(a + x), y]},
  {D[y(mx/(a + x) − s), x], D[y(mx/(a + x) − s), y]}};
MatrixForm[s1]
```

$$\begin{pmatrix} \frac{mxy}{(a+x)^2} - \frac{my}{a+x} - \frac{x\alpha}{k} + (1 - \frac{x}{k})\alpha & -\frac{mx}{a+x} \\ (-\frac{mx}{(a+x)^2} + \frac{m}{a+x})y & -s + \frac{mx}{a+x} \end{pmatrix}$$

Because $\mathbf{J}(E_0)$ has one positive eigenvalue, E_0 is unstable.

```
e0 = s1/.rps[[3]];
MatrixForm[e0]
eigs0 = Eigenvalues[e0]
```

$$\begin{pmatrix} \alpha & 0 \\ 0 & -s \end{pmatrix}$$

$\{-s, \alpha\}$

The stability of E_1 is determined by the sign of $m - s - am/(a+k)$.

```
e1 = s1/.rps[[1]];
MatrixForm[e1]
eigs1 = Eigenvalues[e1]
```

$$\begin{pmatrix} -\alpha & -\frac{km}{a+k} \\ 0 & \frac{km}{a+k} - s \end{pmatrix}$$

$\left\{\frac{km-as-ks}{a+k}, -\alpha\right\}$

The eigenvalues of $\mathbf{J}(E_2)$ are quite complicated.

```
e2 = s1/.rps[[2]];
MatrixForm[e2]
eigs2 = Eigenvalues[e2]
```

$$\begin{pmatrix} -\frac{as\alpha}{k(m-s)} + \left(1 - \frac{as}{k(m-s)}\right)\alpha + \frac{ams\left(akm\alpha - a^2 s\alpha - aks\alpha\right)}{k(m-s)^3\left(a+\frac{as}{m-s}\right)^2} - \frac{m\left(akm\alpha - a^2 s\alpha - aks\alpha\right)}{k(m-s)^2\left(a+\frac{as}{m-s}\right)} & -\frac{ams}{(m-s)\left(a+\frac{as}{m-s}\right)} \\ \frac{\left(-\frac{ams}{(m-s)\left(a+\frac{as}{m-s}\right)^2} + \frac{m}{a+\frac{as}{m-s}}\right)\left(akm\alpha - a^2 s\alpha - aks\alpha\right)}{k(m-s)^2} & -s + \frac{ams}{(m-s)\left(a+\frac{as}{m-s}\right)} \end{pmatrix}$$

$\{\frac{1}{2km(m-s)}(-ams\alpha + kms\alpha - as^2\alpha - ks^2\alpha - \sqrt{}((ams\alpha - kms\alpha + as^2\alpha + ks^2\alpha)^2 - 4(k^2m^4s\alpha - akm^3s^2\alpha - 3k^2m^3s^2\alpha + 2akm^2s^3\alpha + 3k^2m^2s^3\alpha - akms^4\alpha - k^2ms^4\alpha))),$
$\frac{1}{2km(m-s)}(-ams\alpha + kms\alpha - as^2\alpha - ks^2\alpha + \sqrt{}((ams\alpha - kms\alpha + as^2\alpha + ks^2\alpha)^2 - 4(k^2m^4s\alpha - akm^3s^2\alpha - 3k^2m^3s^2\alpha + 2akm^2s^3\alpha + 3k^2m^2s^3\alpha - akms^4\alpha - k^2ms^4\alpha)))\}$

Instead of using the eigenvalues, we compute the characteristic polynomial of $\mathbf{J}(E_2)$, $p(\lambda) = c_2\lambda^2 + c_1\lambda + c_0$, and examine the coefficients. Notice that c_2 is always positive.

```
cpe2 = CharacteristicPolynomial[e2, λ]//Simplify
```

$$\frac{as\alpha(m(-s+\lambda)+s(s+\lambda))+k(m-s)\left(-s\alpha(s+\lambda)+m\left(s\alpha+\lambda^2\right)\right)}{km(m-s)}$$

```
c0 = cpe2/.λ->0//Simplify
```

$$\frac{s(k(m-s)-as)\alpha}{km}$$

```
c1 = Coefficient[cpe2, λ]//Simplify
```

$$\frac{s(k(-m+s)+a(m+s))\alpha}{km(m-s)}$$

```
c2 = Coefficient[cpe2, λ^2]//Simplify
```

1

On the other hand, c_0 and $m - s - am/(a+k)$ have the same sign because

```
c0/eigs1[[1]]//Simplify
```

$$\frac{(a+k)s\alpha}{km}$$

is always positive. In particular, if $m - s - am/(a+k) < 0$, E_1 is stable. Because c_0 is negative, by Descartes' rule of signs, it follows that $p(\lambda)$ will have one positive root and hence E_2 will be unstable.

On the other hand, if $m - s - am/(a+k) > 0$ so that E_1 is unstable, E_2 may be either stable or unstable. To illustrate these two possibilities let $\alpha = K = m = 1$ and $a = 1/10$. We recalculate.

```
α = 1; k = 1; m = 1; a = 1/10;

rps = Solve[{αx(1 − 1/kx) − mxy/(a + x)==0, y(mx/(a + x) − s)==0}, {x, y}]
```

$$\left\{\{x\to 1, y\to 0\}, \left\{x\to -\frac{s}{10(-1+s)}, y\to \frac{10-11s}{100(-1+s)^2}\right\}, \{x\to 0, y\to 0\}\right\}$$

```
s1 = {{D[αx(1 − 1/kx) − mxy/(a + x), x], D[αx(1 − 1/kx) − mxy/(a + x), y]},
    {D[y(mx/(a + x) − s), x], D[y(mx/(a + x) − s), y]}};
MatrixForm[s1]
```

$$\begin{pmatrix} 1-2x+\frac{xy}{\left(\frac{1}{10}+x\right)^2}-\frac{y}{\frac{1}{10}+x} & -\frac{x}{\frac{1}{10}+x} \\ \left(-\frac{x}{\left(\frac{1}{10}+x\right)^2}+\frac{1}{\frac{1}{10}+x}\right)y & -s+\frac{x}{\frac{1}{10}+x} \end{pmatrix}$$

```
eigs1 = Eigenvalues[s1]
```

$\{\frac{1}{2(1+10x)^2}(1 - s + 28x - 20sx + 160x^2 - 100sx^2 - 200x^3 - 10y - \sqrt{}((-1 + s - 28x + 20sx - 160x^2 + 100sx^2 + 200x^3 + 10y)^2 - 4(-s + 10x - 38sx + 280x^2 - 520sx^2 + 2400x^3 - 2800sx^3 + 4000x^4 - 2000sx^4 - 20000x^5 + 20000sx^5 + 10sy + 200sxy + 1000sx^2y))), \frac{1}{2(1+10x)^2}(1 - s + 28x - 20sx + 160x^2 - 100sx^2 - 200x^3 - 10y + \sqrt{}((-1 + s - 28x + 20sx - 160x^2 + 100sx^2 + 200x^3 + 10y)^2 - 4(-s + 10x - 38sx + 280x^2 - 520sx^2 + 2400x^3 - 2800sx^3 + 4000x^4 - 2000sx^4 - 20000x^5 + 20000sx^5 + 10sy + 200sxy + 1000sx^2y)))\}$

```
e2 = s1/.rps[[2]];
cpe2 = CharacteristicPolynomial[e2, λ]//Simplify
```

$\frac{-11s^3+s^2(21-11\lambda)-10\lambda^2+s(-10+9\lambda+10\lambda^2)}{10(-1+s)}$

```
c0 = cpe2/.λ->0//Simplify
```

$s - \frac{11s^2}{10}$

```
c1 = Coefficient[cpe2, λ]//Simplify
```

$\frac{(9-11s)s}{10(-1+s)}$

```
c2 = Coefficient[cpe2, λ^2]//Simplify
```

1

Using `Reduce`, we see that

1. c_0, c_1, and c_2 are positive if $9/11 < s < 10/11$, and
2. c_0 and c_2 are positive and c_1 is negative if $0 < s < 9/11$.

```
Reduce[c0 > 0&&c1 > 0, s]
```

$\frac{9}{11} < s < \frac{10}{11}$

```
Reduce[c0 > 0&&c1 < 0, s]
```

$0 < s < \frac{9}{11}$

In the first situation, E_2 is stable; in the second E_2 is unstable.

Using $s = 19/22$, we graph the direction field associated with the system as well as various solutions in Fig. 6.43 (a). In the plot, notice that all nontrivial solutions approach $E_2 \approx (.63, .27)$; E_2 is stable – a situation that cannot occur with the standard predator–prey equations.

```
rps/.s->19/22//N
```

```
{{x → 1., y → 0.}, {x → 0.633333, y → 0.268889}, {x → 0., y → 0.}}
```

```
Clear[pvf, numgraph, i1, i2]
pvf = StreamPlot[{αx(1 - x/k) - mxy/(a+x), y(mx/(a+x) - 19/22)}, {x, 0, 1},
  {y, 0, 1}, StreamPoints → Fine, StreamStyle → CMYKColor[0, 1, .79, .2]];

Clear[x, y]
numgraph[init_, s_, opts___]:=Module[{numsol},
numsol = NDSolve[{x'[t]==αx[t](1 - 1/kx[t]) - mx[t]y[t]/(a + x[t]),
  y'[t]==y[t](mx[t]/(a + x[t]) - s), x[0]==init[[1]], y[0]==init[[2]]},
  {x[t], y[t]}, {t, 0, 50}];
ParametricPlot[Evaluate[{x[t], y[t]}/.numsol], {t, 0, 50},
  PlotStyle → {{Thickness[.01], CMYKColor[0, 0, 0, .6]}}, opts]]

i1 = Table[numgraph[{1, i}, 19/22], {i, 0, 1, 1/10}];
i2 = Table[numgraph[{i, 1}, 19/22], {i, 0, 1, 1/10}];
Show[i1, i2, pvf, PlotRange->{{0, 1}, {0, 1}}, AspectRatio->Automatic]
```

On the other hand, using $s = 8/11$ (so that E_2 is unstable) in Fig. 6.43 (b) we see that all nontrivial solutions appear to approach a limit cycle.

```
rps/.s->8/11//N
```

```
{{x → 1., y → 0.}, {x → 0.266667, y → 0.268889}, {x → 0., y → 0.}}
```

```
i1 = Table[numgraph[{1, i}, 19/22], {i, 0, 1, 1/10}];
i2 = Table[numgraph[{i, 1}, 19/22], {i, 0, 1, 1/10}];
p1 = Show[i1, i2, pvf, PlotRange->{{0, 1}, {0, 1}}, AspectRatio->Automatic]
```

The limit cycle is shown more clearly in Fig. 6.43 (c).

```
numgraph[{.759, .262}, 8/11, PlotRange->{{0, 1}, {0, 1}}, AspectRatio->Automatic]
```

As we have seen in similar situations, these commands can be collected into a single `Manipulate` command to investigate the situation. See Fig. 6.44.

```
Clear[pvf, numgraph, i1, i2, α, k, m]
Manipulate[
pvf = StreamPlot[{αx(1 - x/k) - mxy/(a+x), y(mx/(a+x) - s)}, {x, 0, 1},
```

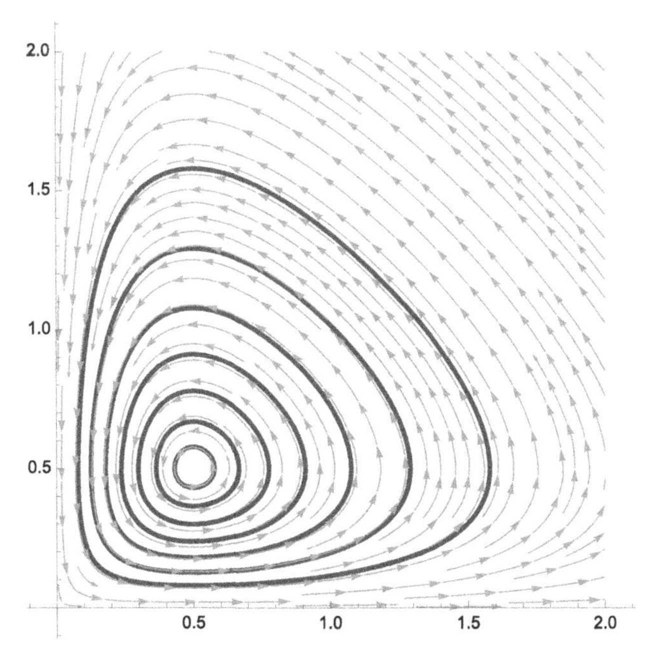

FIGURE 6.41 Multiple limit cycles about the interior rest point. (University of South Dakota colors)

```
  {y, 0, 1}, StreamPoints → Fine, StreamStyle → CMYKColor[0, 1, .79, .2]];
numgraph[init_, s_]:=Module[{numsol},
numsol = NDSolve[{x'[t]==αx[t](1 − 1/kx[t]) − mx[t]y[t]/(a + x[t]),
  y'[t]==y[t](mx[t]/(a + x[t]) − s), x[0]==init[[1]], y[0]==init[[2]]},
  {x[t], y[t]}, {t, 0, 50}];
ParametricPlot[Evaluate[{x[t], y[t]}/.numsol], {t, 0, 50}, PlotPoints → 200,
  PlotStyle → {{Thickness[.01], CMYKColor[0, 0, 0, .6]}}]];
i1 = Table[numgraph[{1, i}, s], {i, 0, 1, 1/10}];
i2 = Table[numgraph[{i, 1}, s], {i, 0, 1, 1/10}];
Show[i1, i2, pvf, PlotRange->{{0, 1}, {0, 1}}, AspectRatio->Automatic],
  {{α, 1}, 0, 5}, {{k, 1}, 0, 5}, {{m, 1}, 0, 5}, {{a, 1/10}, 0, 1}, {{s, 8/11}, 0, 5}]
```

Example 6.32: Van-der-Pol's equation

Also see Example 6.29.

In Example 6.29 we saw that **Van-der-Pol's equation** $x'' + \mu\left(x^2 - 1\right)x' + x = 0$ is equivalent to the system $\begin{cases} x' = y \\ y' = \mu\left(1 - x^2\right)y - x \end{cases}$. Classify the equilibrium points, use `NDSolve` to approximate the solutions to this nonlinear system, and plot the phase plane.

Solution. We find the equilibrium points by solving $\begin{cases} y = 0 \\ \mu\left(1 - x^2\right)y - x = 0 \end{cases}$. From the first equation, we see that $y = 0$. Then, substitution of $y = 0$ into the second equation yields

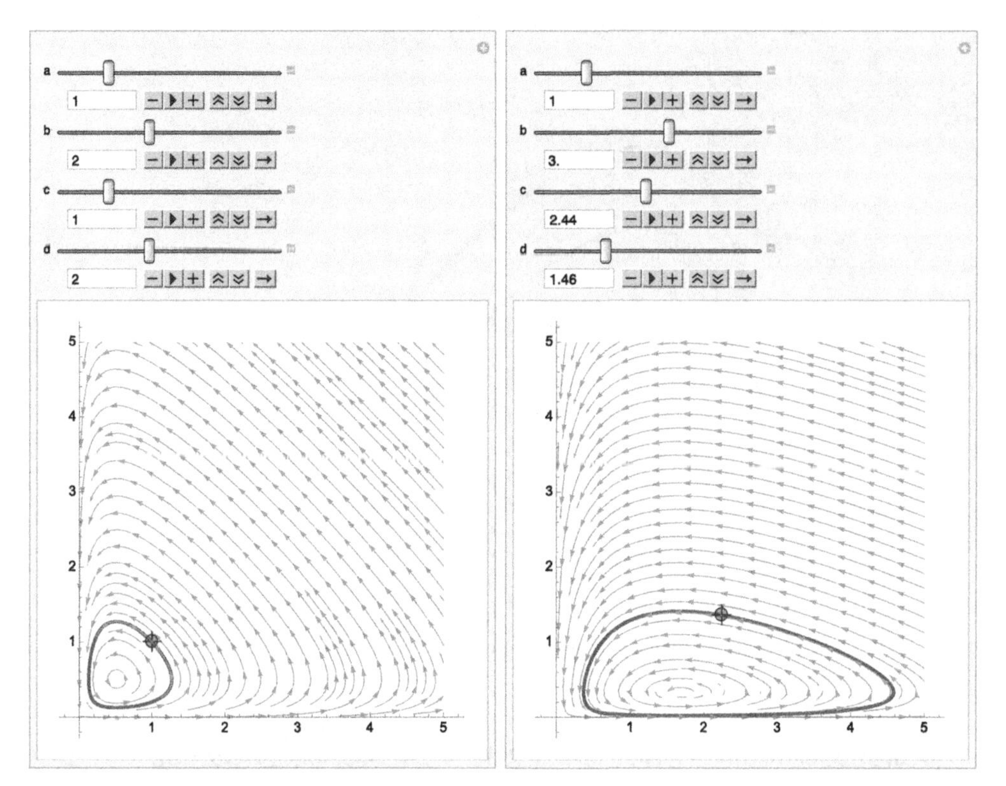

FIGURE 6.42 Multiple limit cycles about the interior rest point.

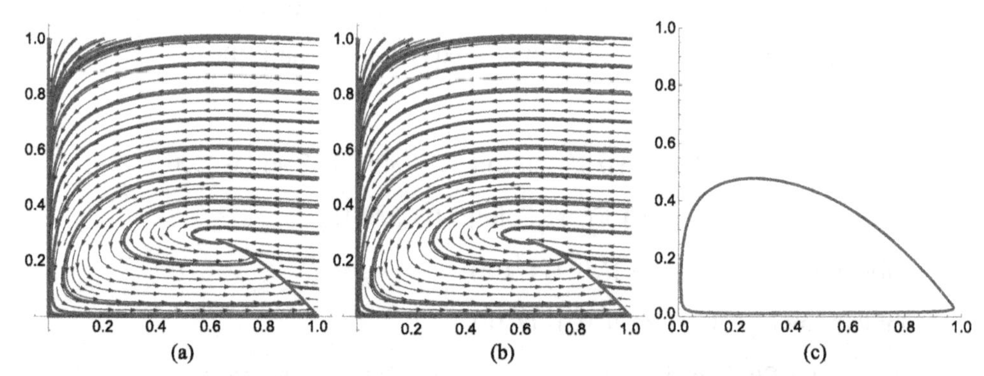

FIGURE 6.43 (a) $s = 19/22$; (b) $s = 8/11$; (c) A better view of the limit cycle without the direction field. (University of Utah colors)

$x = 0$. Therefore, the only equilibrium point is $(0, 0)$. The Jacobian matrix for this system is

$$\mathbf{J}(x, y) = \begin{pmatrix} 0 & 1 \\ -1 - 2\mu x y & -\mu\left(x^2 - 1\right) \end{pmatrix}.$$

The eigenvalues of $\mathbf{J}(0, 0)$ are $\lambda_{1,2} = \frac{1}{2}\left(\mu \pm \sqrt{\mu^2 - 4}\right)$.

```
Clear[f, g, x, y, μ]

f[x_, y_] = y;

g[x_, y_] = −x − μ(x^2 − 1)y;
```

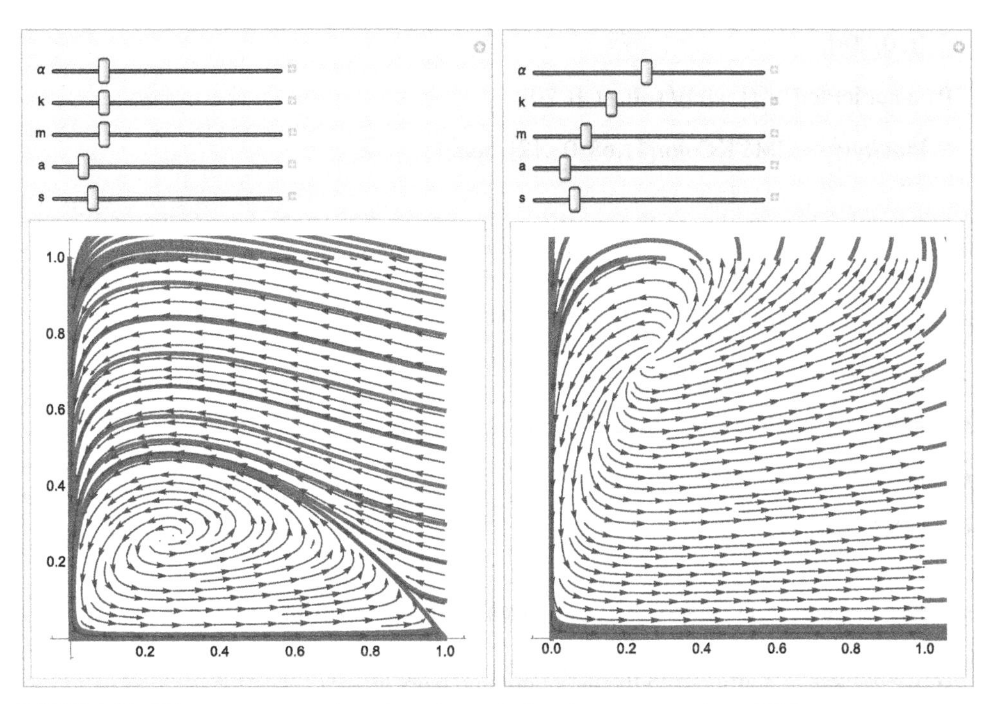

FIGURE 6.44 Using `Manipulate` to investigate the standard predator–prey equations of Kolmogorov type.

$$\text{jac} = \begin{pmatrix} D[f[x,y],x] & D[f[x,y],y] \\ D[g[x,y],x] & D[g[x,y],y] \end{pmatrix};$$

jac/.{x->0, y->0}//Eigenvalues

$\{\frac{1}{2}(\mu - \sqrt{-4+\mu^2}), \frac{1}{2}(\mu + \sqrt{-4+\mu^2})\}$

Notice that if $\mu > 2$, then both eigenvalues are positive and real. Hence, we classify $(0, 0)$ as an **unstable node**. On the other hand, if $0 < \mu < 2$, then the eigenvalues are a complex conjugate pair with a positive real part. Hence, $(0, 0)$ is an **unstable spiral**. (We omit the case $\mu = 2$ because the eigenvalues are repeated.)

We now show several curves in the phase plane that begin at various points for various values of μ. First, we define the function `sol`, which given μ, x_0, and y_0, generates a numerical solution to the initial-value problem

$$\begin{cases} x' = y \\ y' = \mu\left(1 - x^2\right) y - x \\ x(0) = x_0 \,,\, y(0) = y_0 \end{cases}$$

and then parametrically graphs the result for $0 \le t \le 20$.

```
Clear[sol]

sol[μ_, {x0_, y0_}, opts___]:=Module[{eqone, eqtwo, solt},

  eqone = x'[t]==y[t];

  eqtwo = y'[t]==μ(1 − x[t]^2)y[t] − x[t];

  solt = NDSolve[{eqone, eqtwo, x[0]==x0, y[0]==y0}, {x[t], y[t]},
```

```
    {t, 0, 20}];
  ParametricPlot[{x[t], y[t]}/.solt, {t, 0, 20},
    PlotStyle → CMYKColor[1, .68, 0, .12], opts]]
```

We then use `Table` and `Union` to generate a list of ordered pairs `initconds` that will correspond to the initial conditions in the initial-value problem.

```
initconds1 = Table[{0.1Cos[t], 0.1Sin[t]}, {t, 0, 2π, 2π/9}];
initconds2 = Table[{−5, i}, {i, −5, 5, 10/9}];
initconds3 = Table[{5, i}, {i, −5, 5, 10/9}];
initconds4 = Table[{i, 5}, {i, −5, 5, 10/9}];
initconds5 = Table[{i, −5}, {i, −5, 5, 10/9}];

initconds = initconds1 ∪ initconds2 ∪ initconds3 ∪ initconds4 ∪ initconds5;
```

Last, we use `Map` to apply `sol` to the list of ordered pairs in `initconds` for $\mu = 1/2$.

```
somegraphs1 = Map[sol[1/2, #]&, initconds];

phase1 = Show[somegraphs1, PlotRange → {{−5, 5}, {−5, 5}}, AspectRatio → 1,
  Ticks → {{−4, 4}, {−4, 4}}]
```

Similarly, we use `Map` to apply `sol` to the list of ordered pairs in `initconds` for $\mu = 1$, $3/2$, and 3.

```
somegraphs2 = Map[sol[1, #, DisplayFunction->Identity]&, initconds];

phase2 = Show[somegraphs2, PlotRange → {{−5, 5}, {−5, 5}}, AspectRatio → 1,
  Ticks → {{−4, 4}, {−4, 4}}]

somegraphs3 = Map[sol[3/2, #, DisplayFunction->Identity]&, initconds];

phase3 = Show[somegraphs3, PlotRange → {{−5, 5}, {−5, 5}}, AspectRatio → 1,
  Ticks → {{−4, 4}, {−4, 4}}]

somegraphs4 = Map[sol[3, #, DisplayFunction->Identity]&, initconds];

phase4 = Show[somegraphs3, PlotRange → {{−5, 5}, {−5, 5}}, AspectRatio → 1,
  Ticks → {{−4, 4}, {−4, 4}}]
```

All four graphs are shown together in Fig. 6.45. In each figure, we see that all of the curves approach a curve called a *limit cycle*. Physically, the fact that the system has a limit cycle in-

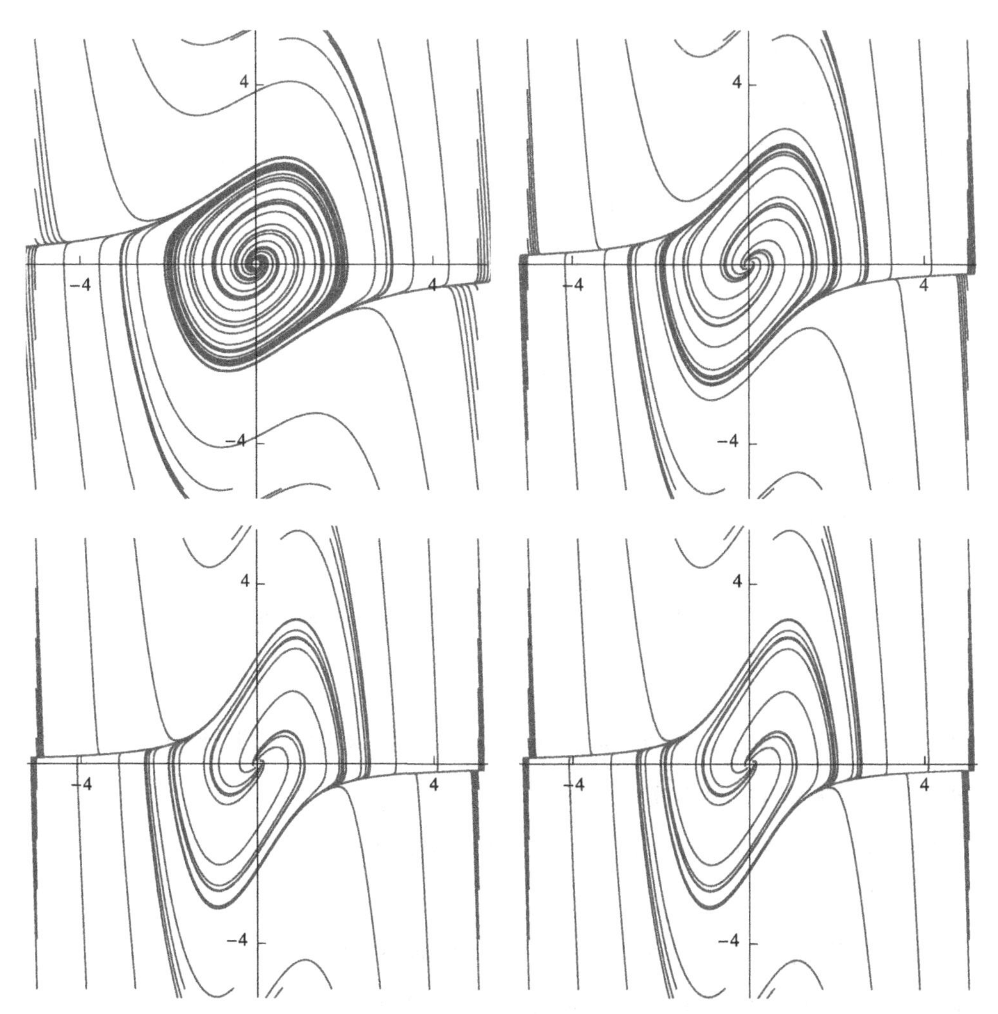

FIGURE 6.45 Solutions to the Van-der-Pol equation for various values of μ. (SUNY colors)

dicates that for all oscillations, the motion eventually becomes periodic, which is represented by a closed curve in the phase plane.

```
Show[GraphicsGrid[{{phase1, phase2}, {phase3, phase4}}]]
```

On the other hand, in Fig. 6.46 we graph the solutions that satisfy the initial conditions $x(0) = 1$ and $y(0) = 0$ parametrically and individually for various values of μ. Notice that for small values of μ the system more closely approximates that of the harmonic oscillator because the damping coefficient is small. The curves are more circular than those for larger values of μ.

```
Clear[x, y, t, s]
graph[μ_]:=Module[{numsol, pp, pxy},
numsol = NDSolve[{x'[t] == y[t], y'[t] == μ(1 − x[t]^2)y[t] − x[t], x[0] == 1, ..
  y[0] == 0}, {x[t], y[t]}, {t, 0, 20}];
pp = ParametricPlot[{x[t], y[t]}/.numsol, {t, 0, 20},
  PlotRange → {{−5, 5}, {−5, 5}}, AspectRatio → 1, Ticks → {{−4, 4}, {−4, 4}},
```

```
    PlotStyle → CMYKColor[0, .06, .95, 0]];
pxy = Plot[Evaluate[{x[t], y[t]}/.numsol], {t, 0, 20},
    PlotStyle → {{CMYKColor[0, .59, .96, 0]}, {CMYKColor[0, .99, .0, 0]}},
    PlotRange → {−5, 5}, AspectRatio → 1, Ticks → {{5, 10, 15}, {−4, 4}},
    DisplayFunction → Identity];
    GraphicsRow[{pxy, pp}]]

graphs = Table[graph[i], {i, 0.25, 3, 2.75/9}];

toshow = Partition[graphs, 2];
Show[GraphicsGrid[toshow]]
```

An alternative to comparing the graphics together is to use `Manipulate` to create an animation of how the μ values affect the solutions of the equation (see Fig. 6.47).

```
Manipulate[
sol[μ_, {x0_, y0_}, opts___]:=Module[{eqone, eqtwo, solt}, eqone = x'[t]==y[t];
    eqtwo = y'[t]==μ (1 − x[t]^2) y[t] − x[t];
    solt = NDSolve[{eqone, eqtwo, x[0]==x0, y[0]==y0}, {x[t], y[t]},
      {t, 0, 20}];
    ParametricPlot[{x[t], y[t]}/.solt, {t, 0, 20}, PlotPoints → 200]];
    initconds1 = Table[{0.1Cos[t], 0.1Sin[t]}, {t, 0, 2π, 2π/9}];
    initconds2 = Table[{−5, i}, {i, −5, 5, 10/9}];
    initconds3 = Table[{5, i}, {i, −5, 5, 10/9}];
    initconds4 = Table[{i, 5}, {i, −5, 5, 10/9}];
    initconds5 = Table[{i, −5}, {i, −5, 5, 10/9}];
    initconds = initconds1 ∪ initconds2 ∪ initconds3 ∪ initconds4 ∪ initconds5;
    somegraphs1 = Map[sol[μ, #, DisplayFunction->Identity,
      PlotStyle → CMYKColor[1, .68, 0, .12]]&, initconds];
    pvf = StreamPlot[{y, μ(1 − x^2)y − x}, {x, −5, 5}, {y, −5, 5},
      StreamPoints → Fine, StreamStyle->CMYKColor[0, .59, .96, 0]];
    phase1 = Show[{somegraphs1, pvf}, PlotRange → {{−5, 5}, {−5, 5}}, AspectRatio → 1,
      Ticks → {{−4, 4}, {−4, 4}}], {{μ, 3}, 0, 6}]
```
□

Although linearization can help you determine local behavior near rest points, the long-term behavior of solutions to nonlinear systems can be quite complicated, even for deceptively simple looking systems.

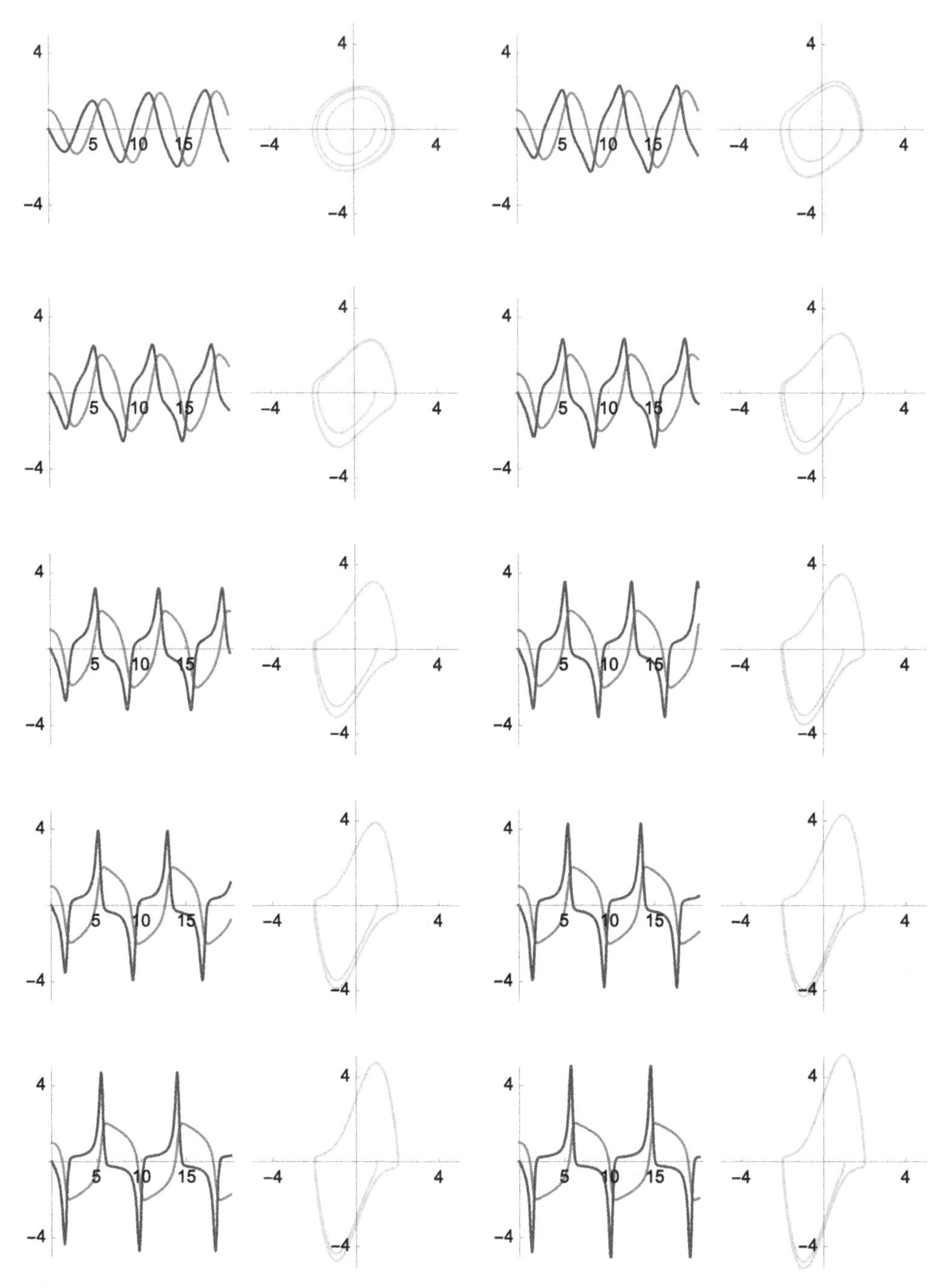

FIGURE 6.46 The solutions to the Van-der-Pol equation satisfying $x(0) = 1$ and $y(0) = 0$ individually for various values of μ.

Example 6.33: Lorenz Equations

The **Lorenz equations** are

$$\begin{cases} dx/dt = a(y - x) \\ dy/dt = bx - y - xz \\ dz/dt = xy - cz \end{cases}$$

See texts like Jordan and Smith's *Nonlinear Ordinary Differential Equations*, [10], for discussions of ways to analyze systems like the Rössler attractor and the Lorenz equations.

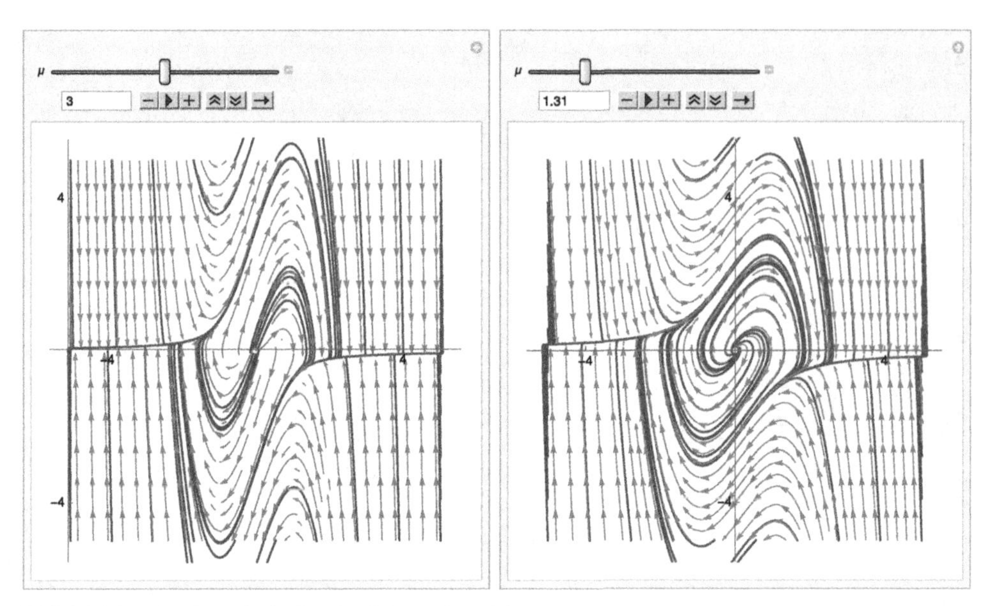

FIGURE 6.47 Varying μ in the Van-der-Pol equation with `Manipulate`.

Graph the solutions to the Lorenz equations if $a = 7$, $b = 28$, and $c = 3$ if the initial conditions are $x(0) = 3$, $y(0) = 4$, and $z(0) = 2$.

Solution. We present three solutions. First, we define the function `lorenzsol` that uses `NDSolve` to solve the Lorenz system for various parameter values and initial conditions. We then use `lorenzsol` to numerically solve the Lorenz system if $a = 7$, $b = 28$, an $c = 3$ and the initial conditions are $x(0) = 3$, $y(0) = 4$, and $z(0) = 2$.

```
lorenzsol[a_, b_, c_][{x0_, y0_, z0_}, ts_:{t, 0, 1000}, opts___]:=
Module[{numsol},
  numsol =
    NDSolve[{x'[t] == -ax[t] + ay[t], y'[t] == bx[t] - y[t] - x[t]z[t],
    z'[t] == x[t]y[t] - cz[t], x[0] == x0, y[0] == y0, z[0] == z0},
    {x[t], y[t], z[t]}, ts, MaxSteps -> 1000000]
  ]

n1 = lorenzsol[7, 28, 3][{3, 4, 2}]
```

{{x[t] → InterpolatingFunction[][t], y[t] → InterpolatingFunction[][t], z[t] →
InterpolatingFunction[][t]}}

The solutions are graphed parametrically in Fig. 6.48 and individually in Fig. 6.49.

```
p1a = ParametricPlot[Evaluate[{x[t], y[t]}/.n1], {t, 700, 1000},
  PlotPoints -> 1000, AspectRatio -> 1, AxesLabel -> {x, y},
  PlotStyle -> CMYKColor[1, .69, .08, .54]];
```

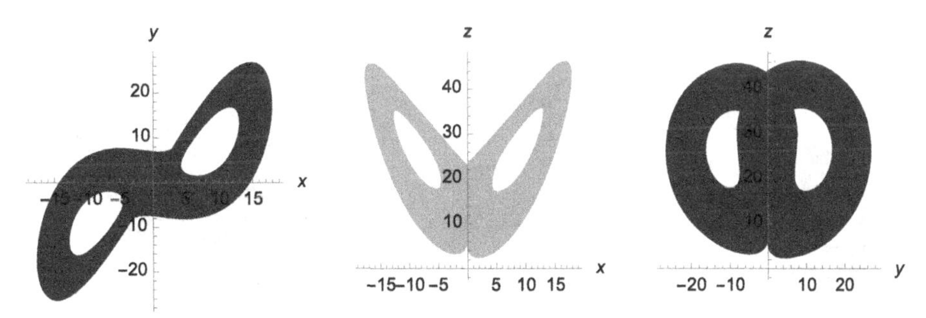

FIGURE 6.48 Plots of x vs y, x vs z, and y vs z. (West Virginia University colors)

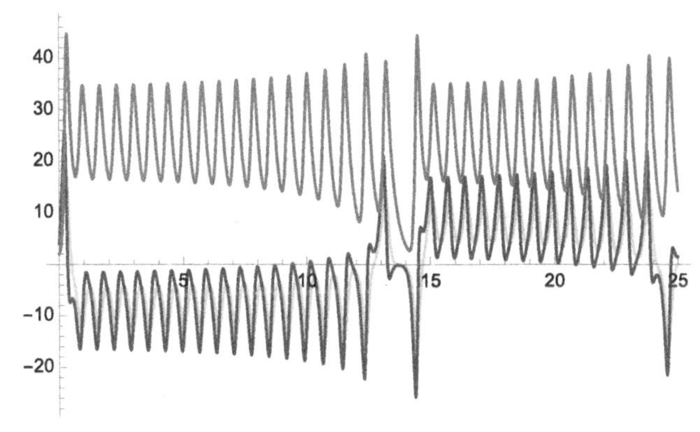

FIGURE 6.49 Plots of x, y, and z as functions of t.

```
p1b = ParametricPlot[Evaluate[{x[t], z[t]}/.n1], {t, 700, 1000},
  PlotPoints → 1000, AspectRatio → 1, AxesLabel → {x, z},
  PlotStyle → CMYKColor[0, .3, 1, .05]];
p1c = ParametricPlot[Evaluate[{y[t], z[t]}/.n1], {t, 700, 1000},
  PlotPoints → 1000, AspectRatio → 1, AxesLabel → {y, z},
  PlotStyle → CMYKColor[1, .69, .08, .54]];
Show[GraphicsRow[{p1a, p1b, p1c}]]

Plot[Evaluate[{x[t], y[t], z[t]}/.n1], {t, 0, 25},
  PlotStyle → {{CMYKColor[.36, .03, .28, .04]},
  {CMYKColor[.03, .91, .86, .12]}, {CMYKColor[0, .76, 1, 0]}},
  PlotPoints → 1000]
```

In `lorenzplot`, we combine `lorenzsol` to also graph the solutions parametrically.

```
lorenzplot[a_, b_, c_][{x0_, y0_, z0_}, ts_:{t, 700, 1000},
opts___]:=Module[{numsol},
numsol =
  NDSolve[{x'[t] == −ax[t] + ay[t], y'[t] == bx[t] − y[t] − x[t]z[t],
```

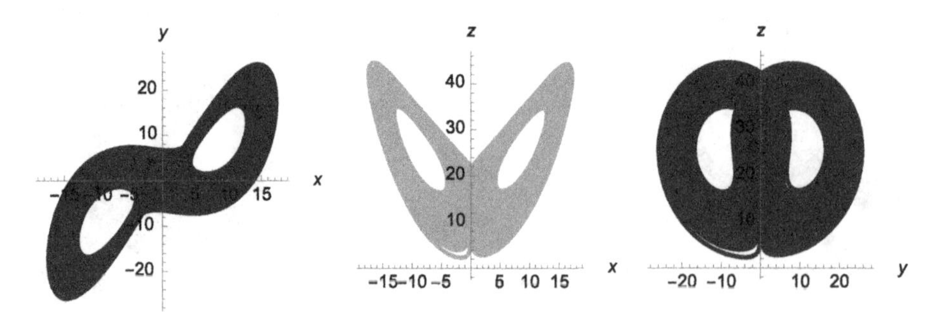

FIGURE 6.50 Plots of x vs y, x vs z, and y vs z.

```
    z'[t] == x[t]y[t] - cz[t], x[0] == x0, y[0] == y0, z[0] == z0},
    {x[t], y[t], z[t]}, ts, MaxSteps -> 1000000];
  p1a = ParametricPlot[Evaluate[{x[t], y[t]}/.numsol], ts,
    PlotPoints -> 1000, AspectRatio -> 1, AxesLabel -> {x, y},
    PlotStyle -> CMYKColor[1, .69, .08, .54]];
  p1b = ParametricPlot[Evaluate[{x[t], z[t]}/.numsol], ts,
    PlotPoints -> 1000, AspectRatio -> 1, AxesLabel -> {x, z},
    PlotStyle -> CMYKColor[0, .3, 1, .05]];
  p1c = ParametricPlot[Evaluate[{y[t], z[t]}/.numsol], ts,
    PlotPoints -> 1000, AspectRatio -> 1, AxesLabel -> {y, z},
    PlotStyle -> CMYKColor[1, .69, .08, .54]];
  Show[GraphicsRow[{p1a, p1b, p1c}, opts]]]

lorenzplot[7, 27.2, 3][{3, 4, 2}]
```

We then use `lorenzsol` to numerically solve the Lorenz system if $a = 7$, $b = 27.2$, an $c = 3$ and the initial conditions are $x(0) = 3$, $y(0) = 4$, and $z(0) = 2$ and graph the results in Fig. 6.50.

So that you can experiment with different parameters and initial-conditions, we use `Manipulate` to solve the Lorenz system using initial conditions $x(0) = x_0$, $y(0) = y_0$, and $z(0) = z_0$ for $950 \le t \le 1000$, generate parametric plots of x versus y, y versus z, x versus z, and x versus y versus z, and display the four resulting plots as a graphics array.

```
Manipulate[
Clear[x, y, z];
lorenzsol[a_, b_, c_][{x0_, y0_, z0_}, ts_:{t, 0, 1000},
opts___]:=Module[{numsol},
numsol =
  NDSolve[{x'[t] == -ax[t] + ay[t],
    y'[t] == bx[t] - y[t] - x[t]z[t],
```

```
    z'[t] == x[t]y[t] - cz[t], x[0] == x0,
    y[0] == y0, z[0] == z0}, {x[t], y[t], z[t]},
    ts, MaxSteps → Infinity]];
n1 = lorenzsol[a, b, c][{x0, y0, z0}];
  p1a = ParametricPlot[Evaluate[{x[t], y[t]}/.n1],
    {t, 500, 1000}, PlotPoints → 3000, AspectRatio → 1,
    PlotStyle → CMYKColor[1, .69, .08, .54],
    AxesLabel → {x, y}];
  p1b = ParametricPlot[Evaluate[{x[t], z[t]}/.n1],
    {t, 500, 1000}, PlotPoints → 3000, AspectRatio → 1,
    PlotStyle → CMYKColor[0, .3, 1, .05],
    AxesLabel → {x, z}];
  p1c = ParametricPlot[Evaluate[{y[t], z[t]}/.n1],
    {t, 500, 1000}, PlotPoints → 3000, AspectRatio → 1,
    PlotStyle → CMYKColor[0, .3, 1, .05],
    AxesLabel → {y, z}];
  p1d = ParametricPlot3D[
    Evaluate[{x[t], y[t], z[t]}/.n1], {t, 500, 1000},
    PlotPoints → 3000, BoxRatios → {1, 1, 1},
    PlotStyle → CMYKColor[1, .69, .08, .54],
    AxesLabel → {x, y, z}];
  Show[GraphicsGrid[{{p1a, p1b}, {p1c, p1d}}]], {{a, 7}, 5, 9},
    {{b, 27.2}, 25, 30}, {{c, 3}, 1, 5}, {{x0, 3}, 0, 5}, {{y0, 4}, 0, 5},
    {{z0, 2}, 0, 5}]
```

Because the behavior of solutions can be quite intricate, we include the option `MaxSteps->Infinity` in the `NDSolve` command to help Mathematica capture the oscillatory behavior in the long-term solution. See Fig. 6.51. You should be aware that numerically solving the Lorenz equations and graphing the solutions involves considerable memory and time even on a very fast computer. Consequently, although the `Manipulate` command works well, you may find it more useful to use `lorenzsol` to investigate the behavior of the Lorenz system, especially if you need three-dimensional plots. □

6.5 SOME PARTIAL DIFFERENTIAL EQUATIONS

6.5.1 The One-Dimensional Wave Equation

Suppose that we pluck a string (like a guitar or violin string) of length p and constant mass density that is fixed at each end. A question that we might ask is: What is the position of

Out[1]=

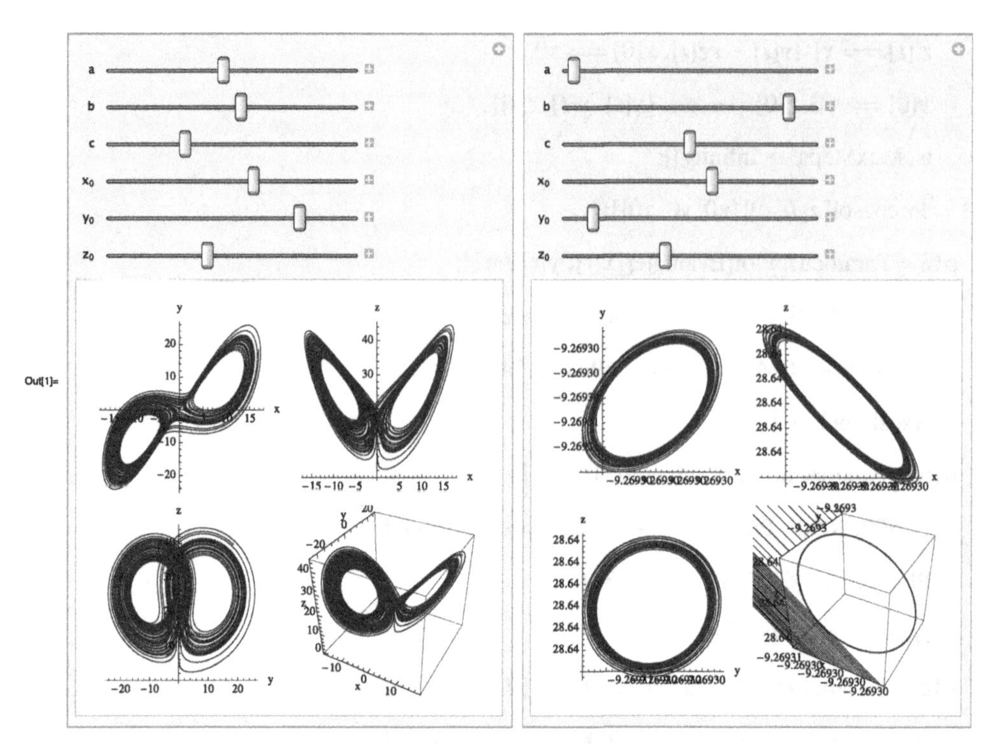

FIGURE 6.51 Comparing a chaotic solution to a non-chaotic solution with `Manipulate`.

the string at a particular instance of time? We answer this question by modeling the physical situation with a partial differential equation, namely the wave equation in one spatial variable:

$$c^2\frac{\partial^2 u}{\partial x^2}=\frac{\partial^2 u}{\partial t^2} \qquad \text{or} \qquad c^2 u_{xx}=u_{tt}. \tag{6.41}$$

In equation (6.41), $c^2=T/\rho$, where T is the tension of the string and ρ is the constant mass of the string per unit length. The solution $u(x,t)$ represents the displacement of the string from the x-axis at time t. To determine u we must describe the boundary and initial conditions that model the physical situation. At the ends of the string, the displacement from the x-axis is fixed at zero, so we use the homogeneous boundary conditions $u(0,t)=u(p,t)=0$ for $t>0$. The motion of the string also depends on the displacement and the velocity at each point of the string at $t=0$. If the initial displacement is given by $f(x)$ and the initial velocity by $g(x)$, we have the initial conditions $u(x,0)=f(x)$ and $u_t(x,0)=g(x)$ for $0\le x\le p$. Therefore, we determine the displacement of the string with the initial-boundary value problem

$$\begin{cases} c^2\dfrac{\partial^2 u}{\partial x^2}=\dfrac{\partial^2 u}{\partial t^2},\ 0<x<p,\ t>0 \\ u(0,t)=u(p,t)=0,\ t>0 \\ u(x,0)=f(x),\ u_t(x,0)=g(x),\ 0<x<p. \end{cases} \tag{6.42}$$

This problem is solved through separation of variables by assuming that $u(x,t)=X(x)T(t)$. Substitution into equation (6.41) yields

λ is a constant.

$$c^2X''T=XT'' \qquad \text{or} \qquad \frac{X''}{X}=\frac{T''}{c^2T}=-\lambda$$

so we obtain the two second-order ordinary differential equations $X''+\lambda X=0$ and $T''+c^2\lambda T=0$. At this point, we solve the equation that involves the homogeneous boundary conditions. The boundary conditions in terms of $u(x,t)=X(x)T(t)$ are $u(0,t)=X(0)T(t)=0$ and $u(p,t)=X(p)T(t)=0$, so we have $X(0)=0$ and $X(p)=0$. Therefore, we determine

$X(x)$ by solving the *eigenvalue problem*

$$\begin{cases} X'' + \lambda X = 0, \; 0 < x < p \\ X(0) = X(p) = 0. \end{cases}$$

The eigenvalues of this problem are $\lambda_n = (n\pi/p)^2$, $n = 1, 3, \ldots$ with corresponding eigenfunctions $X_n(x) = \sin(n\pi x/p)^2$, $n = 1, 3, \ldots$. Next, we solve the equation $T'' + c^2\lambda_n T = 0$. A general solution is

$$T_n(t) = a_n \cos\left(c\sqrt{\lambda_n t}\right) + b_n \sin\left(c\sqrt{\lambda_n t}\right) = a_n \cos\left(\frac{cn\pi t}{p}\right) + b_n \sin\left(\frac{cn\pi t}{p}\right),$$

where the coefficients a_n and b_n must be determined. Putting this information together, we obtain

$$u_n(x,t) = \left(a_n \cos\left(\frac{cn\pi t}{p}\right) + b_n \sin\left(\frac{cn\pi t}{p}\right)\right) \sin\left(\frac{n\pi x}{p}\right),$$

so by the Principle of Superposition, we have

$$u(x,t) = \sum_{n=1}^{\infty} \left(a_n \cos\left(\frac{cn\pi t}{p}\right) + b_n \sin\left(\frac{cn\pi t}{p}\right)\right) \sin\left(\frac{n\pi x}{p}\right).$$

Applying the initial displacement $u(x,0) = f(x)$ yields

$$u(x,0) = \sum_{n=1}^{\infty} a_n \sin\left(\frac{n\pi x}{p}\right) = f(x),$$

so a_n is the *Fourier sine series coefficient* for $f(x)$, which is given by

$$a_n = \frac{2}{p} \int_0^p f(x) \sin\left(\frac{n\pi x}{p}\right) dx, \quad n = 1, 2, \ldots.$$

In order to determine b_n, we must use the initial velocity. Therefore, we compute

$$\frac{\partial u}{\partial t}(x,t) = \sum_{n=1}^{\infty} \left(-a_n \frac{cn\pi}{p} \sin\left(\frac{cn\pi t}{p}\right) + b_n \frac{cn\pi}{p} \cos\left(\frac{cn\pi t}{p}\right)\right) \sin\left(\frac{n\pi x}{p}\right).$$

Then,

$$\frac{\partial u}{\partial t}(x,0) = \sum_{n=1}^{\infty} b_n \frac{cn\pi}{p} \sin\left(\frac{n\pi x}{p}\right) = g(x)$$

so $b_n \frac{cn\pi}{p}$ represents the Fourier sine series coefficient for $g(x)$ which means that

$$b_n = \frac{p}{cn\pi} \int_0^p g(x) \sin\left(\frac{n\pi x}{p}\right) dx, \quad n = 1, 2, \ldots.$$

Example 6.34

Solve $\begin{cases} u_{xx} = u_{tt}, \; 0 < x < 1, \; t > 0 \\ u(0,t) = u(1,t) = 0, \; t > 0 \\ u(x,0) = \sin \pi x, \; u_t(x,0) = 3x + 1, \; 0 < x < 1. \end{cases}$

Solution. The initial displacement and velocity functions are defined first.

```
f[x_] = Sin[πx];
g[x_] = 3x + 1;
```

Next, the functions to determine the coefficients a_n and b_n in the series approximation of the solution $u(x,t)$ are defined. Here, $p = c = 1$.

$$a_1 = 2\int_0^1 f[x]\text{Sin}[\pi x]dx$$

1

$$a_{n\_} = 2\int_0^1 f[x]\text{Sin}[n\pi x]dx$$

$$\frac{2\text{Sin}[n\pi]}{\pi - n^2\pi}$$

$$b_{n\_} = \frac{2\int_0^1 g[x]\text{Sin}[n\pi x]dx}{n\pi} \text{//Simplify}$$

$$\frac{2n\pi - 8n\pi\text{Cos}[n\pi] + 6\text{Sin}[n\pi]}{n^3\pi^3}$$

Because n represents an integer, these results indicate that $a_n = 0$ for all $n \geq 2$, which we confirm with `Simplify` together with the `Assumptions` by instructing Mathematica to assume that n is an integer.

$$\text{Simplify}\left[\frac{2\text{Sin}[n\pi]}{\pi - n^2\pi}, \text{Assumptions} \to \text{Element}[n, \text{Integers}]\right]$$

0

$$\text{Simplify}\left[\frac{2n\pi - 8n\pi\text{Cos}[n\pi] + 6\text{Sin}[n\pi]}{n^3\pi^3},\right.$$
$$\left.\text{Assumptions} \to \text{Element}[n, \text{Integers}]\right]$$

$$\frac{2 - 8(-1)^n}{n^2\pi^2}$$

We use `Table` to calculate the first ten values of b_n.

Table[{n, b_n, b_n//N}, {n, 1, 10}]//TableForm

1	$\frac{10}{\pi^2}$	1.01321
2	$-\frac{3}{2\pi^2}$	−0.151982
3	$\frac{10}{9\pi^2}$	0.112579
4	$-\frac{3}{8\pi^2}$	−0.0379954
5	$\frac{2}{5\pi^2}$	0.0405285
6	$-\frac{1}{6\pi^2}$	−0.0168869
7	$\frac{10}{49\pi^2}$	0.0206778
8	$-\frac{3}{32\pi^2}$	−0.00949886
9	$\frac{10}{81\pi^2}$	0.0125088
10	$-\frac{3}{50\pi^2}$	−0.00607927

The function u defined next computes the nth term in the series expansion. Thus, uapprox determines the approximation of order k by summing the first k terms of the expansion, as illustrated with approx[10].

Notice that we define uapprox[n] so that Mathematica "remembers" the terms uapprox that are computed. That is, Mathematica does not need to recompute uapprox[n-1] to compute uapprox[n] provided that uapprox[n-1] has already been computed.

Clear[u, uapprox]

u[n_] = b_nSin[$n\pi t$]Sin[$n\pi x$];

uapprox[k_]:=uapprox[k] = uapprox[k − 1] + u[k];

uapprox[0] = Cos[πt]Sin[πx];

uapprox[10]

$$\text{Cos}[\pi t]\text{Sin}[\pi x] + \frac{10\text{Sin}[\pi t]\text{Sin}[\pi x]}{\pi^2} - \frac{3\text{Sin}[2\pi t]\text{Sin}[2\pi x]}{2\pi^2} + \frac{10\text{Sin}[3\pi t]\text{Sin}[3\pi x]}{9\pi^2} - \frac{3\text{Sin}[4\pi t]\text{Sin}[4\pi x]}{8\pi^2} + \frac{2\text{Sin}[5\pi t]\text{Sin}[5\pi x]}{5\pi^2} - \frac{\text{Sin}[6\pi t]\text{Sin}[6\pi x]}{6\pi^2} + \frac{10\text{Sin}[7\pi t]\text{Sin}[7\pi x]}{49\pi^2} - \frac{3\text{Sin}[8\pi t]\text{Sin}[8\pi x]}{32\pi^2} + \frac{10\text{Sin}[9\pi t]\text{Sin}[9\pi x]}{81\pi^2} - \frac{3\text{Sin}[10\pi t]\text{Sin}[10\pi x]}{50\pi^2}$$

To illustrate the motion of the string, we graph uapprox[10], the tenth partial sum of the series, on the interval [0, 1] for 16 equally spaced values of t between 0 and 2 in Fig. 6.52.

somegraphs =

Table[Plot[Evaluate[uapprox[10]], {x, 0, 1}, PlotRange → {$-\frac{3}{2}, \frac{3}{2}$},

Ticks → {{0, 1}, {−1, 1}}, PlotStyle → CMYKColor[.53, .72, .77, .57]],

{t, 0, 2, $\frac{2}{15}$}];

toshow = Partition[somegraphs, 4];

Show[GraphicsGrid[toshow]]

If instead we wished to see the motion of the string, we can use Animate. We show a frame from the resulting animation.

uapprox[10]

Animate[

Plot[$\text{Cos}[\pi t]\text{Sin}[\pi x] + \frac{10\text{Sin}[\pi t]\text{Sin}[\pi x]}{\pi^2} - \frac{3\text{Sin}[2\pi t]\text{Sin}[2\pi x]}{2\pi^2} + \frac{10\text{Sin}[3\pi t]\text{Sin}[3\pi x]}{9\pi^2} - \frac{3\text{Sin}[4\pi t]\text{Sin}[4\pi x]}{8\pi^2} + \frac{2\text{Sin}[5\pi t]\text{Sin}[5\pi x]}{5\pi^2} - \frac{\text{Sin}[6\pi t]\text{Sin}[6\pi x]}{6\pi^2} + \frac{10\text{Sin}[7\pi t]\text{Sin}[7\pi x]}{49\pi^2} - \frac{3\text{Sin}[8\pi t]\text{Sin}[8\pi x]}{32\pi^2} + \frac{10\text{Sin}[9\pi t]\text{Sin}[9\pi x]}{81\pi^2} - \frac{3\text{Sin}[10\pi t]\text{Sin}[10\pi x]}{50\pi^2}$**, {$x$, 0, 1},**

PlotRange → {−3/2, 3/2}, Ticks → {{0, 1}, {−1, 1}},

PlotStyle->CMYKColor[0, .24, .94, 0]], {t, 0, 2}]

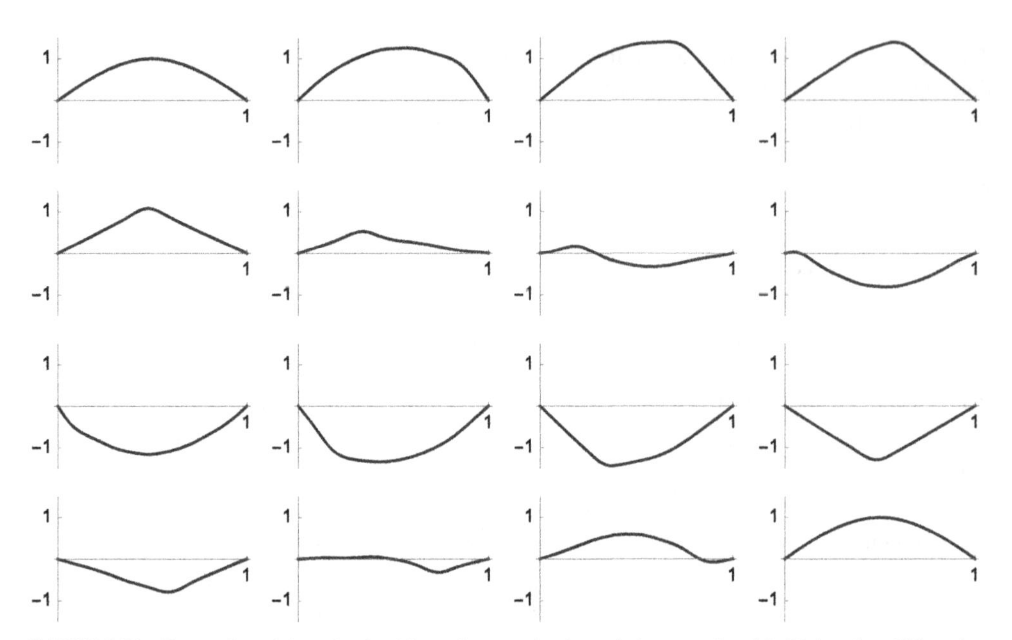

FIGURE 6.52 The motion of the string for 16 equally spaced values of t between 0 and 2. (University of Wyoming colors)

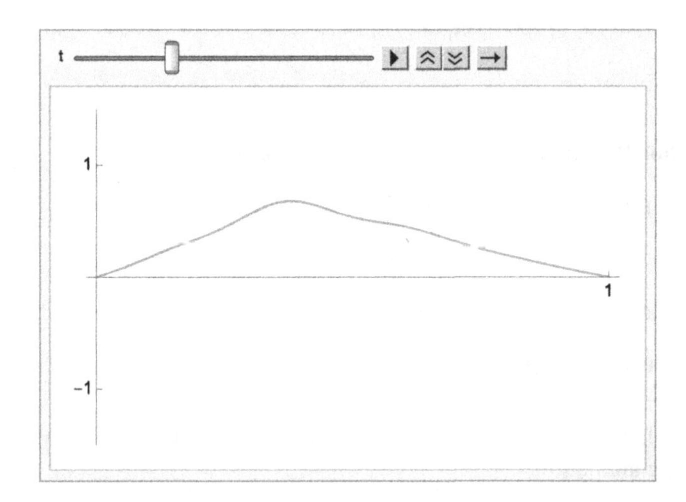

Finally, we remark that `DSolve` can find **D'Alembert's solution** to the wave equation.

Clear[u, c]

DSolve[c^2D[u[x, t], {x, 2}]==D[u[x, t], {t, 2}],

u[x, t], {x, t}]

$\{\{u[x,t] \to C[1][t - \frac{x}{\sqrt{c^2}}] + C[2][t + \frac{x}{\sqrt{c^2}}]\}\}$

DSolve[$c^2\partial_{\{x,2\}}u[x,t]==\partial_{\{t,2\}}u[x,t]$ **, u[x, t], {x, t}]**

$\{\{u[x,t] \to C[1][t - \frac{x}{\sqrt{c^2}}] + C[2][t + \frac{x}{\sqrt{c^2}}]\}\}$ □

6.5.2 The Two-Dimensional Wave Equation

Another interesting problem involving two spatial dimensions (x and y) is the wave equation. The two-dimensional wave equation in a circular region which is radially symmetric (not dependent on θ) with boundary and initial conditions is expressed in polar coordinates as

$$\begin{cases} c^2\left(\dfrac{\partial^2 u}{\partial r^2}+\dfrac{1}{r}\dfrac{\partial u}{\partial r}\right)=\dfrac{\partial^2 u}{\partial t^2},\ 0<r<\rho,\ t>0 \\ u(\rho,t)=0,\ |u(0,t)|<\infty,\ t>0 \\ u(r,0)=f(r),\ \dfrac{\partial u}{\partial t}(r,0)=g(r),\ 0<r<\rho. \end{cases}$$

Notice that the boundary condition $u(\rho,t)=0$ indicates that u is fixed at zero around the boundary; the condition $|u(0,t)|<\infty$ indicates that the solution is bounded at the center of the circular region. Like the wave equation discussed previously, this problem is typically solved through separation of variables by assuming a solution of the form $u(r,t)=F(r)G(t)$. Applying separation of variables yields the solution

$$u(r,t)=\sum_{n=1}^{\infty}\left(A_n\cos ck_n t+B_n\sin ck_n t\right)J_0(k_n r),$$

where $\lambda_n=c\alpha_n/\rho$, and the coefficients A_n and B_n are found through application of the initial displacement and velocity functions. With

α_n represents the nth zero of the Bessel function of the first kind of order zero.

$$u(r,0)=\sum_{n=1}^{\infty}A_n J_0(k_n r)=f(r)$$

and the orthogonality conditions of the Bessel functions, we find that

$$A_n=\frac{\int_0^\rho rf(r)J_0(k_n r)\,dr}{\int_0^\rho r\left[J_0(k_n r)\right]^2 dr}=\frac{2}{\left[J_1(\alpha_n)\right]^2}\int_0^\rho rf(r)J_0(k_n r)\,dr,\ n=1,2,\ldots.$$

Similarly, because

$$\frac{\partial u}{\partial t}(r,0)=\sum_{n=1}^{\infty}\left(-ck_n A_n\sin ck_n t+ck_n B_n\cos ck_n t\right)J_0(k_n r)$$

we have

$$u_t(r,0)=\sum_{n=1}^{\infty}ck_n B_n J_0(k_n r)=g(r).$$

Therefore,

$$B_n=\frac{\int_0^\rho rg(r)J_0(k_n r)\,dr}{ck_n\int_0^\rho r\left[J_0(k_n r)\right]^2 dr}=\frac{2}{ck_n\left[J_1(\alpha_n)\right]^2}\int_0^\rho rg(r)J_0(k_n r)\,dr,\ n=1,2,\ldots.$$

As a practical matter, in nearly all cases, these formulas are difficult to evaluate.

Example 6.35

Solve $$\begin{cases} \dfrac{\partial^2 u}{\partial r^2}+\dfrac{1}{r}\dfrac{\partial u}{\partial r}=\dfrac{\partial^2 u}{\partial t^2},\ 0<r<1,\ t>0 \\ u(1,t)=0,\ |u(0,t)|<\infty,\ t>0 \\ u(r,0)=r(r-1),\ \dfrac{\partial u}{\partial t}(r,0)=\sin\pi r,\ 0<r<1. \end{cases}$$

Solution. In this case, $\rho=1$, $f(r)=r(r-1)$, and $g(r)=\sin\pi r$. The command `BesselJZero[n,k]` represents the kth zero of the Bessel function $J_k(x)$. To obtain an approximation of the number, use `N`.

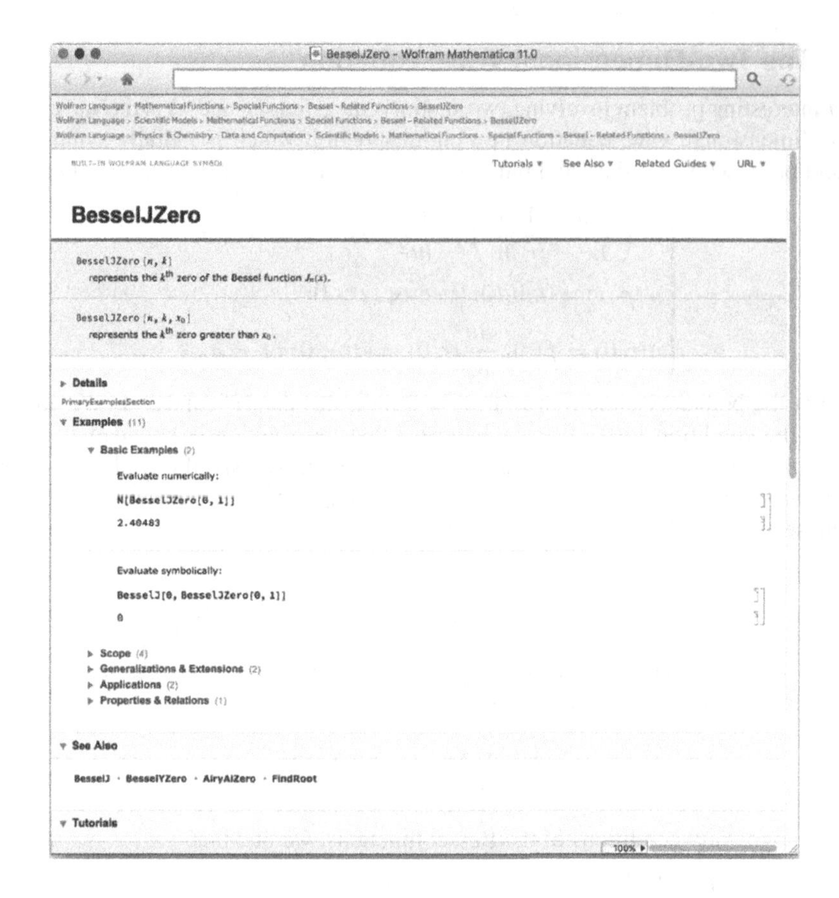

$\alpha_{\text{n\_}}$:=Evaluate[BesselJZero[0, n]//N]

Next, we define the constants ρ and c and the functions $f(r) = r(r-1)$, $g(r) = \sin \pi r$, and $k_n = \alpha_n/\rho$.

$c = 1$;

$\rho = 1$;

f[r_] $= r(r-1)$;

g[r_] = Sin[πr];

$k_{\text{n\_}}$:=$k_n = \frac{\alpha_n}{\rho}$;

The formulas for the coefficients A_n and B_n are then defined so that an approximate solution may be determined. (We use lowercase letters to avoid any possible ambiguity with built-in Mathematica functions.) Note that we use `NIntegrate` to approximate the coefficients and avoid the difficulties in integration associated with the presence of the Bessel function of order zero.

$a_{\text{n\_}}$:=a_n = (2NIntegrate[$r f[r]$BesselJ[0, $k_n r$], {r, 0, ρ}])/BesselJ[1, α_n]^2;

$b_{\text{n\_}}$:=b_n = (2NIntegrate[$r g[r]$BesselJ[0, $k_n r$], {r, 0, ρ}])/
(ck_nBesselJ[1, α_n]^2)

We now compute the first ten values of A_n and B_n. Because a and b are defined using the form $a_{n\_} := a_n = \ldots$ and $b_{n\_} := b_n = \ldots$, Mathematica remembers these values for later use.

Table[{n, a_n, b_n}, {n, 1, 10}]//TableForm

1	1	0.52118
2	0.208466	−0.145776
3	0.00763767	−0.0134216
4	0.0383536	−0.00832269
5	0.00534454	−0.00250503
6	0.0150378	−0.00208315
7	0.00334937	−0.000882012
8	0.00786698	−0.000814719
9	0.00225748	−0.000410202
10	0.00479521	−0.000399219

The nth term of the series solution is defined in u. Then, an approximate solution is obtained in uapprox by summing the first ten terms of u.

```
u[n_, r_, t_]:=(a_n Cos[c k_n t] + b_n Sin[c k_n t])BesselJ[0, k_n r];
uapprox[r_, t_] = Sum_{n=1}^{10} u[n, r, t];
```

We graph uapprox for several values of t in Fig. 6.53.

```
somegraphs =
  Table[ParametricPlot3D[{rCos[θ], rSin[θ], uapprox[r, t]},
  {r, 0, 1}, {θ, −π, π}, Boxed → False, PlotRange → {−1.25, 1.25},
  BoxRatios → {1, 1, 1}, Ticks → {{−1, 1}, {−1, 1}, {−1, 1}}],
  {t, 0, 1.5, 1.5/8}];
toshow = Partition[somegraphs, 3];
Show[GraphicsGrid[toshow]]
```

In order to actually watch the drumhead move, we can use Animate. Be aware, however, that generating many three-dimensional graphics and then animating the results uses a great deal of memory and can take considerable time, even on a relatively powerful computer. We show one frame from the animation that results from the following Animate command.

```
Animate[ParametricPlot3D[{rCos[θ], rSin[θ], uapprox[r, t]},
  {r, 0, 1}, {θ, −π, π}, Boxed → False,
```

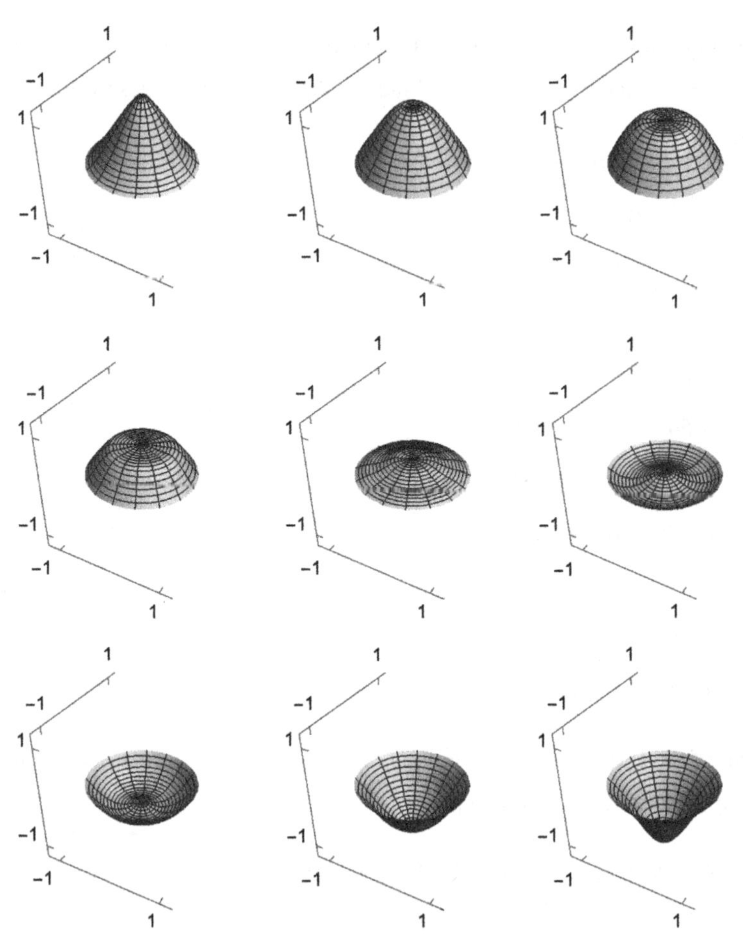

FIGURE 6.53 The drumhead for nine equally spaced values of t between 0 and 1.5.

```
PlotRange → {−1.25, 1.25}, BoxRatios → {1, 1, 1},
Ticks → {{−1, 1}, {−1, 1}, {−1, 1}}],
{t, 0, 1.5}]
```

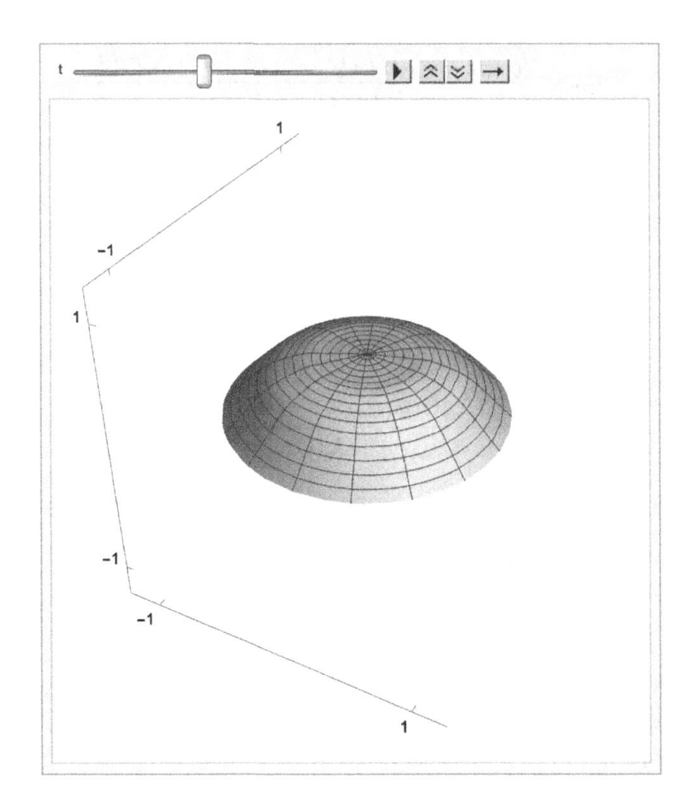

□

If the displacement of the drumhead is not radially symmetric, the problem that describes the displacement of a circular membrane in its general case is

$$\begin{cases} c^2\left(\dfrac{\partial^2 u}{\partial r^2}+\dfrac{1}{r}\dfrac{\partial u}{\partial r}+\dfrac{1}{r^2}\dfrac{\partial^2 u}{\partial \theta^2}\right)=\dfrac{\partial^2 u}{\partial t^2},\ 0<r<\rho,\ -\pi<\theta<\pi,\ t>0 \\ u(\rho,\theta,t)=0,\ |u(0,\theta,t)|<\infty,\ -\pi\le\theta\le\pi,\ t>0 \\ u(r,\pi,t)=u(r,-\pi,t),\ \dfrac{\partial u}{\partial \theta}(r,\pi,t)=\dfrac{\partial u}{\partial \theta}(r,-\pi,t),\ 0<r<\rho,\ t>0 \\ u(r,\theta,0)=f(r,\theta),\ \dfrac{\partial u}{\partial t}(r,\pi,0)=g(r,\theta),\ 0<r<\rho,\ -\pi<\theta<\pi. \end{cases} \tag{6.43}$$

Using separation of variables and assuming that $u(r,\theta,t)=R(t)H(\theta)T(t)$, we obtain that a general solution is given by

$$\begin{aligned} u(r,\theta,t)=&\sum_n a_{0n}J_0(\lambda_{0n}r)\cos(\lambda_{0n}ct)+\sum_{m,n}a_{mn}J_m(\lambda_{mn}r)\cos(m\theta)\cos(\lambda_{mn}ct)+ \\ &\sum_{m,n}b_{mn}J_m(\lambda_{mn}r)\sin(m\theta)\cos(\lambda_{mn}ct)+\sum_n A_{0n}J_0(\lambda_{0n}r)\sin(\lambda_{0n}ct)+ \\ &\sum_{m,n}A_{mn}J_m(\lambda_{mn}r)\cos(m\theta)\sin(\lambda_{mn}ct)+ \\ &\sum_{m,n}B_{mn}J_m(\lambda_{mn}r)\sin(m\theta)\sin(\lambda_{mn}ct), \end{aligned}$$

where J_m represents the mth Bessel function of the first kind, α_{mn} denotes the nth zero of the Bessel function $y=J_m(x)$, and $\lambda_{mn}=\alpha_{mn}/\rho$. The coefficients are given by the following formulas.

$$a_{0n}=\frac{\int_0^{2\pi}\int_0^{\rho}f(r,\theta)J_0(\lambda_{0n}r)\,r\,dr\,d\theta}{2\pi\int_0^{\rho}[J_0(\lambda_{0n}r)]^2\,r\,dr}$$

$$a_{mn} = \frac{\int_0^{2\pi} \int_0^{\rho} f(r,\theta) J_m(\lambda_{mn} r) \cos(m\theta)\, r\, dr\, d\theta}{\pi \int_0^{\rho} [J_m(\lambda_{mn} r)]^2 r\, dr}$$

$$b_{mn} = \frac{\int_0^{2\pi} \int_0^{\rho} f(r,\theta) J_m(\lambda_{mn} r) \sin(m\theta)\, r\, dr\, d\theta}{\pi \int_0^{\rho} [J_m(\lambda_{mn} r)]^2 r\, dr}$$

$$A_{0n} = \frac{\int_0^{2\pi} \int_0^{\rho} g(r,\theta) J_0(\lambda_{0n} r)\, r\, dr\, d\theta}{2\pi \lambda_{0n} c\pi \int_0^{\rho} [J_0(\lambda_{0n} r)]^2 r\, dr}$$

$$A_{mn} = \frac{\int_0^{2\pi} \int_0^{\rho} g(r,\theta) J_m(\lambda_{mn} r) \cos(m\theta)\, r\, dr\, d\theta}{\pi \lambda_{mn} c \int_0^{\rho} [J_m(\lambda_{mn} r)]^2 r\, dr}$$

$$B_{mn} = \frac{\int_0^{2\pi} \int_0^{\rho} g(r,\theta) J_m(\lambda_{mn} r) \sin(m\theta)\, r\, dr\, d\theta}{\pi \lambda_{mn} c \int_0^{\rho} [J_m(\lambda_{mn} r)]^2 r\, dr}$$

Example 6.36

Solve
$$\begin{cases} 10^2 \left(\frac{\partial^2 u}{\partial r^2} + \frac{1}{r}\frac{\partial u}{\partial r} + \frac{1}{r^2}\frac{\partial^2 u}{\partial \theta^2} \right) = \frac{\partial^2 u}{\partial t^2}, \\ 0 < r < 1,\ -\pi < \theta < \pi,\ t > 0 \\ u(1,\theta,t) = 0,\ |u(0,\theta,t)| < \infty,\ -\pi \le \theta \le \pi,\ t > 0 \\ u(r,\pi,t) = u(r,-\pi,t),\ \frac{\partial u}{\partial \theta}(r,\pi,t) = \frac{\partial u}{\partial \theta}(r,-\pi,t). \\ 0 < r < 1,\ t > 0 \\ u(r,\theta,0) = \cos(\pi r/2)\sin\theta, \\ \frac{\partial u}{\partial t}(r,\pi,0) = (r-1)\cos(\pi\theta/2),\ 0 < r < 1,\ -\pi < \theta < \pi \end{cases}$$

Solution. To calculate the coefficients, we will need to have approximations of the zeros of the Bessel functions, so we use `BesselJZero` together with `N` and `Evaluate` to define α_{mn} to be an approximation of the nth zero of $y = J_m(x)$. We illustrate the use of α_{mn} by using it to compute the first five zeros of $y = J_0(x)$.

```
α_{m_,n_}:=α_{m,n} = Evaluate[BesselJZero[m, n]//N]

Table[α_{0,n}, {n, 1, 5}]

{2.40483, 5.52008, 8.65373, 11.7915, 14.9309}
```

The appropriate parameter values as well as the initial condition functions are defined as follows. Notice that the functions describing the initial displacement and velocity are defined as the product of functions. This enables the subsequent calculations to be carried out using `NIntegrate`.

```
Clear[a, f, f1, f2, g1, g2, A, c, g, capa, capb, b]

c = 10;

ρ = 1;

f1[r_] = Cos[πr/2];

f2[θ_] = Sin[θ];

f[r_, θ_]:=f[r, θ] = f1[r]f2[θ];
```

```
g1[r_] = r − 1;
g2[θ_] = Cos[πθ/2];
g[r_, θ_]:=g[r, θ] = g1[r]g2[θ];
```

The coefficients a_{0n} are determined with the function a.

```
Clear[a]
a[n_]:=a[n] =
N[
(NIntegrate[f1[r]BesselJ[0, α0,n r]r, {r, 0, ρ}].
NIntegrate[f2[t], {t, 0, 2π}])/
   (2πNIntegrate[rBesselJ[0, α0,n r]^2, {r, 0, ρ}])];
```

Hence, as represents a table of the first five values of a_{0n}. Chop is used to round off very small numbers to zero.

```
as = Table[a[n]//Chop, {n, 1, 5}]
{0, 0, 0, 0, 0}
```

Because the denominator of each integral formula used to find a_{mn} and b_{mn} is the same, the function bjmn which computes this value is defined next. A table of nine values of this coefficient is then determined.

```
bjmn[m_, n_]:=
bjmn[m, n] = N[NIntegrate[rBesselJ[m, αm,n r]^2, {r, 0, ρ}]]
Table[Chop[bjmn[m, n]], {m, 1, 3}, {n, 1, 3}]
{{0.0811076, 0.0450347, 0.0311763}, {0.0576874, 0.0368243, 0.0270149},
{0.0444835, 0.0311044, 0.0238229}}
```

We also note that in evaluating the numerators of a_{mn} and b_{mn} we must compute $\int_0^\rho r f_1(r) J_m(\alpha_{mn} r)\, dr$. This integral is defined in fbjmn and the corresponding values are found for $n = 1, 2, 3$ and $m = 1, 2, 3$.

```
Clear[fbjmn]
fbjmn[m_, n_]:=fbjmn[m, n] =
   N[NIntegrate[f1[r]BesselJ[m, αm,n r]r, {r, 0, ρ}]]
Table[Chop[fbjmn[m, n]], {m, 1, 3}, {n, 1, 3}]
{{0.103574, 0.020514, 0.0103984}, {0.0790948, 0.0275564, 0.0150381},
{0.0628926, 0.0290764, 0.0171999}}
```

The formula to compute a_{mn} is then defined and uses the information calculated in fbjmn and bjmn. As in the previous calculation, the coefficient values for $n = 1, 2, 3$ and $m = 1, 2, 3$ are determined.

```
a[m_, n_]:=
a[m, n] = N[(fbjmn[m, n]NIntegrate[f2[t]Cos[mt], {t, 0, 2π}])/
  (π bjmn[m, n])];
Table[Chop[a[m, n]], {m, 1, 3}, {n, 1, 3}]
{{0, 0, 0}, {0, 0, 0}, {0, 0, 0}}
```

A similar formula is then defined for the computation of b_{mn}.

```
b[m_, n_]:=b[m, n] =
N[(fbjmn[m, n]NIntegrate[f2[t]Sin[mt], {t, 0, 2π}])/
  (π bjmn[m, n])];
Table[Chop[b[m, n]], {m, 1, 3}, {n, 1, 3}]
{{1.277, 0.455514, 0.333537}, {0, 0, 0}, {0, 0, 0}}
```

Note that defining the coefficients in this manner a[m_,n_]:=a[m,n]=... and b[m_,n_]:=b[m,n]=... so that Mathematica "remembers" previously computed values which reduces computation time. The values of A_{0n} are found similarly to those of a_{0n}. After defining the function capa to calculate these coefficients, a table of values is then found.

```
capa[n_]:=capa[n] =
N[(NIntegrate[g1[r]BesselJ[0, α0,n r]r, {r, 0, ρ}]
NIntegrate[g2[t], {t, 0, 2π}])/
  (2π cα0,n NIntegrate[r BesselJ[0, α0,n r]^2, {r, 0, ρ}])];
Table[Chop[capa[n]], {n, 1, 6}]
{0.00142231, 0.0000542518, 0.0000267596, 6.419764234815093*^-6,
4.958428464118777 5*^-6, 1.8858472721004265*^-6}
```

The value of the integral of the component of g, g1, which depends on r and the appropriate Bessel functions, is defined as gbjmn.

```
gbjmn[m_, n_]:=gbjmn[m, n] = NIntegrate[g1[r]*
  BesselJ[m, αm,n r]r, {r, 0, ρ}]//N
Table[gbjmn[m, n]//Chop, {m, 1, 3}, {n, 1, 3}]
```

{{−0.0743906, −0.019491, −0.00989293}, {−0.0554379, −0.0227976, −0.013039},
{−0.0433614, −0.0226777, −0.0141684}}

Then, A_{mn} is found by taking the product of integrals, `gbjmn` depending on r and one depending on θ. A table of coefficient values is generated in this case as well.

```
capa[m_, n_]:=capa[m, n] =
  N[(gbjmn[m, n]NIntegrate[g2[t]Cos[mt], {t, 0, 2π}])/
  (π α_{m,n} cbjmn[m, n])];
Table[Chop[capa[m, n]], {m, 1, 3}, {n, 1, 3}]
```

{{0.0035096, 0.000904517, 0.000457326}, {−0.00262692, −0.00103252, −0.000583116},
{−0.000503187, −0.000246002, −0.000150499}}

Similarly, the B_{mn} are determined.

```
capb[m_, n_]:=capb[m, n] =
  N[(gbjmn[m, n]NIntegrate[g2[t]Sin[mt], {t, 0, 2π}])/
  (π α_{m,n} cbjmn[m, n])];
Table[Chop[capb[m, n]], {m, 1, 3}, {n, 1, 3}]
```

{{0.00987945, 0.00254619, 0.00128736}, {−0.0147894, −0.00581305, −0.00328291},
{−0.00424938, −0.00207747, −0.00127095}}

Now that the necessary coefficients have been found, we construct an approximate solution to the wave equation by using our results. In the following, `term1` represents those terms of the expansion involving a_{0n}, `term2` those terms involving a_{mn}, `term3` those involving b_{mn}, `term4` those involving A_{0n}, `term5` those involving A_{mn}, and `term6` those involving B_{mn}.

```
Clear[term1, term2, term3, term4, term5, term6]
term1[r_, t_, n_]:=a[n]BesselJ[0, α_{0,n} r]Cos[α_{0,n} ct];
term2[r_, t_, θ_, m_, n_]:=
  a[m, n]BesselJ[m, α_{m,n} r]Cos[mθ]Cos[α_{m,n} ct];
term3[r_, t_, θ_, m_, n_]:=
  b[m, n]BesselJ[m, α_{m,n} r]Sin[mθ]Cos[α_{m,n} ct];
term4[r_, t_, n_]:=capa[n]BesselJ[0, α_{0,n} r]Sin[α_{0,n} ct];
term5[r_, t_, θ_, m_, n_]:=
  capa[m, n]BesselJ[m, α_{m,n} r]Cos[mθ]Sin[α_{m,n} ct];
term6[r_, t_, θ_, m_, n_]:=
```

```
capb[m, n]BesselJ[m, α_{m,n} r]Sin[mθ]Sin[α_{m,n} ct];
```

Therefore, our approximate solution is given as the sum of these terms as computed in `u`.

```
Clear[u]
u[r_, t_, th_]:= Sum_{n=1}^{5} term1[r, t, n] + Sum_{m=1}^{3} Sum_{n=1}^{3} term2[r, t, th, m, n]+
  Sum_{m=1}^{3} Sum_{n=1}^{3} term3[r, t, th, m, n] + Sum_{n=1}^{5} term4[r, t, n]+
  Sum_{m=1}^{3} Sum_{n=1}^{3} term5[r, t, th, m, n] + Sum_{m=1}^{3} Sum_{n=1}^{3} term6[r, t, th, m, n];
uc = Compile[{r, t, th}, u[r, t, th]]
CompiledFunction[]
```

The solution is *compiled* in `uc`. The command `Compile` is used to compile functions. `Compile` returns a `CompiledFunction` which represents the compiled code. Generally, compiled functions take less time to perform computations than uncompiled functions, although compiled functions can only be evaluated for numerical arguments.

Next, we define the function `tplot` which uses `ParametricPlot3D` to produce the graph of the solution for a particular value of t. Note that the x and y coordinates are given in terms of polar coordinates.

```
Clear[tplot]
tplot[t_]:=ParametricPlot3D[{rCos[θ], rSin[θ], uc[r, t, θ]},
  {r, 0, 1}, {θ, −π, π}, PlotPoints → {20, 20},
  BoxRatios → {1, 1, 1}, Axes → False, Boxed → False]
```

A table of nine plots for nine equally spaced values of t from $t = 0$ to $t = 1$ using increments of $1/8$ is then generated. This table of graphs is displayed as a graphics array (grid) in Fig. 6.54.

```
somegraphs = Table[tplot[t], {t, 0, 1, 1/8}];
toshow = Partition[somegraphs, 3];
Show[GraphicsGrid[toshow]]
```

Of course, we can animate the motion of the drumhead with `Animate` as illustrated in Fig. 6.55. Be aware, however, that generating many three-dimensional graphics and then animating the results uses a great deal of memory and can take considerable time, even on a relatively powerful computer.

```
Animate[tplot[t], {t, 0, 1, 1/8}]
```

□

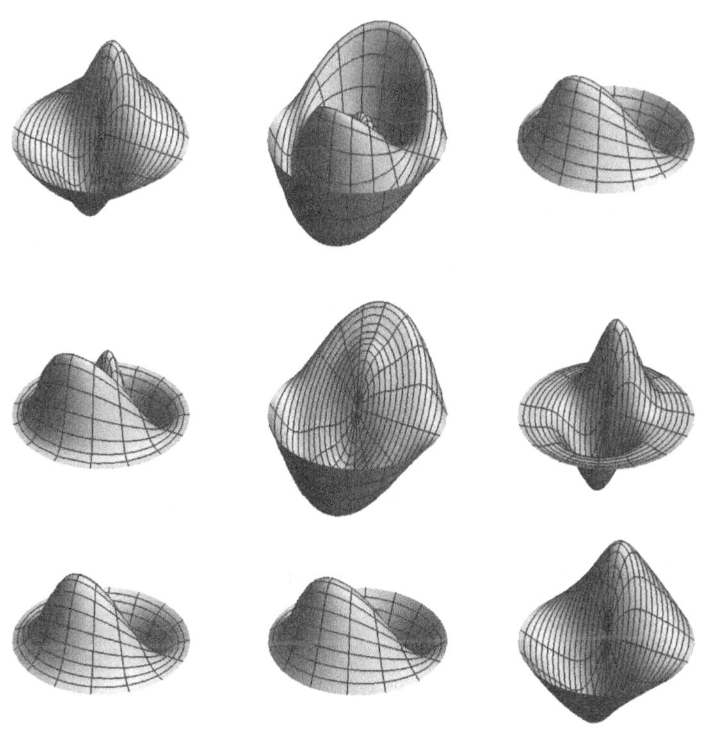

FIGURE 6.54 The drumhead for nine equally spaced values of t from $t = 0$ to $t = 1$.

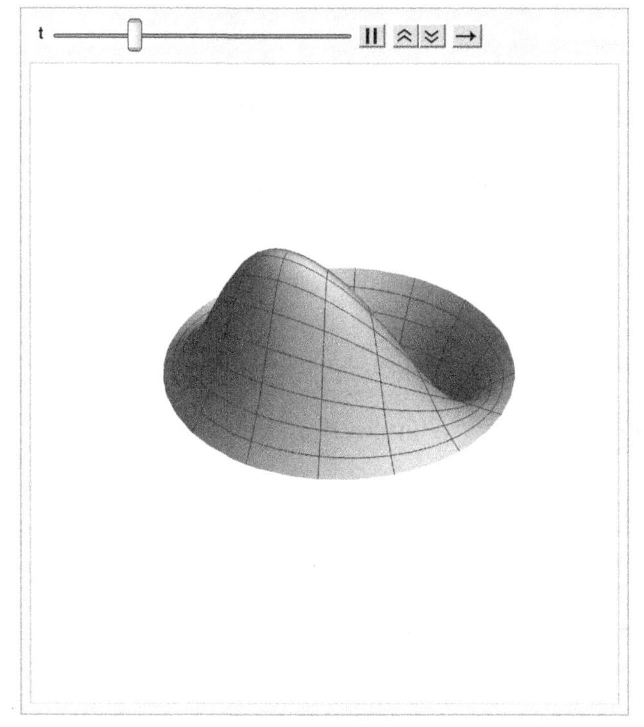

FIGURE 6.55 Using `Animate` to see the motion of the drumhead.

6.5.3 Other Partial Differential Equations

A partial differential equation of the form

$$a(x,y,u)\frac{\partial u}{\partial x}+b(x,y,u)\frac{\partial u}{\partial y}=c(x,y,u) \quad (6.44)$$

is called a **first-order, quasi-linear partial differential equation**. In the case when $c(x,y,u)=0$, equation (6.44) is **homogeneous**; if a and b are independent of u, equation (6.44) is **almost linear**; and when $c(x,y,u)$ can be written in the form $c(x,y,u)=d(x,y)u+s(x,y)$, equation (6.44) is **linear**. Quasi-linear partial differential equations can frequently be solved using the *method of characteristics*.

Example 6.37

Use the *method of characteristics* to solve the initial-value problem $\begin{cases}-3xtu_x+u_t=xt\\ u(x,0)=x.\end{cases}$

Solution. For this problem, the *characteristic system* is

$$\begin{aligned}\partial x/\partial r&=-3xt, & x(0,s)&=s\\ \partial t/\partial r&=1, & t(0,s)&=0\\ \partial u/\partial r&=xt, & u(0,s)&=s.\end{aligned}$$

We begin by using `DSolve` to solve $\partial t/\partial r=1$, $t(0,s)=0$

```
d1 = DSolve[{D[t[r], r]==1, t[0]==0}, t[r], r]
```

$\{\{t[r]\to r\}\}$

and obtain $t=r$. Thus, $\partial x/\partial r=-3xr$, $x(0,s)=s$, which we solve next

```
d2 = DSolve[{D[x[r], r]== - 3x[r]r, x[0]==s}, x[r], r]
```

$\{\{x[r]\to e^{-\frac{3r^2}{2}}s\}\}$

and obtain $x=se^{-3r^2/2}$. Substituting $r=t$ and $x=se^{-3r^2/2}$ into $\partial u/\partial r=xt$, $u(0,s)=s$ and using `DSolve` to solve the resulting equation yields the following result, named `d3`.

```
d3 = DSolve[{D[u[r], r]==E^(-3r^2/2) sr, u[0]==s}, u[r], r]
```

$\{\{u[r]\to \frac{1}{3}e^{-\frac{3r^2}{2}}(-1+4e^{\frac{3r^2}{2}})s\}\}$

To find $u(x,t)$, we must solve the system of equations

$$\begin{cases}t=r\\ x=se^{-3r^2/2}\end{cases}$$

for r and s. Substituting $r=t$ into $x=se^{-3r^2/2}$ and solving for s yields $s=xe^{3t^2}/2$. Thus, the solution is given by replacing the values obtained above in the solution obtained in `d3`. We do this below by using `ReplaceAll` (`/.`) to replace each occurrence of r and s in `d3[[1,1,2]]`, the solution obtained in `d3`, by the values $r=t$ and $s=xe^{3t^2}/2$. The resulting output represents the solution to the initial-value problem.

d3[[1, 1, 2]]/.{r->t, s->xExp[3/2t^2]}//Simplify

$\frac{1}{3}(-1+4e^{\frac{3t^2}{2}})x$

In this example, `DSolve` can also solve this first-order partial differential equation.
Next, we use `DSolve` to find a general solution of $-3xtu_x + u_t = xt$ and name the resulting output `gensol`.

gensol = DSolve[−3xtD[u[x, t], x] + D[u[x, t], t]==xt,

u[x, t], {x, t}]

$\{\{u[x,t] \to \frac{1}{3}(-x+3C[1][\frac{1}{6}(3t^2+2\text{Log}[x])])\}\}$

The output represents an arbitrary function of $-\frac{3}{2}t^2 - \ln x$. The explicit solution is extracted from `gensol` with `gensol[[1,1,2]]`, the same way that results are extracted from the output of `DSolve` commands involving ordinary differential equations.

gensol[[1, 1, 2]]

$\frac{1}{3}(-x+3C[1][\frac{1}{6}(3t^2+2\text{Log}[x])])$

To find the solution that satisfies $u(x,0) = x$ we replace each occurrence of t in the solution by 0.

gensol[[1, 1, 2]]/.t->0

$\frac{1}{3}(-x+3C[1][\frac{\text{Log}[x]}{3}])$

Thus, we must find a function $f(x)$ so that

$$-\frac{1}{2}x + f(\ln x) = x$$
$$f(\ln x) = \frac{3}{2}x.$$

Certainly $f(t) = \frac{4}{3}e^{-t}$ satisfies the above criteria. We define $f(t) = \frac{4}{3}e^{-t}$ and then compute $f(\ln x)$ to verify that $f(\ln x) = \frac{3}{2}x$.

Clear[f]

f[t_] = 4Exp[−t]/3;

f[−Log[x]]

$\frac{4x}{3}$

Thus, the solution to the initial-value problem is given by $-\frac{1}{3}x + f\left(-\frac{3}{2}t^2 - \ln x\right)$ which is computed and named `sol`. Of course, the result returned is the same as that obtained previously.

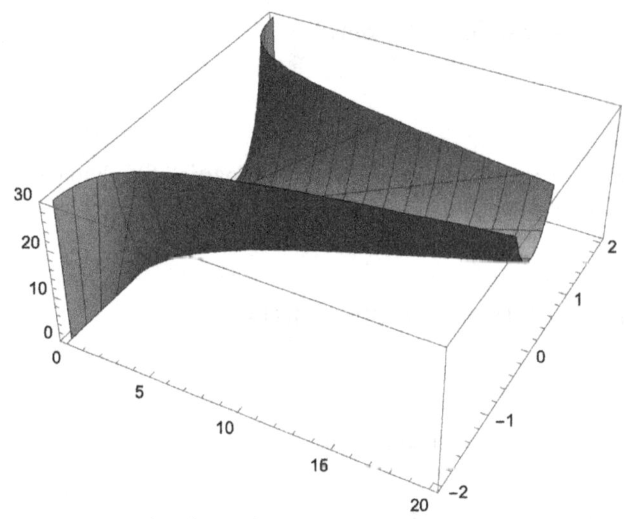

FIGURE 6.56 Plot of $u(x,t) = \frac{1}{3}x\left(4e^{3t^2/2} - 1\right)$.

sol = Simplify[$-\frac{x}{3} + f[-\frac{3t^2}{2} - \text{Log}[x]]$]

$\frac{1}{3}(-1 + 4e^{\frac{3t^2}{2}})x$

Last, we use `Plot3D` to graph `sol` on the rectangle $[0, 20] \times [-2, 2]$ in Fig. 6.56.

Plot3D[sol, {x, 0, 20}, {t, −2, 2}, PlotRange → {0, 30}, PlotPoints → 30,

ClippingStyle → None] □

Bibliography

[1] Elsa Abbena, Simon Salamon, Alfred Gray, Modern Differential Geometry of Curves and Surfaces with Mathematica, 3rd edition, Chapman and Hall/CRC, Boca Raton, FL, 2006.
[2] Martha L. Abell, James P. Braselton, Differential Equations with Mathematica, 4th edition, Academic Press, Burlington, MA, 2016.
[3] Michael F. Barnsley, Fractals Everywhere: New Edition, Dover Books, Mineola, NY, 2012.
[4] James P. Braselton, Martha L. Abell, Lorraine M. Braselton, When is a surface not orientable? Int. J. Math. Educ. Sci. Technol. 33 (4) (2002) 529–541.
[5] C. Henry Edwards, David E. Penney, Calculus, 6th edition, Pearson, New York, 2002.
[6] David P. Feldman, Chaos and Fractals: An Elementary Introduction, Oxford University Press, Oxford, 2012.
[7] Karl F. Graff, Wave Motion in Elastic Solids, Dover Books, Mineola, NY, 1991.
[8] Alfred Gray, Modern Differential Geometry of Curves and Surfaces, 1st edition, CRC Press, Boca Raton, FL, 1993.
[9] John Gray, Mastering Mathematica: Programming Methods and Applications, 1st edition, Academic Press, Burlington, MA, 1994.
[10] Dominic Jordan, Peter Smith, Nonlinear Ordinary Differential Equations: An Introduction for Scientists and Engineers, 4th edition, Oxford University Press, Oxford, 2007.
[11] Ron Larson, Bruce H. Edwards, Calculus, 10th edition, Cengage Learning, Boston, 2013.
[12] Roman E. Maeder, The Mathematica Programmer II, 1st edition, Academic Press, Burlington, MA, 1996.
[13] Roman E. Maeder, Programming in Mathematica, 3rd edition, Addison-Wesley Professional, Boston, MA, 1997.
[14] W.S. Massey, Algebraic Topology: An Introduction, Springer, New York, 1990.
[15] John A. Rafter, Martha L. Abell, James P. Braselton, Statistics with Mathematica, Academic Press, Burlington, MA, 1998.
[16] James Stewart, Calculus, 8th edition, Brooks Cole, Pacific Grove, CA, 2015.
[17] Eric W. Weisstein, The CRC Encyclopedia of Mathematics, 3rd edition, Chapman and Hall/CRC, Boca Raton, FL, 2009.
[18] Paul Wellin, Programming with Mathematica: An Introduction, 1st edition, Cambridge University Press, Cambridge, 2013.
[19] Paul Wellin, Essentials of Programming in Mathematica, 1st edition, Cambridge University Press, Cambridge, 2016.

Index

U

V

W

Y

Printed in the United States
By Bookmasters